T0189801

Lecture Notes in Computer Science   12906

More information about this subseries at http://www.springer.com/series/7412

Marleen de Bruijne · Philippe C. Cattin ·
Stéphane Cotin · Nicolas Padoy ·
Stefanie Speidel · Yefeng Zheng ·
Caroline Essert (Eds.)

Medical Image Computing and Computer Assisted Intervention – MICCAI 2021

24th International Conference
Strasbourg, France, September 27 – October 1, 2021
Proceedings, Part VI

 Springer

Editors
Marleen de Bruijne 🆔
Erasmus MC - University Medical Center
Rotterdam
Rotterdam, The Netherlands

University of Copenhagen
Copenhagen, Denmark

Stéphane Cotin 🆔
Inria Nancy Grand Est
Villers-lès-Nancy, France

Stefanie Speidel 🆔
National Center for Tumor Diseases
(NCT/UCC)
Dresden, Germany

Caroline Essert 🆔
ICube, Université de Strasbourg, CNRS
Strasbourg, France

Philippe C. Cattin 🆔
University of Basel
Allschwil, Switzerland

Nicolas Padoy 🆔
ICube, Université de Strasbourg, CNRS
Strasbourg, France

Yefeng Zheng 🆔
Tencent Jarvis Lab
Shenzhen, China

ISSN 0302-9743   ISSN 1611-3349 (electronic)
Lecture Notes in Computer Science
ISBN 978-3-030-87230-4   ISBN 978-3-030-87231-1 (eBook)
https://doi.org/10.1007/978-3-030-87231-1

LNCS Sublibrary: SL6 – Image Processing, Computer Vision, Pattern Recognition, and Graphics

This Springer imprint is published by the registered company Springer Nature Switzerland AG
The registered company address is: Gewerbestrasse 11, 6330 Cham, Switzerland

Preface

The 24th edition of the International Conference on Medical Image Computing and Computer Assisted Intervention (MICCAI 2021) has for the second time been placed under the shadow of COVID-19. Complicated situations due to the pandemic and multiple lockdowns have affected our lives during the past year, sometimes perturbing the researchers work, but also motivating an extraordinary dedication from many of our colleagues, and significant scientific advances in the fight against the virus. After another difficult year, most of us were hoping to be able to travel and finally meet in person at MICCAI 2021, which was supposed to be held in Strasbourg, France. Unfortunately, due to the uncertainty of the global situation, MICCAI 2021 had to be moved again to a virtual event that was held over five days from September 27 to October 1, 2021. Taking advantage of the experience gained last year and of the fast-evolving platforms, the organizers of MICCAI 2021 redesigned the schedule and the format. To offer the attendees both a strong scientific content and an engaging experience, two virtual platforms were used: Pathable for the oral and plenary sessions and SpatialChat for lively poster sessions, industrial booths, and networking events in the form of interactive group video chats.

These proceedings of MICCAI 2021 showcase all 531 papers that were presented at the main conference, organized into eight volumes in the Lecture Notes in Computer Science (LNCS) series as follows:

- Part I, LNCS Volume 12901: Image Segmentation
- Part II, LNCS Volume 12902: Machine Learning 1
- Part III, LNCS Volume 12903: Machine Learning 2
- Part IV, LNCS Volume 12904: Image Registration and Computer Assisted Intervention
- Part V, LNCS Volume 12905: Computer Aided Diagnosis
- Part VI, LNCS Volume 12906: Image Reconstruction and Cardiovascular Imaging
- Part VII, LNCS Volume 12907: Clinical Applications
- Part VIII, LNCS Volume 12908: Microscopic, Ophthalmic, and Ultrasound Imaging

These papers were selected after a thorough double-blind peer review process. We followed the example set by past MICCAI meetings, using Microsoft's Conference Managing Toolkit (CMT) for paper submission and peer reviews, with support from the Toronto Paper Matching System (TPMS), to partially automate paper assignment to area chairs and reviewers, and from iThenticate to detect possible cases of plagiarism.

Following a broad call to the community we received 270 applications to become an area chair for MICCAI 2021. From this group, the program chairs selected a total of 96 area chairs, aiming for diversity — MIC versus CAI, gender, geographical region, and a mix of experienced and new area chairs. Reviewers were recruited also via an open call for volunteers from the community (288 applications, of which 149 were selected by the program chairs) as well as by re-inviting past reviewers, leading to a total of 1340 registered reviewers.

We received 1630 full paper submissions after an original 2667 intentions to submit. Four papers were rejected without review because of concerns of (self-)plagiarism and dual submission and one additional paper was rejected for not adhering to the MICCAI page restrictions; two further cases of dual submission were discovered and rejected during the review process. Five papers were withdrawn by the authors during review and after acceptance.

The review process kicked off with a reviewer tutorial and an area chair meeting to discuss the review process, criteria for MICCAI acceptance, how to write a good (meta-)review, and expectations for reviewers and area chairs. Each area chair was assigned 16–18 manuscripts for which they suggested potential reviewers using TPMS scores, self-declared research area(s), and the area chair's knowledge of the reviewers' expertise in relation to the paper, while conflicts of interest were automatically avoided by CMT. Reviewers were invited to bid for the papers for which they had been suggested by an area chair or which were close to their expertise according to TPMS. Final reviewer allocations via CMT took account of reviewer bidding, prioritization of area chairs, and TPMS scores, leading to on average four reviews performed per person by a total of 1217 reviewers.

Following the initial double-blind review phase, area chairs provided a meta-review summarizing key points of reviews and a recommendation for each paper. The program chairs then evaluated the reviews and their scores, along with the recommendation from the area chairs, to directly accept 208 papers (13%) and reject 793 papers (49%); the remainder of the papers were sent for rebuttal by the authors. During the rebuttal phase, two additional area chairs were assigned to each paper. The three area chairs then independently ranked their papers, wrote meta-reviews, and voted to accept or reject the paper, based on the reviews, rebuttal, and manuscript. The program chairs checked all meta-reviews, and in some cases where the difference between rankings was high or comments were conflicting, they also assessed the original reviews, rebuttal, and submission. In all other cases a majority voting scheme was used to make the final decision. This process resulted in the acceptance of a further 325 papers for an overall acceptance rate of 33%.

Acceptance rates were the same between medical image computing (MIC) and computer assisted interventions (CAI) papers, and slightly lower where authors classified their paper as both MIC and CAI. Distribution of the geographical region of the first author as indicated in the optional demographic survey was similar among submitted and accepted papers.

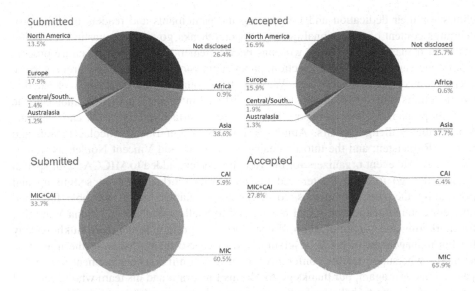

New this year, was the requirement to fill out a reproducibility checklist when submitting an intention to submit to MICCAI, in order to stimulate authors to think about what aspects of their method and experiments they should include to allow others to reproduce their results. Papers that included an anonymous code repository and/or indicated that the code would be made available were more likely to be accepted. From all accepted papers, 273 (51%) included a link to a code repository with the camera-ready submission.

Another novelty this year is that we decided to make the reviews, meta-reviews, and author responses for accepted papers available on the website. We hope the community will find this a useful resource.

The outstanding program of MICCAI 2021 was enriched by four exceptional keynote talks given by Alyson McGregor, Richard Satava, Fei-Fei Li, and Pierre Jannin, on hot topics such as gender bias in medical research, clinical translation to industry, intelligent medicine, and sustainable research. This year, as in previous years, high-quality satellite events completed the program of the main conference: 28 workshops, 23 challenges, and 14 tutorials; without forgetting the increasingly successful plenary events, such as the Women in MICCAI (WiM) meeting, the MICCAI Student Board (MSB) events, the 2nd Startup Village, the MICCAI-RSNA panel, and the first "Reinforcing Inclusiveness & diverSity and Empowering MICCAI" (or RISE-MICCAI) event.

MICCAI 2021 has also seen the first edition of CLINICCAI, the clinical day of MICCAI. Organized by Nicolas Padoy and Lee Swanstrom, this new event will hopefully help bring the scientific and clinical communities closer together, and foster collaborations and interaction. A common keynote connected the two events. We hope this effort will be pursued in the next editions.

We would like to thank everyone who has contributed to making MICCAI 2021 a success. First of all, we sincerely thank the authors, area chairs, reviewers, and session chairs for their dedication and for offering the participants and readers of these proceedings content of exceptional quality. Special thanks go to our fantastic submission platform manager Kitty Wong, who has been a tremendous help in the entire process from reviewer and area chair selection, paper submission, and the review process to the preparation of these proceedings. We also thank our very efficient team of satellite events chairs and coordinators, led by Cristian Linte and Matthieu Chabanas: the workshop chairs, Amber Simpson, Denis Fortun, Marta Kersten-Oertel, and Sandrine Voros; the challenges chairs, Annika Reinke, Spyridon Bakas, Nicolas Passat, and Ingerid Reinersten; and the tutorial chairs, Sonia Pujol and Vincent Noblet, as well as all the satellite event organizers for the valuable content added to MICCAI. Our special thanks also go to John Baxter and his team who worked hard on setting up and populating the virtual platforms, to Alejandro Granados for his valuable help and efficient communication on social media, and to Shelley Wallace and Anna Van Vliet for marketing and communication. We are also very grateful to Anirban Mukhopadhay for his management of the sponsorship, and of course many thanks to the numerous sponsors who supported the conference, often with continuous engagement over many years. This year again, our thanks go to Marius Linguraru and his team who supervised a range of actions to help, and promote, career development, among which were the mentorship program and the Startup Village. And last but not least, our wholehearted thanks go to Mehmet and the wonderful team at Dekon Congress and Tourism for their great professionalism and reactivity in the management of all logistical aspects of the event.

Finally, we thank the MICCAI society and the Board of Directors for their support throughout the years, starting with the first discussions about bringing MICCAI to Strasbourg in 2017.

We look forward to seeing you at MICCAI 2022.

September 2021

Marleen de Bruijne
Philippe Cattin
Stéphane Cotin
Nicolas Padoy
Stefanie Speidel
Yefeng Zheng
Caroline Essert

Organization

General Chair

Caroline Essert   Université de Strasbourg, CNRS, ICube, France

Program Chairs

Marleen de Bruijne   Erasmus MC Rotterdam, The Netherlands,
     and University of Copenhagen, Denmark
Philippe C. Cattin   University of Basel, Switzerland
Stéphane Cotin   Inria, France
Nicolas Padoy   Université de Strasbourg, CNRS, ICube, IHU, France
Stefanie Speidel   National Center for Tumor Diseases, Dresden, Germany
Yefeng Zheng   Tencent Jarvis Lab, China

Satellite Events Coordinators

Cristian Linte   Rochester Institute of Technology, USA
Matthieu Chabanas   Université Grenoble Alpes, France

Workshop Team

Amber Simpson   Queen's University, Canada
Denis Fortun   Université de Strasbourg, CNRS, ICube, France
Marta Kersten-Oertel   Concordia University, Canada
Sandrine Voros   TIMC-IMAG, INSERM, France

Challenges Team

Annika Reinke   German Cancer Research Center, Germany
Spyridon Bakas   University of Pennsylvania, USA
Nicolas Passat   Université de Reims Champagne-Ardenne, France
Ingerid Reinersten   SINTEF, NTNU, Norway

Tutorial Team

Vincent Noblet   Université de Strasbourg, CNRS, ICube, France
Sonia Pujol   Harvard Medical School, Brigham and Women's
     Hospital, USA

Clinical Day Chairs

Nicolas Padoy   Université de Strasbourg, CNRS, ICube, IHU, France
Lee Swanström   IHU Strasbourg, France

Sponsorship Chairs

Anirban Mukhopadhyay   Technische Universität Darmstadt, Germany
Yanwu Xu   Baidu Inc., China

Young Investigators and Early Career Development Program Chairs

Marius Linguraru   Children's National Institute, USA
Antonio Porras   Children's National Institute, USA
Daniel Racoceanu   Sorbonne Université/Brain Institute, France
Nicola Rieke   NVIDIA, Germany
Renee Yao   NVIDIA, USA

Social Media Chairs

Alejandro Granados   King's College London, UK
  Martinez
Shuwei Xing   Robarts Research Institute, Canada
Maxence Boels   King's College London, UK

Green Team

Pierre Jannin   INSERM, Université de Rennes 1, France
Étienne Baudrier   Université de Strasbourg, CNRS, ICube, France

Student Board Liaison

Éléonore Dufresne   Université de Strasbourg, CNRS, ICube, France
Étienne Le Quentrec   Université de Strasbourg, CNRS, ICube, France
Vinkle Srivastav   Université de Strasbourg, CNRS, ICube, France

Submission Platform Manager

Kitty Wong   The MICCAI Society, Canada

Virtual Platform Manager

John Baxter   INSERM, Université de Rennes 1, France

Program Committee

Ehsan Adeli	Stanford University, USA
Iman Aganj	Massachusetts General Hospital, Harvard Medical School, USA
Pablo Arbelaez	Universidad de los Andes, Colombia
John Ashburner	University College London, UK
Meritxell Bach Cuadra	University of Lausanne, Switzerland
Sophia Bano	University College London, UK
Adrien Bartoli	Université Clermont Auvergne, France
Christian Baumgartner	ETH Zürich, Switzerland
Hrvoje Bogunovic	Medical University of Vienna, Austria
Weidong Cai	University of Sydney, Australia
Gustavo Carneiro	University of Adelaide, Australia
Chao Chen	Stony Brook University, USA
Elvis Chen	Robarts Research Institute, Canada
Hao Chen	Hong Kong University of Science and Technology, Hong Kong SAR
Albert Chung	Hong Kong University of Science and Technology, Hong Kong SAR
Adrian Dalca	Massachusetts Institute of Technology, USA
Adrien Depeursinge	HES-SO Valais-Wallis, Switzerland
Jose Dolz	ÉTS Montréal, Canada
Ruogu Fang	University of Florida, USA
Dagan Feng	University of Sydney, Australia
Huazhu Fu	Inception Institute of Artificial Intelligence, United Arab Emirates
Mingchen Gao	University at Buffalo, The State University of New York, USA
Guido Gerig	New York University, USA
Orcun Goksel	Uppsala University, Sweden
Alberto Gomez	King's College London, UK
Ilker Hacihaliloglu	Rutgers University, USA
Adam Harrison	PAII Inc., USA
Mattias Heinrich	University of Lübeck, Germany
Yi Hong	Shanghai Jiao Tong University, China
Yipeng Hu	University College London, UK
Junzhou Huang	University of Texas at Arlington, USA
Xiaolei Huang	The Pennsylvania State University, USA
Jana Hutter	King's College London, UK
Madhura Ingalhalikar	Symbiosis Center for Medical Image Analysis, India
Shantanu Joshi	University of California, Los Angeles, USA
Samuel Kadoury	Polytechnique Montréal, Canada
Fahmi Khalifa	Mansoura University, Egypt
Hosung Kim	University of Southern California, USA
Minjeong Kim	University of North Carolina at Greensboro, USA

Zhong Xue   Shanghai United Imaging Intelligence, China
Xin Yang   Huazhong University of Science and Technology,
                               China
Jianhua Yao   National Institutes of Health, USA
Zhaozheng Yin   Stony Brook University, USA
Yixuan Yuan   City University of Hong Kong, Hong Kong SAR
Liang Zhan   University of Pittsburgh, USA
Tuo Zhang   Northwestern Polytechnical University, China
Yitian Zhao   Chinese Academy of Sciences, China
Luping Zhou   University of Sydney, Australia
S. Kevin Zhou   Chinese Academy of Sciences, China
Dajiang Zhu   University of Texas at Arlington, USA
Xiahai Zhuang   Fudan University, China
Maria A. Zuluaga   EURECOM, France

Reviewers

Alaa Eldin Abdelaal   Chloé Audigier
Khalid Abdul Jabbar   Kamran Avanaki
Purang Abolmaesumi   Angelica Aviles-Rivero
Mazdak Abulnaga   Suyash Awate
Maryam Afzali   Dogu Baran Aydogan
Priya Aggarwal   Qinle Ba
Ola Ahmad   Morteza Babaie
Sahar Ahmad   Hyeon-Min Bae
Euijoon Ahn   Woong Bae
Alireza Akhondi-Asl   Junjie Bai
Saad Ullah Akram   Wenjia Bai
Dawood Al Chanti   Ujjwal Baid
Daniel Alexander   Spyridon Bakas
Sharib Ali   Yaël Balbastre
Lejla Alic   Marcin Balicki
Omar Al-Kadi   Fabian Balsiger
Maximilian Allan   Abhirup Banerjee
Pierre Ambrosini   Sreya Banerjee
Sameer Antani   Shunxing Bao
Michela Antonelli   Adrian Barbu
Jacob Antunes   Sumana Basu
Syed Anwar   Mathilde Bateson
Ignacio Arganda-Carreras   Deepti Bathula
Mohammad Ali Armin   John Baxter
Md Ashikuzzaman   Bahareh Behboodi
Mehdi Astaraki   Delaram Behnami
Angélica Atehortúa   Mikhail Belyaev
Gowtham Atluri   Aicha BenTaieb

Camilo Bermudez
Gabriel Bernardino
Hadrien Bertrand
Alaa Bessadok
Michael Beyeler
Indrani Bhattacharya
Chetan Bhole
Lei Bi
Gui-Bin Bian
Ryoma Bise
Stefano B. Blumberg
Ester Bonmati
Bhushan Borotikar
Jiri Borovec
Ilaria Boscolo Galazzo
Alexandre Bousse
Nicolas Boutry
Behzad Bozorgtabar
Nathaniel Braman
Nadia Brancati
Katharina Breininger
Christopher Bridge
Esther Bron
Rupert Brooks
Qirong Bu
Duc Toan Bui
Ninon Burgos
Nikolay Burlutskiy
Hendrik Burwinkel
Russell Butler
Michał Byra
Ryan Cabeen
Mariano Cabezas
Hongmin Cai
Jinzheng Cai
Yunliang Cai
Sema Candemir
Bing Cao
Qing Cao
Shilei Cao
Tian Cao
Weiguo Cao
Aaron Carass
M. Jorge Cardoso
Adrià Casamitjana
Matthieu Chabanas

Ahmad Chaddad
Jayasree Chakraborty
Sylvie Chambon
Yi Hao Chan
Ming-Ching Chang
Peng Chang
Violeta Chang
Sudhanya Chatterjee
Christos Chatzichristos
Antong Chen
Chang Chen
Cheng Chen
Dongdong Chen
Geng Chen
Hanbo Chen
Jianan Chen
Jianxu Chen
Jie Chen
Junxiang Chen
Lei Chen
Li Chen
Liangjun Chen
Min Chen
Pingjun Chen
Qiang Chen
Shuai Chen
Tianhua Chen
Tingting Chen
Xi Chen
Xiaoran Chen
Xin Chen
Xuejin Chen
Yuhua Chen
Yukun Chen
Zhaolin Chen
Zhineng Chen
Zhixiang Chen
Erkang Cheng
Jun Cheng
Li Cheng
Yuan Cheng
Farida Cheriet
Minqi Chong
Jaegul Choo
Aritra Chowdhury
Gary Christensen

Daan Christiaens
Stergios Christodoulidis
Ai Wern Chung
Pietro Antonio Cicalese
Özgün Çiçek
Celia Cintas
Matthew Clarkson
Jaume Coll-Font
Toby Collins
Olivier Commowick
Pierre-Henri Conze
Timothy Cootes
Luca Corinzia
Teresa Correia
Hadrien Courtecuisse
Jeffrey Craley
Hui Cui
Jianan Cui
Zhiming Cui
Kathleen Curran
Claire Cury
Tobias Czempiel
Vedrana Dahl
Haixing Dai
Rafat Damseh
Bilel Daoud
Neda Davoudi
Laura Daza
Sandro De Zanet
Charles Delahunt
Yang Deng
Cem Deniz
Felix Denzinger
Hrishikesh Deshpande
Christian Desrosiers
Blake Dewey
Neel Dey
Raunak Dey
Jwala Dhamala
Yashin Dicente Cid
Li Ding
Xinghao Ding
Zhipeng Ding
Konstantin Dmitriev
Ines Domingues
Liang Dong

Mengjin Dong
Nanqing Dong
Reuben Dorent
Sven Dorkenwald
Qi Dou
Simon Drouin
Niharika D'Souza
Lei Du
Hongyi Duanmu
Nicolas Duchateau
James Duncan
Luc Duong
Nicha Dvornek
Dmitry V. Dylov
Oleh Dzyubachyk
Roy Eagleson
Mehran Ebrahimi
Jan Egger
Alma Eguizabal
Gudmundur Einarsson
Ahmed Elazab
Mohammed S. M. Elbaz
Shireen Elhabian
Mohammed Elmogy
Amr Elsawy
Ahmed Eltanboly
Sandy Engelhardt
Ertunc Erdil
Marius Erdt
Floris Ernst
Boris Escalante-Ramírez
Maria Escobar
Mohammad Eslami
Nazila Esmaeili
Marco Esposito
Oscar Esteban
Théo Estienne
Ivan Ezhov
Deng-Ping Fan
Jingfan Fan
Xin Fan
Yonghui Fan
Xi Fang
Zhenghan Fang
Aly Farag
Mohsen Farzi

Lina Felsner
Jun Feng
Ruibin Feng
Xinyang Feng
Yuan Feng
Aaron Fenster
Aasa Feragen
Henrique Fernandes
Enzo Ferrante
Jean Feydy
Lukas Fischer
Peter Fischer
Antonio Foncubierta-Rodríguez
Germain Forestier
Nils Daniel Forkert
Jean-Rassaire Fouefack
Moti Freiman
Wolfgang Freysinger
Xueyang Fu
Yunguan Fu
Wolfgang Fuhl
Isabel Funke
Philipp Fürnstahl
Pedro Furtado
Ryo Furukawa
Jin Kyu Gahm
Laurent Gajny
Adrian Galdran
Yu Gan
Melanie Ganz
Cong Gao
Dongxu Gao
Linlin Gao
Siyuan Gao
Yixin Gao
Yue Gao
Zhifan Gao
Alfonso Gastelum-Strozzi
Srishti Gautam
Bao Ge
Rongjun Ge
Zongyuan Ge
Sairam Geethanath
Shiv Gehlot
Nils Gessert
Olivier Gevaert

Sandesh Ghimire
Ali Gholipour
Sayan Ghosal
Andrea Giovannini
Gabriel Girard
Ben Glocker
Arnold Gomez
Mingming Gong
Cristina González
German Gonzalez
Sharath Gopal
Karthik Gopinath
Pietro Gori
Michael Götz
Shuiping Gou
Maged Goubran
Sobhan Goudarzi
Dushyant Goyal
Mark Graham
Bertrand Granado
Alejandro Granados
Vicente Grau
Lin Gu
Shi Gu
Xianfeng Gu
Yun Gu
Zaiwang Gu
Hao Guan
Ricardo Guerrero
Houssem-Eddine Gueziri
Dazhou Guo
Hengtao Guo
Jixiang Guo
Pengfei Guo
Xiaoqing Guo
Yi Guo
Yulan Guo
Yuyu Guo
Krati Gupta
Vikash Gupta
Praveen Gurunath Bharathi
Boris Gutman
Prashnna Gyawali
Stathis Hadjidemetriou
Mohammad Hamghalam
Hu Han

Liang Han
Xiaoguang Han
Xu Han
Zhi Han
Zhongyi Han
Jonny Hancox
Xiaoke Hao
Nandinee Haq
Ali Hatamizadeh
Charles Hatt
Andreas Hauptmann
Mohammad Havaei
Kelei He
Nanjun He
Tiancheng He
Xuming He
Yuting He
Nicholas Heller
Alessa Hering
Monica Hernandez
Carlos Hernandez-Matas
Kilian Hett
Jacob Hinkle
David Ho
Nico Hoffmann
Matthew Holden
Sungmin Hong
Yoonmi Hong
Antal Horváth
Md Belayat Hossain
Benjamin Hou
William Hsu
Tai-Chiu Hsung
Kai Hu
Shi Hu
Shunbo Hu
Wenxing Hu
Xiaoling Hu
Xiaowei Hu
Yan Hu
Zhenhong Hu
Heng Huang
Qiaoying Huang
Yi-Jie Huang
Yixing Huang
Yongxiang Huang

Yue Huang
Yufang Huang
Arnaud Huaulmé
Henkjan Huisman
Yuankai Huo
Andreas Husch
Mohammad Hussain
Raabid Hussain
Sarfaraz Hussein
Khoi Huynh
Seong Jae Hwang
Emmanuel Iarussi
Kay Igwe
Abdullah-Al-Zubaer Imran
Ismail Irmakci
Mobarakol Islam
Mohammad Shafkat Islam
Vamsi Ithapu
Koichi Ito
Hayato Itoh
Oleksandra Ivashchenko
Yuji Iwahori
Shruti Jadon
Mohammad Jafari
Mostafa Jahanifar
Amir Jamaludin
Mirek Janatka
Won-Dong Jang
Uditha Jarayathne
Ronnachai Jaroensri
Golara Javadi
Rohit Jena
Rachid Jennane
Todd Jensen
Won-Ki Jeong
Yuanfeng Ji
Zhanghexuan Ji
Haozhe Jia
Jue Jiang
Tingting Jiang
Xiang Jiang
Jianbo Jiao
Zhicheng Jiao
Amelia Jiménez-Sánchez
Dakai Jin
Yueming Jin

Bin Jing
Anand Joshi
Yohan Jun
Kyu-Hwan Jung
Alain Jungo
Manjunath K N
Ali Kafaei Zad Tehrani
Bernhard Kainz
John Kalafut
Michael C. Kampffmeyer
Qingbo Kang
Po-Yu Kao
Neerav Karani
Turkay Kart
Satyananda Kashyap
Amin Katouzian
Alexander Katzmann
Prabhjot Kaur
Erwan Kerrien
Hoel Kervadec
Ashkan Khakzar
Nadieh Khalili
Siavash Khallaghi
Farzad Khalvati
Bishesh Khanal
Pulkit Khandelwal
Maksim Kholiavchenko
Naji Khosravan
Seyed Mostafa Kia
Daeseung Kim
Hak Gu Kim
Hyo-Eun Kim
Jae-Hun Kim
Jaeil Kim
Jinman Kim
Mansu Kim
Namkug Kim
Seong Tae Kim
Won Hwa Kim
Andrew King
Atilla Kiraly
Yoshiro Kitamura
Tobias Klinder
Bin Kong
Jun Kong
Tomasz Konopczynski

Bongjin Koo
Ivica Kopriva
Kivanc Kose
Mateusz Kozinski
Anna Kreshuk
Anithapriya Krishnan
Pavitra Krishnaswamy
Egor Krivov
Frithjof Kruggel
Alexander Krull
Elizabeth Krupinski
Serife Kucur
David Kügler
Hugo Kuijf
Abhay Kumar
Ashnil Kumar
Kuldeep Kumar
Nitin Kumar
Holger Kunze
Tahsin Kurc
Anvar Kurmukov
Yoshihiro Kuroda
Jin Tae Kwak
Yongchan Kwon
Francesco La Rosa
Aymen Laadhari
Dmitrii Lachinov
Alain Lalande
Tryphon Lambrou
Carole Lartizien
Bianca Lassen-Schmidt
Ngan Le
Leo Lebrat
Christian Ledig
Eung-Joo Lee
Hyekyoung Lee
Jong-Hwan Lee
Matthew Lee
Sangmin Lee
Soochahn Lee
Étienne Léger
Stefan Leger
Andreas Leibetseder
Rogers Jeffrey Leo John
Juan Leon
Bo Li

Chongyi Li
Fuhai Li
Hongming Li
Hongwei Li
Jian Li
Jianning Li
Jiayun Li
Junhua Li
Kang Li
Mengzhang Li
Ming Li
Qing Li
Shaohua Li
Shuyu Li
Weijian Li
Weikai Li
Wenqi Li
Wenyuan Li
Xiang Li
Xiaomeng Li
Xiaoxiao Li
Xin Li
Xiuli Li
Yang Li
Yi Li
Yuexiang Li
Zeju Li
Zhang Li
Zhiyuan Li
Zhjin Li
Gongbo Liang
Jianming Liang
Libin Liang
Yuan Liang
Haofu Liao
Ruizhi Liao
Wei Liao
Xiangyun Liao
Roxane Licandro
Gilbert Lim
Baihan Lin
Hongxiang Lin
Jianyu Lin
Yi Lin
Claudia Lindner
Geert Litjens

Bin Liu
Chi Liu
Daochang Liu
Dong Liu
Dongnan Liu
Feng Liu
Hangfan Liu
Hong Liu
Huafeng Liu
Jianfei Liu
Jingya Liu
Kai Liu
Kefei Liu
Lihao Liu
Mengting Liu
Peng Liu
Qin Liu
Quande Liu
Shengfeng Liu
Shenghua Liu
Shuangjun Liu
Sidong Liu
Siqi Liu
Tianrui Liu
Xiao Liu
Xinyang Liu
Xinyu Liu
Yan Liu
Yikang Liu
Yong Liu
Yuan Liu
Yue Liu
Yuhang Liu
Andrea Loddo
Nicolas Loménie
Daniel Lopes
Bin Lou
Jian Lou
Nicolas Loy Rodas
Donghuan Lu
Huanxiang Lu
Weijia Lu
Xiankai Lu
Yongyi Lu
Yueh-Hsun Lu
Yuhang Lu

Imanol Luengo
Jie Luo
Jiebo Luo
Luyang Luo
Ma Luo
Bin Lv
Jinglei Lv
Junyan Lyu
Qing Lyu
Yuanyuan Lyu
Andy J. Ma
Chunwei Ma
Da Ma
Hua Ma
Kai Ma
Lei Ma
Anderson Maciel
Amirreza Mahbod
S. Sara Mahdavi
Mohammed Mahmoud
Saïd Mahmoudi
Klaus H. Maier-Hein
Bilal Malik
Ilja Manakov
Matteo Mancini
Tommaso Mansi
Yunxiang Mao
Brett Marinelli
Pablo Márquez Neila
Carsten Marr
Yassine Marrakchi
Fabio Martinez
Andre Mastmeyer
Tejas Sudharshan Mathai
Dimitrios Mavroeidis
Jamie McClelland
Pau Medrano-Gracia
Raghav Mehta
Sachin Mehta
Raphael Meier
Qier Meng
Qingjie Meng
Yanda Meng
Martin Menten
Odyssée Merveille
Islem Mhiri

Liang Mi
Stijn Michielse
Abhishek Midya
Fausto Milletari
Hyun-Seok Min
Zhe Min
Tadashi Miyamoto
Sara Moccia
Hassan Mohy-ud-Din
Tony C. W. Mok
Rafael Molina
Mehdi Moradi
Rodrigo Moreno
Kensaku Mori
Lia Morra
Linda Moy
Mohammad Hamed Mozaffari
Sovanlal Mukherjee
Anirban Mukhopadhyay
Henning Müller
Balamurali Murugesan
Cosmas Mwikirize
Andriy Myronenko
Saad Nadeem
Vishwesh Nath
Rodrigo Nava
Fernando Navarro
Amin Nejatbakhsh
Dong Ni
Hannes Nickisch
Dong Nie
Jingxin Nie
Aditya Nigam
Lipeng Ning
Xia Ning
Tianye Niu
Jack Noble
Vincent Noblet
Alexey Novikov
Jorge Novo
Mohammad Obeid
Masahiro Oda
Benjamin Odry
Steffen Oeltze-Jafra
Hugo Oliveira
Sara Oliveira

Arnau Oliver
Emanuele Olivetti
Jimena Olveres
John Onofrey
Felipe Orihuela-Espina
José Orlando
Marcos Ortega
Yoshito Otake
Sebastian Otálora
Cheng Ouyang
Jiahong Ouyang
Xi Ouyang
Michal Ozery-Flato
Danielle Pace
Krittin Pachtrachai
J. Blas Pagador
Akshay Pai
Viswanath Pamulakanty Sudarshan
Jin Pan
Yongsheng Pan
Pankaj Pandey
Prashant Pandey
Egor Panfilov
Shumao Pang
Joao Papa
Constantin Pape
Bartlomiej Papiez
Hyunjin Park
Jongchan Park
Sanghyun Park
Seung-Jong Park
Seyoun Park
Magdalini Paschali
Diego Patiño Cortés
Angshuman Paul
Christian Payer
Yuru Pei
Chengtao Peng
Yige Peng
Antonio Pepe
Oscar Perdomo
Sérgio Pereira
Jose-Antonio Pérez-Carrasco
Fernando Pérez-García
Jorge Perez-Gonzalez
Skand Peri

Matthias Perkonigg
Mehran Pesteie
Jorg Peters
Jens Petersen
Kersten Petersen
Renzo Phellan Aro
Ashish Phophalia
Tomasz Pieciak
Antonio Pinheiro
Pramod Pisharady
Kilian Pohl
Sebastian Pölsterl
Iulia A. Popescu
Alison Pouch
Prateek Prasanna
Raphael Prevost
Juan Prieto
Sergi Pujades
Elodie Puybareau
Esther Puyol-Antón
Haikun Qi
Huan Qi
Buyue Qian
Yan Qiang
Yuchuan Qiao
Chen Qin
Wenjian Qin
Yulei Qin
Wu Qiu
Hui Qu
Liangqiong Qu
Kha Gia Quach
Prashanth R.
Pradeep Reddy Raamana
Mehdi Rahim
Jagath Rajapakse
Kashif Rajpoot
Jhonata Ramos
Lingyan Ran
Hatem Rashwan
Daniele Ravì
Keerthi Sravan Ravi
Nishant Ravikumar
Harish RaviPrakash
Samuel Remedios
Yinhao Ren

Yudan Ren
Mauricio Reyes
Constantino Reyes-Aldasoro
Jonas Richiardi
David Richmond
Anne-Marie Rickmann
Leticia Rittner
Dominik Rivoir
Emma Robinson
Jessica Rodgers
Rafael Rodrigues
Robert Rohling
Michal Rosen-Zvi
Lukasz Roszkowiak
Karsten Roth
José Rouco
Daniel Rueckert
Jaime S. Cardoso
Mohammad Sabokrou
Ario Sadafi
Monjoy Saha
Pramit Saha
Dushyant Sahoo
Pranjal Sahu
Maria Sainz de Cea
Olivier Salvado
Robin Sandkuehler
Gianmarco Santini
Duygu Sarikaya
Imari Sato
Olivier Saut
Dustin Scheinost
Nico Scherf
Markus Schirmer
Alexander Schlaefer
Jerome Schmid
Julia Schnabel
Klaus Schoeffmann
Andreas Schuh
Ernst Schwartz
Christina Schwarz-Gsaxner
Michaël Sdika
Suman Sedai
Anjany Sekuboyina
Raghavendra Selvan
Sourya Sengupta

Youngho Seo
Lama Seoud
Ana Sequeira
Maxime Sermesant
Carmen Serrano
Muhammad Shaban
Ahmed Shaffie
Sobhan Shafiei
Mohammad Abuzar Shaikh
Reuben Shamir
Shayan Shams
Hongming Shan
Harshita Sharma
Gregory Sharp
Mohamed Shehata
Haocheng Shen
Li Shen
Liyue Shen
Mali Shen
Yiqing Shen
Yiqiu Shen
Zhengyang Shen
Kuangyu Shi
Luyao Shi
Xiaoshuang Shi
Xueying Shi
Yemin Shi
Yiyu Shi
Yonghong Shi
Jitae Shin
Boris Shirokikh
Suprosanna Shit
Suzanne Shontz
Yucheng Shu
Alberto Signoroni
Wilson Silva
Margarida Silveira
Matthew Sinclair
Rohit Singla
Sumedha Singla
Ayushi Sinha
Kevin Smith
Rajath Soans
Ahmed Soliman
Stefan Sommer
Yang Song

Youyi Song
Aristeidis Sotiras
Arcot Sowmya
Rachel Sparks
William Speier
Ziga Spiclin
Dominik Spinczyk
Jon Sporring
Chetan Srinidhi
Anuroop Sriram
Vinkle Srivastav
Lawrence Staib
Marius Staring
Johannes Stegmaier
Joshua Stough
Robin Strand
Martin Styner
Hai Su
Yun-Hsuan Su
Vaishnavi Subramanian
Gérard Subsol
Yao Sui
Avan Suinesiaputra
Jeremias Sulam
Shipra Suman
Li Sun
Wenqing Sun
Chiranjib Sur
Yannick Suter
Tanveer Syeda-Mahmood
Fatemeh Taheri Dezaki
Roger Tam
José Tamez-Peña
Chaowei Tan
Hao Tang
Thomas Tang
Yucheng Tang
Zihao Tang
Mickael Tardy
Giacomo Tarroni
Jonas Teuwen
Paul Thienphrapa
Stephen Thompson
Jiang Tian
Yu Tian
Yun Tian

Aleksei Tiulpin
Hamid Tizhoosh
Matthew Toews
Oguzhan Topsakal
Antonio Torteya
Sylvie Treuillet
Jocelyne Troccaz
Roger Trullo
Chialing Tsai
Sudhakar Tummala
Verena Uslar
Hristina Uzunova
Régis Vaillant
Maria Vakalopoulou
Jeya Maria Jose Valanarasu
Tom van Sonsbeek
Gijs van Tulder
Marta Varela
Thomas Varsavsky
Francisco Vasconcelos
Liset Vazquez Romaguera
S. Swaroop Vedula
Sanketh Vedula
Harini Veeraraghavan
Miguel Vega
Gonzalo Vegas Sanchez-Ferrero
Anant Vemuri
Gopalkrishna Veni
Mitko Veta
Thomas Vetter
Pedro Vieira
Juan Pedro Vigueras Guillén
Barbara Villarini
Satish Viswanath
Athanasios Vlontzos
Wolf-Dieter Vogl
Bo Wang
Cheng Wang
Chengjia Wang
Chunliang Wang
Clinton Wang
Congcong Wang
Dadong Wang
Dongang Wang
Haifeng Wang
Hongyu Wang

Hu Wang
Huan Wang
Kun Wang
Li Wang
Liansheng Wang
Linwei Wang
Manning Wang
Renzhen Wang
Ruixuan Wang
Sheng Wang
Shujun Wang
Shuo Wang
Tianchen Wang
Tongxin Wang
Wenzhe Wang
Xi Wang
Xiaosong Wang
Yan Wang
Yaping Wang
Yi Wang
Yirui Wang
Zeyi Wang
Zhangyang Wang
Zihao Wang
Zuhui Wang
Simon Warfield
Jonathan Weber
Jürgen Weese
Dong Wei
Donglai Wei
Dongming Wei
Martin Weigert
Wolfgang Wein
Michael Wels
Cédric Wemmert
Junhao Wen
Travis Williams
Matthias Wilms
Stefan Winzeck
James Wiskin
Adam Wittek
Marek Wodzinski
Jelmer Wolterink
Ken C. L. Wong
Chongruo Wu
Guoqing Wu

Ji Wu
Jian Wu
Jie Ying Wu
Pengxiang Wu
Xiyin Wu
Ye Wu
Yicheng Wu
Yifan Wu
Tobias Wuerfl
Pengcheng Xi
James Xia
Siyu Xia
Wenfeng Xia
Yingda Xia
Yong Xia
Lei Xiang
Deqiang Xiao
Li Xiao
Yiming Xiao
Hongtao Xie
Lingxi Xie
Long Xie
Weidi Xie
Yiting Xie
Yutong Xie
Xiaohan Xing
Chang Xu
Chenchu Xu
Hongming Xu
Kele Xu
Min Xu
Rui Xu
Xiaowei Xu
Xuanang Xu
Yongchao Xu
Zhenghua Xu
Zhoubing Xu
Kai Xuan
Cheng Xue
Jie Xue
Wufeng Xue
Yuan Xue
Faridah Yahya
Ke Yan
Yuguang Yan
Zhennan Yan

Changchun Yang
Chao-Han Huck Yang
Dong Yang
Erkun Yang
Fan Yang
Ge Yang
Guang Yang
Guanyu Yang
Heran Yang
Hongxu Yang
Huijuan Yang
Jiancheng Yang
Jie Yang
Junlin Yang
Lin Yang
Peng Yang
Xin Yang
Yan Yang
Yujiu Yang
Dongren Yao
Jiawen Yao
Li Yao
Qingsong Yao
Chuyang Ye
Dong Hye Ye
Menglong Ye
Xujiong Ye
Jingru Yi
Jirong Yi
Xin Yi
Youngjin Yoo
Chenyu You
Haichao Yu
Hanchao Yu
Lequan Yu
Qi Yu
Yang Yu
Pengyu Yuan
Fatemeh Zabihollahy
Ghada Zamzmi
Marco Zenati
Guodong Zeng
Rui Zeng
Oliver Zettinig
Zhiwei Zhai
Chaoyi Zhang

Daoqiang Zhang
Fan Zhang
Guangming Zhang
Hang Zhang
Huahong Zhang
Jianpeng Zhang
Jiong Zhang
Jun Zhang
Lei Zhang
Lichi Zhang
Lin Zhang
Ling Zhang
Lu Zhang
Miaomiao Zhang
Ning Zhang
Qiang Zhang
Rongzhao Zhang
Ru-Yuan Zhang
Shihao Zhang
Shu Zhang
Tong Zhang
Wei Zhang
Weiwei Zhang
Wen Zhang
Wenlu Zhang
Xin Zhang
Ya Zhang
Yanbo Zhang
Yanfu Zhang
Yi Zhang
Yishuo Zhang
Yong Zhang
Yongqin Zhang
You Zhang
Youshan Zhang
Yu Zhang
Yue Zhang
Yueyi Zhang
Yulun Zhang
Yunyan Zhang
Yuyao Zhang
Can Zhao
Changchen Zhao
Chongyue Zhao
Fenqiang Zhao
Gangming Zhao

He Zhao
Jun Zhao
Li Zhao
Qingyu Zhao
Rongchang Zhao
Shen Zhao
Shijie Zhao
Tengda Zhao
Tianyi Zhao
Wei Zhao
Xuandong Zhao
Yiyuan Zhao
Yuan-Xing Zhao
Yue Zhao
Zixu Zhao
Ziyuan Zhao
Xingjian Zhen
Guoyan Zheng
Hao Zheng
Jiannan Zheng
Kang Zheng
Shenhai Zheng
Yalin Zheng
Yinqiang Zheng
Yushan Zheng
Jia-Xing Zhong
Zichun Zhong

Bo Zhou
Haoyin Zhou
Hong-Yu Zhou
Kang Zhou
Sanping Zhou
Sihang Zhou
Tao Zhou
Xiao-Yun Zhou
Yanning Zhou
Yuyin Zhou
Zongwei Zhou
Dongxiao Zhu
Hancan Zhu
Lei Zhu
Qikui Zhu
Xinliang Zhu
Yuemin Zhu
Zhe Zhu
Zhuotun Zhu
Aneeq Zia
Veronika Zimmer
David Zimmerer
Lilla Zöllei
Yukai Zou
Lianrui Zuo
Gerald Zwettler
Reyer Zwiggelaar

Outstanding Reviewers

Neel Dey   New York University, USA
Monica Hernandez   University of Zaragoza, Spain
Ivica Kopriva   Rudjer Boskovich Institute, Croatia
Sebastian Otálora   University of Applied Sciences and Arts Western
                         Switzerland, Switzerland
Danielle Pace   Massachusetts General Hospital, USA
Sérgio Pereira   Lunit Inc., South Korea
David Richmond   IBM Watson Health, USA
Rohit Singla   University of British Columbia, Canada
Yan Wang   Sichuan University, China

Honorable Mentions (Reviewers)

Mazdak Abulnaga	Massachusetts Institute of Technology, USA
Pierre Ambrosini	Erasmus University Medical Center, The Netherlands
Hyeon-Min Bae	Korea Advanced Institute of Science and Technology, South Korea
Mikhail Belyaev	Skolkovo Institute of Science and Technology, Russia
Bhushan Borotikar	Symbiosis International University, India
Katharina Breininger	Friedrich-Alexander-Universität Erlangen-Nürnberg, Germany
Ninon Burgos	CNRS, Paris Brain Institute, France
Mariano Cabezas	The University of Sydney, Australia
Aaron Carass	Johns Hopkins University, USA
Pierre-Henri Conze	IMT Atlantique, France
Christian Desrosiers	École de technologie supérieure, Canada
Reuben Dorent	King's College London, UK
Nicha Dvornek	Yale University, USA
Dmitry V. Dylov	Skolkovo Institute of Science and Technology, Russia
Marius Erdt	Fraunhofer Singapore, Singapore
Ruibin Feng	Stanford University, USA
Enzo Ferrante	CONICET/Universidad Nacional del Litoral, Argentina
Antonio Foncubierta-Rodríguez	IBM Research, Switzerland
Isabel Funke	National Center for Tumor Diseases Dresden, Germany
Adrian Galdran	University of Bournemouth, UK
Ben Glocker	Imperial College London, UK
Cristina González	Universidad de los Andes, Colombia
Maged Goubran	Sunnybrook Research Institute, Canada
Sobhan Goudarzi	Concordia University, Canada
Vicente Grau	University of Oxford, UK
Andreas Hauptmann	University of Oulu, Finland
Nico Hoffmann	Technische Universität Dresden, Germany
Sungmin Hong	Massachusetts General Hospital, Harvard Medical School, USA
Won-Dong Jang	Harvard University, USA
Zhanghexuan Ji	University at Buffalo, SUNY, USA
Neerav Karani	ETH Zurich, Switzerland
Alexander Katzmann	Siemens Healthineers, Germany
Erwan Kerrien	Inria, France
Anitha Priya Krishnan	Genentech, USA
Tahsin Kurc	Stony Brook University, USA
Francesco La Rosa	École polytechnique fédérale de Lausanne, Switzerland
Dmitrii Lachinov	Medical University of Vienna, Austria
Mengzhang Li	Peking University, China
Gilbert Lim	National University of Singapore, Singapore
Dongnan Liu	University of Sydney, Australia

Bin Lou	Siemens Healthineers, USA
Kai Ma	Tencent, China
Klaus H. Maier-Hein	German Cancer Research Center (DKFZ), Germany
Raphael Meier	University Hospital Bern, Switzerland
Tony C. W. Mok	Hong Kong University of Science and Technology, Hong Kong SAR
Lia Morra	Politecnico di Torino, Italy
Cosmas Mwikirize	Rutgers University, USA
Felipe Orihuela-Espina	Instituto Nacional de Astrofísica, Óptica y Electrónica, Mexico
Egor Panfilov	University of Oulu, Finland
Christian Payer	Graz University of Technology, Austria
Sebastian Pölsterl	Ludwig-Maximilians Universität, Germany
José Rouco	University of A Coruña, Spain
Daniel Rueckert	Imperial College London, UK
Julia Schnabel	King's College London, UK
Christina Schwarz-Gsaxner	Graz University of Technology, Austria
Boris Shirokikh	Skolkovo Institute of Science and Technology, Russia
Yang Song	University of New South Wales, Australia
Gérard Subsol	Université de Montpellier, France
Tanveer Syeda-Mahmood	IBM Research, USA
Mickael Tardy	Hera-MI, France
Paul Thienphrapa	Atlas5D, USA
Gijs van Tulder	Radboud University, The Netherlands
Tongxin Wang	Indiana University, USA
Yirui Wang	PAII Inc., USA
Jelmer Wolterink	University of Twente, The Netherlands
Lei Xiang	Subtle Medical Inc., USA
Fatemeh Zabihollahy	Johns Hopkins University, USA
Wei Zhang	University of Georgia, USA
Ya Zhang	Shanghai Jiao Tong University, China
Qingyu Zhao	Stanford University, China
Yushan Zheng	Beihang University, China

Mentorship Program (Mentors)

Shadi Albarqouni	Helmholtz AI, Helmholtz Center Munich, Germany
Hao Chen	Hong Kong University of Science and Technology, Hong Kong SAR
Nadim Daher	NVIDIA, France
Marleen de Bruijne	Erasmus MC/University of Copenhagen, The Netherlands
Qi Dou	The Chinese University of Hong Kong, Hong Kong SAR
Gabor Fichtinger	Queen's University, Canada
Jonny Hancox	NVIDIA, UK

Nobuhiko Hata	Harvard Medical School, USA
Sharon Xiaolei Huang	Pennsylvania State University, USA
Jana Hutter	King's College London, UK
Dakai Jin	PAII Inc., China
Samuel Kadoury	Polytechnique Montréal, Canada
Minjeong Kim	University of North Carolina at Greensboro, USA
Hans Lamecker	1000shapes GmbH, Germany
Andrea Lara	Galileo University, Guatemala
Ngan Le	University of Arkansas, USA
Baiying Lei	Shenzhen University, China
Karim Lekadir	Universitat de Barcelona, Spain
Marius George Linguraru	Children's National Health System/George Washington University, USA
Herve Lombaert	ETS Montreal, Canada
Marco Lorenzi	Inria, France
Le Lu	PAII Inc., China
Xiongbiao Luo	Xiamen University, China
Dzung Pham	Henry M. Jackson Foundation/Uniformed Services University/National Institutes of Health/Johns Hopkins University, USA
Josien Pluim	Eindhoven University of Technology/University Medical Center Utrecht, The Netherlands
Antonio Porras	University of Colorado Anschutz Medical Campus/Children's Hospital Colorado, USA
Islem Rekik	Istanbul Technical University, Turkey
Nicola Rieke	NVIDIA, Germany
Julia Schnabel	TU Munich/Helmholtz Center Munich, Germany, and King's College London, UK
Debdoot Sheet	Indian Institute of Technology Kharagpur, India
Pallavi Tiwari	Case Western Reserve University, USA
Jocelyne Troccaz	CNRS, TIMC, Grenoble Alpes University, France
Sandrine Voros	TIMC-IMAG, INSERM, France
Linwei Wang	Rochester Institute of Technology, USA
Yalin Wang	Arizona State University, USA
Zhong Xue	United Imaging Intelligence Co. Ltd, USA
Renee Yao	NVIDIA, USA
Mohammad Yaqub	Mohamed Bin Zayed University of Artificial Intelligence, United Arab Emirates, and University of Oxford, UK
S. Kevin Zhou	University of Science and Technology of China, China
Lilla Zollei	Massachusetts General Hospital, Harvard Medical School, USA
Maria A. Zuluaga	EURECOM, France

Contents – Part VI

Image Reconstruction

Two-Stage Self-supervised Cycle-Consistency Network for Reconstruction
of Thin-Slice MR Images.................................... 3
  Zhiyang Lu, Zheng Li, Jun Wang, Jun Shi, and Dinggang Shen

Over-and-Under Complete Convolutional RNN for MRI Reconstruction 13
  Pengfei Guo, Jeya Maria Jose Valanarasu, Puyang Wang,
  Jinyuan Zhou, Shanshan Jiang, and Vishal M. Patel

TarGAN: Target-Aware Generative Adversarial Networks
for Multi-modality Medical Image Translation 24
  Junxiao Chen, Jia Wei, and Rui Li

Synthesizing Multi-tracer PET Images for Alzheimer's Disease Patients
Using a 3D Unified Anatomy-Aware Cyclic Adversarial Network......... 34
  Bo Zhou, Rui Wang, Ming-Kai Chen, Adam P. Mecca, Ryan S. O'Dell,
  Christopher H. Van Dyck, Richard E. Carson, James S. Duncan,
  and Chi Liu

Generalised Super Resolution for Quantitative MRI Using Self-supervised
Mixture of Experts 44
  Hongxiang Lin, Yukun Zhou, Paddy J. Slator, and Daniel C. Alexander

TransCT: Dual-Path Transformer for Low Dose Computed Tomography 55
  Zhicheng Zhang, Lequan Yu, Xiaokun Liang, Wei Zhao, and Lei Xing

IREM: High-Resolution Magnetic Resonance Image Reconstruction via
Implicit Neural Representation 65
  Qing Wu, Yuwei Li, Lan Xu, Ruiming Feng, Hongjiang Wei, Qing Yang,
  Boliang Yu, Xiaozhao Liu, Jingyi Yu, and Yuyao Zhang

DA-VSR: Domain Adaptable Volumetric Super-Resolution
for Medical Images 75
  Cheng Peng, S. Kevin Zhou, and Rama Chellappa

Improving Generalizability in Limited-Angle CT Reconstruction
with Sinogram Extrapolation 86
  Ce Wang, Haimiao Zhang, Qian Li, Kun Shang, Yuanyuan Lyu,
  Bin Dong, and S. Kevin Zhou

Fast Magnetic Resonance Imaging on Regions of Interest: From Sensing
to Reconstruction   97
   Liyan Sun, Hongyu Huang, Xinghao Ding, Yue Huang, Xiaoqing Liu,
   and Yizhou Yu

InDuDoNet: An Interpretable Dual Domain Network for CT Metal Artifact
Reduction   107
   Hong Wang, Yuexiang Li, Haimiao Zhang, Jiawei Chen, Kai Ma,
   Deyu Meng, and Yefeng Zheng

Depth Estimation for Colonoscopy Images with Self-supervised Learning
from Videos   119
   Kai Cheng, Yiting Ma, Bin Sun, Yang Li, and Xuejin Chen

Joint Optimization of Hadamard Sensing and Reconstruction
in Compressed Sensing Fluorescence Microscopy   129
   Alan Q. Wang, Aaron K. LaViolette, Leo Moon, Chris Xu,
   and Mert R. Sabuncu

Multi-contrast MRI Super-Resolution via a Multi-stage
Integration Network.   140
   Chun-Mei Feng, Huazhu Fu, Shuhao Yuan, and Yong Xu

Generator Versus Segmentor: Pseudo-healthy Synthesis   150
   Yunlong Zhang, Chenxin Li, Xin Lin, Liyan Sun, Yihong Zhuang,
   Yue Huang, Xinghao Ding, Xiaoqing Liu, and Yizhou Yu

Real-Time Mapping of Tissue Properties for Magnetic
Resonance Fingerprinting   161
   Yilin Liu, Yong Chen, and Pew-Thian Yap

Estimation of High Framerate Digital Subtraction Angiography Sequences
at Low Radiation Dose   171
   Nazim Haouchine, Parikshit Juvekar, Xin Xiong, Jie Luo, Tina Kapur,
   Rose Du, Alexandra Golby, and Sarah Frisken

RLP-Net: A Recursive Light Propagation Network for 3-D
Virtual Refocusing   181
   Changyeop Shin, Hyun Ryu, Eun-Seo Cho, and Young-Gyu Yoon

Noise Mapping and Removal in Complex-Valued Multi-Channel MRI via
Optimal Shrinkage of Singular Values   191
   Khoi Minh Huynh, Wei-Tang Chang, Sang Hun Chung, Yong Chen,
   Yueh Lee, and Pew-Thian Yap

Self Context and Shape Prior for Sensorless Freehand 3D
Ultrasound Reconstruction   201
 Mingyuan Luo, Xin Yang, Xiaoqiong Huang, Yuhao Huang, Yuxin Zou,
 Xindi Hu, Nishant Ravikumar, Alejandro F. Frangi, and Dong Ni

Universal Undersampled MRI Reconstruction   211
 Xinwen Liu, Jing Wang, Feng Liu, and S. Kevin Zhou

A Neural Framework for Multi-variable Lesion Quantification Through
B-Mode Style Transfer   222
 SeokHwan Oh, Myeong-Gee Kim, Youngmin Kim, Hyuksool Kwon,
 and Hyeon-Min Bae

Temporal Feature Fusion with Sampling Pattern Optimization
for Multi-echo Gradient Echo Acquisition and Image Reconstruction.   232
 Jinwei Zhang, Hang Zhang, Chao Li, Pascal Spincemaille,
 Mert Sabuncu, Thanh D. Nguyen, and Yi Wang

Dual-Domain Adaptive-Scaling Non-local Network for CT Metal
Artifact Reduction.   243
 Tao Wang, Wenjun Xia, Yongqiang Huang, Huaiqiang Sun, Yan Liu,
 Hu Chen, Jiliu Zhou, and Yi Zhang

Towards Ultrafast MRI via Extreme *k*-Space Undersampling
and Superresolution.   254
 Aleksandr Belov, Joël Stadelmann, Sergey Kastryulin,
 and Dmitry V. Dylov

Adaptive Squeeze-and-Shrink Image Denoising for Improving Deep
Detection of Cerebral Microbleeds   265
 Hangfan Liu, Tanweer Rashid, Jeffrey Ware, Paul Jensen,
 Thomas Austin, Ilya Nasrallah, Robert Bryan, Susan Heckbert,
 and Mohamad Habes

3D Transformer-GAN for High-Quality PET Reconstruction   276
 Yanmei Luo, Yan Wang, Chen Zu, Bo Zhan, Xi Wu, Jiliu Zhou,
 Dinggang Shen, and Luping Zhou

Learnable Multi-scale Fourier Interpolation for Sparse View CT
Image Reconstruction   286
 Qiaoqiao Ding, Hui Ji, Hao Gao, and Xiaoqun Zhang

U-DuDoNet: Unpaired Dual-Domain Network for CT Metal
Artifact Reduction.   296
 Yuanyuan Lyu, Jiajun Fu, Cheng Peng, and S. Kevin Zhou

Task Transformer Network for Joint MRI Reconstruction
and Super-Resolution.   307
  Chun-Mei Feng, Yunlu Yan, Huazhu Fu, Li Chen, and Yong Xu

Conditional GAN with an Attention-Based Generator and a 3D
Discriminator for 3D Medical Image Generation   318
  Euijin Jung, Miguel Luna, and Sang Hyun Park

Multimodal MRI Acceleration via Deep Cascading Networks
with Peer-Layer-Wise Dense Connections   329
  Xiao-Xin Li, Zhijie Chen, Xin-Jie Lou, Junwei Yang, Yong Chen,
  and Dinggang Shen

Rician Noise Estimation for 3D Magnetic Resonance Images Based
on Benford's Law   340
  Rosa Maza-Quiroga, Karl Thurnhofer-Hemsi,
  Domingo López-Rodríguez, and Ezequiel López-Rubio

Deep J-Sense: Accelerated MRI Reconstruction via Unrolled
Alternating Optimization   350
  Marius Arvinte, Sriram Vishwanath, Ahmed H. Tewfik,
  and Jonathan I. Tamir

Label-Free Physics-Informed Image Sequence Reconstruction
with Disentangled Spatial-Temporal Modeling   361
  Xiajun Jiang, Ryan Missel, Maryam Toloubidokhti, Zhiyuan Li,
  Omar Gharbia, John L. Sapp, and Linwei Wang

High-Resolution Hierarchical Adversarial Learning for OCT Speckle
Noise Reduction   372
  Yi Zhou, Jiang Li, Meng Wang, Weifang Zhu, Yuanyuan Peng,
  Zhongyue Chen, Lianyu Wang, Tingting Wang, Chenpu Yao, Ting Wang,
  and Xinjian Chen

Self-supervised Learning for MRI Reconstruction with a Parallel
Network Training Framework   382
  Chen Hu, Cheng Li, Haifeng Wang, Qiegen Liu, Hairong Zheng,
  and Shanshan Wang

Acceleration by Deep-Learnt Sharing of Superfluous Information
in Multi-contrast MRI   392
  Sudhanya Chatterjee, Suresh Emmanuel Joel, Ramesh Venkatesan,
  and Dattesh Dayanand Shanbhag

Sequential Lung Nodule Synthesis Using Attribute-Guided Generative
Adversarial Networks   402
  Sungho Suh, Sojeong Cheon, Dong-Jin Chang, Deukhee Lee,
  and Yong Oh Lee

A Data-Driven Approach for High Frame Rate Synthetic Transmit Aperture
Ultrasound Imaging 412
  Yinran Chen, Jing Liu, Jianwen Luo, and Xiongbiao Luo

Interpretable Deep Learning for Multimodal Super-Resolution
of Medical Images 421
  Evaggelia Tsiligianni, Matina Zerva, Iman Marivani, Nikos Deligiannis,
  and Lisimachos Kondi

MRI Super-Resolution Through Generative Degradation Learning 430
  Yao Sui, Onur Afacan, Ali Gholipour, and Simon K. Warfield

Task-Oriented Low-Dose CT Image Denoising 441
  Jiajin Zhang, Hanqing Chao, Xuanang Xu, Chuang Niu, Ge Wang,
  and Pingkun Yan

Revisiting Contour-Driven and Knowledge-Based Deformable Models:
Application to 2D-3D Proximal Femur Reconstruction from X-ray Images . . . 451
  Christophe Chênes and Jérôme Schmid

Memory-Efficient Learning for High-Dimensional MRI Reconstruction 461
  Ke Wang, Michael Kellman, Christopher M. Sandino, Kevin Zhang,
  Shreyas S. Vasanawala, Jonathan I. Tamir, Stella X. Yu,
  and Michael Lustig

SA-GAN: Structure-Aware GAN for Organ-Preserving Synthetic
CT Generation 471
  Hajar Emami, Ming Dong, Siamak P. Nejad-Davarani,
  and Carri K. Glide-Hurst

Clinical Applications - Cardiac

Distortion Energy for Deep Learning-Based Volumetric Finite Element
Mesh Generation for Aortic Valves 485
  Daniel H. Pak, Minliang Liu, Theodore Kim, Liang Liang,
  Raymond McKay, Wei Sun, and James S. Duncan

Ultrasound Video Transformers for Cardiac Ejection Fraction Estimation 495
  Hadrien Reynaud, Athanasios Vlontzos, Benjamin Hou, Arian Beqiri,
  Paul Leeson, and Bernhard Kainz

EchoCP: An Echocardiography Dataset in Contrast Transthoracic
Echocardiography for Patent Foramen Ovale Diagnosis 506
  Tianchen Wang, Zhihe Li, Meiping Huang, Jian Zhuang, Shanshan Bi,
  Jiawei Zhang, Yiyu Shi, Hongwen Fei, and Xiaowei Xu

Transformer Network for Significant Stenosis Detection in CCTA
of Coronary Arteries 516
  Xinghua Ma, Gongning Luo, Wei Wang, and Kuanquan Wang

Training Automatic View Planner for Cardiac MR Imaging
via Self-supervision by Spatial Relationship Between Views. 526
  Dong Wei, Kai Ma, and Yefeng Zheng

Phase-Independent Latent Representation for Cardiac Shape Analysis 537
  Josquin Harrison, Marco Lorenzi, Benoit Legghe, Xavier Iriart,
  Hubert Cochet, and Maxime Sermesant

Cardiac Transmembrane Potential Imaging with GCN Based Iterative
Soft Threshold Network. 547
  Lide Mu and Huafeng Liu

AtrialGeneral: Domain Generalization for Left Atrial Segmentation
of Multi-center LGE MRIs. 557
  Lei Li, Veronika A. Zimmer, Julia A. Schnabel, and Xiahai Zhuang

TVnet: Automated Time-Resolved Tracking of the Tricuspid Valve
Plane in MRI Long-Axis Cine Images with a Dual-Stage Deep
Learning Pipeline 567
  Ricardo A. Gonzales, Jérôme Lamy, Felicia Seemann, Einar Heiberg,
  John A. Onofrey, and Dana C. Peters

Clinical Applications - Vascular

Deep Open Snake Tracker for Vessel Tracing. 579
  Li Chen, Wenjin Liu, Niranjan Balu, Mahmud Mossa-Basha,
  Thomas S. Hatsukami, Jenq-Neng Hwang, and Chun Yuan

MASC-Units:Training Oriented Filters for Segmenting
Curvilinear Structures 590
  Zewen Liu and Timothy Cootes

Vessel Width Estimation via Convolutional Regression 600
  Rui-Qi Li, Gui-Bin Bian, Xiao-Hu Zhou, Xiaoliang Xie, Zhen-Liang Ni,
  Yan-Jie Zhou, Yuhan Wang, and Zengguang Hou

Renal Cell Carcinoma Classification from Vascular Morphology. 611
  Rudan Xiao, Eric Debreuve, Damien Ambrosetti, and Xavier Descombes

Correction to: TVnet: Automated Time-Resolved Tracking of the Tricuspid
Valve Plane in MRI Long-Axis Cine Images with a Dual-Stage Deep
Learning Pipeline C1
  Ricardo A. Gonzales, Jérôme Lamy, Felicia Seemann, Einar Heiberg,
  John A. Onofrey, and Dana C. Peters

Author Index 623

Image Reconstruction

Two-Stage Self-supervised Cycle-Consistency Network for Reconstruction of Thin-Slice MR Images

Zhiyang Lu[1], Zheng Li[1], Jun Wang[1,2], Jun Shi[1,2(✉)], and Dinggang Shen[3,4(✉)]

[1] Key Laboratory of Specialty Fiber Optics and Optical Access Networks, Joint International Research Laboratory of Specialty Fiber Optics and Advanced Communication, School of Communication and Information Engineering, Shanghai University, Shanghai, China
junshi@shu.edu.cn

[2] Shanghai Institute for Advanced Communication and Data Science, School of Communication and Information Engineering, Shanghai University, Shanghai, China

[3] School of Biomedical Engineering, ShanghaiTech University, Shanghai, China
Dinggang.Shen@gmail.com

[4] Shanghai United Imaging Intelligence Co., Ltd., Shanghai 200230, China

Abstract. The thick-slice magnetic resonance (MR) images are often structurally blurred in coronal and sagittal views, which causes harm to diagnosis and image post-processing. Deep learning (DL) has shown great potential to reconstruct the high-resolution (HR) thin-slice MR images from those low-resolution (LR) cases, which we refer to as the slice interpolation task in this work. However, since it is generally difficult to sample abundant paired LR-HR MR images, the classical fully supervised DL-based models cannot be effectively trained to get robust performance. To this end, we propose a novel Two-stage Self-supervised Cycle-consistency Network (TSCNet) for MR slice interpolation, in which a two-stage self-supervised learning (SSL) strategy is developed for unsupervised DL network training. The paired LR-HR images are synthesized along the sagittal and coronal directions of input LR images for network pretraining in the first-stage SSL, and then a cyclic interpolation procedure based on triplet axial slices is designed in the second-stage SSL for further refinement. More training samples with rich contexts along all directions are exploited as guidance to guarantee the improved interpolation performance. Moreover, a new cycle-consistency constraint is proposed to supervise this cyclic procedure, which encourages the network to reconstruct more realistic HR images. The experimental results on a real MRI dataset indicate that TSCNet achieves superior performance over the conventional and other SSL-based algorithms, and obtains competitive qualitative and quantitative results compared with the fully supervised algorithm.

Keywords: Magnetic resonance imaging · Thin-slice magnetic resonance image · Slice interpolation · Self-supervised learning · Cycle-consistency constraint

© Springer Nature Switzerland AG 2021
M. de Bruijne et al. (Eds.): MICCAI 2021, LNCS 12906, pp. 3–12, 2021.
https://doi.org/10.1007/978-3-030-87231-1_1

1 Introduction

Magnetic resonance imaging (MRI) is widely used for diagnosis of various diseases. Due to the issues of scanning time and signal-noise ratio [1], the thick-slice magnetic resonance (MR) images are generally acquired in clinical routine, which have low between-slice resolution along the axial direction [2]. Since the MR images with high spatial resolution are desirable to provide detailed visual information and facilitate further image post-processing [3], it is important to reconstruct high-resolution (HR) thin-slice MR images from these low-resolution (LR) thick-slice cases, which we refer to as the slice interpolation task in this work.

Deep learning (DL) has gained great reputation for image super-resolution (SR) in recent years [4–6]. Some pioneering works have also applied DL to reconstruct thin-slice MR image [7–10], which typically build a convolutional neural network (CNN) based model to perform slice interpolation with paired LR and HR images. However, due to the difficulty of collecting HR MR images in clinical practice, there are generally few datasets with sufficient paired samples for model training. Consequently, the performance of these DL-based slice interpolation models degrades seriously under the classical full voxel-to-voxel supervision.

Self-supervised learning (SSL), a commonly used approach for unsupervised learning, is one plausible solution to the lack of HR images, which can generate paired data only based on input data and enables network training without ground truth [11]. SSL has been successfully applied to various image reconstruction tasks, including image SR and video frame interpolation [12–14]. Motivated by the SSL-based SR strategy on 2D images, Zhao et al. proposed to downsample the axial slices of input LR images to form the LR-HR slice pairs for training an SR network, which was then applied to improve slice resolutions along the other two directions [15–17]. However, this algorithm ignores the contextual information between the axial slices during interpolation, which causes anatomical distortion in reconstructed axial slices. On the other hand, although the SSL-based video frame interpolation methods can be directly applied to MR slice interpolation by taking the consecutive axial slices as video frames [14], they cannot exploit the information from the dense slices along other directions to enhance interpolation performance. Therefore, the existing SSL-based approaches still cannot achieve improved reconstruction for slice interpolation in MRI.

To this end, we propose a novel Two-stage Self-supervised Cycle-consistency Network (TSCNet), in which a new two-stage SSL strategy is developed to train an interpolation network in an unsupervised manner to perform MRI slice interpolation. The paired LR-HR images are synthesized from the sagittal and coronal slices of input LR images for preliminary network training in the first-stage SSL. Then, since the central slice of the input triplet axial slices can be estimated from the two intermediate slices interpolated between each two adjacent slices of the triplet, a new cycle-consistency constraint is developed to supervise this cyclic interpolation procedure in the second-stage SSL, which can further refine the network and boost the interpolation performance. The experimental results on a real MRI dataset indicate its qualitatively and quantitatively comparable performance with the fully supervised algorithm.

The main contributions of this work are two-fold as follows:

(1) A novel TSCNet is proposed to learn MRI slice interpolation with a two-stage SSL strategy only using LR images. The paired LR-HR images are synthesized based on sagittal and coronal directions for pretraining in the first-stage SSL, whereas the sparse axial slices can be exploited in the triplet manner as useful guidance for network refinement in the second-stage SSL. Training data with rich contexts about anatomical structures from all of the three directions can thus be exploited by the two-stage SSL to guarantee improved interpolation performance.

(2) A new cycle-consistency constraint strategy is developed to supervise the cyclic interpolation procedure based on the input triplet axial slices in the second-stage SSL to further refine the pretrained interpolation network. This strategy allows for SSL free from unwanted axial downsampling and can effectively encourage the network to interpolate more realistic intermediate slices required in HR images.

2   Method

Figure 1 illustrates the framework of the proposed TSCNet. An interpolation network trained by the strategy of two-stage SSL will interpolate the intermediate slices between the input two adjacent slices of thick-slice LR images. The general pipeline of the two-stage SSL is described as follows:

(1) In the first-stage SSL, the input LR image is downsampled into synthesized LR-HR data based on the dense slices along its sagittal and coronal directions. Sufficient paired samples can thus be generated for pretraining the interpolation network to learn the rich knowledge about anatomical structures

(2) In the second-stage SSL, the central slice of input triplet axial slices is estimated in a cyclic interpolation procedure for network refinement, which is supervised by the powerful cycle-consistency constraint to implicitly force the interpolated intermediate slices needed in reconstructed HR images to be more realistic.

(3) In the testing stage, the well-trained interpolation network interpolates the slices along the axial direction of input LR images to reconstruct HR thin-slice images.

2.1   Interpolation Network

TSCNet provides an effective two-stage SSL strategy for training the interpolation network with any structure. Specifically, the network structure used in this work is shown in Fig. 2. Two paralleled feature extraction modules are utilized to extract high-level features respectively from the input adjacent slices of LR images. The generated two sets of features that contain rich contextual information are combined by concatenation into joint feature representations, which are then refined through a slice reconstruction module to reconstruct the intermediate slice.

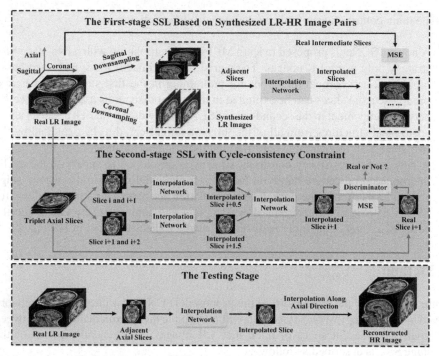

Fig. 1. Framework of the Two-stage Self-supervised Cycle-consistency Network. The slice $i +$ 0.5 denotes the interpolated slices between the i-th and $(i + 1)$-th axial slices of the LR image.

The residual dense network (RDN) [18] is utilized to construct each of the three modules in the interpolation network. Specifically, the RDN contains cascaded residual dense blocks (RDB), which is a powerful convolutional block that takes advantage of both residual and dense connections to fully aggregate hierarchical features, the cascaded RDB can thus provide sufficient capability of feature extraction and refinement, which enables the interpolation network to get improved reconstruction performance. More details about the structure of RDN can be referred to [18].

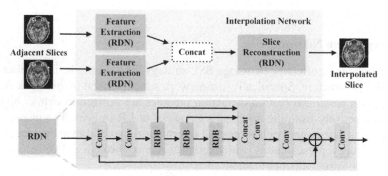

Fig. 2. Architecture of the interpolation network and the residual dense network (RDN).

2.2 The First-Stage SSL Based on Synthesized LR-HR Image Pairs

Due to the absence of real HR images, SSL requires to synthesize the paired LR-HR training data from the real LR images to train the interpolation network. One plausible method of SSL is to downsample the LR images along the axial directions. However, since the axial slices are originally sparse, this strategy can only get the paired data that have severe discrepancy between the input adjacent slices, which easily misguides the network for coarse interpolation. We thus synthesize LR-HR pairs by downsampling LR images on the dense slices along the sagittal and coronal directions for network pretraining in the first-stage SSL.

Given a real LR image $I \in \mathbb{R}^{X \times Y \times Z}$, we downsample the image by factor 2 along its sagittal and coronal directions, respectively, to get two synthesized LR images $I_{\downarrow sag} \in \mathbb{R}^{(X/2) \times Y \times Z}$ and $I_{\downarrow cor} \in \mathbb{R}^{X \times (Y/2) \times Z}$, and regard I as the HR version of these two images. Subsequently, the adjacent two sagittal slices in $I_{\downarrow sag}$ are extracted as the input data of training, and the corresponding real intermediate slices are derived from I as the ground truth, constructing a training dataset $\{S^i_{\downarrow sag}, S^{i+1}_{\downarrow sag}, S^{2i}_{sag}\}_{i=1}^{(X/2)-1}$, in which $S^i_{\downarrow sag}$ denotes the i-th sagittal slice in $I_{\downarrow sag}$ and S^{2i}_{sag} is the $2i$-th sagittal slice in I. With the same operation, we can also get $\{S^i_{\downarrow cor}, S^{i+1}_{\downarrow cor}, S^{2i}_{cor}\}_{i=1}^{(Y/2)-1}$ as training data based on the coronal slices of $I_{\downarrow cor}$ and I.

When two adjacent slices of the synthesized LR image are fed into the interpolation network, the missing intermediate slice can be generated. To train this model, we utilize the MSE loss function to enforce the pixel-wise consistency between the interpolated slice and the real intermediate slice, which can be formulated as:

$$l_{mse} = \frac{1}{X-2} \sum_{i=1}^{(X/2)-1} \| S^{2i}_{sag} - \mathcal{F}\left(S^i_{\downarrow sag}, S^{i+1}_{\downarrow sag}\right) \|^2 + \frac{1}{Y-2} \sum_{i=1}^{(Y/2)-1} \| S^{2i}_{cor} - \mathcal{F}\left(S^i_{\downarrow cor}, S^{i+1}_{\downarrow cor}\right) \|^2 \tag{1}$$

where $\mathcal{F}(\bullet)$ denotes the mapping function of the interpolation network.

2.3 The Second-Stage SSL with Cycle-Consistency Constraint

The contexts of coronal and sagittal slices in the original LR images have been exploited within the synthesized dataset to guide the training of interpolation network in the first-stage SSL. Thus, the pretrained interpolation network can already perform coarse slice interpolation. However, the contexts along the axial direction are still ignored in training, and the used MSE loss function easily introduces image smoothness and artifacts, resulting in the reconstruction of unrealistic axial slices. We thus design a cyclic interpolation procedure using triplet axial slices and develops a cycle-consistence constraint to supervise this procedure, which further refine the interpolation network to achieve more realistic interpolation results in the second-stage SSL.

Given a real LR image $I \in \mathbb{R}^{X \times Y \times Z}$, a set of consecutive triplet axial slices can be extracted from I as $\{S^i_{axi}, S^{i+1}_{axi}, S^{i+2}_{axi}\}_{i=1}^{Z-2}$, in which S^i_{axi} denotes its i-th axial slice. Then, the two sets of adjacent two slices in a triplet are fed into the interpolation network, respectively, to interpolate the intermediate slices, which can be formulated as:

$$\overline{S}^{i+0.5}_{axi} = \mathcal{F}\left(S^i_{axi}, S^{i+1}_{axi}\right) \tag{2}$$

$$\overline{S}_{axi}^{i+1.5} = \mathcal{F}\left(S_{axi}^{i+1}, S_{axi}^{i+2}\right) \tag{3}$$

in which $\overline{S}_{axi}^{i+0.5}$ denotes the interpolated slice between S_{axi}^{i} and S_{axi}^{i+1}, whereas $\overline{S}_{axi}^{i+1.5}$ is the interpolated slice between S_{axi}^{i+1} and S_{axi}^{i+2}.

Given $\overline{S}_{axi}^{i+0.5}$ and $\overline{S}_{axi}^{i+1.5}$, the central slice S_{axi}^{i+1} of the original input triplet can be reversely estimated through the interpolation network, which is formulated as:

$$\overline{S}_{axi}^{i+1} = \mathcal{F}\left(\overline{S}_{axi}^{i+0.5}, \overline{S}_{axi}^{i+1.5}\right) \tag{4}$$

We can thus adopt S_{axi}^{i+1} as the ground truth to supervise this cyclic procedure through a cycle-consistency constraint, which implicitly enforces the model to reconstruct realistic $\overline{S}_{axi}^{i+0.5}$ and $\overline{S}_{axi}^{i+1.5}$. Specifically, in addition to the MSE loss function, we add an adversarial loss to constrain the similarity between S_{axi}^{i+1} and $\overline{S}_{axi}^{i+1}$, which aims to overcome the issue of image smoothness in reconstructed slices.

To this end, we adopt the discriminator in SRGAN to distinguish between the estimated $\overline{S}_{axi}^{i+1}$ and the real S_{axi}^{i+1}, whose objective function is defined as follows [19]:

$$l_D = \frac{1}{Z-2}\sum_{i=1}^{Z-2}\left[-logD\left(S_{axi}^{i+1}\right) - log\left(1 - \overline{S}_{axi}^{i+1}\right)\right] \tag{5}$$

where $D(\bullet)$ denotes the function of the discriminator.

Meanwhile, the cycle-consistency constraint is formulated as:

$$l_{cyc} = \frac{1}{2Z-4}\sum_{i=1}^{Z-2}\|S_{axi}^{i+1} - \overline{S}_{axi}^{i+1}\|^2 - \frac{\lambda}{Z-2}\sum_{i=1}^{Z-2}\left[log\left(\overline{S}_{axi}^{i+1}\right)\right] \tag{6}$$

where λ is the parameter to balance the MSE and adversarial loss.

Finally, we utilize l_{mse} together with l_{cyc} to guarantee the stability of training, and the total loss of the second-stage SSL for network refinement is defined as:

$$L = l_{mse} + l_{cyc} \tag{7}$$

3   Experiments

3.1   Dataset

The proposed TSCNet algorithm was evaluated on 64 T1 MR brain images selected from the publicly available Alzheimer's Disease Neuroimaging Initiative (ADNI) dataset [20]. All of the MR images were sampled at $1 \times 1 \times 1$ mm^3 and zero-padded to the voxel size of $256 \times 256 \times 256$ as the real HR thin-slice images. Then we downsampled the isotropic volumes by factors of 2 along the axial direction to generate the LR thick-slice images with the voxel size of $256 \times 256 \times 128$.

3.2 Experimental Design

To evaluate the performance of our proposed TSCNet algorithm, we compared it with the following algorithms for slice interpolation:

1) Trilinear Interpolation: A conventional algorithm to interpolate 3D images.
2) EDSSR [15]: An SSL-based DL algorithm for MR slice interpolation. It generated paired LR-HR axial slices from LR images to train an SR network, which was then used to perform SR on coronal and sagittal slices to reconstruct HR images with the technique of Fourier Burst Accumulation. It is worth mentioning that a data augmentation operation was applied on EDSSR to form the SMORE (3D) algorithm proposed in the more recent paper [17]. However, we did not conduct any similar operations in our experiments.
3) Full-supervised Interpolation Network (FSIN): A fully supervised algorithm based on the interpolation network, which used the adjacent axial slices of real LR images as input and the corresponding intermediate slices in real HR images as ground truth, and is trained with the common MSE loss function.

An ablation experiment was also performed to compare our proposed TSCNet algorithm with the following two variants of TSCNet:

1) TSCNet with Adapted First-stage SSL (TSCNet with AFS): This algorithm modified the first-stage SSL based on the common methods for video frame interpolation to evaluate the effectiveness of this stage, which downsampled real LR images only along the axial direction to synthesize paired LR-HR training samples.
2) TSCNet without Second-stage SSL (TSCNet w/o SS): This algorithm only adopted the first-stage SSL to train the interpolation network, and removed the second-stage SSL to validate the effectiveness of proposed cycle-consistency constraint.

We performed the 4-fold cross-validation to evaluate the performance of different algorithms. The commonly used peak signal-to-noise ratio (PSNR) and structural similarity index (SSIM) were adopted as the evaluation indices. The results of all folds were reported with the format of the mean $\pm$ SD (standard deviation).

3.3 Implementation Details

In our implementations, all real images were normalized to voxel values between 0 and 1 as the network inputs. Each module in TSCNet removed the upsampling operation in original RDN and has three RDB with five layers. For training TSCNet, the first-stage strategy was conduct for 100 epochs, and the discriminator and the interpolation network were optimized alternately for another 150 epochs in the second-stage, and the parameter λ in l_{cyc} was set to 0.1. The batch size was set to 6 for the two adjacent coronal or sagittal slices, and 3 for triplet axial slices. All of the DL-based algorithms were trained by an Adam optimizer with a learning rate of 0.0001.

3.4 Experimental Results

Figure 3 shows the interpolated slices along the axial directions in the HR thin-slice images reconstructed by different slice interpolation algorithms. It can be found that our proposed TSCNet can reconstruct more realistic slices with reference to the ground truth compared with the conventional and existing SSL-based algorithms. Trilinear interpolation generally reconstructs blurred results, and EDSSR generates inaccurate tissue structures. TSCNet also achieves superior performance over its two variants with clearer tissue boundaries and less artifacts in ablation study, which validates the effectiveness of each stage in the two-stage SSL.

Moreover, TSCNet can get visually comparable performance with the fully-supervised FSIN, and even better results in the third and fourth rows in Fig. 3, which demonstrates its robust capability for interpolating realistic slices.

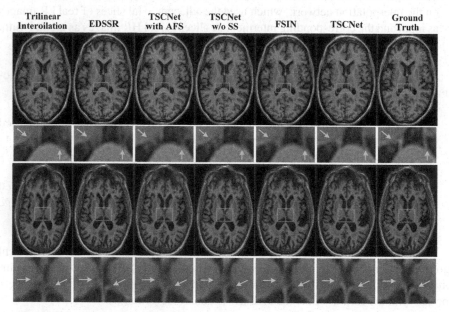

Fig. 3. Visual comparison of different slice interpolation algorithms on ADNI dataset.

Table 1 gives the qualitative results of experiments, which indicate that TSCNet outperforms the conventional trilinear interpolation and the SSL-based EDSSR with the PSNR of 36.14 ± 0.43 dB and SSIM of 0.9419 ± 0.0040. Specifically, TSCNet improves by 0.7 dB and 0.0075 on the two indices compared with EDSSR, which demonstrates that more contextual information related to the missing intermediate slices can be learned through the proposed two-stage SSL strategy for superior performance. TSCNet also gets an improvement by at least 0.22 dB and 0.0025 on PSNR and SSIM over its two variants in ablation study, which validates the effectiveness of each stage in proposed SSL strategy. It can be concluded that more training samples can be obtained in TSCNet to enhance the robustness of interpolation network compared with TSCNet with AFS, whereas the cycle-consistency constraint utilized in the second-stage of TSCNet can benefit to reconstruct more realistic HR images.

It is worth mentioning that TSCNet achieves close quantitative performance to the fully-supervised algorithm. Besides, it takes advantage of SSL to avoid the requirement for the expensive HR ground truth, and can even reconstruct visually better HR images as shown in the qualitative results.

Table 1. Quantitative results of different algorithms for slice interpolation.

Algorithm	Category	PSNR (dB)	SSIM
Trilinear interpolation	Convention	34.03 ± 0.37	0.9139 ± 0.0037
EDSSR	SSL	35.44 ± 0.18	0.9344 ± 0.0012
TSCNet with AFS	SSL	35.86 ± 0.43	0.9364 ± 0.0036
TSCNet w/o SS	SSL	35.92 ± 0.44	0.9394 ± 0.0043
TSCNet (ours)	SSL	36.14 ± 0.43	0.9419 ± 0.0040
FSIN	Full supervision	36.53 ± 0.39	0.9447 ± 0.0039

4 Conclusion

In summary, a novel TSCNet algorithm is proposed to address the slice interpolation problem for reconstruction of HR thin-slice MR images. It develops an SSL-based strategy with a novel cycle-consistency constraint to exploit contextual information along all directions for training the interpolation network only based on LR thick-slice input, which enables improved reconstruction free from the guidance of the clinically rare HR ground truth. The experimental results validate the effectiveness of TSCNet by demonstrating its superior performance over the traditional and other SSL-based algorithms, and comparable performance with fully supervised algorithms.

In future work, TSCNet will be combined with the fully supervised algorithm to further enhance the reconstruction performance. The novel structure of the interpolation network will also be explored to boost the interpolation performance in TSCNet.

Acknowledgements. This work is supported by National Natural Science Foundation of China (81830058) and the 111 Project (D20031).

References

1. Plenge, E., Poot, D.H., Bernsen, M., et al.: Super-resolution methods in MRI: can they improve the trade-off between resolution, signal-to-noise ratio, and acquisition time? Magn. Reson. Med. **68**(6), 1983–1993 (2012)

2. Jia, Y., Gholipour, A., He, Z., et al.: A new sparse representation framework for reconstruction of an isotropic high spatial resolution MR volume from orthogonal anisotropic resolution scans. IEEE Trans. Med. Imag. **36**(5), 1182–1193 (2017)
3. Van Reeth, E., Tham, I.W., Tan, C.H., et al.: Super-resolution in magnetic resonance imaging: a review. Concepts Magn. Reson. Part A **40**(6), 306–325 (2012)
4. Yang, W., Zhang, X., Tian, Y., et al.: Deep learning for single image super-resolution: a brief review. IEEE Trans. Multimedia **21**(12), 3106–3121 (2019)
5. Shi, J., Li, Z., Ying, S., et al.: MR image super-resolution via wide residual networks with fixed skip connection. IEEE J. Biomed. Health inform. **23**(3), 1129–1140 (2018)
6. Li, Z., et al.: A two-stage multi-loss super-resolution network for arterial spin labeling magnetic resonance imaging. In: Shen, D., et al. (eds.) MICCAI 2019. LNCS, vol. 11766, pp. 12–20. Springer, Cham (2019). https://doi.org/10.1007/978-3-030-32248-9_2
7. Li, Z., Wang, Y., Yu, J.: Reconstruction of thin-slice medical images using generative adversarial network. In: Wang, Q., Shi, Y., Suk, H.-I., Suzuki, K. (eds.) MLMI 2017. LNCS, vol. 10541, pp. 325–333. Springer, Cham (2017). https://doi.org/10.1007/978-3-319-67389-9_38
8. Chen, Y., Xie, Y., Zhou, Z., et al.: Brain MRI super resolution using 3D deep densely connected neural networks. In: ISBI, pp. 739–742 (2018)
9. Du, J., He, Z., Wang, L., et al.: Super-resolution reconstruction of single anisotropic 3D MR images using residual convolutional neural network. Neurocomputing **392**, 209–220 (2020)
10. Peng, C., Lin, W.A., Liao, H., et al.: SAINT: spatially aware interpolation network for medical slice synthesis. In: CVPR, pp. 7750–7759 (2020)
11. Jing, L., Tian, Y.: Self-supervised visual feature learning with deep neural networks: a survey. IEEE Trans. Pattern Anal. Mach. Intell. (2020)
12. Huang, J.B., Singh, A., Ahuja, N.: Single image super-resolution from transformed self-exemplars. In: CVPR, pp. 5197–5206 (2015)
13. Yuan, Y., Liu, S., Zhang, J., et al.: Unsupervised image super-resolution using cycle-in-cycle generative adversarial networks. In: CVPR, pp. 701–710 (2018)
14. Reda, F.A., Sun, D., Dundar, A., et al.: Unsupervised video interpolation using cycle consistency. In: ICCV, pp. 892–900 (2019)
15. Zhao, C., Carass, A., Dewey, B.E., et al.: Self super-resolution for magnetic resonance images using deep networks. In: ISBI, pp. 365–368 (2018)
16. Zhao, C., et al.: A deep learning based anti-aliasing self super-resolution algorithm for MRI. In: Frangi, A.F., Schnabel, J.A., Davatzikos, C., Alberola-López, C., Fichtinger, G. (eds.) MICCAI 2018. LNCS, vol. 11070, pp. 100–108. Springer, Cham (2018). https://doi.org/10.1007/978-3-030-00928-1_12
17. Zhao, C., Dewey, B.E., Pham, D.L., et al.: SMORE: a self-supervised anti-aliasing and super-resolution algorithm for MRI using deep learning. IEEE Trans. Med. Imag. **40**(3), 805–817 (2020)
18. Zhang, Y., Tian, Y., Kong, Y., et al.: Residual dense network for image super-resolution. In: CVPR, pp. 2472–2481 (2018)
19. Ledig, C., Theis, L., Huszár, F., et al.: Photo-realistic single image super-resolution using a generative adversarial network. In: CVPR, pp. 4681–4690 (2017)
20. Jack, Jr., C.R., Bernstein, M.A., Fox, N.C., et al.: The Alzheimer's disease neuroimaging initiative (ADNI): MRI methods. J. Magn. Reson. Imag. **27**(4), 685–691 (2008)

Over-and-Under Complete Convolutional RNN for MRI Reconstruction

Pengfei Guo[1]([✉]), Jeya Maria Jose Valanarasu[2], Puyang Wang[2],
Jinyuan Zhou[3], Shanshan Jiang[3], and Vishal M. Patel[1,2]

[1] Department of Computer Science, Johns Hopkins University, Baltimore, MD, USA
pguo4@jhu.edu
[2] Department of Electrical and Computer Engineering, Johns Hopkins University,
Baltimore, MD, USA
[3] Department of Radiology, Johns Hopkins University, Baltimore, MD, USA

Abstract. Reconstructing magnetic resonance (MR) images from under-sampled data is a challenging problem due to various artifacts introduced by the under-sampling operation. Recent deep learning-based methods for MR image reconstruction usually leverage a generic auto-encoder architecture which captures low-level features at the initial layers and high-level features at the deeper layers. Such networks focus much on global features which may not be optimal to reconstruct the fully-sampled image. In this paper, we propose an **O**ver-and-**U**nder **C**omplete Convolutional **R**ecurrent Neural Network (OUCR), which consists of an overcomplete and an undercomplete Convolutional Recurrent Neural Network (CRNN). The overcomplete branch gives special attention in learning local structures by restraining the receptive field of the network. Combining it with the undercomplete branch leads to a network which focuses more on low-level features without losing out on the global structures. Extensive experiments on two datasets demonstrate that the proposed method achieves significant improvements over the compressed sensing and popular deep learning-based methods with less number of trainable parameters.

Keywords: Convolutional RNN · MRI reconstruction · Deep learning

1 Introduction

Magnetic resonance imaging (MRI) is a noninvasive medical imaging approach that provides various tissue contrast mechanisms for visualizing anatomical structures and functions. Due to the hardware constraint, one major limitation of MRI is relatively slow data acquisition process, which subsequently causes higher imaging cost and patients' discomfort in many clinical applications [3].

Electronic supplementary material The online version of this chapter (https://doi.org/10.1007/978-3-030-87231-1_2) contains supplementary material, which is available to authorized users.

© Springer Nature Switzerland AG 2021
M. de Bruijne et al. (Eds.): MICCAI 2021, LNCS 12906, pp. 13–23, 2021.
https://doi.org/10.1007/978-3-030-87231-1_2

While the exploitation of advanced hardware and parallel imaging [20] can mitigate such issue, a common approach is to shorten the image acquisition time by under-sampling k-space (also known as Compressed Sensing (CS)) [9,19]. However, reconstructing an image directly from partial k-space data results in a suboptimal image with aliasing artifacts. To deal with this issue, nonlinear recovery algorithms based on ℓ_0, ℓ_1 or total variation minimization are often used to recover the image from incomplete k-space data. More advanced CS-based image reconstruction algorithms have been combined with parallel imaging [15], low-rank constraint terms [16], and dictionary learning [18]. Unfortunately, though CS image reconstruction algorithms are able to recover images, they lack noise-like textures and as a result many physicians find CS reconstructed images as "artificial". Moreover, when large errors are not reduced during optimization, high-frequency oscillatory artifacts cannot be properly removed [23]. Thus, the acceleration factors of CS-based algorithms are generally limited between 2.5 and 3 for typical MR images [23].

Recent advances in deep neural networks open a new possibility to solve the inverse problem of MR image reconstruction in an efficient manner [8]. Artificial neural network-based image reconstruction methods have been shown to provide much better MR image quality than conventional CS-based methods [1,2,4,7, 12,14,21,25,31]. Most DL-based methods are convolutional which consist of a set of convolution and down/up sampling layers for efficient learning. The main intuition behind this kind of network architecture is that at the initial layers, the receptive field of the filters is smaller, so low-level features (e.g., edges) are captured. At deeper layers, the receptive field of the filters is larger, so high-level features (e.g., the interpretation of input) are captured. Using such generic architecture for MR image reconstruction might not be optimal, since low-level vision tasks are mainly concerned with extracting descriptions from input rather than the interpretation of input [5]. Prior to DL era, overcomplete representations were explored for dealing with noisy observations in the vision and image processing tasks [13,30]. In an overcomplete architecture, the increase of receptive field is restricted through the network, which forces the filters to focus on low-level features [32]. Recently, overcomplete representations have been explored for image segmentation [29] and image restoration [32].

In this paper, a novel **O**ver-and-**U**nder **C**omplete Convolutional **R**ecurrent Neural Network (OUCR) is proposed that can recover a fully-sampled image from the under-sampled k-space data for accelerated MR image reconstruction. To summarize, the following are our key contributions: **1.** An over-complete convolutional RNN architecture is explored for MR image reconstruction. **2.** To recover finer details better, the proposed OUCR consists of two branches that can leverage the features of both undercomplete and overcomplete CRNN. **3.** Extensive experiments are conducted on two datasets and it is demonstrated that the proposed method achieves significant improvements over CS-based as well as popular DL-based methods and is more parameter efficient.

2   Methodology

Overcomplete Networks. Overcomplete representations were first explored for signal representation where overcomplete bases were used such that the number of basis functions are more than the input signal samples [13]. This enabled a high flexibility at expressing the structure of the data. In neural networks, overcomplete fully connected networks were observed to be better feature extractors and hence perform well at denoising [30]. Recently, overcomplete convolutional networks are noted to be better at extracting local features because of restraining the receptive field when compared to the generic encoder-decoder networks [27–29,32]. Rather than using max-pooling layers at the encoder like an undercomplete convolutional network, in an overcomplete convolutional network, one uses upsampling layers. Figure 1(a) and (b) visually explains this concept in CNNs. As can be seen from Fig. 1(c) and (d), the overcomplete networks focus on fine structures and learning local features from an image as the receptive field is constrained even at deeper layers. More examples of comparison between over and under complete networks can be found in the supplementary material.

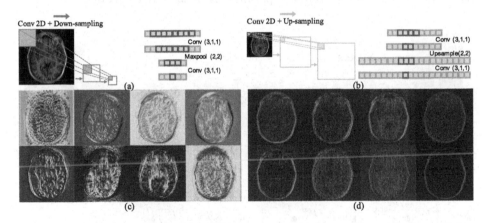

Fig. 1. *Top row*: Explanation for the receptive field change in (a) undercomplete and (b) overcomplete networks. Conv (3, 1, 1) represents a convolution layer with kernel size as 3, stride as 1 and padding as 1. Upsample (2, 2) represents a nearest neighbor upsampling layer with upsampling coefficients as (2, 2). Maxpool (2, 2) represents maxpooling layer with kernel size as 2 and stride as 2. The red pixels denote receptive field. It can be noted that the receptive field is constrained in overcomplete network compared to undercomplete network. *Bottom row*: Visualization of filter responses (feature maps from ResBlock) for (c) undercomplete and (d) overcomplete CRNN. By restricting the size of receptive field, the overcomplete CRNN is able to focus on low-level features. (Color figure online)

MR Image Reconstruction. Let $x \in \mathbb{C}^N$ denote the observed under-sampled k-space data, $y \in \mathbb{C}^M$ is the fully-sampled image that we want to reconstruct. To obtain a regularized solution, the optimization problem of MRI reconstruction can be formulated as follows [22,25]:

$$\min_{y} R(y) + \lambda \|x - F_D y\|_2^2. \tag{1}$$

Here, $R(y)$ is the regularization term and λ controls the contribution of second term. F_D represents the undersampling Fourier encoding matrix that is defined as the multiplication of the Fourier transform matrix with a binary undersampling mask D. The ratio of the amount of k-space data required for a fully-sampled image to the amount collected in an accelerated acquisition is controlled by the acceleration factor (AF). The approximate fully-sampled image $\bar{y}$ can be measured from the observed under-sampled k-space data x via an optimization process. One can solve the objective function (Eq. 1) based on iterative optimization methods, such as gradient descent. A CRNN [22,31] is capable of modeling the iterative optimization process in Eq. 1 as $\bar{y} = \text{CRNN}(\bar{x}, x, D, \Theta)$, where $\bar{y}$ is the reconstructed MR image from CRNN model, $\bar{x}$ is the zero-filled image and Θ denotes the trainable parameters of the CRNN model.

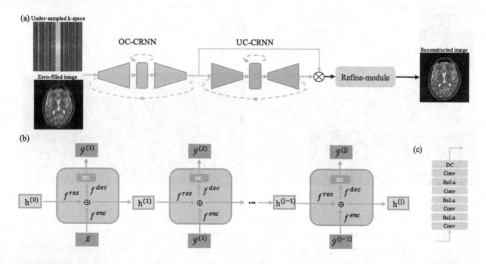

Fig. 2. (a) An overview of the proposed OUCR. Here, $\otimes$ denotes the channel-wise concatenation. We denote overcomplete and undercomplete CRNN as OC-CRNN and UC-CRNN, respectively. (b) A schematic of unrolled CRNN iterations. Here, $\oplus$ denotes the element-wise addition. (c) The network configuration of the refine-module. DC represents the data consistency layer.

OUCR. To have special attention in learning low-level feature structures while not losing out on the global structures and inspired by previous CRNN methods [22,31], we propose a novel OUCR network as shown in Fig. 2. OUCR consists of two CRNN modules with different receptive fields to reconstruct MR images (i.e. OC-CRNN and UC-CRNN in Fig. 2(a)) and an refine-module (RM). In a CRNN module f_i, let f_i^{enc}, f_i^{dec}, and f_i^{res} denote the encoder, decoder, and ResBlock, respectively (Fig. 2(b)). A data consistency (DC) layer is added at the end of each module to reinforce the data consistency in k-space. The iterations of a CRNN module can be unrolled as follows:

$$
\begin{aligned}
\bar{y}_i^{(j+1)} &= \text{DC}(f_i(\bar{y}_i^{(j)}, h_i^{(j)}), x, D), \\
&= F^{-1}[Dx + (1 - D)F[f_i(\bar{y}_i^{(j)}, h_i^{(j)})]], \\
&= F^{-1}[Dx + (1 - D)F[f_i^{\text{dec}}(f_i^{\text{res}}(h_i^{(j)}) + f_i^{\text{enc}}(\bar{y}_i^{(j)}))]],
\end{aligned}
\tag{2}
$$

where h_i^j is the hidden state after iteration j and F^{-1} denotes the inverse Fourier transform. After J iterations of the two CRNN modules, the final reconstructed MR image $\bar{y}$ is formulated as follows:

$$
\begin{aligned}
\bar{y}_{\text{oc}} &= \text{OC-CRNN}(\bar{x}, x, D, \Theta_{\text{oc}}), \\
\bar{y}_{\text{uc}} &= \text{UC-CRNN}(\bar{y}_{\text{oc}}, x, D, \Theta_{\text{uc}}), \\
\bar{y} &= \text{DC}(\text{RM}(\bar{y}_{\text{oc}} \otimes \bar{y}_{\text{uc}}, \Theta_{\text{rm}}), x, D),
\end{aligned}
\tag{3}
$$

where $\otimes$ denotes the channel-wise concatenation and $\Theta_{\text{oc}}, \Theta_{\text{uc}}, \Theta_{\text{rm}}$ denote the parameters of the overcomplete, undercomplete CRNN and RM network, respectively. The intuition behind using both OC-CRNN and UC-CRNN is to make use of both local and global features. While we focus more on the local features using the OC-CRNN, the global features are not neglected altogether as they still have meaningful information for proper reconstruction. In each CRNN module, we have two convolutional blocks in both encoder and decoder. Each convolutional block in the encoder has a 2D convolutional layer followed by an upsampling layer in OC-CRNN or a max-pooling layer in UC-CRNN. In the decoder, each convolutional block has a 2D convolutional layer followed by max-pooling layer in OC-CRNN or upsampling layer in UC-CRNN. More details regarding the network configuration can be found in the supplementary material.

3 Experiments and Results

Evaluation and Implementation Details. The following two datasets are used for conducting experiments – **fastMRI** [11] and **HPKS** [6,10]. The fastMRI dataset consists of single-coil coronal proton density-weighted knee images corresponding to 1172 subjects. In particular, 973 subjects' data is used for training, and 199 subjects' data (fastMRI validation dataset) is used for testing. For each subject, there are approximately 35 knee images that contain tissues. The **HPKS** dataset is collected by an anonymous medical center from post-treatment patients with malignant glioma. T_1-weighted images from 144 subjects are used, where 102 subjects' data are used for training, 14 subjects' data set are used for validation, and 28 subjects' data are used for testing. For each subject, 15 axial cross-sectional images that contain brain tissues are provided in this dataset. We simulated k-space measurements using the same

sampling mask function as the fastMRI challenge [11] with 4× and 8× accelerations. All models were trained using the ℓ_1 loss with Adam optimizer by the
following hyperparameters: initial learning rate of 1.5×10^{-4} then reduced by a
factor of 0.9 every 5 epochs; 50 maximum epochs; batch size of 4; the number
of CRNN iteration J of 5. SSIM and PSNR are used as the evaluation metrics
for comparison.

MR Image Reconstruction Results. Table 1 shows the results correspond-
ing to seven different methods evaluated on the HPKS and fastMRI datasets.
The performance of the proposed model was compared with compressed sensing
(CS) [26], UNet [24], KIKI-Net [4], Kiu-net [29], D5C5 [25], and PC-RNN [31].
For a fair comparison, UNet [24] and Kiu-net [29] are modified for data with
real and imaginary channels and a DC layer is added at the end of the networks.
KIKI-Net [4] which conducts interleaved convolution operation on image and
k-space domains achieves better performance than UNet [24] on HPKS. Kiu-
net [29] which is an overcomplete variant architecture of UNet [24] outperforms
UNet [24] and KIKI-Net [4]. By leveraging the cascade of convolutional neural
networks, D5C5 [25] outperforms the Kiu-net [29] in both HPKS and fastMRI
datasets. PC-RNN [31] which learns the mapping in an iterative way by CRNN
from three different scales achieves the second best performance. As it can be
seen from Table 1, the proposed OUCR outperforms other methods by leveraging the overcomplete architecture. Figure 4 shows the qualitative results of two
datasets with 4× and 8× accelerations. It can be observed that the proposed
OUCR yields reconstructed images with remarkable visual similarity to the reference images compared to the others (see the last column of each sub-figure
in Fig. 4) in two datasets with different modalities. The reported improvements
achieved by OUCR are statistically significant ($p < 10^{-5}$). The computational
details of different methods and statistical significance test are provided in Supplementary Table 1 and 2, respectively.

Table 1. Quantitative results on the HPKS and fastMRI dataset. Param denotes the
number of parameters.

Method	Param	HPKS				fastMRI			
		PSNR		SSIM		PSNR		SSIM	
		4×	8×	4×	8×	4×	8×	4×	8×
CS	–	29.94	24.96	0.8705	0.7125	29.54	26.99	0.5736	0.4870
UNet	8.634 M	34.47	29.47	0.9155	0.8249	31.88	29.78	0.7142	0.6424
KIKI-Net	1.790 M	35.35	29.86	0.9363	0.8436	31.87	29.27	0.7172	0.6355
Kiu-Net	7.951 M	35.35	30.18	0.9335	0.8467	32.06	29.86	0.7228	0.6456
D5C5	2.237 M	37.51	30.40	0.9595	0.8623	32.25	29.65	0.7256	0.6457
PC-RNN	1.482 M	38.36	31.54	0.9696	0.8965	32.37	30.17	0.7281	0.6585
OUCR	**1.192 M**	**39.33**	**32.14**	**0.9747**	**0.9044**	**32.61**	**30.59**	**0.7354**	**0.6634**

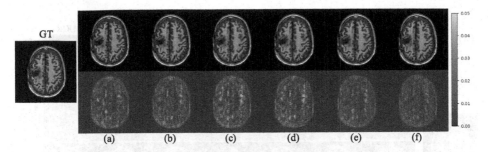

Fig. 3. Qualitative results and error maps of ablation study on HPKS with AF = 4. (a) UC-CRNN. (b) UC-CRNN + RM. (c) OC-CRNN. (d) OC-CRNN + RM. (e) UC-CRNN + OC-CRNN. (f) UC-CRNN + OC-CRNN + RM (proposed OUCR).

In the k-space domain, the center frequencies determine the overall image contrast, brightness, and general shape. The peripheral area of k-space contains high spatial frequency information that controls edges, details, sharp transitions [17]. To further analyze the performance of different methods in low and high spatial frequency, we carry out the k-space analysis in Fig. 5. We reconstruct an MR image from partially masked k-space and compare it with the reference image that is applied same mask on k-space, as shown in Fig. 5 top row. The reported PSNR and SSIM are presented by Boxplot in Fig. 5 bottom row. It can be seen that the proposed OUCR exhibits better reconstruction performance than other methods for both low and high frequency information.

Ablation Study. We conduct a detailed ablation study to separately evaluate the effectiveness of using OC-CRNN, UC-CRNN, and RM in the proposed framework. The results are shown in Table 2. We start with only using OC-CRNN and UC-CRNN. It can be noted that the performance of OC-CRNN is lesser than UC-CRNN, since even though OC-CRNN captures the low-level features properly it does not capture most high-level features like UC-CRNN. Then, we show that adding RM with each individual module can improve the reconstruction quality. Finally, combing both CRNN networks with RM (proposed OUCR) results in the best performance. Figure 3 illustrates the qualitative improvements after adding each major block, which is consistent with the results reported in Table 2. Moreover, we observe that increasing the number of CRNN iterations can further improve the performance of the proposed OUCR, but consequently leads to lower computational efficiency. Due to space constraint, an ablation study regarding CRNN iterations, k-space analysis on fastMRI dataset, and more visualizations are provided in the supplementary material.

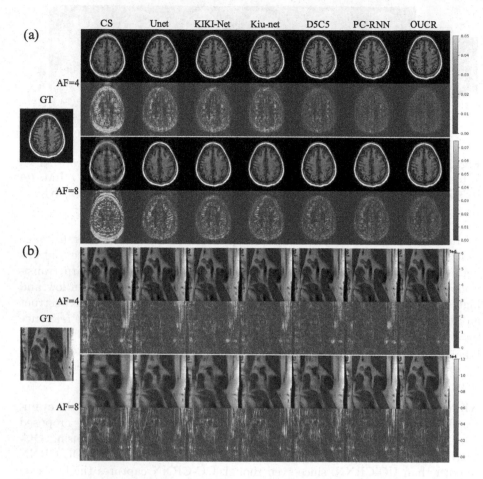

Fig. 4. Qualitative comparison of different methods on (a) HPKS and (b) fastMRI dataset. The second row of each subplot shows the corresponding error maps.

Table 2. Ablation study of designed modules in term of reconstruction quality on HPKS with AF = 4.

Modules	PSNR	SSIM
UC-CRNN	37.87	0.9644
UC-CRNN + RM	38.02	0.9658
OC-CRNN	36.97	0.9564
OC-CRNN + RM	37.34	0.9600
UC-CRNN + OC-CRNN	38.98	0.9719
UC-CRNN + OC-CRNN + RM (OUCR)	**39.33**	**0.9747**

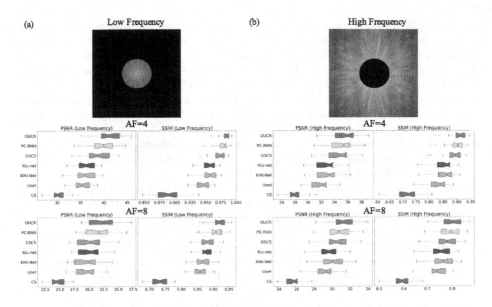

Fig. 5. K-space analysis on (a) low frequency and (b) high frequency. *Top row*: the examples of masked k-space image. *Bottom row*: the Boxplot of reconstruction performance by different methods on HPKS dataset.

4   Discussion and Conclusion

We proposed a novel over-and-under complete convolutional RNN (OUCR) for MR image reconstruction. The purposed method leverages an overcomplete network to specifically capture low-level features, which are typically missed out in the other MR image reconstruction methods. Moreover, we incorporate an undercomplete CRNN, which results in an effective learning of low and high level information. The proposed method achieves better performance on two datasets and has less numbers of trainable parameters as compared to the CS and popular DL-based methods, including UNet [24], KIKI-Net [4], Kiu-net [29], D5C5 [25], and PC-RNN [31]. This study demonstrates the potential of using overcomplete networks in MR image reconstruction task.

Acknowledgment. This work was supported by grants from the National Science Foundation (1910141) and the National Institutes of Health (R37CA248077).

References

1. Akçakaya, M., Moeller, S., Weingärtner, S., Uğurbil, K.: Scan-specific robust artificial-neural-networks for k-space interpolation (RAKI) reconstruction: database-free deep learning for fast imaging. Magn. Reson. Med. **81**(1), 439–453 (2019)

2. Chen, E.Z., Chen, T., Sun, S.: MRI image reconstruction via learning optimization using neural ODEs. In: Martel, A.L., et al. (eds.) MICCAI 2020. LNCS, vol. 12262, pp. 83–93. Springer, Cham (2020). https://doi.org/10.1007/978-3-030-59713-9_9

3. Edmund, J.M., Nyholm, T.: A review of substitute CT generation for MRI-only radiation therapy. Radiat. Oncol. **12**(1), 1–15 (2017)

4. Eo, T., Jun, Y., Kim, T., Jang, J., Lee, H.J., Hwang, D.: Kiki-net: cross-domain convolutional neural networks for reconstructing undersampled magnetic resonance images. Magn. Reson. Med. **80**(5), 2188–2201 (2018)

5. Fisher, R.B.: Cvonline: The evolving, distributed, non-proprietary, on-line compendium of computer vision (2008). https://homepages.inf.ed.ac.uk/rbf/CVonline. Accessed 28 Jan 2006

6. Guo, P., Wang, P., Yasarla, R., Zhou, J., Patel, V.M., Jiang, S.: Anatomic and molecular MR image synthesis using confidence guided CNNS. IEEE Trans. Med. Imaging, 1 (2020). https://doi.org/10.1109/TMI.2020.3046460

7. Guo, P., Wang, P., Zhou, J., Jiang, S., Patel, V.M.: Multi-institutional collaborations for improving deep learning-based magnetic resonance image reconstruction using federated learning. In: Proceedings of the IEEE/CVF Conference on Computer Vision and Pattern Recognition (CVPR), pp. 2423–2432 (2021)

8. Guo, P., Wang, P., Zhou, J., Patel, V.M., Jiang, S.: Lesion mask-based simultaneous synthesis of anatomic and molecular MR Images Using a GAN. In: Martel, A.L., et al. (eds.) MICCAI 2020. LNCS, vol. 12262, pp. 104–113. Springer, Cham (2020). https://doi.org/10.1007/978-3-030-59713-9_11

9. Haldar, J.P., Hernando, D., Liang, Z.P.: Compressed-sensing MRI with random encoding. IEEE Trans. Med. Imaging **30**(4), 893–903 (2010)

10. Jiang, S., et al.: Identifying recurrent malignant glioma after treatment using amide proton transfer-weighted MR imaging: a validation study with image-guided stereotactic biopsy. Clin. Cancer Res. **25**(2), 552–561 (2019)

11. Knoll, F., et al.: fastmri: A publicly available raw k-space and DICOM dataset of knee images for accelerated MR image reconstruction using machine learning. Radiol. Artif. Intell. **2**(1), e190007 (2020)

12. Lee, D., Yoo, J., Tak, S., Ye, J.C.: Deep residual learning for accelerated MRI using magnitude and phase networks. IEEE Trans. Biomed. Eng. **65**(9), 1985–1995 (2018)

13. Lewicki, M.S., Sejnowski, T.J.: Learning overcomplete representations. Neural Comput. **12**(2), 337–365 (2000)

14. Liang, D., Cheng, J., Ke, Z., Ying, L.: Deep magnetic resonance image reconstruction: inverse problems meet neural networks. IEEE Sign. Process. Mag. **37**(1), 141–151 (2020)

15. Liang, D., Liu, B., Wang, J., Ying, L.: Accelerating sense using compressed sensing. Magn. Reson. Med. Offic. J. Int. Soc. Magn. Reson. Med. **62**(6), 1574–1584 (2009)

16. Majumdar, A.: Improving synthesis and analysis prior blind compressed sensing with low-rank constraints for dynamic MRI reconstruction. Magn. Reson. Imaging **33**(1), 174–179 (2015)

17. Mezrich, R.: A perspective on k-space. Radiology **195**(2), 297–315 (1995)

18. Patel, V.M., Chellappa, R.: Sparse representations, compressive sensing and dictionaries for pattern recognition. In: The First Asian Conference on Pattern Recognition, pp. 325–329. IEEE (2011)

19. Patel, V.M., Maleh, R., Gilbert, A.C., Chellappa, R.: Gradient-based image recovery methods from incomplete fourier measurements. IEEE Trans. Image Process. **21**(1), 94–105 (2011)

20. Pruessmann, K.P., Weiger, M., Scheidegger, M.B., Boesiger, P.: Sense: sensitivity encoding for fast MRI. Magn. Reson. Med. Offic. J. Int. Soc. Magn. Reson. Med. **42**(5), 952–962 (1999)
21. Putzky, P., Welling, M.: Invert to learn to invert. arXiv preprint arXiv:1911.10914 (2019)
22. Qin, C., et al.: Convolutional recurrent neural networks for dynamic MR image reconstruction. IEEE Trans. Med. Imaging **38**(1), 280–290 (2019). https://doi.org/10.1109/TMI.2018.2863670
23. Ravishankar, S., Bresler, Y.: MR image reconstruction from highly undersampled k-space data by dictionary learning. IEEE Trans. Med. Imaging **30**(5), 1028–1041 (2010)
24. Ronneberger, O., Fischer, P., Brox, T.: U-Net: convolutional networks for biomedical image segmentation. In: Navab, N., Hornegger, J., Wells, W.M., Frangi, A.F. (eds.) MICCAI 2015. LNCS, vol. 9351, pp. 234–241. Springer, Cham (2015). https://doi.org/10.1007/978-3-319-24574-4_28
25. Schlemper, J., et al.: A deep cascade of convolutional neural networks for dynamic MR image reconstruction. IEEE Trans. Med. Imaging **37**(2), 491–503 (2017)
26. Tamir, J.I., Ong, F., Cheng, J.Y., Uecker, M., Lustig, M.: Generalized magnetic resonance image reconstruction using the Berkeley advanced reconstruction toolbox. In: ISMRM Workshop on Data Sampling & Image Reconstruction, Sedona, AZ (2016)
27. Valanarasu, J.M.J., Patel, V.M.: Overcomplete deep subspace clustering networks. In: Proceedings of the IEEE/CVF Winter Conference on Applications of Computer Vision, pp. 746–755 (2021)
28. Valanarasu, J.M.J., Sindagi, V.A., Hacihaliloglu, I., Patel, V.M.: Kiu-net: overcomplete convolutional architectures for biomedical image and volumetric segmentation. arXiv preprint arXiv:2010.01663 (2020)
29. Valanarasu, J.M.J., Sindagi, V.A., Hacihaliloglu, I., Patel, V.M.: KiU-Net: towards accurate segmentation of biomedical images using over-complete representations. In: Martel, A.L., et al. (eds.) MICCAI 2020. LNCS, vol. 12264, pp. 363–373. Springer, Cham (2020). https://doi.org/10.1007/978-3-030-59719-1_36
30. Vincent, P., Larochelle, H., Bengio, Y., Manzagol, P.A.: Extracting and composing robust features with denoising autoencoders. In: Proceedings of the 25th International Conference on Machine Learning, pp. 1096–1103 (2008)
31. Wang, P., Chen, E.Z., Chen, T., Patel, V.M., Sun, S.: Pyramid convolutional RNN for MRI reconstruction. arXiv preprint arXiv:1912.00543 (2019)
32. Yasarla, R., Valanarasu, J.M.J., Patel, V.M.: Exploring overcomplete representations for single image deraining using CNNS. IEEE J. Select. Top. Sign. Process. **15**(2), 229–239 (2020)

TarGAN: Target-Aware Generative Adversarial Networks for Multi-modality Medical Image Translation

Junxiao Chen[1], Jia Wei[1(✉)], and Rui Li[2]

[1] School of Computer Science and Engineering, South China University of Technology, Guangzhou, China
cs_xiao@mail.scut.edu.cn, csjwei@scut.edu.cn
[2] Golisano College of Computing and Information Sciences, Rochester Institute of Technology, Rochester, NY 14623, USA
rxlics@rit.edu

Abstract. Paired multi-modality medical images, can provide complementary information to help physicians make more reasonable decisions than single modality medical images. But they are difficult to generate due to multiple factors in practice (e.g., time, cost, radiation dose). To address these problems, multi-modality medical image translation has aroused increasing research interest recently. However, the existing works mainly focus on translation effect of a whole image instead of a critical target area or Region of Interest (ROI), e.g., organ and so on. This leads to poor-quality translation of the localized target area which becomes blurry, deformed or even with extra unreasonable textures. In this paper, we propose a novel target-aware generative adversarial network called **TarGAN**, which is a generic multi-modality medical image translation model capable of (1) learning multi-modality medical image translation without relying on paired data, (2) enhancing quality of target area generation with the help of target area labels. The generator of TarGAN jointly learns mapping at two levels simultaneously—whole image translation mapping and target area translation mapping. These two mappings are interrelated through a proposed crossing loss. The experiments on both quantitative measures and qualitative evaluations demonstrate that TarGAN outperforms the state-of-the-art methods in all cases. Subsequent segmentation task is conducted to demonstrate effectiveness of synthetic images generated by TarGAN in a real-world application. Our code is available at https://github.com/cs-xiao/TarGAN.

Keywords: Multi-modality translation · GAN · Abdominal organs · 2D

Electronic supplementary material The online version of this chapter (https://doi.org/10.1007/978-3-030-87231-1_3) contains supplementary material, which is available to authorized users.

1   Introduction

Medical imaging, a powerful diagnostic and research tool creating visual representations of anatomy, has been widely available for disease diagnosis and surgery planning [2]. In current clinical practice, Computed Tomography (CT) and Magnetic Resonance Imaging (MRI) are most commonly used. Since CT and multiple MR imaging modalities provide complementary information, an effective integration of these different modalities can help physicians make more informative decisions.

Since it is difficult and costly to obtain paired multi-modality images in clinical practice, there is a growing demand for developing multi-modality image translations to assist clinical diagnosis and treatment [17].

Existing works can be categorized into two types. One is crossing-modality medical image translation between two modalities, which has scalability issues to the increasing number of modalities [18,19], since these methods have to train $n(n-1)$ generator models in order to learn all mappings between n modalities. The other is multi-modality image translation [1,7,16,17]. In this category, some methods [7,17] rely on paired data, which is hard to acquire in clinical reality. Other methods [1,16] can learn from unpaired data, however, they tend to lead to deformation in target area without prior knowledge, as concluded by Zhang et al. [19]. As demonstrated in Fig. 1, the state-of-the-art multi-modality image translation methods give rise to poor quality local translations. The translated target area (For example, Liver, in red curves) is blurry, deformed or perturbed with redundant unreasonable textures. Comparing to them, our method can not only perform whole image translation in competitive quality but also achieve significantly better local translation for the target area.

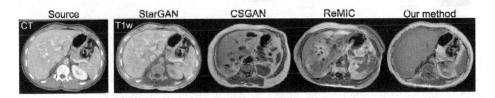

Fig. 1. Translation results (CT to T1w) of different methods are shown here. The target area (i.e., liver) is contoured in red. (Color figure online)

To address the above issues, we present a novel unified general-purpose multi-modality medical image translation method named "Target-Aware Generative Adversarial Networks" (TarGAN). We incorporate target labels to enable the generator to focus on local translation of target area. The generator has two input-output streams. One stream translates a whole image from source modality to target modality, the other focuses on translating a target area. In particular, we combine the cycle-consistency loss [21] and the backbone of StarGAN [1] to learn the generator, which enables our model to scale up to modality increase without relying on paired data. Then, the untraceable constraint [20] is employed to further improve translation quality of synthetic images. To avoid the deformation of output images caused by untraceable constraint, we construct a shape-consistency loss [3] with an auxiliary network, namely shape controller. We further propose a novel crossing loss to allow the generator to focus on the target area when translating the whole image to target modality. Trained in an end-to-end fashion, TarGAN can not only accomplish multi-modality translation but also properly retain the target area information in the synthetic images.

Overall, the Contributions of This Work Are: (1) We propose TarGAN to generate multi-modality medical images with high-quality local translation on target areas by integrating global and local mappings with a crossing loss. (2) We show qualitative and quantitative performance evaluations on multi-modality medical image translation tasks with CHAOS2019 dataset [12], demonstrating our method's superiority over the state-of-the-art methods. (3) We further use the synthetic images generated from TarGAN to improve the performance of a segmentation task, which indicates that the synthetic images generated by TarGAN achieve the improvement by enriching the information of source images.

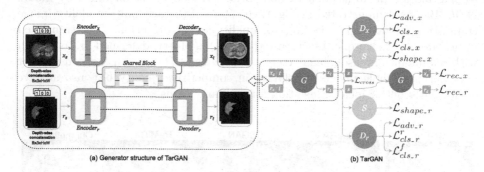

Fig. 2. The illustration of TarGAN. As in **(b)**, TarGAN consists of four modules (G, S, D_x, D_r). The generator G translates a source whole image x_s and a source target area image r_s to a target whole image x_t and a target area image r_t. The detailed structure of G is shown in **(a)**. The shape controller S preserves the invariance of anatomy structures. The discriminators D_x and D_r distinguish whether a whole image and its target area are real or fake and determine which modalities the source images come from.

2   Methods

2.1   Proposed Framework

Given an image x_s from source modality s and its corresponding target area label y, we specify a target area image r_s which only contains the target area by binarization operation $y \cdot x_s$. Given any target modality t, our goal is to train a single generator G that can translate any input image x_s of source modality s to the corresponding output image x_t of target modality t, and translate the input target area image r_s of source modality s to the corresponding output target area image r_t of target modality t simultaneously, denoted as $G(x_s, r_s, t) \rightarrow (x_t, r_t)$. Figure 2 illustrates the architecture of TarGAN, which is composed of four modules described below.

To achieve the aforementioned goal, we design a double input-output streams **generator** G consisting of a shared middle block and two pairs of encoder-decoder. Combining with the shared middle block, both encoder-decoder pairs translate an input image into an output image of the target modality t. One stream's input is the whole image x_s, and the other's input only includes the target area r_s. The shared middle block is designed to implicitly enable G to focus on target area in whole image translation. Note that target area label y of x_s is not available in test phase, so the input block $Encoder_r$ and output block $Decoder_r$ are not used at that time.

Given a synthetic image x_t or r_t from G, **the shape controller** S generates a binary mask which can represent the foreground area of the synthetic image.

Lastly, we use **two discriminators denoted as D_x and D_r** corresponding to two output streams of G. The probability distributions inferred by D_x distinguish whether the whole image is real or fake, and determine which modality the whole image comes from. Similarly, the D_r distinguish whether the target area image is real or fake, and to determine which modality the target area image comes from.

2.2   Training Objectives

Adversarial Loss. To minimize the difference between the distributions of generated images and real images, we define the adversarial loss as

$$
\begin{aligned}
\mathcal{L}_{adv\_x} &= \mathbb{E}_{x_s}[\log D_{src\_x}(x_s)] + \mathbb{E}_{x_t}[\log(1 - D_{src\_x}(x_t))], \\
\mathcal{L}_{adv\_r} &= \mathbb{E}_{r_s}[\log D_{src\_r}(r_s)] + \mathbb{E}_{r_t}[\log(1 - D_{src\_r}(r_t))].
\end{aligned}
\tag{1}
$$

Here, D_{src_x} and D_{src_r} represent the probability distributions of real or fake over input whole images and target area images.

Modality Classification Loss. To assign the generated image to their target modality t, we impose the modality classification loss on G, D_x and D_r. The loss consists of two terms: modality classification loss of real images which is used to optimize D_x and D_r, denoted as $\mathcal{L}^r_{cls\_(x/r)}$, and modality classification loss of fake images which is used to optimize G, denoted as $\mathcal{L}^f_{cls\_(x/r)}$. In addition, to eliminate synthetic images' style features from source modalities, the untraceable constraint [20] is combined into $\mathcal{L}^r_{cls\_(x/r)}$ as:

$$
\begin{aligned}
\mathcal{L}^r_{cls\_x} &= \mathbb{E}_{x_s,s}[-\log D_{cls\_x}(s|x_s)] + \lambda_u \mathbb{E}_{x_t,s'}[-\log D_{cls\_x}(s'|x_t)], \\
\mathcal{L}^r_{cls\_r} &= \mathbb{E}_{r_s,s}[-\log D_{cls\_r}(s|r_s)] + \lambda_u \mathbb{E}_{r_t,s'}[-\log D_{cls\_r}(s'|r_t)].
\end{aligned}
\tag{2}
$$

Here, D_{cls_x} and D_{cls_r} represent the probability distributions over modality labels and input images. s' indicates whether an input image is fake, and is translated from a source modality s [20]. Besides, we define $\mathcal{L}^f_{cls\_(x/r)}$ as

$$\mathcal{L}^f_{cls\_x} = \mathbb{E}_{x_t,t}[-\log D_{cls\_x}(t|x_t)], \quad \mathcal{L}^f_{cls\_r} = \mathbb{E}_{r_t,t}[-\log D_{cls\_r}(t|r_t)]. \quad (3)$$

Shape Consistency Loss. Since the untraceable constraint can affect the shape of anatomy structures in synthetic images by causing structure deformation, we correct it by adding a shape consistency loss [3] to G with shape controller S as

$$\mathcal{L}_{shape\_x} = \mathbb{E}_{x_t,b^x}[||b^x - S(x_t)||_2^2], \quad \mathcal{L}_{shape\_r} = \mathbb{E}_{r_t,b^r}[||b^r - S(r_t)||_2^2], \quad (4)$$

where b^x and b^r are the binarizations (with 1 indicating foreground pixels and 0 otherwise) of x_s and r_s. S constrains G to focus on the multi-modality mapping in a content area.

Reconstruction Loss. To allow G to preserve the modality-invariant characteristics of the whole image x_s and its target area image r_s, we employ a cycle consistency loss [21] as

$$\mathcal{L}_{rec\_x} = \mathbb{E}_{x_s,x'_s}[||x_s - x'_s||_1], \quad \mathcal{L}_{rec\_r} = \mathbb{E}_{r_s,r'_s}[||r_s - r'_s||_1]. \quad (5)$$

Note that x'_s and r'_s are from $G(x_t, r_t, s)$. Given the paired synthetic image (x_t, r_t) and the source modality s, G tries to reconstruct the input images (x_s, r_s).

Crossing Loss. To enforce G to focus on a target area when generating a whole image x_t, we directly regularize G with a crossing loss defined as

$$\mathcal{L}_{cross} = \mathbb{E}_{x_t,r_t,y}[||x_t \cdot y - r_t||_1], \quad (6)$$

where y is the target area label corresponding to x_s. By minimizing the crossing loss, G can jointly learn from double input-output streams and share information between them.

Complete Objective. By combining the proposed losses together, our complete objective functions are as follows:

$$\mathcal{L}_{D_{(x/r)}} = -\mathcal{L}_{adv\_(x/r)} + \lambda^r_{cls} \mathcal{L}^r_{cls\_(x/r)}, \quad (7)$$

$$\mathcal{L}_G = \mathcal{L}_{adv\_(x/r)} + \lambda^f_{cls} \mathcal{L}^f_{cls\_(x/r)} + \lambda_{rec} \mathcal{L}_{rec\_(x/r)} + \lambda_{cross} \mathcal{L}_{crossing}, \quad (8)$$

$$\mathcal{L}_{G,S} = \mathcal{L}_{shape\_(x/r)}, \quad (9)$$

where λ^r_{cls}, λ^f_{cls}, λ_{rec}, λ_{cross} and λ_u (Eqs. (2)) are hyperparameters to control the relative importance of each loss.

Table 1. Quantitative evaluations on synthetic images of different methods. (↑ denotes higher is better, while ↓ denotes lower is better)

Method	FID↓			S-score(%)↑		
	CT	T1w	T2w	CT	T1w	T2w
StarGAN [1]	0.0488	0.1179	0.2615	42.89	29.23	42.17
CSGAN [19]	0.0484	0.1396	0.4819	56.72	45.67	69.09
ReMIC [16]	0.0912	0.1151	0.5925	51.03	32.00	69.58
Our method	**0.0418**	**0.0985**	**0.2431**	**57.13**	**65.79**	**69.63**

3 Experiments and Results

3.1 Settings

Dataset. We use 20 patients' data in each modality (CT, T1-weighted and T2-weighted). They are from the Combined Healthy Abdominal Organ Segmentation (CHAOS) Challenge [11]. Detailed imaging parameters are shown in supplementary material. We resize all slices as 256×256 uniformly. 50% data from each modality are randomly selected as training data, while the rest as test data. Because CT scans only have liver labels, we set liver as the target area.

Baseline Methods. Translation results comparisons are conducted against the state-of-the-art translation methods, StarGAN [1], CSGAN [19] and ReMIC [16]. Note that we implement an unsupervised ReMIC because of the lack of ground-truth images.

Target segmentation performances are also evaluated against the above methods. We train and test models using only real images of each modality, denoted as **Single**. We use the mean results of two segmentation models of each modality from CSGAN and use the segmentation model G_s from ReMIC. As for StarGAN and TarGAN, inspired by *'image enrichment'* [5], we extend every single modality to multiple modalities and concatenate multiple modalities within each sample, as [CT] → [CT, synthetic T1w, synthetic T2w].

Evaluation Metrics. In the translation tasks, due to the lack of ground-truth images, we can not use the common metrics like PSNR, SSIM, etc. So we evaluate both the visual quality and the integrity of target area structures of generated images using Frechét inception distance (**FID**) [6] and segmentation score (**S-score**) [19]. We compute FID and S-score for each modality and report their average values. The details on above metrics are further described in supplementary material.

In the segmentation tasks, dice coefficient (**DICE**) and relative absolute volume difference (**RAVD**) are used as metrics. We compute each metric on every modality, and report their average values and standard deviations.

Implementation Details. We use U-net [15] as the backbone of G and S. In G, only half of the channels are used for every skip connection. As for D_x and D_r, we implement the backbone with PatchGAN [9]. Details of above networks are included in the supplementary material. All the liver segmentation experiments are conducted with nnU-Net [8] except CSGAN and ReMIC.

To stabilize the training process, we adopt Wasserstein GAN loss with a gradient penalty [4,14] using $\lambda_{gp} = 10$ and two-timescale update rule (TTUR) [6] for G and D. The learning rates for G, S are set to 10^{-4}, while that of D is set to 3×10^{-4}. We set $\lambda_{cls}^r = 1$, $\lambda_{cls}^f = 1$, $\lambda_{rec} = 1$, $\lambda_{cross} = 50$ and $\lambda_u = 0.01$. The batch size and training epoch are set to 4 and 50, respectively. We use the Adam optimizer [13] with momentum parameters $\beta_1 = 0.5$ and $\beta_2 = 0.9$. All images are normalized to $[-1, 1]$ prior to the training and test. We use exponential moving averages over parameters [10] of G during test, with a decay of 0.999. Our implementation is trained on an NVIDIA GTX 2080Ti with PyTorch.

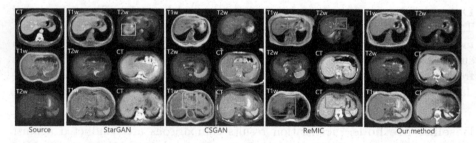

Fig. 3. Multi-modality medical image translation results. Red boxes highlight the redundant textures, and blue boxes indicate the deformed structures. (Color figure online)

3.2 Results and Analyses

Image Translation. Figure 3 shows qualitative results on each pair of modal image translation. As shown, StarGAN fails to translate image from CT to T1w and produces many artifacts in MRI to CT translation. CSGAN sometimes adds redundant textures (marked by the red boxes) in the target area while retaining the shape of target. ReMIC tends to generate relatively realistic synthetic images while deforming the structure of target area in most cases (marked by the blue boxes). Comparing to above methods, TarGAN generates translation results in higher visual quality and properly preserves the target structures. Facilitated by the proposed crossing loss, TarGAN can jointly learn the mappings of the target area and the whole image among different modalities, and further make G focus on the target areas to improve their quality. Furthermore, as shown in Table 1, TarGAN outperforms all the baselines in terms of FID and S-score, which suggests TarGAN produces the most realistic medical images, and the target area integrity of synthetic images derived from TarGAN is significantly better.

Table 2. Liver segmentation results (mean ± standard deviation) on different medical modalities.

Method	DICE(%)↑			RAVD(%)↓		
	CT	T1w	T2w	CT	T1w	T2w
Single	96.29 ± 0.74	93.53 ± 2.43	89.24 ± 8.18	3.31 ± 1.80	3.81 ± 3.49	11.68 ± 14.37
StarGAN [1]	96.65 ± 0.34	92.71 ± 1.66	86.38 ± 4.95	3.07 ± 1.53	5.40 ± 2.87	15.71 ± 9.85
CSGAN [19]	96.08 ± 2.05	87.47 ± 5.97	86.35 ± 6.29	4.47 ± 3.94	15.74 ± 14.18	**8.23 ± 8.55**
ReMIC [16]	93.81 ± 1.43	86.33 ± 8.50	82.70 ± 4.36	5.33 ± 3.55	8.06 ± 8.80	10.62 ± 6.80
Our method	**97.06 ± 0.62**	**94.02 ± 2.00**	**90.94 ± 6.28**	**2.33 ± 1.60**	**3.50 ± 1.82**	9.92 ± 11.17

Liver Segmentation. The quantitative segmentation results are shown in Table 2. Our method achieves better performance than all other methods on most of the metrics. This suggests TarGAN can not only generate realistic images for every modality, but also properly retain liver structure in synthetic images. The high-quality local translation for the target areas plays a key role in the improvement of liver segmentation performance. By jointly learning from real and synthetic images, the segmentation models can incorporate more information on the liver areas within each sample.

Ablation Test. We conduct an ablation test to validate effectiveness of different parts of TarGAN in terms of preserving target area information. For ease of presentation, we denote **shape controller**, **target area translation mapping** and **crossing loss** as **S**, **T** and **C**, respectively. As shown in Table 3, **TarGAN without (w/o) S, T, C** is closely similar to StarGAN except using our implementation. The proposed crossing loss plays a key role in TarGAN, which increases the mean of S-score from **TarGAN w/o C** 51.03% to 64.18%.

Table 3. Ablation study on different components of TarGAN. Note that **TarGAN w/o S, T** and **TarGAN w/o T** don't exist, since **T** is the premise of **C**.

Method	S-score(%)			
	CT	T1w	T2w	Mean
TarGAN w/o S, T, C	30.64	35.05	67.45	44.38
TarGAN w/o S, C	39.78	29.96	67.47	45.74
TarGAN w/o T, C	37.42	38.33	68.85	48.20
TarGAN w/o C	43.00	38.83	71.27	51.03
TarGAN w/o S	56.69	59.37	**71.89**	62.65
TarGAN	**57.13**	**65.79**	69.63	**64.18**

4 Conclusion

In this paper, we propose a novel general-purpose method TarGAN to mainly address two challenges in multi-modality medical image translation: learning multi-modality medical image translation without relying on paired data, and improving the quality of local translation on target area. A novel translation mapping mechanism is introduced to enhance the target area quality during generating the whole image. Additionally, by using the shape controller to alleviate the deformation problem caused by the untraceable constraint and combining a novel crossing loss in generator G, TarGAN addresses both challenges within a unified framework. Both the quantitative and qualitative evaluations show the superiority of TarGAN in comparison with the state-of-the-art methods. We further conduct a segmentation task to demonstrate effectiveness of synthetic images generated by TarGAN in a real application.

Acknowledgments. This work is supported in part by the Natural Science Foundation of Guangdong Province (2017A030313358, 2017A030313355, 2020A15 15010717), the Guangzhou Science and Technology Planning Project (201704030051), the Fundamental Research Funds for the Central Universities (2019MS073), NSF-1850492 (to R.L.) and NSF-2045804 (to R.L.).

References

1. Choi, Y., Choi, M., Kim, M., Ha, J.W., Kim, S., Choo, J.: Stargan: unified generative adversarial networks for multi-domain image-to-image translation. In: Proceedings of the IEEE Conference on Computer Vision and Pattern Recognition, pp. 8789–8797 (2018)
2. Ernst, P., Hille, G., Hansen, C., Tönnies, K., Rak, M.: A CNN-based framework for statistical assessment of spinal shape and curvature in whole-body MRI images of large populations. In: Shen, D., et al. (eds.) MICCAI 2019. LNCS, vol. 11767, pp. 3–11. Springer, Cham (2019). https://doi.org/10.1007/978-3-030-32251-9_1
3. Fu, C., et al.: Three dimensional fluorescence microscopy image synthesis and segmentation. In: Proceedings of the IEEE Conference on Computer Vision and Pattern Recognition Workshops, pp. 2221–2229 (2018)
4. Gulrajani, I., Ahmed, F., Arjovsky, M., Dumoulin, V., Courville, A.C.: Improved training of Wasserstein GANS. In: Advances in Neural Information Processing Systems, pp. 5767–5777 (2017)
5. Gupta, L., Klinkhammer, B.M., Boor, P., Merhof, D., Gadermayr, M.: GAN-based image enrichment in digital pathology boosts segmentation accuracy. In: Shen, D., et al. (eds.) MICCAI 2019. LNCS, vol. 11764, pp. 631–639. Springer, Cham (2019). https://doi.org/10.1007/978-3-030-32239-7_70
6. Heusel, M., Ramsauer, H., Unterthiner, T., Nessler, B., Hochreiter, S.: GANS trained by a two time-scale update rule converge to a local Nash equilibrium. In: Advances in Neural Information Processing Systems, pp. 6626–6637 (2017)
7. Huang, P.U., et al.: CoCa-GAN: common-feature-learning-based context-aware generative adversarial network for glioma grading. In: Shen, D., et al. (eds.) MICCAI 2019. LNCS, vol. 11766, pp. 155–163. Springer, Cham (2019). https://doi.org/10.1007/978-3-030-32248-9_18
8. Isensee, F., Jaeger, P.F., Kohl, S.A., Petersen, J., Maier-Hein, K.H.: NNU-net: a self-configuring method for deep learning-based biomedical image segmentation. Nat. Meth. **18**(2), 203–211 (2021)

9. Isola, P., Zhu, J.Y., Zhou, T., Efros, A.A.: Image-to-image translation with conditional adversarial networks. In: IEEE Conference on Computer Vision and Pattern Recognition (2017)

10. Karras, T., Aila, T., Laine, S., Lehtinen, J.: Progressive growing of GANS for improved quality, stability, and variation. In: International Conference on Learning Representations (2018)

11. Kavur, A.E., et al.: Chaos challenge-combined (CT-MR) healthy abdominal organ segmentation. Med. Image Anal. **69**, 101950 (2021)

12. Kavur, A.E., Selver, M.A., Dicle, O., Barış, M., Gezer, N.S.: Chaos-combined (CT-MR) healthy abdominal organ segmentation challenge data. In: Proceedings of IEEE International Symposium Biomedical Image (ISBI) (2019)

13. Kingma, D.P., Ba, J.: Adam: a method for stochastic optimization. In: ICLR (Poster) (2015)

14. Martin Arjovsky, S., Bottou, L.: Wasserstein generative adversarial networks. In: Proceedings of the 34th International Conference on Machine Learning, Sydney, Australia (2017)

15. Ronneberger, O., Fischer, P., Brox, T.: U-Net: convolutional networks for biomedical image segmentation. In: Navab, N., Hornegger, J., Wells, W.M., Frangi, A.F. (eds.) MICCAI 2015. LNCS, vol. 9351, pp. 234–241. Springer, Cham (2015). https://doi.org/10.1007/978-3-319-24574-4_28

16. Shen, L., et al.: Multi-domain image completion for random missing input data. IEEE Trans. Med. Imaging **40**(4), 1113–1122 (2020)

17. Xin, B., Hu, Y., Zheng, Y., Liao, H.: Multi-modality generative adversarial networks with tumor consistency loss for brain MR image synthesis. In: 2020 IEEE 17th International Symposium on Biomedical Imaging (ISBI), pp. 1803–1807. IEEE (2020)

18. Yu, B., Zhou, L., Wang, L., Shi, Y., Fripp, J., Bourgeat, P.: EA-GANS: edge-aware generative adversarial networks for cross-modality MR image synthesis. IEEE Trans. Med. Imaging **38**(7), 1750–1762 (2019)

19. Zhang, Z., Yang, L., Zheng, Y.: Translating and segmenting multimodal medical volumes with cycle-and shape-consistency generative adversarial network. In: Proceedings of the IEEE Conference on Computer Vision and Pattern Recognition, pp. 9242–9251 (2018)

20. Zhu, D., et al.: UGAN: Untraceable GAN for multi-domain face translation. arXiv preprint arXiv:1907.11418 (2019)

21. Zhu, J.Y., Park, T., Isola, P., Efros, A.A.: Unpaired image-to-image translation using cycle-consistent adversarial networks. In: Proceedings of the IEEE International Conference on Computer Vision, pp. 2223–2232 (2017)

Synthesizing Multi-tracer PET Images for Alzheimer's Disease Patients Using a 3D Unified Anatomy-Aware Cyclic Adversarial Network

Bo Zhou[1]([⊠]), Rui Wang[2,3], Ming-Kai Chen[2], Adam P. Mecca[4], Ryan S. O'Dell[4], Christopher H. Van Dyck[4], Richard E. Carson[1,2], James S. Duncan[1,2], and Chi Liu[1,2]

[1] Department of Biomedical Engineering, Yale University, New Haven, CT, USA
bo.zhou@yale.edu
[2] Department of Radiology and Biomedical Imaging, Yale University, New Haven, CT, USA
[3] Department of Engineering Physics, Tsinghua University, Beijing, China
[4] Department of Psychiatry, Yale University, New Haven, CT, USA

Abstract. Positron Emission Tomography (PET) is an important tool for studying Alzheimer's disease (AD). PET scans can be used as diagnostics tools, and to provide molecular characterization of patients with cognitive disorders. However, multiple tracers are needed to measure glucose metabolism (^{18}F-FDG), synaptic vesicle protein (^{11}C-UCB-J), and β-amyloid (^{11}C-PiB). Administering multiple tracers to patient will lead to high radiation dose and cost. In addition, access to PET scans using new or less-available tracers with sophisticated production methods and short half-life isotopes may be very limited. Thus, it is desirable to develop an efficient multi-tracer PET synthesis model that can generate multi-tracer PET from single-tracer PET. Previous works on medical image synthesis focus on one-to-one fixed domain translations, and cannot simultaneously learn the feature from multi-tracer domains. Given 3 or more tracers, relying on previous methods will also create heavy burden on the number of models to be trained. To tackle these issues, we propose a 3D unified anatomy-aware cyclic adversarial network (UCAN) for translating multi-tracer PET volumes with one unified generative model, where MR with anatomical information is incorporated. Evaluations on a multi-tracer PET dataset demonstrate the feasibility that our UCAN can generate high-quality multi-tracer PET volumes, with NMSE less than 15% for all PET tracers. Our code is available at https://github.com/bbbbbbzhou/UCAN.

Keywords: Brain PET · Alzheimer's disease · Multi-tracer synthesis

Electronic supplementary material The online version of this chapter (https://doi.org/10.1007/978-3-030-87231-1_4) contains supplementary material, which is available to authorized users.

M. de Bruijne et al. (Eds.): MICCAI 2021, LNCS 12906, pp. 34–43, 2021.
https://doi.org/10.1007/978-3-030-87231-1_4

1   Introduction

Alzheimer's Disease (AD) is a progressive neurodegenerative disorder and the most common cause of dementia. Positron Emission Tomography (PET), as a functional neuroimaging technique, is commonly used in biomedical research for studying the brain's metabolic, biochemical, and other physiological alterations, thus providing essential information for understanding underlying pathology and helping early diagnosis/differential diagnosis of AD [1]. Several important radiotracers have been developed for studying AD's biomarkers, including ^{18}F-Fluorodeoxyglucose (^{18}F-FDG) [2], ^{11}C-UCB-J [3], and ^{11}C-Pittsburgh Compound-B (^{11}C-PiB) [4]. Specifically, ^{18}F-FDG provides the metabolic marker that assesses regional cerebral metabolism, and previous studies have revealed that cerebral glucose hypometabolism on ^{18}F-FDG is a downstream marker of neuronal injury and neurodegeneration [5]. On the other hand, synapse loss measured by synaptic vesicle protein (SV2A) is another important feature of neurodegeneration [6,7]. The quantification of SV2A using ^{11}C-UCB-J serves as another essential marker for synaptic density. Similarly, recent studies have found that the early stage of AD can be characterized by the presence of asymptomatic β-amyloidosis or increased β-amyloid burden [8]. Thus, the quantification of β-amyloid using ^{11}C-PiB or related tracers is also an indispensable marker for AD studies. Multi-tracer brain PET imaging can provide AD patient with comprehensive brain evaluations on various pathophysiological aspects, enabling more accurate early diagnosis and differential diagnosis as compared to a single tracer study.

However, multi-tracer brain PET is challenging to deploy in real-world scenarios due to the radiation dose to both patient and healthcare providers, as well as the increasing cost when more tracer studies are involved. The ability to synthesize multi-tracer PET images from single-tracer PET images is paramount in terms of providing more information for assessing AD with minimal radiation dose and cost. As ^{11}C-UCB-J reflects the neural synaptic density, ^{18}F-FDG reflects the cell metabolism (high activities at the neural synapse) and ^{11}C-PiB reflecting the neural growth/repair activities, the PET images from these three tracers are correlated which post the possibility to predict one tracer image from another tracer image. With the recent advances in deep learning based image translation techniques [9,10], these techniques have been widely utilized in medical imaging for translation between individual modalities and individual acquisition protocols, such as MR-PET [11], DR-DE [12], and PET-PET [13]. Even though similar strategies can be utilized for synthesizing multi-tracer PET images from single-tracer PET image, it will result in the need for training large amount of one-to-one domain translation model ($\mathcal{A}_n^2 = n \times (n-1)$) and leaving features learned from multiple domain under-utilized. Let ^{18}F-FDG, ^{11}C-UCB-J, and ^{11}C-PiB be denoted as PET tracer A, B, and C, respectively. Given the three PET tracers scenario, traditional methods require $\mathcal{A}_3^2 = 6$ translation models for translating $A \rightarrow B$ / $A \rightarrow C$ / $B \rightarrow A$ / $B \rightarrow C$ / $C \rightarrow A$ / $C \rightarrow B$. In addition, brain PET tracer images, as functional images, are challenging to translate between tracer domain without prior knowledge of anatomy.

To tackle these limitations, we developed a 3D unified anatomy-aware cyclic adversarial network (UCAN) for translating multi-tracer PET volumes with one unified generative model, where MR with anatomical information is incorporated. Our UCAN is a paired multi-domain 3D translation model based on the previous design of unpaired 2D model of StarGAN [14]. The general pipeline of UCAN is illustrated in Fig. 1. Our UCAN uses training data of multiple tracer domains along with anatomical MR to learn the mapping between all available tracer domains using one single generator. Rather than learning a one-to-one fixed translation, our generator takes tracer volume/MR volume/tracer domain information as input, and learns to flexibly translate the tracer volume into the target tracer domain. Specifically, we use a customized label, a three-channel binary volume with one-hot-channel, for representing tracer domain information. Given trio of tracer volumes, we randomly sample a target tracer domain label and train the generator to translate an input tracer volume into the target tracer domain. During inference stage, we can simply set the tracer domain label and map the original tracer volume into the other tracer domain, hence producing multi-tracer PET volumes from single-tracer PET volume using one generator. Moreover, we proposed to use a 3D Dual Squeeze-and-Excitation Network (DuSE-Net) as our UCAN's generator to better fuse the multiple input information for synthesizing tracer volumes. Extensive experiments on human brain data demonstrate that our UCAN with DuSE-Net can efficiently produce high-quality multi-tracer PET volumes from a single-tracer PET volume using single generator.

2 Methods

The framework and training pipeline of our UCAN is illustrated in Fig. 1. Our UCAN consists of one generator G and one discriminator D for multi-tracer-domain translations. We aim to obtain a single G that can map between tracer domains. Specifically, we train our G to map an input tracer volume X_{pet} into an output tracer volume Y_{pet} conditioned on both the target tracer domain label M and MR volume X_{mr}, i.e. $Y_{pet} = G(X_{pet}, X_{mr}, M)$. Specifically, M is a three-channel binary volume with one-hot-channel for target domain labeling, and can be written as

$$M = [c_1, c_2, c_3] \tag{1}$$

where [] is the channel-wise concatenation operator, and c_n is a one channel binary volume for domain labeling, i.e. $c_1 = 1, c_2 = 0, c_3 = 0$ for tracer A. M is randomly generated over the training process. In parallel, we deploy a discriminator D with auxiliary classifier design [15] for discrimination of source domain (i.e. real/fake) and class domain (i.e. tracer A/B/C).

Loss Function. To ensure the generated tracer volume lies in target domain and being realistic, we propose four loss components as follows:

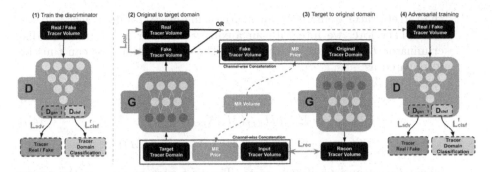

Fig. 1. Our UCAN consists of one generator G (red block) and one discriminator D (blue block) with detailed structures demonstrated in our supplemental materials. **(1)** D is trained to discriminate real/fake tracer volumes and classify the real volume's tracer type. **(2)** G inputs channel-wise concatenation of original tracer volume, MR volume, and target tracer domain label for generation of the target tracer volume. **(3)** Given predicted target tracer volume, G inputs it along with MR volume and original tracer domain label and tries to reconstruct the original tracer volume. **(4)** G learns to synthesize fake target tracer volumes that are indistinguishable from real tracer volumes and can be classified into target domain by D. The four loss components are marked in red. (Color figure online)

(1) Pair Loss: Given trios of tracer volumes and corresponding MR volume, we randomly sample pairs of tracer volumes along with their MR volume for our pair training. We use a pair loss:

$$\mathcal{L}_{pair} = \mathbb{E}[||G(X_{pet}, X_{mr}, M) - Y_{pet}^{GT}||_1] \qquad (2)$$

where G generates a tracer volume $G(X_{pet}, X_{mr}, M) \rightarrow Y_{pet}$ with input of X_{pet} and conditioned on both the MR volume X_{mr} and the target tracer domain label M. Here, our $\mathcal{L}_{pair}$ aims to minimize the L_1 difference between ground truth target tracer volume Y_{pet}^{GT} and predicted target tracer volume Y_{pet}.

(2) Adversarial Loss: To ensure the synthesized tracer volume is perceptually indistinguishable from the real tracer volume, we utilize an adversarial loss:

$$\mathcal{L}_{adv} = -\mathbb{E}[log D_{gan}(X_{pet})] - \mathbb{E}[log(1 - D_{gan}(G(X_{pet}, X_{mr}, M)))] \qquad (3)$$

where G tries to generate realistic tracer volume $G(X_{pet}, X_{mr}, M) \rightarrow Y_{pet}$ that can fool our discriminator D_{gan}, while D_{gan} tries to classify if the input tracer volume is real or fake. That is, D aims to minimize the above equation while G aims to maximize it. In Fig. 1, D_{gan} is one of the distribution output from D.

(3) Classification Loss: While D_{gan} ensures visually plausible tracer volume, given the input volume X_{pet} and a target tracer domain label M, the predicted target tracer volume Y_{pet} also needs to be properly classified into

the target tracer domain M for synthesizing class-specific tracer appearance. Therefore, we integrate an auxiliary classifier D_{clsf} into our discriminator D, as illustrated in Fig. 1. The classification loss can thus be formulated as:

$$\mathcal{L}_{clsf} \sim \{\mathcal{L}^r_{clsf}, \mathcal{L}^f_{clsf}\} \tag{4a}$$

$$\mathcal{L}^r_{clsf} = -\mathbb{E}[log D_{clsf}(M^{\dagger}|X_{pet})] \tag{4b}$$

$$\mathcal{L}^f_{clsf} = -\mathbb{E}[log D_{clsf}(M|G(X_{pet}, X_{mr}, M)) \tag{4c}$$

where $\mathcal{L}_{clsf}$ consists of two parts. The first part of $\mathcal{L}^r_{clsf}$ is the tracer domain classification loss of real tracer volumes for optimizing D, which encourages D to learn to classify a real tracer volume X_{pet} into its corresponding original tracer domain $M^{\dagger}$. The second part of $\mathcal{L}^f_{clsf}$ is the loss for the tracer domain classification of fake tracer volumes, which encourages G to generate tracer volumes that can be classified into the target tracer domain M.

(4) Cyclic-Reconstruction Loss: With the adversarial loss and classification loss, G learns to generate tracer volumes that are both realistic and lie in its correct target tracer domain. With the pair loss, G receives direct supervision for synthesizing target tracer volume to learn to preserve content while only alternating the domain-related characteristics. To reinforce the preservation of content over the translation process, we add a cyclic-reconstruction loss:

$$\mathcal{L}_{rec} = \mathbb{E}[||X_{pet} - G(G(X_{pet}, X_{mr}, M), X_{mr}, M^{\dagger})||_1] \tag{5}$$

where G inputs the predicted target tracer volume $G(X_{pet}, X_{mr}, M)$, the same MR volume X_{mr}, and the original tracer domain label $M^{\dagger}$ for generating the cyclic-reconstructed volume $G(G(X_{pet}, X_{mr}, M), X_{mr}, M^{\dagger})$ with G utilized twice here. Then, we aim to minimize the L_1 difference between original tracer volume X_{pet} and cyclic-reconstructed volume.

The full objective function consists of four loss components and can be written as:

$$\mathcal{L}_G = \mathcal{L}_{pair} + \alpha_{rec}\mathcal{L}_{rec} + \alpha_{adv}\mathcal{L}_{adv} + \alpha_{clsf}\mathcal{L}^f_{clsf} \tag{6}$$

$$\mathcal{L}_D = \alpha_{adv}\mathcal{L}_{adv} + \alpha_{clsf}\mathcal{L}^r_{clsf} \tag{7}$$

where α_{clsf}, α_{adv}, and α_{rec} are weighting parameters for classification loss, adversarial loss, and cyclic-reconstruction loss, respectively. In our experiments, we empirically set $\alpha_{clsf} = 0.1$, $\alpha_{adv} = 0.1$, and $\alpha_{rec} = 0.5$. During training, we updated G and D alternatively by: optimizing D using $\mathcal{L}_D$ with G fixed, then optimizing G using $\mathcal{L}_G$ with D fixed. To make the above objective function fully optimized, our G needs to generate realistic target tracer volume with low values on all four-loss components.

Sub-networks Design. Our discriminator consists of 6 convolutional layers. The first 4 sequential layers extract the features from the input. Then, the feature are simultaneously fed into the last two convolutional layers for classification of real/fake and classification of tracer domain. On the other hand, our generator is a 3D Dual Squeeze-and-Excitation Network (DuSE-Net), and the architecture is based on a 3D U-Net [16] with enhancements on 1.) spatial-wise and channel-wise feature re-calibration [17,18] and 2.) feature transformation in the latent space. Specifically, to better fuse the multi-channel input consisting of tracer domain volume, MR volume, and tracer volume as illustrated in Fig. 1, we propose to use a 3D dual squeeze-and-excitation block at each 3D U-Net's outputs, such that channel-wise and spatial-wise features are iteratively re-calibrated and fused. In the latent space, we also utilize 3 sequential residual convolutional blocks to further transform the feature representation. Network architecture and implementation details are summarized in our supplementary.

2.1   Evaluation with Human Data

We collected 35 human studies in our evaluation, including 24 patients diagnosed with AD and 11 patients as healthy control. Each study contains paired brain scans with ^{18}F-FDG PET scan (tracer A) for glucose metabolism, ^{11}C-UCB-J PET scan (tracer B) for synaptic density, and ^{11}C-PiB PET scan (tracer C) for amyloid, and corresponding MR scan for anatomy. All PET scans were acquired on an ECAT HRRT (high-resolution research tomograph) scanner, which is dedicated to PET brain studies, and were performed using 90-min dynamic acquisitions after tracer injections. Given the difference in kinetics for each tracer, we generated standard uptake value (SUV) images using 60–90 min post-injection data for ^{18}F-FDG, 40–60 min post-injection data for ^{11}C-UCB-J, and 50–70 min post-injection data for ^{11}C-PiB. All PET sinograms were reconstructed into $256 \times 256 \times 207$ volume size with the voxel size of $1.219 \times 1.219 \times 1.231$ mm^3. For each patient, the three tracer PET scans and MR imaging were performed on different days, and the PET images were registered to the MR. The volume size after registration is $176 \times 240 \times 256$ with a voxel size of $1.2 \times 1.055 \times 1.055$ mm^3. We normalized our data by first dividing the maximal value, and then multiplying it by 2, and then minus 1 to ensure the intensity lies in $[-1, 1]$. During the evaluation, the data are denormalized.

We performed five-fold cross validation, where we divided our dataset into a training set of 28 studies and test set of 7 studies in each fold validation. The evaluation was performed on the all 35 studies with all tracer domain translations evaluated. For quantitative evaluation, the performance were assessed using Normalized Mean Square Error (NMSE) and Structural Similarity Index (SSIM) by comparing the predicted tracer volumes and the ground-truth tracer volumes. To evaluate the bias in important brain ROIs, we calculated each ROI's bias using:

$$Bias = \frac{\sum_{m \in R} Y_m^{pred} - \sum_{m \in R} Y_m^{gt}}{\sum_{m \in R} Y_m^{gt}} \tag{8}$$

where R is the specific ROI chosen in the brain. Y^{pred} and Y^{gt} are the predicted tracer volume and ground-truth tracer volume, respectively. For comparative study, we compared our results against the previous translation methods, including cGAN [10] and StarGAN [14].

3   Results

The qualitative results of our UCAN is shown in Fig. 2. As we can see, given any single tracer volume, our UCAN can generate reconstructions for the remaining two tracers. Because glucose metabolism (measured by tracer A) is positively correlated to the amount of the SV2A (measured by tracer B), the general appearance is similar between image from tracer A and tracer B, except in regions such as thalamus (blue arrows). On the other hand, tracer C (^{11}C-PiB) has a low correlation with tracer A (^{18}F-FDG) and tracer B (^{11}C-UCB-J), the translations are more challenging for $C \rightarrow A$, $C \rightarrow B$, $A \rightarrow C$, and $B \rightarrow C$. While the synthetic results are still not quite consistent with the ground truth globally, we can observe that consistent tracer distribution can be generated in some regions of tracer C (e.g. green arrows).

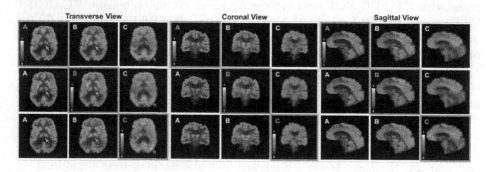

Fig. 2. Multi-tracer synthesis results using **UCAN with MR**. Input PET volume (blue text/box) and corresponding synthesized PET volumes (white text/box on the same row) are visualization in transverse, coronal, and sagittal 2D views. All volumes are displayed with the same window/level. A: ^{18}F-FDG, B: ^{11}C-UCB-J, C: ^{11}C-PiB. (Color figure online)

Table 1 outlines the quantitative comparison of different methods. Specifically, we compared our UCAN with the classical one-to-one fixed translation method (cGAN [10]), and the unified translation method (StarGAN [14]). Both NMSE and SSIM are evaluated for any two tracers' translations. Our UCAN without MR (UCAN-MR) achieves slightly better performance compared to the conventional one-to-one cGAN, while using only one model. Our UCAN-MR also demonstrates superior performance as compared to the previous unified StarGAN for all the translation paths. Incorporating anatomy information into UCAN significantly improves the translation performance for all the translation paths. For the most challenging translation paths, such as $C \to A$ and $C \to B$, we reduced the NMSE from 19.13 to 10.23 and 16.81 to 8.63, respectively.

Table 1. Quantitative comparison of tracer synthesis results using SSIM and NMSE. The optimal and second optimal results are marked in red and **blue**. A: ^{18}F-FDG, B: ^{11}C-UCB-J, C: ^{11}C-PiB. † means the difference between ours and cGAN are significant at $p < 0.05$.

SSIM/NMSE ($\times 100\%$)	$A \to B$	$A \to C$	$B \to A$	$B \to C$	$C \to A$	$C \to B$
cGAN [10]×6	.767/16.91	.714/19.13	.797/19.21	.689/23.77	.752/24.26	.779/18.81
StarGAN [14]	.701/36.83	.585/40.13	.601/43.23	.574/38.99	.609/45.20	.607/37.16
UCAN-MR	.789/14.76	.728/18.82	.814/15.81	.691/19.98	.823/19.13	.802/16.81
UCAN+MR	.841/8.94†	.764/14.58†	.899/8.21†	.773/13.64†	.871/10.23†	.821/8.63†

Table 2. Ablation study on our UCAN with different configurations. ✓ and ✗ means module used and not used in our UCAN.

SSIM/NMSE	$\mathcal{L}_{pair}$	DuSE	$A \to B$	$A \to C$	$B \to A$	$B \to C$	$C \to A$	$C \to B$
UCAN-MR	✗	✗	.710/34.41	.589/39.36	.608/41.64	.592/38.68	.621/43.41	.612/35.00
	✓	✗	.779/16.99	.718/25.63	.811/18.32	.620/25.12	.804/26.24	.771/21.20
	✗	✓	.774/17.59	.611/39.19	.810/20.85	.600/38.11	.647/36.55	.747/28.01
	✓	✓	.789/14.76	.728/18.82	.814/15.81	.691/19.98	.823/19.13	.802/16.81
UCAN+MR	✗	✗	.766/22.12	.647/24.61	.807/23.98	.640/22.56	.821/21.57	.800/17.91
	✓	✗	.798/10.44	.724/18.84	.866/9.03	.742/17.63	.847/12.96	.808/12.89
	✗	✓	.778/16.91	.653/23.67	.811/19.31	.649/21.57	.828/20.02	.802/16.97
	✓	✓	.841/8.94	.764/14.58	.899/8.21	.773/13.64	.871/10.23	.821/8.63

We performed ablation studies on our UCAN with different configurations, including with or without pair loss, 3D DuSE block, and MR information. The quantitative results are summarized in Table 2. UCAN without pair loss is difficult to generate the correct tracer image with low correlation, such as $C \to A/B$. Adding 3D DuSE block along with pair loss allows UCAN to better fuse the input information, thus generating better synthesis results. UCAN with pair loss, DuSE block, and MR information provides the best synthesis results. The corresponding visualization comparison are included in our supplementary.

In parallel, we evaluated the bias of our UCAN+MR results in 9 important ROIs, and the results are summarized in Fig. 3. From the results, the mean bias and associated standard deviation are no more than 15% for all the ROIs, including hippocampus and entorhinal cortex which are two of the first areas impaired by AD and is an important region for AD early diagnosis.

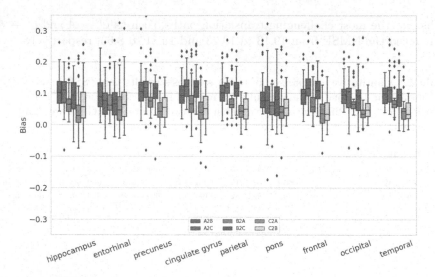

Fig. 3. Quantitative evaluation of bias in 9 ROIs for UCAN+MR. Mean and standard deviation of bias in 9 AD-related brain ROIs is illustrated for the synthesis between all three tracers. A: ^{18}F-FDG, B: ^{11}C-UCB-J, C: ^{11}C-PiB.

4   Conclusion

In this work, we proposed a 3D unified anatomy-aware cyclic adversarial network (UCAN), a framework for translating multi-tracer PET volumes with one unified generative model. Our UCAN consisting of four key loss components and domain label volume allows us to take tracer volume/MR volume/tracer domain label as input, and learns to flexibly translate the tracer volume into the target tracer domain. Moreover, the DuSE-Net in our UCAN allows us to better fuse multiple input information for the unified synthesis tasks. Preliminary evaluation using human studies suggested the feasibility that our method is able to generate high-quality multi-tracer PET volumes with acceptable bias in important AD-related ROIs.

References

1. Valotassiou, V., et al.: SPECT and PET imaging in Alzheimer's disease. Ann. Nucl. Med. **32**(9), 583–593 (2018)
2. Cohen, A.D., Klunk, W.E.: Early detection of Alzheimer's disease using PiB and FDG pet. Neurobiol. Dis. **72**, 117–122 (2014)
3. Finnema, S.J., et al.: Imaging synaptic density in the living human brain. Sci. Transl. Med. **8**(348), 348ra96–348ra96 (2016)
4. Klunk, W.E., et al.: Imaging brain amyloid in Alzheimer's disease with Pittsburgh compound-B. Ann. Neurol. Off. J. Am. Neurol. Assoc. Child Neurol. Society **55**(3), 306–319 (2004)

5. Márquez, F., Yassa, M.A.: Neuroimaging biomarkers for Alzheimer's disease. Mol. Neurodegener. **14**(1), 21 (2019)
6. Chen, Z., Brodie, M.J., Liew, D., Kwan, P.: Treatment outcomes in patients with newly diagnosed epilepsy treated with established and new antiepileptic drugs: a 30-year longitudinal cohort study. JAMA Neurol. **75**(3), 279–286 (2018)
7. Nabulsi, N., et al.: [11c] UCB-J: a novel pet tracer for imaging the synaptic vesicle glycoprotein 2A (SV2A). J. Nucl. Med. **55**(supplement 1), 355–355 (2014)
8. Sperling, R.A., et al.: Toward defining the preclinical stages of Alzheimer's disease: recommendations from the national institute on aging-alzheimer's association workgroups on diagnostic guidelines for alzheimer's disease. Alzheimer's Dementia **7**(3), 280–292 (2011)
9. Ronneberger, O., Fischer, P., Brox, T.: U-Net: convolutional networks for biomedical image segmentation. In: Navab, N., Hornegger, J., Wells, W.M., Frangi, A.F. (eds.) MICCAI 2015. LNCS, vol. 9351, pp. 234–241. Springer, Cham (2015). https://doi.org/10.1007/978-3-319-24574-4_28
10. Isola, P., Zhu, J.Y., Zhou, T., Efros, A.A.: Image-to-image translation with conditional adversarial networks. In: Proceedings of the IEEE Conference on Computer Vision and Pattern Recognition, pp. 1125–1134 (2017)
11. Sikka, A., Peri, S.V., Bathula, D.R.: MRI to FDG-PET: cross-modal synthesis using 3D U-Net for multi-modal Alzheimer's classification. In: Gooya, A., Goksel, O., Oguz, I., Burgos, N. (eds.) SASHIMI 2018. LNCS, vol. 11037, pp. 80–89. Springer, Cham (2018). https://doi.org/10.1007/978-3-030-00536-8_9
12. Zhou, B., Lin, X., Eck, B., Hou, J., Wilson, D.: Generation of virtual dual energy images from standard single-shot radiographs using multi-scale and conditional adversarial network. In: Jawahar, C.V., Li, H., Mori, G., Schindler, K. (eds.) ACCV 2018. LNCS, vol. 11361, pp. 298–313. Springer, Cham (2019). https://doi.org/10.1007/978-3-030-20887-5_19
13. Wang, R., et al.: Generation of synthetic pet images of synaptic density and amyloid from 18 F-FDG images using deep learning. Med. Phys. (2021)
14. Choi, Y., Choi, M., Kim, M., Ha, J.W., Kim, S., Choo, J.: StarGAN: unified generative adversarial networks for multi-domain image-to-image translation. In: Proceedings of the IEEE Conference on Computer Vision and Pattern Recognition, pp. 8789–8797 (2018)
15. Odena, A., Olah, C., Shlens, J.: Conditional image synthesis with auxiliary classifier GANs. In: Proceedings of the 34th International Conference on Machine Learning-Volume 70, pp. 2642–2651. JMLR. org (2017)
16. Çiçek, Ö., Abdulkadir, A., Lienkamp, S.S., Brox, T., Ronneberger, O.: 3D U-Net: learning dense volumetric segmentation from sparse annotation. In: Ourselin, S., Joskowicz, L., Sabuncu, M.R., Unal, G., Wells, W. (eds.) MICCAI 2016. LNCS, vol. 9901, pp. 424–432. Springer, Cham (2016). https://doi.org/10.1007/978-3-319-46723-8_49
17. Roy, A.G., Navab, N., Wachinger, C.: Recalibrating fully convolutional networks with spatial and channel "squeeze and excitation" blocks. IEEE Trans. Med. Imaging **38**(2), 540–549 (2018)
18. Hu, J., Shen, L., Sun, G.: Squeeze-and-excitation networks. In: Proceedings of the IEEE Conference on Computer Vision and Pattern Recognition, pp. 7132–7141 (2018)

Generalised Super Resolution for Quantitative MRI Using Self-supervised Mixture of Experts

Hongxiang Lin[1,2,3]($\boxtimes$), Yukun Zhou[2,4,5], Paddy J. Slator[2,3], and Daniel C. Alexander[2,3]

[1] Research Center for Healthcare Data Science, Zhejiang Lab, Hangzhou, China
hxlin@zhejianglab.edu.cn
[2] Centre for Medical Image Computing, University College London, London, UK
[3] Department of Computer Science, University College London, London, UK
[4] Department of Medical Physics and Biomedical Engineering, UCL, London, UK
[5] NIHR Biomedical Research Centre at Moorfields Eye Hospital, London, UK

Abstract. Multi-modal and multi-contrast imaging datasets have diverse voxel-wise intensities. For example, quantitative MRI acquisition protocols are designed specifically to yield multiple images with widely-varying contrast that inform models relating MR signals to tissue characteristics. The large variance across images in such data prevents the use of standard normalisation techniques, making super resolution highly challenging. We propose a novel self-supervised mixture-of-experts (SS-MoE) paradigm for deep neural networks, and hence present a method enabling improved super resolution of data where image intensities are diverse and have large variance. Unlike the conventional MoE that automatically aggregates expert results for each input, we explicitly assign an input to the corresponding expert based on the predictive pseudo error labels in a self-supervised fashion. A new gater module is trained to discriminate the error levels of inputs estimated by Multi-scale Quantile Segmentation. We show that our new paradigm reduces the error and improves the robustness when super resolving combined diffusion-relaxometry MRI data from the Super MUDI dataset. Our approach is suitable for a wide range of quantitative MRI techniques, and multi-contrast or multi-modal imaging techniques in general. It could be applied to super resolve images with inadequate resolution, or reduce the scanning time needed to acquire images of the required resolution. The source code and the trained models are available at https://github.com/hongxiangharry/SS-MoE.

Keywords: Self supervision · Mixture of experts · Quantitative MRI · Generalised super resolution · Pseudo labels

---

H. Lin and Y. Zhou contributed equally.

Electronic supplementary material The online version of this chapter (https://doi.org/10.1007/978-3-030-87231-1_5) contains supplementary material, which is available to authorized users.

M. de Bruijne et al. (Eds.): MICCAI 2021, LNCS 12906, pp. 44–54, 2021.
https://doi.org/10.1007/978-3-030-87231-1_5

1   Introduction

Quantitative Magnetic Resonance Imaging (qMRI) can measure and map chemical, physical, and physiological values that strongly relate to underlying tissue structure and function. Such measurements have the potential to improve diagnosis, prognosis and monitoring of a wide variety of diseases. However, qMRI has not yet been widely used in the clinic, due to long acquisition times and noise sensitivity. Super-resolution (SR) reconstruction techniques enable images with the same spatial resolution to be acquired within reduced scanning times and with improved signal-to-noise ratios [30]. Improved SR techniques can hence increase the likelihood of clinical adoption of qMRI, as well as similar multi-modal or multi-contrast imaging techniques, such as multi-contrast X-ray [32] and multi-modal functional imaging [29].

Deep learning based SR for medical imaging has demonstrated significant improvements over existing techniques [4,5,18,19,33,34]. However, the data normalisation required for deep learning SR hinders its application to multi-modal or multi-contrast techniques such as qMRI, as such imaging datasets have diverse voxel-wise intensities leading to large variances that prevent the use of standard normalisation techniques. In conventional MRI SR, intensity normalisation is performed within single images using a method such as Z-score, fussy C-mean, or Gaussian mixture model [22]. These approaches can be applied to individual qMRI images sequentially to normalise the intensity scale, but this affects the relationship between voxelwise intensities and MR sequence parameters, biasing downstream analyses that interpret these relationships to estimate underlying tissue properties. An alternative, used by most state-of-the-art deep learning architectures in computer vision and medical imaging, is batch normalisation [13]. However, similarly to intensity normalisation, the reconstruction accuracy degrades rapidly when the training batches have a large variance [23].

To generalise SR to data with large underlying variance, such as qMRI, we propose a self-supervised mixture-of-experts (SS-MoE) paradigm that can augment any encoder-decoder network backbone. The conventional mixture of experts automatically aggregates expert results for each input; see [14] and its recent extensions in [8,24,35]. Unlike this, our proposed SS-MoE discriminates an input data by predicting the error class in a self-supervised fashion, to make hard assignments to the corresponding expert decoder network [8]. This divide-and-conquer strategy predictively clusters input data belonging to the same error level, thereby reducing the variance in each cluster. As such, the resulting outputs of the multi-hand networks formulate a predictive distribution with multiple peaks with respect to error clusters, rather than a single peak; see a similar analysis developed in [31]. In this paper, we select a U-Net variant as an encoder-decoder network backbone [11,17], which was employed in the Image Quality Transfer framework, a patch-based machine learning approach used to enhance the resolution and/or contrast in diffusion MRI and structural MRI [2,3,17,26,27]. We apply our method to Super MUDI challenge [21] combined diffusion-relaxometry MRI data; this is a challenging qMRI dataset for SR

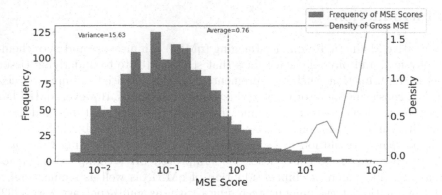

Fig. 1. (Blue block) Histogram of mean-squared-error (MSE) between interpolated LR volumes and HR volumes in Subject cdmri0015 in the first anisotropic Super MUDI Dataset. All MSE scores were clustered into 30 bins within log scale. (Red line) Density of the gross MSE scores over all volumes. Cubic spline method was used to interpolate LR volume. (Color figure online)

as the simultaneous inversion recovery, multi-echo gradient echo, and diffusion sequences all combine to yield large intensity variance in voxels across volumes.

Our main contributions are: 1) To reduce population variance, we separately assign the inputs into multiple expert networks based on the predictive pseudo error labels. 2) We train a new gater module to predict the pseudo error labels from the extracted high-level perceptual features from the baseline network; pseudo error labels are typically estimated by unsupervised or heuristic ways, and are used to train the gater. Our overall paradigm is non-end-to-end so is potentially extendable to most other encoder-decoder network architectures.

2 Method

2.1 Data Description

We perform SR on the publicly available Super MUDI dataset [1,21], which comprises combined diffusion-relaxometry brain scans on 5 healthy subjects using the ZEBRA technique [12]. Each subject comprises: original isotropic data with $2.5 \times 2.5 \times 2.5$ mm high-resolution (HR) voxels, corresponding 2×-downsampled anisotropic data with $2.5 \times 2.5 \times 5$ mm low-resolution (LR) voxels, and 2×-downsampled isotropic data with $5 \times 5 \times 5$ mm LR voxels. Thus, two super resolution tasks can be defined: 2× through-plane SR and 2× isotropic SR. We split each kind of data to have 6720 3D volumes to enable an error analysis across the volumes on a subject. We observed that the majority of reconstruction errors are concentrated in regions with small errors, whilst the largest errors contribute most to the overall error; see Fig. 1.

Next, to establish the training paired patches for the two tasks, we first randomly cropped N patches of the shape $(16, 16, 16)$ whose non-background voxels account for over 50% of patch volume from the original resolution data to serve as the ground-truth HR patch y_i, where $i = 1, \cdots, N$. To form the corresponding LR half x_i, we cropped the same field of view, respectively from the two downsampled data.

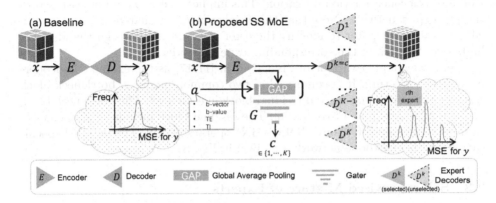

Fig. 2. Conceptual comparison of (a) the baseline network and (b) the proposed self-supervised mixture-of-expert (SS-MoE) network. The two networks are commonly built on an encoder-decoder (E-D) architecture. The gater G infers the pseudo error label c by inputting a combination of the acquisition parameter and GAP-encoded features from the input LR patch x. (a) The HR patch y is predicted from a single network that is trained on all data with diverse intensity levels. The predicted HR patch will be subject to a distribution with single peak centred at the average MSE score. (b) The network performs in two stages: we first infer the pseudo error label c for the input LR patch x. Then the HR patch outputs via the particular expert decoder D^c identified by the error class c. Under this framework, the output HR patches demonstrate multi-peak distribution which satisfies the need of quantitative super resolution with diverse intensities.

2.2 Backbone Network Architecture

Figure 2(a) and (b) show two digraphs of the proposed backbone network, where its nodes represent the block of neural network layers and the edges are directed. Let adjacency matrices E, D, and G be the encoder, decoder, gater branches, respectively, and let $E^O := O \circ E$ output the last activation in the encoder where O denotes an adjacency matrix used to operate Global Average Pooling (GAP) to the last activation node of the encoder. The nodes in E and D consist of regular convolutional neural networks with the down- or up-sampling operations, whereas the nodes in G comprise a feedforward network with a softmax activation at the end. Given the input LR patch x a combination of E and D outputs the HR patches y:

$$y = D \circ E(x). \tag{1}$$

Moreover, we can incorporate the additional condition of the MR acquisition parameter a into the encoder features, making the error class estimation more robust in terms of the scanning process. A combination of E^O and G outputs a predictive pseudo error label c:

$$c = G \circ [E^O(x), a], \tag{2}$$

where $[\cdot]$ is a concatenation operation. This implies that x should be assigned to the cth expert network. Note that the weights of E are shared in Eqs. 1 and 2, which enables the gater to identify the error class and then assign the identical high-level features to the corresponding expert decoder.

Here, we adopt a variant of SR U-Net in [11,17] as the exampled encoder-decoder architecture. It is comprised of two common modules, bottleneck block right before concatenation and residual block in each level. The bottleneck block has a similar design as FSRCNN [6]. The residual block includes a skip connection over a number of "Conv+ReLU+BN" operations [10]. The detailed specifications of the backbone network are given in Fig. S1.

2.3   Self-supervised Mixture of Experts

Training Phase One: Estimate Pseudo Error Labels via Baseline Model. We first train a baseline model based on Eq. 1. Given N training LR-HR patch pairs $\{(x_i, y_i)\}_{i=1}^{N}$, we optimise the weights θ, φ in terms of the baseline encoder and decoder by minimising the mean-squared-error (MSE) loss function $\mathcal{L}$:

$$\theta^*, \varphi^* = \arg \min_{\theta, \varphi} \sum_{i=1}^{N} \mathcal{L}(y_i, D_\varphi \circ E_\theta(x_i)). \tag{3}$$

The trained encoder and decoder can be used to calculate the baseline MSE score: $e_i = \mathcal{L}(y_i, D_{\varphi^*} \circ E_{\theta^*}(x_i))$.

Next, we estimate K clusters from the obtained error scores. Multiscale quantile segmentation (MQS) is a way to partition the error scores into K clusters at $K-1$ quantiles [15]. Here, we adopted MQS, denoted by $\mathcal{J}$, to presumably identify pseudo error labels $c_i \in \{1, \cdots, K\}$ mapped from e_i, i.e.

$$c_i = \mathcal{J}(e_i) = \mathcal{J}\left(\mathcal{L}(y_i, D_{\varphi^*} \circ E_{\theta^*}(x_i))\right). \tag{4}$$

The estimated c_i will be used as the ground-truth labels when training the gater G in Phase Two. We also tested on other alternative segmenters such as the empirical rule[1] or K-means clustering, but observed that the overall approach performed best using MQS.

Training Phase Two: Train Gater to Classify Input Patch for Expert Network. We adopt a supervised way to train the gater G. The detailed architecture of G is specified in the supplementary material. Given the trained encoder

---

[1] The rule empirically selects an equispaced grid along a power-law distribution as class boundaries.

E_{θ^*}, the input LR patch x_i, the pseudo error class c_i, and the acquisition parameter a_i, we optimise the weights ψ of the gater by minimising the cross-entropy loss $\mathcal{L}_{CE}$:

$$\psi^* = \arg\min_{\psi} \sum_{i=1}^{N} \mathcal{L}_{CE}(c_i, G_{\psi} \circ [E_{\theta^*}^{O}(x_i), a_i]). \tag{5}$$

Then we calculate the predictive pseudo error labels $\tilde{c}_i$ by Eq. 2, that is, $\tilde{c}_i = G_{\psi^*} \circ [E_{\theta^*}^{O}(x_i), a_i]$.

We finally group up LR-HR patch pairs into K subsets according to their error class indices. We denote the kth subset as $S^{(k)}$, where $k = 1, \cdots, K$, and rename the indices of LR-HR patch pairs into the kth subset as $S^{(k)} = \{(x_i^{(k)}, y_i^{(k)}) : i = 1, \cdots, N^{(k)}\}$. Each subset will be used to train an expert network in Phase Three.

Training Phase Three: Train Multiple Expert Networks with Assigned Training Patches. We freeze the encoder section and train multiple decoders with respect to the aforementioned split training subsets. Given any $S^{(k)}$ for $k = 1, \cdots, K$, we optimise the weights $\varphi^{(k)}$ of the kth decoder $D^{(k)}$ in a way similar to Eq. 3:

$$\varphi^{(k)*} = \arg\min_{\varphi^{(k)}} \sum_{i=1}^{N^{(k)}} \mathcal{L}(y_i^{(k)}, D_{\varphi^{(k)}}^{(k)} \circ E_{\theta^*}(x_i^{(k)})). \tag{6}$$

Usually, we can train the decoder $D^{(k)}$ from scratch. However, when one subset has a relatively small number of data, we choose to initialise the decoder weights with the pre-trained baseline decoder weights φ^*, and then continuously train on the subset $S^{(k)}$.

Test Phase. At the test phase, we need to first predict the pseudo error label by the test LR patch $\hat{x}$ and its acquisition parameter $\hat{a}$, and then assign $\hat{x}$ to the corresponding expert network to predict the output HR patch $\hat{y}$. Specifically, we predict the pseudo error class label $\hat{c}$ by Eq. 2:

$$\hat{c} = G_{\psi^*} \circ [E_{\theta^*}^{O}(\hat{x}), \hat{a}], \tag{7}$$

and then predict the HR output by Eq. 1:

$$\hat{y} = D_{\varphi^{\hat{c}}}^{\hat{c}} \circ E_{\theta^*}(\hat{x}). \tag{8}$$

3 Experiments

3.1 Implementation Details

The overall method was implemented by Tensorflow 2.0. Our program is required to run on an Nvidia GPU with at least 12 GB memory. For training, we used

ADAM [16] as the optimiser with a starting learning rate of 10^{-3} and a decay of 10^{-6}. We set the batch size as 64. We initialised the network weights with Glorot normal initialiser [7]. All networks required 100/20/20 training epochs respectively from Phase One to Three. All networks at Phases One and Two were trained on uniformly sampled patch pairs of around $270k^2$, while at Phase Three, the expert networks fine-tuned through randomly sampled 100k pairs for speedup. MSE was used as both the loss function and the evaluation metric. We employed 5-fold cross validation to evaluate the proposed method on the Super MUDI datasets. Specifically, one of the subjects containing 1344 volumes was used for validation on a fold, and we randomly sampled 100k patch pairs out of the remaining 5376 volumes for training. We employed the statistics, such as average, variance, and quartiles, to summarise of the distribution of MSE scores over all volumes. We used a two-tailed Wilcoxon signed-rank test to determine statistical significance of the performance between any two compared methods.

3.2   Results

Since the results comprised 1344 MSE scores we used descriptive statistics to characterise the distribution of MSE scores. We conducted the comparative study over cubic spline interpolation, SR U-Net [17] as a backbone, Hard MoE [8] as a baseline model, and the proposed SS-MoE. All the neural networks had comparable model capacity. In Table 1, we observe that SS-MoE had the best performance over the others measured by average, variance, and median, and significantly reduced maximal MSE score; The full MSE score distributions are shown in Fig. S2. With Table S1, we further confirmed that SS-MoE boosted the performance of the SR U-Net backbone, outperformed over nearly all the rest methods with statistical significance ($p < 0.001$), and reduced the distribution variance. To visualise SR performance in an individual volume map, Fig. 3 compares the coronal views of different reconstructed results on the 119th volume of the last subject for 2× through-plane SR task. We observe that our proposed method enhanced resolution and showed lower error score in a zoomed region.

We further analysed the effect of hyper-parameters by increasing the number of pseudo error classes as shown in Table 2. We observe a large improvement over all statistics when increasing from 2 to 4, and a modest improvement going from 4 to 8. The computational cost grew several times with more pseudo error classes since the number of network weights increased. Considering the cost-performance ratio, we recommend choosing 4 pseudo error classes for SS-MoE in this context. We infer from Fig. S3 that the mis-classified labels were mostly concentrated around the hard boundary, which implied that clusters of predictive error labels may be overlapping but may not largely degrade the performance of SS-MoE.

---

[2] Uniformly crop patches by the function `extract_patches` in scikit-learn 0.22.

Table 1. 5-fold cross-validation results for cubic spline interpolation, SR U-Net, Hard MoE and the proposed SS-MoE both with the SR U-Net backbone. Two SR tasks were performed on the anisotropic-voxel (Aniso.) and the isotropic-voxel (Iso.) Super MUDI datasets. For each fold, we evaluated MSE scores on 1344 volumes and used their statistics (Stats.) to characterise the distribution. The mean and std of the average statistics over the 5 cross-validation folds were computed.

Dataset	Stats.	Cubic spline interpolationk	SR U-Net [17] (Baseline)	Hard MoE [8] (MoE baseline)	SS-MoE (Proposed)
Aniso.	Average	0.744 ± 0.010	0.363 ± 0.049	0.383 ± 0.042	**0.305 ± 0.048**
	Variance	14.899 ± 1.496	3.607 ± 1.105	4.230 ± 0.857	**2.765 ± 0.941**
	Median	0.104 ± 0.002	0.048 ± 0.006	0.045 ± 0.008	**0.041 ± 0.006**
	Max	97.28 ± 12.02	50.45 ± 6.77	55.24 ± 4.64	**43.59 ± 5.94**
Iso.	Average	1.583 ± 0.014	0.658 ± 0.075	0.717 ± 0.088	**0.648 ± 0.036**
	Variance	62.905 ± 4.2393	13.927 ± 3.893	17.289 ± 5.353	**13.829 ± 4.064**
	Median	0.2351 ± 0.0029	0.0781 ± 0.010	0.083 ± 0.009	**0.075 ± 0.010**
	Max	194.89 ± 18.67	96.64 ± 11.12	108.91 ± 14.27	**96.09 ± 11.01**

Table 2. Statistics of MSE-score distribution on the isotropic-voxel Super MUDI dataset v.s. the number of pseudo error classes (#Classes) in SS-MoE. The number of network Weights (#Weights) are given. All the experiments were validated on the setup of the first cross-validation fold that was used to train the SS-MoE model and predict on the same 1344 volumes.

#Classes	#Weights	Average	Variance	Min	Q1	Median	Q3	Max
2	4.42×10^6	0.5593	9.8079	0.0032	0.0230	0.0664	0.2297	85.0411
4	8.76×10^6	0.5537	9.5210	0.0032	0.0228	0.0653	0.2294	83.8385
8	1.74×10^7	0.5527	9.4552	0.0031	0.0228	0.0649	0.2282	83.3759

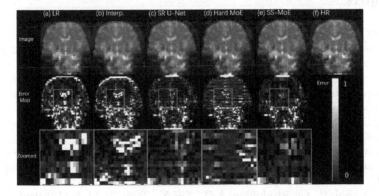

Fig. 3. Visual comparison of the coronal views by (a) LR, (b) Interpolation (Interp.), (c) the variant SR U-Net backbone, (d) the Hard MoE baseline, (e) the proposed SS-MoE, and (f) HR images on the 119th volume of the last subject for 2× through-plane SR task. The error maps are normalised square difference between the reconstructed volumes and the HR volumes for each voxel. Zoomed regions of the error maps are highlighted.

4   Discussion and Conclusion

We propose a novel SS-MoE paradigm for SR of multi-modal or multi-contrast imaging datasets that have diverse intensities and large variance, such as [9, 28,29,32]. Our SS-MoE approach can append to any baseline encoder-decoder network, allowing incorporation of state-of-the-art SR networks; in this paper we utilised the best deep neural network in the leaderboard of the Super MUDI challenge. We demonstrate that our approach reduces both errors and variances when super resolving combined diffusion-relaxometry qMRI data. The proposed SS-MoE performed better than MoE due to convex loss function enabling robust training and memory footprint independent to the number of experts [20].

The SS-MoE paradigm also provides a way for future improvement and application. First, the pseudo error labels are estimated and then predicted in a self-supervised way, and hence the gater used to predict them may highly depend on the segmenters, such as MQS, and how good the baseline model is. This may limit the entire performance of SS-MoE, especially for the generalisablility of the gater; see the supplementary material. Automatically discriminating the inputs without using the baseline model like the idea in [25] will be valuable to explore. On the other hand, our method has the potential to super resolve a variety of qMRI data types, ultimately accelerating the acquisition process and increasing clinical viability. In future work, we will investigate if super resolved images offer better visualisation of pathologies, such as lesions or tumours.

Acknowledgements. This work was supported by EPSRC grants EP/M020533/1, EP/R014019/1, and EP/V034537/1 as well as the NIHR UCLH Biomedical Research Centre.

References

1. CDMRI super MUDI challenge 2020. https://www.developingbrain.co.uk/data/
2. Alexander, D.C., Zikic, D., Ghosh, A., et al.: Image quality transfer and applications in diffusion MRI. Neuroimage **152**, 283–298 (2017)
3. Blumberg, S.B., Tanno, R., Kokkinos, I., Alexander, D.C.: Deeper image quality transfer: training low-memory neural networks for 3D images. In: Frangi, A.F., Schnabel, J.A., Davatzikos, C., Alberola-López, C., Fichtinger, G. (eds.) MICCAI 2018. LNCS, vol. 11070, pp. 118–125. Springer, Cham (2018). https://doi.org/10.1007/978-3-030-00928-1_14
4. Chen, G., Dong, B., Zhang, Y., Lin, W., Shen, D., Yap, P.T.: XQ-SR: joint x-q space super-resolution with application to infant diffusion MRI. Med. Image Anal. **57**, 44–55 (2019)
5. Chen, Y., Shi, F., Christodoulou, A.G., Xie, Y., Zhou, Z., Li, D.: Efficient and accurate MRI super-resolution using a generative adversarial network and 3D multi-level densely connected network. In: Frangi, A.F., Schnabel, J.A., Davatzikos, C., Alberola-López, C., Fichtinger, G. (eds.) MICCAI 2018. LNCS, vol. 11070, pp. 91–99. Springer, Cham (2018). https://doi.org/10.1007/978-3-030-00928-1_11

6. Dong, C., Loy, C.C., Tang, X.: Accelerating the super-resolution convolutional neural network. In: Leibe, B., Matas, J., Sebe, N., Welling, M. (eds.) ECCV 2016. LNCS, vol. 9906, pp. 391–407. Springer, Cham (2016). https://doi.org/10.1007/978-3-319-46475-6_25

7. Glorot, X., Bengio, Y.: Understanding the difficulty of training deep feedforward neural networks. In: Proceedings of the Thirteenth International Conference on Artificial Intelligence and Statistics, pp. 249–256 (2010)

8. Gross, S., Ranzato, M., Szlam, A.: Hard mixtures of experts for large scale weakly supervised vision. In: 2017 IEEE Conference on Computer Vision and Pattern Recognition (CVPR), pp. 5085–5093. IEEE (2017)

9. Grussu, F., et al.: Multi-parametric quantitative in vivo spinal cord MRI with unified signal readout and image denoising. Neuroimage **217**, 116884 (2020)

10. He, K., Zhang, X., Ren, S., Sun, J.: Deep residual learning for image recognition. In: 2016 IEEE Conference on Computer Vision and Pattern Recognition (CVPR), pp. 770–778. IEEE (2016)

11. Heinrich, L., Bogovic, J.A., Saalfeld, S.: Deep learning for isotropic super-resolution from non-isotropic 3D electron microscopy. In: Descoteaux, M., Maier-Hein, L., Franz, A., Jannin, P., Collins, D.L., Duchesne, S. (eds.) MICCAI 2017. LNCS, vol. 10434, pp. 135–143. Springer, Cham (2017). https://doi.org/10.1007/978-3-319-66185-8_16

12. Hutter, J., Slator, P.J., Christiaens, D., et al.: Integrated and efficient diffusion-relaxometry using ZEBRA. Sci. Rep. **8**(1), 1–13 (2018)

13. Ioffe, S., Szegedy, C.: Batch normalization: accelerating deep network training by reducing internal covariate shift. In: Proceedings of the 32nd International Conference on Machine Learning. Proceedings of Machine Learning Research, vol. 37, pp. 448–456. PMLR, Lille (2015)

14. Jacobs, R.A., Jordan, M.I., Nowlan, S.E., Hinton, G.E.: Adaptive mixture of experts (1991)

15. Jula Vanegas, L., Behr, M., Munk, A.: Multiscale quantile segmentation. J. Am. Stat. Assoc., 1–14 (2021)

16. Kingma, D.P., Ba, J.: Adam: a method for stochastic optimization. arXiv preprint arXiv:1412.6980 (2014)

17. Lin, H., et al.: Deep learning for low-field to high-field MR: image quality transfer with probabilistic decimation simulator. In: Knoll, F., Maier, A., Rueckert, D., Ye, J.C. (eds.) MLMIR 2019. LNCS, vol. 11905, pp. 58–70. Springer, Cham (2019). https://doi.org/10.1007/978-3-030-33843-5_6

18. Lyu, Q., et al.: Multi-contrast super-resolution MRI through a progressive network. IEEE Trans. Med. Imaging **39**(9), 2738–2749 (2020)

19. Ma, J., Yu, J., Liu, S., et al.: PathSRGAN: multi-supervised super-resolution for cytopathological images using generative adversarial network. IEEE Trans. Med. Imaging **39**(9), 2920–2930 (2020)

20. Makkuva, A., Viswanath, P., Kannan, S., Oh, S.: Breaking the gridlock in mixture-of-experts: consistent and efficient algorithms. In: Proceedings of the 36th International Conference on Machine Learning. Proceedings of Machine Learning Research, vol. 97, pp. 4304–4313. PMLR (2019)

21. Pizzolato, M., Palombo, M., Bonet-Carne, E., et al.: Acquiring and predicting multidimensional diffusion (MUDI) data: an open challenge. In: Gyori, N., Hutter, J., Nath, V., Palombo, M., Pizzolato, M., Zhang, F. (eds.) Computational Diffusion MRI, pp. 195–208. Springer, Cham (2020). https://doi.org/10.1007/978-3-030-73018-5

22. Reinhold, J.C., Dewey, B.E., Carass, A., Prince, J.L.: Evaluating the impact of intensity normalization on MR image synthesis. In: Medical Imaging 2019: Image Processing, p. 126. SPIE (2019)
23. Shen, S., Yao, Z., Gholami, A., Mahoney, M., Keutzer, K.: PowerNorm: rethinking batch normalization in transformers. In: Proceedings of the 37th International Conference on Machine Learning. Proceedings of Machine Learning Research, vol. 119, pp. 8741–8751. PMLR (2020)
24. Shi, Y., Siddharth, N., Paige, B., Torr, P.: Variational mixture-of-experts autoencoders for multi-modal deep generative models. In: Wallach, H., Larochelle, H., Beygelzimer, A., d'Alché-Buc, F., Fox, E., Garnett, R., (eds.) Advances in Neural Information Processing Systems, vol. 32, Curran Associates, Inc., (2019). https://proceedings.neurips.cc/paper/2019/file/0ae775a8cb3b499ad1fca944e6f5c836-Paper.pdf
25. Tanno, R., Arulkumaran, K., Alexander, D., Criminisi, A., Nori, A.: Adaptive neural trees. In: Proceedings of the 36th International Conference on Machine Learning. Proceedings of Machine Learning Research, vol. 97, pp. 6166–6175. PMLR (2019)
26. Tanno, R., et al.: Bayesian image quality transfer with CNNs: exploring uncertainty in dMRI super-resolution. In: Descoteaux, M., Maier-Hein, L., Franz, A., Jannin, P., Collins, D.L., Duchesne, S. (eds.) MICCAI 2017. LNCS, vol. 10433, pp. 611–619. Springer, Cham (2017). https://doi.org/10.1007/978-3-319-66182-7_70
27. Tanno, R., Worrall, D.E., Kaden, E., et al.: Uncertainty modelling in deep learning for safer neuroimage enhancement: demonstration in diffusion MRI. NeuroImage **225**, 117366 (2020)
28. Tong, Q., et al.: Multicenter dataset of multi-shell diffusion MRI in healthy traveling adults with identical settings. Sci. Data **7**(1), 1–7 (2020)
29. Van Essen, D.C., Smith, S.M., Barch, D.M., Behrens, T.E., Yacoub, E., Ugurbil, K.: The WU-MINN human connectome project: an overview. Neuroimage **80**, 62–79 (2013)
30. Van Steenkiste, G., Poot, D.H., Jeurissen, B., et al.: Super-resolution T1 estimation: quantitative high resolution T1 mapping from a set of low resolution T1-weighted images with different slice orientations. Magn. Reson. Med. **77**(5), 1818–1830 (2017)
31. Wilson, A.G., Izmailov, P.: Bayesian deep learning and a probabilistic perspective of generalization. In: Advances in Neural Information Processing Systems, vol. 33, pp. 4697–4708. Curran Associates, Inc. (2020)
32. Zhang, R., Garrett, J., Ge, Y., Ji, X., Chen, G.H., Li, K.: Design, construction, and initial results of a prototype multi-contrast X-ray breast imaging system. In: Medical Imaging 2018: Physics of Medical Imaging, vol. 176, p. 31. SPIE (2018)
33. Zhang, Y., Yap, P.T., Chen, G., Lin, W., Wang, L., Shen, D.: Super-resolution reconstruction of neonatal brain magnetic resonance images via residual structured sparse representation. Med. Image Anal. **55**, 76–87 (2019)
34. Zhao, C., Shao, M., Carass, A., et al.: Applications of a deep learning method for anti-aliasing and super-resolution in MRI. Magn. Reson. Imaging **64**, 132–141 (2019)
35. Zheng, Z., et al.: Self-supervised mixture-of-experts by uncertainty estimation. Proc. AAAI Conf. Artif. Intell. **33**, 5933–5940 (2019)

TransCT: Dual-Path Transformer for Low Dose Computed Tomography

Zhicheng Zhang[1(✉)], Lequan Yu[1,2], Xiaokun Liang[1], Wei Zhao[1], and Lei Xing[1]

[1] Department of Radiation Oncology, Stanford University, Stanford, USA
zzc623@stanford.edu

[2] Department of Statistics and Actuarial Science, The University of Hong Kong, Hong Kong, People's Republic of China

Abstract. Low dose computed tomography (LDCT) has attracted more and more attention in routine clinical diagnosis assessment, therapy planning, *etc.*, which can reduce the dose of X-ray radiation to patients. However, the noise caused by low X-ray exposure degrades the CT image quality and then affects clinical diagnosis accuracy. In this paper, we train a transformer-based neural network to enhance the final CT image quality. To be specific, we first decompose the noisy LDCT image into two parts: high-frequency (HF) and low-frequency (LF) compositions. Then, we extract content features (X_{L_c}) and latent texture features (X_{L_t}) from the LF part, as well as HF embeddings (X_{H_f}) from the HF part. Further, we feed X_{L_t} and X_{H_f} into a modified transformer with three encoders and decoders to obtain well-refined HF texture features. After that, we combine these well-refined HF texture features with the pre-extracted X_{L_c} to encourage the restoration of high-quality LDCT images with the assistance of piecewise reconstruction. Extensive experiments on Mayo LDCT dataset show that our method produces superior results and outperforms other methods.

1 Introduction

Computed tomography (CT) system, as noninvasive imaging equipment, has been widely used for medical diagnosis and treatment [16,18]. However, concerns about the increase of X-ray radiation risk have become an unavoidable problem for all CT vendors and medical institutions [2]. Since x-ray imaging is mainly based on a photon-noise-dominated process [27], lowering the X-ray dose will result in degraded CT images. Therefore, on the premise of ensuring CT image quality, how to reduce the X-ray radiation dose as far as possible becomes a promising and significant research topic [2].

Compared to sparse or limited-view CT [33] and other hardware-based strategies [34], lowering single X-ray exposure dose [11,22] is the most convenient and affordable method. To obtain high-quality LDCT images, previous works can be mainly classified into two categories: model-based and data-driven methods. The key to model-based methods is to use a mathematical model for the description of each process of CT imaging: noise characteristics in the sinogram

© Springer Nature Switzerland AG 2021
M. de Bruijne et al. (Eds.): MICCAI 2021, LNCS 12906, pp. 55–64, 2021.
https://doi.org/10.1007/978-3-030-87231-1_6

domain [15,30], image prior information in the image domain, such as sparsity in gradient domain [13] and low rank [3], as well as defects in CT hardware systems [32]. This kind of methods are independent of a large training dataset, while the accuracy of the model depiction limits its performance.

With the development of deep learning in medical image reconstruction and analysis [29], many data-driven works have been proposed to reconstruct LDCT images with convolution neural network (CNN) [25]. Kang *et al.* proposed a CNN-based neural network with the assistance of directional wavelets, suggesting the potential of deep learning technique in LDCT. Similarly, Chen *et al.* employed residual learning to extract noise in the LDCT images and obtain superior performance [5]. However, these methods need FBP-reconstructed LDCT images as the inputs, which belong to image post-processing. To get rid of the influence of traditional analytic algorithms (*e.g.* FBP), Zhu *et al.* suggested that 'AUTOMAP' was a direct reconstruction method from the measurement data to the final image [35]. Then again, the first fully-connected layer as domain transform has a huge memory requirement, which makes AUTOMAP unavailable for large-scale CT reconstruction [24]. Besides, many works with the combination of iterative reconstruction and deep learning have been proposed as deep unrolled approaches. This kind of method used CNNs as special regularizations plugged into conventional iterative reconstruction. They not only inherit the advantages of the convenient calculation of system matrix in conventional algorithms but also get rid of the complicated manual design regularization [7,10,11].

Despite the success of CNNs in LDCT reconstruction, CNN-based methods heavily rely on cascaded convolution layers to extract high-level features since the convolution operation has its disadvantage of a limited receptive field that only perceives local areas. Moreover, this disadvantage makes it difficult for CNN-based methods to make full of the similarity across large regions [26,31], which makes CNN-based methods less efficient in modeling various structural information in CT images [14]. To overcome this limitation, Transformers [23], which solely depend on attention mechanisms instead, have emerged as a powerful architectures in many fields, such as natural language processing (NLP) [8], image segmentation [6],image recognition [9], *etc.* In addition to these high-level tasks, Transformer has also been tentatively investigated for some lower-level tasks [4,28], which can model all pairwise interactions between image regions and capture long-range dependencies by computing interactions between any two positions, regardless of their positional distance.

For image denoising, noise is mainly contained in the high-frequency subband. Moreover, the remaining low-frequency sub-band not only contains the main image content, but also contains the weakened image textures, which are noise-free. These weakened image textures can be used to help noise removal in the high-frequency sub-band. Inspired by this observation, in this paper, we present the first work, TransCT, to explore the potential of transformers in LDCT imaging. Firstly, we decompose the noisy LDCT image into high-frequency (HF) and low-frequency (LF) parts. To remove the image noise on the premise of retaining the image content, we extract the corresponding content features (X_{L_c}) and latent texture features (X_{L_t}) from the LF part. Simultaneously, we extract the corresponding embeddings (X_{H_f}) from the HF part. Since transformers can only use sequences as input, we then convert X_{L_t} and X_{H_f} into separated sequences as the input of transformer encoder and decoder, respectively. To preserve the fine details of the final LDCT images, we integrate the output of the transformer decoder and some specific features from the LF part and then piecewise reconstruct high-quality and high-resolution LDCT images by stages. Extensive experiments on Mayo LDCT dataset demonstrate the superiority of our method over other methods.

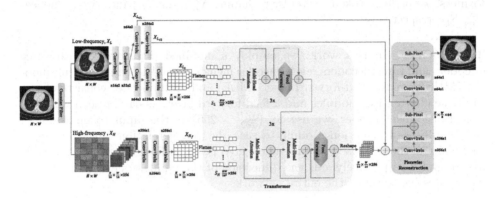

Fig. 1. The overall architecture of the proposed TransCT. '$n64s2$' means the convolution layer has 64 kernels with stride 2. Sub-Pixel layer is the upsampling layer [21].

2   Method

Figure 1 illustrates the overview of our proposed framework. For image denoising, an intuitive solution is to decompose the noisy image into HF and LF parts, and then the noise is mainly left in the HF part, which also contains plenty of image textures. However, noise removal only in the HF part breaks the relationship between the HF and LF parts since there are also weakened latent textures in the LF part with reduced noise. Therefore, we can remove the noise in the HF part with the assistance of the latent textures from the LF part. In this work, given the noisy LDCT image X with the size of $H \times W$, we first use a Gaussian filter with a standard deviation of 1.5 to decompose the LDCT image into two compositions: HF part X_H and LF part X_L.

$$X = X_H + X_L \tag{1}$$

To use the latent textures in X_L, we firstly extract the corresponding content features X_{L_c} and texture features X_{L_t} from X_L using shallow two CNNs. Further, we use these texture features and embeddings from X_H to train a transformer and get high-level features of X_H, combined with content features from X_L to reconstruct the final high-quality LDCT image.

2.1   TransCT

Sequence. Similar with what other works have done [6], we firstly employ two convolution layers with stride 2 to obtain low-resolution features from X_L, and then set two paths to extract content features $X_{L_{c_1}}$ ($\frac{H}{8} \times \frac{W}{8} \times 64$), $X_{L_{c_2}}$ ($\frac{H}{16} \times \frac{W}{16} \times 256$) and latent texture feature X_{L_t} ($\frac{H}{32} \times \frac{W}{32} \times 256$), respectively. For X_H, we employ sub-pixel layer to make X_H to be low-resolution images ($\frac{H}{16} \times \frac{W}{16} \times 256$), and final high-level features X_{H_f} can be obtained with three convolution layers. The goal is to get a sequence of moderate dimensions eventually. To take advantage of the characteristic of long-range dependencies of transformers, we perform tokenization by reshaping X_{L_t} and X_{H_f} into two sequences S_L, S_H, respectively.

Transformer. In this work, we employ a modified transformer with three encoders and three decoders, each encoder includes a multi-head attention module (MHSA) and a feed-forward layer (MLP) and each decoder consists of two multi-head attention modules and a feed-forward layer, as can be seen in Fig. 1. For transformer encoder, we use S_L ($\frac{WH}{1024} \times 256$) as the input token, followed by a multi-head attention module to seek the global relationship across large regions, and then we use two fully-connected layers (whose number of the node are $8c$ and c, respectively. c is the dimension of the input sequence) to increase the expressive power of the entire network.

$$
\begin{aligned}
Z &= MHSA(S_L^{i-1}) + S_L^{i-1} \\
S_L^i &= MLP(Z) + Z \\
s.t. \quad & i \in \{1, 2, 3\}
\end{aligned}
\tag{2}
$$

After acquiring the latent texture features S_L^3 from X_L, we feed S_H ($\frac{WH}{256} \times 256$) into the first multi-head attention module and treat S_L^3 as the key and value of each transformer decoder in the second multi-head attention module.

$$
\begin{aligned}
Z &= MHSA(S_H^{i-1}) + S_H^{i-1} \\
Z &= MHSA(Z, S_L^3, S_L^3) + Z \\
S_H^i &= MLP(Z) + Z \\
s.t. \quad & i \in \{1, 2, 3\}
\end{aligned}
\tag{3}
$$

Piecewise Reconstruction. Since the transformer only output features Y, we combine Y with $X_{L_{c_1}}$, $X_{L_{c_2}}$ to piecewise reconstruct the final high-quality LDCT images. In our work, the output of the transformer has the size of $\frac{H}{16} \times \frac{W}{16} \times 256$. Here, we reconstruct the high-resolution LDCT image piecewise. In the first step, we add Y and $X_{L_{c_2}}$ and then feed the output into a ResNet with two 'Conv2d + Leaky-ReLU(lrelu)' layers, followed by a sub-pixel layer which results in higher-resolution features with size of $\frac{H}{8} \times \frac{W}{8} \times 64$. Similarly, we add these higher-resolution features and $X_{L_{c_1}}$. After another ResNet with two 'Conv2d + lrelu' layers and sub-pixel layer, we can get the final output with the size of $H \times W$.

2.2 Loss Function

The MSE measures the difference between the output and normal dose CT images (NDCT), which reduces the noise in the input LDCT images. Formally, the MSE is defined as follows:

$$\min_{\theta} \mathcal{L} = ||I_{ND} - F_{\theta}(I_{LD})||_2^2 \qquad (4)$$

Where I_{ND} is the NDCT image and I_{LD} is the LDCT image, F is the proposed model and θ denotes the network parameters.

2.3 Implementation

In this work, the proposed framework was implemented in python based on Tensorflow [1] library. We used the Adam [12] optimizer to optimize all the parameters of the framework. We totally trained 300 epochs with a mini-batch size of 8. The learning rate was set as 0.0001 in the first 180 epochs and then reduced to 0.00001 for the next 120 epochs. The configuration of our computational platform is Intel(R) Core(Tm) i7-7700K CPU @4.20 GHz, 32 GB RAM, and a GeForce GTX TITAN X GPU with 12 GB RAM. We initialized all the variations with xavier initialization. Our code is publicly available at https://github.com/zzc623/TransCT.

3 Experiments

Datasets. In this work, we used a publicly released dataset for *the 2016 NIH-AAPM-Mayo Clinic Low-Dose CT Grand Challenge*[1] [17]. In this dataset, normal-dose abdominal CT images of $1mm$ slice thickness were taken from 10 anonymous patients and the corresponding quarter-dose CT images were simulated by inserting Poisson noise into the projection data. To better train the proposed TransCT, we divided the original 10 training patient cases into 7/1/2 cases, related to the training/validation/testing datasets, respectively. Before network training, we converted CT value of each pixel into its corresponding attenuation value under the assumption that the x-ray source was monochromatic at 60 keV.

Comparison with Other Methods. We compared our method with baseline methods: Non-local Mean (NLM), RED-CNN [5], MAP-NN [19], which are the high-performance LDCT methods. NLM can be found in the scikit-image library[2]. Since there is no public well-trained model for RED-CNN and MAP-NN, we re-train these methods with the same dataset.

Figure 2 shows the results randomly selected from the testing dataset. As compared to LDCT (Fig. 2(B)), NLM and all the DL-based methods can remove

---

[1] https://www.aapm.org/GrandChallenge/LowDoseCT/.
[2] https://scikit-image.org/.

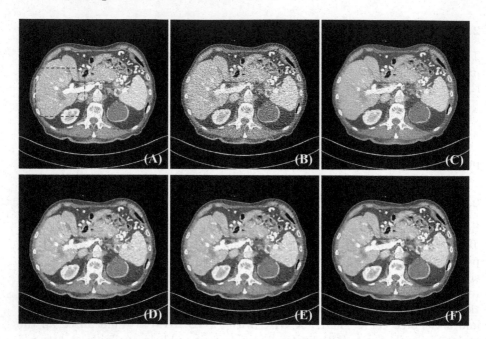

Fig. 2. Visual comparisons from Mayo testing dataset. (A) NDCT, (B) LDCT, (C) NLM, (D) RED-CNN, (E) MAP-NN, (F) TransCT. The display window is [−160, 240] *HU*.

noise to a certain extent, while our proposed TransCT is more close to NDCT. By investigating the local region in Fig. 3, we can see that the blood vessels (red arrows) are not obvious with NLM in (Fig. 3(C)). RED-CNN and MAP-NN generate, more or less, some additional light tissues (yellow arrow in (Fig. 3(D))) and shadows (green arrow in (Fig. 3(E))), respectively.

Quantitative Analysis. To quantitatively compare all the related methods, we conducted 5-fold cross-validation for all methods on Mayo dataset and employed Root Mean Square Error (RMSE), Structural Similarity (SSIM), and Visual Information Fidelity (VIF) [20] as image quality metrics. Among the three metrics, RMSE and SSIM mainly focus on pixel-wise similarity, and VIF uses natural statistics models to evaluate psychovisual features of the human visual system. From Table 1, we can see that all the related methods improve the image quality on all three metrics. To be specific, Red-CNN is superior to MAP-NN at the pixel-wise level while inferior to MAP-NN in terms of VIF. As compared to LDCT, our TransCT can decrease RMSE by 40.5%, improve SSIM by 12.3%, and VIF by 93.7%. For clinical evaluation, limited by clinical ethics, we evaluated all the methods on clinical CBCT images from a real pig head. The tube current was: 80 mA for NDCT and 20 mA for LDCT. From Table 1, our method outperforms others with superior robustness.

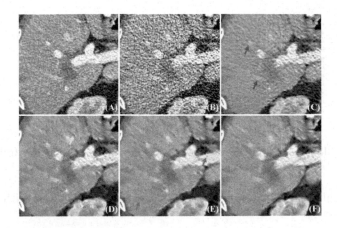

Fig. 3. The zoomed regions marked by the red box in Fig. 2(A). (A) NDCT, (B) LDCT, (C) NLM, (D) RED-CNN, (E) MAP-NN, (F) TransCT. The display window is $[-160, 240]$ HU. (Color figure online)

3.1 Ablation Study

On the Influence of Piecewise Reconstruction. In this work, after the output of transformer decoder, we used two resnet blocks and two sub-pixel layers to piecewise reconstruct the high-quality high-resolution LDCT image. The goal is to restore image detail more finely. To evaluate the influence of piecewise reconstruction, we modified the proposed TransCT and removed the piecewise reconstruction. After the output of the third transformer decoder, we used a sub-pixel layer to directly reconstruct the noise-free high-resolution HF texture, and then we added this HF texture and X_L to obtain the final LDCT image. Specifically, we have removed six convolution layers, including the path of content extraction ($X_{L_{c1}}$ and $X_{L_{c2}}$) and four convolution layers in the final two resnet blocks. Figure 4(a) shows the RMSE value on the validation dataset at each epoch. We can see that in about the first 20 epochs, the RMSE from modified TransCT decreases faster since its model scale is smaller than our

Table 1. Quantitative results (MEAN ± SDs) associated with different methods. Red and blue indicate the best and the second-best results, respectively.

Dataset		LDCT	NLM	RED-CNN	MAP-NN	TransCT
Mayo	RMSE	37.167 ± 7.245	25.115 ± 4.54	22.204 ± 3.89	22.492 ± 3.897	$\mathbf{22.123 \pm 3.784}$
	SSIM	0.822 ± 0.053	0.908 ± 0.031	0.922 ± 0.025	0.921 ± 0.025	$\mathbf{0.923 \pm 0.024}$
	VIF	0.079 ± 0.032	0.133 ± 0.037	0.152 ± 0.037	0.150 ± 0.038	$\mathbf{0.153 \pm 0.039}$
Pig	RMSE	50.776 ± 3.7	42.952 ± 5.971	37.551 ± 5.334	37.744 ± 4.883	$\mathbf{36.999 \pm 5.25}$
	SSIM	0.701 ± 0.02	0.799 ± 0.043	0.861 ± 0.03	0.86 ± 0.027	$\mathbf{0.87 \pm 0.029}$
	VIF	0.023 ± 0.002	0.040 ± 0.004	0.066 ± 0.006	0.063 ± 0.006	$\mathbf{0.069 \pm 0.007}$

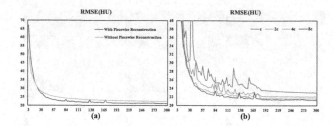

Fig. 4. RMSE results on the validation dataset during the network trainings.

TransCT, while the convergence was inferior to our TransCT with piecewise reconstruction.

On the Influence of Model Size. Generally, larger network size will lead to stronger neural network learning ability. In terms of each transformer encoder and decoder, which includes a two-layer feed-forward network, respectively, when the dimension of the input sequence is fixed, the dimension of the hidden layer in the feed-forward network will determine the network size. Here, we adjusted the dimension of the hidden layer $\{c, 2c, 4c\}$ to investigate the influence of model size. From Fig. 4(b), we can see that the smaller the dimension of the hidden layer is, the larger the fluctuation of the convergence curve is, the larger the final convergent value will be. Therefore, we conclude that larger model results in a better performance. In this work, we set the dimension of the hidden layer in the feed-forward network at $8c$.

Ablation Studies on Transformer Module and Dual-Path Module. To investigate the effectiveness of the transformer module and dual-path module, we conducted two additional experiments. First, we used a revised module ("Conv+3×ResNet blocks") to replace the transformer module. We concatenated X_{H_f} and the output from the fourth Conv layer (n128s2, before X_{L_t}) and then inputted it into the revised module. As for the dual-path module, we discarded the HF path and inputted the $X_{L_{c_2}}$ into 3 transformer encoders, whose output will be combined with $X_{L_{c_1}}$ and $X_{L_{c_2}}$ in the piecewise reconstruction stage. The results on the validation dataset were shown in Table 2, we can see that our TransCT with transformer module and dual-path module can obtain better performance.

4 Conclusion

Inspired by the internal similarity of the LDCT image, we present the first transformer-based neural network for LDCT, which can explore large-range dependencies between LDCT pixels. To ease the impact of noise on high-frequency texture recovery, we employ a transformer encoder to further excavate the low-frequency part of the latent texture features and then use these texture

Table 2. Ablation studies on transformer module and dual-path module conducted on the validation dataset.

	RMSE	SSIM	VIF
w/o transformer module	22.62 ± 2.068	0.927 ± 0.013	0.13 ± 0.023
w/o dual-path module	21.711 ± 1.997	0.931 ± 0.012	0.14 ± 0.025
TransCT	21.199 ± 2.054	0.933 ± 0.012	0.144 ± 0.025

features to restore the high-frequency features from noisy high-frequency parts of LDCT image. The final high-quality LDCT image can be piecewise reconstructed with the combination of low-frequency content and high-frequency features. In the future, we will further explore the learning ability of TransCT and introduce self-supervised learning to lower the need for the training dataset.

Acknowledgements. This work was partially supported by NIH (1 R01CA227713) and a Faculty Research Award from Google Inc.

References

1. Abadi, M., Barham, P., et al.: TensorFlow: a system for large-scale machine learning. In: OSDI, pp. 265–283 (2016)
2. Brenner, D.J., Hall, E.J.: Computed tomography–an increasing source of radiation exposure. N. Engl. J. Med. **357**(22), 2277–2284 (2007)
3. Cai, J.F., Jia, X., et al.: Cine cone beam CT reconstruction using low-rank matrix factorization: algorithm and a proof-of-principle study. IEEE Trans. Med. Imag. **33**(8), 1581–1591 (2014)
4. Chen, H., Wang, Y., et al.: Pre-trained image processing transformer. In: CVPR, pp. 12299–12310 (2021)
5. Chen, H., Zhang, Y., et al.: Low-dose CT with a residual encoder-decoder convolutional neural network. IEEE Trans. Med. Imag. **36**(12), 2524–2535 (2017)
6. Chen, J., Lu, Y., et al.: TransuNet: transformers make strong encoders for medical image segmentation. arXiv:2102.04306 (2021)
7. Chun, I.Y., Zheng, X., Long, Y., Fessler, J.A.: BCD-net for low-dose CT reconstruction: acceleration, convergence, and generalization. In: Shen, D., et al. (eds.) MICCAI 2019. LNCS, vol. 11769, pp. 31–40. Springer, Cham (2019). https://doi.org/10.1007/978-3-030-32226-7_4
8. Devlin, J., Chang, M.W., et al.: BERT: pre-training of deep bidirectional transformers for language understanding. arXiv:1810.04805 (2018)
9. Dosovitskiy, A., Beyer, L., et al.: An image is worth 16x16 words: transformers for image recognition at scale. arXiv:2010.11929 (2020)
10. Gupta, H., Jin, K.H., et al.: CNN-based projected gradient descent for consistent CT image reconstruction. IEEE Trans. Med. Imag. **37**(6), 1440–1453 (2018)
11. He, J., Yang, Y., et al.: Optimizing a parameterized plug-and-play ADMM for iterative low-dose CT reconstruction. IEEE Trans. Med. Imag. **38**(2), 371–382 (2018)
12. Kingma, D.P., Ba, J.: Adam: a method for stochastic optimization. arXiv:1412.6980 (2014)

13. LaRoque, S.J., Sidky, E.Y., Pan, X.: Accurate image reconstruction from few-view and limited-angle data in diffraction tomography. JOSA A **25**(7), 1772–1782 (2008)
14. Li, M., Hsu, W., et al.: SACNN: self-attention convolutional neural network for low-dose CT denoising with self-supervised perceptual loss network. IEEE Trans. Med. Imag. **39**(7), 2289–2301 (2020)
15. Manduca, A., Yu, L., et al.: Projection space denoising with bilateral filtering and CT noise modeling for dose reduction in CT. Med. Phys. **36**(11), 4911–4919 (2009)
16. Mathews, J.P., Campbell, Q.P., et al.: A review of the application of x-ray computed tomography to the study of coal. Fuel **209**, 10–24 (2017)
17. McCollough, C.H., Bartley, A.C., et al.: Low-dose CT for the detection and classification of metastatic liver lesions: results of the 2016 low dose CT grand challenge. Med. Phys. **44**(10), e339–e352 (2017)
18. Seeram, E.: Computed tomography: physical principles, clinical applications, and quality control. Elsevier Health Sciences (2015)
19. Shan, H., Padole, A., et al.: Competitive performance of a modularized deep neural network compared to commercial algorithms for low-dose CT image reconstruction. Nat. Mach. Intell. **1**(6), 269–276 (2019)
20. Sheikh, H.R., Bovik, A.C.: Image information and visual quality. IEEE Trans. Image Process. **15**(2), 430–444 (2006)
21. Shi, W., Caballero, J., et al.: Real-time single image and video super-resolution using an efficient sub-pixel convolutional neural network. In: CVPR, pp. 1874–1883 (2016)
22. Tian, Z., Jia, X., et al.: Low-dose CT reconstruction via edge-preserving total variation regularization. Phys. Med. Biol. **56**(18), 5949 (2011)
23. Vaswani, A., Shazeer, N., et al.: Attention is all you need. arXiv:1706.03762 (2017)
24. Wang, G., Ye, J.C., et al.: Image reconstruction is a new frontier of machine learning. IEEE Trans. Med. Imag. **37**(6), 1289–1296 (2018)
25. Wang, G., Ye, J.C., De Man, B.: Deep learning for tomographic image reconstruction. Nat. Mach. Intell. **2**(12), 737–748 (2020)
26. Wang, X., Girshick, R., et al.: Non-local neural networks. In: CVPR, pp. 7794–7803 (2018)
27. Xu, Q., Yu, H., et al.: Low-dose x-ray CT reconstruction via dictionary learning. IEEE Trans. Med. Imag. **31**(9), 1682–1697 (2012)
28. Yang, F., Yang, H., et al.: Learning texture transformer network for image super-resolution. In: CVPR, pp. 5791–5800 (2020)
29. Yu, L., Zhang, Z., et al.: Deep sinogram completion with image prior for metal artifact reduction in CT images. IEEE Trans. Med. Imag. **40**(1), 228–238 (2020)
30. Yu, L., Manduca, A., et al.: Sinogram smoothing with bilateral filtering for low-dose CT. In: Medical Imaging 2008: Physics of Medical Imaging, vol. 6913, p. 691329
31. Zhang, H., Goodfellow, I., et al.: Self-attention generative adversarial networks. In: ICML, pp. 7354–7363 (2019)
32. Zhang, Z., Yu, L., et al.: Modularized data-driven reconstruction framework for non-ideal focal spot effect elimination in computed tomography. Med. Phys. (2021)
33. Zhang, Z., Liang, X., et al.: A sparse-view CT reconstruction method based on combination of DenseNet and deconvolution. IEEE Trans. Med. Imag. **37**(6), 1407–1417 (2018)
34. Zhang, Z., Yu, S., et al.: A novel design of ultrafast micro-CT system based on carbon nanotube: a feasibility study in phantom. Phys. Med. **32**(10), 1302–1307 (2016)
35. Zhu, B., Liu, J.Z., et al.: Image reconstruction by domain-transform manifold learning. Nature **555**(7697), 487–492 (2018)

IREM: High-Resolution Magnetic Resonance Image Reconstruction via Implicit Neural Representation

Qing Wu[1], Yuwei Li[1], Lan Xu[1], Ruiming Feng[2], Hongjiang Wei[2], Qing Yang[3], Boliang Yu[1], Xiaozhao Liu[1], Jingyi Yu[1(✉)], and Yuyao Zhang[1,4(✉)]

[1] School of Information Science and Technology, ShanghaiTech University, Shanghai, China
{yujingyi,zhangyy8}@shanghaitech.edu.cn
[2] School of Biomedical Engineering, Shanghai Jiao Tong University, Shanghai, China
[3] Institute of Brain-Intelligence Technology, Zhangjiang Laboratory, Shanghai, China
[4] Shanghai Engineering Research Center of Intelligent Vision and Imaging, ShanghaiTech University, Shanghai, China

Abstract. For collecting high-quality high-resolution (HR) MR image, we propose a novel image reconstruction network named IREM, which is trained on multiple low-resolution (LR) MR images and achieve an arbitrary up-sampling rate for HR image reconstruction. In this work, we suppose the desired HR image as an implicit continuous function of the 3D image spatial coordinate, and the thick-slice LR images as several sparse discrete samplings of this function. Then the super-resolution (SR) task is to learn the continuous volumetric function from a limited observation using a fully-connected neural network combined with Fourier feature positional encoding. By simply minimizing the error between the network prediction and the acquired LR image intensity across each imaging plane, IREM is trained to represent a continuous model of the observed tissue anatomy. Experimental results indicate that IREM succeeds in representing high-frequency image features, and in real scene data collection, IREM reduces scan time and achieves high-quality high-resolution MR imaging in terms of SNR and local image detail.

Keywords: MRI · High resolution · Super resolution · Implicit neural representation

1 Introduction

High Resolution (HR) medical images provide rich structural details to facilitate early and accurate diagnosis. In MRI, unpredictable patient movements result in difficulties to collect artifact-free high-resolution 3D images at sub-millimeter within a single scan. Thus, multiple anisotropic low-resolution (LR) MR scans with thick slice are always conducted at different scan orientations with a relatively short scan time to eliminate motion artifacts. MRI Super-Resolution (SR)

© Springer Nature Switzerland AG 2021
M. de Bruijne et al. (Eds.): MICCAI 2021, LNCS 12906, pp. 65–74, 2021.
https://doi.org/10.1007/978-3-030-87231-1_7

techniques are then used to enhance the image spatial resolution from one or more LR images. According to the number of processed images, related works can be divided into two categories: multiple image super-resolution (MISR) [2,19] or single image super-resolution (SISR) [17].

MRI SR techniques were originally proposed as MISR algorithms [14]. Multiple thick-sliced LR images of the same subject were acquired to reconstruct an HR image. Gholipour et al. [5] developed a model-based MISR technique (always known as SRR) that enabled isotropic volumetric image reconstruction from arbitrarily oriented slice acquisitions. SRR achieved great success and was broadly applied for motion-prone MR acquisitions (i.e., fetus brain construction [4]). Jia et al. [9] proposed a Sparse Representation based SR approach using only the multiple LR scans to build the over-complete dictionary. Previous studies [5,21,22,24] showed that for MISR methods, in a given acquisition time, when the alignment among multiple frames is known or correctly estimated, the SR reconstructed image presented higher SNR (thick-slice improves SNR in single voxel) when comparing with the isotropic image acquired with similar spatial resolution. But it is challenging for MISR methods to build submillimeter-level image detail, such as in brain cerebellum.

On the other hand, SISR is attracting attentions due to its advantage of shorter scan time to further reduce the motion artifacts. Recently, the performance of SISR methods were significantly improved by Deep convolution neural network (CNN) because of its non-linearity to imitate transformation between LR and HR images. Various network architectures for MRI SISR have been proposed [1,3,11,12,15]. However, there have to exist a large HR image dataset as ground truth to train the SR network and the learned model lacks generalization for the input images with different qualities, e.g., different spatial resolutions, SNRs. As a result, this kind of methods cannot handle SR tasks when no HR ground truth is available (i.e., fetus brain reconstruction).

In this paper, we propose IREM, which is short for *Implicit REpresentation for MRI*. Our work intends to further improve MISR reconstruction performance via deep neural network. Instead of CNN architecture, we train a fully-connected deep network (MLP) to learn and represent the desired HR image. Specifically, we suppose the HR image as an implicit continuous function of the 3D image spatial coordinate, and the 2D thick-slice LR images as several sparse discrete samplings of this function. Then the SR task is to learn the continuous volumetric function from a limited 2D observations using an extended MLP. The MLP takes the positional encoded 3D coordinate (x, y, z) as input and outputs the image intensity $I(x, y, z)$ of the voxel at that location. The positional encoding is connected via Fourier feature mapping [20] to project the input 3D coordinates into a higher dimensional space, for enabling the network to learn sufficient high frequency image features from the LR image stacks. By minimizing the error between the network prediction and the acquired LR image intensity across each LR imaging plane, IREM is trained to reconstruct a continuous model of the observed tissue anatomy. We demonstrate that our resulting HR MRI reconstruction method quantitatively and qualitatively outperforms state-of-the-art MISR methods. The major advantage of IREM is summarized as below:

- No need for large amount of HR data set for training the network;
- No constraints for input LR image resolution and up-sampling scale;
- Comparing with the realistic HR image acquisition with similar total scan time, the SR image provides equivalent level of image quality and higher SNR.

2   Method

Inspired by a very recent implicit scene representation work [13], we build a deep learning model that learn to represent an image function $I(x, y, z)$ from input voxel location (x, y, z). In model training, the spatial coordinate of the LR images are input into a fully-connected network with positional encoding, while the model is trained to output the image intensity $I(x, y, z)$ of the voxel at that location with minimizing error comparing with the LR image intensity. Then, the fine-trained network reconstruct the desired HR image with arbitrary up-sampling SR rate (with arbitrary input (x, y, z)). The spatial locations that are not observed in the LR images are thus the "test data" generated by the proposed SR model.

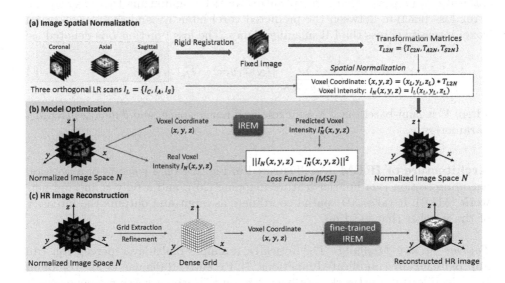

Fig. 1. An overview of the proposed IREM approach.

An overview of IREM is depicted in Fig. 1, which is presented in three stages: a) Image Spatial Normalization; b) Model Optimization; c) HR Image Reconstruction.

2.1   Image Spatial Normalization

As demonstrated in Fig. 1(a), we first build a set of anisotropic LR (thick-sliced MR scanning) image stacks $I_L = \{I_C, I_A, I_S\}$. For simplifying, we use three orthogonal scanning orientations (coronal, axial, sagittal) to represent the LR stacks. Then we randomly select one from the three LR scans as fixed image and rigidly register all the LR scans to the fixed image to generate transformation matrices $T_{L2N} = \{T_{C2N}, T_{A2N}, T_{S2N}\}$. Finally, we utilize transformation matrices T_{L2N} to transfer each LR image from its original space to the normalized image space N.

In the normalized space, the image intensities from different LR image stacks at same coordinate (x, y, z) represents coherent observations of the same tissue anatomy in different imaging orientations. Note that the pair of voxel coordinate and intensity in the normalized space N: $(x, y, z) \rightarrow I_N(x, y, z)$ are transformed from LR scans at position $(x_L, y_L, z_L) \rightarrow I_L(x_L, y_L, z_L)$ by transformation matrix (i.e., $(x, y, z) = (x_L, y_L, z_L) * T_{L2N}$ and $I_N(x, y, z) = I_L(x_L, y_L, z_L)$).

2.2   Model Optimization

The process of model optimization is illustrated in Fig. 1(b). After the image spatial normalization, the training data set for building IREM model is prepared. We feed the voxel coordinate (x, y, z) into IREM to compute the predicted voxel intensity $I_N^*(x, y, z)$. Then, we optimize IREM by minimizing the mean square error loss function between the predicted voxel intensity and the real observed voxel intensity across the LR imaging planes. The loss function L is denoted as:

$$L(\theta) = \frac{1}{\mathcal{K}} \sum_{(x,y,z) \in N} \|I_N(x, y, z) - I_N^*(x, y, z)\|^2 \tag{1}$$

where $\mathcal{K}$ is mini-batch size, N is normalized image space, and θ denotes IREM's parameters.

Architecture of IREM. As illustrated in Fig. 2, IREM consists of a position encoding section (via Fourier feature mapping [20]), and a fully-connected network (MLP). It takes 3D spatial coordinate as input and outputs the intensity of the voxel at that location.

Fourier Feature Mapping. Theoretically, a simple multi-layer perceptron can approximate any complicated functions [7]. Recently, Rahaman et al. [16] found that deep learning networks are biased towards learning lower frequency functions in practical training. To this end, Mildenhall et al. proposed Fourier feature mapping [20] that maps the low-dimensional input to a higher dimensional space and thus enable network to learn higher frequency image feature. In IREM, we perform Fourier feature mapping [20] to map the 3D voxel coordinates to a higher

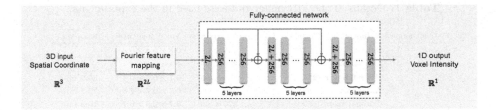

Fig. 2. Architecture of IREM.

dimensional space $\mathbb{R}^{2L} (2L > 3)$ before passing them to the fully-connected network. Let $\gamma (\cdot)$ denotes Fourier feature mapping from the space $\mathbb{R}^3$ to $\mathbb{R}^{2L}$ and it is calculated by

$$\gamma (\mathcal{P}) = [\cos (2\pi \mathcal{B}\mathcal{P}), \sin (2\pi \mathcal{B}\mathcal{P})]^T \qquad (2)$$

where $\mathcal{P} = (x, y, z) \in \mathbb{R}^3$ and each element in $\mathcal{B} \in \mathbb{R}^{L \times 3}$ is sampled from gaussian distribution $\mathcal{N} (0, 1)$.

Fully-Connected Network. The network has eighteen fully-connected layers. Each fully-connected layer is followed by a batch normalization layer [8] and a ReLU activation [6]. In order to ease the difficulty of optimizing IREM, we include two skip connections that concatenate the input of the fully-connected network to the 6th layer's activation and the 12th layer's activation, respectively. And except that there are $2L$, $2L + 256$, $2L + 256$ neurons in the 1st, 7th, and 13th fully-connected layers. Other fully-connected layers have all 256 neurons.

2.3 HR Image Reconstruction

The sketch map of an HR image reconstruction by a fine-trained IREM is shown in Fig. 1(c). We build a dense grid (i.e., a coordinate set with more voxels than in LR stacks) in the normalized image space N, with isotropic image resolution. Then, we pass each voxel coordinate (x, y, z) in the grid into IREM to produce predicted the voxel intensity $I_N^* (x, y, z)$. An HR image thus is reconstructed.

3 Experiments

3.1 Data

We conduct experiments on three datasets. The details are shown in Table 1. Dataset #A consists of T1-weighted (T1w) HR brain MR scans from five healthy adults on a 7T MR scanner and #B consists of 3T T2 flair images from two patients with lesions in brain white matter. Dataset #C consists of four T1w brain MR scans (An HR scan used for reference and three orthogonal LR scans used for input images) from the same volunteer.

Table 1. Details of the MR image datasets used in the experiments.

Dataset name	# Image	Number of slices	Matrix size	Voxel size (mm^3)	Scan time
#A	Sub1-sub5	244	244 × 244	0.8 × 0.8 × 0.8	–
#B	Sub1-sub2	64	256 × 256	1 × 1 × 2	–
#C	HR reference	254	368 × 345	0.7 × 0.7 × 0.7	30.9 min
	Axial scan	80	320 × 320	0.7 × 0.7 × 2.8	10.3 min
	Coronal scan	80	320 × 320	0.7 × 2.8 × 0.7	10.3 min
	Sagittal scan	64	320 × 320	2.8 × 0.7 × 0.7	8.2 min

Evaluation for the Accuracy of Implicit HR Image Representation on Simulation Data. For dataset #A and #B, we employ the original MR image as ground truth (GT) HR image and downsample the image by the factor of {4, 8} in the three image dimensions, to simulate three orthogonal LR scans. We then use the built LR stacks to train our model, and compare the reconstructed HR image with the GT to evaluate the ability of IREM to learn the high frequency image content in the implicit HR image.

Evaluation for the Performance in Real Data Collection Protocol. We conduct a real data collection scene using IREM on dataset #C. In this case, no actual GT can be built. We scan a HR T1w HR brain image with 0.7 × 0.7 × 0.7 mm³ isotropic in 30.9 min as HR reference. Then three anisotropic LR images are scanned in coronal, axial, and sagittal orientations in about 10 min each, so the total scan time of 3 LR stacks is about 2 min shorter than the HR scan.

3.2   Implementation Details

We adopt Adam optimizer [10] to train IREM through back-propagation with a mini-batch size of 2500, and the hyperparameters of the Adam are set as follows: $\beta_1 = 0.9$, $\beta_2 = 0.999$, $\varepsilon = 10^{-8}$. The learning rate starts from 10^{-4} and decays by factor 0.5 every 500 epochs. For fair comparison, we implement two MISR methods, including Super-Resolution Reconstruction [4] (SRR) and B-spline interpolation, to compare with proposed work. We assess quantitatively the performance of the three methods in terms of peak signal-to-noise ratio (PSNR), structural similarity (SSMI) [23].

3.3   Results

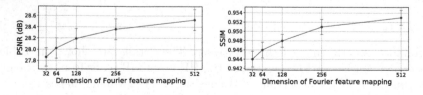

Fig. 3. Evolution of the performance of IREM on the different dimensions $2L$ of Fourier feature mapping [20] for dataset #A. Here the factor k of down-sampling is 8.

Effect of the Fourier Feature Mapping. Based on dataset #A, we investigate the effect of Fourier feature mapping [20] on the performance of IREM. As shown in Fig. 3, the performance of IREM improves with the dimension $2L$ of Fourier feature mapping [20], but its growth rate gradually decreases, which is consistent with the previous work [20]. To achieve the balance between efficiency and accuracy, the dimension $2L$ is set as 256 in all the experiments below.

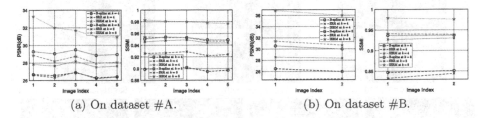

(a) On dataset #A.   (b) On dataset #B.

Fig. 4. Quantitative results (PSNR(dB)/SSMI) of B-spline interpolation, SRR [4], and IREM on dataset #A and #B. Here k denotes the factor of down-sampling.

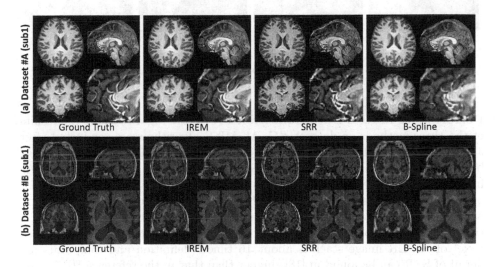

Fig. 5. Qualitative results of B-spline interpolation, SRR [4], and IREM on dataset #A (sub1) and #B (sub1). Here the factor k of down-sampling is 4.

Performance on Simulation Validation. On dataset #A, the quantitative evaluation results of B-spline interpolation, SRR [4], and IREM are shown in Fig. 4(a). We indicate that IREM consistently outperforms the two baselines at all factors k of down-sampling in terms of all metrics. Figure 5(a) demonstrates that the representative slices of an HR image (#A sub1) reconstructed from the three methods, and the results from IREM visually are closest to GT HR image.

On dataset #B, the quantitative and qualitative evaluation results of the three methods are shown in Fig. 4(b) and Fig. 5(b), respectively. We can see that IREM achieves the best performance, which is consistent with the experiments on dataset #A.

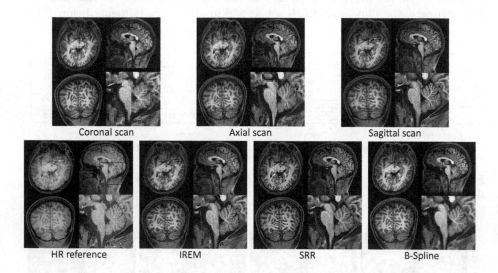

Coronal scan   Axial scan   Sagittal scan

HR reference   IREM   SRR   B-Spline

Fig. 6. Results of B-spline interpolation, SRR [4], and IREM on dataset #C.

Performance in Real Data Collection Protocol. Figure 6 shows the results of the three methods on dataset #C. All MISR algorithms are conducted after a unique image spatial normalization step. Comparing with IREM, the image built from B-Spline is more blurry and SRR [4] yields an image with artifacts. As indicated in the enlarged part, the image built from IREM achieves equivalent qualitative image details comparing with the HR reference in the cerebellum, which is one of the most complicated anatomy in human brain. Besides, benefitting from the multiple anisotropic thick slice scanning strategy, the SNR in each voxel of the LR image stack is about 16 times (definition and computational detail of SNR can be found in [18]) higher than that in the reference HR image. Thus the image contrast between white matter and gray matter in the reconstructed image is better than that in the reference HR image. The result suggests that IREM is a more effective and reasonable pipeline to achieve high-quality HR image comparing with scanning directly an isotropic HR image.

4   Conclusion

In this paper, we proposed IREM, a novel implicit representation based deep learning framework to improve the performance of multiple image super-resolution task. In IREM, we combined spatial encoding and fully connected neural network, then trained a powerful model that precisely predict MR image intensity from input spatial coordinates. The HR reconstruction result on two set of simulated image data indicated the ability of IREM to accurately approach the implicit HR space. While the real scene data collection demonstrated that based on multiple low-resolution MR scans with thicker slice, IREM reconstructed HR image with both improved anatomy detail and image contrast.

Acknowledgements. This study is supported by the National Natural Science Foundation of China (No. 62071299, 61901256).

References

1. Chen, Y., Christodoulou, A.G., Zhou, Z., Shi, F., Xie, Y., Li, D.: MRI super-resolution with GAN and 3d multi-level DenseNet: smaller, faster, and better. arXiv preprint arXiv:2003.01217 (2020)
2. Chen, Y., Shi, F., Christodoulou, A., Zhou, Z., Xie, Y., Li, D.: Efficient and Accurate MRI Super-Resolution Using a Generative Adversarial Network and 3D Multi-level Densely Connected Network, pp. 91–99, September 2018. https://doi.org/10.1007/978-3-030-00928-1_11
3. Delannoy, Q., et al.: SegSRGAN: super-resolution and segmentation using generative adversarial networks-application to neonatal brain MRI. Comput. Biol. Med. **120**, 103755 (2020)
4. Ebner, M., et al.: An automated framework for localization, segmentation and super-resolution reconstruction of fetal brain MRI. NeuroImage **206**, 116324 (2019). https://doi.org/10.1016/j.neuroimage.2019.116324
5. Gholipour, A., Estroff, J., Warfield, S.: Robust super-resolution volume reconstruction from slice acquisitions: application to fetal brain MRI. IEEE Trans. Med. Imag. **29**, 1739–58 (2010). https://doi.org/10.1109/TMI.2010.2051680
6. He, K., Zhang, X., Ren, S., Sun, J.: Delving deep into rectifiers: surpassing human-level performance on ImageNet classification. In: IEEE International Conference on Computer Vision (ICCV 2015), vol. 1502, February 2015. https://doi.org/10.1109/ICCV.2015.123
7. Hornik, K., Stinchcomb, M., White, H.: Multilayer feedforward networks are universal approximator. IEEE Trans. Neural Netw. **2**, 359–366 (1989)
8. Ioffe, S., Szegedy, C.: Batch normalization: accelerating deep network training by reducing internal covariate shift, February 2015
9. Jia, Y., He, Z., Gholipour, A., Warfield, S.: Single anisotropic 3-d MR image upsampling via overcomplete dictionary trained from in-plane high resolution slices. IEEE J. Biomed. Health Inform. **20** (2015). https://doi.org/10.1109/JBHI.2015.2470682
10. Kingma, D., Ba, J.: Adam: a method for stochastic optimization. In: International Conference on Learning Representations, December 2014
11. Lyu, Q., et al.: Multi-contrast super-resolution MRI through a progressive network. IEEE Trans. Med. Imaging **39**(9), 2738–2749 (2020)
12. Lyu, Q., Shan, H., Wang, G.: MRI super-resolution with ensemble learning and complementary priors. IEEE Trans. Comput. Imag. **6**, 615–624 (2020)
13. Mildenhall, B., Srinivasan, P., Tancik, M., Barron, J., Ramamoorthi, R., Ng, R.: NeRF: representing scenes as neural radiance fields for view synthesis, pp. 405–421, November 2020. https://doi.org/10.1007/978-3-030-58452-8_24

14. Peled, S., Yeshurun, Y.: Superresolution in MRI: application to human white matter fiber tract visualization by diffusion tensor imaging. Magn. Reson. Med. Off. J. Soc. Magn. Reson. Med. Soc. Magn. Reson. Med. **45**, 29–35 (2001). https://doi.org/10.1002/1522-2594(200101)45:18216;29::aid-mrm10058217;3.0.co;2-z

15. Pham, C.H., Ducournau, A., Fablet, R., Rousseau, F.: Brain MRI super-resolution using deep 3d convolutional networks. In: 2017 IEEE 14th International Symposium on Biomedical Imaging (ISBI 2017), pp. 197–200. IEEE (2017)

16. Rahaman, N., et al.: On the spectral bias of neural networks. In: International Conference on Machine Learning, pp. 5301–5310. PMLR (2019)

17. Rueda, A., Malpica, N., Romero, E.: Single-image super-resolution of brain MR images using overcomplete dictionaries. Med. Image Anal. **17**, 113–132 (2012). https://doi.org/10.1016/j.media.2012.09.003

18. Scherrer, B., Gholipour, A., Warfield, S.K.: Super-resolution reconstruction to increase the spatial resolution of diffusion weighted images from orthogonal anisotropic acquisitions. Med. Image Anal. **16**(7), 1465–1476 (2012)

19. Shi, F., Cheng, J., Wang, L., Yap, P.T., Shen, D.: LRTV: MR image super-resolution with low-rank and total variation regularizations. IEEE Trans. Med. Imag. **34** (2015). https://doi.org/10.1109/TMI.2015.2437894

20. Tancik, M., et al.: Fourier features let networks learn high frequency functions in low dimensional domains. arXiv preprint arXiv:2006.10739 (2020)

21. Ur Rahman, S., Wesarg, S.: Combining short-axis and long-axis cardiac MR images by applying a super-resolution reconstruction algorithm. Fraunhofer IGD 7623, March 2010. https://doi.org/10.1117/12.844356

22. Van Reeth, E., Tan, C.H., Tham, I., Poh, C.L.: Isotropic reconstruction of a 4-d MRI thoracic sequence using super-resolution. Magn. Reson. Med. Off. J. Soc. Magn. Reson. Med. Soc. Magn. Reson. Med. **73** (2015). https://doi.org/10.1002/mrm.25157

23. Wang, Z., Bovik, A., Sheikh, H., Member, S., Simoncelli, E.: Image quality assessment: from error measurement to structural similarity. IEEE Trans. Imag. Process. **13** (2003)

24. Yang, J., Wright, J., Yu, L.: Image super-resolution via sparse representation. Image Process. IEEE Trans. **19**, 2861–2873 (2010). https://doi.org/10.1109/TIP.2010.2050625

DA-VSR: Domain Adaptable Volumetric Super-Resolution for Medical Images

Cheng Peng[1], S. Kevin Zhou[2,3,4], and Rama Chellappa[1(✉)]

[1] Johns Hopkins University, Baltimore, MD, USA
{cpeng26,rchella4}@jhu.edu
[2] Medical Imaging, Robotics, and Analytic Computing Laboratory and Engineering
(MIRACLE) Group, Beijing, China
[3] School of Biomedical Engineering and Suzhou Institute for Advance Research,
University of Science and Technology of China, Suzhou 215123, China
[4] Key Lab of Intelligent Information Processing of Chinese Academy of Sciences
(CAS), Institute of Computing Technology, CAS, Beijing 100190, China

Abstract. Medical image super-resolution (SR) is an active research area that has many potential applications, including reducing scan time, bettering visual understanding, increasing robustness in downstream tasks, etc. However, applying deep-learning-based SR approaches for clinical applications often encounters issues of domain inconsistency, as the test data may be acquired by different machines or on different organs. In this work, we present a novel algorithm called domain adaptable volumetric super-resolution (DA-VSR) to better bridge the domain inconsistency gap. DA-VSR uses a unified feature extraction backbone and a series of network heads to improve image quality over different planes. Furthermore, DA-VSR leverages the in-plane and through-plane resolution differences on the test data to achieve a self-learned domain adaptation. As such, DA-VSR combines the advantages of a strong feature generator learned through supervised training and the ability to tune to the idiosyncrasies of the test volumes through unsupervised learning. Through experiments, we demonstrate that DA-VSR significantly improves super-resolution quality across numerous datasets of different domains, thereby taking a further step toward real clinical applications.

1 Introduction

Medical imaging such as Magnetic Resonance Imaging (MRI) and Computed Tomography (CT) are crucial to clinical diagnosis. To facilitate faster and less costly acquisitions, it is routine to acquire a few high-resolution cross sectional images in CT/MRI, leading to a low through-plane resolution when the acquired images are organized into an anisotropic volume. The anisotropic volumes lead to

Electronic supplementary material The online version of this chapter (https://doi.org/10.1007/978-3-030-87231-1_8) contains supplementary material, which is available to authorized users.

© Springer Nature Switzerland AG 2021
M. de Bruijne et al. (Eds.): MICCAI 2021, LNCS 12906, pp. 75–85, 2021.
https://doi.org/10.1007/978-3-030-87231-1_8

difficulties in understanding the patient's anatomy both for physicians and automated algorithms [12,17]. One way to address this is through super-resolution (SR) algorithms [26,27], which upsample along the axis with a low resolution. SR has witnessed great improvement in the image domain with Convolutional Neural Network (CNN)-based algorithms [4,8,9,20], where the formulation typically involves supervised learning between a low resolution (LR) image and its paired high resolution (HR) groundtruth. Various improvements have been made to reduce computation [5,14], enhance feature extraction efficiency [10,22,23], and improve robustness [15,21].

Volumetric SR for medical images poses unique challenges. Firstly, the high dimensional and anisotropic nature of volumetric images lead to difficulties in computational cost and learning efficiency. While there exists many work on 2D medical image SR [3,6,11,18,19,25], few directly tackle 3D medical image SR due to high computational cost and limited sample size. Chen et al. [2] apply a DenseNet-based CNN algorithms called mDCSRN on volumetric data with 3D kernels. Wang et al. [17] ease the 3DCNN memory bottleneck by using a more efficiently implemented DenseNet with residual connections. These methods still require patch-by-patch inference on large-size volume, which can lead to undesirable border artifacts and inefficiency. Peng et al. [13] propose SAINT, which can super-resolve volumetric images with multiple upsampling factors with a single network. Furthermore, it addresses the memory constraint at inference by performing an ensemble of 2D SR operations on through-plane images.

Another challenge arises from the need for high reliability. Under a supervised learning framework, if a test image comes from a distribution not well represented in training, e.g. of a different body part or by a different machine, performance often degrades in unexpected ways. Therefore, semi-supervised or self-supervised SR methods provide distinct advantages if they can learn directly from test datasets. Zhao et al. [24] propose SMORE, a self-supervised SR algorithm that leverages the high in-plane resolution to create LR-HR pairs for learning, and applies the learned model on lower through-plane resolution. Implicitly, SMORE assumes that the in-plane and through-plane images are from same or similar distributions, which may not be true for many cases.

To address the issue of robustly super-resolving volumetric medical images, we propose a novel algorithm named Domain Adaptable Volumetric Super-Resolution (DA-VSR). DA-VSR follows SAINT's thinking in addressing volumetric SR based on a series of slice-wise SR. DA-VSR uses a single feature extraction backbone and assigns small, task-specific network heads for upsampling and fusion. Inspired by SMORE, DA-VSR leverages the resolution differences across dimensions as a self-supervised signal for domain adaptation at *test time*. Specifically, DA-VSR designs an additional self-supervised network head that can help align features on test images through in-plane super-resolution. As a result, DA-VSR enjoys the benefit of a strong feature backbone obtained by supervised training, and the ability to adapt to various distributions through unsupervised training. To summarize,

- We design a slice-based volumetric SR network called DA-VSR. DA-VSR uses a Unified Feature Extraction (UFE) backbone and a series of lightweight network heads to perform super-resolution.
- We propose an in-plane SR head that propagates gradients to the UFE backbone both in training and testing. As such, DA-VSR is more robust, and can adapt its features to the test data distribution.
- We experiment with a diverse set of medical imaging data on different parts of the organ, and find large quantitative and visual improvement in SR quality on datasets out of the training distribution.

2   Domain Adaptable Volumetric Super-Resolution

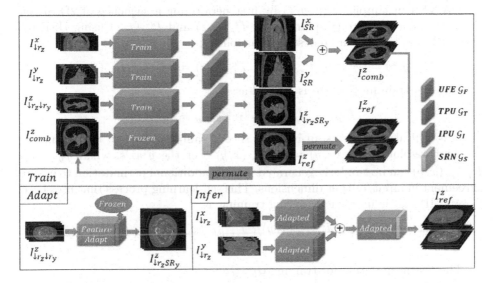

Fig. 1. The overall pipeline of Domain Adaptable Volumetric Super-Resolution (DA-VSR). DA-VSR contains three stages. The network parameters are first trained in supervised setting. An additional adaptation stage is proposed to fit to test data. Finally inference is done through an adapted feature backbone and network heads. Networks of the same color share weights.

Consider $I(x, y, z) \in \mathbb{R}^{X \times Y \times Z}$ as a densely sampled volumetric medical image. Following the notations in [13], we refer to x, y, and z as the sagittal, coronal, and axial axis, and $I^x(y, z)$, $I^y(x, z)$, and $I^z(x, y)$ as the sagittal, coronal, and axial slices. The task of super-resolution seeks to recover $I(x, y, z)$ from its partially observed, downsampled version $I_{\downarrow}(x, y, z)$. This work focuses on finding a transformation $\mathcal{F}: \mathbb{R}^{X \times Y \times \frac{Z}{r_z}} \to \mathbb{R}^{X \times Y \times Z}$ that super-resolves an axially sparse volume $I_{\downarrow r_z}(x, y, z)$ to $I(x, y, z)$, where r_z is the sparsity factor in the axial axis. The super-resolving function $\mathcal{F}$ is most popularly approximated through a CNN

$\mathcal{F}_{\theta_S}$, where θ_S denotes the network parameters learned from a training set $\mathcal{S}$ that contains LR-HR pairs $\{I^{\mathcal{S}}_{\downarrow r_z}(x,y,z), I^{\mathcal{S}}(x,y,z)\}$ for supervised learning. While $\mathcal{F}_{\theta_S}$ may be near-optimal for $\mathcal{S}$, its performance degrades when used on a test set $\mathcal{T}$ from a different distribution. Hence, we would like to approximate a better $\mathcal{F}_{\theta_T}$ based on θ_S and $I^{\mathcal{T}}_{\downarrow r_z}(x,y,z)$. As shown in Fig. 1, DA-VSR consists of a supervised training stage that yields an $\mathcal{F}_{\theta_S}$, and a self-supervised adaptation stage that adjusts its network parameters to fit with $\mathcal{T}$.

2.1 Network Structure

DA-VSR consists of four components: a Unified Feature Extraction (UFE) backbone and three lightweight network heads called Through-Plane Upsampler (TPU), In-Plane Upsampler (IPU), and Slice Refinement Net (SRN), which are denoted as $\mathcal{G}_F$, $\mathcal{G}_T$, $\mathcal{G}_I$, and $\mathcal{G}_S$, respectively.

DA-VSR upsamples $I_{\downarrow}(x,y,z)$ by first performing a sequence of 2D SR on the low resolution through-plane slices $I^{x}_{\downarrow r_z}(y,z)$ and $I^{y}_{\downarrow r_z}(x,z)$ using UFE $\mathcal{G}_F$ and TPU $\mathcal{G}_T$, which can be described as follows:

$$I^{x}_{SR}(y,z) = \mathcal{G}_T \circ \mathcal{G}_F(I^{x}_{\downarrow r_z}(y,z)), \quad I^{y}_{SR}(x,z) = \mathcal{G}_T \circ \mathcal{G}_F(I^{y}_{\downarrow r_z}(x,z)). \quad (1)$$

The loss for training $\mathcal{G}_T \circ \mathcal{G}_F$ is defined as:

$$\mathcal{L}_{tpu} = \|I^{x}_{SR} - I^{x}_{gt}\|_1 + \|I^{y}_{SR} - I^{y}_{gt}\|_1. \quad (2)$$

DA-VSR also performs an in-plane slice upsampling by first downsampling the high resolution $I^{z}_{\downarrow r_z}(x,y)$ to $I^{z}_{\downarrow r_z \downarrow r_y}(x,y)$ over the y axis, where $r_y = r_z$. $I^{z}_{\downarrow r_z \downarrow r_x}(x,y)$ can be similarly generated with an additional permutation to ensure agreement in input/output dimensions. This self-learning process that uses UFE $\mathcal{G}_F$ and IPU $\mathcal{G}_I$ is introduced during training as:

$$I^{z}_{\downarrow r_z SR_x}(x,y) = \mathcal{G}_I \circ \mathcal{G}_F(I^{z}_{\downarrow r_z \downarrow r_x}(x,y)), \quad I^{z}_{\downarrow r_z SR_y}(x,y) = \mathcal{G}_I \circ \mathcal{G}_F(I^{z}_{\downarrow r_z \downarrow r_y}(x,y)). \quad (3)$$

The loss formulation for training $\mathcal{G}_I \circ \mathcal{G}_F$ is:

$$\mathcal{L}_{ipu} = \|I^{z}_{\downarrow r_z SR_x} - I^{z}_{\downarrow r_z}\|_1 + \|I^{z}_{\downarrow r_z SR_y} - I^{z}_{\downarrow r_z}\|_1. \quad (4)$$

During training, the overall loss is defined as $\mathcal{L}_{main} = \lambda_{tpu} * \mathcal{L}_{tpu} + \lambda_{ipu} * \mathcal{L}_{ipu}$, where λ_{tpu} and λ_{ipu} are selected as 2 and 1 empirically. After $\mathcal{G}_F$, $\mathcal{G}_T$, and $\mathcal{G}_I$ are trained to convergence, we <u>freeze</u> their network parameters to obtain $\mathcal{G}_F^{fro}$, $\mathcal{G}_T^{fro}$, and $\mathcal{G}_I^{fro}$, respectively. The super-resolved slices I^{x}_{SR} and I^{y}_{SR} are reformatted into volumes and averaged to yield a single volume $I_{comb}(x,y,z) = \frac{1}{2}(I^{x}_{SR}(x,y,z) + I^{y}_{SR}(x,y,z))$. We then feed the axial slices $I^{z}_{comb}(x,y)$ into a frozen UFE $\mathcal{G}_F^{fro}$ and a SRN $\mathcal{G}_S$. As such, we reuse the already well-trained feature extraction from a deep $\mathcal{G}_F$ and a lightweight $\mathcal{G}_S$ to perform axial refinement in $I^{z}_{comb}(x,y)$. Note that this is different from SAINT, which trains an independent, relatively shallow network from scratch to perform axial refinement. The forward process and training loss are:

$$I^{z}_{ref}(x,y) = \mathcal{G}_S^{fro} \circ \mathcal{G}_F(I^{z}_{comb}(x,y)), \mathcal{L}_{ref} = \|I^{z}_{ref} - I^{z}_{gt}\|_1. \quad (5)$$

2.2 Self-supervised Adaptation

After all networks are trained to convergence with dataset $\mathcal{S}$, DA-VSR uses a simple yet effective adaptation stage before inference on test set $\mathcal{T}$, as inspired by SMORE [24]. We seek to update network parameters in $\mathcal{G}_F$ to fit to $\mathcal{T}$ through the common task of in-plane super-resolution. In particular, we <u>freeze</u> $\mathcal{G}_I$ during adaptation and only allow backpropagating gradients to modify parameters in $\mathcal{G}_F$ to form $\mathcal{G}_F^{adp}$. This process is described as:

$$I_{\downarrow r_z SR_x}^z(x,y) = \mathcal{G}_I^{adp} \circ \mathcal{G}_F^{fro}(I_{\downarrow r_z \downarrow r_x}^z(x,y)), \quad I_{\downarrow r_z SR_y}^z(x,y) = \mathcal{G}_I^{adp} \circ \mathcal{G}_F^{fro}(I_{\downarrow r_z \downarrow r_y}^z(x,y)), \tag{6}$$

where $\mathcal{G}_I^{fro}$ denotes a frozen, pretrained $\mathcal{G}_I$. The loss formulation is similar to Eq. (4). We find that this approach effectively prevents overfitting of adaptation to the test data $\mathcal{T}$. If $\mathcal{G}_I$ is not frozen, the composition of $\mathcal{G}_I \circ \mathcal{G}_G$ effectively becomes a SMORE setup after tuning. With $\mathcal{G}_T$ and $\mathcal{G}_I$ both learned through the training data, $\mathcal{G}_I$ serves as proxy that constrains features generated by $\mathcal{G}_F$ to not veer far from the effective range of $\mathcal{G}_T$. As shown in Fig. 1, after self-supervised adaptation, inference can be done straigtforwardly. The LR through-plane slices $I_{\downarrow r_z}^x(y,z)$ and $I_{\downarrow r_z}^y(x,z)$ are upsampled by $\mathcal{G}_T^{adp} \circ \mathcal{G}_F^{fro}$, combined together as a volume through averaging, and fed to $\mathcal{G}_S^{adp} \circ \mathcal{G}_F^{fro}$ as a series of axial slices. The refined slices I_{ref}^z are formatted to form the final upsampled volume.

3 Experiments

3.1 Implementation Details

To ensure a fair comparison, we implement all compared models to have similar number of parameters, as shown in Table 1 and Table 2. For DA-VSR, we use Residual Dense Blocks (RDBs) [23] as the building block for UFE. Specifically, we use six RDBs, each of which has eight convolution layers and a growth rate of thirty-two. For TPU, IPU and SRN, we use a lightweight design of three convolution layers. TPU and IPU use additional pixel shuffling layers to upsample the spatial dimension on the axial axis. Following [13], the LR input image is formatted as three consecutive slices.

SAINT [13] is similarly implemented with a RDB-based network. 3DRDN is implemented with two RDBs, each containing eight convolutional layers and with a growth rate of thirty-two. 3DRCAN is implemented with three residual groups, each containing three residual blocks and with a feature dimension of sixty-four.

3.2 Dataset

For training and validation, we use 890 CT scans from the publicly available LIDC-IDRI [1] dataset, which are taken on lungs. We use 810 volumes for training, 30 volumes for validation, and 50 volumes for testing. Additionally, we use

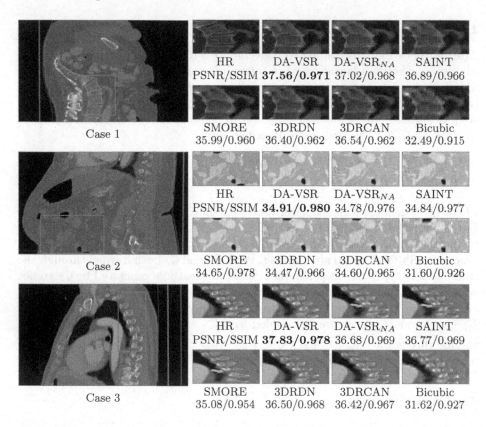

Fig. 2. Visual comparisons of DA-VSR and other state-of-the-art implementations from the sagittal plane, highlight regions are contrast-adjusted. Case 1 is from the Colon dataset, Case 2 and 3 are from the Kidney dataset [7]. In particular, Case 3 is a lung region cropped from a kidney-containing image. Please refer to the supplemental material for more visual comparisons.

a slew of CT datasets that are acquired for different organs, including liver [16], colon [16], and kidney [7] for testing. For consistency and following previous literature [2,17], all volumes are selected with slice thickness between 1 mm to 3.5 mm, and interpolated to 2.5 mm. We then discard volumes with less than 128 axial slices. After pre-processing, we obtain 120 liver volumes, 30 colon volumes, and 59 kidney volumes for testing. Due to large memory cost at inference time for 3D-kernel baselines, in-plane pixel resolution is downsampled from 512×512 to 256×256. For training, 3D-kernel baselines use patches of size $64 \times 64 \times Z$.

3.3   Ablation Study

We examine the effectiveness of through-plane SR in DA-VSR as compared to other implementations. Specifically, we compare the full DA-VSR to:

- DA-VSR$_{SMORE}$: DA-VSR that only uses test data to train, implemented similar to SMORE [24].
- DA-VSR$_{SAINT}$: DA-VSR without a self-supervised $\mathcal{G}_I$, similar to SAINT [13].
- DA-VSR^{NA}: DA-VSR without a self-supervised adaptation stage.
- DA-VSR$^A_{nofro}$: DA-VSR with a self-supervised adaptation stage, where $\mathcal{G}_I$ is not frozen during adaptation.

Table 1. Quantitative ablation study of through-plane upsampling for DA-VSR in terms of PSNR and SSIM, measured based on $I^x_{SR}(y, z)$. The best results are in **bold**, and the second best results are underlined. All baselines achieved similar performance compared to the original papers.

Scale	Method	Param	Lung	Liver	Colon	Kidney
X4	DA-VSR$_{SMORE}$	2.8M	38.25/0.971	39.05/0.981	40.00/0.986	37.18/0.975
	DA-VSR$_{SAINT}$	2.8M	**39.77/0.976**	39.11/0.980	40.32/0.987	37.53/0.975
	DA-VSR^{NA}	2.9M	39.67/0.976	39.29/0.981	40.43/0.987	37.60/0.976
	DA-VSR$^A_{nofro}$	2.9M	N/A	38.94/0.981	39.89/0.983	36.91/0.974
	DA-VSR	2.9M	N/A	**39.51/0.982**	**40.60/0.988**	**38.07/0.978**

The performances are summarized in Table 1. Improvements can be seen from DA-VSR^{NA} to the full DA-VSR when self-supervised adaptation is applied. We also observe that if $\mathcal{G}_I$ is not frozen during adaptation, performance is severely degraded after adaptation. Note for adaptation, DA-VSR$^A_{nofro}$ and DA-VSR are both trained to convergence, which typically takes five to ten epochs. In comparison, adaptation on DA-VSR with a frozen $\mathcal{G}^{fro}_I$ is stable and performance does not degrade even if trained over many epochs. This can be helpful as no scheme for early stopping is required. We also observe that quantitatively DA-VSR^{NA} performs slightly better over unseen datasets than DA-VSR$_{SAINT}$, despite slightly worse performance over lung. This may be attributed to DA-VSR's self-supervised in-plane upsampling process, as it forces the network observe a more diverse data distribution. Finally, DA-VSR$_{SMORE}$ performs well over unseen datasets and in some instances is nearly on par with supervised methods, e.g. over the Liver dataset. The performance of DA-VSR$_{SMORE}$ fluctuates depending on (1) how similar in-plane and through-plane statistics are, and (2) the sample size of the test dataset. In comparison, DA-VSR is less reliant on these factors, as its feature extractor is trained with supervision and on a large dataset.

3.4 Quantitative Evaluation

As shown in Table 2, we compare the full DA-VSR pipeline against other state-of-the-art SR implementations: a 3D-kernel variant of RDN [20], a 3D-kernel variant of RCAN [22], and SAINT [13]. We find that 3D RDN and RCAN is not as efficient as a slice-based volumetric SR approach like SAINT under similar network size. As an overall pipeline, DA-VSR performs slightly better than SAINT on unseen dataset, and significantly better when adaptation is applied. We find that by using a unified feature backbone to generate features, DA-VSR's slice refinement stage converges much faster than SAINT's despite using less parameters; please refer to the supplemental material for illustration.

Table 2. Quantitative <u>volume-wise</u> evaluation of DA-VSR against SoTA SR implementations in terms of PSNR and SSIM. The best results are in **bold**, and the second best results are <u>underlined</u>. All baselines achieved similar performance compared to the original papers. Please refer to supplementary material for significance tests.

Scale	Method	Param	Lung	Liver	Colon	Kidney
X4	Bicubic	N/A	33.72/0.941	34.56/0.955	35.23/0.964	33.37/0.948
	3D RCAN	2.9M	39.30/0.975	38.95/0.979	40.11/0.986	37.41/0.975
	3D RDN	2.9M	39.39/0.976	39.04/0.980	40.22/0.986	37.46/0.975
	SAINT	2.9M	**40.01/0.977**	39.30/0.981	40.51/0.987	37.70/0.976
	DA-VSR^{NA}	3.0M	<u>39.90/0.977</u>	<u>39.48/0.981</u>	<u>40.68/0.988</u>	<u>37.82/0.977</u>
	DA-VSR	3.0M	N/A	**39.74/0.983**	**40.83/0.988**	**38.28/0.979**
X6	Bicubic	N/A	31.47/0.913	32.55/0.935	32.98/0.943	31.31/0.924
	3D RCAN	3.0M	36.56/0.962	36.62/0.970	37.19/0.975	35.04/0.962
	3D RDN	3.3M	36.73/0.963	36.67/0.970	37.23/0.976	35.10/0.963
	SAINT	2.9M	**37.23/0.966**	36.76/0.971	37.37/0.977	35.26/0.964
	DA-VSR^{NA}	3.0M	<u>37.14/0.965</u>	<u>36.86/0.972</u>	<u>37.44/0.977</u>	<u>35.31/0.964</u>
	DA-VSR	3.0M	N/A	**37.18/0.973**	**37.78/0.979**	**35.55/0.966**

While metrics like PSNR and SSIM are useful to understand performance in aggregate, they can often be too coarse. Domain drift does not happen uniformly on a CT image. Some patches do not suffer as much since similar patches can be observed in the training set, leading to similar overall PSNR metrics; however, some patches suffer heavily due to lack of observations. We provide visualization on SR results, as shown in Fig. 2, to better show where improvements are most often observed. We observe that supervised techniques without adaptation can lead to significant overfitting issues over unseen test sets and create unfaithful details, as seen in Case 2 of Fig. 2 and indicated by the orange arrows. Compared to supervised methods, SMORE produces results that are smoother and more similar to the groundtruth if those patterns are seen in axial slices. For organs that exhibit different patterns between axial and other axes, such as the sagittal spinal structure, SMORE can generate unreliable or overly smooth patterns, as shown in Case 1 of Fig. 2. In comparison, since DA-VSR goes through supervised training on a lung dataset, which contains both LR and HR spine patterns, we observe that it performs much better than SMORE even with adaptation. Interestingly, we also observe improvements over the few unseen cases where a region of lung is included, e.g. in Case 3, by using our proposed adaptation stage. Despite being trained on the lung dataset, other supervised methods still experience local discontinuity over small scale bone structures. In this case, SMORE generates smoother but less structurally reliable details. DA-VSR takes the advantages of both approaches and generate smoother and more reliable details under this challenging case. As no two individuals are the same, DA-VSR's ability to reduce minor distribution differences can be valuable in real SR applications.

4 Conclusion

We propose a Domain-Adaptable Volumetric Super-Resolution (DA-VSR). Inspired by SAINT [13] and SMORE [24], DA-VSR leverages the advantages in supervised and self-supervised learning. Specifically, DA-VSR uses supervised training to learn a strong feature generator with various task-specific network heads, and a self-supervised domain adaptation stage to better fit to unseen test sets. We carefully evaluate our approach between training and testing CT datasets that are acquired on different organs. We find that DA-VSR produce consistent improvements in quantitative measurements and visual quality. Our approach is conceptually straightforward and can be implemented with different network structures. Future work includes investigating the effect of anisotropic resolution to self-supervised adaptation and the effect of DA-VSR over other types of domain gap, e.g. cross machine, cross modality, etc.

References

1. Armato, S.G., III., et al.: The lung image database consortium (LIDC) and image database resource initiative (IDRI): a completed reference database of lung nodules on CT scans. Med. Phys. **38**(2), 915–931 (2011)
2. Chen, Y., Shi, F., Christodoulou, A.G., Xie, Y., Zhou, Z., Li, D.: Efficient and accurate MRI super-resolution using a generative adversarial network and 3D multi-level densely connected network. In: Frangi, A.F., Schnabel, J.A., Davatzikos, C., Alberola-López, C., Fichtinger, G. (eds.) MICCAI 2018, Part I. LNCS, vol. 11070, pp. 91–99. Springer, Cham (2018). https://doi.org/10.1007/978-3-030-00928-1_11
3. Cherukuri, V., Guo, T., Schiff, S.J., Monga, V.: Deep MR brain image super-resolution using spatio-structural priors. IEEE Trans. Image Process. **29**, 1368–1383 (2020)
4. Dong, C., Loy, C.C., He, K., Tang, X.: Image super-resolution using deep convolutional networks. CoRR abs/1501.00092 (2015). http://arxiv.org/abs/1501.00092
5. Dong, C., Loy, C.C., Tang, X.: Accelerating the super-resolution convolutional neural network. CoRR abs/1608.00367 (2016). http://arxiv.org/abs/1608.00367
6. Georgescu, M., Ionescu, R.T., Verga, N.: Convolutional neural networks with intermediate loss for 3d super-resolution of CT and MRI scans. IEEE Access **8**, 49112–49124 (2020)
7. Heller, N., et al.: The kits19 challenge data: 300 kidney tumor cases with clinical context, CT semantic segmentations, and surgical outcomes (2019)

8. Kim, J., Lee, J.K., Lee, K.M.: Accurate image super-resolution using very deep convolutional networks. CoRR abs/1511.04587 (2015). http://arxiv.org/abs/1511.04587

9. Kim, J., Lee, J.K., Lee, K.M.: Deeply-recursive convolutional network for image super-resolution. In: 2016 IEEE Conference on Computer Vision and Pattern Recognition, CVPR 2016, Las Vegas, NV, USA, 27–30 June 2016, pp. 1637–1645 (2016)

10. Liu, J., Zhang, W., Tang, Y., Tang, J., Wu, G.: Residual feature aggregation network for image super-resolution. In: 2020 IEEE/CVF Conference on Computer Vision and Pattern Recognition, CVPR 2020, Seattle, WA, USA, 13–19 June 2020, pp. 2356–2365. IEEE (2020)

11. Park, J., Hwang, D., Kim, K.Y., Kang, S.K., Kim, Y.K., Lee, J.S.: Computed tomography super-resolution using deep convolutional neural network. Phys. Med. Biol. **63**(14), 145011 (2018)

12. Peng, C., Lin, W.A., Liao, H., Chellappa, R., Zhou, S.K.: Deep slice interpolation via marginal super-resolution, fusion and refinement (2019)

13. Peng, C., Lin, W., Liao, H., Chellappa, R., Zhou, S.K.: SAINT: spatially aware interpolation network for medical slice synthesis. In: 2020 IEEE/CVF Conference on Computer Vision and Pattern Recognition, CVPR 2020, Seattle, WA, USA, 13–19 June 2020, pp. 7747–7756. IEEE (2020)

14. Shi, W., et al.: Real-time single image and video super-resolution using an efficient sub-pixel convolutional neural network (2016)

15. Shocher, A., Cohen, N., Irani, M.: "zero-shot" super-resolution using deep internal learning. In: 2018 IEEE Conference on Computer Vision and Pattern Recognition, CVPR 2018, Salt Lake City, UT, USA, 18–22 June 2018, pp. 3118–3126. IEEE Computer Society (2018)

16. Simpson, A.L., et al.: A large annotated medical image dataset for the development and evaluation of segmentation algorithms (2019)

17. Wang, J., Chen, Y., Wu, Y., Shi, J., Gee, J.: Enhanced generative adversarial network for 3d brain MRI super-resolution. In: IEEE Winter Conference on Applications of Computer Vision, WACV 2020, Snowmass Village, CO, USA, 1–5 March 2020, pp. 3616–3625. IEEE (2020)

18. You, C., et al.: CT super-resolution GAN constrained by the identical, residual, and cycle learning ensemble (GAN-circle). IEEE Trans. Med. Imaging **39**(1), 188–203 (2020). https://doi.org/10.1109/TMI.2019.2922960

19. Yu, H., et al.: Computed tomography super-resolution using convolutional neural networks. In: 2017 IEEE International Conference on Image Processing, ICIP 2017, Beijing, China, 17–20 September 2017, pp. 3944–3948. IEEE (2017)

20. Zhang, K., Zuo, W., Gu, S., Zhang, L.: Learning deep CNN denoiser prior for image restoration. In: 2017 IEEE Conference on Computer Vision and Pattern Recognition, CVPR 2017, Honolulu, HI, USA, 21–26 July 2017, pp. 2808–2817 (2017)

21. Zhang, K., Zuo, W., Zhang, L.: Deep plug-and-play super-resolution for arbitrary blur kernels. In: IEEE Conference on Computer Vision and Pattern Recognition, CVPR 2019, Long Beach, CA, USA, 16–20 June 2019, pp. 1671–1681. Computer Vision Foundation / IEEE (2019)

22. Zhang, Y., Li, K., Li, K., Wang, L., Zhong, B., Fu, Y.: Image super-resolution using very deep residual channel attention networks. In: Ferrari, V., Hebert, M., Sminchisescu, C., Weiss, Y. (eds.) ECCV 2018, Part VII. LNCS, vol. 11211, pp. 294–310. Springer, Cham (2018). https://doi.org/10.1007/978-3-030-01234-2_18

23. Zhang, Y., Tian, Y., Kong, Y., Zhong, B., Fu, Y.: Residual dense network for image super-resolution. CoRR abs/1802.08797 (2018). http://arxiv.org/abs/1802.08797
24. Zhao, C., Dewey, B.E., Pham, D.L., Calabresi, P.A., Reich, D.S., Prince, J.L.: SMORE: a self-supervised anti-aliasing and super-resolution algorithm for MRI using deep learning. IEEE Trans. Med. Imag. **40**(3), 805–817 (2021)
25. Zhao, X., Zhang, Y., Zhang, T., Zou, X.: Channel splitting network for single MR image super-resolution. IEEE Trans. Image Process. **28**(11), 5649–5662 (2019)
26. Zhou, S.K., et al.: A review of deep learning in medical imaging: imaging traits, technology trends, case studies with progress highlights, and future promises. In: Proceedings of the IEEE (2021)
27. Zhou, S.K., Rueckert, D., Fichtinger, G.: Handbook of Medical Image Computing and Computer Assisted Intervention. Academic Press, San Diego (2019)

Improving Generalizability in Limited-Angle CT Reconstruction with Sinogram Extrapolation

Ce Wang[1,2], Haimiao Zhang[3], Qian Li[1,2], Kun Shang[4], Yuanyuan Lyu[5], Bin Dong[6], and S. Kevin Zhou[1,7(✉)]

[1] Key Lab of Intelligent Information Processing of Chinese Academy of Sciences (CAS), Institute of Computing Technology, CAS, Beijing, China
[2] Suzhou Institute of Intelligent Computing Technology, Chinese Academy of Sciences, Suzhou, China
[3] Institute of Applied Mathematics, Beijing Information Science and Technology University, Beijing, China
[4] Research Center for Medical AI, Shenzhen Institutes of Advanced Technology, Chinese Academy of Sciences, Shenzhen, China
[5] Z^2Sky Technologies Inc., Suzhou, China
[6] Beijing International Center for Mathematical Research, Peking University, Beijing, China
[7] Medical Imaging, Robotics, and Analytic Computing Laboratory and Engineering (MIRACLE) School of Biomedical Engineering and Suzhou Institute for Advanced Research, University of Science and Technology of China, Suzhou 215123, China

Abstract. Computed tomography (CT) reconstruction from X-ray projections acquired within a limited angle range is challenging, especially when the angle range is extremely small. Both analytical and iterative models need more projections for effective modeling. Deep learning methods have gained prevalence due to their excellent reconstruction performances, but such success is mainly limited within the same dataset and does not generalize across datasets with different distributions. Hereby we propose ExtraPolationNetwork for limited-angle CT reconstruction via the introduction of a sinogram extrapolation module, which is theoretically justified. The module complements extra sinogram information and boots model generalizability. Extensive experimental results show that our reconstruction model achieves state-of-the-art performance on NIH-AAPM dataset, similar to existing approaches. More importantly, we show that using such a sinogram extrapolation module significantly improves the generalization capability of the model on unseen datasets (e.g., COVID-19 and LIDC datasets) when compared to existing approaches.

Keywords: Limited-angle CT reconstruction · Sinogram extrapolation · Model generalizability

1 Introduction and Motivation

In healthcare, Computed Tomography (CT) based on X-ray projections is an indispensable imaging modality for clinical diagnosis. Limited-angle (LA) CT is a common type of acquisition in many scenarios, such as to reduce radiation dose in low-dose CT or forced to take projections in a restricted range of angles in C-arm CT [17] and dental CT. However, the deficiency of projection angles brings significant challenge to image reconstruction and may lead to severe artifacts in the reconstructed images.

Many CT image reconstruction algorithms have been proposed in the literature to improve image quality, which can be categorized as model-based and deep-learning-based methods. For example, Filtered Back Projection (FBP) [20], as a representative analytical method, is widely used for reconstructing a high-quality image efficiently. However, FBP prefers acquisition with full-ranged views which makes using it for LACT sub-optimal. The (some times extreme) reduction on the range of projection angles decreases the effectiveness of the commercial CT reconstruction algorithms. To overcome such challenge, iterative regularization-based algorithms [6,12,15,18,21,27] are proposed to leverage prior-knowledge on the image to be reconstructed and achieve better reconstruction performance for LACT. Notice that those iterative algorithms are often computationally expensive and require careful case-by-case hyperparameter tuning.

Currently, deep learning (DL) techniques have been widely adopted in CT and demonstrate promising reconstruction performance [2,8,24,28,32,33]. By further combining the iterative algorithms with DL, a series of iterative frameworks with the accordingly designed neural-network-based modules are proposed [1,3,5,7,13,19,29]. ADMMNet [25] introduces a neural-network-based module in reconstruction problem and achieves remarkable performance. Furthermore, DuDoNet [10,11], ADMM-CSNet [26] and LEARN++ [31] improve reconstruction results with an enhancement module in the projection domain, which inspires us to fuse dual-domain learning in our model design.

Although deep-learning-based algorithms have achieved state-of-the-art performance, they are also known to easily over-fit on training data, which is not expected in practice. MetaInvNet [30] is then proposed to improve the reconstruction performance with sparse-view projections, demonstrating good model generalizability. They attempt to find better initialization for an iterative HQS-CG [6] model with a U-Net [16] and achieve better generalization performance in such scenarios. But they still focus on the case with a large range of acquired projections, which limits the application of their model in practice. How to obtain a highly generalizable model when learning from practical data is still difficult.

To retain model generalizability in LACT reconstruction, we propose a model, called ExtraPolationNetwork (EPNet), for recovering high-quality CT images. In this model, we utilize dual-domain learning to emphasize data consistency between image domain and projection domain, and introduce an extrapolation module. The proposed extrapolation module helps complement missed information in the projection domain and provides extra details for reconstruction. Extensive experimental results show that the model achieves state-of-the-art performance on the NIH-AAPM dataset [14]. Furthermore, we also achieve better generalization performance on additional datasets, COVID-19 and LIDC [4]. This empirically verifies the effectiveness of the proposed extrapolation module. We make our implementation available at https://github.com/mars11121/EPNet.

2   Problem Formulation

CT reconstruction aims to reconstruct clean image u from the projection data Y with unknown noise n, whose mathematical formulation is:

$$Y = Au + n,$$

where A is the Radon transform. For LACT, the projection data Y is incomplete as a result of the decrease of angle range (the view angle $\alpha \in [0, \alpha_{max}]$ with $\alpha_{max} < 180°$). The reduced sinogram information limits the performance of current reconstruction methods. Therefore, the estimation of a more complete sinogram $\widetilde{Y}$ is necessary to enhance model reconstruction performance. To yield such an accurate estimation, the consistency between incomplete projection Y and complete projection $\widetilde{Y}$ is crucial. We assume Y is obtained by some operations (e.g. downsampling operation) from $\widetilde{Y}$. Besides, $\widetilde{Y}$ and clean image u should also be consistent under the corresponding transformation matrix $\widetilde{A}$. Consequently, we propose the following constraints:

$$\widetilde{A}u = \widetilde{Y}, \quad P\widetilde{Y} = Y, \tag{1}$$

where P is the downsampling matrix.

In this way, the final model becomes the following optimization problem:

$$\min_{u} \quad \frac{1}{2}\|Au - Y\|^2 + R(u)$$
$$s.t. \quad \widetilde{A}u = \widetilde{Y}, \quad P\widetilde{Y} = Y, \tag{2}$$

where $R(\cdot)$ is a regularization term incorporating image priors.

3   Proposed Method

In this section, we introduce full details on the proposed method, which is depicted in Fig. 1. Our model is built by unrolling the HQS-CG [6] algorithm with N iterations. The HQS-CG algorithm is briefly introduced in Sect. 3.1. Specifically, we utilize the Init-CNN module [30] to search for a better initialization for Conjugate Gradient (CG) algorithm in each iteration. The input of the module is composed of reconstructed images from the image domain and projection domain. In the image domain, we retrain the basic HQS-CG model and use the CG module for reconstruction. In the projection domain, we first use our proposed Extrapolation Layer (EPL) to estimate extra sinograms. Then, we use Sinogram Enhancement Network (SENet) to inpaint the extrapolated sinograms and reconstruct them with Radon Inversion Layer (RIL) [10], which is capable of backpropagating gradients to the previous layer. Section 3.2 introduces the details of the involved modules.

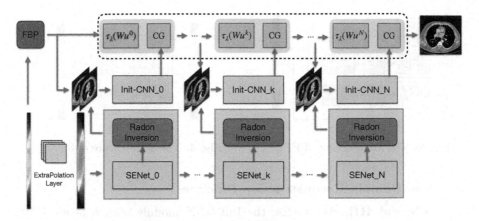

Fig. 1. The framework of our proposed EPNet. It reconstructs CT image via two parallel pipelines in the image domain and sinogram domain. In the sinogram domain, we propose an extrapolation layer before SENet to take extra prior information of image details. The reconstructed images of sinogram domain and image domain are then concatenated followed by an "Init-CNN" module to provide a more accurate initialization estimation for the Conjugate-Gradient algorithm.

3.1 HQS-CG Algorithm

Traditionally, there exist many effective algorithms to solve objective (2). One such algorithm is Half Quadratic Splitting (HQS) [6], which solves the following:

$$\min_{u,z} \quad \frac{1}{2}\|Au-Y\|^2 + \lambda\|z\|_1 + \frac{1}{2}\sum_{i=1}^{M}\gamma_i\|W_iu-z_i\|^2 + \beta_1\|P\widetilde{Y}-Y\|^2 + \beta_2\|\widetilde{A}u-\widetilde{Y}\|^2,$$

$$(3)$$

where $W = (W_1, W_2, \ldots, W_M)$ is a $M-$channel operator, $z = (z_1, z_2, \ldots, z_M)$, $\lambda > 0$, $\beta_1 > 0$, $\beta_2 > 0$, and $\gamma = (\gamma_1, \gamma_2, \ldots, \gamma_M)$ with $\{\gamma_i\}_{i=1}^{M} > 0$. The operator W is chosen as the highpass components of the piecewise linear tight wavelet frame transform. With alternating optimization among $\widetilde{Y}$, u, and z, the final closed-form solution could be derived as follows:

$$\widetilde{Y}^{k+1} = \left(\beta_1 P^T P + \beta_2\right)^{-1}\left[\beta_1 P^T Y + \beta_2 \widetilde{A} u^k\right],$$

$$u^{k+1} = \left(A^T A + \sum_{i=1}^{M}\gamma_i W_i^T W_i + 2\beta_2 \widetilde{A}^T \widetilde{A}\right)^{-1}\left[A^T Y + \sum_{i=1}^{M}\gamma_i W_i^T z_i^{\ k} + 2\beta_2 \widetilde{A}^T \widetilde{Y}^{k+1}\right],$$

$$z_i^{k+1} = \tau_{\lambda/\gamma_i}(W_i u^{k+1}), i = 1, \ldots, M,$$

where $\tau_\lambda(x) = \mathrm{sgn}(x)\max\{\|x\| - \lambda, 0\}$ is the soft-thresholding operator.

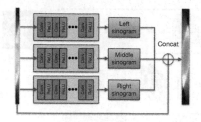   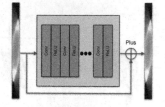

Fig. 2. The architecture of EPL.   **Fig. 3.** The architecture of SENet.

3.2 Dual-Domain Reconstruction Pipelines

Init-CNN and RIL. We realize the Init-CNN module with a heavy U-Net architecture with skip connection, which stabilizes the training. Besides, the heavy U-Net shares parameters across different steps, which is proved more powerful for the final reconstruction. The Radon Inversion Layer (RIL) is first introduced in DuDoNet [10], which builds dual-domain learning for reconstruction. We here use the module to obtain the reconstructed image from the projection domain.

EPL. As introduced, the reduction of angle range is the main bottleneck in the limited-angle scenario. Besides, usual interpolation techniques are not suitable in this case. But few researchers consider extrapolating sinograms with CNNs, which provides more details of images since sinograms contain both spatial and temporal (or view angle) information of the corresponding images. Compared with the image domain difference, sinograms from different data distributions also have similarities in the temporal dimension. To utilize such an advantage, we propose a module called "Extrapolation Layer (EPL)" to extrapolate sinograms before SENet.

As shown in Fig. 2, the EPL module is composed of three parallel convolutional neural networks, where the left and the right networks are used to predict neighboring sinograms of the corresponding sides and the middle one is used to denoise the input. The outputs of the three networks are then concatenated, followed with the proposed supervision defined as follows:

$$\mathcal{L}_{EPL} = (1 + mask) \times \|Y_{out} - Y_{gt}\|_1 + \|RIL(Y_{out}) - RIL(Y_{gt})\|_1, \qquad (4)$$

where Y_{out} is the predicted sinogram, Y_{gt} is the corresponding ground-truth, and $mask$ is a binary matrix to emphasize the bilateral prediction. Here, we utilize RIL to realize a dual-domain consistency for the prediction, which makes the module estimation more accurate when embedded into the whole model.

SENet. With extrapolated sinograms, we then use SENet to firstly enhance the quality of sinograms, which is designed as a light CNN as in Fig. 3. At last, the enhanced sinograms are mapped to the image domain via RIL, which would help decrease the different optimization directions in our dual-domain learning. The objective for SENet is as follows:

$$\mathcal{L}_{SE} = \|Y_{se} - Y_{gt}\|_1 + \|RIL(Y_{se}) - u_{gt}\|_1, \qquad (5)$$

Table 1. Quantitative results of models on different testing datasets. The best performances in each row are in **bold**.

Ablation study	EPL30		EPL30$_{re}$		EPL60		EPL120		DuDoNet		DuDoEPL30	
	PSNR	SSIM	PSNR	SSIM	PSNR	SSIM	PSNR	SSIM	PSNR	SSIM	PSNR	SSIM
AAPM-test	27.63	0.882	**28.03**	**0.886**	25.41	0.867	27.87	0.885	23.71	0.716	22.63	0.842
COVID-test	**19.59**	**0.713**	18.11	0.649	18.68	0.674	16.90	0.584	5.59	0.252	6.94	0.295
LIDC-test	**19.80**	**0.726**	18.81	0.670	19.19	0.690	18.07	0.627	5.45	0.246	6.93	0.304

where Y_{se} is the enhanced sinogram, Y_{gt} and u_{gt} are the corresponding ground-truth sinogram and image, respectively.

Loss function. With the above modules, the full objective function of EPNet is defined by:

$$\mathcal{L} = \sum_{i=1}^{N} \|u_i - u_{gt}\|_2 + \mu\mathcal{L}_{ssim}(u_N, u_{gt}) + \mathcal{L}_{EPL} + \mathcal{L}_{SE}, \tag{6}$$

where N is the total iterations of unrolled back-bone HQS-CG model, $\{u\}_{i=1}^{N}$ is the reconstructed image of each iteration, and $\mathcal{L}_{ssim}$ is the SSIM loss.

4 Experimental Results

4.1 Datasets and Experimental Settings

Datasets. We first train and test models on the "2016 NIH-AAPM-Mayo Clinic Low Dose CT Grand Challenge" dataset [14]. Specifically, we choose 1,746 slices of five patients for training and 1,716 slices of another five patients for testing. To further show our models' generalization capability, we test our models on 1,958 slices of four patients chosen from the COVID-19 dataset, and 1,635 slices of six patients from the LIDC dataset [4]. The two datasets are also composed of chest CT images but from different scenarios and machines, which constitutes good choices for testing the generalization capability. All the experiments are conducted with Fan-Beam Geometry and the number of detector elements is set to 800. Besides, we add mixed noise, composed of 5% Gaussian noise and Poisson noise with an intensity of $5e^6$, to all simulated sinograms [30].

Implementations and Training Settings. All the compared models are trained and tested with the corresponding angle number (15, 30, 60, 90) except MetaInvNet_ori, which is trained with 180 angle number as Zhang et al. [30] do. Our models are implemented using the PyTorch framework. We use the Adam optimizer [9] with $(\beta_1, \beta_2) = (0.9, 0.999)$ to train these models. The learning rate starts from 0.0001. Models are all trained on a Nvidia 3090 GPU card for 10 epochs with a batch size of 1.

Evaluation Metric. Quantitative results are measured by the multi-scale structural similarity index (SSIM) (with level = 5, Gaussian kernel size = 11, and standard deviation = 1.5) [23] and peak signal-to-noise ratio (PSNR) [22] (Table 2).

Table 2. Quantitative results on AAPM-test, COVID-test and LIDC-test sets. We also test the computation time when sinogram number is fixed as 60.

AAPM-test	$\alpha_{max} = 15$		$\alpha_{max} = 30$		$\alpha_{max} = 60$		$\alpha_{max} = 90$		Time (s)
	PSNR	SSIM	PSNR	SSIM	PSNR	SSIM	PSNR	SSIM	
FBP	7.12	0.399	8.61	0.495	12.63	0.643	14.98	0.726	0.0123
HQS-CG	18.66	0.621	19.95	0.675	21.65	0.744	23.96	0.800	15.7620
DuDoNet	20.04	0.633	21.11	0.661	23.71	0.716	24.95	0.753	1.5083
FBPCovNet	20.38	0.802	22.04	0.831	23.69	0.855	28.09	0.897	1.3901
MetaInvNet_ori	17.60	0.800	18.99	0.822	21.47	0.856	24.04	0.884	1.6528
MetaInvNet	**21.99**	**0.819**	**24.03**	**0.845**	**28.21**	**0.887**	**30.05**	**0.902**	1.5885
EPNet	**21.92**	**0.820**	**23.65**	**0.842**	27.63	0.882	30.40	0.906	1.8859
COVID-test	PSNR	SSIM	PSNR	SSIM	PSNR	SSIM	PSNR	SSIM	Time (s)
FBP	8.24	0.4073	9.31	0.4516	11.19	0.5177	12.98	0.5688	0.0101
HQS-CG	**17.84**	**0.611**	**19.29**	**0.679**	**21.18**	**0.782**	**23.41**	**0.824**	16.6641
DuDoNet	4.15	0.214	4.29	0.231	5.59	0.252	6.26	0.268	1.4262
FBPCovNet	3.96	0.273	3.88	0.279	4.91	0.296	6.60	0.328	1.4226
MetaInvNet_ori	**16.03**	**0.565**	16.80	0.584	18.45	0.644	20.86	0.719	1.8310
MetaInvNet	13.05	0.339	15.30	0.298	15.31	0.531	18.37	0.591	1.8199
EPNet	15.56	0.508	**17.31**	**0.616**	**19.59**	**0.713**	**21.57**	**0.745**	1.9004
LIDC-test	PSNR	SSIM	PSNR	SSIM	PSNR	SSIM	PSNR	SSIM	Time (s)
FBP	8.57	0.439	9.57	0.492	11.52	0.551	13.48	0.595	0.0116
HQS-CG	**18.71**	**0.602**	**20.30**	**0.673**	**22.27**	**0.787**	**24.46**	**0.837**	16.50
DuDoNet	4.12	0.205	4.19	0.223	5.45	0.246	6.24	0.266	1.4423
FBPCovNet	3.89	0.276	3.80	0.284	4.89	0.303	6.74	0.343	1.4328
MetaInvNet_ori	**16.59**	**0.561**	17.40	0.581	18.74	0.634	20.75	0.698	1.7403
MetaInvNet	13.57	0.367	15.66	0.517	15.45	0.547	18.33	0.583	1.7616
EPNet	16.07	0.564	**18.02**	**0.630**	19.80	0.726	22.18	0.751	1.9180

4.2   Ablation Study

To investigate the effectiveness of different modules and used hyperparameters for models, we firstly conduct an ablation study with the following configurations, where the number of the input sinogram angle is fixed to $\alpha_{max} = 60$:

a) EPL30: our model with pretrained EPL fixed and extrapolate 30 angles,

b) EPL30$_{re}$: our model with pretrained EPL not fixed and extrapolate 30 angles,

c) EPL60: our model with pretrained EPL fixed and extrapolate 60 angles,

d) EPL120: our model with pretrained EPL fixed and extrapolate 120 angles,

e) DuDoEPL30: DuDoNet with our proposed EPL and extrapolate 30 angles.

The quantitative comparison is shown in Table 1. When comparing (a) and (b), retraining parameters of the pretrained EPL module reduces the generalizability of our model, so we fix the EPL module parameters in later experiments. Besides, we investigate the most suitable extrapolated angle number for EPL. Comparing models (a) (c) (d), when increasing the number of extrapolated angles from 30 to 120, the reconstruction performance on AAPM-test is not affected, but the generalization performance gradually reduces. Therefore, we fix the number of extrapolated angles as 30 in all later experiments. Besides, we also insert the module in DuDoNet [10], the reconstruction performance drops a lot, but the module also improves generalization result by about 1.5 dB.

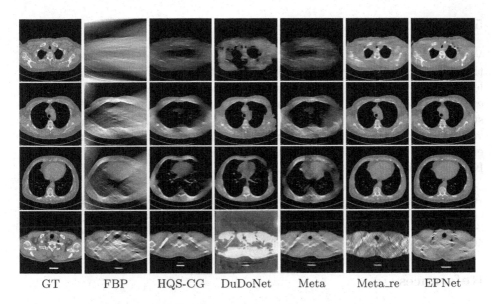

GT   FBP   HQS-CG   DuDoNet   Meta   Meta_re   EPNet

Fig. 4. The visualization of compared methods. The first three rows are results on AAPM-test set with 30, 60, and 90 input sinograms. The last row shows results on COVID-test set with 90 sinograms.

4.3   Quantitative and Qualitative Results Comparison

Quantitative Results Comparison. Then, we quantitatively compare our models with model-based and data-driven models. Results on the AAPM-test set show that the performance of our models and retrained MetaInvNet [30] are the best. Besides, the original training setting of MetaInvNet has achieved a better generalization performance on COVID-test and LIDC-test sets, but they need more projections to train the model and our models have also achieved better generalizability results than it except when $\alpha_{max} = 15$, which is due to the extremely limited sinogram information fed into extrapolation layer. On the other hand, HQS-CG has kept their performance across different data distributions, however the prior knowledge modeling limits their reconstruction performance on AAPM-test set, and the tuning and computation time is too expensive.

Qualitative Results Comparison. We also visualize the reconstruction results of these methods on AAPM-test and COVID-test datasets. As in the first three rows of Fig. 4, the reconstructed images from ours and retrained MetaInvNet show the best visualization quality on AAPM-test set across different angle numbers. Besides, our results show sharper details with the additional utilization of $\mathcal{L}_{SE}$ in the projection domain. When testing the reconstructed image on the COVID-test set, our result also gives sharper details but with more artifacts since the data distribution is very different. Although HQS-CG has achieved better quantitative results on the COVID-test dataset, the reconstructed image of their model in the fourth row is even smoother than FBP.

5   Conclusion

We propose the novel EPNet for limited-angle CT image reconstruction and the model achieves exciting generalization performance. We utilize dual-domain learning for data consistency in two domains and propose an EPL module to estimate extra sinograms, which provide useful information for the final reconstruction. Quantitative and qualitative comparisons with competing methods verify the reconstruction performance and the generalizability of our model. The effectiveness encourages us to further explore designing a better architecture for EPL in the future.

Acknowledgements. This work was supported in part by the National Natural Science Foundation of China (NSFC) under Grant 11831002, in part by the Beijing Natural Science Foundation under Grant 180001, in part by the NSFC under Grant 12090022, and in part by the Beijing Academy of Artificial Intelligence (BAAI).

References

1. Adler, J., Öktem, O.: Learned primal-dual reconstruction. IEEE Trans. Med. Imaging **37**(6), 1322–1332 (2018)
2. Chen, H., et al.: Low-dose CT with a residual encoder-decoder convolutional neural network. IEEE Trans. Med. Imaging **36**(12), 2524–2535 (2017)
3. Cheng, W., Wang, Y., Li, H., Duan, Y.: Learned full-sampling reconstruction from incomplete data. IEEE Trans. Comput. Imag. **6**, 945–957 (2020)
4. Clark, K., et al.: The cancer imaging archive (TCIA): maintaining and operating a public information repository. J. Digit. Imaging **26**(6), 1045–1057 (2013)
5. Ding, Q., Chen, G., Zhang, X., Huang, Q., Ji, H., Gao, H.: Low-dose CT with deep learning regularization via proximal forward backward splitting. Phys. Med. Biol. **65**, 125009 (2020)
6. Geman, D., Yang, C.: Nonlinear image recovery with half-quadratic regularization. IEEE Trans. Image Process. **4**(7), 932–946 (1995)
7. Gupta, H., Jin, K.H., Nguyen, H.Q., McCann, M.T., Unser, M.: CNN-based projected gradient descent for consistent CT image reconstruction. IEEE Trans. Med. Imaging **37**(6), 1440–1453 (2018)
8. Jin, K.H., McCann, M.T., Froustey, E., Unser, M.: Deep convolutional neural network for inverse problems in imaging. IEEE Trans. Image Process. **26**(9), 4509–4522 (2017)
9. Kingma, D.P., Ba, J.: Adam: a method for stochastic optimization. arXiv preprint arXiv:1412.6980 (2014)
10. Lin, W.A., et al.: DuDoNet: dual domain network for CT metal artifact reduction. In: Proceedings of the IEEE/CVF Conference on Computer Vision and Pattern Recognition, pp. 10512–10521 (2019)
11. Lyu, Y., Lin, W.-A., Liao, H., Lu, J., Zhou, S.K.: Encoding metal mask projection for metal artifact reduction in computed tomography. In: Martel, A.L., et al. (eds.) MICCAI 2020. LNCS, vol. 12262, pp. 147–157. Springer, Cham (2020). https://doi.org/10.1007/978-3-030-59713-9_15
12. Mahmood, F., Shahid, N., Skoglund, U., Vandergheynst, P.: Adaptive graph-based total variation for tomographic reconstructions. IEEE Signal Process. Lett. **25**(5), 700–704 (2018)

13. Mardani, M., et al.: Neural proximal gradient descent for compressive imaging. arXiv preprint arXiv:1806.03963 (2018)
14. McCollough, C.: TU-FG-207A-04: overview of the low dose CT grand challenge. Med. Phys. **43**(6Part35), 3759–3760 (2016)
15. Rantala, M., et al.: Wavelet-based reconstruction for limited-angle x-ray tomography. IEEE Trans. Med. Imaging **25**(2), 210–217 (2006)
16. Ronneberger, O., Fischer, P., Brox, T.: U-Net: convolutional networks for biomedical image segmentation. In: Navab, N., Hornegger, J., Wells, W.M., Frangi, A.F. (eds.) MICCAI 2015. LNCS, vol. 9351, pp. 234–241. Springer, Cham (2015). https://doi.org/10.1007/978-3-319-24574-4_28
17. Schafer, S., et al.: Mobile C-arm cone-beam CT for guidance of spine surgery: image quality, radiation dose, and integration with interventional guidance. Med. Phys. **38**(8), 4563–4574 (2011)
18. Sidky, E.Y., Pan, X.: Image reconstruction in circular cone-beam computed tomography by constrained, total-variation minimization. Phys. Med. Biol. **53**(17), 4777 (2008)
19. Solomon, O., et al.: Deep unfolded robust PCA with application to clutter suppression in ultrasound. IEEE Trans. Med. Imaging **39**(4), 1051–1063 (2019)
20. Wang, G., Zhang, Y., Ye, X., Mou, X.: Machine Learning for Tomographic Imaging. IOP Publishing, Bristol (2019)
21. Wang, T., Nakamoto, K., Zhang, H., Liu, H.: Reweighted anisotropic total variation minimization for limited-angle CT reconstruction. IEEE Trans. Nucl. Sci. **64**(10), 2742–2760 (2017)
22. Wang, Z., Bovik, A.C., Sheikh, H.R., Simoncelli, E.P.: Image quality assessment: from error visibility to structural similarity. IEEE Trans. Image Process. **13**(4), 600–612 (2004)
23. Wang, Z., Simoncelli, E.P., Bovik, A.C.: Multiscale structural similarity for image quality assessment. In: The Thirty-Seventh Asilomar Conference on Signals, Systems & Computers, vol. 2, pp. 1398–1402. IEEE (2003)
24. Yang, Q., et al.: Low-dose CT image denoising using a generative adversarial network with Wasserstein distance and perceptual loss. IEEE Trans. Med. Imaging **37**(6), 1348–1357 (2018)
25. Yang, Y., Sun, J., Li, H., Xu, Z.: Deep ADMM-Net for compressive sensing MRI. Adv. Neural. Inf. Process. Syst. **29**, 10–18 (2016)
26. Yang, Y., Sun, J., Li, H., Xu, Z.: ADMM-CSNet: a deep learning approach for image compressive sensing. IEEE Trans. Pattern Anal. Mach. Intell. **42**(3), 521–538 (2018)
27. Zeng, D., et al.: Spectral CT image restoration via an average image-induced nonlocal means filter. IEEE Trans. Biomed. Eng. **63**(5), 1044–1057 (2015)
28. Zhang, H.M., Dong, B.: A review on deep learning in medical image reconstruction. J. Oper. Res. Soc. China, 1–30 (2020)
29. Zhang, H., Dong, B., Liu, B.: JSR-Net: a deep network for joint spatial-radon domain CT reconstruction from incomplete data. In: ICASSP 2019-2019 IEEE International Conference on Acoustics, Speech and Signal Processing (ICASSP), pp. 3657–3661. IEEE (2019)
30. Zhang, H., Liu, B., Yu, H., Dong, B.: MetaInv-Net: meta inversion network for sparse view CT image reconstruction. IEEE Trans. Med. Imaging **40**(2), 621–634 (2021)
31. Zhang, Y., et al.: LEARN++: recurrent dual-domain reconstruction network for compressed sensing CT. arXiv preprint arXiv:2012.06983 (2020)

32. Zhou, S.K., et al.: A review of deep learning in medical imaging: imaging traits, technology trends, case studies with progress highlights, and future promises. In: Proceedings of the IEEE (2021)
33. Zhou, S.K., Rueckert, D., Fichtinger, G.: Handbook of Medical Image Computing and Computer Assisted Intervention. Academic Press, San Diego (2019)

Fast Magnetic Resonance Imaging on Regions of Interest: From Sensing to Reconstruction

Liyan Sun[1], Hongyu Huang[1], Xinghao Ding[1(✉)], Yue Huang[1], Xiaoqing Liu[2], and Yizhou Yu[2,3]

[1] School of Informatics, Xiamen University, Xiamen, China
dxh@xmu.edu.cn
[2] Deepwise AI Lab, Beijing, China
[3] The University of Hong Kong, Pokfulam, Hong Kong

Abstract. Magnetic Resonance Imaging (MRI) in a specific Region Of Interest (ROI) is valuable in detecting biomarkers for diagnosis, treatment, prognosis accurately. However, few existing methods study ROI in both data acquisition and image reconstruction when accelerating MRI by partial k-space measurements. Aiming at utilizing limited sampling resources efficiently on most relevant and desirable imaging contents in fast MRI, we propose a deep network framework called *ROICSNet*. With a learnable k-space sampler, an ad-hoc sampling pattern is adapted to a certain type of ROI organ. A cascaded Convolutional Neural Network (CNN) is used as the MR image reconstructor. By using a ROI prediction branch and a three-phase training strategy, the reconstructor is better guided onto the regions where radiologists hope to look closer. Experiments are performed on T1-modality abdominal MRI to demonstrate its state-of-the-art reconstruction accuracy compared with recent general and ROI-based fast MRI approaches. Our model achieves accurate imaging on fine details in ROI under a high accelerator factor and showed promise in real-world MRI application.

Keywords: Compressed sensing MRI · Regions of interest · Deep convolutional neural networks

1 Introduction

To overcome the major drawback of MRI in slow data acquisition, compressed sensing enables fast MRI by only measuring partial k-space data [12,17]. The fast MRI techniques alleviate the discomforts of scanned subjects and reduce artifacts caused by respiratory motion in abdominal imaging [7]. With the cutting-edge

Electronic supplementary material The online version of this chapter (https://doi.org/10.1007/978-3-030-87231-1_10) contains supplementary material, which is available to authorized users.

© Springer Nature Switzerland AG 2021
M. de Bruijne et al. (Eds.): MICCAI 2021, LNCS 12906, pp. 97–106, 2021.
https://doi.org/10.1007/978-3-030-87231-1_10

CNN models, researches on fast MRI reconstruction based on deep learning are thriving. The pioneering work [14] utilized a vanilla deep network for CS-MRI. Some deep networks were built by unrolling an iterative optimization. For example, a Deep Cascaded CNN (DC-CNN) [11] was proposed by unrolling optimization with learnable convolutional transform. Some models enriched fine image details by generative adversarial networks (GAN) [10]. Most existing deep networks reconstruct a MRI using k-space measurements acquired by a pre-fixed sampling trajectory. In SparseMRI by Lustig et al. [8], an under-sampling trajectory with incoherence was proposed using a probabilistic map which overlooked imaging contents. Recently, under-sampling trajectory was inferred from training data and merged with following image reconstruction into a joint learning framework under a deep network architecture [1].

In the previous Compressed Sensing MRI (CS-MRI) pipeline, the correlation of partial k-space under-sampling and image reconstruction on specific image contents has not been paid enough attentions. It is often the case that only a sub-region is of the interest for radiologists despite rich yet redundant contextual information is also measured. A weighted sparsity-promoting term was imposed to focus on ROI [6,9,16]. Contrast to the sparsity-based CS-MRI, a deep network was proposed to incorporate ROI finetune training using a ROI-weighted loss called ROIRecNet [13]. Although this work outperformed sparsity-driven ROI CS-MRI with fixed sparse transform domains, it failed to exploit k-space sampling resources in the frontend.

With the purpose to accelerate MRI while maintaining high imaging quality on ROI, we propose a deep network called *ROICSNet* designed for ROI-driven fast MRI from sensing to reconstruction with a ROI prediction branch and a multi-phase training. We enable k-space sampler to be learnable and use a DC-CNN model as the backbone for MRI reconstruction. We validate our model using the T1 MR images from the Combined Healthy Abdominal Organ Segmentation (CHAOS) datasets [2–4]. The contributions are summarized as follows:

- To our knowledge, we propose the first deep network for joint learning of k-space sampling and image reconstruction towards ROI.
- We proposed a ROI prediction branch under a multi-phase training strategy to guide deep MRI reconstructor with the model designs in both loss function and feature learning.
- We compared our model with state-of-the-art CS-MRI approaches. We showed the ROICSNet model is capable of recovering fine structures like vessels in an organ under low under-sampling ratios, which is promising in real-world MR imaging where high acceleration factor is crucial.

2   Methods

We show the architecture of the ROICSNet in Fig. 1 with a k-space sampler and MRI reconstructor. We describe the adaptive k-space sampling and cascaded CNN as backbone for reconstruction and how they are combined in a pipeline.

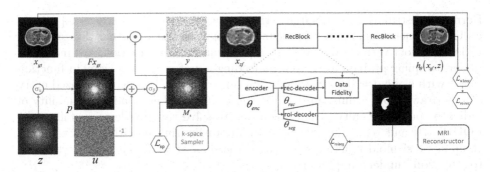

Fig. 1. The flowchart of the proposed ROICSNet where k-space sampler and MRI reconstructor are optimized jointly. We present k-space sampler in blue and MRI reconstructor in purple. We denote loss functions in red. The dashed line represents the re-training using ROI reconstruction loss. (Color figure online)

A multi-phase training strategy is proposed including ROI region prediction, whole image reconstruction and ROI targeting to empower the network with ROI semantics.

2.1 Problem Statement

The problem of compressed sensing MRI can be expressed as

$$x = \underset{x}{\operatorname{argmin}} \frac{\lambda}{2} \|F_u x - y\|_2^2 + g_\theta(x), \tag{1}$$

where the first term called data fidelity imposes k-space consistency in reconstruction and the second term is a prior regularizing solution space for the ill-posed problem. Deep networks extract rich MRI semantics as a deep prior $g_\theta(x)$ with learnable parameters θ. We denote a full Fourier matrix as $F \in \mathbb{C}^{P \times P}$, and a under-sampling Fourier matrix is $F_u \in \mathbb{C}^{P \times P}$ is identical to a full Fourier matrix F except for rows of un-sampled k-space positions are zero-filled. A under-sampling matrix F_u maps a MR image $x \in \mathbb{C}^{P \times 1}$ to a zero-filling partial k-space measurements $y \in \mathbb{C}^{P \times 1}$ and the data fidelity term presents such a constraint. The parameter λ balances the two terms. We further rewrite Eq. 1 into

$$x = \underset{x}{\operatorname{argmin}} \frac{\lambda}{2} \|MF(x - x_{gt})\|_2^2 + g_\theta(x). \tag{2}$$

The operator $M \in \mathbb{C}^{P \times P}$ denotes a diagonal under-sampling matrix where zeroes are filled to un-sampled k-space positions. The x_{gt} represents a latent full-sampled MRI to which a CS-MRI model approximates.

2.2 Adaptive Sampler

We assume each diagonal entry m_i of M is drawn from a Bernoulli distribution with the probability p_i denoted as $m_i \sim \text{Bern}(p_i)$. We realize a sampling of

Bernoulli distribution by reparametrization $m_i = \mathbb{1}\{u_i \leq p_i\}$ where $\mathbb{1}\{\cdot\}$ is an indicator function mapping a boolean-true input to one and otherwise zero, and u_i is a random variable drawn from a $[0, 1]$ uniform distribution. We model an under-sampling probability map p by feeding a vector z into a sigmoid function $\sigma_\alpha(\cdot)$ with slope $\alpha = 0.25$ to squeeze its range into $[0, 1]$. Each initialized entry of z is drawn from a uniform distribution $[-15, 15]$. The nondifferentiability of indicator poses difficulty in optimization. To solve this problem, we follow the LOUPE [1] and relax the indicator by approximating it with a differentiable steep-slope sigmoid function $\sigma_\beta(\cdot)$ with the slope $\beta = 12$. Thus we have a surrogate "soft" under-sampling mask

$$M_s = \sigma_\beta\left(\sigma_\alpha\left(z\right) - u\right), \quad u \sim U(0, 1). \tag{3}$$

With a "soft" under-sampling mask M_s, we can have a corresponding under-sampled k-space measurements $y = M_s F x_{gt}$ and zero-filled MR image $x_{zf} = F^H M_s F x_{gt}$. The differentiability of the "soft" mask enables adaptation to different ROI according to practical clinical needs.

2.3 Deep Reconstructor

Motivated by the success of deep networks [11] in compressed sensing MRI, we model the prior term $g_\theta(x)$ in Eq. 2 using a deep CNN $g_{\text{cnn}}(\cdot|\theta)$

$$x = \underset{x}{\arg\min}\,\frac{\lambda}{2}\left\|MF\left(x - x_{gt}\right)\right\|_2^2 + \frac{1}{2}\left\|x - g_{\text{cnn}}\left(x_{zf}|\theta\right)\right\|_2^2. \tag{4}$$

The optimization in Eq. 4 admits an approximate closed solution under noiseless condition and leads to a data fidelity layer l_{df} [11,13] producing a k-space corrected MR image,

$$l_{df}\left(x_{in}\right) = F^H\left(MFx_{gt} + \left(I - M\right)Fx_{in}\right). \tag{5}$$

The input of l_{df} is $x_{in} = g_{\text{cnn}}(x_{zf}|\theta)$.

We propose a cascaded deep reconstructor with blocks. A block denoted "RecBlock" consists of a reconstruction branch and ROI prediction branch. The RecBlocks are stacked to mimic the unrolling of optimization-based CS-MRI motivated by the DC-CNN model [11]. The reconstruction branch approximates the full-sampled MRI and ROI prediction branch segments the image into its ROI annotation. The reconstruction and ROI prediction branch share an encoder θ_{enc} made up of convolution and down-sampling operations based on the fact that shallow layers extracts sharable low-level image features [15]. A reconstruction decoder (rec-decoder) θ_{rec} takes as input the encoder features. In the data fidelity layer, we replace the fixed binary under-sampling mask M in Eq. 5 with the learnable "soft" under-sampling mask M_s as

$$l_{df}\left(x_{in}\right) = F^{-1}\left(M_s F x_{gt} + \left(I - M_s\right)F x_{in}\right). \tag{6}$$

In doing so, M_s can also be optimized in each data fidelity layer in different depths of the deep reconstructor.

In the ROI prediction branch, a ROI decoder (roi-decoder) θ_{seg} predicts the ROI label. The rec-decoder and roi-decoder shares the similar network architecture except the output layer. The output of rec-decoder is activated by a identity mapping while the output of roi-decoder is squeezed between 0 and 1 using a sigmoid function for binary image segmentation. The MRI deep reconstructor and k-space sampler can be jointly optimized due to the differentiability of the two module in a unified pipeline. The network architecture of the shared encoder, rec-decoder and roi-decoder is detailed in supplementary materials.

2.4 Training Strategy

We next describe how the ROI is incorporated into the joint sensing and reconstruction deep framework under a three-phase training strategy including (1) pre-training for ROI label prediction; (2) training for whole image reconstruction; (3) re-training for ROI reconstruction.

In this first phase, the parameters of ROI prediction branch in each RecBlock θ_{enc} and θ_{seg} are updated using full-sampled ROI-annotated images with the loss function $\mathcal{L}_{\text{roiseg}}$ combining binary cross entropy loss and Dice coefficient loss,

$$
\mathcal{L}_{\text{roiseg}} \left(\left\{ x_{gt}^k, b^k \right\}_{k=1}^{K}, \theta_{enc}, \theta_{seg} \right)
$$
$$
= -\frac{1}{K} \sum\nolimits_{k=1}^{K} \left(\beta b^k \log s^k + (1 - b^k) \log(1 - s^k) \right) \tag{7}
$$
$$
+ \alpha (1 - \frac{2 \sum_{k=1}^{K} b^k s^k}{\sum_{k=1}^{K} b^k + \sum_{k=1}^{K} s^k}),
$$

where b^k is the binary ROI label, and s^k is the predicted ROI region. We suppose K samples in a training batch. We set α and β to 0.1 and 3, respectively. The parameter β is set larger for the foreground ROI. The pre-training for ROI region prediction initializes the encoder and roi-decoder towards ROI by the semantic image segmentation. No ROI region is needed in testing phase, which is practical in clinical use.

In the second phase, the parameters of roi-decoder θ_{seg} are frozen and the parameters of encoder θ_{enc} and rec-decoder θ_{rec} are optimized for whole MR images reconstruction with the loss function $\mathcal{L}_{\text{whrec}}$, the "soft" sampling mask M_s is also updated. A loss function $\mathcal{L}_{\text{whrec}}$ for θ_{enc} and θ_{rec} is

$$
\mathcal{L}_{\text{whrec}} \left(\left\{ x_{gt}^k \right\}_{k=1}^{K}, \theta_{enc}, \theta_{rec}, z \right) = (1 - \lambda_{\text{sp}}) \frac{1}{K} \sum\nolimits_{k=1}^{K} \left\| h_\theta \left(x_{zf}^k, z \right) - x_{gt}^k \right\|_1 \tag{8}
$$
$$
+ \lambda_{\text{sp}} \mathcal{L}_{\text{sp}},
$$

where $h_{\theta,\omega} \left(x_{zf}^k, z \right)$ is the unified mapping from a vector z and a zero-filled MR image x_{zf}^k to the reconstructed MR image. By training for whole image reconstruction, we have a good initialization on an under-sampling mask by guiding a k-space sampler towards the data distribution of whole MR images, which reduces the complexity in further search for accurate imaging on ROI.

Table 1. The averaged ROI-PSNR (dB) results on liver, kidney and spleen with 2D random sampling patterns under 20% sampling ratio.

ROI region	DC-CNN	ROIRecNet	LOUPE	ROI-LOUPE	ROICSNet
Liver	27.77	28.69	30.25	31.10	**32.89**
Kidney	27.03	28.28	29.96	31.45	**34.68**
Spleen	28.15	29.75	31.24	32.98	**35.86**

In the third phase, we apply a ROI reconstruction loss $\mathcal{L}_{\text{roirec}}$ and ROI prediction loss $\mathcal{L}_{\text{roiseg}}$ to jointly re-train the whole model. The ROI reconstruction loss is

$$
\mathcal{L}_{\text{roirec}}\left(\left\{x_{gt}^k, b^k\right\}_{k=1}^K, \theta_{enc}, \theta_{rec}, \theta_{seg}\right) =
$$
$$
(1 - \lambda_{\text{sp}})\frac{1}{K}\sum_{k=1}^K \left\|b^k\left(h_\theta\left(x_{zf}^k, z\right) - x_{gt}^k\right)\right\|_1 + \lambda_{\text{sp}}\mathcal{L}_{\text{sp}} + \lambda_{\text{roiseg}}\mathcal{L}_{\text{roiseg}}. \tag{9}
$$

In Eq. 9, we observe the ROI reconstruction loss $\mathcal{L}_{\text{roirec}}$ leads to network training both in sensing and reconstruction, in comparison with the prior CS-MRI work ROIRecNet [13] where the ROI loss only works on reconstruction parts with a fixed sampling pattern. We show the re-training in Fig. 1 using dashed lines. The parameter λ_{roiseg} is set 0.01.

In the second and third training phases, we follow [1] and impose the regularization on sparse under-sampling by applying an $\mathcal{L}_1$ norm on the soft under-sampling mask M_s denoted as $\mathcal{L}_{\text{sp}} = \|M_s\|_1$. We empirically set λ_{sp} to 0.01.

Binarizing Soft Sampling Mask. In order to obey physical restriction when apply compressed sensing MRI, we binarize trained "soft" under-sampling mask M_s to indicate sampling or not on a certain k-space position. If total number of Q measurements are to be sampled, we sort values of M_s in a descending order and activate top Q positions to ones and suppress the others to zeros, resulting a physically feasible under-sampling mask of the sampling ratio $(Q/P)\%$. This binarization is justified by the fact that a diagonal entry of M_s closer to 1 indicates a higher probability that this position should be sampled. We then re-match the binarized under-sampling mask with the image reconstruction by fine-tuning the deep reconstructor.

3   Implementations

We adopt Adam [5] as the optimizer. The pre-training of ROI region prediction takes 100 training epochs, the training of whole image reconstruction 50 epochs and the retraining of ROI reconstruction 200 epochs. The fine-tuning to match a binarized mask to the MRI reconstructor takes 10 epochs. We set batchsize 4 in our work. We implement our model on a graphic card GTX 1080 Ti using Tensorflow as platform. We use the metric ROI-PSNR [13] as the metric which only evaluates Peak Signal-to-Noise Ratio (PSNR) in ROI within an image.

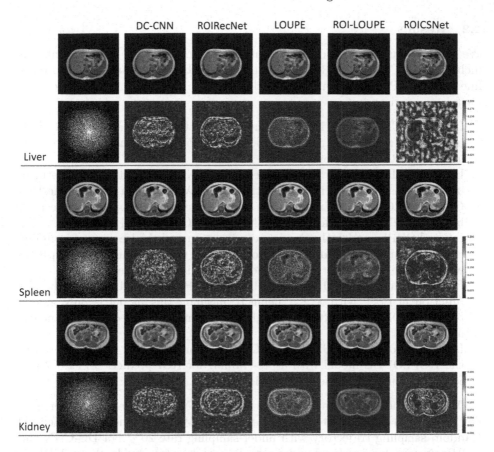

Fig. 2. We show reconstruction results of compared models on 2D 20% under-sampling masks with liver, spleen, kidney being ROI, respectively. In the leftmost column we show the full-sampled ground truth MR images and learned under-sampling patterns from ROICSNet. We also give reconstruction error maps for better visualization.

4 Experiments

4.1 Data

We test our model the Combined Healthy Abdominal Organ Segmentation (CHAOS) datasets[1]. We use publicized 40 scans from T1-DUAL in phase MRI sequence. The data sets are acquired by a 1.5T Philips MRI having a resolution of 256 × 256. In this dataset, segmentation labels for liver, kidney and spleen are provided. We split the datasets into training and testing according to the ratio 80%:20%.

---

[1] https://chaos.grand-challenge.org/.

4.2 Results

We compare our ROICSNet model with several recent state-of-the-art models including DC-CNN [11][2], ROIRecNet [13], LOUPE [1][3] and one proposed baseline ROI-LOUPE. The number of stack blocks for all the compared models is 1 for fair comparison.

– In the DC-CNN model, k-space sampling trajectories are pre-fixed and whole images are targeted for reconstruction. This model serves as a baseline deep-learning MR image reconstructor.
– In the ROIRecNet model, a trained DC-CNN model is further fine-tuned using the loss function in Eq. 9 with λ_{sp} and λ_{roiseg} being 0. Both the DC-CNN and ROIRecNet are based on the pre-fixed under-sampling trajectory, but ROIRecNet outperforms DC-CNN via a ROI-weighted loss.
– In the LOUPE model, the image reconstruction and the learning of sampling trajectory are optimized simultaneously using the loss function in Eq. 8 for whole image reconstruction. However, no ROI is emphasized.
– In the baseline ROI-LOUPE model, the LOUPE is re-trained with the loss function in Eq. 9 with λ_{roiseg} being 0, meaning this model is only guided towards ROI by the ROI-weighted loss but no ROI prediction branch is applied.
– In our proposed ROICSNet, the ROI prediction branch is incorporated into the ROI-LOUPE model with a pre-training for ROI prediction in Eq. 7 and full fine-tuning using the loss function in Eq. 9.

We show their comparative results in ROI-PSNR in Table 1 with liver, kidney and spleen as designated ROI, respectively. We adopt the widely-used 2D Random sampling trajectory with under-sampling rate 20%. For DC-CNN and ROIRecNet, we use pre-fixed under-sampling masks generated by SparseMRI [8]. For LOUPE, ROI-LOUPE and ROICSNet, under-sampling trajectories of the identical sampling ratio are inferred from data. All the compared models are adjusted to the optimal performance.

In Table 1, we observe ROIRecNet outperforms DC-CNN by more than 1 dB in ROI-PSNR despite degradation in backgrounds, which is of little significance in providing meaningful visual information. By a joint learning of k-space sampling and MRI reconstructor, a plain LOUPE achieves better reconstruction than ROIRecNet over 1 dB, proving adaptive k-space sampling is able to collaborate with following image reconstruction. Furthermore, by the ad hoc design on both k-space sensing and image reconstruction towards ROI, our simple ROI-LOUPE outperforms LOUPE by more than 1 dB in liver, 3 dB in kidney and 2 dB in spleen in PSNR. In Fig. 2, we show some qualitative results of compared methods and learned specified sampling masks. The visual results are consistent with the quantitative results. We observe the proposed ROICSNet with adaptive k-space sampling and high-level supervision recovers fine details like vessels in liver, which is critical in real-world MRI.

---

[2] https://github.com/js3611/Deep-MRI-Reconstruction.
[3] https://github.com/cagladbahadir/LOUPE.

We increase the number of cascaded blocks from 1 to 3 for all the compared models. Under the 2D 20% sampling pattern with liver being ROI, the DC-CNN, ROIRecNet, LOUPE and LOUPE-ROI achieves the ROI-PSNR value at 29.84 dB, 32.07 dB, 35.64 dB and 36.02 dB. In contrast, the ROICSNet achieves the ROI-PSNR value at 44.99 dB. The ROICSNet maintains advantages over compared methods with more cascaded blocks. We also experiment with 1D Cartesian line-based sampling trajectory, we learn the sampling probability of each line according to modified Eq. 3. Taking liver as ROI, we observe the DC-CNN, ROIRecNet, LOUPE, ROI-LOUPE and ROICSNet achieve the ROI-PSNR value at 28.62 dB, 30.96 dB, 31.21 dB, 32.15 dB and 33.89 dB on 1D 20% Cartesian sampling pattern. The results demonstrate our ROICSNet model outperforms other state-of-the-art models in the line-based sampling pattern.

5 Conclusions

In this work, we proposed a compressed sensing MRI model from sensing to reconstruction called ROICSNet targeting on specific organs to achieve fast and accurate imaging. We trained an adaptive k-space sampler and an image reconstructor jointly. The experiments demonstrate the state-of-the-art performance of the proposed model. It is noteworthy that our model recovers fine yet important image details like vessels within ROI liver at the accelerator factor 5×, which shows its promising application in abdominal MRI for fast high-resolution imaging. In the future research, we plan to evaluate the effect of bias in manual labeling of ROI on the ROICSNet model and extend it to multi-coil MRI.

Acknowledgements. This work was supported in part by National Natural Science Foundation of China under Grants U19B2031, 61971369, in part by Fundamental Research Funds for the Central Universities 20720200003, in part by the Science and Technology Key Project of Fujian Province, China (No. 2019HZ020009).

References

1. Bahadir, C.D., Wang, A.Q., Dalca, A.V., Sabuncu, M.R.: Deep-learning-based optimization of the under-sampling pattern in MRI. IEEE Trans. Comput. Imaging **6**, 1139–1152 (2020)
2. Kavur, A.E., et al.: CHAOS challenge-combined (CT-MR) healthy abdominal organ segmentation. Med. Image Anal. **69**, 101950 (2021)
3. Kavur, A.E., et al.: Comparison of semi-automatic and deep learning-based automatic methods for liver segmentation in living liver transplant donors. Diagn. Interv. Radiol. **26**(1), 11 (2020)
4. Kavur, A.E., Selver, M.A., Dicle, O., Bariş, M., Gezer, N.S.: CHAOS - combined (CT-MR) healthy abdominal organ segmentation challenge data, April 2019. https://doi.org/10.5281/zenodo.3362844
5. Kingma, D.P., Ba, J.: Adam: a method for stochastic optimization. In: International Conference on Learning Representations (2015)

6. Konar, A.S., Vajuvalli, N.N., Rao, R., Jain, D., Babu, D.R., Geethanath, S.: Accelerated dynamic contrast enhanced MRI based on region of interest compressed sensing. Magn. Reson. Imaging **67**, 18–23 (2020)

7. Li, Z., et al.: Expiration-phase template-based motion correction of free-breathing abdominal dynamic contrast enhanced MRI. IEEE Trans. Biomed. Eng. **62**(4), 1215–1225 (2014)

8. Lustig, M., Donoho, D., Pauly, J.M.: Sparse MRI: the application of compressed sensing for rapid MR imaging. Magn. Reson. Med. **58**(6), 1182–1195 (2007)

9. Oh, H., Lee, S.: Visually weighted reconstruction of compressive sensing MRI. Magn. Reson. Imaging **32**(3), 270–280 (2014)

10. Quan, T.M., Nguyen-Duc, T., Jeong, W.K.: Compressed sensing MRI reconstruction using a generative adversarial network with a cyclic loss. IEEE Trans. Med. Imaging **37**(6), 1488–1497 (2018)

11. Schlemper, J., Caballero, J., Hajnal, J.V., Price, A., Rueckert, D.: A deep cascade of convolutional neural networks for MR image reconstruction. In: Niethammer, M., et al. (eds.) IPMI 2017. LNCS, vol. 10265, pp. 647–658. Springer, Cham (2017). https://doi.org/10.1007/978-3-319-59050-9_51

12. Sun, J., Li, H., Xu, Z., et al.: Deep ADMM-Net for compressive sensing MRI. In: Advances in Neural Information Processing Systems, pp. 10–18 (2016)

13. Sun, L., Fan, Z., Ding, X., Huang, Y., Paisley, J.: Region-of-interest undersampled MRI reconstruction: a deep convolutional neural network approach. Magn. Reson. Imaging **63**, 185–192 (2019)

14. Wang, S., et al.: Accelerating magnetic resonance imaging via deep learning. In: International Symposium on Biomedical Imaging, pp. 514–517. IEEE (2016)

15. Zeiler, M.D., Fergus, R.: Visualizing and understanding convolutional networks. In: Fleet, D., Pajdla, T., Schiele, B., Tuytelaars, T. (eds.) ECCV 2014. LNCS, vol. 8689, pp. 818–833. Springer, Cham (2014). https://doi.org/10.1007/978-3-319-10590-1_53

16. Zhang, C., van de Giessen, M., Eisemann, E., Vilanova, A.: User-guided compressed sensing for magnetic resonance angiography. In: Annual International Conference of the IEEE Engineering in Medicine and Biology Society, pp. 2416–2419. IEEE (2014)

17. Zhou, B., Zhou, S.K.: DuDoRNet: learning a dual-domain recurrent network for fast MRI reconstruction with deep T1 prior. In: Proceedings of the IEEE/CVF Conference on Computer Vision and Pattern Recognition, pp. 4273–4282 (2020)

InDuDoNet: An Interpretable Dual Domain Network for CT Metal Artifact Reduction

Hong Wang[1,2], Yuexiang Li[2], Haimiao Zhang[3], Jiawei Chen[2], Kai Ma[2],
Deyu Meng[1,4(✉)], and Yefeng Zheng[2(✉)]

[1] Xi'an Jiaotong University, Xi'an, Shaan'xi, People's Republic of China
hongwang01@stu.xjtu.edu.cn, dymeng@mail.xjtu.edu.cn
[2] Tencent Jarvis Lab, Shenzhen, People's Republic of China
{vicyxli,kylekma,yefengzheng}@tencent.com
[3] Beijing Information Science and Technology University, Beijing,
People's Republic of China
hmzhang@bistu.edu.cn
[4] Macau University of Science and Technology, Taipa, Macau,
People's Republic of China

Abstract. For the task of metal artifact reduction (MAR), although deep learning (DL)-based methods have achieved promising performances, most of them suffer from two problems: 1) the CT imaging geometry constraint is not fully embedded into the network during training, leaving room for further performance improvement; 2) the model interpretability is lack of sufficient consideration. Against these issues, we propose a novel interpretable dual domain network, termed as InDuDoNet, which combines the advantages of model-driven and data-driven methodologies. Specifically, we build a joint spatial and Radon domain reconstruction model and utilize the proximal gradient technique to design an iterative algorithm for solving it. The optimization algorithm only consists of simple computational operators, which facilitate us to correspondingly unfold iterative steps into network modules and thus improve the interpretablility of the framework. Extensive experiments on synthesized and clinical data show the superiority of our InDuDoNet. Code is available in https://github.com/hongwang01/InDuDoNet.

Keywords: Metal artifact reduction · Imaging geometry · Physical interpretability · Multi-class segmentation · Generalization ability

1 Introduction

Computed tomography (CT) images reconstructed from X-ray projections play an important role in clinical diagnosis and treatment planning. However, due to

Electronic supplementary material The online version of this chapter (https://doi.org/10.1007/978-3-030-87231-1_11) contains supplementary material, which is available to authorized users.

M. de Bruijne et al. (Eds.): MICCAI 2021, LNCS 12906, pp. 107–118, 2021.
https://doi.org/10.1007/978-3-030-87231-1_11

the metallic implants within patients, CT images are always adversely affected by undesirable streaking and shading artifacts, which may consequently affect the clinical diagnosis [3,18]. Hence, metal artifact reduction (MAR), as a potential solution, gains increasing attention from the community. Various traditional hand-crafted methods [2,10,16,17] have been proposed for the MAR task. Driven by the significant success of deep learning (DL) in medical image reconstruction and analysis [9,20,21], researchers began to apply the convolutional neural network (CNN) for MAR in recent years [12,13,15,28,32].

Existing deep-learning-based MAR methods can be grouped into three research lines, *i.e.,* sinogram enhancement, image enhancement, and dual enhancement (joint sinogram and image). Concretely, the sinogram-enhancement-based approaches adopt deep networks to directly repair metal-corrupted sinogram [5,11,18] or utilize the forward projection (FP) of a prior image to correct the sinogram [6,32]. For the image enhancement line, researchers exploit the residual learning [8] or adversarial learning [12,25] on CT images only for metal artifact reduction. The dual enhancement of sinogram and image is a recently-emerging direction for MAR. The mutual learning between the sinogram and CT image proposed by recent studies [13,15,28] significantly boosts the performance of MAR. Nevertheless, these deep-learning-based MAR techniques share some common drawbacks. The most evident one is that most of them regard MAR as the general image restoration problem and neglect the inherent physical geometry constraints during network training. Yet such constraints are potentially helpful to further boost the performance of MAR. Besides, due to the nature of almost black box, the existing approaches relying on the off-the-shelf deep networks are always lack of sufficient model interpretability for the specific MAR task, making them difficult to analyze the intrinsic role of network modules.

To alleviate these problems, we propose a novel interpretable dual domain network, termed as InDuDoNet, for the MAR task, which sufficiently embeds the intrinsic imaging geometry model constraints into the process of mutual learning between spatial (image) and Radon (sinogram) domains, and is flexibly integrated with the dual-domain-related prior learning. Particularly, we propose a concise dual domain reconstruction model and utilize the proximal gradient technique [1] to design an optimization algorithm. Different from traditional solvers [30] for the model containing heavy operations (*e.g.,* matrix inversion), the proposed algorithm consists of only simple computations (*e.g.,* point-wise multiplication) and thus facilitates us to easily unfold it as a network architecture. The specificity of our framework lies in the exact step-by-step corresponding relationship between its modules and the algorithm operations, naturally resulting in its fine physical interpretability. Comprehensive experiments on synthetic and clinical data substantiate the effectiveness of our method.

2   Method

In this section, we first theoretically formulate the optimization process for dual domain MAR, and then present the InDuDoNet which is constructed by correspondingly unfolding the optimization process into network modules in details.

Formulation of Dual Domain Model. Given the observed metal-affected sinogram $Y \in \mathbb{R}^{N_b \times N_p}$, where N_b and N_p are the number of detector bins and projection views, respectively, traditional iterative MAR is formulated as:

$$\min_{X} \|(1 - Tr) \odot (\mathcal{P}X - Y)\|_F^2 + \lambda g(X), \tag{1}$$

where $X \in \mathbb{R}^{H \times W}$ is the clean CT image (*i.e.*, spatial domain); H and W are the height and width of the CT image, respectively; $\mathcal{P}$ is the Radon transform (*i.e.*, forward projection); Tr is the binary metal trace; $\odot$ is the point-wise multiplication; $g(\cdot)$ is a regularizer for delivering the prior information of X and λ is a trade-off parameter. For the spatial and Radon domain mutual learning, we further execute the joint regularization and transform the problem (1) to:

$$\min_{S,X} \|\mathcal{P}X - S\|_F^2 + \alpha \|(1 - Tr) \odot (S - Y)\|_F^2 + \lambda_1 g_1(S) + \lambda_2 g_2(X), \tag{2}$$

where S is the clean sinogram (*i.e.*, Radon domain); α is a weight factor balancing the data consistency between spatial and Radon domains; $g_1(\cdot)$ and $g_2(\cdot)$ are regularizers embedding the priors of the to-be-estimated S and X, respectively.

Clearly, correcting the normalized metal-corrupted sinogram is easier than directly correcting the original metal-affected sinogram, since the former profile is more homogeneous [17,30]. We thus rewrite the sinogram S as:

$$S = \widetilde{Y} \odot \widetilde{S}, \tag{3}$$

where $\widetilde{Y}$ is normalization coefficient, usually set as the FP of a prior image $\widetilde{X}$, *i.e.*, $\widetilde{Y} = \mathcal{P}\widetilde{X}$;[1] $\widetilde{S}$ is the normalized sinogram. By substituting Eq. (3) into Eq. (2), we can derive the dual domain reconstruction problem as:

$$\min_{\widetilde{S},X} \left\|\mathcal{P}X - \widetilde{Y} \odot \widetilde{S}\right\|_F^2 + \alpha \left\|(1 - Tr) \odot (\widetilde{Y} \odot \widetilde{S} - Y)\right\|_F^2 + \lambda_1 g_1(\widetilde{S}) + \lambda_2 g_2(X). \tag{4}$$

As presented in Eq. (4), our goal is to jointly estimate $\widetilde{S}$ and X from Y. In the traditional prior-based MAR methods, regularizers $g_1(\cdot)$ and $g_2(\cdot)$ are manually formulated as explicit forms [30], which cannot always capture complicated and diverse metal artifacts. Owning to the sufficient and adaptive prior fitting capability of CNN [23,26], we propose to automatically learn the dual-domain-related priors $g_1(\cdot)$ and $g_2(\cdot)$ from training data using network modules in the following. Similarly, adopting such a data-driven strategy to learn implicit models has been applied in other vision tasks [22,24,29].

---

[1] We utilize a CNN to flexibly learn $\widetilde{X}$ and $\widetilde{Y}$ from training data as shown in Fig. 1.

2.1 Optimization Algorithm

Since we want to construct an interpretable deep unfolding network for solving the problem (4) efficiently, it is critical to build an optimization algorithm with possibly simple operators that can be transformed to network modules easily. Traditional solver [30] for the dual domain model (4) contains complex operations, *e.g.*, matrix inversion, which are hard for such unfolding transformation. We thus prefer to build a new solution algorithm for problem (4), which only involves simple computations. Particularly, $\widetilde{S}$ and X are alternately updated as:

Updating $\widetilde{S}$: The normalized sinogram $\widetilde{S}$ can be updated by solving the quadratic approximation [1] of the problem (4) about $\widetilde{S}$, written as:

$$\min_{\widetilde{S}} \frac{1}{2} \left\| \widetilde{S} - \left(\widetilde{S}_{n-1} - \eta_1 \nabla f\left(\widetilde{S}_{n-1} \right) \right) \right\|_F^2 + \lambda_1 \eta_1 g_1(\widetilde{S}), \tag{5}$$

where $\widetilde{S}_{n-1}$ is the updated result after $(n-1)$ iterations; η_1 is the stepsize parameter; and $f\left(\widetilde{S}_{n-1} \right) = \left\| \mathcal{P}X_{n-1} - \widetilde{Y}\widetilde{S}_{n-1} \right\|_F^2 + \alpha \left\| (1 - Tr)(\widetilde{Y}\widetilde{S}_{n-1} - Y) \right\|_F^2$ (note that we omit $\odot$ used in Eq. (4) for simplicity). For general regularization terms [4], the solution of Eq. (5) is:

$$\widetilde{S}_n = \mathrm{prox}_{\lambda_1 \eta_1} \left(\widetilde{S}_{n-1} - \eta_1 \nabla f\left(\widetilde{S}_{n-1} \right) \right). \tag{6}$$

By substituting $\nabla f\left(\widetilde{S}_{n-1} \right) = \widetilde{Y}\left(\widetilde{Y}\widetilde{S}_{n-1} - \mathcal{P}X_{n-1} \right) + \alpha(1 - Tr)$ $\widetilde{Y}\left(\widetilde{Y}\widetilde{S}_{n-1} - Y \right)$ into Eq. (6), the updating rule of $\widetilde{S}$ is:

$$\begin{aligned} \widetilde{S}_n &= \mathrm{prox}_{\lambda_1 \eta_1} \left(\widetilde{S}_{n-1} - \eta_1 \left(\widetilde{Y}\left(\widetilde{Y}\widetilde{S}_{n-1} - \mathcal{P}X_{n-1} \right) + \alpha(1 - Tr)\widetilde{Y}\left(\widetilde{Y}\widetilde{S}_{n-1} - Y \right) \right) \right) \\ &\triangleq \mathrm{prox}_{\lambda_1 \eta_1} \left(\widehat{S}_{n-1} \right), \end{aligned}$$

$$\tag{7}$$

where $\mathrm{prox}_{\lambda_1 \eta_1}(\cdot)$ is the proximal operator related to the regularizer $g_1(\cdot)$. Instead of fixed hand-crafted image priors [30,31], we adopt a convolutional network module to automatically learn $\mathrm{prox}_{\lambda_1 \eta_1}(\cdot)$ from training data (detailed in Sect. 2.2).

Updating X: Also, the image X can be updated by solving the quadratic approximation of Eq. (4) with respect to X:

$$\min_X \frac{1}{2} \| X - (X_{n-1} - \eta_2 \nabla h\left(X_{n-1} \right)) \|_F^2 + \lambda_2 \eta_2 g_2(X), \tag{8}$$

where $\nabla h\left(X_{n-1} \right) = \mathcal{P}^T \left(\mathcal{P}X_{n-1} - \widetilde{Y}\widetilde{S}_n \right)$. Thus, the updating formula of X is:

$$X_n = \mathrm{prox}_{\lambda_2 \eta_2} \left(X_{n-1} - \eta_2 \mathcal{P}^T \left(\mathcal{P}X_{n-1} - \widetilde{Y}\widetilde{S}_n \right) \right) \triangleq \mathrm{prox}_{\lambda_2 \eta_2} \left(\widehat{X}_{n-1} \right), \tag{9}$$

where $\text{prox}_{\lambda_2\eta_2}(\cdot)$ is dependent on $g_2(\cdot)$. Using the iterative algorithm (Eqs. (7) and (9)), we can correspondingly construct the deep unfolding network in Sect. 2.2.

2.2   Overview of InDuDoNet

Recent studies [23,26] have demonstrated the excellent interpretability of unfolding models. Motivated by these, we propose a deep unfolding framework, namely InDuDoNet, specifically fitting the MAR task. The pipeline of our framework is illustrated in Fig. 1, which consists of Prior-net, N-stage $\widetilde{S}$-net, and N-stage X-net with parameters θ_{prior}, $\theta_{\widetilde{s}}^{(n)}$, and $\theta_x^{(n)}$, respectively. Note that $\widetilde{S}$-net and X-net are step-by-step constructed based on the updating rules as expressed in Eqs. (7) and (9), which results in a specific physical interpretability of our framework. All the parameters including θ_{prior}, $\{\theta_{\widetilde{s}}^{(n)}, \theta_x^{(n)}\}_{n=1}^N$, η_1, η_2, and α can be automatically learned from the training data in an end-to-end manner.

Prior-Net. Prior-net in Fig. 1 is utilized to learn $\widetilde{Y}$ from the concatenation of metal-affected image X_{ma} and linear interpolation (LI) corrected image X_{LI} [10]. Our Prior-net has a similar U-shape architecture [20] to the PriorNet in [28].

$\widetilde{S}$-Net and X-Net. With $\widetilde{Y}$ generated by Prior-net, the framework reconstructs the artifact-reduced sinogram $\widetilde{S}$ and the CT image X via sequential updates of $\widetilde{S}$-net and X-net. As shown in Fig. 1(a), N stages are involved in our framework, which correspond to N iterations of the algorithm for solving (4). Each stage shown in Fig. 1(b) is constructed by unfolding the updating rules Eqs. (7) and (9), respectively. Particularly, for the n-th stage, $\widehat{S}_{n-1}$ is firstly computed based on Eq. (7) and then fed to a deep network $\text{proxNet}_{\theta_{\widetilde{s}}^{(n)}}(\cdot)$ to execute the operator $\text{prox}_{\lambda_1\eta_1}(\cdot)$. Then, we obtain the updated normalized sinogram: $\widetilde{S}_n - \text{proxNet}_{\theta_{\widetilde{s}}^{(n)}}\left(\widehat{S}_{n-1}\right)$. Similar operation is taken to process $\widehat{X}_{n-1}$ computed based on Eq. (9) and the updated artifact-reduced

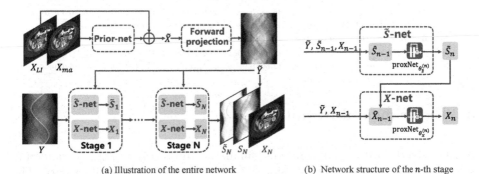

(a) Illustration of the entire network   (b) Network structure of the n-th stage

Fig. 1. (a) The proposed network architecture consists of a Prior-net, N-stage $\widetilde{S}$-net, and N-stage X-net. It outputs the normalized sinogram $\widetilde{S}_N$, sinogram S_N, and image X_N. (b) The detailed structure at the n-th stage, in which $\widetilde{S}_n$ and X_n are successively updated by $\widetilde{S}$-net and X-net, respectively, based on the algorithm in Eqs. (7) (9).

image is: $X_n = \text{proxNet}_{\theta_x^{(n)}}\left(\widehat{X}_{n-1}\right)$. $\text{proxNet}_{\theta_{\widetilde{s}}^{(n)}}(\cdot)$ and $\text{proxNet}_{\theta_x^{(n)}}(\cdot)$ have the same structure—four $[Conv+BN+ReLU+Conv+BN+Skip\ Connection]$ residual blocks [7]. After N stages of optimization, the framework can well reconstruct the normalized sinogram $\widetilde{S}_N$, and therefore yield the final sinogram S_N by $\widetilde{Y} \odot \widetilde{S}_N$ (refer to Eq. (3)), and the CT image X_N.

Remark: Our network is expected to possess both the advantages of the model-driven and data-driven methodologies. Particularly, compared with traditional prior-based methods, our network can flexibly learn sinogram-related and image-related priors through $\text{proxNet}_{\theta_{\widetilde{s}}^{(n)}}(\cdot)$ and $\text{proxNet}_{\theta_x^{(n)}}(\cdot)$ from training data. Compared with deep MAR methods, our framework incorporates both CT imaging constraints and dual-domain-related priors into the network architecture.

Training Loss. We adopt the mean square error (MSE) for the extracted sinogram $\widetilde{Y} \odot \widetilde{S}_n$ and image X_n at every stage as the training objective function:

$$\mathcal{L} = \sum_{n=0}^{N} \beta_n \|X_n - X_{gt}\|_F^2 \odot (1 - M) + \gamma \left(\sum_{n=1}^{N} \beta_n \left\|\widetilde{Y} \odot \widetilde{S}_n - Y_{gt}\right\|_F^2\right), \quad (10)$$

where X_{gt} and Y_{gt} are ground truth image and metal-free sinogram, respectively. We simply set $\beta_N = 1$ to make the outputs at the final stage play a dominant role, and $\beta_n = 0.1$ $(n = 0, \cdots, N - 1)$ to supervise each middle stage. γ is a hyperparamter to balance the weight of different loss items and we empirically set it as 0.1. We initialize X_0 by passing X_{LI} through a proximal network $\text{proxNet}_{\theta_x^{(0)}}$.

Table 1. Effect of the total stage number N on the performance of the proposed InDuDoNet on synthesized data with PSNR (dB) and SSIM.

N	Large metal $\longrightarrow$ Small metal					Average
$N = 0$	28.91/0.9280	30.42/0.9400	34.45/0.9599	36.72/0.9653	37.18/0.9673	33.54/0.9521
$N = 1$	34.10/0.9552	35.91/0.9726	38.48/0.9820	39.94/0.9829	40.39/0.9856	37.76/0.9757
$N = 3$	33.46/0.9564	37.14/0.9769	40.33/0.9868	42.55/0.9896	42.68/0.9908	39.23/0.9801
$N = 6$	34.59/0.9764	38.95/0.9890	42.28/0.9941	44.09/0.9945	45.09/0.9953	41.00/<u>0.9899</u>
$N = 10$	36.74/0.9801	39.32/0.9896	41.86/0.9931	44.47/0.9942	45.01/0.9948	<u>41.48</u>/**0.9904**
$N = 12$	36.52/0.9709	40.01/0.9896	42.66/0.9955	44.17/0.9960	44.84/0.9967	**41.64**/0.9897

3   Experimental Results

Synthesized Data. Following the simulation protocol in [28], we randomly select a subset from the DeepLesion [27] to synthesize metal artifact data. The metal masks are from [32], which contain 100 metallic implants with different shapes and sizes. We choose 1,000 images and 90 metal masks to synthesize the training samples, and pair the additional 200 CT images from 12 patients with the remaining 10 metal masks to generate 2,000 images for testing. The sizes of the 10 metallic implants for test data are: [2061, 890, 881, 451, 254, 124, 118, 112, 53, 35] in pixels. Consistent to [13,15], we simply put the adjacent sizes into one group when reporting MAR performance. We adopt the procedures widely used by existing studies [12,13,15,28,32] to simulate Y and X_{ma}. All the CT images are resized to 416×416 pixels and 640 projection views are uniformly spaced in $360°$. The resulting sinograms are of the size $N_b \times N_p$ as 641×640.

Clinical Data. We further assess the feasibility of the proposed InDuDoNet on a clinical dataset, called CLINIC-metal [14], for pelvic fracture segmentation. The dataset includes 14 testing volumes labeled with multi-bone, *i.e.*, sacrum, left hip, right hip, and lumbar spine. The clinical images are resized and processed using the same protocol to the synthesized data. Similar to [12,28], the clinical metal masks are segmented with a thresholding (2,500 HU).

Evaluation Metrics. The peak signal-to-noise ratio (PSNR) and structured similarity index (SSIM) with the code from [32] are adopted to evaluate the performance of MAR. Since we perform the downstream multi-class segmentation on the CLINIC-metal dataset to assess the improvement generated by different MAR approaches to clinical applications, the Dice coefficient (DC) is adopted as the metric for the evaluation of segmentation performance.

Training Details. Based on a NVIDIA Tesla V100-SMX2 GPU, we implement our network with PyTorch [19] and differential operations $\mathcal{P}$ and $\mathcal{P}^T$ in ODL library.[2] We adopt the Adam optimizer with $(\beta_1, \beta_2) = (0.5, 0.999)$. The initial learning rate is 2×10^{-4} and divided by 2 every 40 epochs. The total epoch is 100 with a batch size of 1. Similar to [28], in each training iteration, we randomly select an image and a metal mask to synthesize a metal-affected sample.

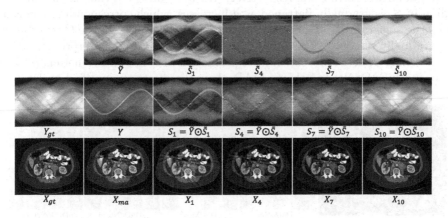

Fig. 2. The recovery normalization coefficient $\widetilde{Y}$, normalized sinogram $\widetilde{S}_n$, sinogram S_n, and image X_n at different stages ($N = 10$). The red pixels stand for metallic implant. (Color figure online)

---

[2] https://github.com/odlgroup/odl.

Table 2. PSNR (dB) and SSIM of different MAR methods on synthesized data.

Methods	Large metal $\longrightarrow$ Small metal					Average
Input	24.12/0.6761	26.13/0.7471	27.75/0.7659	28.53/0.7964	28.78/0.8076	27.06/0.7586
LI [10]	27.21/0.8920	28.31/0.9185	29.86/0.9464	30.40/0.9555	30.57/0.9608	29.27/0.9347
NMAR [17]	27.66/0.9114	28.81/0.9373	29.69/0.9465	30.44/0.9591	30.79/0.9669	29.48/0.9442
CNNMAR [32]	28.92/0.9433	29.89/0.9588	30.84/0.9706	31.11/0.9743	31.14/0.9752	30.38/0.9644
DuDoNet [13]	29.87/0.9723	30.60/0.9786	31.46/0.9839	31.85/0.9858	31.91/0.9862	31.14/0.9814
DSCMAR [28]	34.04/0.9343	33.10/0.9362	33.37/0.9384	32.75/0.9393	32.77/0.9395	33.21/0.9375
DuDoNet++ [15]	36.17/0.9784	38.34/0.9891	40.32/0.9913	41.56/0.9919	42.08/0.9921	39.69/0.9886
InDuDoNet (ours)	**36.74/0.9801**	**39.32/0.9896**	**41.86/0.9931**	**44.47/0.9942**	**45.01/0.9948**	**41.48/0.9904**

3.1   Ablation Study

Table 1 lists the performance of our framework under different stage number N. The $N = 0$ entry means that the initialization X_0 is directly regarded as the reconstruction result. Taking $N = 0$ as the baseline, we can find that with only one stage ($N = 1$), the MAR performance yielded by our proposed InDuDoNet is already evidently improved, which validates the essential role of the mutual learning between $\tilde{S}$-net and X-net. When $N = 12$, the SSIM is slightly lower than that of $N = 10$. The underlying reason is that the more stages cause a deeper network and may suffer from gradient vanishing. Hence, for better performance and fewer network parameters, we choose $N = 10$ in all our experiments.[3]

Model Verification. We conduct a model verification experiment to present the mechanism underlying the network modules ($\tilde{S}$-net and X-net). The evaluation results are shown in Fig. 2. The normalized sinogram $\tilde{S}_n$, sinogram S_n, and CT image X_n generated at different stages ($n = 1, 4, 7, 10$) are presented on the first, second and third rows, respectively. It can be observed that the metal trace region in $\tilde{S}_n$ is gradually flattened as n increases, which correspondingly ameliorates the sinogram S_n. Thus, the metal artifacts contained in the CT image X_n are gradually removed. The results verify the design of our interpretable iterative learning framework—the mutual promotion of $\tilde{S}$-net and X-net enables the proposed InDuDoNet to achieve MAR along the direction specified by Eq. (4).

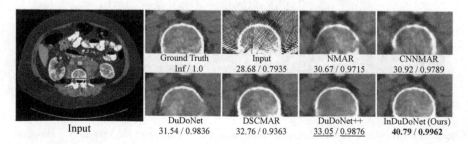

Fig. 3. Comparison for medium metallic implants. PSNR (dB)/SSIM below is for reference. The display window is $[-175, 275]$ HU. The red pixels stand for metallic implants. (Color figure online)

---

[3] More analysis on network parameter and testing time are in *supplementary material*.

Table 3. The Dice coefficient (DC) results on the downstream segmentation task.

Bone	Input	LI	NMAR	CNNMAR	DuDoNet	DSCMAR	DuDoNet++	InDuDoNet
Sacrum	0.9247	0.9086	0.9151	0.9244	0.9326	0.9252	**0.9350**	0.9348
Left hip	0.9543	0.9391	0.9427	0.9485	0.9611	0.9533	0.9617	**0.9630**
Right hip	0.8747	0.9123	0.9168	0.9250	0.9389	0.9322	0.9379	**0.9421**
Lumbar spine	0.9443	0.9453	0.9464	0.9489	0.9551	0.9475	**0.9564**	0.9562
Average DC	0.9245	0.9263	0.9303	0.9367	0.9469	0.9396	0.9478	**0.9490**

3.2 Performance Evaluation

Synthesized Data. We compare the proposed InDuDoNet with current state-of-the-art (SOTA) MAR approaches, including traditional LI [10] and NMAR [17], DL-based CNNMAR [32], DuDoNet [13], DSCMAR [28], and DuDoNet++ [15]. For LI, NMAR, and CNNMAR, we directly use the released code and model. We re-implement DuDoNet, DSCMAR, and DuDoNet++, since there is no official code. Table 2 reports the quantitative comparison. We can observe that most of DL-based methods consistently outperform the conventional LI and NMAR, showing the superiority of data-driven deep CNN for MAR. The dual enhancement approaches (*i.e.*, DuDoNet, DSCMAR, and DuDoNet++) achieve higher PSNR than the sinogram-enhancement-only CNN-MAR. Compared to DuDoNet, DSCMAR, and DuDoNet++, our dual-domain method explicitly embeds the physical CT imaging geometry constraints into the mutual learning between spatial and Radon domains, *i.e.*, jointly regularizing the sinogram and CT image recovered at each stage. Hence, our method achieves the highest PSNRs and SSIMs for all metal sizes as listed. The visual comparisons are shown in Fig. 3.[4]

Clinical Data. We further evaluate all MAR methods on clinical downstream pelvic fracture segmentation task using the CLINIC-metal dataset. A U-Net is firstly trained using the clinical metal-free dataset (CLINIC [14]) and then tested on the metal-artifact-reduced CLINIC-metal CT images generated by different MAR approaches. The segmentation accuracy achieved by the metal-free-trained U-Net is reported in Table 3. We can observe that in average, our method finely outperforms other SOTA approaches. This comparison fairly demonstrates that our network generalizes well for clinical images with unknown metal materials and geometries and is potentially useful for clinical applications (See Footnote 4).

4 Conclusion

In this paper, we have proposed a joint spatial and Radon domain reconstruction model for the metal artifact reduction (MAR) task and constructed an interpretable network architecture, namely InDuDoNet, by unfolding an iterative optimization algorithm with only simple computations involved. Extensive

---

[4] More comparisons of MAR and bone segmentation are in *supplementary material*.

experiments were conducted on synthesized and clinical data. The experimental results demonstrated the effectiveness of our dual-domain MAR approach as well as its superior interpretability beyond current SOTA deep MAR networks.

Acknowledgements. This research was supported by National Key R&D Program of China (2020YFA0713900), the Macao Science and Technology Development Fund under Grant 061/2020/A2, Key-Area Research and Development Program of Guangdong Province, China (No. 2018B010111001), the Scientific and Technical Innovation 2030-"New Generation Artificial Intelligence" Project (No. 2020AAA0104100), the China NSFC projects (62076196, 11690011, 61721002, U1811461).

References

1. Beck, A., Teboulle, M.: A fast iterative shrinkage-thresholding algorithm for linear inverse problems. SIAM J. Imaging Sci. **2**(1), 183–202 (2009)
2. Chang, Z., Ye, D.H., Srivastava, S., Thibault, J.B., Sauer, K., Bouman, C.: Prior-guided metal artifact reduction for iterative X-ray computed tomography. IEEE Trans. Med. Imaging **38**(6), 1532–1542 (2018)
3. De Man, B., Nuyts, J., Dupont, P., Marchal, G., Suetens, P.: Metal streak artifacts in X-ray computed tomography: a simulation study. IEEE Trans. Nuclear Sci. **46**(3), 691–696 (1999)
4. Donoho, D.L.: De-noising by soft-thresholding. IEEE Trans. Inf. Theory **41**(3), 613–627 (1995)
5. Ghani, M.U., Karl, W.C.: Fast enhanced CT metal artifact reduction using data domain deep learning. IEEE Trans. Comput. Imaging **6**, 181–193 (2019)
6. Gjesteby, L., Yang, Q., Xi, Y., Zhou, Y., Zhang, J., Wang, G.: Deep learning methods to guide CT image reconstruction and reduce metal artifacts. In: Medical Imaging 2017: Physics of Medical Imaging, vol. 10132, p. 101322W. International Society for Optics and Photonics (2017)
7. He, K., Zhang, X., Ren, S., Sun, J.: Deep residual learning for image recognition. In: Proceedings of the IEEE Conference on Computer Vision and Pattern Recognition, pp. 770–778 (2016)
8. Huang, X., Wang, J., Tang, F., Zhong, T., Zhang, Y.: Metal artifact reduction on cervical CT images by deep residual learning. Biomed. Eng. Online **17**(1), 1–15 (2018)
9. Ji, W., et al.: Learning calibrated medical image segmentation via multi-rater agreement modeling. In: Proceedings of the IEEE/CVF Conference on Computer Vision and Pattern Recognition, pp. 12341–12351 (2021)
10. Kalender, W.A., Hebel, R., Ebersberger, J.: Reduction of CT artifacts caused by metallic implants. Radiology **164**(2), 576–577 (1987)
11. Liao, H., et al.: Generative mask pyramid network for CT/CBCT metal artifact reduction with joint projection-sinogram correction. In: Shen, D., et al. (eds.) MICCAI 2019. LNCS, vol. 11769, pp. 77–85. Springer, Cham (2019). https://doi.org/10.1007/978-3-030-32226-7_9
12. Liao, H., Lin, W.A., Zhou, S.K., Luo, J.: ADN: artifact disentanglement network for unsupervised metal artifact reduction. IEEE Trans. Med. Imaging **39**(3), 634–643 (2019)
13. Lin, W.A., et al.: DuDoNet: dual domain network for CT metal artifact reduction. In: Proceedings of the IEEE/CVF Conference on Computer Vision and Pattern Recognition, pp. 10512–10521 (2019)

14. Liu, P., et al.: Deep learning to segment pelvic bones: large-scale CT datasets and baseline models. arXiv preprint arXiv:2012.08721 (2020)

15. Lyu, Y., Lin, W.-A., Liao, H., Lu, J., Zhou, S.K.: Encoding metal mask projection for metal artifact reduction in computed tomography. In: Martel, A.L., et al. (eds.) MICCAI 2020. LNCS, vol. 12262, pp. 147–157. Springer, Cham (2020). https://doi.org/10.1007/978-3-030-59713-9_15

16. Mehranian, A., Ay, M.R., Rahmim, A., Zaidi, H.: X-ray CT metal artifact reduction using wavelet domain l_0 sparse regularization. IEEE Trans. Med. Imaging **32**(9), 1707–1722 (2013)

17. Meyer, E., Raupach, R., Lell, M., Schmidt, B., Kachelrieß, M.: Normalized metal artifact reduction (NMAR) in computed tomography. Med. Phys. **37**(10), 5482–5493 (2010)

18. Park, H.S., Lee, S.M., Kim, H.P., Seo, J.K., Chung, Y.E.: CT sinogram-consistency learning for metal-induced beam hardening correction. Med. Phys. **45**(12), 5376–5384 (2018)

19. Paszke, A., et al.: Automatic differentiation in PyTorch (2017)

20. Ronneberger, O., Fischer, P., Brox, T.: U-Net: convolutional networks for biomedical image segmentation. In: Navab, N., Hornegger, J., Wells, W.M., Frangi, A.F. (eds.) MICCAI 2015. LNCS, vol. 9351, pp. 234–241. Springer, Cham (2015). https://doi.org/10.1007/978-3-319-24574-4_28

21. Wang, G., Ye, J.C., Mueller, K., Fessler, J.A.: Image reconstruction is a new frontier of machine learning. IEEE Trans. Med. Imaging **37**(6), 1289–1296 (2018)

22. Wang, H., et al.: Structural residual learning for single image rain removal. Knowl.-Based Syst. **213**, 106595 (2021)

23. Wang, H., Xie, Q., Zhao, Q., Meng, D.: A model-driven deep neural network for single image rain removal. In: Proceedings of the IEEE/CVF Conference on Computer Vision and Pattern Recognition, pp. 3103–3112 (2020)

24. Wang, H., Yue, Z., Xie, Q., Zhao, Q., Zheng, Y., Meng, D.: From rain generation to rain removal. In: Proceedings of the IEEE/CVF Conference on Computer Vision and Pattern Recognition, pp. 14791–14801 (2021)

25. Wang, J., Zhao, Y., Noble, J.H., Dawant, B.M.: Conditional generative adversarial networks for metal artifact reduction in CT images of the ear. In: Frangi, A.F., Schnabel, J.A., Davatzikos, C., Alberola-López, C., Fichtinger, G. (eds.) MICCAI 2018. LNCS, vol. 11070, pp. 3–11. Springer, Cham (2018). https://doi.org/10.1007/978-3-030-00928-1_1

26. Xie, Q., Zhou, M., Zhao, Q., Xu, Z., Meng, D.: MHF-Net: An interpretable deep network for multispectral and hyperspectral image fusion. IEEE Trans. Pattern Anal. Mach. Intell. (2020). https://doi.org/10.1109/TPAMI.2020.3015691

27. Yan, K., et al.: Deep lesion graphs in the wild: relationship learning and organization of significant radiology image findings in a diverse large-scale lesion database. In: Proceedings of the IEEE Conference on Computer Vision and Pattern Recognition, pp. 9261–9270 (2018)

28. Yu, L., Zhang, Z., Li, X., Xing, L.: Deep sinogram completion with image prior for metal artifact reduction in CT images. IEEE Trans. Med. Imaging **40**(1), 228–238 (2020)

29. Yue, Z., Yong, H., Zhao, Q., Zhang, L., Meng, D.: Variational image restoration network. arXiv preprint arXiv:2008.10796 (2020)

30. Zhang, H., Dong, B., Liu, B.: A reweighted joint spatial-radon domain CT image reconstruction model for metal artifact reduction. SIAM J. Imaging Sci. **11**(1), 707–733 (2018)

31. Zhang, H., Wang, L., Li, L., Cai, A., Hu, G., Yan, B.: Iterative metal artifact reduction for X-ray computed tomography using unmatched projector/backprojector pairs. Med. Phys. **43**(6Part1), 3019–3033 (2016)
32. Zhang, Y., Yu, H.: Convolutional neural network based metal artifact reduction in X-ray computed tomography. IEEE Trans. Med. Imaging **37**(6), 1370–1381 (2018)

Depth Estimation for Colonoscopy Images with Self-supervised Learning from Videos

Kai Cheng[1], Yiting Ma[1], Bin Sun[3], Yang Li[3], and Xuejin Chen[1,2(✉)]

[1] National Engineering Laboratory for Brain-inspired Intelligence Technology and Application, University of Science and Technology of China, Hefei, China
xjchen99@ustc.edu.cn
[2] Institute of Artificial Intelligence, Hefei, China
[3] The First Affiliated Hospital of Anhui Medical University, Hefei, China

Abstract. Depth estimation in colonoscopy images provides geometric clues for downstream medical analysis tasks, such as polyp detection, 3D reconstruction, and diagnosis. Recently, deep learning technology has made significant progress in monocular depth estimation for natural scenes. However, without sufficient ground truth of dense depth maps for colonoscopy images, it is significantly challenging to train deep neural networks for colonoscopy depth estimation. In this paper, we propose a novel approach that makes full use of both synthetic data and real colonoscopy videos. We use synthetic data with ground truth depth maps to train a depth estimation network with a generative adversarial network model. Despite the lack of ground truth depth, real colonoscopy videos are used to train the network in a self-supervision manner by exploiting temporal consistency between neighboring frames. Furthermore, we design a masked gradient warping loss in order to ensure temporal consistency with more reliable correspondences. We conducted both quantitative and qualitative analysis on an existing synthetic dataset and a set of real colonoscopy videos, demonstrating the superiority of our method on more accurate and consistent depth estimation for colonoscopy images.

Keywords: Colonoscopy · Depth estimation · Self-supervised learning · Videos · Temporal consistency

1 Introduction

Colorectal cancer is recently reported as the third most prevalent malignancy and the fourth most common cause of cancer-associated death worldwide [1, 13, 15]. Colonoscopy is an effective technique for the prevention and treatment of

Electronic supplementary material The online version of this chapter (https://doi.org/10.1007/978-3-030-87231-1_12) contains supplementary material, which is available to authorized users.

M. de Bruijne et al. (Eds.): MICCAI 2021, LNCS 12906, pp. 119–128, 2021.
https://doi.org/10.1007/978-3-030-87231-1_12

colon cancer. Many approaches have been proposed for colorectal polyp detection and diagnosis in colonoscopy images and videos [5,9,20,21]. While geometric features, e.g. location, size, and shape of polyps, are critical for colorectal polyp diagnosis, depth estimation from colonoscopy images could help a lot in deriving 3D geometric information of the intestinal environment.

Many efforts have been put into depth estimation and 3D reconstruction of intestinal environments from colonoscopy videos. Low-level geometric clues are utilized in earlier model-based approaches. Hong et al. [3] estimate depths from colon fold contours. Zhao et al. [22] combine structure-from-motion (SfM) and shape-from-shading (SfS) techniques for surface reconstruction. These approaches suffer from reflection and low texture of colonoscopy images, resulting in inconsistency and great sparseness of the estimated depth maps. In order to enhance surface textures for more robust SfM, Widya et al. [17–19] use chromoendoscopy with the surface dyed by indigo carmine. However, chromoendoscopy is not very common, which leads to limitations in application. Besides, these approaches require dense feature extraction and global optimization, which is computationally expensive.

Deep-learning-based approaches have achieved remarkable performance in general depth estimation recently. Compared with natural scenes where ground-truth depth can be obtained using depth cameras or LiDARs, acquiring the ground-truth depth for colonoscopy videos is arduous. Ma et al. [7] use an SfM approach [12] to generate sparse colonoscopy depth maps as ground-truth to train a depth estimation network. However, due to the inherited limitation of SfM in low-quality reconstruction for textureless and non-Lambertian surfaces, it is challenging to obtain accurate dense depth maps for supervised learning. Assuming temporal consistency between frames in videos, unsupervised depth estimation has also been studied [2,6,23]. Liu et al. [6] propose a self-supervised depth estimation method for monocular endoscopic images using depth consistency check between adjacent frames with camera poses estimated by SfM. Freedman et al. [2] propose a calibration-free unsupervised method by predicting depth, camera pose, and intrinsics simultaneously. However, for colonoscopy videos with weak illumination in complex environments, these unsupervised approaches face significant challenges posed by frequent occlusions between colon folds and non-Lambertian surfaces.

Many works use synthetic data to produce precise ground truth depth for network training. Mahmood et al. [8] train a joint convolutional neural network-conditional random field framework on synthetic data and transfer real endoscopy images to synthetic style using a transformer network. Rau et al. [11] train an image translation network pix2pix [4] with synthetic image-and-depth pairs to directly translate a colonoscopy image into a depth map. In order to reduce the domain gap between synthetic data and real images, the GAN loss also involves the depth maps predicted from real colonoscopy images but L_1 loss is not computed since no ground truth is available for real images. By doing so, the generator is expected to learn to predict realistic-looking depth maps from real images. However, without accurate supervision on the details in the predicted depth map, it is non-trivial for the generator to precisely predict depth for unseen textures in real colonoscopy images that deviate from synthetic data.

In this paper, we not only utilize synthetic data with ground truth depth to help the network learn fine appearance features for depth estimation but also exploit the temporal consistency between neighboring frames to make full use of unlabeled real colonoscopy videos for self-supervision. Moreover, we design a masked gradient warping loss to filter out non-reliable correspondence caused by occlusions or reflections. A more powerful image translation model [16] is also employed in our framework to enhance the quality of depth estimation. We evaluate our method on the synthetic dataset [11] and our real colonoscopy videos. The results show that our method achieves more accurate and temporally consistent depth estimation for colonoscopy images.

2 Methodology

Given a single colonoscopy image $\mathbf{F}$, our goal is to train a deep neural network DepthNet G that directly generates a depth map $\mathbf{D}$ as $\mathbf{D} = G(\mathbf{F})$. In order to train the DepthNet G, we leverage both the synthetic data for full supervision and real colonoscopy videos for self-supervision via temporal consistency. The framework of our approach is shown in Fig. 1. First, we adopt a high-resolution image translation model to train DepthNet in an adversarial manner with synthetic data. Second, we introduce self-supervision during the network training by enforcing temporal consistency between the predicted depths of neighboring frames of real colonoscopy videos.

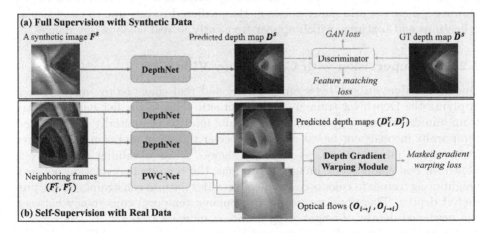

Fig. 1. Overview of our approach. (a) We first train DepthNet as a conditional GAN with synthetic image-and-depth pairs. (b) The DepthNet is then finetuned with self-supervision by checking the temporal consistency between neighboring frames.

(a) Colonoscopy image   (b) Deconv   (c) Upsample+Conv   (d) Self-supervised

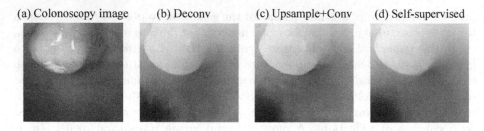

Fig. 2. Checkerboard artifacts. From a colonoscopy image (a), the original pix2pixHD model produces a depth map with checkerboard artifacts (b). Checkerboard artifact is alleviated by replacing the deconvolution layers in the generator with upsampling and convolution layers (c). Smoother depth is generated with our self-supervised model (d).

2.1 Training Baseline Model with Synthetic Data

We adopt the high-resolution image translation network pix2pixHD [16] as our baseline model to translate a colonoscopy image to a depth map. It consists of a coarse-to-fine generator and a multi-scale discriminator in order to produce high-resolution images. The network is trained in an adversarial manner with a GAN loss and feature matching loss [16] on the synthetic dataset [11] which contains paired synthetic colonoscopy images and the corresponding depth maps. However, the original pix2pixHD model produces results with checkerboard artifacts [10], as Fig. 2(b) shows. In order to alleviate this effect, we replace the deconvolution layers in the generator with upsampling and convolutional layers, similar to [6]. Figure 2(c) shows that the checkerboard effect is alleviated by replacing the deconvolutional layers with upsampling and convolutional layers. However, there are still many noises in the predicted results due to the specular reflections and textures, which appear frequently in real colonoscopy images.

2.2 Self-supervision with Colonoscopy Videos

Due to the domain gap between synthetic and real colonoscopy images, when applying the DepthNet trained on the synthetic data to predict depth directly from clinical colonoscopy images, the results tend to be spatially jumping and temporally inconsistent because of the specular reflection and complex textures in intestinal environments, as Fig. 2(c) shows. While obtaining ground-truth depth for real colonoscopy images is arduous, the temporal correlation between neighboring frames in colonoscopy videos provides natural constraints on the predicted depths. Therefore, we propose to enforce temporal consistency between the predicted depths of neighboring frames in network training.

For two neighboring frames in a real colonoscopy video $\mathbf{F}_i^r$ and $\mathbf{F}_j^r$, the Depth-Net estimates two depth maps $\mathbf{D}_i^r$ and $\mathbf{D}_j^r$ respectively. In order to check the consistency between these two depth maps, a typical way is to warp one frame to the other according to the camera pose and intrinsic, which are not easy to obtain. In order to avoid camera calibration, we propose a calibration-free warping module that finds pixel correspondences from optical flows. A pre-trained

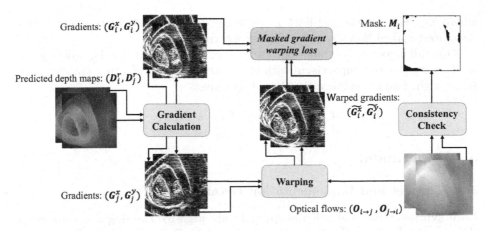

Fig. 3. Depth gradient warping module to check the temporal structural consistency of the predicted depth maps of two neighboring frames.

network PWC-Net [14] is employed to infer optical flows. Due to self-occlusions and reflections in colons, brightness consistency is not guaranteed so that errors in optical flow estimation are inevitable. In order to filter out the optical flow noises, we estimate optical flows $\mathbf{O}_{i \to j}$ and $\mathbf{O}_{j \to i}$ in two directions. Then we check if a pixel $\mathbf{p}$ can be warped back to the same position from frame i to frame j by $\mathbf{O}_{i \to j}$ then from frame j to frame i by $\mathbf{O}_{j \to i}$. If not, the pixel $\mathbf{p}$ is filtered out when checking temporal consistency. Therefore, we compute a mask $\mathbf{M}_i$ for frame i as

$$\mathbf{M}_i(\mathbf{p}) = \begin{cases} 0, & |\mathbf{O}_{i \to j}(\mathbf{p}) + \mathbf{O}_{j \to i}(\mathbf{q})| > \varepsilon \\ 1, & otherwise \end{cases} \qquad (1)$$

where $\mathbf{q}$ is the corresponding location in frame $\mathbf{F}_j^r$ of the pixel $\mathbf{p}$ in frame $\mathbf{F}_i^r$ according to the estimated optical flow $\mathbf{q} = \mathbf{p} + \mathbf{O}_{i \to j}(\mathbf{p})$. Note that we use bilinear interpolation of $\mathbf{O}_{j \to i}(\mathbf{q})$ for a subpixel $\mathbf{q}$. ε is a threshold for the forward-backward warping distance check. We set $\varepsilon = 1$ in our experiments.

However, the camera shifts at two neighboring frames. As a result, the absolute depth values of the correspondence pixels in two neighboring frames are not equal. Instead of comparing the depth values directly, we encourage the structural consistency between two depth maps by comparing the gradients of two depth maps through the depth gradient warping module. As Fig. 3 shows, we compute the gradients $(\mathbf{G}_i^x, \mathbf{G}_i^y)$, $(\mathbf{G}_j^x, \mathbf{G}_j^y)$ of the two predicted depth maps $\mathbf{D}_i^r$ and $\mathbf{D}_j^r$ in x and y direction. Then we check the consistency between the depth gradients of two neighboring frames with the mask $\mathbf{M}_i$ to calculate the masked gradient warping loss for self-supervision:

$$L_{MGW} = \frac{1}{|\mathbf{M}_i|} \sum_{\mathbf{p} \in \mathbf{F}_i^r} \mathbf{M}_i(\mathbf{p}) \left(\left| \mathbf{G}_i^x(\mathbf{p}) - \tilde{\mathbf{G}}_i^x(\mathbf{p}) \right| + \left| \mathbf{G}_i^y(\mathbf{p}) - \tilde{\mathbf{G}}_i^y(\mathbf{p}) \right| \right), \qquad (2)$$

where $\widetilde{\mathbf{G}}_i^x, \widetilde{\mathbf{G}}_i^y$ are the gradient maps warped from $\mathbf{G}_j^x, \mathbf{G}_j^y$ according to the estimated optical flow $\mathbf{O}_{j \to i}$ by bilinear interpolation.

Our full objective combines both self-supervision with masked gradient warping loss L_{MGW} and supervision with GAN loss L_{GAN} and feature matching loss L_{FM} with α and γ balance the three loss terms:

$$L = \alpha L_{MGW} + \gamma L_{FM} + L_{GAN}. \tag{3}$$

3   Experiments

3.1   Dataset and Implementation Details

Both synthetic and real colonoscopy data are used for training and evaluation. We use the UCL synthetic dataset published by Rau *et al.* [11]. The dataset consists of 16,016 pairs of synthetic endoscopic images and the corresponding depth maps. Following their split strategy, the dataset is divided randomly into training, validation, and test set by 6:1:3. We also collect 57 clinical colonoscopy videos from different patients. In the training stage, we use neighboring frames from each video at different intervals. Trading off overlap and interval between frame pairs, we choose four intervals including 1, 4, 8, and 16 frames. The final dataset of real colonoscopy data contains 6,352 training pairs and 4,217 test pairs.

Both the synthetic images and real images are resized to 512×512. We train our network in two steps. In the first step, we train our model on the synthetic data only. In the second step, we finetune the model with self-supervision on real colonoscopy frames. The batch size of synthetic images and real images for the first step and second step is set 8 and 4 respectively. We employ Adam optimizer with $\beta_1 = 0.5$ and $\beta_2 = 0.999$. The learning rate starts with $5e - 5$ and linearly decays. We update the generator every iteration while update the discriminator every 5 iterations. The framework is implemented in PyTorch 1.4 and trained on 4 Nvidia Titan XP GPUs. The first step training takes 70 epochs and we add the second step finetuning with real data at the last 10 epochs. The weight for the masked gradient warping loss $\alpha = 5$ initially and linearly increases by 2 in the second step. The weight of feature matching loss $\gamma = 2$.

3.2   Quantitative Evaluation

In order to quantitatively evaluate the performance of our method on depth estimation of colonoscopy images, we compare our method with previous approaches on the UCL synthetic dataset [11]. We adopt the same three metrics including the absolute L_1 distance, the relative error, and the root-mean-squared-error $RMSE$ between the ground truth and prediction. The results are reported in Table 1.

Table 1. Quantitative evaluation on the UCL synthetic dataset (* in cm, ** in %).

Method	Mean L_1-error*	Mean relative L_1-error**	Mean RMSE*
Pix2pix [11]	0.232 ± 0.046	8.2 ± 2.0	0.236
Extended pix2pix [11]	0.171 ± 0.034	6.4 ± 1.7	0.175
Our baseline	0.032 ± 0.011	1.5 ± 2.0	0.056
Ours	0.033 ± 0.012	1.6 ± 2.2	0.057

Our baseline model is only trained with synthetic data. It shows that a better conditional GAN model (pix2pixHD instead of pix2pix) brings great performance improvement. While only tested on synthetic data, our model that is fine-tuned with real colonoscopy videos does not make further improvement on synthetic data. This is reasonable because the self-supervision between neighboring frames leverages temporal consistency for more depth data from real colonoscopy video but it does not bring more information for the synthetic data.

Although the self-supervision with temporal consistency does not bring gain on the mean accuracy, it significantly improves the temporal consistency between the estimated depths on both the synthetic and real colonoscopy data. We quantify the temporal consistency by the masked gradient warping loss L_{MGW}, which reflects the structural consistency between the estimated depth maps of two neighboring frames. Table 2 demonstrates that our method reduces the masked gradient warping loss on both the synthetic data and real data.

Table 2. Masked gradient warping loss on synthetic and real colonoscopy datasets.

Method	Synthetic dataset (cm/pixel)	Real dataset (cm/pixel)
Baseline	0.024 ± 0.003	0.025 ± 0.020
Ours	**0.020 ± 0.003**	**0.012 ± 0.011**

3.3 Qualitative Evaluation on Real Data

Without ground-truth depths for quantitative comparison on real colonoscopy data, we evaluate our method qualitatively by comparing the depth prediction results with other methods. First, we compare our method with Rau *et al.* [11] and show some examples in Fig. 4. For the first three examples, we observe the wrongly predicted location of the lumen, missed polyps, and misinterpreted geometry of the lumen respectively in the results generated by Rau *et al.*. For the last three examples, we can see that our method generates more accurate predictions, proving that our model better captures geometric structure details.

We also verify our model in regards to the temporal consistency of the depth estimation. As shown in Fig. 5, without supervision by temporal consistency, the baseline model tends to predict discontinuous depths on the polyp surface due

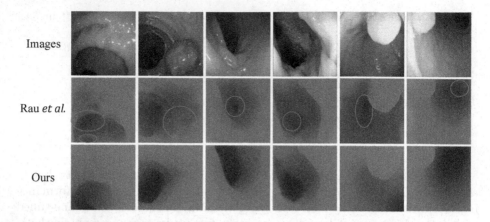

Fig. 4. Comparison of our method with Rau *et al.* [11]. The red ellipses highlight the inaccurate depth predictions such as wrong locations of the lumen, missed polyp, and misinterpreted geometry of the lumen.

to the specular reflection in the colonoscopy frames. These depth noises also lead to the discontinuity between neighboring frames. In comparison, the depths predicted by our fine-tuned model are more spatially smooth and temporally consistent, avoiding the interruption by specular reflections and textures.

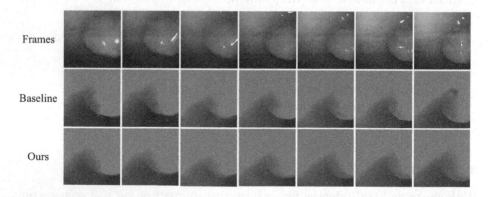

Fig. 5. Depth estimation for adjacent frames in a real colonoscopy video. Compared with the results generated by the baseline model, our model produces more consistent results avoiding the noises caused by specular reflections and textures.

4   Conclusion

We propose a novel depth estimation approach for colonoscopy images that makes full use of both synthetic and real data. Considering the depth estimation task as an image translation problem, we employ a conditional generative network as the backbone model. While the synthetic dataset which contains image-and-depth pairs provides precise supervision on the depth estimation network, we exploit unlabeled real colonoscopy videos for self-supervision. We designed a masked gradient warping loss to ensure the temporal consistency of the estimated depth maps of two neighboring frames during network training. The experimental results demonstrate that our method produces more accurate and temporally consistent depth estimation for both synthetic and real colonoscopy videos. The robust depth estimation will facilitate the accuracy of many downstream medical analysis tasks, such as polyp diagnosis and 3D reconstruction, and assist colonoscopists in polyp localization and removal in the future.

Acknowledgments. We acknowledge funding from National Natural Science Foundation of China under Grants 61976007 and 62076230.

References

1. Arnold, M., Sierra, M.S., Laversanne, M., Soerjomataram, I., Jemal, A., Bray, F.: Global patterns and trends in colorectal cancer incidence and mortality. Gut **66**(4), 683–691 (2017). https://doi.org/10.1136/gutjnl-2015-310912
2. Freedman, D., Blau, Y., Katzir, L., Aides, A., Shimshoni, I., Veikherman, D., Golany, T., Gordon, A., Corrado, G., Matias, Y., Rivlin, E.: Detecting deficient coverage in colonoscopies. IEEE Trans. Med. Imag. **39**(11), 3451–3462 (2020). https://doi.org/10.1109/TMI.2020.2994221
3. Hong, D., Tavanapong, W., Wong, J., Oh, J., de Groen, P.C.: 3D reconstruction of virtual colon structures from colonoscopy images. Comput. Med. Imag. Graph. **38**(1), 22–33 (2014). https://doi.org/10.1016/j.compmedimag.2013.10.005
4. Isola, P., Zhu, J., Zhou, T., Efros, A.A.: Image-to-image translation with conditional adversarial networks. In: IEEE Conference on Computer Vision and Pattern Recognition, pp. 5967–5976 (2017). https://doi.org/10.1109/CVPR.2017.632
5. Itoh, H., et al.: Towards automated colonoscopy diagnosis: Binary polyp size estimation via unsupervised depth learning. In: Medical Image Computing and Computer Assisted (MICCAI 2018), pp. 611–619 (2018)
6. Liu, X., Sinha, A., Ishii, M., Hager, G.D., Reiter, A., Taylor, R.H., Unberath, M.: Dense depth estimation in monocular endoscopy with self-supervised learning methods. IEEE Trans. Med. Imag. **39**(5), 1438–1447 (2020). https://doi.org/10.1109/TMI.2019.2950936
7. Ma, R., Wang, R., Pizer, S., Rosenman, J., McGill, S.K., Frahm, J.M.: Real-time 3D reconstruction of colonoscopic surfaces for determining missing regions. In: Medical Image Computing and Computer Assisted Intervention, pp. 573–582 (2019). https://doi.org/10.1007/978-3-030-32254-0_64
8. Mahmood, F., Durr, N.J.: Deep learning and conditional random fields-based depth estimation and topographical reconstruction from conventional endoscopy. Med. Image Anal. **48**, 230–243 (2018). https://doi.org/10.1016/j.media.2018.06.005
9. Nadeem, S., Kaufman, A.: Depth reconstruction and computer-aided polyp detection in optical colonoscopy video frames. arXiv preprint arXiv:1609.01329 (2016)
10. Odena, A., Dumoulin, V., Olah, C.: Deconvolution and checkerboard artifacts. Distill (2016). https://doi.org/10.23915/distill.00003

11. Rau, A., Edwards, P.E., Ahmad, O.F., Riordan, P., Janatka, M., Lovat, L.B., Stoyanov, D.: Implicit domain adaptation with conditional generative adversarial networks for depth prediction in endoscopy. Int. J. Comput. Assist. Radiol. Surg. **14**(7), 1167–1176 (2019). https://doi.org/10.1007/s11548-019-01962-w
12. Schonberger, J.L., Frahm, J.: Structure-from-motion revisited. In: IEEE Conference on Computer Vision and Pattern Recognition, pp. 4104–4113 (2016). https://doi.org/10.1109/CVPR.2016.445
13. Stark, U.A., Frese, T., Unverzagt, S., Bauer, A.: What is the effectiveness of various invitation methods to a colonoscopy in the early detection and prevention of colorectal cancer? protocol of a systematic review. Syst. Rev. **9**(1), 1–7 (2020). https://doi.org/10.1186/s13643-020-01312-x
14. Sun, D., Yang, X., Liu, M., Kautz, J.: PWC-net: CNNs for optical flow using pyramid, warping, and cost volume. In: 2018 IEEE/CVF Conference on Computer Vision and Pattern Recognition, pp. 8934–8943 (2018). https://doi.org/10.1109/CVPR.2018.00931
15. Waluga, M., Zorniak, M., Fichna, J., Kukla, M., Hartleb, M.: Pharmacological and dietary factors in prevention of colorectal cancer. J. Physiol. Pharmacol. **69**(3) (2018). https://doi.org/10.26402/jpp.2018.3.02
16. Wang, T., Liu, M., Zhu, J., Tao, A., Kautz, J., Catanzaro, B.: High-resolution image synthesis and semantic manipulation with conditional GANs. In: IEEE Conference on Computer Vision and Pattern Recognition, pp. 8798–8807 (2018). https://doi.org/10.1109/CVPR.2018.00917
17. Widya, A.R., Monno, Y., Okutomi, M., Suzuki, S., Gotoda, T., Miki, K.: Stomach 3D reconstruction based on virtual chromoendoscopic image generation. In: The 42nd Annual International Conference of the IEEE Engineering in Medicine Biology Society (EMBC), pp. 1848–1852 (2020). https://doi.org/10.1109/EMBC44109.2020.9176016
18. Widya, A.R., Monno, Y., Okutomi, M., Suzuki, S., Gotoda, T., Miki, K.: Whole stomach 3D reconstruction and frame localization from monocular endoscope video. IEEE J. Trans. Eng. Health Med. **7**, 1–10 (2019). https://doi.org/10.1109/JTEHM.2019.2946802
19. Widya, A.R., Monno, Y., Okutomi, M., Suzuki, S., Gotoda, T., Miki, K.: Self-supervised monocular depth estimation in gastroendoscopy using GAN-augmented images. In: Medical Imaging 2021: Image Processing. vol. 11596, p. 1159616 (2021). https://doi.org/10.1117/12.2579317
20. Yu, L., Chen, H., Dou, Q., Qin, J., Heng, P.A.: Integrating online and offline three-dimensional deep learning for automated polyp detection in colonoscopy videos. IEEE J. Biomed. Health Inform. **21**(1), 65–75 (2017). https://doi.org/10.1109/JBHI.2016.2637004
21. Zhang, R., Zheng, Y., Poon, C.C., Shen, D., Lau, J.Y.: Polyp detection during colonoscopy using a regression-based convolutional neural network with a tracker. Patt. Recogn. **83**, 209–219 (2018). https://doi.org/10.1016/j.patcog.2018.05.026
22. Zhao, Q., Price, T., Pizer, S., Niethammer, M., Alterovitz, R., Rosenman, J.: The endoscopogram: A 3D model reconstructed from endoscopic video frames. In: Medical Image Computing and Computer-Assisted Intervention (MICCAI 2016), pp. 439–447 (2016). https://doi.org/10.1007/978-3-319-46720-7_51
23. Zhou, T., Brown, M., Snavely, N., Lowe, D.G.: Unsupervised learning of depth and ego-motion from video. In: IEEE Conference on Computer Vision and Pattern Recognition, pp. 6612–6619 (2017). https://doi.org/10.1109/CVPR.2017.700

Joint Optimization of Hadamard Sensing and Reconstruction in Compressed Sensing Fluorescence Microscopy

Alan Q. Wang[1]([✉]), Aaron K. LaViolette[2]([✉]), Leo Moon[2], Chris Xu[2], and Mert R. Sabuncu[1]

[1] School of Electrical and Computer Engineering, Cornell University, Ithaca, USA
aw847@cornell.edu
[2] School of Applied and Engineering Physics, Cornell University, Ithaca, USA

Abstract. Compressed sensing fluorescence microscopy (CS-FM) proposes a scheme whereby less measurements are collected during sensing and reconstruction is performed to recover the image. Much work has gone into optimizing the sensing and reconstruction portions separately. We propose a method of jointly optimizing both sensing and reconstruction end-to-end under a total measurement constraint, enabling learning of the optimal sensing scheme concurrently with the parameters of a neural network-based reconstruction network. We train our model on a rich dataset of confocal, two-photon, and wide-field microscopy images comprising of a variety of biological samples. We show that our method outperforms several baseline sensing schemes and a regularized regression reconstruction algorithm. Our code is publicly-available at https://github.com/alanqrwang/csfm.

Keywords: Fluorescence microscopy · Compressed sensing · Joint optimization

1 Introduction

Fluorescence microscopy (FM) is an imaging modality widely used in biological research [12]. FM follows a Poissonian-Gaussian noise model which determines the signal-to-noise ratio (SNR) [6,33]. One common solution to increase SNR involves taking multiple measurements and averaging them; however, this invariably leads to an increase in the total number of photons used to generate an image, which can be undesirable when factors like photo-damage, photobleaching, and viability of the sample are considered [8,14]. Therefore, the trade-off between image quality and the total number of measurements collected is important to optimize [11,24,26].

Compressed sensing (CS) seeks to address this by collecting fewer measurements during sensing [7,16,20,27] and performing reconstruction on the noisy measurements to recover the image. Typically, reconstruction involves iterative optimization [2,3] or data-driven techniques [28,30].

M. de Bruijne et al. (Eds.): MICCAI 2021, LNCS 12906, pp. 129–139, 2021.
https://doi.org/10.1007/978-3-030-87231-1_13

Recently, researchers in multiple domains have explored the idea of jointly optimizing both the sensing and reconstruction parts end-to-end on training data, thus allowing for improved performance compared to separate optimization [1,21,28]. In this paper, we build on this idea and propose a method of jointly optimizing sensing and reconstruction in CS-FM.

Our contributions are as follows. We design a stochastic sensing model which is capable of learning the relative importance of each coefficient in the Hadamard sensing basis. Simultaneously, we train the parameters of a neural network which reconstructs the noisy measurements from the sensing model. Both parts are optimized end-to-end on training data, subject to a loss which maximizes reconstruction quality under a total measurement constraint. We evaluate both our sensing and reconstruction models and show a performance gain over CS-FM benchmarks.

2   Background

2.1   Fluorescence Microscopy and Hadamard Sensing

Let $x \in \mathbb{R}^N$ be a vector of N pixels in the field-of-view (FOV), where x_i is the signal from pixel i for a fixed acquisition time. In FM, a popular basis for collecting measurements of x is the Hadamard basis, which is well-studied, computationally-efficient, and easily realized via a digital micromirror device (DMD) in the imaging setup [19,20,22,31].

In Hadamard sensing, measurements correspond to the Hadamard coefficients. Mathematically, these coefficients are given by a linear transformation of x by a Hadamard matrix $H \in \{-1,1\}^{N \times N}$, where each row H_i is a Hadamard pattern. To physically measure each coefficient, H is reformulated as the difference between two binary, complementary matrices $H^+, H^- \in \{0,1\}^{N \times N}$:

$$H = \frac{1}{2}\left(J + H\right) - \frac{1}{2}\left(J - H\right) := H^+ - H^-, \tag{1}$$

where J is a matrix of ones. To obtain the measurement associated with the ith Hadamard coefficient, two sequential acquisitions are made by illuminating the sample according to the two binary patterns corresponding to rows H_i^+ and H_i^-. All the light for each acquisition is collected at one detector, and the Hadamard coefficient is computed by subtracting the two measurements[1].

For noise considerations, we adopt a Poissonian-Gaussian noise model [6,15, 33]. Incorporating this noise model with the above complementary measurement scheme, each Hadamard coefficient y_i follows a distribution

$$y_i \sim p(y_i \mid x) = \left[a\mathcal{P}\left(\frac{1}{a}H_i^+ x\right) + \mathcal{N}(0,b)\right] - \left[a\mathcal{P}\left(\frac{1}{a}H_i^- x\right) + \mathcal{N}(0,b)\right]. \tag{2}$$

---

[1] Note we are not considering the effects of a point-spread function here.

Here, $\mathcal{P}$ is a Poisson distribution which models signal-dependent uncertainty, while $\mathcal{N}$ is a normal distribution which models signal-independent uncertainty. Note that $a > 0$ is a conversion parameter (e.g. one detected photon corresponds to a signal of a). All noise components are assumed to be independent.

To increase SNR, we consider the case where multiple measurements are collected and then averaged. In particular, for S_i i.i.d. measurements[2] of the ith coefficient $\boldsymbol{y}_i^{(1)}, ..., \boldsymbol{y}_i^{(S_i)} \sim p(\boldsymbol{y}_i \mid \boldsymbol{x})$,

$$
\begin{aligned}
\boldsymbol{y}_i^{avg} &= \frac{1}{S_i} \sum_{s=1}^{S_i} \boldsymbol{y}_i^{(s)} \\
&\sim \frac{1}{S_i} \left[a\mathcal{P} \left(\frac{S_i}{a} H_i^+ \boldsymbol{x} \right) + \mathcal{N}(0, S_i b) \right] - \frac{1}{S_i} \left[a\mathcal{P} \left(\frac{S_i}{a} H_i^- \boldsymbol{x} \right) + \mathcal{N}(0, S_i b) \right].
\end{aligned}
\tag{3}
$$

It follows that while the mean is invariant to S_i, the variance decreases as S_i increases:

$$
\mathbb{E}[\boldsymbol{y}_i^{avg}] = H_i^+ \boldsymbol{x} - H_i^- \boldsymbol{x} = H_i \boldsymbol{x},
\tag{4}
$$

$$
Var(\boldsymbol{y}_i^{avg}) = \frac{1}{S_i} \left[aH_i^+ \boldsymbol{x} + b \right] - \frac{1}{S_i} \left[aH_i^- \boldsymbol{x} + b \right].
\tag{5}
$$

Note that the number of photons coming from the sample to produce the image is proportional to the sum of the S_i's, i.e. the total number of measurements. In this work, we pose the problem of optimizing the SNR-photon tradeoff as one of maximizing image quality subject to a constraint on the total measurements.

2.2   Sensing and Reconstruction Optimization

Prior work has explored the idea of jointly optimizing the sensing and reconstruction portions of computational imaging in different domains. The underlying intuition is that optimizing both parts concurrently leads to better performance than optimizing separately. Deterministic sensing patterns have been jointly optimized with neural-network based reconstruction schemes in designing the color multiplexing pattern of camera sensors [4], the LED illumination pattern in Fourier ptychography [10], the optical parameters of a camera lens [18], and microscopy cell detection [28]. More recently, stochastic sampling of sensing patterns have been explored in the context of CS magnetic resonance imaging [1,32] and very-long-baseline-interferometry (VLBI) array design [21]. In this paper, we extend the idea of joint optimization of sensing strategies and neural-network based reconstruction to CS-FM.

---

[2] Note we are defining one measurement to be the computation of one Hadamard coefficient without averaging, despite the fact that this is a two-step process in the physical imaging setup.

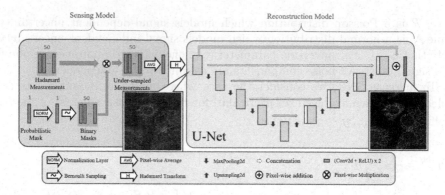

Fig. 1. Proposed model. Both sensing and reconstruction are trained jointly and end-to-end according to Eq. (7).

3   Proposed Method

Suppose S measurements are made at each of the N Hadamard coefficients. In the unconstrained case, this corresponds to NS total measurements. At each coefficient, the S measurements are averaged and follow a distribution according to Eq. (3). Given a compression ratio of $\alpha \in [0,1]$ which restricts the total number of measurements to αNS, the question we pose is as follows: how do we learn the optimal *allocation* of these αNS measurements across Hadamard bases, while simultaneously learning the parameters of a neural network reconstruction model, such that we maximize reconstruction quality?

3.1   End-to-End Sensing and Reconstruction Scheme

Let x denote a vectorized ground truth image from a dataset $\mathcal{D}$ and let $y_i^{(1)}, ..., y_i^{(S)} \sim p(y_i \mid x)$ denote S measurements at the ith Hadamard coefficient. Define $P \in [0,1]^N$ as a *probabilistic mask*, normalized such that $\frac{1}{N}\|P\|_1 = \alpha$.

For each Hadamard coefficient, we draw S i.i.d. samples from a Bernoulli distribution with parameter P_i, i.e. $M_i^{(1)}, ..., M_i^{(S)} \sim \mathcal{B}(P_i)$. In particular, $M_i^{(s)} \in \{0,1\}$ indicates if measurement $s \in \{1, ..., S\}$ should be included in the average. As such, we multiply $M_i^{(s)}$ with measurement $y_i^{(s)}$. Finally, we take the average over all included measurements to arrive at the per-coefficient measurement z_i:

$$z_i = \frac{\sum_{s=1}^{S} M_i^{(s)} y_i^{(s)}}{\sum_{t=1}^{S} M_i^{(t)}}, \quad \text{where } M_i^{(s)} \sim \mathcal{B}(P_i) \text{ and } y_i^{(s)} \sim p(y_i \mid x). \quad (6)$$

If we collect all z_i's in a vector z and collect all $M_i^{(s)}$'s in M, then we can summarize the sensing model by a sensing function $z = g(M, x)$. Note that $M \sim \prod_{s=1}^{S} \prod_{i=1}^{N} \mathcal{B}(P_i)$.

z is subsequently the input to the reconstruction model, which first transforms z to image space via the Hadamard transform and then passes the result into a neural network f_θ with parameters θ. In this work, we use a U-Net architecture [17]. Thus, the final reconstruction is $\hat{x} = f_\theta(Hz)$.

The proposed model is illustrated in Fig. 1.

3.2 Loss Function

The end-to-end sensing and reconstruction scheme is optimized jointly, according to a loss that maximizes image quality under the normalization constraint:

$$\min_{P,\theta} \mathbb{E}_{M \sim \prod_s \prod_i \mathcal{B}(P_i)} \sum_{x \in \mathcal{D}} \|\hat{x} - x\|_2^2 \quad \text{s.t.} \quad \hat{x} = f_\theta(Hg(M,x)) \quad \text{and} \quad \frac{1}{N}\|P\|_1 = \alpha. \quad (7)$$

A higher value of the optimized P_i^* implies that more measurements for the ith coefficient will be averaged, resulting in a higher SNR for that coefficient. Thus, P_i^* effectively encodes the "relative importance" of each location in Hadamard space for reconstruction. During imaging, one would draw S Bernoulli realizations with parameter P_i^* to determine the number of measurements to collect at basis i. Note the total required measurements will be αNS on average.

3.3 Implementation

To allow gradient computation through the Bernoulli sampling operation, we use the straight-through Gumbel-Softmax approximation [9,13], which samples from a Bernoulli distribution in the forward pass and computes gradients in the backward pass via a differentiable relaxation of the sampling procedure. A temperature hyperparameter τ controls the degree of relaxation.

To enforce the normalization constraint in Eq. (7), we add a closed-form differentiable normalization layer to the reconstruction network [1,32]. Let $\tilde{P}$ be an unconstrained probabilistic mask, and let $\bar{p} = \frac{1}{N}\|\tilde{P}\|_1$ be the average value of $\tilde{P}$. Note that $1 - \bar{p}$ is the average value of $1 - \|\tilde{P}\|_1$. Then,

$$P = \begin{cases} \frac{\alpha}{\bar{p}}\tilde{P}, & \text{if } \bar{p} \geq \alpha, \\ 1 - \frac{1-\alpha}{1-\bar{p}}(1 - \tilde{P}), & \text{otherwise.} \end{cases} \quad (8)$$

It can be shown that Eq. (8) yields $P \in [0,1]^N$ and $\frac{1}{N}\|P\|_1 = \alpha$, thus enforcing the hard constraint in Eq. (7).

4   Experiments

Training Details. In this work, we used a U-Net architecture [17] for f_θ, but alternative architectures which directly incorporate the forward model may be used as well, such as unrolled architectures [5,29] and physics-based neural networks [10]. In particular, our U-Net has 64 hidden channels per encoder layer and

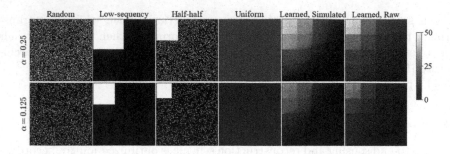

Fig. 2. Baseline and learned masks for $\alpha = 0.25$ (top) and $\alpha = 0.125$ (bottom). All masks are displayed in 2D-sequence order.

intermediate ReLU activations. The network was trained with ADAM optimizer. We set the temperature hyperparameter $\tau = 0.8$ via grid-search on validation loss. All training and testing experiments in this paper were performed on a machine equipped with an Intel Xeon Gold 6126 processor and an NVIDIA Titan Xp GPU. All models were implemented in Pytorch.

Baselines. For reconstruction, we compare our U-Net based method against a widely used regularized regression reconstruction algorithm (TV-W) [23][3]. The algorithm minimizes a least-squares cost function with two regularization terms: total variation and L1-penalty on wavelet coefficients. The weights of these terms was established via grid search on validation performance.

As is common in the CS-FM literature, we implemented baseline binary sensing schemes that collect either all (e.g., $S = 50$) measurements or no measurement at each basis. This can be denoted with a binary mask $\boldsymbol{P} \in \{0, 1\}^N$ where 1 indicates the basis is sampled and 0 indicates the basis is not sampled. Binary baseline masks include random (R), low-sequency (LS), and half-half (HH) [20] Hadamard masks, shown in Fig. 2. Non-binary stochastic masks include uniform (U) and learned masks. For U, on average α-percent of all available (S) measurements were used at each basis, realized via Bernoulli sampling; i.e., $\boldsymbol{P}$ is a constant vector of α's. We binarize the learned stochastic masks for the TV-W algorithm by taking the top α-percentile of basis patterns. This cannot be achieved for uniform masks, so we only evaluate them on U-Net models.

Data. To illustrate the computational utility of our proposed method, models were trained and tested on a combination of in-house data and the publicly-available Fluorescence Microscopy Denoising (FMD) dataset, comprising of confocal, two-photon, and wide-field microscopy images [33]. Our combined dataset consists of a variety of representative biological samples, namely blood vessels, neurons, and lymph nodes. Each FOV has $S = 50$ noisy pixel-wise acquisitions,

---

[3] We used publicly-available code from: https://github.com/alanqrwang/HQSNet.

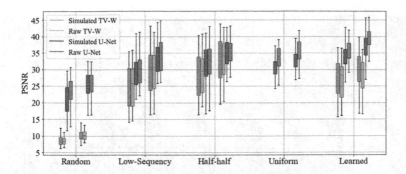

Fig. 3. PSNR for all masks, $\alpha = 0.125$ (left) and $\alpha = 0.25$ (right).

which consist of raw measurements at constant excitation power. Ground truth images $\boldsymbol{x}$ were obtained by pixel-wise averaging over all S frames.

To obtain $\boldsymbol{y}_i^{(s)}$ input Hadamard measurements, we implemented two approaches: simulated and raw. In the simulated approach, we applied random noise to the ground truth $\boldsymbol{x}$ following Eq. (2). Values of a and b were obtained from [33][4]. In the raw approach, we applied the Hadamard transform to raw pixel-wise measurements in a FOV. Although this is not exact since individual noise contributions are not additive, it is a reasonable approximation as we are interested in the relative importance of each basis pattern.

All images were cropped to fixed grid sizes of 256×256. The final training, validation, and test splits were 1000, 100, and 500 cropped images, respectively; and included non-overlapping independent samples. Reconstructions were evaluated against ground-truth images on peak signal-to-noise ratio (PSNR) and structural similarity index measure (SSIM) [25].

4.1 Masks

Figure 2 shows baseline and learned masks, displayed in 2D-sequence order [7]. We notice that the learned masks prioritize low-sequence Hadamard patterns but also spread out across higher-sequence patterns. We reason that the sensing and reconstruction scheme learns to provide the U-Net with mostly low-sequence information, and then subsequently "assigns" the U-Net the task of imputing the missing high-sequence information.

4.2 Reconstruction Methods

Figure 3 shows boxplots of PSNR values for both $\alpha = 0.25$ and $\alpha = 0.125$. U-Net reconstruction models outperform TV-W for all conditions. The pattern holds for both simulated and raw measurements. Similarly, learned masks outperform the baseline masks consistently for U-Net reconstruction (which the sensing was

---

[4] Negative values of b were replaced with 0 following the paper's experiments [33].

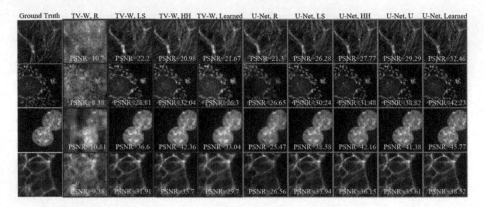

Fig. 4. Representative slices for $\alpha = 0.25$. Left is ground truth, blue is TV-W reconstructions, red is U-Net reconstructions. Masks are indicated in titles. (Color figure online)

Table 1. Performance for $\alpha = 0.25$ and $\alpha = 0.125$. Mean $\pm$ standard deviation across test cases. Simulated/Raw.

α	Method	Mask	PSNR	SSIM
0.25	TV-W	R	$10.40 \pm 2.066/10.44 \pm 1.847$	$0.275 \pm 0.127/0.286 \pm 0.126$
		LS	$29.15 \pm 6.516/29.15 \pm 6.405$	$0.887 \pm 0.094/0.886 \pm 0.095$
		HH	$33.20 \pm 6.544/33.42 \pm 5.948$	$0.905 \pm 0.094/0.913 \pm 0.084$
		Learned	$29.87 \pm 5.525/28.20 \pm 5.080$	$0.917 \pm 0.049/0.901 \pm 0.051$
	U-Net	R	$25.43 \pm 3.932/25.46 \pm 3.565$	$0.692 \pm 0.124/0.692 \pm 0.135$
		LS	$33.86 \pm 4.593/34.30 \pm 4.866$	$0.904 \pm 0.060/0.917 \pm 0.058$
		HH	$35.04 \pm 3.882/35.80 \pm 3.942$	$0.918 \pm 0.043/0.927 \pm 0.042$
		U	$31.44 \pm 6.264/35.66 \pm 3.360$	$0.841 \pm 0.125/0.924 \pm 0.024$
		Learned	$36.38 \pm 4.061/39.38 \pm 3.254$	$0.932 \pm 0.044/0.953 \pm 0.018$
0.125	TV-W	R	$8.646 \pm 1.499/8.631 \pm 1.316$	$0.216 \pm 0.108/0.218 \pm 0.101$
		LS	$24.91 \pm 5.984/24.90 \pm 6.042$	$0.805 \pm 0.148/0.801 \pm 0.146$
		HH	$28.39 \pm 6.439/28.27 \pm 6.129$	$0.843 \pm 0.137/0.847 \pm 0.131$
		Learned	$27.21 \pm 5.365/26.29 \pm 5.320$	$0.873 \pm 0.069/0.862 \pm 0.070$
	U-Net	R	$21.31 \pm 4.700/22.94 \pm 4.527$	$0.539 \pm 0.203/0.613 \pm 0.150$
		LS	$29.49 \pm 4.957/30.00 \pm 4.720$	$0.813 \pm 0.129/0.826 \pm 0.125$
		HH	$31.93 \pm 5.089/32.40 \pm 5.146$	$0.871 \pm 0.084/0.877 \pm 0.083$
		U	$29.26 \pm 5.867/33.57 \pm 3.301$	$0.806 \pm 0.121/0.894 \pm 0.042$
		Learned	$33.98 \pm 3.173/36.01 \pm 3.224$	$0.895 \pm 0.062/0.923 \pm 0.033$

optimized for), while half-half masks perform best for TV-W reconstruction. We conjecture that this is due to the U-Net model being better at imputing the missing information in the learned mask. Table 1 shows metrics for all models on PSNR and SSIM. Figure 4 shows representative reconstructions for raw data, which allows us to qualitatively appreciate the superior quality afforded by the learned mask and U-Net reconstruction.

5   Conclusion

We presented a novel sensing and reconstruction model for CS-FM, which is jointly trained end-to-end. We show that our scheme effectively learns the optimal sensing and reconstruction, and outperforms established baselines. Future work will include evaluating the proposed strategy on prospectively-collected data.

Acknowledgments. This work was, in part, supported by NIH R01 grants (R01LM012719 and R01AG053949, to MRS), the NSF NeuroNex grant (1707312, to MRS), an NSF CAREER grant (1748377, to MRS), and an NSF NeuroNex Hub grant (DBI-1707312, to CX).

References

1. Bahadir, C.D., Wang, A.Q., Dalca, A.V., Sabuncu, M.R.: Deep-learning-based optimization of the under-sampling pattern in MRI. IEEE Trans. Comput. Imaging **6**, 1139–1152 (2020)
2. Beck, A., Teboulle, M.: A fast iterative shrinkage-thresholding algorithm for linear inverse problems. SIAM J. Imaging Sci. **2**(1), 183–202 (2009)
3. Boyd, S., Parikh, N., Chu, E., Peleato, B., Eckstein, J.: Distributed optimization and statistical learning via the alternating direction method of multipliers. Found. Trends Mach. Learn. **3**(1), 1–122 (2011)
4. Chakrabarti, A.: Learning sensor multiplexing design through back-propagation. In: Proceedings of the 30th International Conference on Neural Information Processing Systems, NIPS 2016, pp. 3089–3097. Curran Associates Inc., Red Hook (2016)
5. Diamond, S., Sitzmann, V., Heide, F., Wetzstein, G.: Unrolled optimization with deep priors. CoRR abs/1705.08041 (2017)
6. Foi, A., Trimeche, M., Katkovnik, V., Egiazarian, K.: Practical poissonian-gaussian noise modeling and fitting for single-image raw-data. IEEE Trans. Image Process. **17**(10), 1737–1754 (2008)
7. Gibson, G.M., Johnson, S.D., Padgett, M.J.: Single-pixel imaging 12 years on: a review. Opt. Express **28**(19), 28190–28208 (2020)
8. Hopt, A., Neher, E.: Highly nonlinear photodamage in two-photon fluorescence microscopy. Biophys. J . **80**(4), 2029–2036 (2001)
9. Jang, E., Gu, S., Poole, B.: Categorical reparametrization with gumble-softmax. In: Proceedings of the International Conference on Learning Representations 2017. OpenReviews.net, April 2017

10. Kellman, M., Bostan, E., Chen, M., Waller, L.: Data-driven design for Fourier ptychographic microscopy. In: 2019 IEEE International Conference on Computational Photography (ICCP), pp. 1–8 (2019)
11. Lee, S., Negishi, M., Urakubo, H., Kasai, H., Ishii, S.: Mu-net: multi-scale U-net for two-photon microscopy image denoising and restoration. Neural Netw. **125**, 92–103 (2020)
12. Lichtman, J., Conchello, J.: Fluorescence microscopy. Nat. Methods **2**, 910–919 (2005)
13. Maddison, C.J., Mnih, A., Teh, Y.W.: The concrete distribution: a continuous relaxation of discrete random variables. In: International Conference on Learning Representations (2017)
14. Magidson, V., Khodjakov, A.: Circumventing photodamage in live-cell microscopy. In: Sluder, G., Wolf, D.E. (eds.) Digital Microscopy, Methods in Cell Biology, vol. 114, pp. 545–560. Academic Press (2013)
15. Makitalo, M., Foi, A.: Optimal inversion of the generalized Anscombe transformation for Poisson-Gaussian noise. IEEE Trans. Image Process. **22**(1), 91–103 (2013)
16. Parot, V.J., et al.: Compressed Hadamard microscopy for high-speed optically sectioned neuronal activity recordings. J. Phys. D Appl. Phys. **52**(14), 144001 (2019)
17. Ronneberger, O., Fischer, P., Brox, T.: U-Net: convolutional networks for biomedical image segmentation. In: Navab, N., Hornegger, J., Wells, W.M., Frangi, A.F. (eds.) MICCAI 2015. LNCS, vol. 9351, pp. 234–241. Springer, Cham (2015). https://doi.org/10.1007/978-3-319-24574-4_28
18. Sitzmann, V., et al.: End-to-end optimization of optics and image processing for achromatic extended depth of field and super-resolution imaging. ACM Trans. Graph. **37**(4), 1–13 (2018)
19. Streeter, L., Burling-Claridge, G.R., Cree, M.J., Künnemeyer, R.: Optical full Hadamard matrix multiplexing and noise effects. Appl. Opt. **48**(11), 2078–2085 (2009)
20. Studer, V., Bobin, J., Chahid, M., Mousavi, H.S., Candes, E., Dahan, M.: Compressive fluorescence microscopy for biological and hyperspectral imaging. Proc. Natl. Acad. Sci. **109**(26), E1679–E1687 (2012)
21. Sun, H., Dalca, A.V., Bouman, K.L.: Learning a probabilistic strategy for computational imaging sensor selection. In: 2020 IEEE International Conference on Computational Photography (ICCP), pp. 1–12 (2020)
22. Sun, M.J., Meng, L.T., Edgar, M.: A Russian Dolls ordering of the Hadamard basis for compressive single-pixel imaging. Sci. Rep. **7** (2017). Article number: 3464
23. Wang, A.Q., Dalca, A.V., Sabuncu, M.R.: Neural network-based reconstruction in compressed sensing MRI without fully-sampled training data. In: Deeba, F., Johnson, P., Würfl, T., Ye, J.C. (eds.) MLMIR 2020. LNCS, vol. 12450, pp. 27–37. Springer, Cham (2020). https://doi.org/10.1007/978-3-030-61598-7_3
24. Wang, H., Rivenson, Y., Jin, Y.: Deep learning enables cross-modality super-resolution in fluorescence microscopy. Nat. Methods **16**, 103–110 (2019)
25. Wang, Z., Bovik, A.C., Sheikh, H.R., Simoncelli, E.P.: Image quality assessment: from error visibility to structural similarity. Trans. Image Process. **13**(4), 600–612 (2004)
26. Weigert, M., Schmidt, U., Boothe, T.: Content-aware image restoration: pushing the limits of fluorescence microscopy. Nat. Methods **15**(12), 1090–1097 (2018)
27. Wijesinghe, P., Escobet-Montalbán, A., Chen, M., Munro, P.R.T., Dholakia, K.: Optimal compressive multiphoton imaging at depth using single-pixel detection. Opt. Lett. **44**(20), 4981 (2019)

28. Xue, Y., Bigras, G., Hugh, J., Ray, N.: Training convolutional neural networks and compressed sensing end-to-end for microscopy cell detection. IEEE Trans. Med. Imaging **38**(11), 2632–2641 (2019)
29. Yang, Y., Sun, J., Li, H., Xu, Z.: Deep ADMM-Net for compressive sensing MRI. In: Lee, D., Sugiyama, M., Luxburg, U., Guyon, I., Garnett, R. (eds.) Advances in Neural Information Processing Systems, vol. 29. Curran Associates, Inc. (2016)
30. Yao, R., Ochoa, M., Yan, P.: Net-FLICS: fast quantitative wide-field fluorescence lifetime imaging with compressed sensing - a deep learning approach. Light Sci. Appl. **8** (2019). Article number: 26
31. Yu, X., Yang, F., Gao, B., Ran, J., Huang, X.: Deep compressive single pixel imaging by reordering Hadamard basis: a comparative study. IEEE Access **8**, 55773–55784 (2020)
32. Zhang, J., et al.: Extending LOUPE for K-space under-sampling pattern optimization in multi-coil MRI. In: Deeba, F., Johnson, P., Würfl, T., Ye, J.C. (eds.) MLMIR 2020. LNCS, vol. 12450, pp. 91–101. Springer, Cham (2020). https://doi.org/10.1007/978-3-030-61598-7_9
33. Zhang, Y., et al.: A Poisson-Gaussian denoising dataset with real fluorescence microscopy images. In: 2019 IEEE/CVF Conference on Computer Vision and Pattern Recognition (CVPR), pp. 11702–11710 (2019)

Multi-contrast MRI Super-Resolution via a Multi-stage Integration Network

Chun-Mei Feng[1,2], Huazhu Fu[2], Shuhao Yuan[2], and Yong Xu[1(✉)]

[1] Shenzhen Key Laboratory of Visual Object Detection and Recognition,
Harbin Institute of Technology, Shenzhen, China
[2] Inception Institute of Artificial Intelligence, Abu Dhabi, UAE
https://github.com/chunmeifeng/MINet

Abstract. Super-resolution (SR) plays a crucial role in improving the image quality of magnetic resonance imaging (MRI). MRI produces multi-contrast images and can provide a clear display of soft tissues. However, current super-resolution methods only employ a single contrast, or use a simple multi-contrast fusion mechanism, ignoring the rich relations among different contrasts, which are valuable for improving SR. In this work, we propose a multi-stage integration network (*i.e.,* MINet) for multi-contrast MRI SR, which explicitly models the dependencies between multi-contrast images at different stages to guide image SR. In particular, our MINet first learns a hierarchical feature representation from multiple convolutional stages for each of different-contrast image. Subsequently, we introduce a multi-stage integration module to mine the comprehensive relations between the representations of the multi-contrast images. Specifically, the module matches each representation with all other features, which are integrated in terms of their similarities to obtain an enriched representation. Extensive experiments on fastMRI and real-world clinical datasets demonstrate that 1) our MINet outperforms state-of-the-art multi-contrast SR methods in terms of various metrics and 2) our multi-stage integration module is able to excavate complex interactions among multi-contrast features at different stages, leading to improved target-image quality.

Keywords: Magnetic resonance imaging · Super-resolution · Multi-contrast

1 Introduction

Magnetic resonance imaging (MRI) is a popular technique in medical imaging. Unlike other modalities, such as computed tomography (CT) or nuclear imaging, MRI can provide clear information about tissue structure and function without inducing ionizing radiation. However, due to the complex data acquisition process, it is difficult to obtain high-resolution (HR) MRI images [6,7].

This work was done during the internship of C.-M. Feng at Inception Institute of Artificial Intelligence.

© Springer Nature Switzerland AG 2021
M. de Bruijne et al. (Eds.): MICCAI 2021, LNCS 12906, pp. 140–149, 2021.
https://doi.org/10.1007/978-3-030-87231-1_14

Since the super-resolution (SR) can improve the image quality without changing the MRI hardware, this post-processing tool has been widely used to overcome the challenge of obtaining HR MRI scans [8]. Bicubic and b-spline interpolation are two basic SR interpolation methods [5,6]; however, they inevitably lead to blurred edges and blocking artifacts. Relying on the inherent redundancy of the transformation domain, iterative deblurring algorithms [11,19], low rank [18], and dictionary learning methods [1] have made significant progress in MRI SR. Recently, deep learning which offers higher resolution, has also become widely used for the task [2,4,9,10,15–17]. For example, Akshay et al. applied a 3D residual network to generate thin-slice MR images of knees [2]. Chen et al. used a densely connected SR network to restore HR details from a single low-resolution (LR) input image [4]. Lyu et al. used a generative adversarial network (GAN) framework for CT denoising and then transferred it to MRI SR [14]. However, the above methods focus on mono-contrast acquisition to restore HR images, ignoring complementary multi-contrast information.

MRI produces multi-contrast images under different settings but with the same anatomical structure, e.g., T1 and T2 weighted images (T1WI and T2WI), as well as proton density and fat-suppressed proton density weighted images (PDWI and FS-PDWI), which can provide complementary information to each other. For example, T1WI describe morphological and structural information, while T2WI describe edema and inflammation. Further, PDWI provide information on structures such as articular cartilage, and have a high signal-to-noise ratio (SNR) for tissues with little difference in PDWI, while FS-PDWI can inhibit fat signals and highlight the contrast of tissue structures such as cartilage ligaments [3]. In clinical settings, T1WI have shorter repetition time (TR) and echo time (TE) than T2WI, while PDWI usually take shorter than FS-PDWI in the scanning process. Therefore, the former are easier to acquire than the latter. These can be employed to add further information to a single LR image. For instance, relevant HR information from T1WI or PDWI can be used to assist the generation of SR T2WI or FS-PDWI [21]. Recently, several methods for multi-contrast SR have been proposed [13,24,27,28]. For example, Zheng et al. used the local weight similarity and relation model of the gradient value between images of different contrast to restore an SR image from its counterpart LR image [27,28]. Zeng et al. proposed a deep convolutional neural network to simultaneously obtain single- and multi-contrast SR images [24]. Lyu et al. introduced a progressive network, which employs a composite loss function for multi-contrast SR [13]. However, though significant progress has been made, existing methods do not explore the interaction between different modalities in various stages, which is especially critical for the target-contrast restoration.

To explore the correlations among hierarchical stages of different contrast, we propose a multi-stage integration network (MINet). In our method, the features of each stage are interacted and weighted. Specifically, the complementary features are fused with the corresponding target features, to obtain a comprehensive feature, which can guide the learning of the target features. Our main contributions are as follows: 1) We design a novel multi-contrast SR network to guide the

SR restoration of the target contrast through the HR auxiliary contrast features. 2) We explore the response of multi-contrast fusion at different stages, obtain the dependency relationship between the fused features, and improve their representation ability. 3) We perform extensive experiments on fastMRI and clinical datasets, demonstrating that our method can obtain superior results compared with the current state of the arts.

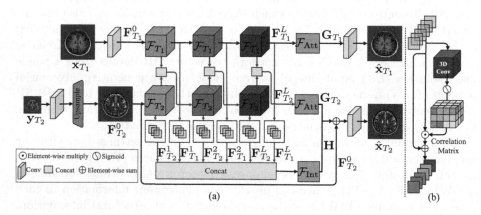

Fig. 1. (a) Network architecture of the proposed multi-contrast MRI SR model. (b) Details of the channel-spatial attention module (*i.e.*, $\mathcal{F}_{\text{Att}}$).

2   Methodology

2.1   Overall Architecture

Given an LR image $\mathbf{y} \in \mathbb{R}^{N \times N}$, we aim to learn a neural network that can provide an HR image $\mathbf{x} \in \mathbb{R}^{M \times M}$ ($M > N$). While mainstream efforts address this task by directly restoring $\mathbf{x}$ from $\mathbf{y}$, we propose to solve it from a novel perspective by introducing an extra HR contrast image with the same structural information, which provides more effective results. As shown in Fig. 1(a), we can employ the HR T1WI as an auxiliary to improve the resolution of the LR T2WI. More specifically, our MINet accepts two images as input, *i.e.*, an HR T1WI $\mathbf{x}_{T_1}$ and an LR T2WI $\mathbf{y}_{T_2}$. Each image is processed by two independent convolutional branches to obtain multi-stage feature representations. To take advantage of the complementary information between the two images of different contrast, we fuse the representations at the same stage in the two branches so that the anatomical structure features of the auxiliary branch can be fully propagated to the target SR branch. Moreover, we propose a multi-stage integration component to explore the dependencies of multi-contrast features in different stages and allocate appropriate attention weights to them, yielding a more powerful holistic feature representation for SR.

Multi-stage Feature Representation. In order to obtain the anatomical structure of the HR T1WI, we apply a 3×3 convolutional layer to obtain an initial representation $\mathbf{F}_{T_1}^0$. Similarly, the LR T2WI is also processed by a convolutional layer, followed by a sub-pixel convolution to restore the scale of the representation (*i.e.*, $\mathbf{F}_{T_2}^0$).

Subsequently, multi-stage feature representations are learned through a cascade of residual groups [25]. In particular, at the lth stage, we obtain the intermediate features as follows:

$$\mathbf{F}_{T_1}^l = \mathcal{F}_{T_1}^l(\mathbf{F}_{T_1}^{l-1}), \tag{1}$$

$$\mathbf{F}_{T_2}^l = \mathcal{F}_{T_2}^l([\mathbf{F}_{T_1}^{l-1}, \mathbf{F}_{T_2}^{l-1}]). \tag{2}$$

Here, $\mathcal{F}_{T_1}^l$ and $\mathcal{F}_{T_2}^l$ represent residual groups at the lth stage of each respective branch, and $[,]$ denotes the concatenation operation. Through Eq. (2), our model leverages T1WI features to guide the learning of the T2WI at each stage, enabling progressive feature fusion. Different from most existing methods [12,13], which only use the representations from the final stage for SR, our model stores all the intermediate features at multiple stages:

$$\mathbf{F} = [\mathbf{F}_{T_1}^1, \mathbf{F}_{T_2}^1, \mathbf{F}_{T_1}^2, \mathbf{F}_{T_2}^2, \cdots \mathbf{F}_{T_1}^L, \mathbf{F}_{T_2}^L]. \tag{3}$$

Here, $\mathbf{F}$ denotes a multi-stage feature representation obtained by concatenating all the intermediate features, and L indicates the number of residual groups.

Multi-Contrast Feature Enhancement. While Eq. (3) aggregates multi-stage features to achieve a more comprehensive representation, it treats each feature independently and does not fully explore the relations among them. To this end, we propose a multi-stage integration module $\mathcal{F}_{\text{Int}}$ (§2.2) which learns the relations among features and uses them to modulate each feature. Formally, we have: $\mathbf{H} = \mathcal{F}_{\text{Int}}(\mathbf{F})$, where $\mathbf{H}$ represents the enriched feature representation.

In addition, we employ a channel-spatial attention module $\mathcal{F}_{\text{Att}}$ to modulate the feature $\mathbf{F}_{T_2}^L$ as $\mathbf{G}_{T_2} = \mathcal{F}_{\text{Att}}(\mathbf{F}_{T_2}^L)$. Inspired by the spatial [20] and channel attention mechanisms [25], we design the channel-spatial attention module, which reveals responses from all dimensions of the feature maps. As shown in Fig. 1(b), the features output by the residual groups $\mathbf{F}_{T_2}^L$ are sent into a 3D convolutional layer to generate attention map $\mathbf{A}_{tt}$ by capturing joint channels and spatial features. This can be written as

$$\mathbf{G}_{T_2} = \texttt{sigmoid}(\mathbf{A}_{tt}) \odot \mathbf{F}_{T_2}^L + \mathbf{F}_{T_2}^L, \tag{4}$$

where $\texttt{sigmoid}$ is the sigmoid operation, and $\odot$ denotes element-wise multiplication. As a result, Eq. (4) provides a feasible way to enrich $\mathbf{F}_{T_2}^L$ by making use of the context within the feature. We also use a residual layer to effectively preserve information from the original feature map. Note that $\mathbf{G}_{T_1}$ is also calculated using Eq. (4).

Image Reconstruction. Since our framework is a pre-upsampling SR, we use a convolutional layer $\mathcal{F}_{\mathrm{Rec}}$ to obtain the final restored image $\hat{\mathbf{x}}_{T_2}$, which can be written as:

$$\hat{\mathbf{x}}_{T_2} = \mathcal{F}_{\mathrm{Rec}}(\mathbf{F}^0_{T_2} \oplus \mathbf{G}_{T_2} \oplus \mathbf{H}), \tag{5}$$

where $\oplus$ denotes element-wise summation. In addition, as shown in Fig. 1, we also compute the reconstruction of the T1WI using a convolutional layer with input $\mathbf{G}_{T_1}$. Note that the reconstruction procedure is only used as an auxiliary.

Loss Function. We simply use the L_1 loss to evaluate the reconstruction results of the T1WI and T2WI. The final loss function is

$$L = \frac{1}{N} \sum_{n=1}^{N} \alpha \left\| \hat{\mathbf{x}}^n_{T_2} - \mathbf{x}^n_{T_2} \right\|_1 + \beta \left\| \hat{\mathbf{x}}^n_{T_1} - \mathbf{x}^n_{T_1} \right\|_1, \tag{6}$$

where α and β weight the trade-off between the T1WI and T2WI reconstruction.

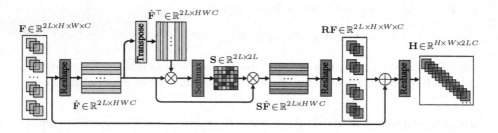

Fig. 2. Architecture of the proposed multi-stage integration module.

2.2 Multi-stage Integration Module

Although we have aggregated the T1WI and T2WI features in the intermediate stages for T2WI SR, the correlation among different stages has been missed, especially for MRI in different contrasts. In multi-contrast SR, different contrasts can be considered as different classes, and features from different stages can be considered as responses to specific classes. Therefore, we design a multi-stage integration module. As shown in Fig. 2, by obtaining the dependencies between features of the T1WI in each stage with other stages as well as the T2WI in each stage, the network can allocate different attention weights to the features of different contrasts and stages. Thus, while the network pays attention to the more informative stages in the T2WI, it also emphasizes the stages in the T1WI that can guide the SR of the T2WI.

Recall that $\mathbf{F} \in \mathbb{R}^{2L \times H \times W \times C}$ (Eq. (3)) concatenates all intermediate representations together. The multi-stage integration module aims to learn the correlations between each pair of features. In particular, $\mathbf{F}$ is first flattened into a matrix representation $\hat{\mathbf{F}} \in \mathbb{R}^{2L \times HWC}$ for computational convenience. Then, we find the correspondence between each pair of features $\hat{\mathbf{F}}_i \in \mathbb{R}^{HWC}$ and $\hat{\mathbf{F}}_j \in \mathbb{R}^{HWC}$ using the following bilinear model:

$$\begin{aligned} \mathbf{S} &= \mathtt{softmax}(\hat{\mathbf{F}}\hat{\mathbf{F}}^\top) \\ &= \mathtt{softmax}([\hat{\mathbf{F}}_1, \cdots, \hat{\mathbf{F}}_{2L}][\hat{\mathbf{F}}_1, \cdots, \hat{\mathbf{F}}_{2L}]^\top) \in [0,1]^{2L \times 2L}. \end{aligned} \tag{7}$$

Here, $\mathbf{S}$ denotes the affinity matrix which stores similarity scores corresponding to each pair of features in $\hat{\mathbf{F}}$, $i.e.$, the $(i,j)^{th}$ element of $\hat{\mathbf{F}}$ gives the similarity between $\hat{\mathbf{F}}_i$ and $\hat{\mathbf{F}}_j$. $\mathtt{softmax}(\cdot)$ normalizes each row of the input. Next, attention summaries are computed as $\mathbf{S}\hat{\mathbf{F}} \in \mathbb{R}^{2L \times HWC}$ and used to generate an enhanced representation in the following residual form:

$$\mathbf{H} = \mathtt{reshape}(\mathbf{RF} + \mathbf{F}) \in \mathbb{R}^{H \times W \times 2LC}, \tag{8}$$

where $\mathbf{RF}$ is reshaped by $\mathbf{S}\hat{\mathbf{F}}$. In this way, our model is able to learn a more comprehensive and holistic feature representation $\mathbf{H}$ by exploring the relations between features of multi-contrast images at multiple stages.

3   Experiments

Datasets and Baselines. We evaluate our approach on three datasets: 1) **fastMRI** [23] is the largest open-access MRI dataset. Following [22], we filter out 227 and 24 pairs of PD and FS-PDWI volumes for training and validation. 2) **SMS** was constructed using a 3T Siemens Magnetom Skyra system on 155 patients. Each MRI scan includes a T1WI and a T2WI with full k-space sampling (TR$_{T1}$ = 2001 ms, TE$_{T1}$ = 10.72 ms, slice thickness = 5 mm, matrix = $320 \times 320 \times 20$, field of view (FOV) = 230×200 mm^2, TR$_{T2}$ = 4511 ms, TE$_{T2}$ − 112.86 ms). 3) **uMR** was constructed using a 3T whole body scanner (United Imaging Healthcare uMR 790[1].) on 50 patients. Each MRI scan includes a T1WI and a T2WI with full k-space sampling (TR$_{T1}$ = 2001 ms, TE$_{T1}$ = 10.72 ms, slice thickness = 4 mm, matrix = $320 \times 320 \times 19$, FOV = 220×250 mm^2, TR$_{T1}$ = 4511 ms, TE$_{T1}$ = 112.86 ms). The collection of the clinical datasets was approved by the Institutional Review Board. We split **SMS** and **uMR** patient-wise into training/validation/test sets with a ratio of 7:1:2. For **fastMRI**, we ues the PDWI to guide the SR of FS-PDWI, while for all other datasets we use the T1WI to guide the T2WI. We compare our method with two single-contrast SR methods (EDSR [12], SMORE [26]) and two multi-contrast methods (Zeng et al. [13], Lyu et al. [24]).

Experimental Setup. We implement our model in PyTorch with two NVIDIA Tesla V100 GPUs and 32 GB of memory per card. Our model is trained using

---

[1] Provided by United Imaging Healthcare, Shanghai, China United Imaging Healthcare, Shanghai, China

the Adam optimizer with a learning rate of 1e–5, for 50 epochs. The parameters α and β are empirically set to 0.3 and 0.7, respectively. We use $L = 6$ residual groups in out network. All the compared methods are retrained using their default parameter settings.

Quantitative Results. Table 1 reports the average SSIM and PSNR scores with respect to different datasets under 2× and 4× enlargement. As can be seen, our approach yields the best results on all datasets. This demonstrates that our model can effectively fuse the two contrasts, which is beneficial to the restoration of the target contrast. Notably, the single-contrast SR methods, *e.g.*,EDSR [12] and SMORE [26], are far less effective than the multi-contrast models. More importantly, however, the multi-contrast SR models, *i.e.*, Zeng *et al.* [13] and Lyu *et al.* [24], are also less effective than our model, since they do not mine fused features of different contrast or the interaction between different modes at each stage. In particular, when the scaling factor is 2× on **fastMRI**, we improve the PSNR from 29.484 dB to 31.769 dB, and SSIM from 0.682 to 0.709, as compared to the current best approach, *i.e.*, Lyu *et al.* [24]. Although it is more difficult to restore images under 4× enlargement than 2×, our model still outperforms previous methods in extremely challenging settings, which can be attributed to its strong capability in multi-contrast image restoration.

Table 1. Quantitative results on three datasets with different enlargement scales, in terms of SSIM and PSNR. The best and second-best results are marked in red and blue, respectively.

Dataset	fastMRI [23]				SMS				uMR			
Scale	2×		4×		2×		4×		2×		4×	
Metrics	PSNR	SSIM	PSNR	SSIM	PSNR	SSIM	PSNR	SSIM	PSNR	SSIM	PSNR	SSIM
Bicubic	16.571	0.459	13.082	0.105	21.548	0.780	19.508	0.706	21.107	0.730	19.072	0.720
EDSR [12]	26.669	0.512	18.363	0.208	36.415	0.962	31.484	0.886	35.394	0.965	31.165	0.907
SMORE [26]	28.278	0.667	21.813	0.476	38.106	0.972	32.091	0.901	36.547	0.972	31.971	0.918
Zeng *et al.* [13]	28.870	0.670	23.255	0.507	38.164	0.973	32.484	0.912	36.435	0.971	31.859	0.921
Lyu *et al.* [24]	29.484	0.682	28.219	0.574	39.194	0.978	33.667	0.931	37.139	0.977	32.231	0.929
MINet	31.769	0.709	29.819	0.601	40.549	0.983	35.032	0.948	37.997	0.980	34.219	0.956

Qualitative Evaluation. Figure 3 provides the 2× and 4× enlargement of target-contrast images and their corresponding error maps on the **fastMRI** and **SMS** datasets, respectively. The more obvious the texture in the error map, the worse the restoration. As can be seen, the multi-contrast methods outperform the single-contrast methods. However, our restoration exhibits less of a chessboard effect and fewer structural losses compared to other methods, which is owed to the fact that our model can effectively learn aggregated features from multiple stages. More importantly, the error maps for different enlargement scales demonstrate the robustness of our method across various datasets.

Table 2. Ablation study on the SMS dataset with 2× enlargement. The best and second-best results are marked in red and blue, respectively

Variant	NMSE↓	PSNR↑	SSIM↑
$w/o\ \mathcal{F}_{\text{Aux}}$	0.0037 ± 0.012	38.5731 ± 0.015	0.9700 ± 0.009
$w/o\ \mathcal{F}_{\text{Int}}$	0.0025 ± 0.002	39.2001 ± 0.108	0.9776 ± 0.025
$w/o\ \mathcal{F}_{\text{Att}}$	0.0022 ± 0.236	39.2654 ± 0.001	0.9791 ± 0.037
MINet	0.0018 ± 0.004	40.5491 ± 0.014	0.9830 ± 0.009

Ablation Study. In this section, we evaluate the effectiveness of the key components of our model through an ablation study. We construct three models: $w/o\ \mathcal{F}_{\text{Aux}}$, which is our model without auxiliary contrast, $w/o\ \mathcal{F}_{\text{Int}}$, which is our model without $\mathcal{F}_{\text{Int}}$, and $w/o\ \mathcal{F}_{\text{Att}}$, which is our model without $\mathcal{F}_{\text{Att}}$. We summarize the 2× enlargement results on **SMS** in Table 2. From this table, we observe that single-contrast $\mathcal{F}_{\text{Aux}}$ performs the worst, which is consistent with our conclusion that auxiliary contrast can provide supplementary information for the SR of the target contrast. Since $\mathcal{F}_{\text{Int}}$ reflects the interaction between two contrasts at different stages, the results of $w/o\ \mathcal{F}_{\text{Int}}$ are not optimal. Further, $w/o\ \mathcal{F}_{\text{Att}}$ outperforms $w/o\ \mathcal{F}_{\text{Int}}$ because $\mathcal{F}_{\text{Att}}$ only enhances the features of a single-contrast, and cannot learn the interaction between two different contrasts. Finally, our full MINet, which enhances both the single-contrast feature from $\mathcal{F}_{\text{Att}}$ and the fused feature from $\mathcal{F}_{\text{Int}}$, yields the best results, demonstrating its powerful capability in mining crucial information to guide the target-contrast restoration.

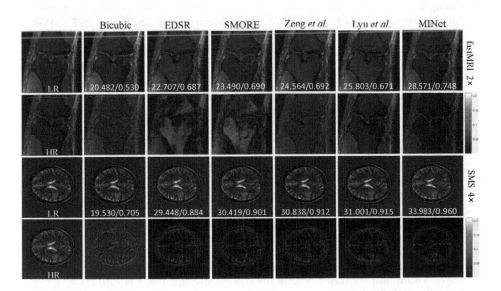

Fig. 3. Visual reconstruction results and error maps of different methods.

4   Conclusion

We have proposed a multi-stage integration network for multi-contrast MRI super-resolution, with the aim of effectively restoring the target-contrast image with guidance from a complementary contrast. Specifically, our model explores modality-specific properties within each modality, as well as the interaction between multi-contrast images. $\mathcal{F}_{Int}$ and $\mathcal{F}_{Att}$ are collaboratively applied to multi-level and multi-contrast features, helping to capture more informative features for the target-contrast image restoration. This work provides a potential guide for further research into the interaction between multi-contrast images for MRI super-resolution.

References

1. Bhatia, K.K., Price, A.N., Shi, W., Hajnal, J.V., Rueckert, D.: Super-resolution reconstruction of cardiac MRI using coupled dictionary learning. In: 2014 IEEE 11th International Symposium on Biomedical Imaging (ISBI), pp. 947–950. IEEE (2014)
2. Chaudhari, A.S., et al.: Super-resolution musculoskeletal MRI using deep learning. Magn. Reson. Med. **80**(5), 2139–2154 (2018)
3. Chen, W., et al.: Accuracy of 3-T MRI using susceptibility-weighted imaging to detect meniscal tears of the knee. Knee Surg. Sports Traumatol. Arthrosc. **23**(1), 198–204 (2015)
4. Chen, Y., Xie, Y., Zhou, Z., Shi, F., Christodoulou, A.G., Li, D.: Brain MRI super resolution using 3D deep densely connected neural networks. In: 2018 IEEE 15th International Symposium on Biomedical Imaging (ISBI 2018), pp. 739–742. IEEE (2018)
5. Dong, C., Loy, C.C., He, K., Tang, X.: Image super-resolution using deep convolutional networks. IEEE Trans. Patt. Anal. Mach. Intell. **38**(2), 295–307 (2015)
6. Feng, C.M., Wang, K., Lu, S., Xu, Y., Li, X.: Brain mri super-resolution using coupled-projection residual network. Neurocomputing (2021)
7. Feng, C.M., Yan, Y., Chen, G., Fu, H., Xu, Y., Shao, L.: Accelerated multi-modal MRI imaging with transformers (2021)
8. Feng, C.M., Yan, Y., Fu, H., Chen, L., Xu, Y.: Task transformer network for joint MRI reconstruction and super-resolution. In: International Conference on Medical Image Computing and Computer Assisted Intervention (MICCAI) (2021)
9. Feng, C.M., Yang, Z., Chen, G., Xu, Y., Shao, L.: Dual-octave convolution for accelerated parallel MR image reconstruction. In: Proceedings of the 35th AAAI Conference on Artificial Intelligence (AAAI) (2021)
10. Feng, C.M., Yang, Z., Fu, H., Xu, Y., Yang, J., Shao, L.: Donet: dual-octave network for fast MR image reconstruction. IEEE Trans. Neural Netw. Learn. Syst. (2021)
11. Hardie, R.: A fast image super-resolution algorithm using an adaptive wiener filter. IEEE Trans. Image Process. **16**(12), 2953–2964 (2007)
12. Lim, B., Son, S., Kim, H., Nah, S., Mu Lee, K.: Enhanced deep residual networks for single image super-resolution. In: Proceedings of the IEEE Conference on Computer Vision and Pattern Recognition Workshops, pp. 136–144 (2017)
13. Lyu, Q., et al.: Multi-contrast super-resolution MRI through a progressive network. IEEE Trans. Med. Imag. (2020)

14. Lyu, Q., You, C., Shan, H., Wang, G.: Super-resolution MRI through deep learning. arXiv preprint arXiv:1810.06776 (2018)
15. McDonagh, S., et al.: Context-sensitive super-resolution for fast fetal magnetic resonance imaging. In: Molecular Imaging, Reconstruction and Analysis of Moving Body Organs, and Stroke Imaging and Treatment, pp. 116–126. Springer (2017). https://doi.org/10.1007/978-3-319-67564-0
16. Oktay, O., et al.: Multi-input cardiac image super-resolution using convolutional neural networks. In: Ourselin, S., Joskowicz, L., Sabuncu, M.R., Unal, G., Wells, W. (eds.) MICCAI 2016. LNCS, vol. 9902, pp. 246–254. Springer, Cham (2016). https://doi.org/10.1007/978-3-319-46726-9_29
17. Pham, C.H., Ducournau, A., Fablet, R., Rousseau, F.: Brain MRI super-resolution using deep 3d convolutional networks. In: 2017 IEEE 14th International Symposium on Biomedical Imaging (ISBI 2017), pp. 197–200. IEEE (2017)
18. Shi, F., Cheng, J., Wang, L., Yap, P.T., Shen, D.: Lrtv: MR image super-resolution with low-rank and total variation regularizations. IEEE Trans. Med. Imag. **34**(12), 2459–2466 (2015)
19. Tourbier, S., Bresson, X., Hagmann, P., Thiran, J.P., Meuli, R., Cuadra, M.B.: An efficient total variation algorithm for super-resolution in fetal brain MRI with adaptive regularization. NeuroImage **118**, 584–597 (2015)
20. Woo, S., Park, J., Lee, J.-Y., Kweon, I.S.: CBAM: convolutional block attention module. In: Ferrari, V., Hebert, M., Sminchisescu, C., Weiss, Y. (eds.) ECCV 2018. LNCS, vol. 11211, pp. 3–19. Springer, Cham (2018). https://doi.org/10.1007/978-3-030-01234-2_1
21. Xiang, L., et al..: Deep-learning-based multi-modal fusion for fast MR reconstruction. IEEE Trans. Biomed. Eng. **66**(7), 2105–2114 (2018)
22. Xuan, K., Sun, S., Xue, Z., Wang, Q., Liao, S.: Learning MRI k-space subsampling pattern using progressive weight pruning. In: Martel, A.L., et al. (eds.) MICCAI 2020. LNCS, vol. 12262, pp. 178–187. Springer, Cham (2020). https://doi.org/10.1007/978-3-030-59713-9_18
23. Zbontar, J., Knoll, F., Sriram, A., Muckley, M.J., Bruno, M., Defazio, A., Parente, M., Geras, K.J., Katsnelson, J., Chandarana, H., et al.: fastmri: An open dataset and benchmarks for accelerated mri. arXiv preprint arXiv:1811.08839 (2018)
24. Zeng, K., Zheng, H., Cai, C., Yang, Y., Zhang, K., Chen, Z.: Simultaneous single- and multi-contrast super-resolution for brain MRI images based on a convolutional neural network. Comput. Biol. Med. **99**, 133–141 (2018)
25. Zhang, Y., Li, K., Li, K., Wang, L., Zhong, B., Fu, Y.: Image super-resolution using very deep residual channel attention networks. In: Proceedings of the European Conference on Computer Vision (ECCV), pp. 286–301 (2018)
26. Zhao, C., et al.: A deep learning based anti-aliasing self super-resolution algorithm for MRI. In: Frangi, A.F., Schnabel, J.A., Davatzikos, C., Alberola-López, C., Fichtinger, G. (eds.) MICCAI 2018. LNCS, vol. 11070, pp. 100–108. Springer, Cham (2018). https://doi.org/10.1007/978-3-030-00928-1_12
27. Zheng, H., et al.: Multi-contrast brain magnetic resonance image super-resolution using the local weight similarity. BMC Med. Imag. **17**(1), 6 (2017)
28. Zheng, H., et al.: Multi-contrast brain MRI image super-resolution with gradient-guided edge enhancement. IEEE Access **6**, 57856–57867 (2018)

Generator Versus Segmentor: Pseudo-healthy Synthesis

Yunlong Zhang[1], Chenxin Li[1], Xin Lin[1], Liyan Sun[1], Yihong Zhuang[1], Yue Huang[1(✉)], Xinghao Ding[1], Xiaoqing Liu[2], and Yizhou Yu[2]

[1] School of Informatics, Xiamen University, Xiamen, China
yhuang2010@xmu.edu.cn
[2] Deepwise AI Lab, Beijing, China

Abstract. This paper investigates the problem of pseudo-healthy synthesis that is defined as synthesizing a subject-specific pathology-free image from a pathological one. Recent approaches based on Generative Adversarial Network (GAN) have been developed for this task. However, these methods will inevitably fall into the trade-off between preserving the subject-specific identity and generating healthy-like appearances. To overcome this challenge, we propose a novel adversarial training regime, Generator versus Segmentor (GVS), to alleviate this trade-off by a divide-and-conquer strategy. We further consider the deteriorating generalization performance of the segmentor throughout the training and develop a pixel-wise weighted loss by muting the well-transformed pixels to promote it. Moreover, we propose a new metric to measure how healthy the synthetic images look. The qualitative and quantitative experiments on the public dataset BraTS demonstrate that the proposed method outperforms the existing methods. Besides, we also certify the effectiveness of our method on datasets LiTS. Our implementation and pre-trained networks are publicly available at https://github.com/Au3C2/Generator-Versus-Segmentor.

Keywords: Pseudo-healthy synthesis · Adversarial training · Medical images segmentation

1 Introduction

Pseudo-healthy synthesis is defined as synthesizing a subject-specific pathology-free image from a pathological one [15]. Generating such images has been proven to be valuable for a variety of medical image analysis tasks [15], such as segmentation [1,5,6,12,16], detection [13], and providing additional diagnostic information for pathological analysis [3,12]. By definition, *a perfect pseudo-healthy image should maintain both healthiness (i.e., the pathological regions are indistinguishable from healthy ones in synthetic images) and subject identity (i.e., belonging*

Y. Zhang and C. Li—Equal contribution.

© Springer Nature Switzerland AG 2021
M. de Bruijne et al. (Eds.): MICCAI 2021, LNCS 12906, pp. 150–160, 2021.
https://doi.org/10.1007/978-3-030-87231-1_15

to the same subject as the input). Note that both of them are essential and indispensable. The importance of the former is self-explanatory, and the latter is also considerable since generating another healthy counterpart is meaningless.

In this paper, we focus on promoting the pseudo-healthy synthesis from both above-mentioned aspects: healthiness and subject identity. *The existing GAN-based methods attained promising results but still existed the trade-off between changing the entire appearance towards a healthy counterpart and keeping visual similarity.* Thus, we utilize a divide-and-conquer strategy to alleviate this trade-off. Concretely, we divide an image into healthy/pathological regions and apply the individual constraint for each of them. The first constraint is keeping visual consistency for healthy pixels before and after synthesis, and the second one is mapping pathological pixels into pixel-level healthy distribution (i.e., the distribution of healthy pixels). Furthermore, to measure the distributional shift between healthy and pathological pixels, a segmentor is introduced into the adversarial training.

Contributions. Our contributions are summarized as three-fold. (1) We introduce a segmentor as the 'discriminator' by originality. The zero-sum game between it and the generator contributes to synthesize better pseudo-healthy images by alleviating the above-mentioned trade-off. (2) We further consider the persistent degradation of generalization performance of the segmentor. To alleviate this issue, we propose a pixel-wise weighted loss by muting the well-transformed pixels. (3) The only gold standard to measure the healthiness of synthetic images is the subjective assessment. However, it is time-consuming and costly, while being subject to inter- and intra-observer variability. Hence, it deviates from reproducibility. Inspired by the study of label noise, we propose a new metric to measure the healthiness.

Related Work. According to various clinical scenarios, a series of methods for pseudo-healthy synthesis were proposed and mainly included the *pathology-deficiency* (i.e., lacking pathological images in the training phase) [7,10,11] and *pathology-sufficiency* based methods (i.e., having plenty of pathological images in the training phase) [3,12,15]. The pathology-deficiency based methods aimed to learn the normative distribution from plenty of healthy images. Concretely, they adopt the VAE [10], AAE [7], and GAN [11] to reconstruct healthy images. In the testing phase, the out-of-distribution regions (i.e., lesions) cannot be reconstructed and were transformed into healthy-like ones. In contrast, the pathology-sufficiency based methods introduced pathological images along with *image-level* [3] or *pixel-level* [12,15] labeling. The existing methods aimed to translate pathological images into normal ones by the GAN. Besides the adversarial loss that aligns the distributions between pathological and synthetic images, the VA-GAN [3] introduced $\mathcal{L}_1$ loss to assure the visual consistency. In the process of applying Cycle-GAN [18] to pseudo-healthy synthesis, the ANT-GAN proposed two improvements, which are the shortcut to simplify the optimization and the masked L2 loss to better preserve the normal regions. The PHS-GAN considered the one-to-many problem when applying the Cycle-GAN into pseudo-healthy synthesis. It disentangled the pathology information from what seems to be a

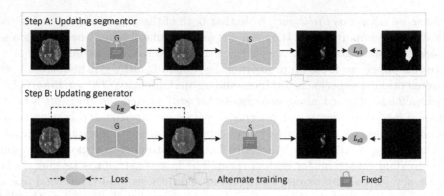

Fig. 1. Training workflow. The model is optimized by iteratively alternating Step A and Step B. In Step A, we fix the generator **G** and update the segmentor **S** with L_{s1}. In Step B, we fix the segmentor **S** and update the generator **G** with $L_{s2} + \lambda L_R$.

healthy image part and then combined the disentangled information with the pseudo-healthy images to reconstruct the pathological images.

2   Methods

In this section, the proposed GVS method for pathology-sufficiency pseudo-healthy synthesis with pixel-level labeling is introduced. Assume a set of pathological image x_p and corresponding pixel-level lesion annotations y_t are given.

2.1   Basic GVS Flowchart

The training workflow of the proposed GVS is shown in Fig. 1. The generator gradually synthesizes healthy-like images by iteratively alternating Step A and Step B. The specific steps are as follows.

Step A. As shown in Fig. 1, we fix the generator **G** and update the segmentor **S** to segment the lesions. The lesion annotation y_t is adopted, and the loss is:

$$\mathcal{L}_{s1} = \mathcal{L}_{ce}(\mathbf{S}(\mathbf{G}(x_p)), y_t), \tag{1}$$

where L_{ce} denotes the cross-entropy loss.

Step B. In this step, we fix the segmentor **S** and update the generator **G**, aiming to remove the lesions and preserve the identity of pathological images. Specifically, on the one hand, it is expected that the generator **G** can synthesize healthy-like appearances that do not contain lesions. Therefore, an adversarial loss is used:

$$\mathcal{L}_{s2} = \mathcal{L}_{ce}(\mathbf{S}(\mathbf{G}(x_p))), y_h), \tag{2}$$

where y_h denotes the zero matrix with the same size as y_t. To deceive the segmentor, the generator further compensates the distributional difference between pathological and healthy regions. On the other hand, the synthetic images should be visually consistent with the pathological ones [3,12,15]. Therefore, the generator $\mathbf{G}$ is trained with a residual loss:

$$\mathcal{L}_R = \mathcal{L}_{mse}(x_p, \mathbf{G}(x_p)), \tag{3}$$

where $\mathcal{L}_{mse}$ denotes pixel-wise $\mathcal{L}_2$ loss. The total training loss of $\mathbf{G}$ is:

$$\mathcal{L}_G = \mathcal{L}_{s2} + \lambda\mathcal{L}_R, \tag{4}$$

where λ denotes a hyperparameter that represents the trade-offs between the healthiness and identity, and it is subject to $\lambda > 0$.

2.2 Improved Residual Loss

This section further considers whether it is reasonable to set the same visual similarity in normal and pathological regions. Apparently, it is reasonable to keep visual consistency in normal regions between pathological and synthetic images. However, it is contradictory to remove lesions and keep pixel values within pathological regions. In order to alleviate this contradiction, we assume that potential normal tissues for lesion regions and normal tissues in the same pathological image have similar pixel values. Based on this, we improve the residual loss as follows:

$$\mathcal{L}_{R+} = \mathcal{L}_{mse}((1-y_t)\odot x_p, (1-y_t)\odot \mathbf{G}(x_p)) + \lambda_1\mathcal{L}_{mse}(y_t\odot \overline{x}_{pn}, y_t\odot \mathbf{G}(x_p)), \tag{5}$$

where $\odot$ represents the pixel-wise multiplication, and $\overline{x}_{pn}$ denotes a matrix filled with the average value of normal tissue in the same image x_p and has the same size as x_p; λ_1 denotes a hyperparameter that controls the power of visual consistency in the lesion regions, and $0 < \lambda_1 < 1$ since the potential normal tissues are close but not equal to the average value of normal tissues.

2.3 Training a Segmentor with Strong Generalization Ability

The generalization ability of segmentor is further considered. During the training, the pathological regions are gradually transformed into healthy-like ones. As shown in Fig. 2(b), the major part of the lesion region has been well transformed, so these pixels should be labeled as 'healthy'. However, the basic GVS still views all the pixels in this region as lesions. In other words, the well-transformed parts are forced to fit the false label, which weakens the generalization capacity of neural networks [17]. As shown in Fig. 2(b), we observe that predictions of segmentor substantially deviate from labels, which suggests the poor generalization of the segmentor. To meet this challenge, we present a novel pixel-level weighted cross-entropy loss for lesion segmentation. The difference map between pathological

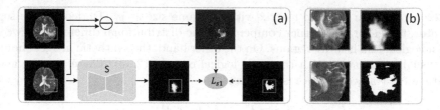

Fig. 2. (a) Framework of the pixel-level weighted cross-entropy loss. (b) The blue, yellow, green, and red boxes denote zoom-in views corresponding to the pathological image, synthetic image, segmentation prediction, and lesion annotation, respectively. (Color figure online)

and synthetic images can be used as an indicator to measure the transformation degree:

$$\mathcal{L}_{wce} = \frac{1}{N} \sum_{i=1}^{N} w(i) y_t(i) log(\mathbf{S}(\mathbf{G}(x_p))(i)), \tag{6}$$

where N denotes the number of pixels. The weight w associated with difference maps are defined as:

$$w = \begin{cases} 0.1, & 1 - m < 0.1, \\ 1 - m, & \text{Otherwise}, \end{cases} \tag{7}$$

where $m = \text{Normalization}(x_p - \mathbf{G}(x_p))$ denotes the normalized difference map. In this work, $w[w < 0.1] = 0.1$ because the minimum value does not represent perfect-transformation, and it is necessary to keep a subtle penalty.

The complete GVS is developed by replacing $\mathcal{L}_R$ with $\mathcal{L}_{R+}$, and then replacing $\mathcal{L}_{s1}$ with $\mathcal{L}_{wce}$. Please note that the GVS refers to the complete GVS in the next subsections.

3 Experiments

3.1 Implementation Details

Data. Our method is mainly evaluated on BraTS2019 dataset [2,8], which contains 259 GBM (i.e., glioblastoma) and 76 LGG (i.e., lower-grade glioma) volumes. Here, we utilize the T2-weighted volumes of GBM, and they are split into training (234 volumes) and test sets (25 volumes). To verify the potential to different modalities and organs, we also present some visual results on the LiTS [4], which contains 131 CT scans of the liver. The slice resolution is 512×512. The datasets are divided into training (118 scans) and test sets (13 scans). The intensity of images is rescaled to the range of $[0, 1]$.

Network. Both the generator and the segmentor adopt the 2D U-Net architectures [9], which consist of an encoder-decoder architecture with symmetric skip connections. They both downsample and upsample four times and adapt the bilinear upsampling method. Furthermore, unlike the generator, the segmentor contains the instance normalization and softmax layer.

Training Details. The proposed method is implemented on Pytorch and an NVIDIA TITAN XP GPU. We use Adam optimizer, with an initial learning rate of 0.001 and decrease it by 0.1 after $0.8 * total\_epoch$. The $total\_epoch$ is set to 20. The batch size is set 8 for the BraTS and 4 for the LiTS. Lastly, λ and λ_1 are set to 1.0 and 0.1.

3.2 Evaluation Metrics

In this section, two metrics, $\mathbb{S}_{dice}$ and iD, are introduced to evaluate the healthiness and identity, respectively.

Zhang et al. [17] revealed an interesting phenomenon that convergence time on the false/noisy labels increases by a constant factor compared with that on the true labels. Similar to this, aligning well-transformed pixels (i.e., these pixels can be viewed as healthy ones) and lesion annotations is counterfactual and hampers the convergence. Thus, how healthy the synthetic images look is negatively related to the convergence time. Inspired by this, we present a new metric to assess the healthiness, which is defined as the accumulated dice score throughout the training process: $\mathbb{S}_{dice} = \sum_{e=1}^{epochs} dice_e$, where $epochs$ denotes the number of training epochs, and $dice_e$ denotes the dice evaluated on the training data at the e-th epoch.

Here, whether the $\mathbb{S}_{dice}$ can correctly assess the healthiness of synthetic images is testified. To this end, we calculate the $\mathbb{S}_{dice}$ of GVS(0.1), GVS(0.4), GVS(0.7), GVS(1.0), and original images. Note that the GVS(a) denotes the synthetic images generated by the GVS trained on $a\%$ training data. Generally, the GVS(a) should have more healthy appearances than the GVS(b) when $a > b$. The results show that the $\mathbb{S}_{dice}$ can correctly reflect the healthiness in order. In addition, we also discover that the $\mathbb{S}_{dice}$ is unstable in the small value area, which may result in false results when synthetic images have similar healthy appearances. One possible reason for this situation is that the random parameter initialization may lead to different results at each trial (Fig. 3).

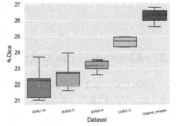

Fig. 3. The $\mathbb{S}_{dice}$ evaluated on different synthetic images. The GVS(1.0) represents the synthetic images generated by the GVS that is trained on 100% of training data.

Following Xia et. [15], the identity is expressed as the structural similarity calculated on normal pixels, which is defined as $iD = $ MS-SSIM$[(1 - y_t) \odot \mathbf{G}(x_p), (1 - y_t) \odot x_p]$, where the $\odot$ denotes the pixel-wise multiplication, and the MS-SSIM() representes a masked Multi-Scale Structural Similarity Index [14].

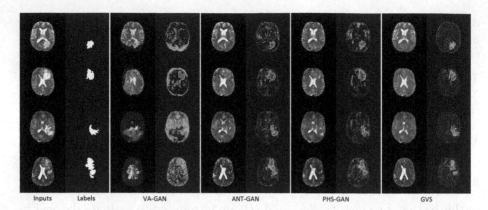

Fig. 4. The qualitative results of synthetic images. The first and second columns show inputs and corresponding labels, respectively. The next eight columns show the synthetic images and difference maps of the VA-GAN, ANT-GAN, PHS-GAN, and GVS.

Table 1. Evaluation results of the VA-GAN, PHS-GAN, ANT-GAN, GVS and its variants, as well as the baseline. The performance of the baseline was evaluated on original images. The average values and standard deviations of the evaluation results were calculated over three runs. The best mean value for each metric is shown in bold.

	VA-GAN	PHS-GAN	ANT-GAN	GVS	w/o $\mathcal{L}_{R+}$	w/o $\mathcal{L}_{wce}$	Baseline
$iD \uparrow$	$0.74_{0.03}$	$0.97_{0.03}$	$0.96_{0.02}$	$\mathbf{0.99}_{0.01}$	$\mathbf{0.99}_{0.01}$	$\mathbf{0.99}_{0.01}$	$1.00_{0.00}$
$\mathbb{S}_{dice} \downarrow$	–	$23.32_{0.45}$	$23.11_{0.41}$	$\mathbf{21.75}_{0.13}$	$22.11_{0.14}$	$23.66_{0.15}$	$26.23_{0.17}$

3.3 Comparisons with Other Methods

We compare our GVS with existing pathology-deficiency based methods, including the VA-GAN [3], ANT-GAN [12], and PHS-GAN [15]. The VA-GAN[1] and PHS-GAN[2] are implemented using their official codes, and the ANT-GAN is implemented on the code provided by the authors. Note that the VA-GAN uses *image-level* labels, and the ANT-GAN and PHS-GAN utilize *pixel-level* labels. Our experimental setting keeps *consistent* with the last two. Next, we analyze the performance of all methods qualitatively and quantitatively.

Qualitative results are shown in Fig. 4. We first analyze the identity by comparing the reconstruction performance in the normal regions. The proposed method achieves higher reconstruction quality than the other methods (better see the difference maps). For example, compared with the existing methods, our method better preserves the high-frequency details (i.e., edges). Overall, the VA-GAN could not keep the subject identity and loses a part of lesion regions in some cases. The PHS-GAN and ANT-GAN preserve the major normal regions but lose some details. Among all the methods, the proposed method achieves

[1] https://github.com/baumgach/vagan-code.
[2] https://github.com/xiat0616/pseudo-healthy-synthesis.

the best subject identity. Then, we further analyze the healthiness, which can be judged by comparing whether pathological and normal regions are harmonious. The synthetic images generated by VA-GAN are visually unhealthy due to the poor reconstruction. The PHS-GAN and ANT-GAN remove most of the lesions, but some artifacts still remain. The GVS achieves the best healthiness due to indistinguishable pathological and normal regions.

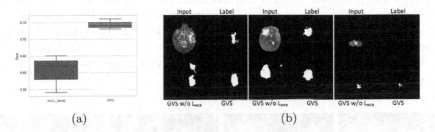

(a)   (b)

Fig. 5. (a) Generalization ability of the segmentors in the GVS and GVS w/o $\mathcal{L}_{wce}$. (b) Visual comparison of synthetic images of the GVS and GVS w/o $\mathcal{L}_{wce}$. Three blocks show three examples. Each block contain the input, label, predictions of the GVS without $\mathcal{L}_{R+}$ and GVS.

We report quantitative results in Table 1. The first metric, iD, is used to assess the identity. Our GVS achieves a iD value of 0.99, outperforming the second place, PHS-GAN, by 0.02, which is an evident improvement when the SSIM is large. The proposed $\mathbb{S}_{dice}$ is used to assess healthiness. Since the VA-GAN can not reconstruct the normal tissue, as shown in the third column in Fig. 4), its $\mathbb{S}_{dice}$ value is meaningless and not considered. In addition, compared with baseline (i.e., original images), the $\mathbb{S}_{dice}$ of the ANT-GAN and PHS-GAN decline from 26.23 to 23.32 and 23.11, respectively. The proposed method further improves the $\mathbb{S}_{dice}$ to 21.75.

3.4   Ablation Study

To verify the claim that the $\mathcal{L}_{wce}$ can alleviate the poor generalization of the segmentor, we respectively calculate the segmentation performance of the segmentors trained by the GVS and GVS w/o $\mathcal{L}_{wce}$. As shown in Fig. 5(a), the GVS achieves a higher average dice score and lower variance compared to the GVS w/o $\mathcal{L}_{wce}$, which confirms that the $\mathcal{L}_{wce}$ could effectively improve the generalization ability of the segmentor. Similar conclusions are also derived from the visual examples in Fig. 5(b). Predictions of segmentor trained by GVS w/o $\mathcal{L}_{wce}$ deviate from the labels severely. After adding the $\mathcal{L}_{wce}$, the predictions are more accurate. Note that only relying on the $\mathcal{L}_{wce}$ cannot solve the generalization problem of the segmentor entirely (see the third example in Fig. 5(b)). Hence, this problem needs further exploration in the future. Furthermore, benefiting from the better generalization of the segmentor, the healthiness attains further improvement ($23.66 \rightarrow 21.75$ in Table 1). We also conduct the ablation study on the improved residual loss, and results are shown in Table 1. We observe that the $\mathbb{S}_{dice}$ increases from 21.75 to 22.11 after replacing $\mathcal{L}_R$ with $\mathcal{L}_{R+}$.

3.5   Results on LiTS Dataset

The proposed GVS is also evaluated on a public CT dataset, LiTS. The results show that the proposed method can maintain the identity and transform the high-contrast lesions well, as shown in the top row in Fig. 6, However, the synthetic results of low-contrast images still exist subtle artifacts in the pathological regions, as shown in the bottom row in Fig. 6. We conjecture that the reason may be that the segmentor cannot accurately detect low-contrast lesions, which further results in poor transformation.

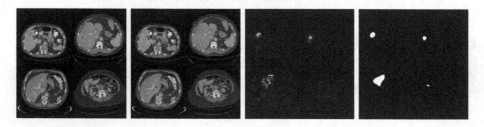

Fig. 6. The visual results on the LiTS dataset. The blocks from left to right represent the pathological images, synthetic images, difference maps, and lesion annotations.

4   Conclusions

This paper proposes an adversarial training framework by iteratively training the generator and segmentor to synthesize pseudo-healthy images. Also, taking the rationality of residual loss and the generalization ability of the segmentor into account, the improved residual loss and pixel-wise weighted cross-entropy loss are introduced. In experiments, the effectiveness of the proposed scheme is verified on two public datasets, BraTS and LiTS.

Limitations and Future Work. One limitation of the proposed GVS is requiring densely labeled annotations. In clinical application, huge amounts of accurate segmentation labels are hardly available. Hence, it is necessary to relax the demand for accurate pixel-level annotations in the next step. Another concern is the instability of the proposed $\mathbb{S}_{dice}$. In the future, we plan to improve this by further exploiting the more characteristics of noise labels.

Acknowledgments. This work was supported in part by National Key Research and Development Program of China (No. 2019YFC0118101), in part by National Natural Science Foundation of China under Grants U19B2031, 61971369, in part by Fundamental Research Funds for the Central Universities 20720200003, in part by the Science and Technology Key Project of Fujian Province, China (No. 2019HZ020009).

References

1. Andermatt, S., Horváth, A., Pezold, S., Cattin, P.: Pathology segmentation using distributional differences to images of healthy origin. In: Crimi, A., Bakas, S., Kuijf, H., Keyvan, F., Reyes, M., van Walsum, T. (eds.) BrainLes 2018. LNCS, vol. 11383, pp. 228–238. Springer, Cham (2019). https://doi.org/10.1007/978-3-030-11723-8_23
2. Bakas, S., et al.: Advancing the cancer genome atlas glioma MRI collections with expert segmentation labels and radiomic features. Sci. Data **4**, 170117 (2017)
3. Baumgartner, C.F., Koch, L.M., Can Tezcan, K., Xi Ang, J., Konukoglu, E.: Visual feature attribution using Wasserstein GANs. In: Proceedings of the IEEE Conference on Computer Vision and Pattern Recognition, pp. 8309–8319 (2018)
4. Bilic, P., et al.: The liver tumor segmentation benchmark (LiTs). arXiv preprint arXiv:1901.04056 (2019)
5. Bowles, C., et al.: Pseudo-healthy image synthesis for white matter lesion segmentation. In: Tsaftaris, S.A., Gooya, A., Frangi, A.F., Prince, J.L. (eds.) SASHIMI 2016. LNCS, vol. 9968, pp. 87–96. Springer, Cham (2016). https://doi.org/10.1007/978-3-319-46630-9_9
6. Bowles, C., et al.: Brain lesion segmentation through image synthesis and outlier detection. NeuroImage Clin. **16**, 643–658 (2017)
7. Chen, X., Konukoglu, E.: Unsupervised detection of lesions in brain MRI using constrained adversarial auto-encoders. arXiv preprint arXiv:1806.04972 (2018)
8. Menze, B.H., et al.: The multimodal brain tumor image segmentation benchmark (BRATS). IEEE Trans. Med. Imaging **34**(10), 1993–2024 (2014)
9. Ronneberger, O., Fischer, P., Brox, T.: U-net: convolutional networks for biomedical image segmentation. In: Navab, N., Hornegger, J., Wells, W.M., Frangi, A.F. (eds.) MICCAI 2015. LNCS, vol. 9351, pp. 234–241. Springer, Cham (2015). https://doi.org/10.1007/978-3-319-24574-4_28
10. Sato, D., et al.: A primitive study on unsupervised anomaly detection with an autoencoder in emergency head CT volumes. In: Medical Imaging 2018: Computer-Aided Diagnosis, vol. 10575, p. 105751P. International Society for Optics and Photonics (2018)
11. Schlegl, T., Seeböck, P., Waldstein, S.M., Langs, G., Schmidt-Erfurth, U.: f-AnoGAN: fast unsupervised anomaly detection with generative adversarial networks. Med. Image Anal. **54**, 30–44 (2019)
12. Sun, L., Wang, J., Huang, Y., Ding, X., Greenspan, H., Paisley, J.: An adversarial learning approach to medical image synthesis for lesion detection. IEEE J. Biomed. Health Inf. (2020)
13. Tsunoda, Y., Moribe, M., Orii, H., Kawano, H., Maeda, H.: Pseudo-normal image synthesis from chest radiograph database for lung nodule detection. In: Kim, Y.S., Ryoo, Y.J., Jang, M., Bae, Y.-C. (eds.) Advanced Intelligent Systems. AISC, vol. 268, pp. 147–155. Springer, Cham (2014). https://doi.org/10.1007/978-3-319-05500-8_14
14. Wang, Z., Simoncelli, E.P., Bovik, A.C.: Multiscale structural similarity for image quality assessment. In: The Thrity-Seventh Asilomar Conference on Signals, Systems & Computers, 2003, vol. 2, pp. 1398–1402. IEEE (2003)
15. Xia, T., Chartsias, A., Tsaftaris, S.A.: Pseudo-healthy synthesis with pathology disentanglement and adversarial learning. Med. Image Anal. **64**, 101719 (2020)

16. Ye, D.H., Zikic, D., Glocker, B., Criminisi, A., Konukoglu, E.: Modality propagation: coherent synthesis of subject-specific scans with data-driven regularization. In: Mori, K., Sakuma, I., Sato, Y., Barillot, C., Navab, N. (eds.) MICCAI 2013. LNCS, vol. 8149, pp. 606–613. Springer, Heidelberg (2013). https://doi.org/10.1007/978-3-642-40811-3_76
17. Zhang, C., Bengio, S., Hardt, M., Recht, B., Vinyals, O.: Understanding deep learning requires rethinking generalization. arXiv preprint arXiv:1611.03530 (2016)
18. Zhu, J.Y., Park, T., Isola, P., Efros, A.A.: Unpaired image-to-image translation using cycle-consistent adversarial networks. In: Proceedings of the IEEE International Conference on Computer Vision, pp. 2223–2232 (2017)

Real-Time Mapping of Tissue Properties for Magnetic Resonance Fingerprinting

Yilin Liu[1], Yong Chen[3]($\boxtimes$), and Pew-Thian Yap[1,2]($\boxtimes$)

[1] Department of Computer Science, University of North Carolina at Chapel Hill, Chapel Hill, USA
ptyap@med.unc.edu
[2] Department of Radiology and Biomedical Research Imaging Center (BRIC), University of North Carolina at Chapel Hill, Chapel Hill, USA
[3] Department of Radiology, Case Western Reserve University, Cleveland, USA
yxc235@case.edu

Abstract. Magnetic resonance fingerprinting (MRF) is a relatively new multi-parametric quantitative imaging method that involves a two-step process: (i) reconstructing a series of time frames from highly-undersampled non-Cartesian spiral k-space data and (ii) pattern matching using the time frames to infer tissue properties (e.g., T_1 and T_2 relaxation times). In this paper, we introduce a novel end-to-end deep learning framework to seamlessly map the tissue properties directly from spiral k-space MRF data, thereby avoiding time-consuming processing such as the non-uniform fast Fourier transform (NUFFT) and the dictionary-based fingerprint matching. Our method directly consumes the non-Cartesian k-space data, performs adaptive density compensation, and predicts multiple tissue property maps in one forward pass. Experiments on both 2D and 3D MRF data demonstrate that quantification accuracy comparable to state-of-the-art methods can be accomplished within 0.5 s, which is 1,100 to 7,700 times faster than the original MRF framework. The proposed method is thus promising for facilitating the adoption of MRF in clinical settings.

Keywords: Magnetic resonance fingerprinting · End-to-End learning · Non-Cartesian MRI reconstruction · Deep learning

1 Introduction

Magnetic resonance fingerprinting (MRF) [12] is a new quantitative imaging paradigm that allows fast and parallel measurement of multiple tissue properties in a single acquisition, unlike conventional methods that quantify one specific tissue property at a time. MRF randomizes multiple acquisition parameters to

---

This work was supported in part by United States National Institutes of Health (NIH) grant EB006733.

M. de Bruijne et al. (Eds.): MICCAI 2021, LNCS 12906, pp. 161–170, 2021.
https://doi.org/10.1007/978-3-030-87231-1_16

generate unique signal evolutions, called "fingerprints", that encode information of multiple tissue properties of interest. 1000–3000 time points are usually acquired and one image is reconstructed for each time point. Dictionary matching (DM) is then used to match the fingerprint at each pixel to a pre-defined dictionary of fingerprints associated with a wide range of tissue properties.

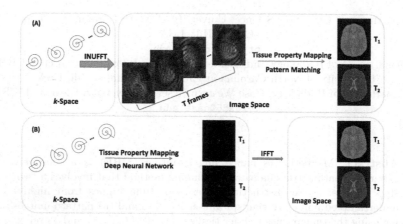

Fig. 1. (A) The original MRF framework. (B) The proposed framework.

To improve the clinical feasibility of MRF, many studies have investigated replacing DM with deep neural networks to accelerate tissue mapping [2–4,8,9]. However, these methods, similar to DM, operate on the reconstructed MRF images and are therefore still limited by the speed and computational efficiency of conventional reconstruction methods. Particularly, since MRF employs a spiral k-space sampling trajectory for robustness to motion [12], the reconstruction is non-trivial and more time-consuming than the Cartesian case.

A major challenge is that the computationally efficient inverse fast Fourier transform (FFT) cannot be directly applied to non-Cartesian data. Besides, the density of the samples varies along the non-Cartesian trajectory and must be compensated for to ensure high-quality reconstruction. Most existing non-Cartesian MRI reconstruction methods thus consist of independent steps that are *not optimized end-to-end*, relying heavily on Non-Uniform Fast Fourier Transform (NUFFT) [5].

Fewer deep learning based reconstruction methods focus on non-Cartesian sampling [6,17] than on Cartesian sampling [16,18,19,21]. AUTOMAP [22] attempts to use a fully-connected network (FNN) to learn the full mapping from raw k-space data to images, including the Fourier transform (FT). Although FNN makes no assumptions on the sampling pattern and aligns with the nature of FT, the network size is quadratic with the image size $N \times N$ ($O(N^4)$), incurring immense memory costs that limit scalability to large images, especially to high-dimensional MRF data with thousands of time frames. Moreover, MRF involves an additional tissue mapping that may require a tailored network architecture based on convolutional neural networks (CNNs) [3,4] or recurrent neural network (RNNs) [9] for optimal performance.

Our aim in this paper is to introduce a framework for real-time tissue quantification directly from non-Cartesian k-space MRF data using only regular 2D convolutions and FFT, providing a computationally more feasible solution to high-dimensional MR data reconstruction and allowing greater flexibility in network design. Mimicking DFT directly with locally-connected CNNs is not effective since every point in the k-space has a global effect on the image. Our approach is inspired by the gridding process in NUFFT [1,5]. However, instead of explicitly incorporating the memory-costly gridding kernel of NUFFT as in [6,17], we show for the first time that gridding and tissue mapping can be performed seamlessly in a single mapping. Experiments on 2D and 3D MRF data demonstrate that our completely end-to-end framework achieves results *on par* with state-of-the-art methods that use more complicated reconstruction schemes while being orders of magnitude faster. To the best of our knowledge, no prior methods have demonstrated the feasibility of end-to-end non-Cartesian MRI reconstruction *for data as high-dimensional as MRF* in a single framework dealing with both reconstruction and tissue mapping simultaneously without the need for NUFFT.

2   Methods

2.1   Problem Formulation

With the MRF sequence employed in this study, only 1/48-th of the full data, i.e., a single spiral, is collected for each time frame for significant acceleration. The original MRF framework first reconstructs an image from each spiral of length n using NUFFT, leading to T highly-aliased images. The image series $\mathbb{C}^{M \times M \times T}$ are then mapped to the corresponding tissue property T1 and T2 maps with image dimensions $M \times M$.

In contrast, our approach directly maps the highly-undersampled spiral k-space MRF data $\mathbb{C}^{n \times T}$ to the Cartesian k-space of the T_1 or T_2 map, and finally to the image space of T_1 or T_2 map simply via inverse FFT (Fig. 1).

Let each data point in k-space be represented as a location vector $p_i \in \mathbb{R}^2$ and a signal value $f_i \in \mathbb{C}$. To grid the signal $S(q)$, convolution is applied via weighted summation of the signal contributions of K neighboring sampled data points of q:

$$S(q) = \sum_{i=1}^{K} f_i g(p_i - q)d_i,\qquad(1)$$

where $g(\cdot)$ denotes the gridding kernel centered at q, and d_i is the density compensation factor for data point p_i. Points in sparsely sampled regions are associated with greater compensation factor. Density compensation is required because in non-Cartesian imaging, the central k-space (low-frequency components) is usually more densely sampled than the outer k-space (high-frequency components).

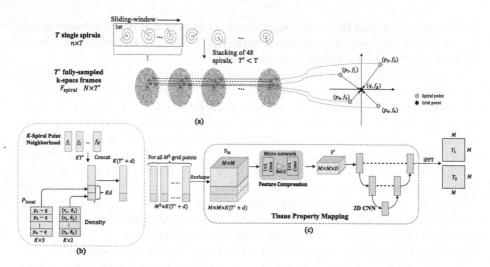

Fig. 2. Illustration of the proposed method. The point distribution features ($d = 5$) consist of the relative Cartesian offsets of p_i with respect to q, the radial distance of p_i with respect to q, and the density of p_i represented as the polar coordinates of p_i with respect to the k-space center.

2.2   Proposed Framework

Instead of performing gridding in k-space and tissue mapping in image space separately as in most existing methods [2,4], we propose to perform tissue mapping *directly from k-space*. This allows gridding and tissue mapping for thousands of time frames to be performed simultaneously via a single CNN, which is key to achieving real-time tissue quantification. Applying CNNs to non-Cartesian data, however, is not straightforward. Here, without actually interpolating each grid point on the Cartesian k-space of MRF frames, our key idea is to directly use the signal time courses of K nearest neighboring spiral points of a target grid point q to infer the corresponding tissue properties (Fig. 2(a)), based on their relative positions to the target grid point and their densities (Fig. 2(b)). K Nearest-Neighbor (KNN) search and density estimation only need to be computed once for each trajectory and pre-stored; therefore the required time cost is negligible. Individual components of the proposed framework are described next.

2.3   Sliding-Window Stacking of Spirals

In MRF, data are typically sampled incoherently in the temporal dimension via a series of rotated spirals. Each MRF time frame is highly undersampled with one spiral. Here, we combine every 48 temporally consecutive spirals in a sliding-window fashion for full k-space coverage (Fig. 2(a, Left)). This reduces the number of time frames from T to $T' = T/48$ and allows each spiral point to be associated with a T' dimensional feature vector f_i. The input to the network thus become: $F_{\text{spiral}} = \{f_1, f_2, ..., f_N\}, f_i \in \mathbb{C}^{T'}$. From Eq. (1), sampled points are gridded based only on their relative positions to the target grid point, i.e., $p_i - q$. Thus, as exemplified in Fig. 2(a, Right), different f_i contributes differently according to its spatial proximity with respect to the center grid point q in a K-point local neighborhood.

2.4 Learned Density Compensation

In non-Cartesian imaging, measurement density varies in k-space and is typically dense at the center and sparse at the peripheral of k-space. Density compensation (DC) can thus be viewed as a function of data location on a k space sampling trajectory with respect to the k-space center. This is different from gridding using local weighting with respect to a target grid point. Thus, we propose to parameterize the DC function using 2D polar coordinates of the sampled points:

$$d_i = f_{\text{dc}}(r_i, \theta_i), \tag{2}$$

where $r_i \geq 0$ and $0 \leq \theta_i < 2\pi$. Straightforward choices of f_{dc} are $d_i = \frac{\theta_i}{r_i}$ and $d_i = r_i\theta_i$. However, rather than fixing and handcrafting f_{dc}, we learn the DC function to adapt to different sampling trajectories via a network that is sensitive to sample locations. This is achieved by directly concatenating the polar coordinates with the spiral point features, inspired by "CoordConv" [10]. By simply giving the convolution access to the input coordinates, the network can adaptively decide where to compensate by learning different weights for the density features associated with different spiral points. This is unlike conventional methods where DC weighting functions are computed analytically [7] or iteratively [14]. See [10] for more information on translation-variant CoordConv.

2.5 Tissue Mapping via Agglomerated Neighboring Features

The features for each target grid point are agglomerated from its K nearest neighbors from a stack of T' spirals. This transforms the spiral data F_{spiral} to a grid, allowing regular 2D convolutions to be applied directly. Concatenating point features with additional d-dimensional point distribution information required by gridding and density compensation leads to input $F_{\text{in}} \in \mathbb{R}^{M \times M \times K(T'+d)}$. Since our framework does not emphasize on and is not limited to a certain network architecture, we extend an existing U-Net [15] based MRF tissue quantification network [4] to make it fully end-to-end, mapping the agglomerated features F_{in} directly to the corresponding tissue property maps. To improve computational efficiency, a micro-network is employed preceding the quantification network to reduce the dimensionality of each target grid feature vector $f \in \mathbb{R}^{K(T'+d)}$ by a shared linear transformation $W \in \mathbb{R}^{K(T'+d) \times D}$, implemented as an 1×1 convolution:

$$f'_j = \text{ReLU}(\text{Conv}([f_1; f_2; ...; f_K])), \tag{3}$$

where $[\cdot;\cdot]$ denotes concatenation, $f'_j \in \mathbb{R}^D$, and $j \in \{1, 2, \ldots, M \times M\}$. The resulting feature map $F' \in \mathbb{R}^{M \times M \times D}$ is then fed to the quantification network.

Network Parameters and Training. Our network backbone consists of a micro-network and a 2D U-Net, which is $\sim 10^5$ lighter than AUTOMAP [22]. AUTOMAP is computationally expensive when applied to MRF ($\sim 5 \times 10^{12}$ params). The micro-network is composed of four 1×1 convolutional layers, each followed by batch normalization and ReLU. The number of output channels of all 1×1 convolutions is D ($D = 64$ for T1, and $D = 164$ for T2, chosen by cross validation). The input channel number of the micro-network is $K(T' + d)$, where $K = 4$. The network was trained in batches of 2 samples and optimized via ADAM with an initial learning rate of 0.0002, which was decayed by 99% after each epoch. Following [4], relative-L1 was used as the objective function. Two GPUs (TITAN X, 12G) were used for training.

3   Experiments and Results

Datasets. 2D MRF datasets were acquired from six normal subjects, each consisting of 10 to 12 scans. For each scan, a total of $2,304$ MRF time points were acquired and each contains only one spiral readout of length $n = 2,052$. Two 3D MRF datasets were used for evaluation. The first 3D MRF dataset with a spatial resolution of $1.2 \times 1.2 \times 3$ mm^3 were collected from three subjects, each covering 64 slices. A total of 576 time points were acquired for each scan. The second 3D MRF datasets were acquired from six volunteers with a high isotropic resolution of 1 mm, each covering 144 slices. 768 time points were collected for each scan. For both 3D datasets, FFT was first applied in the slice-encoding direction, and then the data of each subject were processed slice-by-slice, just as in the 2D case. All MRI measurements were performed on a Siemens 3T scanner with a 32-channel head coil. Real and imaginary parts of the complex-valued MRF signals are concatenated. For acceleration, only the first 25% time frames in each 2D MRF scan and the first 50% in each 3D MRF scan were used for training. The training data size for *each* 2D and 3D scan is $32 \times 2,052 \times 576$ and $32 \times 2,052 \times 288$ (or 384) (# coils$\times$ # spiral readouts $\times$ # time frames), respectively. The ground-truth T_1 and T_2 maps with 256×256 voxels were obtained via dictionary matching using all 2,304 time frames in 2D MRF and all 576 (or 768) time frames in 3D MRF.

Experimental Setup. 1) We compared our end-to-end approach with four state-of-the-art MRF methods: a U-Net based deep learning method (**SCQ**) [4], dictionary matching (**DM**) [12], SVD-compressed dictionary matching (**SDM**) [13], and a low-rank approximation method (**Low-Rank**) [11]. Note that these competing methods require first reconstructing the image for each time frame using NUFFT [5]. Leave-one-out cross validation was employed. 2) We also compared our adaptive gridding with typical handcrafted gridding methods, and investigated the effects of including the relative positions and density features. 3) As a proof of concept, we applied our method on the additional high-resolution 3D MRF dataset for qualitative evaluation.

Table 1. Quantitative comparison on the 2D MRF dataset with 4× under-sampling. MAE is computed relative to the ground truth (unit: %). Times reported for reconstruction and pattern matching are per-slice averages.

Method	MAE		SSIM		NRMSE		Recon. (s)	Patt. match. (s)	Total (s)
	T1	T2	T1	T2	T1	T2			
DM	**2.42**	10.06	**0.994**	0.954	**0.0150**	0.0421	467	25	492
SDM	2.42	10.05	0.994	0.954	0.0150	0.0421	467	10	477
Low-Rank	2.87	8.17	0.991	0.960	0.0156	**0.0302**	3133	25	3158
SCQ	4.87	7.53	0.992	0.968	0.0217	0.0309	9.73	0.12	9.85
Ours	4.24	**7.00**	0.986	**0.972**	0.0258	0.0335	–	–	**0.41**

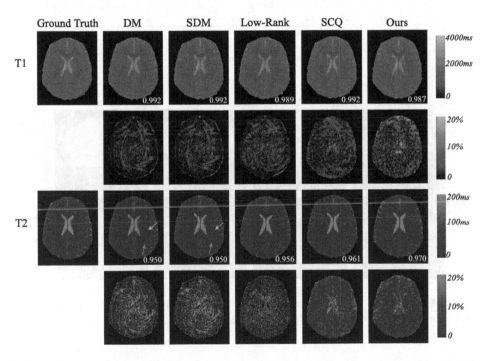

Fig. 3. Example 2D MRF results and the associated error maps with 4× under-sampling. Artifacts are indicated by arrows. SSIM is reported at the bottom right.

Results and Discussion. As shown in Table 1 and Table 2, our method performs overall best in T2 quantification accuracy and achieves competitive accuracy in T1 quantification with processing speed 24 times faster than a CNN method and 1,100 to 7,700 times faster than DM methods. Particularly, for 3D

MRF, our method performs best for most metrics. Qualitative results are shown in Fig. 3 and Fig. 4. The higher T1 than T2 quantification accuracy is consistent with previous findings [4,20]. Due to the sequence used in this study, the early portion of the MRF time frames, which were used for training, contain more information on T1 than T2. Hence, all methods are more accurate in T1 quantification. DM methods exhibit significant artifacts in T2 as indicated by the arrows in Fig. 3. Representative results for the additional high-resolution 3D MRF data are shown in Fig. 5. In the ablation study shown in Table 3, our adaptive gridding performs better than typical handcrafted gridding methods.

Table 2. Quantitative comparison using the first 3D MRF data with 2× under-sampling and a resolution of $1.2 \times 1.2 \times 3\,mm^3$. MAE is computed relative to the ground truth (unit: %). Times reported for reconstruction and pattern matching are per-slice averages.

Method	MAE		SSIM		NRMSE		Recon. (s)	Patt. match. (s)	Total (s)
	T1	T2	T1	T2	T1	T2			
DM	**5.89**	12.19	**0.996**	**0.968**	0.0415	0.0521	140.56	17.01	157.57
SCQ	16.58	16.74	0.933	0.919	0.0652	0.0479	8.58	0.11	8.69
Ours	9.14	**11.78**	0.980	**0.968**	**0.0389**	**0.0323**	–	–	**0.33**

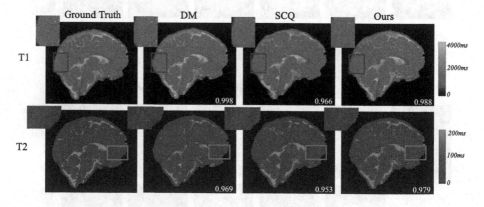

Fig. 4. Example 3D MRF results with 2× under-sampling and $1.2 \times 1.2 \times 3\,mm^3$ resolution. SSIM is reported at the bottom right.

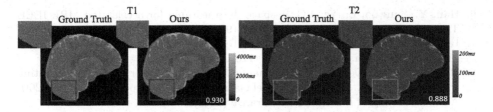

Fig. 5. Example high-resolution 3D MRF results with 2× under-sampling and 1 mm isotropic resolution. SSIM is reported at the bottom right.

Table 3. Comparison of our adaptive gridding method with typical handcrafted gridding methods, and effects of including relative positions and density features.

Gridding method	Average	Bilinear	Gaussian	Ours			
				No xy/density	xy	Density	xy+density
MAE (%) T1	5.59	5.27	5.53	5.24	4.34	4.48	**4.24**
T2	7.74	8.48	7.95	9.05	8.43	7.37	**7.09**

4 Conclusion

In this paper, we introduced a novel and scalable end-to-end framework for direct tissue quantification from non-Cartesian MRF data in milliseconds. With 0.5 s per slice, 120 slices for whole-brain coverage can be processed in one minute, allowing timely re-scan decisions to be made in clinical settings without having to reschedule additional patient visits. It should be noted that the U-Net based network backbone can be replaced with a more advanced architecture to further boost quantification accuracy. Our framework is also agnostic to the data sampling pattern, and thus can be potentially adapted to facilitate other non-Cartesian MRI reconstruction tasks. We believe that our work will improve the clinical feasibility of MRF, and spur the development of fast, accurate and robust reconstruction techniques for non-Cartesian MRI.

References

1. Bernstein, M.A., King, K.F., Zhou, X.J.: Handbook of MRI Pulse Sequences. Elsevier, Amsterdam (2004)
2. Cohen, O., Zhu, B., Rosen, M.S.: MR fingerprinting deep reconstruction network (DRONE). Magn. Reson. Med. **80**(3), 885–894 (2018)
3. Fang, Z., Chen, Y., Hung, S.C., Zhang, X., Lin, W., Shen, D.: Submillimeter MR fingerprinting using deep learning-based tissue quantification. Magn. Reson. Med. **84**(2), 579–591 (2020)
4. Fang, Z., et al.: Deep learning for fast and spatially constrained tissue quantification from highly accelerated data in magnetic resonance fingerprinting. IEEE Trans. Med. Imaging **38**(10), 2364–2374 (2019)
5. Fessler, J.A., Sutton, B.P.: Nonuniform fast fourier transforms using min-max interpolation. IEEE Trans. Signal Process. **51**(2), 560–574 (2003)

6. Han, Y., Sunwoo, L., Ye, J.C.: k-space deep learning for accelerated MRI. IEEE Trans. Med. Imaging **39**(2), 377–386 (2019)
7. Hoge, R.D., Kwan, R.K., Bruce Pike, G.: Density compensation functions for spiral MRI. Magn. Reson. Med. **38**(1), 117–128 (1997)
8. Hoppe, E., et al.: Deep learning for magnetic resonance fingerprinting: accelerating the reconstruction of quantitative relaxation maps. In: Proceedings of the 26th Annual Meeting of ISMRM, Paris, France (2018)
9. Hoppe, E., et al.: RinQ fingerprinting: recurrence-informed quantile networks for magnetic resonance fingerprinting. In: Shen, D., et al. (eds.) MICCAI 2019. LNCS, vol. 11766, pp. 92–100. Springer, Cham (2019). https://doi.org/10.1007/978-3-030-32248-9_11
10. Liu, R., et al.: An intriguing failing of convolutional neural networks and the coordconv solution. arXiv preprint arXiv:1807.03247 (2018)
11. Ma, D., et al.: Applications of low rank modeling to fast 3D MRF. In: Proceedings of the International Society for Magnetic Resonance in Medicine, vol. 25, p. 129 (2017)
12. Ma, D., et al.: Magnetic resonance fingerprinting. Nature **495**(7440), 187–192 (2013)
13. McGivney, D.F., et al.: SVD compression for magnetic resonance fingerprinting in the time domain. IEEE Trans. Med. Imaging **33**(12), 2311–2322 (2014)
14. Pipe, J.G., Menon, P.: Sampling density compensation in MRI: rationale and an iterative numerical solution. Magn. Reson. Med. Off. J. Int. Soc. Magn. Reson. Med. **41**(1), 179–186 (1999)
15. Ronneberger, O., Fischer, P., Brox, T.: U-net: convolutional networks for biomedical image segmentation. In: Navab, N., Hornegger, J., Wells, W.M., Frangi, A.F. (eds.) MICCAI 2015. LNCS, vol. 9351, pp. 234–241. Springer, Cham (2015). https://doi.org/10.1007/978-3-319-24574-4_28
16. Schlemper, J., Caballero, J., Hajnal, J.V., Price, A.N., Rueckert, D.: A deep cascade of convolutional neural networks for dynamic MR image reconstruction. IEEE Trans. Med. Imaging **37**(2), 491–503 (2017)
17. Schlemper, J., et al.: Nonuniform variational network: deep learning for accelerated nonuniform MR image reconstruction. In: Shen, D., et al. (eds.) MICCAI 2019. LNCS, vol. 11766, pp. 57–64. Springer, Cham (2019). https://doi.org/10.1007/978-3-030-32248-9_7
18. Sriram, A., Zbontar, J., Murrell, T., Zitnick, C.L., Defazio, A., Sodickson, D.K.: GrappaNet: combining parallel imaging with deep learning for multi-coil MRI reconstruction. In: Proceedings of the IEEE/CVF Conference on Computer Vision and Pattern Recognition, pp. 14315–14322 (2020)
19. Zhang, Z., Romero, A., Muckley, M.J., Vincent, P., Yang, L., Drozdzal, M.: Reducing uncertainty in undersampled MRI reconstruction with active acquisition. In: Proceedings of the IEEE Conference on Computer Vision and Pattern Recognition, pp. 2049–2058 (2019)
20. Zhao, B., Setsompop, K., Ye, H., Cauley, S.F., Wald, L.L.: Maximum likelihood reconstruction for magnetic resonance fingerprinting. IEEE Trans. Med. Imaging **35**(8), 1812–1823 (2016)
21. Zhou, B., Zhou, S.K.: DuDoRNet: learning a dual-domain recurrent network for fast MRI reconstruction with deep T1 prior. In: Proceedings of the IEEE/CVF Conference on Computer Vision and Pattern Recognition, pp. 4273–4282 (2020)
22. Zhu, B., Liu, J.Z., Cauley, S.F., Rosen, B.R., Rosen, M.S.: Image reconstruction by domain-transform manifold learning. Nature **555**(7697), 487–492 (2018)

Estimation of High Framerate Digital Subtraction Angiography Sequences at Low Radiation Dose

Nazim Haouchine[1,2]([⊠]), Parikshit Juvekar[1,2], Xin Xiong[2,3], Jie Luo[1,2], Tina Kapur[1,2], Rose Du[1,2], Alexandra Golby[1,2], and Sarah Frisken[1,2]

[1] Harvard Medical Shcool, Boston, MA, USA
[2] Brigham and Women's Hospital, Boston, MA, USA
[3] Columbia University, New York City, NY, USA

Abstract. Digital Subtraction Angiography (DSA) provides high resolution image sequences of blood flow through arteries and veins and is considered the gold standard for visualizing cerebrovascular anatomy for neurovascular interventions. However, acquisition frame rates are typically limited to 1–3 fps to reduce radiation exposure, and thus DSA sequences often suffer from stroboscopic effects. We present the first approach that permits generating high frame rate DSA sequences from low frame rate acquisitions eliminating these artifacts without increasing the patient's exposure to radiation. Our approach synthesizes new intermediate frames using a phase-aware Convolutional Neural Network. This network accounts for the non-linear blood flow progression due to vessel geometry and initial velocity of the contrast agent. Our approach outperforms existing methods and was tested on several low frame rate DSA sequences of the human brain resulting in sequences of up to 17 fps with smooth and continuous contrast flow, free of flickering artifacts.

Keywords: Biomedical image synthesis · Digital subtraction angiography · Video interpolation · Convolutional neural networks

1 Introduction and Related Work

Cerebrovascular diseases disrupt the circulation of blood in the brain and include aneurysms, developmental venous angiomas and arteriovenous malformations (AVMs). These diseases may cause blood vessels to rupture that can result in brain hemorrhage with complications including severe headache, seizures, hydrocephalus, brain damage and death. Surgical removal, which is curative, is a preferred treatment [8]. However, it is not without risks. The main risk is post-surgical deficit due to hemorrhage [10]. These risks depend on the malformation, clot size, proximity to the eloquent cortex, the presence of diffuse feeding arteries and deep venous drainage (in the case of AVMs). Risks can be mitigated with meticulous surgical planning and prior understanding of blood flow circulation by careful study of preoperative imaging, which often includes digital subtraction

© Springer Nature Switzerland AG 2021
M. de Bruijne et al. (Eds.): MICCAI 2021, LNCS 12906, pp. 171–180, 2021.
https://doi.org/10.1007/978-3-030-87231-1_17

angiography (DSA), computed tomography angiography or magnetic resonance angiography.

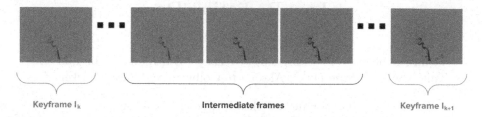

Fig. 1. Our method produces new intermediate images $\{\hat{\mathbf{I}}_{(k,k+1)}\}_N$ between each pair of input images $\mathbf{I}_k$ and $\mathbf{I}_{k+1}$ from a low frame rate DSA sequence to generate a new high frame rate DSA sequence.

DSA is considered the "gold standard" imaging method for evaluating cerebral malformations. It provides high resolution (0.3 mm pixels), dynamic imaging of blood flow through the brain during arterial filling and venous drainage [9]. To visualize blood flow and vascular malformations, a contrast agent/dye is injected into a large artery progressing through the smaller arteries, the capillaries, and then out through the veins (and back to the heart). DSA acquisition protocols involve several parameters, including radiation dose and frame rate [6]. Because higher frame rates acquisitions expose patients to more radiation, an adequate trade-off between acquisition frame rate and radiation dose must be made to reduce patients exposure to radiation [11]. Therefore, current best practice routinely limits the frame rate to 1–3 fps, with frame rates of up to 7.5 fps in some cases [6]. Unfortunately, low frame rate DSA produces stroboscopic (or flicker) effects [11] that make it more challenging for clinicians to interpret the complex dynamic cerebrovascular blood flow, especially in the presence of complex pathology such as arteriovenous malformations.

To eliminate flicker effects, a higher frame rate is required. Because this would require an unsafe radiation dose using conventional methods, we propose a new approach that synthesizes intermediate images to increase the frame rate of low frame rate DSA sequences without increasing radiation dose. Synthesizing images has been studied for the purpose of video interpolation [2] and slow-motion generation [3], for registration of medical images [5] and for graphical animations and rendering [7]. Our solution follows these approaches while preserving coherent blood hemodynamics in intermediate images and does not alter images from the original sequence. Because the progression of contrast agents in the neurovascular structure is non-linear, our method first decomposes the sequence into three phases: arterial, capillary and venous. We extract from these phases an estimate of the volume of the contrast agent that, combined with the original images, will be fed to a Convolutional Neural Network (CNN) to generate new images. The optimized loss function is designed to focus on the regions with high entropy to preserve details of the predicted images. The final sequence is assembled by iteratively interleaving the intermediate images with the original images (see Fig. 1).

We believe that our method will reduce the challenges of interpreting DSA, particularly in the presence of complex cerebrovascular malformations, thus helping to improve preoperative planning and surgical monitoring with the goal of reducing surgical complication rates.

2   Methods

As illustrated in Fig. 1, given two input images $\mathbf{I}_k$ and $\mathbf{I}_{k+1}$, our goal is to predict the intermediate image $\widehat{\mathbf{I}}_{(k,k+1)}$. Inspired by recent work in video frame interpolation, we rely on CNNs to predict the intermediate image. Because we want to control the contribution of the input images $\mathbf{I}_k$ and $\mathbf{I}_{k+1}$ on $\widehat{\mathbf{I}}_{(k,k+1)}$, we take into account the non-linear progression of the contrast through the vascular network. We estimate the contrast agent volume for each image by decomposing the DSA sequence into arterial, capillary and venous phases. In addition, we constrain the model to learn features in regions of the image with rich vessel information. The final composition includes the input and intermediate images and can be repeated iteratively, resulting in a high frame rate DSA sequence.

2.1   Phase Decomposition Using Independent Component Analysis

In order to obtain an estimate of the contrast progression per image, we start by decomposing the DSA sequence into arterial, capillary and venous phases as illustrated in Fig. 2. Because blood flow propagation behaves differently during these three phases, this decomposition permits us to adapt the interpolation mechanism to each phase. We perform the decomposition using Independent Component Analysis (ICA) [4]. ICA is a statistical method that extracts subcomponents of a 1D-signal under the assumption that these subcomponents are independent. The DSA images composing a sequence are first stacked and vectorized to produce a unique 1D-signal. Using ICA on this signal (with three classes) will generate three separate signals corresponding to the arterial, capillary and venous phases. These signals are transformed back to 2D to obtain three distincts images which encode signal information, as well as noise and outliers. To clearly separate meaningful information from the noise and outliers, we use image-base binary thresholding on the histogram distribution. The three thresholded images are used as binary masks to estimate time-density curves (TDCs) [4]. Using the TDCs, we can estimate contrast agent volume that runs through the vascular network between two consecutive images. This amount is normalized between 0 and 1 w.r.t the phase's peak volume. Assuming a sequence $\{\mathbf{I}_i\}_{i=1}^N$ of N DSA images, we can now define the function $v_{(i,j)} = p(\{\mathbf{I}_i\}_{i=1}^N, i, j)$ that estimates for any consecutive images $\mathbf{I}_i$ and $\mathbf{I}_j$, the contrast volume $v_{i,j}$, with $j \in (1, N)$ and $j > i$. This volume will be used to control the contribution of the input images to the intermediate image.

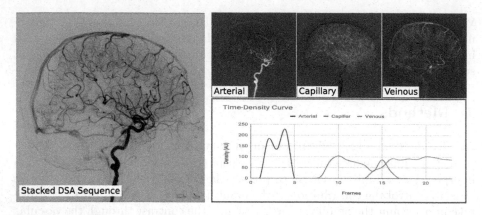

Fig. 2. Decomposition of DSA sequence into arterial, capillary and venous phases using ICA. After stacking and vectorizing all images into a 1D-signal, we can separate the signal to three distinct components and obtain a time-density curve.

2.2 Training and Optimization

Given the training set $\mathcal{T}^e = \{(\mathbf{I}_j, \mathbf{L}_j)\}_j$ composed of DSA images $\mathbf{I}_j$ and their corresponding binary labels $\mathbf{L}_j$, with $j \in (1, N_{\mathcal{T}^e})$, we first start by training a region-of-interest extractor $e(\mathbf{I}; \theta_r)$ with θ_r being the learned parameters for network e. Because a binary segmentation may discard vessels with small diameters and information about the vessels/background boundaries, we use the extractor to generate a per-pixel entropy map $\mathbf{M}$ that will associate with each pixel of an image $\mathbf{I}$ the probability of that pixel being part of the vessels or the background (see Fig. 3). The higher the probability the richer the information around the pixel. We used a patch-based method [1] that optimizes a mean-square error loss function over the parameters θ_r of the network e.

Then given the training set $\mathcal{T}^g = \{(\mathbf{X}_j, \mathbf{Y}_j)\}_j$, we train a generator network $g(\mathbf{X}; \mathbf{M}; \theta_g)$ over the parameters θ_g to interpolate new images. $\mathbf{X}_j$ is composed of a pair of images $\mathbf{I}_k$ and $\mathbf{I}_{k+2}$ and the contrast agent volume $v_{(k,k+2)}$ with $k \in (1, N_{\mathcal{T}^g})$. $\mathbf{Y}_j$ corresponds to the output $\mathbf{I}_{k+1}$ that consists of a skipped image that represents the true intermediate image. Finally, $\mathcal{M} = \{\mathbf{M}_k\}_{k=1}^{N_{\mathcal{T}^g}}$ is the set of entropy maps. Note that $\mathbf{M}$ is not considered an input to train g but will be used in the optimization loss function.

Given an intermediate image $\mathbf{I}_k$ and our predicted intermediate image $\widehat{\mathbf{I}}_k$, the optimization loss function over the parameters θ_g of the network g is as follows:

$$\mathcal{L}(\theta_g; \mathcal{M}; \mathcal{T}^g) = \mathcal{L}_{\text{reconstruction}} + \mathcal{L}_{\text{perceptual}}$$

$$= \frac{\alpha}{N_{\mathcal{T}^g}} \sum_{k=1}^{N_{\mathcal{T}^g}} \mathbf{M}_k \sqrt{\|\widehat{\mathbf{I}}_k - \mathbf{I}_k\|^2 + \epsilon^2} + \frac{\beta}{N_{\mathcal{T}^g}} \sum_{k=1}^{N_{\mathcal{T}^g}} \|\phi(\widehat{\mathbf{I}}_k) - \phi(\mathbf{I}_k)\|_2 \tag{1}$$

where α and β are meta-parameters to control the interaction of the loss function components. The *reconstruction loss* models how good the generation of the image is. We opted for a Charbonnier loss function and confirmed the observations in [17] that it performs better than an ℓ_1 loss function. We penalize this loss using the entropy maps $\mathcal{M}$ to enforce the network to focus on rich vessel information. Moreover, using the reconstruction error alone might produce blur in the predictions. We thus use a *perceptual loss* [14] to make interpolated images sharper and preserve their details, where ϕ denote the conv4_3 features of an ImageNet pretrained VGG16 model [19].

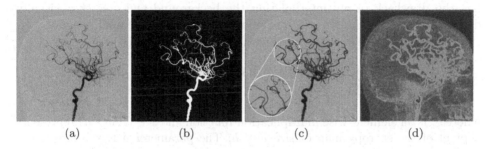

(a)   (b)   (c)   (d)

Fig. 3. Segmentation and entropy maps generation: (a) input DSA image, (b) binary segmentation, (c) extracted contours and (d) entropy map. Although binary segmentation gives accurate geometry, it may discard useful information leading to discrepancies. We use entropy maps instead of binary images to preserve small vessels and information about the vessels/background boundaries.

2.3   Network Details

We adopt a U-Net architecture [18] for both e and g networks. The network e architecture is straightforward and built upon a ResNet34 pretrained model [13] and a Softmax final activation function to generate the entropy map. The network g consists of 6 hierarchies in the encoder, composed of two convolutional and one ReLU layers. Each hierarchy except the last one is followed by an average pooling layer with a stride of 2 to decrease the spatial dimension. There are 6 hierarchies in the decoder part. A bilinear upsampling layer is used at the beginning of each hierarchy, doubling the spatial dimension, followed by two convolutional and ReLU layers. The input consists of a stacked pair of images, while the contrast volume scalar is pointwise added as a 2D features array at the bottom of the network where features are the most dense and spatial dimensions are the least.

For the parameters of the loss function in Eq. 1, we empirically chose, using a validation set, $\alpha = 1$ and $\beta = 0.01$. We optimize the loss using gradient descent [15] for 80 epochs using mini-batches of size 16 and with a learning rate of 0.001.

2.4 Final Composition

To increase the framerate of the DSA sequence $\{\mathbf{I}_i\}_{i=1}^{N}$ we infer an intermediate image for each pair of successive images following:

$$\widehat{\mathbf{I}}_{(i,i+1)} \leftarrow g\big(\{\mathbf{I}_i, \mathbf{I}_{i+1}, p(\{\mathbf{I}_i\}_{i=1}^{N}, i, i+1)\}; e(\mathbf{I}_{i+1}; \widehat{\theta}_r); \widehat{\theta}_e\big) \tag{2}$$

where $(\widehat{\theta}_r, \widehat{\theta}_e)$ are the network parameters found during the training. Using the entropy to enforce learning features on the regions with rich vessel information has the drawback of mis-interpolating the background and creates a visually disturbing effect. Thus, we add a blending step to build the final image $\widehat{\mathbf{I}}_{(i,i+1)}^{f}$ that linearly interpolates the pixels with low entropy as follows:

$$\widehat{\mathbf{I}}_{(i,i+1)}^{f} = \begin{cases} \delta \odot \mathbf{I}_i + (1-\delta) \odot \mathbf{I}_{i+1}, & \text{if } e(\widehat{\mathbf{I}}_{(i,i+1)}; \widehat{\theta}_r)(x) \leq \eta \\ \widehat{\mathbf{I}}_{(i,i+1)}, & \text{otherwise} \end{cases} \tag{3}$$

where $\odot$ is an element-wise multiplication function and $x \in (1, w \times h)$ represents a pixel of the entropy map of size $w \times h$. The parameter δ that controls the interpolation is set empirically to $\delta = 0.6$ while η, a threshold parameter to discard regions with high entropy is set to $\eta = 0.1$.

3 Results

Dataset: Our dataset is composed of 32 DSA sequences for a total of 3216 DSA images (after geometric transformations augmentation). These images were randomly split and 25% of the images were kept for validation. All images were acquired on human subjects using a standard clinical protocol with a biplane General Electric imaging system. We used both frontal and lateral image sequences. The sequences were acquired at 1–3 fps and captured the full cycle of contrast inflow and washout after injection. The dataset $\mathcal{T}^e$ is composed of the totality of images with their corresponding labeled images while the dataset $\mathcal{T}^g$ is composed of pair of images with a leave-one-image-out strategy, skipping every other image to be used as output for the model.

Table 1. Effectiveness of different components of our model.

	woP-MSE	P-MSE	P-eMSE	P-Ch	P-eCh (full model)
PSNR (db)	38.31	38.80	39.90	40.10	**40.40**
IE	11.93	10.90	8.76	8.60	**8.00**

Ablation Study on the Impact of Different Components Design: We first perform an ablation study to analyze the contribution of each component of our approach. We test the impact of having a phase-constrained model, the impact of using a Charbonnier loss instead of a mean-square loss (MSE) and finally the impact of using the entropy maps. To this end, we train five variants of our model: **woP-MSE** which is a traditional U-Net without phase constraints and optimized using an MSE loss function, **P-MSE** and **P-eMSE** which are phase-constrained models using MSE loss function with and without the entropy maps respectively and finally, **P-Ch** and **P-eCh** (full model) which are phase-contrained models using Charbonnier loss function with and without the entropy maps respectively. To quantify the accuracy of our method we use the Peak Signal-to-Noise Ratio (PSNR) as a measure of corruption and noise and the interpolation error (IE), which is defined as the root-mean-squared difference between the ground-truth image and the interpolated image, as a measure of accuracy.

We can observe from Table 1 that removing the phase knowledge from the model harms performance, particularly the interpolation error, while using the Charbonnier loss slightly improves the results w.r.t MSE loss. We also verify that adding the entropy maps improves the predictions, which validate our hypothesis to enforce a learning on regions with rich vessel information.

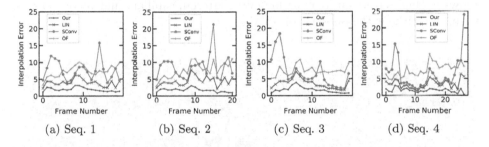

(a) Seq. 1   (b) Seq. 2   (c) Seq. 3   (d) Seq. 4

Fig. 4. Comparison with state-of-the-art methods.

Comparison with State-of-the-Art Methods: We then compare our approach with state-of- the-art methods including neural and non-neural approaches. In addition to our approach (Our), we include a simple bilinear interpolation method (Lin), an optical-flow based interpolation (OF) [12] and a CNN-based method (SConv) [16] for slow motion video generation. We conduct the comparison on four 7.5 fps DSA sequences that are not used during the training. For each triplet of images we leave the middle image out to serve as ground truth. We report the interpolation error in Fig. 4 that shows that our model achieves the best performance on all sequences, on all images. We can notice that our method is only slightly impacted by the contrast agent progression over time, as opposed to the other methods. The performance of our model validates the generalization ability of our approach. Furthermore, in addition to the quantitative measurements, Fig. 5 shows the visual differences between our and others methods and highlights its efficiency.

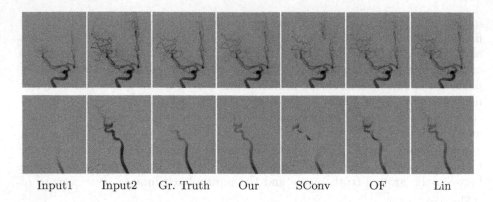

Input1   Input2   Gr. Truth   Our   SConv   OF   Lin

Fig. 5. Samples from the DSA sequences used in our experiments.

Iterative Interpolation: Finally, using our solution we successfully produce high frame rate sequences of up to 17 fps from 3-fps DSA sequences by iteratively interpolating intermediate images from previously predicted images. Figure 6 shows examples from our dataset including arterial, capillary and venous phases with and without the presence of AVMs.

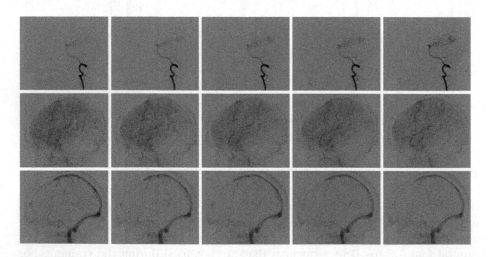

Fig. 6. Samples of high frame rate sequences generated using our method. The first and last column represents the input successive images, and the middle columns represent the estimated intermediate images. Row 1 is the arterial phase (with an AVM), row 2 is the capillary phase and row 3 is the venous phase.

4 Conclusion

We have presented a solution to generate high frame rate DSA sequences at low radiation dose from low frame rate DSA sequences. Using our method, we can increase the framerate of DSA sequences to obtain a continuous blood and contrast flow progression through the cerebral vasculature. The presented approach is clinically practical and can be used with commercially available systems to help clinicians understand the complex dynamic cerebrovascular blood flow, especially in the presence of complex malformations. Our solution is applicable to different organs and procedures, although our experiments involved neurovascular imaging, there is no actual technical limitation for the use of our method in the diagnosis of pulmonary embolisms, renal artery stenosis or any treatment of arterial and venous occlusions.

Our current method is limited to single-frame interpolation and could be extended to produce variable-length multi-image video, where multiple intermediate images are interpolated simultaneously. In addition, future work will investigate ways to estimate 3D high frame rate DSA with the ultimate goal of improving the understanding and diagnosis of arteriovenous malformations.

Acknowledgement. The authors were supported by the following funding bodies and grants: NIH: R01 EB027134-01, NIH: R03 EB032050 and BWH Radiology Department Research Pilot Grant Award.

References

1. Meng, C., et al.: Multiscale dense convolutional neural network for DSA cerebrovascular segmentation. Neurocomputing **373**, 123–134 (2020)
2. Herbst, F., et al.: Occlusion reasoning for temporal interpolation using optical flow. Microsoft technical report (2009)
3. Jiang, H., et al.: Super SloMo: high quality estimation of multiple intermediate frames for video interpolation. In: IEEE CVPR, pp. 9000–9008 (2018). https://doi.org/10.1109/CVPR.2018.00938
4. Hong, J.S., et al.: Validating the automatic independent component analysis of DSA. Am. J. Neuroradiol. **40**(3), 540–542 (2019). https://doi.org/10.3174/ajnr.A5963
5. Leng, J., et al.: Medical image interpolation based on multi-resolution registration. Comput. Math. Appl. **66**(1), 1–18 (2013)
6. Pearl, M., et al.: Practical techniques for reducing radiation exposure during cerebral angiography procedures. J. Neurointerventional Surg. **7**, 141–145 (2014). https://doi.org/10.1136/neurintsurg-2013-010982
7. Narita, R., et al.: Optical flow based line drawing frame interpolation using distance transform to support inbetweenings. In: IEEE ICIP, pp. 4200–4204 (2019)
8. Starke, R., et al.: Treatment guidelines for cerebral arteriovenous malformation microsurgery. Neurosurgery **23**(4), 376–386 (2009)
9. Chng, S.M., et al.: Arteriovenous malformations of the brain and spinal cord. In: Neurology and Clinical Neuroscience, pp. 595–608. Mosby (2007)

10. Thana, T., et al.: Microsurgery for cerebral arteriovenous malformations: postoperative outcomes and predictors of complications in 264 cases. Neurosurg. Focus **37**(3), E10 (2014)

11. Balter, S.: Practical techniques for reducing radiation exposure during cerebral angiography procedures. JAJR Am. J. Roentgenol. **3**, 234–236 (2014). https://doi.org/10.2214/AJR.13.11041

12. Brox, T., Malik, J.: Large displacement optical flow: descriptor matching in variational motion estimation. IEEE Trans. Pattern Anal. Mach. Intell. **33**, 500–513 (2011). https://doi.org/10.1109/TPAMI.2010.143

13. He, K., Zhang, X., Ren, S., Sun, J.: Deep residual learning for image recognition. In: 2016 IEEE Conference on Computer Vision and Pattern Recognition (CVPR), pp. 770–778 (2016). https://doi.org/10.1109/CVPR.2016.90

14. Johnson, J., Alahi, A., Fei-Fei, L.: Perceptual losses for real-time style transfer and super-resolution. In: Leibe, B., Matas, J., Sebe, N., Welling, M. (eds.) ECCV 2016. LNCS, vol. 9906, pp. 694–711. Springer, Cham (2016). https://doi.org/10.1007/978-3-319-46475-6_43

15. Kingma, D.P., Ba, J.: Adam: a method for stochastic optimization. In: Bengio, Y., LeCun, Y. (eds.) 3rd International Conference on Learning Representations, ICLR 2015, San Diego, CA, USA, 7–9 May 2015, Conference Track Proceedings (2015)

16. Niklaus, S., Mai, L., Liu, F.: Video frame interpolation via adaptive separable convolution. In: IEEE International Conference on Computer Vision (2017)

17. Park, J., Ko, K., Lee, C., Kim, C.-S.: BMBC: bilateral motion estimation with bilateral cost volume for video interpolation. In: Vedaldi, A., Bischof, H., Brox, T., Frahm, J.-M. (eds.) ECCV 2020. LNCS, vol. 12359, pp. 109–125. Springer, Cham (2020). https://doi.org/10.1007/978-3-030-58568-6_7

18. Ronneberger, O., Fischer, P., Brox, T.: U-Net: convolutional networks for biomedical image segmentation. In: Navab, N., Hornegger, J., Wells, W.M., Frangi, A.F. (eds.) MICCAI 2015. LNCS, vol. 9351, pp. 234–241. Springer, Cham (2015). https://doi.org/10.1007/978-3-319-24574-4_28

19. Simonyan, K., Zisserman, A.: Very deep convolutional networks for large-scale image recognition. In: International Conference on Learning Representations (2015)

RLP-Net: A Recursive Light Propagation Network for 3-D Virtual Refocusing

Changyeop Shin, Hyun Ryu, Eun-Seo Cho, and Young-Gyu Yoon[✉]

School of Electrical Engineering, KAIST, Daejeon, South Korea
{scey26,ryuhyun1905,eunseo.cho,ygyoon}@kaist.ac.kr

Abstract. High-speed optical 3-D fluorescence microscopy is an essential tool for capturing the rapid dynamics of biological systems such as cellular signaling and complex movements. Designing such an optical system is constrained by the inherent trade-off among resolution, speed, and noise which comes from the limited number of photons that can be collected. In this paper, we propose a recursive light propagation network (RLP-Net) that infers the 3-D volume from two adjacent 2-D wide-field fluorescence images via virtual refocusing. Specifically, we propose a recursive inference scheme in which the network progressively predicts the subsequent planes along the axial direction. This recursive inference scheme reflects that the law of physics for the light propagation remains spatially invariant and therefore a fixed function (i.e., a neural network) for a short distance light propagation can be recursively applied for a longer distance light propagation. Experimental results show that the proposed method can faithfully reconstruct the 3-D volume from two planes in terms of both quantitative measures and visual quality. The source code used in the paper is available at https://github.com/NICALab/rlpnet.

Keywords: Fluorescence microscopy · 3-D volume estimation · Recursive inference · Recursive neural network · Virtual refocusing

1 Introduction

With the recent advancement of genetically encoded fluorescent reporters of biological events [4,5,16,19], the need for high-speed fluorescence microscopy methods that can image such dynamics *in vivo* in 3-D has been consistently increasing [1,3,6,17]. Unfortunately, fluorescence microscopy has an intrinsic performance limit that is set by the excitation light power that the specimen can bear which determines the maximum number of photons that can be collected from the

---

C. Shin and H. Ryu—Equal contributions.

Electronic supplementary material The online version of this chapter (https://doi.org/10.1007/978-3-030-87231-1_18) contains supplementary material, which is available to authorized users.

© Springer Nature Switzerland AG 2021
M. de Bruijne et al. (Eds.): MICCAI 2021, LNCS 12906, pp. 181–190, 2021.
https://doi.org/10.1007/978-3-030-87231-1_18

sample in a given time interval [14]. As the resolution, speed, and noise – the most important performance metrics of fluorescent microscopy – compete for the number of photons against the other metrics, improving one of the metrics inevitably comes with compromising the others [25].

For decoupling the inherent trade-off among the performance metrics of fluorescence microscopy, deep learning has recently emerged as a promising solution [13,21–23]. There are diverse applications and implementations, but the underlying philosophy is the same: minimize the burden of the data acquisition process and predict extensive information from the limited data by exploiting the prior knowledge of the distribution of the data. A notable success was demonstrated in Deep-Z [23], where the authors showed that it is possible to digitally increase the depth-of-field by 20 fold by feeding a single focal plane to a neural network that can generate digitally refocused images. This technique, based on a generative adversarial neural network (GAN) [7,12], takes an input 2-D wide-field fluorescence image concatenated with a matrix that defines the axial distance to the target plane from the input image in a pixel-by-pixel manner. While such a framework provides an opportunity to overcome the inherent limitation of 3-D fluorescence microscopy, there are a number of aspects that can be improved further.

First, Deep-Z [23] relied on the axial asymmetry of the point-spread-function (PSF) to determine the direction of light propagation from the image. However, the axial asymmetry of the PSF is often too small to provide enough directional cues – especially when the object is not sparse in space – which limits the accuracy of the digital refocusing. Second, it would be desirable to reflect the physical properties of light propagation into the network which would help the network learn the general rules.

In this paper, we propose recursive light propagation network (RLP-Net) that infers the 3-D volume from two adjacent 2-D wide-field fluorescence images as shown in Fig. 1a. The framework is inspired by how light propagates in free space and generates an image at each plane as illustrated in Fig. 1b. We exploit the fact that the light propagation law is spatially invariant which leads us to the recursive network design. In addition, in order to cope with the inherent ambiguity of the light direction, we set the goal as inferring a 3-D volume from two adjacent images. Lastly, a step-wise training strategy is devised to achieve a robust performance for the network. The results obtained by refocusing fluorescence microscopy images of fluorescent beads and *Caenorhabditis elegans* demonstrate the effectiveness of the proposed method.

2   Proposed Method

2.1   Recursive Light Propagation Network (RLP-Net)

Network for Modeling Light Propagation. First, let's consider an ideal light propagation function $f_d(\mathbf{I})$ where $\mathbf{I}$ is the input image and d is the light propagation distance. Due to the physical properties of the light propagation (i.e., spatial invariance and continuity) in free space, the function has to meet the following conditions.

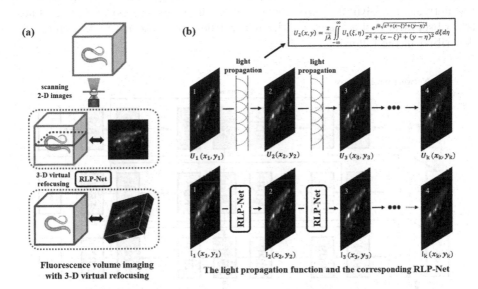

Fig. 1. (a) Illustration of the fluorescence volume imaging with 3-D virtual refocusing. Our proposed method generates a 3-D volume from two 2-D wide-field fluorescence images through virtual refocusing. (b) Light propagation function and the corresponding RLP-Net. The light propagation function is substituted with the proposed network with the recursive inference scheme.

- The function over a very short distance is nearly an identity function: $\lim_{d \to 0} f_d(\mathbf{I}) = \mathbf{I}$.
- The function is commutative: $\forall l, d \in \mathbb{R}, \ f_l \circ f_d(\mathbf{I}) = f_d \circ f_l(\mathbf{I})$.
- The function is additive with respect to the propagation distance:
  $\forall l, d \in \mathbb{R}, \ f_l \circ f_d(\mathbf{I}) = f_{l+d}(\mathbf{I})$.

The first condition indicates that a network with a relatively low capacity would suffice to approximate $f_d(\mathbf{I})$ if d is small. The second and third conditions suggest that $f_d(\mathbf{I})$ can be represented as a composite function of the light propagation function for a unit distance. Therefore, it is natural to approximate the light propagation function as a recursive network (i.e., $f_d(\mathbf{I}) \approx (NN)^d(\mathbf{I})$) as opposed to treating x and d as independent input variables of the network (i.e., $f_d(\mathbf{I}) \approx NN(\mathbf{I}, d)$) where NN denotes a neural network. This recursive design not only allows the network to dynamically adjust its capacity depending on the propagation distance, but it also enforces a proper relation among the light propagation functions for different propagation distances.

Second, another insight can be obtained by looking into the Huygens-Fresnel principle which dictates the light propagation in free space. While the principle relates the input and the output wavefronts which are complex fields, the light propagation function we need to approximate operates on images which are real fields. This suggest that more information, than just one image, should be fed to the network for faithful approximation of the light propagation function. These two observations shed light on the natural design of the network architecture described in the following section.

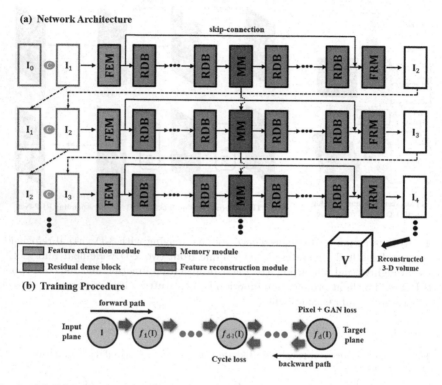

Fig. 2. (a) RLP-Net architecture. The network takes two adjacent images and produces the subsequent plane in a single step. The 3-D volume is reconstructed through recursive inferences. (b) RLP-Net training procedure. The loss function is evaluated based on the inference results where the number of recursive inference steps is gradually increased so that the network can progressively learn to propagate the image further.

Network Architecture. The architecture of RLP-Net is illustrated in Fig. 2a which was designed to reconstruct the 3-D volume through recursive inference of 2-D images. The network takes two adjacent images, $\mathbf{I}_{k-1}$ and $\mathbf{I}_k$, and generates the subsequent image $\mathbf{I}_{k+1}$ where k is an integer number that represents the axial location of the image. Note that the concatenation order of two inputs determines the inference direction. The reconstruction procedure is represented as follows:

$$\mathbf{I}_{k+1}, \mathbf{F}_{k+1} = \Phi(\mathbf{I}_{k-1}, \mathbf{I}_k, \mathbf{F}_k; \Theta), \tag{1}$$

where Φ denotes the approximated light propagation function (i.e., single step RLP-Net), $\mathbf{F}$ is the hidden state feature from the memory module and Θ is the network parameters. A feature extraction module, which is a single convolution layer, was employed at the beginning of the network, followed by 8 residual and dense blocks (RDBs) [20] that were cascaded for global context aggregation. In addition, we introduced a memory module in the middle of the network to fully exploit the extracted information from the earlier steps. As the memory module, Conv-LSTM [24] was used to hold long-term contextual information. A long skip connection [10] was employed to connect multi-scale features across the convolution layers. Lastly, the feature reconstruction module, which is another single convolution layer, generated the target image by aggregating the last feature maps. All the convolution layers had 32 channels and their kernel size was 3×3.

2.2 Training RLP-Net

Loss Function. We used loss function $\mathcal{L}_T$ which consists of three complementary loss functions as follows to ensure that the recursive network faithfully approximates the light propagation function:

$$\mathcal{L}_T = \mathcal{L}_{pixel} + \alpha \mathcal{L}_{cycle} + \beta \mathcal{L}_{GAN}, \tag{2}$$

where α and β are the balancing parameters. First, the pixel loss, $\mathcal{L}_{pixel} = \sum_d ||f_d(\mathbf{I}) - \Phi^d(\mathbf{I})||_1$, which directly compares the reconstructed images and the target images at different distances, was employed where $\Phi(\mathbf{I})$ is a simplified notation of $\Phi(\mathbf{I}_k, \mathbf{I}_{k+1}, \mathbf{F}_{k+1}; \Theta)$. Second, the cycle loss [26], $\mathcal{L}_{cycle} = \sum_{l,d} ||f_{d-l}(\mathbf{I}) - \Phi^{-l}(\Phi^d(\mathbf{I}))||_1$, was introduced to exploit the aforementioned additive property of the light propagation function. Third, the GAN loss [7] $\mathcal{L}_{GAN}$ was added to reflect the natural image statistics of fluorescence microscopy images. We alternatively optimized the generator and discriminator as in [11]. PatchGAN architecture [8] was used as the discriminator network.

Step-Wise Training Strategy. We employed a step-wise training strategy which gradually increases the number of recursive steps, $d = \pm 1, \pm 2, \cdots$, every 50 epochs for calculating the loss in (2). This strategy allowed the network to start with learning easy tasks and move on to difficult ones [2]. The number of recursive steps used for training was capped at 7. We note that the maximum propagation distance was not limited by the number of recursive steps used for training. In addition, the backward step size l was randomly determined to enforce the additive property at every propagation distance.

3 Experiments

3.1 Fluorescence Microscopy Dataset

We used 20 volume images of fluorescent beads and 150 volume images of C. *elegans* for the experimental validation of the proposed method. The fluorescent

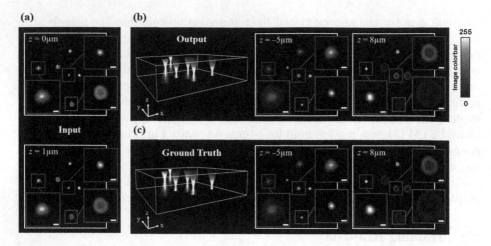

Fig. 3. Virtual refocusing of a fluorescent bead image. (a) Two adjacent images applied as the input. (b) Refocusing result obtained with RLP-Net. A 3-D view of the reconstructed volume (left). Refocused to $z = -5\,\mu m$. The insets below zoom in on the three boxes (center). Refocused to $z = 8\,\mu m$. The insets below zoom in on the boxes (right). (c) As in **b**, but for the ground truth image. Scale bars, $2\,\mu m$.

bead samples were made by randomly distributing fluorescent beads with a diameter of $1\,\mu m$ in 2% agarose. The 3-D images were taken using an epi-fluorescence microscope with a $16\times$ 0.8 NA objective lens. For *C. elegans* imaging, we used nematodes expressing NLS-GCaMP5K under the unc-31 promoter [18] that were mildly paralyzed using 5 mM tetramisole. The 3-D images were taken using an epi-fluorescence microscope with a $40\times$ 1.15 NA objective lens. The axial step size of all images was $1\,\mu m$. We used 70% of the data for training, 15% for validation, and 15% for testing. The images of different samples were used for training and testing to strictly separate the test set from the training set.

3.2   Training Details

All computations were performed on a single GPU (NVIDIA RTX 3090) and the network was implemented using Pytorch [15]. The total loss function, with the balancing parameters $\alpha = 0.01$, $\beta = 0.001$, was minimized using the Adam optimizer [9]. The learning rate was initialized as 10^{-4} for the generator and 2.5×10^{-5} for the discriminator. All weights of convolutional filters were initialized using a normal distribution and the batch size was set as 4. As the unit propagation distance of RLP-Net, $z = 1\,\mu m$ was used. We randomly selected two adjacent lateral planes in the volume images as the input. The data was augmented by random cropping, flipping (horizontal and vertical), and rotating ($90°$, $180°$, $270°$). The network was trained for 600 epochs in total. During the first 100 epochs, only the pixel loss was used for training.

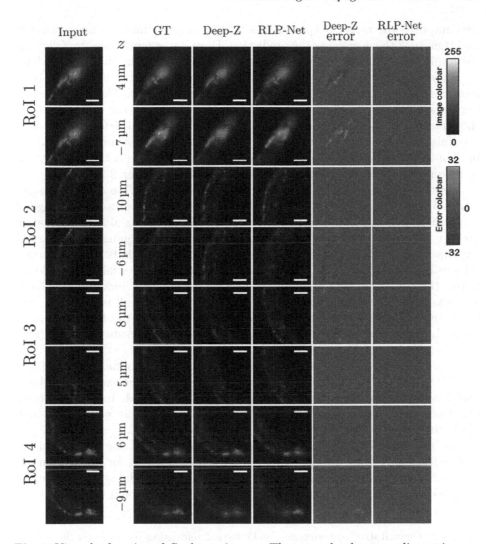

Fig. 4. Virtual refocusing of *C. elegans* images. The network takes two adjacent images as the input (1st column). The ground truth images (2nd column) are compared with the refocusing results from Deep-Z (3rd column) and RLP-Net (4th column). The refocusing errors in Deep-Z (5th column) and RLP-Net (6th column) are shown. Regions-of-interest 1–4 show head ganglia, body (close to head), body (close to tail) and tail ganglia, respectively. Scale bars, 20 μm.

3.3   Evaluation Results

For the objective assessment of the refocusing accuracy, the peak signal-to-noise ratio (PSNR) and structural similarity (SSIM) were measured for the results obtained with Deep-Z, RLP-Net, and RLP-Net without the memory module. For a fair comparison with Deep-Z [23] which takes only one image as the input, we computed the target output images using two input images, independently, and then took the average as the prediction. Deep-Z was reproduced by following the implementation and training details reported in the paper, but with our data.

Figure 3 shows the test results obtained by refocusing the fluorescent bead images (Fig. 3a) to different axial locations using RLP-Net. The inference was performed over an axial range of $\pm 10\,\mu m$ to reconstruct the 3-D volume as shown in Fig. 3b. Not only the overall appearance but also the defocusing pattern was faithfully recovered (Fig. 3c) even though the same network was used for both inference directions, where the inference direction was determined by the concatenation order of two input images as described in Sect. 2.1.

Next, we applied our method to the *C. elegans* dataset. Two adjacent images were applied to RLP-Net to reconstruct the 3-D volume over an axial range of $\pm 10\,\mu m$ (Fig. 4 and Supplementary Video 1). RLP-Net was able to accurately predict the neurons at other axial planes even in the head ganglia and the tail ganglia where neurons were tightly packed, which posed a challenge to accurate refocusing (Fig. S1 and Fig. S2 in the supplementary material). It should be noted that, in each refocusing result, some neurons were brought into focus whereas others were out of focus in a way that matches well with the ground truth image. Deep-Z was also able to accurately predict the images in most cases, but it occasionally refocused the image to the opposite direction (e.g., $z = -6\,\mu m$ plane of RoI 2 in Fig. 4).

The objective measures of the refocusing accuracy are summarized in Table 1. RLP-Net with the memory module showed the highest performance on the bead and *C. elegans* datasets in terms of both PSNR and SSIM. In terms of computational costs, reconstruction of a volume with $128 \times 128 \times 21$ voxels took approximately 0.1 s using RLP-Net and 0.08 s using Deep-Z.

We note that RLP-Net is based on an assumption that the object remains nearly stationary while taking two input images. Therefore, for imaging and 3-D reconstruction of a sample with fast dynamics, a fast axial scanning method (e.g., electrically tunable lens) and a high speed camera should be employed.

Table 1. Quantitative comparison of 3-D virtual refocusing methods on fluorescent beads and *C. elegans* images in terms of PSNR and SSIM.

Dataset	Metric	PSNR (dB)					SSIM				
	Axial distance (μm)	3	6	9	12	Avg	3	6	9	12	Avg
Beads	Deep-Z	29.20	31.17	31.29	32.45	31.03	0.9138	0.9262	0.9267	0.8528	0.9049
	RLP-Net w/o MM	30.50	35.24	34.17	32.98	33.22	0.9801	0.9803	0.9609	0.9034	0.9562
	RLP-Net w MM	**30.77**	**36.84**	**39.36**	**37.37**	**36.09**	**0.9807**	**0.9862**	**0.9750**	**0.9366**	**0.9696**
C-elegans	Deep-Z	39.31	38.21	36.65	35.45	37.41	0.9588	0.9501	0.9383	0.9246	0.9430
	RLP-Net w/o MM	44.21	41.53	38.91	36.37	40.25	0.9853	0.9741	0.9559	0.9251	0.9601
	RLP-Net w MM	**46.72**	**44.48**	**41.10**	**36.55**	**42.21**	**0.9864**	**0.9807**	**0.9714**	**0.9534**	**0.9730**

4 Conclusion

In this paper, we proposed RLP-Net, a new deep-learning method for virtual refocusing of fluorescence microscopy images. RLP-Net was inspired by the light propagation property in which a short distance propagation can be recursively applied for a longer distance propagation. Similarly, our network for unit distance propagation can be applied recursively for a longer distance propagation. We demonstrate the comprehensive results obtained by refocusing experimental fluorescence microscopy images. RLP-Net achieves the state-of-the-art performance in terms of both objective measures and the perceptual quality.

Acknowledgements. This research was supported by National Research Foundation of Korea (2020R1C1C1009869) and the Korea Medical Device Development Fund grant funded by the Korea government (202011B21-05).

References

1. Abrahamsson, S., et al.: Fast multicolor 3D imaging using aberration-corrected multifocus microscopy. Nat. Methods **10**, 60–63 (2013)
2. Bengio, Y., Louradour, J., Collobert, R., Weston, J.: Curriculum learning. In: Proceedings of the 26th Annual International Conference on Machine Learning, pp. 41–48 (2009)
3. Bouchard, M.B., et al.: Swept confocally-aligned planar excitation (SCAPE) microscopy for high-speed volumetric imaging of behaving organisms. Nat. Photonics **9**(2), 113–119 (2015)
4. Chen, Q., et al.: Imaging neural activity using Thy1-GCaMP transgenic mice. Neuron **76**(2), 297–308 (2012)
5. Chen, T.-W., et al.: Ultrasensitive fluorescent proteins for imaging neuronal activity. Nature **499**(7458), 295–300 (2013)
6. Cong, L., et al.: Rapid whole brain imaging of neural activity in freely behaving larval zebrafish (Danio rerio). eLife **6**, e28158 (2017)
7. Goodfellow, I., et al.: Generative adversarial nets. In: Advances in Neural Information Processing Systems, pp. 2672–2680 (2014)
8. Isola, P., Zhu, J.Y., Zhou, T., Efros, A.A.: Image-to-image translation with conditional adversarial networks. In: Proceedings of the IEEE Conference on Computer Vision and Pattern Recognition, pp. 1125–1134. IEEE (2017)
9. Kingma, D.P., Ba, J.: Adam: a method for stochastic optimization. In: Proceedings of the 3rd International Conference on Learning Representations, pp. 1–15 (2015)
10. Mao, X., Shen, C., Yang, Y.-B.: Image restoration using very deep convolutional encoder-decoder networks with symmetric skip connections, pp. 2810–2818 (2016)
11. Mechrez, R., Talmi, I., Shama, F., Zelnik-Manor, L.: Maintaining natural image statistics with the contextual loss. arXiv:1803.04626 (2018)
12. Mirza, M., Osindero, S.: Conditional generative adversarial nets. arXiv:1411.1784 (2014)
13. Ounkomol, C., Seshamani, S., Maleckar, M.M., Collman, F., Johnson, G.R.: Label-free prediction of three-dimensional fluorescence images from transmitted-light microscopy. Nat. Methods **15**, 917–920 (2018). https://doi.org/10.1038/s41592-018-0111-2

14. Pawley, J.: Handbook of Biological Confocal Microscopy, 3rd edn. Springer, Boston (2006). https://doi.org/10.1007/978-0-387-45524-2

15. Paszke, A., et al.: PyTorch: an imperative style, high-performance deep learning library. In: Advances in Neural Information Processing Systems, pp. 8026–8037 (2019)

16. Piatkevich, K.D., et al.: A robotic multidimensional directed evolution of proteins: development and application to fluorescent voltage reporters. Nat. Chem. Biol. **14**, 352–360 (2017)

17. Prevedel, R., et al.: Simultaneous whole-animal 3D imaging of neuronal activity using light-field microscopy. Nat. Methods **11**, 727–730 (2014)

18. Schrödel, T., Prevedel, R., Aumayr, K., Zimmer, M., Vaziri, A.: Brain-wide 3D imaging of neuronal activity in Caenorhabditis elegans with sculpted light. Nat. Methods **10**, 1013–1020 (2013). https://doi.org/10.1038/nmeth.2637

19. Tian, L., et al.: Imaging neural activity in worms, flies and mice with improved GCaMP calcium indicators. Nat. Methods **6**(12), 875–881 (2009)

20. Wang, X., et al.: ESRGAN: enhanced super-resolution generative adversarial networks. In: Leal-Taixé, L., Roth, S. (eds.) ECCV 2018. LNCS, vol. 11133, pp. 63–79. Springer, Cham (2018). https://doi.org/10.1007/978-3-030-11021-5_5

21. Wang, Z., et al.: Real-time volumetric reconstruction of biological dynamics with light-field microscopy and deep learning. Nat. Methods (2021). https://doi.org/10.1038/s41592-021-01058-x

22. Weigert, M., Royer, L., Jug, F., Myers, G.: Isotropic reconstruction of 3D fluorescence microscopy images using convolutional neural networks. In: Descoteaux, M., Maier-Hein, L., Franz, A., Jannin, P., Collins, D., Duchesne, S. (eds.) MICCAI 2017. LNCS, vol. 10434, pp. 126–134. Springer, Cham (2017). https://doi.org/10.1007/978-3-319-66185-8_15

23. Wu, Y., et al.: Three-dimensional virtual refocusing of fluorescence microscopy images using deep learning. Nat. Methods **16**, 1323–1331 (2019). https://doi.org/10.1038/s41592-019-0622-5

24. Xingjian, S., Chen, Z., Wang, H., Yeung, D.Y., Wong, W.K., Woo, W.C.: Convolutional LSTM network: a machine learning approach for precipitation nowcasting. In: Advances in Neural Information Processing Systems, pp. 802–810 (2015)

25. Yoon, Y.-G., et al.: Sparse decomposition light-field microscopy for high speed imaging of neuronal activity. Optica **7**, 1457–1468 (2020)

26. Zhu, J.Y., Park, T., Isola, P., Efros, A.A.: Unpaired image-to-image translation using cycle-consistent adversarial networks. In: Proceedings of the IEEE International Conference on Computer Vision, pp. 2223–2232. IEEE (2017)

Noise Mapping and Removal in Complex-Valued Multi-Channel MRI via Optimal Shrinkage of Singular Values

Khoi Minh Huynh[1], Wei-Tang Chang[2], Sang Hun Chung[1], Yong Chen[3], Yueh Lee[1,2], and Pew-Thian Yap[1,2(✉)]

[1] Department of Biomedical Engineering, University of North Carolina, Chapel Hill, USA
ptyap@med.unc.edu
[2] Department of Radiology and Biomedical Research Imaging Center, University of North Carolina, Chapel Hill, USA
[3] Department of Radiology, Case Western Reserve University, Cleveland, USA

Abstract. In magnetic resonance imaging (MRI), noise is a limiting factor for higher spatial resolution and a major cause of prolonged scan time, owing to the need for repeated scans. Improving the signal-to-noise ratio is therefore key to faster and higher-resolution MRI. Here we propose a method for mapping and reducing noise in MRI by leveraging the inherent redundancy in complex-valued multi-channel MRI data. Our method leverages a provably optimal strategy for shrinking the singular values of a data matrix, allowing it to outperform state-of-the-art methods such as Marchenko-Pastur PCA in noise reduction. Our method reduces the noise floor in brain diffusion MRI by 5-fold and remarkably improves the contrast of spiral lung ^{19}F MRI. Our framework is fast and does not require training and hyper-parameter tuning, therefore providing a convenient means for improving SNR in MRI.

Keywords: Magnetic resonance imaging · Noise removal · Optimal shrinkage

1 Introduction

MRI is a trade-off between spatial resolution and noise—an increase in spatial resolution is associated with a proportional decrease in signal-to-noise ratio (SNR) [1]. While higher resolution is conducive to preserving fine image details, the accompanying increase in the level of noise eventually overwhelms the details, counteracting the benefits of increased spatial resolution [2]. Increasing SNR is therefore essential for high-resolution MRI.

SNR improvement can be achieved with better RF coils and higher magnetic field strength [3]. However, coil improvement has plateaued and higher magnetic

This work was supported in part by United States National Institutes of Health (NIH) grants MH125479 and EB006733.

© Springer Nature Switzerland AG 2021
M. de Bruijne et al. (Eds.): MICCAI 2021, LNCS 12906, pp. 191–200, 2021.
https://doi.org/10.1007/978-3-030-87231-1_19

field strength is not widely available nor suitable for clinical use [4–6]. Another approach is to repeat acquisitions. Since SNR is proportional to the square-root of the number of averages [7], a two-fold increase in spatial resolution in three dimensions will require a $2^3 = 8$-fold increase in SNR or $8^2 = 64$ repetitions. This approach is costly, uncomfortable for patients, and can be complicated by motion and physiological artifacts.

Many post-acquisition techniques have been proposed to improve SNR in MRI. A popular approach is based on random matrix theory (RMT) [1]. A full-rank noisy data matrix can be reduced to a lower-rank noise-reduced matrix by fitting a curve to the Marchenko-Pastur (MP) [8] distribution of the matrix's eigenvalues and discarding noise-related eigenvalues. Ma et al. [9] add a variance-stabilizing transformation (VST) framework to Gaussianize Rician noise in magnitude MR images so that assumptions in RMT hold. Although effective, these two state-of-the-art methods work on channel-combined magnitude data, neglecting the additional information available in multi-channel complex-valued data.

In this paper, we introduce an automated framework for noise mapping and removal of multi-channel complex-valued data using optimal shrinkage of singular values [10]. Our method effectively handles background phase contamination and utilizes multi-channel information, thus yielding results superior to state-of-the-art methods, significantly lowering the noise floor in diffusion MRI and drastically improving the contrast of fast ^{19}F spiral MRI.

2   Methods

2.1   Redundancy in MR Data

Modern MRI data are usually acquired with multi-channel coil. The information between different channels, volumes (e.g., q-space samples in diffusion MRI (dMRI), timepoints in functional MRI (fMRI), repeated scans, or scans from alternative modalities), neighboring voxels share some degree of similarity. Information is redundant because we are studying the same brain, just from different vantage points [1]. It has been shown in [11,12] using redundant information from multiple channels would improve denoising results.

2.2   Optimal Shrinkage of Singular Value and Noise Estimation

A 3-D signal tensor S_{XCV} with indices $x = 1, \ldots, X$, $c = 1, \ldots, C$, and $v = 1, \ldots, V$ corresponding to voxels, channels, and volumes, respectively, can be reshaped as an $M \times N$ matrix S_{MN} with $MN = XCV$. To utilize RMT [1,10,13], we ensure $0 < \frac{M}{N} \leq 1$ and M as large as possible. We take the largest among X, C, and V and compute the product of the other two. We then set N to the larger of the two and M to the other. Due to redundancy, matrix S_{MN} is inherently low rank. With noise, matrix S_{MN} is however full rank. We shrink the singular values of the noisy full-rank matrix S_{MN} to estimate a noise-free low-rank matrix $\hat{S}_{MN}$.

This is achieved by employing optimal shrinkage singular value decomposition (OS-SVD) [10] to shrink the singular values while minimizing the nuclear norm $\|S_{MN} - \hat{S}_{MN}\|_*$. We choose OS-SVD for its theoretically provable optimality [10] and the nuclear norm as the loss function for preservation of structural information [14]. For any singular value s_i of decomposition

$$S_{MN} = \sum_{i=1}^{M} s_i v_i \tilde{v}_i^*, \tag{1}$$

where v_i and $\tilde{v}_i$ are the left and right singular vectors corresponding to s_i, the asterisk denotes the conjungate transpose. The optimal shrinkage $\eta(s_i)$ minimizing the nuclear norm is

$$\eta(s) = \begin{cases} \frac{1}{z^2 y}(z^4 - \delta - \sqrt{\delta}zy), & \text{if } z^4 \geq \delta + \sqrt{\delta}zy, \\ 0, & \text{if } z^4 < \delta + \sqrt{\delta}zy, \end{cases} \tag{2}$$

where $y = \frac{s}{\sigma}$ with noise standard deviation σ, $\delta = \frac{M}{N}$, and

$$z(y) = \frac{1}{\sqrt{2}}\sqrt{y^2 - \delta - 1 + \sqrt{(y^2 - \delta - 1)^2 - 4\delta}}. \tag{3}$$

A robust estimator of σ [10] is

$$\hat{\sigma} = \frac{s_{\text{med}}}{\sqrt{N \cdot \mu_\delta}}, \tag{4}$$

where s_{med} is the median singular value of S_{MN} and μ_δ is the median of the MP distribution, which is the unique solution for $\delta_- \leq x \leq \delta_+$ to the equation

$$\int_{\delta_-}^{x} \frac{\sqrt{(\delta_+ - t)(t - \delta_-)}}{2\pi t} dt = \frac{1}{2}, \tag{5}$$

where $\delta_\pm = (1 \pm \sqrt{\delta})^2$. The estimated noise-reduced data matrix is then

$$\hat{S}_{MN} = \hat{\sigma} \sum_{i=1}^{M} \eta(s_i) v_i \tilde{v}_i^*. \tag{6}$$

2.3   Noise Mapping and Removal

We propose an automated framework for noise mapping and removal with four main steps: (i) channel decorrelation, (ii) phase unwinding, (iii) noise removal, and (iv) channel combination.

Channel Decorrelation: The RMT assumption that noise is independent and identically distributed does not always hold because noise can be highly correlated across channels [15]. We perform channel decorrelation using Mahalanobis whitening transformation:

$$S \leftarrow (\Psi^{-1/2})S, \quad \Psi^{-1/2} = U\psi^{-1/2}U^*, \tag{7}$$

where $\psi = U^*\Psi U$ is the diagonal matrix of eigenvalues of the noise covariance matrix Ψ, estimated by acquiring channel noise without excitation.

Phase Unwinding: The complex-valued MR signal S with magnitude I and phase φ is

$$S = Ie^{i\varphi}. \tag{8}$$

The phase $\varphi = \varphi_{BG} + \varphi_n$ can be decomposed into two parts: random noise phase φ_n and background phase φ_{BG} that contains unwanted information from physiological effects and coil variations [16–18]. We estimate the background phase by performing OS-SVD separately on the real and imaginary parts of the noisy data. Briefly, for each voxel, a local $3 \times 3 \times 3$ block ($X = 27$) across all channels and volumes is used to form S_{XCV}, which is then reshaped to S_{MN}. We then use (6) block-wise to denoise the data and then compute the background phase φ_{BG}. We unwind the phase to obtain phase-corrected signal S_{corr} as

$$S_{\mathrm{corr}} \approx Ie^{i(\varphi - \varphi_{BG})}. \tag{9}$$

After phase unwinding, the imaginary part contains only pure noise and is discarded. Only the real part of S_{corr} is used in subsequent steps.

Noise Mapping and Removal: We employ a sliding-block approach for noise mapping and removal. Similar to the previous step, for each voxel j of the real part of S_{corr}, a local $3 \times 3 \times 3$ block ($X = 27$) is used to form S_{MN}. Then the noise standard deviation $\sigma(j)$ is estimated using (4) and the noise-free signal is recovered using (6). Since a voxel can be in multiple blocks, the final noise-removed signal $\hat{S}(j)$ is estimated as

$$\hat{S}(j) = \frac{\sum_{l=1}^{L(j)} w_l(j)\hat{S}_l(j)}{\sum_{l=1}^{L(j)} w_l(j)}, \tag{10}$$

where $L(j)$ is the number of blocks containing voxel j, $\hat{S}_l(j)$ is the estimated noise-free signal of voxel j in l-th block, and w_l is the corresponding weight. To reduce Gibbs artifacts, we assign a higher weight to a lower-rank block [9]:

$$w_l(j) = \frac{1}{1 + R_l(j)}, \tag{11}$$

where $R_l(j)$ is the rank of $\hat{S}_l(j)$.

Phase Rewinding (Optional): After noise removal, $\hat{S}$ contains the noise-removed real part of S_{corr} with zero phase. If necessary (e.g., for simultaneous multislice (SMS) reconstruction [2]), we rewind the phase by φ_{BG}.

Channel Combination: Multi-channel real-valued noise-free data $\hat{S}$ can be combined using sum-of-squares, spatial matched filter, or coil sensitivity profiles [2, 19].

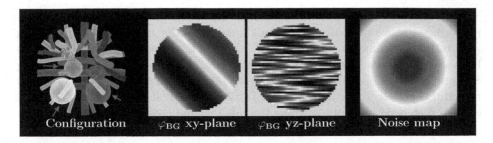

Fig. 1. Data simulation. Configuration of phantom with red arrows pointing to CSF-like regions (left), background phase (middle), and non-stationary noise map (right). (Color figure online)

3 Experiments

For all experiments, we compare our framework with two state-of-the-art approaches and one in-house method based on RMT:

1. RMT denoising of channel-combined magnitude data using MP curve fitting, implemented as 'dwidenoise' in MRtrix3 [1],
2. RMT denoising of channel-combined magnitude data using OS-SVD and VST [9], denoted as 'VST-Mag', and
3. In-house RMT denoising of multi-channel complex-valued (MCC) data using MP curve fitting as in [11], denoted as 'MCC-MP'.

For our method and MCC-MP, a $3 \times 3 \times 3$ block is used. For dwidenoise and VST-Mag a $5 \times 5 \times 5$ block is used as in [1,9].

3.1 Numerical Validation

We simulated noise-free dMRI data with Phantomas [20]. 24, 48, and 96 diffusion-weighted images (DWIs) for $b = 1000, 2000, 3000\,\mathrm{s\,mm^{-2}}$, respectively, and five non-diffusion-weighted images (NDWIs, $b = 0\,\mathrm{s\,mm^{-2}}$) were generated. The noisy signal of channel c, out of 32 channels, is given as

$$S_c = Se^{i\varphi_{\mathrm{BG}}} + \mathcal{N}_c^{\mathrm{R}}(0,\sigma) + i\mathcal{N}_c^{\mathrm{I}}(0,\sigma), \tag{12}$$

where S is the simulated noise-free signal, $\mathcal{N}_c^{\mathrm{R}}$ and $\mathcal{N}_c^{\mathrm{I}}$, are respectively the noise added to the real and imaginary parts of channel c, respectively. The background phase φ_{BG} was created with a bi-dimensional sinusoidal wave in the xy-plane with random initial shifts in the yz-plane to mimic the smooth appearance within each slice but vigorous transition across slices [18]. A non-stationary noise map σ was generated with $\mathrm{SNR} = 2\text{--}15$ (Fig. 1).

Visual Inspection: Our method recovers the signal from the corrupted noisy data, showing structural details similar to the noise-free ground truth (Fig. reffig:phantomdenoise). MCC-MP also yields comparable results as it also harnesses the redundancy of multi-channel data. Both dwidenoise and VST-Mag can remove some of the noise, but they do not perform as well as the proposed method, especially at higher b values. These two methods use channel-combined data and thus do not fully leverage the inherent redundancy across channels.

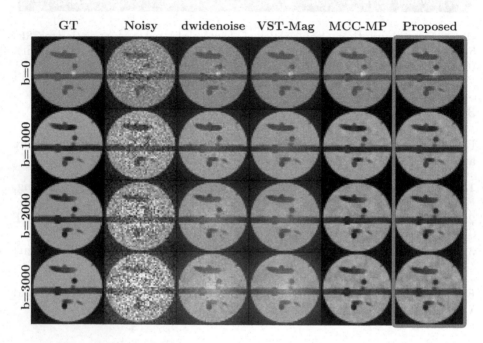

Fig. 2. Denoising of simulated data. Noise-free ground truth (GT) and denoised images given by various methods.

Mean Relative Error: We computed the mean relative errors of the noisy and denoised data with respect to the ground truth. We tested the denoising algorithms with varying V: 26 (1/8 of the total DWIs and 5 NDWIs), 47 (1/4 of the total DWIs and 5 NDWIs), 89 (1/2 of the total DWIs and 5 NDWIs), and 173 (all DWIs and NDWIs). For the different values of V, our method achieves the lowest mean relative errors, followed by MCC-MP (Fig. 3 left). For these two methods, the error decreases with V. A larger V implies a larger M, thus more singular values for shrinkage or MP curve fitting. VST-Mag and dwidenoise are significantly worse than the proposed method.

Noise Floor: We utilized the free-water diffusion signals in cerebrospinal fluid (CSF) regions (red arrows in Fig. 1) to quantify the effects of denoising on the noise floor. Our method reduces the noise floor by 5-fold (Fig. 3 - Right), twice

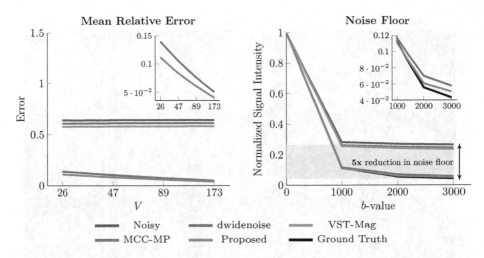

Fig. 3. Mean relative error and noise floor. Left: The error with respect to the number of image volumes. Right: Normalized CSF signal across b-values. Zoomed-in views are provided for closer comparison. The shaded red region marks the noise floor of the noisy data. (Color figure online)

better than MCC-MP. Our method significantly outperforms dwidenoise and VST-Mag.

3.2 In-Vivo High-Resolution Diffusion MRI

We evaluated our method with the diffusion MRI data of a healthy subject: Single-band acquisition, 32-channel head coil, 1 mm isotropic resolution, a total of 65 gradient directions for 4 b-values of 500, 1000, 2000, 3000 s mm^{-2} and an NDWI ($C = 32$, $V = 66$). Our approach is more effective in removing noise by giving cleaner images, especially at high b-values, compared with other methods (Fig. 4). MP curve fitting is less effective than OS-SVD when V is small, causing slightly worse results for MCC-MP than our method, notably in the background.

3.3 In-Vivo Human Lung MRI

We also tested our method with a ^{19}F contrast-enhanced MRI lung dataset, acquired with 4 spiral arms, 8 channels, and 12 repetitions ($C = 8$, $V = 12$), TR $= 11$ ms, and TE $= 0.48$ ms. The acquisition time is 0.4 s compared with 16 s using conventional ^{19}F MRI, reducing the SNR by a factor of $\sqrt{16/0.4} \approx 6.3$. Denoising is hence essential to recover the contrast. Our framework yields strikingly good contrast with clear details compared with other methods, which are unable to remove the noise (Fig. 5). MCC-MP does not yield acceptable results, mainly because of the smaller C and V, leading to bad MP curve-fitting. This observation highlights the limitation of MCC-MP when applied to data with only a small number of channels and volumes. This is due to the reliance of MCC-MP on a sufficient number of eigenvalues for reliable MP curve-fitting.

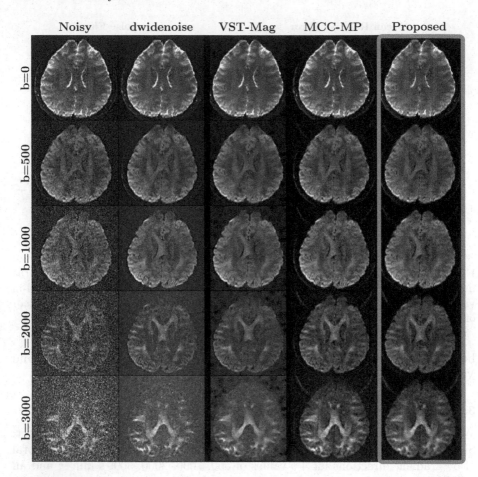

Fig. 4. Diffusion-weighted images with different b-values. Noisy images and denoising results given by the different methods.

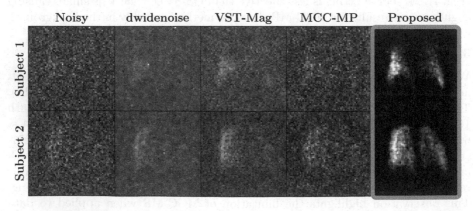

Fig. 5. In-Vivo ^{19}F Lung MRI. Noisy images and denoising results given by the different methods.

4   Conclusion

We have presented a denoising method for fast, accurate, and effective recovery of the MRI signal. The synergistic combination of multi-channel information, phase correction, noise estimation, and optimal signal recovery allows our method to outperform state-of-the-art approaches with remarkably lower noise floor and greater structural visibility. Our method does not require training data and hyper-parameter tuning, and is hence a convenient means for improving SNR to improve the utility of MRI.

References

1. Veraart, J., Fieremans, E., Novikov, D.S.: Diffusion MRI noise mapping using random matrix theory. Magn. Reson. Med. **76**(5), 1582–1593 (2016)
2. Chang, W.T., Huynh, K.M., Yap, P.T., Lin, W.: Navigator-free submillimeter diffusion imaging using multishot-encoded simultaneous multi-slice (MUSIUM). arXiv preprint arXiv:2012.00664 (2020)
3. Budinger, T.F., et al.: Toward 20 T magnetic resonance for human brain studies: opportunities for discovery and neuroscience rationale. Magn. Reson. Mater. Phys. Biol. Med. **29**(3), 617–639 (2016)
4. Fan, Q., et al.: MGH-USC human connectome project datasets with ultra-high B-value diffusion MRI. Neuroimage **124**, 1108–1114 (2016)
5. Kraff, O., Quick, H.H.: 7T: physics, safety, and potential clinical applications. J. Magn. Reson. Imaging **46**(6), 1573–1589 (2017)
6. Ocali, O., Atalar, E.: Ultimate intrinsic signal-to-noise ratio in MRI. Magn. Reson. Med. **39**(3), 462–473 (1998)
7. Haldar, J.P., Liu, Y., Liao, C., Fan, Q., Setsompop, K.: Fast submillimeter diffusion MRI using gSlider-SMS and SNR-enhancing joint reconstruction. Magn. Reson. Med. **84**(2), 762–776 (2020)
8. Marchenko, V.A., Pastur, L.A.: Distribution of eigenvalues for some sets of random matrices. Matematicheskii Sbornik **114**(4), 507–536 (1967)
9. Ma, X., Uğurbil, K., Wu, X.: Denoise magnitude diffusion magnetic resonance images via variance-stabilizing transformation and optimal singular-value manipulation. NeuroImage 116852 (2020)
10. Gavish, M., Donoho, D.L.: Optimal shrinkage of singular values. IEEE Trans. Inf. Theory **63**(4), 2137–2152 (2017)
11. Lemberskiy, G., Baete, S., Veraart, J., Shepherd, T.M., Fieremans, E., Novikov, D.S.: Achieving sub-MM clinical diffusion MRI resolution by removing noise during reconstruction using random matrix theory. In: Proceedings of the ISMRM 27th Annual Meeting, vol. 27 (2019)
12. Lemberskiy, G., Baete, S., Veraart, J., Shepherd, T.M., Fieremans, E., Novikov, D.S.: MRI below the noise floor. In: Proceedings of the ISMRM 28th Annual Meeting, vol. 28 (2020)
13. Cordero-Grande, L., Christiaens, D., Hutter, J., Price, A.N., Hajnal, J.V.: Complex diffusion-weighted image estimation via matrix recovery under general noise models. Neuroimage **200**, 391–404 (2019)
14. Daducci, A., Van De Ville, D., Thiran, J.P., Wiaux, Y.: Sparse regularization for fiber ODF reconstruction: from the suboptimality of l_2 and l_1 priors to l_0. Med. Image Anal. **18**(6), 820–833 (2014)

15. Pruessmann, K.P., Weiger, M., Scheidegger, M.B., Boesiger, P.: SENSE: sensitivity encoding for fast MRI. Magn. Reson. Med. **42**(5), 952–962 (1999)
16. Eichner, C., et al.: Real diffusion-weighted MRI enabling true signal averaging and increased diffusion contrast. Neuroimage **122**, 373–384 (2015)
17. Pizzolato, M., Gilbert, G., Thiran, J.P., Descoteaux, M., Deriche, R.: Adaptive phase correction of diffusion-weighted images. NeuroImage **206**, 116274 (2020)
18. Pizzolato, M., Fick, R., Boutelier, T., Deriche, R.: Noise floor removal via phase correction of complex diffusion-weighted images: influence on DTI and q-space metrics. In: Fuster, A., Ghosh, A., Kaden, E., Rathi, Y., Reisert, M. (eds.) MICCAI 2016. MV, pp. 21–34. Springer, Cham (2017). https://doi.org/10.1007/978-3-319-54130-3_2
19. Dietrich, O., Raya, J.G., Reeder, S.B., Ingrisch, M., Reiser, M.F., Schoenberg, S.O.: Influence of multichannel combination, parallel imaging and other reconstruction techniques on MRI noise characteristics. Magn. Reson. Imaging **26**(6), 754–762 (2008)
20. Caruyer, E., Daducci, A., Descoteaux, M., Houde, J.C., Thiran, J.P., Verma, R.: Phantomas: a flexible software library to simulate diffusion MR phantoms. In: ISMRM (2014)

Self Context and Shape Prior for Sensorless Freehand 3D Ultrasound Reconstruction

Mingyuan Luo[1,2,3], Xin Yang[1,2,3], Xiaoqiong Huang[1,2,3], Yuhao Huang[1,2,3], Yuxin Zou[1,2,3], Xindi Hu[4], Nishant Ravikumar[5,6], Alejandro F. Frangi[1,5,6,7], and Dong Ni[1,2,3(✉)]

[1] National-Regional Key Technology Engineering Laboratory for Medical Ultrasound, School of Biomedical Engineering, Health Science Center, Shenzhen University, Shenzhen, China
nidong@szu.edu.cn
[2] Medical Ultrasound Image Computing (MUSIC) Lab, Shenzhen University, Shenzhen, China
[3] Marshall Laboratory of Biomedical Engineering, Shenzhen University, Shenzhen, China
[4] School of Biomedical Engineering and Information, Nanjing Medical University, Nanjing , China
[5] Centre for Computational Imaging and Simulation Technologies in Biomedicine (CISTIB), University of Leeds, Leeds, UK
[6] Leeds Institute of Cardiovascular and Metabolic Medicine, University of Leeds, Leeds, UK
[7] Medical Imaging Research Center (MIRC), KU Leuven, Leuven, Belgium

Abstract. 3D ultrasound (US) is widely used for its rich diagnostic information. However, it is criticized for its limited field of view. 3D freehand US reconstruction is promising in addressing the problem by providing broad range and freeform scan. The existing deep learning based methods only focus on the basic cases of skill sequences, and the model relies on the training data heavily. The sequences in real clinical practice are a mix of diverse skills and have complex scanning paths. Besides, deep models should adapt themselves to the testing cases with prior knowledge for better robustness, rather than only fit to the training cases. In this paper, we propose a novel approach to sensorless freehand 3D US reconstruction considering the complex skill sequences. Our contribution is three-fold. First, we advance a novel online learning framework by designing a differentiable reconstruction algorithm. It realizes an end-to-end optimization from section sequences to the reconstructed volume. Second, a self-supervised learning method is developed to explore the context information that reconstructed by the testing data itself, promoting the perception of the model. Third, inspired by the effectiveness of shape prior, we also introduce adversarial training to strengthen the learning of anatomical shape prior in the reconstructed volume. By mining the context and structural cues of the testing data, our online

---

M. Luo and X. Yang—Contribute equally to this work.

© Springer Nature Switzerland AG 2021
M. de Bruijne et al. (Eds.): MICCAI 2021, LNCS 12906, pp. 201–210, 2021.
https://doi.org/10.1007/978-3-030-87231-1_20

learning methods can drive the model to handle complex skill sequences. Experimental results on developmental dysplasia of the hip US and fetal US datasets show that, our proposed method can outperform the start-of-the-art methods regarding the shift errors and path similarities.

Keywords: Self context · Shape prior · Freehand 3D ultrasound

1   Introduction

Ultrasound (US) imaging is one of the main diagnostic tools in clinical due to its safety, portability, and low-cost. 3D US is increasingly used in clinical diagnosis [6,8,9] because of its rich context informations which are not offered in 2D US. However, 3D US probe is constrained by the limited field of view and poor operability. Thus, exploring 3D freehand US reconstruction from a series of 2D frames has great application benefits [11]. However, this kind of reconstruction is non-trivial due to the complex in-plane and out-plane shifts among adjacent frames, which are caused by the diverse scan skills and paths. In this paper, as the illustration in Fig. 1, besides the basic linear scan, we also consider three typical scan skills and their hybrid cases which are rarely handled in previous studies, including loop scan, fast-and-slow scan, and sector scan.

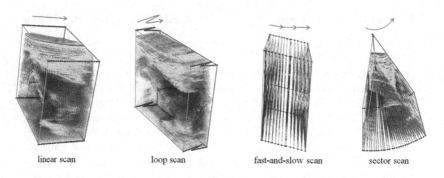

linear scan   loop scan   fast-and-slow scan   sector scan

Fig. 1. Complex scan sequences considered in this work.

External sensor-based solutions can build tracking signal for 3D reconstruction [5,7]. But they are facing magnetic influences, optical occlusion, and expensive costs. Therefore, US volume reconstruction from sensorless freehand scans plays an important role and has tremendous potential applications. Most previous researches are based on the non-deep learning methods [10,12]. Prevost et al. [15] pioneered the deep learning based estimation of relative motion between US images, and later they extended their works by introducing the extra optical flow [2] and integrating a sensor source [14]. Guo et al. [3] proposed a deep contextual learning network for reconstruction, which utilizes 3D convolutions over US video clips for feature extraction. However, these methods mainly handle the basic cases, rather than the complex scan sequences, such as the loop scan and fast-slow scan as shown in Fig. 1. In addition, previous models rely on the training data heavily and ignore the reconstruction robustness on the complex sequences during the test phase. Recently though, self-supervised learning (SSL) and adversarial learning (ADL) strategies in deep learning have been widely proven to learn an enriched representation from image itself and improve method robustness [1, 4, 18].

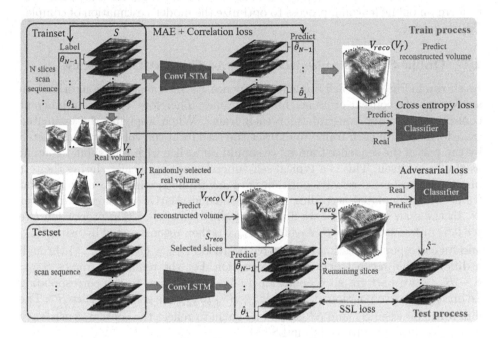

Fig. 2. Overview of our proposed online learning framework.

For the limitations described above, we propose a novel approach to effectively address the difficulty of sensorless freehand 3D US reconstruction under complex skill sequences. Our contribution is three-fold. First, a novel online learning framework is advanced by devising a differentiable algorithm to approximate the reconstruction process. It enables an end-to-end, recurrent optimization from section sequence to the whole reconstructed volume. Second, based on the differentiable design, a SSL scheme is developed to explore the context cues contained in the reconstructed volumes as pseudo supervisions, and hence regularize the prediction of future frames. Third, online ADL is introduced to strengthen the representation learning of anatomical shape priors, and thereby avoid irregular reconstructions. SSL and ADL in the online form are general. They can enable our method to generate plausible visual reconstructions when facing the difficult but practical scan cases that are shown in Fig. 1.

2   Methodology

Figure 2 shows the proposed online learning framework, including the recurrent Convolutional LSTM (ConvLSTM) backbone [16] and the online learning framework to adapt to complex skill sequences. During the training process, the N-length sequence $S = \{s_i | i = 1, \ldots, N\}$ is input into the ConvLSTM to predict the 3D relative transformation parameters $\widehat{\Theta} = \{\widehat{\theta}_i | i = 1, \ldots, N-1\}$ among all adjacent frames, where $\widehat{\theta}_i$ (or θ_i) represents a rigid transformation containing 3 translation and 3 rotation degrees. During the test, SSL and ADL are combined to form an online learning process to optimize the model's estimation of complex skill sequences.

2.1   Online Self-supervised Learning for Context Consistency

As shown in Fig. 2, for each frame in a sequence, its 3D transformation estimation aggregates the context of its neighboring frames. Therefore, there is an inherent context consistency constraint for each frame. When applying the estimated transformation of each frame to extract the slice from the volume reconstructed by the rest of the sequence frames, we should get a slice with very similar content as the frame itself. This is a typical self-supervision. Specifically, during the test phase, the estimated relative position $\widehat{p}_i$ of any frame s_i with respect to the first frame s_1 is calculated according to the estimated relative transform parameter $\widehat{\Theta}$. In the scan sequence S, we firstly uniformly select a certain proportion (we set 0.5) of frames $S_{reco} \subsetneq S$. A volume V_{reco} is then reconstructed by using S_{reco} and its corresponding estimated relative position $\widehat{P}_{reco} = \{\widehat{p}_i | s_i \in S_{reco}\}$ through a differentiable reconstruction approximation. For the remaining frames $S^- = S - S_{reco}$, we perform slice operation in V_{reco} according to the corresponding estimated relative position $\widehat{P}^- = \{\widehat{p}_i | s_i \in S^-\}$ to get the generated slice $\widehat{S}^-$. The difference between S^- and $\widehat{S}^-$ should be small to reflect the context consistency and is used to optimize the ConvLSTM.

2.2   Online Adversarial Learning for Shape Constraint

Shape prior is a strong constraint to regularize the volume reconstruction. Exposing the models to shape prior under the online learning scheme, rather than the offline training, provides us more chances to better explore the constraint for more generalizability, especially in handling the complex scan sequences where unplausible reconstructions often occur. Specifically, as shown in Fig. 2, we propose to encode the shape prior with an adversarial module. The volume reconstruction V_r from the training set is randomly selected as a sample with real anatomical structures. While the volume V_f reconstructed from the testing sequence S with relative estimated position parameters $\widehat{P}$ is taken as a sample of fake structure. A classifier C pre-trained to distinguish between the volume V_r and V_f reconstructed from all training sequences serves as the adversarial discriminator. It adversarially tunes the structure contained in the reconstructed volume via the proposed differentiable reconstruction approximation.

2.3   Differentiable Reconstruction Approximation

The end-to-end optimization of our online learning framework is inseparable from the differentiable reconstruction. In general, most of the unstructured interpolation operations in the reconstruction algorithm (such as the Delaunay algorithm) are not differentiable and make our online learning blocked. As a foundation of our online design, we firstly propose a differentiable reconstruction approximation to mimic the interpolation. As shown in Fig. 3, the volume V is reconstructed using the N slices $\{s_j | j = 1, \ldots, N\}$. For any pixel $v_i \in V$, the distance d_{ij} from any slice s_j and the gray value G_{ij} of the projection points on any slice s_j are calculated. Then the reconstructed gray value G_{v_i} at pixel v_i is calculated as:

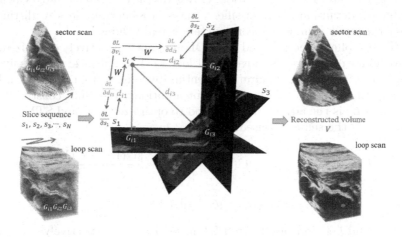

Fig. 3. Two reconstruction examples of differentiable reconstruction approximation.

$$G_{v_i} = \sum_j W(d_{ij})G_{ij} \tag{1}$$

Among them, $W(\cdot)$ is a weight function. Its purpose is to encourage small d_{ij}. The core formulation is softmax operation on the reciprocal of d_{ij}:

$$W(d_{ij}) = \frac{\exp(1/(d_{ij} + \epsilon))}{\sum_j \exp(1/(d_{ij} + \epsilon))}, \tag{2}$$

where ϵ prevents division by 0. If only the nearest slice is weighted, the approximation can be expressed as follows:

$$W(d_{ij}) = \frac{1}{Z}\left(\frac{1}{d_{ij} + \epsilon} \frac{\exp(1/(d_{ij} + \epsilon))}{\sum_j \exp(1/(d_{ij} + \epsilon))}\right), \tag{3}$$

where $1/Z$ is the normalization coefficient such that $\sum_j W(d_{ij}) = 1$.

Figure 3 shows two reconstruction examples. The left volume of each example directly puts the slice into the volume according to the position without interpolation, and the right is the result of differentiable reconstruction approximation.

2.4   Loss Function

In the training phase, the ConvLSTM predicts relative transform parameters $\widehat{\Theta}$ between all adjacent frames in the skill sequence. Its loss function includes two items, the first item is the mean absolute error (MAE) loss (i.e., L1 normalization), and the second item is the case-wise correlation loss from [3], which is beneficial to improve the generalization performance.

$$L = \|\widehat{\Theta} - \Theta\|_1 + (1 - \frac{\mathbf{Cov}(\widehat{\Theta}, \Theta)}{\sigma(\widehat{\Theta})\sigma(\Theta)}) \tag{4}$$

where $\|\cdot\|_1$ indicates L1 normalization, **Cov** gives the covariance, and σ calculates the standard deviation. The classifier C uses cross-entropy loss to distinguish whether the input volume is reconstructed or real volume.

In the test phase, the SSL and the ADL are jointly iteratively optimized. The first two items of L_d are the adversarial loss used to optimize the classifier C, and the third item is the quadratic potential divergence from [17], which helps to prevent the gradient vanishing without additional Lipschitz constraint. The first term of L_g is the adversarial loss used to optimize the ConvLSTM, and the second term is the self-supervised loss.

$$L_d = \mathbb{E}_{V_f \sim \mathbb{P}_{V_f}, V_r \sim \mathbb{P}_{V_r}}[C(V_f) - C(V_r) + \frac{\|C(V_f) - C(V_r)\|_2^2}{2\|V_f - V_r\|_1}] \tag{5}$$

$$L_g = -\mathbb{E}_{V_f \sim \mathbb{P}_{V_f}}[C(V_f)] + \|\widehat{S}^- - S^-\|_1 \tag{6}$$

where $\|\cdot\|_1$ and $\|\cdot\|_2$ indicate L1 and L2 normalization, respectively.

3   Experiments

Materials and Implementation. Our experiments involve two datasets, i.e., 3D developmental dysplasia of the hip (DDH) US dataset and 3D fetus US dataset. The DDH dataset contains 101 US volumes from 14 volunteers. Its average volume size is 370 × 403 × 398 with voxel resolution 0.1 × 0.1 × 0.1 mm. The fetal dataset contains 78 fetal US volumes from 78 volunteers. The gestational age ranges from 10 to 14 weeks. Its average volume size is 402 × 535 × 276, the voxel resolution is 0.3 × 0.3 × 0.3 mm. All data collection is anonymous. The collection and use of data are approved by the local IRB.

It is time-consuming and expensive to collect a large number of real freehand sequences with complex scan skills. In order to verify the proposed framework on the complex skill sequences, the real 3D volumes are used to generate massive complex skill sequences for our training and evaluation. Multiple complex scan sequences are dynamically simulated with diverse scan skills from each US volume to form our final corpus. Specifically, the scan sequences are a complex combination of loop scan, fast-and-slow scan and sector scan with the aim to simulate the loop movement, uneven speed, and anisotropic movement of probe.

For the 3D DDH and fetal US dataset, we randomly divide the dataset into 85/15 and 65/13 volumes for training/testing according to the patients, and the sequence length is 120 and 90, respectively. The number of generated sequences is 100 and 10 for training and testing, respectively. All generated slices are with size 300×300 pixel. In the training phase, the Adam optimizer is used to iteratively optimize our ConvLSTM. The epoch and batchsize are 200 and 4, respectively. The learning rate is 10^{-3}, and the learning rate is reduced by half every 30 epochs. The ConvLSTM has 1 layer and convolutional kernel size is 3. For classifier C, the epoch, batchsize and learning rate is 50, 1 and 10^{-4}, respectively. During the test phase, for each testing sequence, we iterate the online learning with adversarial and self-supervised losses for 30 iterations, and the learning rate is set as 10^{-6}. All codes are implemented in PyTorch [13].

Table 1. The mean (std) results of different methods on the DDH sequences.

Methods	FDR (%)	ADR (%)	MD (mm)	SD (mm)	HD (mm)
CNN [14]	16.25 (10.45)	44.58 (8.71)	21.14 (4.11)	1798.71 (433.18)	21.10 (4.10)
DCL-Net [3]	58.43 (23.73)	101.09 (32.81)	32.73 (11.25)	2060.74 (780.65)	24.62 (10.39)
ConvLSTM	11.15 (3.97)	22.08 (9.03)	7.61 (2.99)	498.84 (197.32)	6.63 (2.77)
ConvLSTM+SSL	10.89 (4.05)	21.93 (9.03)	7.52 (2.94)	493.44 (197.70)	6.58 (2.76)
ConvLSTM+ADL	6.21 (3.09)	13.89 (6.95)	4.70 (2.30)	291.01 (142.08)	4.10 (1.88)
ConvLSTM+SSL+ADL	5.44 (3.03)	13.47 (6.47)	4.44 (2.29)	274.14 (136.89)	3.91 (1.89)

Table 2. The mean (std) results of different methods on the fetus sequences.

Methods	FDR (%)	ADR (%)	MD (mm)	SD (mm)	HD (mm)
CNN [14]	16.59 (9.67)	31.82 (14.02)	24.50 (10.43)	1432.89 (698.88)	19.16 (8.21)
DCL-Net [3]	49.71 (36.57)	71.13 (46.67)	45.63 (30.75)	1993.35 (1339.98)	43.82 (30.76)
ConvLSTM	16.71 (5.22)	31.33 (12.68)	20.57 (8.33)	938.42 (391.39)	15.79 (6.78)
ConvLSTM+SSL	15.86 (5.10)	30.78 (12.20)	20.01 (8.27)	909.95 (382.43)	15.59 (6.53)
ConvLSTM+ADL	10.49 (4.86)	28.15 (9.86)	17.49 (7.68)	762.79 (292.92)	14.18 (5.99)
ConvLSTM+SSL+ADL	9.94 (4.41)	27.08 (9.31)	16.84(7.56)	730.11 (303.20)	14.12 (5.89)

Quantitative and Qualitative Analysis. The current commonly used evaluation indicator is the final drift [3,14], which is the drift of final frame of a sequence, and the drift is the distance between the center points of the frame according to the real relative position and the estimated relative position. On this basis, a series of indicators are used to evaluate the accuracy of our proposed framework in estimating the relative transform parameters among adjacent frames. Final drift rate (FDR) is the final drift divided by the sequence length. Average drift rate (ADR) is the cumulative drift of all frames divided by the length from the frame to the starting point of the sequence, and finally, the average value is calculated. Maximum drift (MD) is the maximum accumulated drift of all frames. Sum of drift (SD) is the sum of accumulated drift of all frames. The bidirectional Hausdorff distance (HD) emphasizes the worst distances between the predicted positions (accumulatively calculated by the relative transform parameters) and the real positions of all the frames in the sequence.

Table 1 and 2 summarize the overall comparison of the proposed online learning framework with other existing methods and ablation frameworks on simulated DDH or fetus complex skill sequences, respectively. "CNN" [14] and "DCL-Net" [3] are considered in this study for comparison. "SSL" and "ADL" are our proposed online self-supervised learning and adversarial learning method, respectively. As can be seen from Table 1 and 2, our proposed methods get consistent and better improvements both on DDH and fetal complex skill sequences. The comparison with the ablation frameworks fully proves the necessity of introducing online learning in the test phase. In particular, our proposed online learning methods help the ConvLSTM improve the accuracy by 5.71%/2.72 mm for DDH sequences, and 6.77%/1.67 mm for fetal sequences in terms of the FDR/HD indicators. Under the complementary effects of mining context consistency and shape constraint, the proposed framework has significantly robust the estimation accuracy of complex skill sequences.

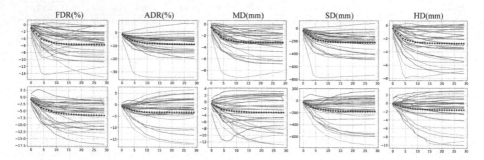

Fig. 4. Indicator declining curves for online iterative optimization on DDH (the first row) and fetus (the second row) US datasets.

Figure 4 shows the indicator declining curves for online iterative optimization of the ConvLSTM on DDH and fetus US datasets. In each subfigure of Fig. 4, the abscissa and ordinate represent the number of iterations and the extent of indicators decline, respectively. The blue dot curve is the average indicator declining curve of all data. It is obvious that the indicators curve of almost all single sequences is declining significantly, and the downward trend is from rapid decline to gentle convergence. Among the various indicators, FDR has the largest decline, which is reduced to 48.78% and 59.48% of the ConvLSTM on the DDH and fetus US dataset, respectively. Figure 5 shows four specific cases of online iterative optimization on DDH and fetus complex skill sequences, respectively. In each subfigure of Fig. 5, the orange and red box represent the first and final frame respectively and the pipeline that change color represent the scan path. It can be observed that through online iterative optimization, the final reconstruction

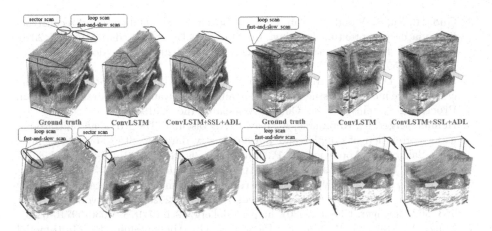

Fig. 5. Specific case of online iterative optimization. The first row and second row shows the DDH and fetus cases, and the yellow arrows indicate the anatomic structure. (Color figure online)

result is significantly improved compared with the ConvLSTM, with smaller drift and more stable trajectory estimation.

4   Conclusion

In this paper, we proposed the first research about freehand complex scan 3D US volume reconstruction. Benefiting from self context, shape prior and differentiable reconstruction approximation, the proposed online learning framework can effectively solve the challenge of 3D ultrasound reconstruction under complex skill sequences. Experiments on two different US datasets prove the effectiveness and versatility of the proposed framework. Future research will focus on extending this framework to more scanning targets and even different modalities.

Acknowledgements. This work was supported by the National Key R&D Program of China (No. 2019YFC0118300), Shenzhen Peacock Plan (No. KQTD20160-53112051497, KQJSCX20180328095606003), Royal Academy of Engineering under the RAEng Chair in Emerging Technologies (CiET1919/19) scheme, EPSRC TUSCA (EP/V04799X/1) and the Royal Society CROSSLINK Exchange Programme (IES/NSFC/201380).

References

1. Chen, L., Bentley, P., Mori, K., Misawa, K., Fujiwara, M., Rueckert, D.: Self-supervised learning for medical image analysis using image context restoration. Med. Image Anal. **58**, 101539 (2019)
2. Farnebäck, G.: Two-frame motion estimation based on polynomial expansion. In: Bigun, J., Gustavsson, T. (eds.) SCIA 2003. LNCS, vol. 2749, pp. 363–370. Springer, Heidelberg (2003). https://doi.org/10.1007/3-540-45103-X_50

3. Guo, H., Xu, S., Wood, B., Yan, P.: Sensorless freehand 3D ultrasound reconstruction via deep contextual learning. In: Martel, A.L., et al. (eds.) MICCAI 2020. LNCS, vol. 12263, pp. 463–472. Springer, Cham (2020). https://doi.org/10.1007/978-3-030-59716-0_44

4. Hendrycks, D., Mazeika, M., Kadavath, S., Song, D.: Using self-supervised learning can improve model robustness and uncertainty. In: Advances in Neural Information Processing Systems, vol. 32. Curran Associates, Inc. (2019)

5. Hennersperger, C., Karamalis, A., Navab, N.: Vascular 3D+T freehand ultrasound using correlation of doppler and pulse-oximetry data. In: Stoyanov, D., Collins, D.L., Sakuma, I., Abolmaesumi, P., Jannin, P. (eds.) IPCAI 2014. LNCS, vol. 8498, pp. 68–77. Springer, Cham (2014). https://doi.org/10.1007/978-3-319-07521-1_8

6. Huang, Y., et al.: Searching collaborative agents for multi-plane localization in 3D ultrasound. In: Martel, A.L., et al. (eds.) MICCAI 2020. LNCS, vol. 12263, pp. 553–562. Springer, Cham (2020). https://doi.org/10.1007/978-3-030-59716-0_53

7. Lang, A., Mousavi, P., Gill, S., Fichtinger, G., Abolmaesumi, P.: Multi-modal registration of speckle-tracked freehand 3D ultrasound to CT in the lumbar spine. Med. Image Anal. **16**(3), 675–686 (2012)

8. Liu, S., et al.: Deep learning in medical ultrasound analysis: a review. Engineering **5**, 261–275 (2019)

9. Looney, P., et al.: Fully automated, real-time 3D ultrasound segmentation to estimate first trimester placental volume using deep learning. JCI Insight **3**(11) (2018)

10. Mercier, L., Langø, T., Lindseth, F., Collins, D.L.: A review of calibration techniques for freehand 3-D ultrasound systems. Ultrasound Med. Biol. **31**, 449–471 (2005)

11. Mohamed, F., Siang, C.V.: A survey on 3D ultrasound reconstruction techniques. Artif. Intell. Appl. Med. Biol. (2019)

12. Mozaffari, M.H., Lee, W.S.: Freehand 3-D ultrasound imaging: a systematic review. Ultrasound Med. Biol. **43**, 2099–2124 (2017)

13. Paszke, A., et al.: Pytorch: An imperative style, high-performance deep learning library. In: Wallach, H., Larochelle, H., Beygelzimer, A., d'Alché-Buc, F., Fox, E., Garnett, R. (eds.) Advances in Neural Information Processing Systems, vol. 32. Curran Associates, Inc. (2019)

14. Prevost, R., et al.: 3D freehand ultrasound without external tracking using deep learning. Med. Image Anal. **48**, 187–202 (2018)

15. Prevost, R., Salehi, M., Sprung, J., Ladikos, A., Bauer, R., Wein, W.: Deep learning for sensorless 3D freehand ultrasound imaging. In: Descoteaux, M., Maier-Hein, L., Franz, A., Jannin, P., Collins, D.L., Duchesne, S. (eds.) MICCAI 2017. LNCS, vol. 10434, pp. 628–636. Springer, Cham (2017). https://doi.org/10.1007/978-3-319-66185-8_71

16. Shi, X., Chen, Z., Wang, H., Yeung, D.Y., Wong, W.K., Woo, W.C.: Convolutional LSTM network: a machine learning approach for precipitation nowcasting. In: Cortes, C., Lawrence, N., Lee, D., Sugiyama, M., Garnett, R. (eds.) Advances in Neural Information Processing Systems, vol. 28. Curran Associates, Inc. (2015)

17. Su, J.: GAN-QP: A novel GAN framework without gradient vanishing and lipschitz constraint (2018)

18. Yi, X., Walia, E., Babyn, P.: Generative adversarial network in medical imaging: a review. Med. Image Anal. **58**, 101552 (2019)

Universal Undersampled MRI Reconstruction

Xinwen Liu[1]($\boxtimes$)(iD), Jing Wang[2], Feng Liu[1], and S. Kevin Zhou[3,4]

[1] School of Information Technology and Electrical Engineering,
The University of Queensland, Brisbane, Australia
xinwen.liu@uq.net.au
[2] The Commonwealth Scientific and Industrial Research Organisation,
Canberra, Australia
[3] Medical Imaging, Robotics, and Analytic Computing Laboratory and Engineering
(MIRACLE), School of Biomedical Engineering and Suzhou Institute for Advanced
Research, University of Science and Technology of China, Suzhou, China
[4] Key Lab of Intelligent Information Processing of Chinese Academy of Sciences
(CAS), Institute of Computing Technology, CAS, Beijing, China

Abstract. Deep neural networks have been extensively studied for undersampled MRI reconstruction. While achieving state-of-the-art performance, they are trained and deployed specifically for one anatomy with limited generalization ability to another anatomy. Rather than building multiple models, a universal model that reconstructs images across different anatomies is highly desirable for efficient deployment and better generalization. Simply mixing images from multiple anatomies for training a single network does not lead to an ideal universal model due to the statistical shift among datasets of various anatomies, the need to retrain from scratch on all datasets with the addition of a new dataset, and the difficulty in dealing with imbalanced sampling when the new dataset is further of a smaller size. In this paper, for the first time, we propose a framework to **learn a universal deep neural network for undersampled MRI reconstruction**. Specifically, anatomy-specific instance normalization is proposed to compensate for statistical shift and allow easy generalization to new datasets. Moreover, the universal model is trained by distilling knowledge from available independent models to further exploit representations across anatomies. Experimental results show the proposed universal model can reconstruct both brain and knee images from NYU fastMRI dataset with high image quality. Also, it is easy to adapt the trained model to new datasets of smaller size, i.e., abdomen, cardiac and prostate, with little effort and superior performance.

Keywords: Deep learning · MRI reconstruction · Universal model

Electronic supplementary material The online version of this chapter (https://doi.org/10.1007/978-3-030-87231-1_21) contains supplementary material, which is available to authorized users.

1   Introduction

Magnetic Resonance Imaging (MRI) is a non-invasive imaging modality with superior soft-tissue visualization, but it is slow due to the sequential acquisition scheme. One direct approach to accelerating MRI is to sample less k-space data; however, the images directly reconstructed from the undersampled data are blurry and full of artifacts. Conventionally, compressed sensing (CS) [14] has been exploited to recover high-quality images from the undersampled measurements. Recently, deep learning (DL) [1,4,11,17,19,21,22,24–26,30] has been extensively studied for higher accuracy and faster speed.

While DL achieves state-of-the-art reconstruction performance [12,18], such success is often within the same dataset; in other words, the network trained on a dataset specific to one anatomy hardly generalize to datasets of other anatomies. To reconstruct images of various anatomies, the existing DL-based reconstruction framework usually needs to train an individual model for each dataset separately; and then multiple models are deployed on machines for application [21]. There are drawbacks to this approach. First, numerous models lead to a large number of parameters, which makes it hard to deploy on commercial MRI machines. Second, datasets of various anatomies contain shared prior knowledge [14]; separately trained networks do not exploit such common knowledge, which may limit the effectiveness of the trained models. While authors in [16] attempted to generalize DL-based reconstruction, the network is trained purely on natural images with limited anatomical knowledge. Therefore, it is highly desirable to train a universal model that exploits cross-anatomy common knowledge and conducts reconstruction for various datasets, even when some datasets are very small.

A straightforward way to have one model for all anatomies is by simply mixing various datasets during training. While this reduces the number of parameters, it unfortunately faces a variety of challenges. As demonstrated in [6,29], various datasets have different distributions, that is, there exist statistical shifts among datasets. Training a network with datasets of varying distributions degrades performance; therefore, simply mixing all data for training does not lead to an ideal universal model. In addition, when a dataset of new anatomy becomes available later, the model needs to be retrained from scratch on all datasets. It is time-consuming and storage-hungering for training and storing all the datasets. Finally, while there exist large-scale datasets, the newly collected datasets are normally small-scale. Mixing datasets of varying sizes for training causes imbalanced sampling [29], so the network may not be sufficiently optimized and the performance is questionable.

To address the issues mentioned above, for the first time, we propose a framework to train a **universal network**, which can reconstruct images of different anatomies and is generalizable to new anatomies easily. The base of the universal model is a convolutional network regularly used for MRI reconstruction, capturing commonly shared knowledge. To compensate for statistical shift and capture the knowledge specific to each dataset, we design an **Anatomy-SPecific Instance Normalization (ASPIN)** module to accompany the base model for reconstructing multiple anatomies. When a new anatomy dataset is available, a new set of ASPIN parameters is inserted and trained with the base of universal model fixed. The new ASPIN only needs few parameters and can be trained fast. Moreover, we propose to distil information from available models, trained purely on one anatomy, to the universal model. **Model distillation** allows the universal network to absorb multi-anatomy knowledge. In this study, we used public large-scale brain and knee data as currently available datasets. We took the abdomen, cardiac and prostate datasets as the new-coming small datasets. The experimental results show that the universal model can recover brain and knee images with better performance and fewer parameters than separately trained models. When generalizing to the new cardiac dataset, and much smaller abdomen and prostate datasets, the universal model also presents superior image qualities to the individually trained models.

2 Methods

2.1 The Overall Framework

Reconstructing an image $\mathbf{x} \in \mathbb{C}^N$ from undersampled measurement $\mathbf{y} \in \mathbb{C}^M$ ($M \ll N$) is an inverse problem, which is formulated as [19,30]:

$$\min_{\mathbf{x}} \quad \mathcal{R}(\mathbf{x}) + \lambda \|\mathbf{y} - \mathbf{F}_u \mathbf{x}\|_2^2, \tag{1}$$

where $\mathcal{R}$ is the regularisation term on $\mathbf{x}$, $\lambda \in \mathbb{R}$ denotes the balancing coefficient between $\mathcal{R}$ and data consistency (DC) term, $\mathbf{F}_u \in \mathbb{C}^{M \times N}$ is an undersampled Fourier encoding matrix. In model-based learning, Eq. (1) is incorporated in the network architecture with $\mathcal{R}$ approximated by the convolutional layers. Taking D5C5 in reference [19] as an example, Eq. (1) is unrolled into a feed-forward network path, cascading five inter-leaved DC blocks and learnable convolutional blocks that capture the prior knowledge of the training data.

Let $\{d_1, d_2, ..., d_A\}$ be A datasets of different anatomies to be reconstructed. Based d_a for an anatomy a, a convolutional network f_a is trained to recover its image. For example, in D5C5, f_a comprises $\{$CNN-t; t=1:5$\}$, where CNN-t is a learnable convolutional block. Conventionally, to recovery images of A anatomies, different networks $\{f_1, f_2, ..., f_A\}$ are trained, one for each anatomy. In this paper, we aim to train a universal network f_u that simultaneously recovers the images of various anatomies, with each convolutional block captures both common and anatomy-specific knowledge. We name such a block as a universal CNN (uCNN); overall f_u is composed of $\{$uCNN-t; t=1:5$\}$.

The universal network f_u in this work is designed to achieve two aims: (A1) it can recover images of different anatomies exemplified by the currently available datasets, which are brain and knee datasets in this study; (A2) when the dataset of new anatomy is available, the trained f_u can be generalized to the new dataset by adding a very few number of parameters.

Figure 1 depicts the overall framework of training f_u, which is comprised of four steps. (S1) We train two D5C5 networks (f_1, f_2) separately for brain

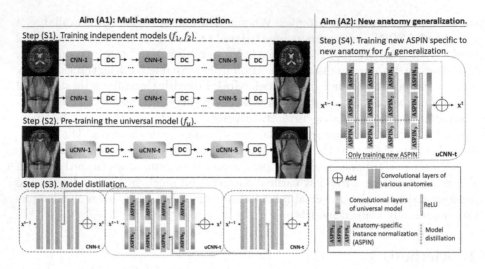

Fig. 1. The overall framework for universal undersampled MRI reconstruction.

and knee image recovery, and name them as independent models. These models are trained following [19], where CNN-t contains five convolutional layers with four ReLU layers and the residual connection. (S2) We pre-train f_u with both datasets. As shown in Fig. 1 Step (S2), CNN-t is replaced with uCNN-t, where two ASPIN modules are inserted, one for each anatomy. Although multi-anatomy knowledge is captured in this step, representation of single anatomy has not yet been fully utilized. (S3) We further optimize the pre-trained f_u with model distillation. As illustrated in Fig. 1 Step (S3), the knowledge captured in the third layer of CNN-t in f_1 and f_2 is transferred to that of uCNN-t. After this step, the universal model can perform reconstruction on both datasets. (S4) When the model is needed for a new anatomy, we insert additional ASPIN to adapt the trained f_u to the new dataset. The new dataset is only used to optimize the parameters of the newly inserted ASPIN, whereas all the other parameters are fixed. The generalized universal model learn to reconstruct new anatomy without forgetting the knowledge of old anatomies, capable of multi-anatomy reconstruction. Overall, Steps (S1–S3) achieves aim (A1), and Step (S4) achieves aim (A2).

2.2 Anatomy-SPecific Instance Normalization (ASPIN)

One issue of mixing data from all anatomies as the training dataset is the statistical shift among datasets of various anatomies. To compensate for the distribution shift, we design the ASPIN module in the network.

Instance normalization (IN), proposed in [7,23], uses the statistics of an instance to normalize features. Let the output activation of a given layer be $h \in \mathbb{R}^{C \times H \times W}$, where C is the number of channels, H is the height of the feature map, and W is the width of the feature map. Here, we ignore the batch dimension for simplicity. The IN function is defined as:

$$IN(h) = \gamma \left(\frac{h - \mu(h)}{\sigma(h)} \right) + \beta, \tag{2}$$

where γ and β are the learnable affine parameters, $\mu(h)$ and $\sigma(h)$ are the mean and standard deviation of the feature map computed across spatial dimensions $(H \times W)$ independently for each channel.

ASPIN is designed by stacking multiple parallel IN functions, each of which is reserved for one anatomy. Specifically, for anatomy d_a, anatomy-specific affine parameters γ_a and β_a are learnt during the training process. The specific ASPIN function for anatomy a is defined as:

$$ASPIN_a(h) = \gamma_a \left(\frac{h - \mu(h)}{\sigma(h)} \right) + \beta_a, \ a = 1, 2, ..., A. \tag{3}$$

where a is the index of anatomy corresponding to the current input image. It is expected that ASPIN captures anatomy-specific knowledge by learning the separate affine parameters for each anatomy. ASPIN allows the antecedent convolutional layer to capture anatomy-invariant knowledge, and the anatomy-specific information is presented by exploiting the captured statistics for the given anatomy. In this way, ASPIN is designed to compensate for the statistical difference of dataset distributions from various anatomies; and the sharable knowledge across anatomies is preserved with the shared convolutional layers.

2.3   Model Distillation

Independent models trained purely on a single anatomy capture more precise representation specifically for one dataset. Transferring the information from the fully-trained independent models to the universal model enables the network to capture more knowledge for each anatomy. To exploit the independent models' representations, we propose model distillation to further optimize the universal model.

The trained f_1 and f_2 in Step (S1) capture the knowledge of one anatomy. As in Fig. 1 Step (S3), we distil the knowledge learnt in the third layer of each CNN-t in f_1 and f_2 to that of each uCNN-t in the pre-trained f_u. The distillation is conducted by minimizing the attention transfer loss [15] between the independent models and the universal model. Let $m_c \in \mathbb{R}^{H \times W}, c = 1, 2, ..., C$ be the cth channel of the output activation h. The spatial attention map of h is defined as $O = \sum_{c=1}^{C} |m_c|^2$. In each cascade t, the attention transfer loss is expressed as:

$$l_{AT}^t = || \frac{O_I^t}{||O_I^t||_2} - \frac{O_U^t}{||O_U^t||_2} ||_1, \tag{4}$$

where O_I^t and O_U^t represent the spatial attention map of the third layer in CNN-t in independent models (f_1 or f_2) and in uCNN-t in f_u, respectively. Overall, the attention transfer loss for D5C5 is $L_{AT} = \sum_{t=1}^{5} l_{AT}^t$.

2.4   Network Training Pipeline

Individual models in Step (S1) are trained with mean-absolute-error (MAE) loss (L_{MAE}) on each of the datasets separately, to fully learn anatomy-specific knowledge. In Step (S2), the universal model is pre-trained on both datasets with L_{MAE}. During the training, we sample a batch of data from the same anatomy in a round-robin fashion in each iteration, and the model with the ASPIN specific to the batch's anatomy is optimized. This training strategy allows each dataset to contribute to the anatomy-invariant representation learning and anatomy-specific representation learning. The pre-trained universal model is further fine-tuned in Step (S3) with combined L_{MAE} and L_{AT} as its loss function: $L = L_{MAE} + \omega L_{AT}$, where ω denotes a balancing coefficient between two terms. In Step (S4), when the new anatomy is available, the newly inserted ASPIN is trained with L_{MAE} while freezing all the other parameters.

3   Experimental Results

3.1   Datasets and Network Configuration

We used five public datasets to evaluate the proposed method. The brain and knee datasets are large-scale DICOM images from the NYU fastMRI[1] database [13,27]. These datasets are used to train the individual base models in Step (S1) and the universal model in Step (2–3). Three small-scale datasets are used to evaluate the generalization ability on new anatomies in Step (S4). They are cardiac dataset from ACDC challenge [2], abdomen dataset from CHAOS challenge [8–10] and prostate dataset from Medical Segmentation Decathlon challenge [20]. In total we have 310,798 brain slices, 274,079 knee slices, 25,251 cardiac slices, 1,268 abdomen slices and 904 prostate slices. For each dataset, 80% of the volumes are randomly extracted for training, 10% for validation, and 10% for testing.

In this study, all the data are emulated single-coil magnitude-only images. We cropped (brain and knee datasets) or resized (cardiac, abdomen and prostate datasets) the images into 320 × 320. Then, we retrospectively undersampled the Cartesian k-space with 1D Gaussian sampling pattern. The evaluation is conducted under acceleration rates of 4× and 6× using structural similarity index (SSIM) and peak signal-to-noise ratio (PSNR). We calculated and reported the mean value of the metrics for reconstructed test images.

All the experiments were carried out on NVIDIA SXM-2 T V100 Accelerator Units, and the network was implemented on Pytorch. We trained networks using Adam optimizer with weight decay to prevent over-fitting. The size of all convolutional kernels is 3 × 3, the number of feature maps in CNN-t and uCNN-t are 32, and ω in the loss function of Step (S3) is empirically chosen to be 0.0001.

---

[1] fastmri.med.nyu.edu.

3.2   Algorithm Comparison

We compared the proposed universal model with two cases. In the first case, we trained five D5C5 models ("Independent") separately for each anatomy. This is the conventional approach that has good performance but uses the most parameters. In the second case, we mixed brain and knee datasets to train a single D5C5 model ("Shared") and then tested its generalization ability to all datasets.

Table 1. Quantitative comparison under acceleration rates of 4× and 6×.

		PSNR (dB)						SSIM (%)					
		Brain	Knee	Card.	Abdo.	Pros.	Avg.	Brain	Knee	Card.	Abdo.	Pros.	Avg.
4×	Undersampled	31.60	30.45	29.20	33.52	26.28	–	78.69	76.47	80.62	75.58	71.55	–
	Independent	40.85	35.46	37.05	40.81	30.91	37.02	97.30	89.84	96.07	96.79	86.03	93.21
	Shared	40.78	35.43	35.18	41.15	30.30	36.57	97.30	89.87	95.15	96.59	84.86	92.75
	Proposed	**43.16**	**36.39**	**37.30**	**42.76**	**31.61**	**38.24**	**98.28**	**91.39**	**96.51**	**98.20**	**87.72**	**94.42**
	w/o MD	42.80	36.25	37.10	42.59	31.51	38.05	98.17	91.19	96.42	98.14	87.51	94.29
6×	Undersampled	21.58	23.25	22.09	24.24	21.55	–	58.47	60.90	61.87	51.66	56.57	–
	Independent	38.04	32.57	33.32	37.97	28.51	34.08	95.48	85.01	92.76	94.43	79.33	89.40
	Shared	37.51	32.31	32.09	37.90	27.74	33.51	94.90	84.55	91.05	92.08	77.98	88.11
	Proposed	**40.03**	**33.66**	**33.72**	**39.67**	**29.26**	**35.27**	**97.02**	**87.25**	**93.53**	**96.37**	**81.80**	**91.19**
	w/o MD	39.49	33.36	33.50	39.48	29.13	34.99	96.63	86.71	93.23	96.08	81.41	90.81

The quantitative results are shown in Table 1. Comparing along the columns, we observed the proposed method has the highest PSNR and SSIM on all datasets under both acceleration rates. The results of "Shared" is the lowest on all cases, which confirms that simply mixing all the datasets does not lead to an ideal model. The proposed method overcomes the issue and improved about 1.7 dB in PSNR on average under both acceleration rates. The improvement of the proposed method over "Shared" is also observed in SSIM, which records around 1.7% (4×) and 3% (6×) increase on average. The "Independent" model, optimised on each anatomy, achieved moderate performance on all datasets. The proposed method absorbed the knowledge from individual base models, and it outperformed "Independent" with considerable margin on brain and knee datasets. It is worth noting that, on the new cardiac, abdomen and prostate datasets, the proposed method also achieved superior performance to the "Independent" models. This indicates that the learned base of the universal model captures the anatomy-common knowledge and using ASPIN further allows adaptation to a new anatomy even with small-scale data. The effectiveness of the large-scale data is also seen in [3,5,16].

We visualized the reconstructed images of different anatomies under 6× acceleration in Fig. 2. Both "Independent" and "Shared" reduced the artifacts presented in the undersampled images. "Proposed" further improved the reconstructed images, with clearer details and reduced error maps observed.

3.3   Ablation Study

To examine the effectiveness of the proposed ASPIN and model distillation, we further conducted ablation study. We removed Step (S3) and trained a network only with Step (S2) for the multi-anatomy reconstruction, named as "w/o MD". In other words, "w/o MD" improves Shared with ASPIN. When generalizing on the new anatomies, the newly inserted ASPIN in "w/o MD" still needs to be trained as in Step (S4).

The effectiveness of ASPIN is illustrated by comparing "Shared" with "w/o MD" in Table 1. We observed that "w/o MD" improves around 2 dB in PSNR on brain dataset and about 1 dB on knee dataset, respectively. This shows the significance of ASPIN in compensating for statistical shift when the training dataset contains data of various anatomies. Comparing "Proposed" with "w/o MD", we observed model distillation improved the quantitative results for brain and knee datasets on both acceleration rates. The improvement means incorporating the knowledge of individual base models in the universal model can further improve the performance. The increase in PSNR and SSIM is also consistently reported on cardiac, abdomen and prostate datasets, which further illustrates the importance of knowledge from individual base models. More ablation studies on different layer for distillation are provided in the supplementary materials, showing network with model distillation consistently outperformed "w/o MD".

In addition, we used the ASPIN trained on brain images to test cardiac data, and PSNR decreased by 1.4 dB. We have done this for each ASPIN and dataset,

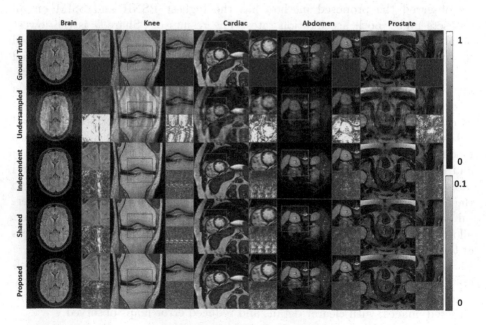

Fig. 2. Visualization of reconstructed images under acceleration rates of 6×. The zoomed-in area and the error maps are presented beside the images.

and the performance always drops. This proves each dataset needs the ASPIN corresponding to its own distribution. Thus, the importance of ASPIN is further validated.

3.4 Model Complexity

A D5C5 model has 144,650 parameters, and we take it as an unit, or $1\times$. "Shared" uses one D5C5 model, so it has least of parameters. For "Independent", reconstructing both brain and knee needs to train two D5C5 networks, leading to $2\times$ parameters. To reconstruct a new anatomy, "Independent" needs another 144,650 parameters. In total, "Independent" needs $5\times$ parameters for five anatomies. On the other hand, the proposed method only has additional parameters on ASPIN for each anatomy. One set of ASPIN for one anatomy has 1,280 parameters, so the proposed method only needs **less than** 1% of additional parameters for each anatomy. Overall, the proposed method has superior performance and only needs few additional parameters to reconstruct multiple anatomies.

4 Conclusion

We propose a novel universal deep neural network to reconstruct MRI of multiple anatomies. Experimental results demonstrate the proposed approach can reconstruct both brain and knee images, and it is easy to adapt to new small datasets while outperforming the individually trained models. While the current design is based on the popular D5C5 architecture, the proposed framework is extendable to other network architectures [28] for improved results. The current study is based on the single-coil magnitude-only images for proof-of-concept. In the future, the network can be adapted to multi-coil complex-valued datasets.

References

1. Aggarwal, H.K., Mani, M.P., Jacob, M.: MoDL: model-based deep learning architecture for inverse problems. IEEE Trans. Med. Imaging **38**(2), 394–405 (2018)
2. Bernard, O., Lalande, A., Zotti, C., Cervenansky, F., Yang, X., Heng, P.A., et al.: Deep learning techniques for automatic MRI cardiac multi-structures segmentation and diagnosis: is the problem solved? IEEE Trans. Med. Imaging **37**(11), 2514–2525 (2018)
3. Dar, S.U.H., Özbey, M., Çatlı, A.B., Çukur, T.: A transfer-learning approach for accelerated MRI using deep neural networks. Magn. Reson. Med. **84**(2), 663–685 (2020)
4. Hammernik, K., et al.: Learning a variational network for reconstruction of accelerated MRI data. Magn. Reson. Med. **79**(6), 3055–3071 (2018)
5. Han, Y., Yoo, J., Kim, H.H., Shin, H.J., Sung, K., Ye, J.C.: Deep learning with domain adaptation for accelerated projection-reconstruction MR. Magn. Reson. Med. **80**(3), 1189–1205 (2018)

6. Huang, C., Han, H., Yao, Q., Zhu, S., Zhou, S.K.: 3D U2-Net: a 3D universal u-net for multi-domain medical image segmentation. In: International Conference on Medical Image Computing and Computer-Assisted Intervention, pp. 291–299. Springer (2019)

7. Huang, X., Belongie, S.: Arbitrary style transfer in real-time with adaptive instance normalization. In: Proceedings of the IEEE International Conference on Computer Vision, pp. 1501–1510 (2017)

8. Kavur, A.E., Gezer, N.S., Barış, M., Aslan, S., Conze, P.H., Groza, V., et al.: CHAOS challenge-combined (CT-MR) healthy abdominal organ segmentation. Med. Image Anal. **69**, 101950 (2021)

9. Kavur, A.E., Gezer, N.S., Barış, M., Şahin, Y., Özkan, S., Baydar, B., et al.: Comparison of semi-automatic and deep learning-based automatic methods for liver segmentation in living liver transplant donors. Diagn. Interv. Radiol. **26**(1), 11 (2020)

10. Kavur, A.E., Selver, M.A., Dicle, O., BarÄśÅ§, M., Gezer, N.S.: CHAOS - Combined (CT-MR) healthy abdominal organ segmentation challenge data, April 2019. https://doi.org/10.5281/zenodo.3362844

11. Knoll, F., Hammernik, K., Kobler, E., Pock, T., Recht, M.P., Sodickson, D.K.: Assessment of the generalization of learned image reconstruction and the potential for transfer learning. Magn. Reson. Med. **81**(1), 116–128 (2019)

12. Knoll, F., et al.: Advancing machine learning for MR image reconstruction with an open competition: overview of the 2019 fastMRI challenge. Magn. Reson. Med. **84**(6), 3054–3070 (2020)

13. Knoll, F., Zbontar, J., Sriram, A., Muckley, M.J., Bruno, M., Defazio, A., et al.: fastMRI: a publicly available raw k-space and DICOM dataset of knee images for accelerated MR image reconstruction using machine learning. Radiol. Artif. Intell. **2**(1), e190007 (2020)

14. Lustig, M., Donoho, D., Pauly, J.M.: Sparse MRI: the application of compressed sensing for rapid MR imaging. Magn. Reson. Med. Offc. J. Int. Soc. Magn. Reson. Med. **58**(6), 1182–1195 (2007)

15. Murugesan, B., Vijayarangan, S., Sarveswaran, K., Ram, K., Sivaprakasam, M.: KD-MRI: a knowledge distillation framework for image reconstruction and image restoration in MRI workflow. In: Medical Imaging with Deep Learning, pp. 515–526. PMLR (2020)

16. Ouyang, C., et al.: Generalising deep learning MRI reconstruction across different domains. arXiv preprint arXiv:1902.10815 (2019)

17. Qin, C., Schlemper, J., Caballero, J., Price, A.N., Hajnal, J.V., Rueckert, D.: Convolutional recurrent neural networks for dynamic MR image reconstruction. IEEE Trans. Med. Imaging **38**(1), 280–290 (2018)

18. Recht, M.P., et al.: Using deep learning to accelerate knee MRI at 3T: results of an interchangeability study. Am. J. Roentgenol. **215**(6), 1421–1429 (2020)

19. Schlemper, J., Caballero, J., Hajnal, J.V., Price, A.N., Rueckert, D.: A deep cascade of convolutional neural networks for dynamic MR image reconstruction. IEEE Trans. Med. Imaging **37**(2), 491–503 (2017)

20. Simpson, A.L., Antonelli, M., Bakas, S., Bilello, M., Farahani, K., Van Ginneken, B., et al.: A large annotated medical image dataset for the development and evaluation of segmentation algorithms. arXiv preprint arXiv:1902.09063 (2019)

21. Sriram, A., et al.: End-to-end variational networks for accelerated MRI reconstruction. In: Martel, A.L., et al. (eds.) MICCAI 2020. LNCS, vol. 12262, pp. 64–73. Springer, Cham (2020). https://doi.org/10.1007/978-3-030-59713-9_7

22. Sriram, A., Zbontar, J., Murrell, T., Zitnick, C.L., Defazio, A., Sodickson, D.K.: GrappaNet: combining parallel imaging with deep learning for multi-coil MRI reconstruction. In: Proceedings of the IEEE/CVF Conference on Computer Vision and Pattern Recognition, pp. 14315–14322 (2020)
23. Ulyanov, D., Vedaldi, A., Lempitsky, V.: Instance normalization: the missing ingredient for fast stylization. arXiv preprint arXiv:1607.08022 (2016)
24. Wang, S., et al.: Accelerating magnetic resonance imaging via deep learning. In: 2016 IEEE 13th International Symposium on Biomedical Imaging, pp. 514–517. IEEE (2016)
25. Yang, G., Yu, S., Dong, H., Slabaugh, G., Dragotti, P.L., Ye, X., et al.: DAGAN: deep de-aliasing generative adversarial networks for fast compressed sensing MRI reconstruction. IEEE Trans. Med. Imaging **37**(6), 1310–1321 (2017)
26. Yang, Y., Sun, J., Li, H., Xu, Z.: Deep ADMM-Net for compressive sensing MRI. In: Proceedings of the 30th International Conference on Neural Information Processing Systems, pp. 10–18 (2016)
27. Zbontar, J., Knoll, F., Sriram, A., Murrell, T., Huang, Z., Muckley, M.J., et al.: fastMRI: an open dataset and benchmarks for accelerated MRI. arXiv preprint arXiv:1811.08839 (2018)
28. Zhou, B., Zhou, S.K.: DuDoRNet: learning a dual-domain recurrent network for fast MRI reconstruction with deep T1 prior. In: Proceedings of the IEEE/CVF Conference on Computer Vision and Pattern Recognition, pp. 4273–4282 (2020)
29. Zhou, S.K., et al.: A review of deep learning in medical imaging: image traits, technology trends, case studies with progress highlights, and future promises. arXiv preprint arXiv:2008.09104 (2020)
30. Zhou, S.K., Rueckert, D., Fichtinger, G.: Handbook of Medical Image Computing and Computer Assisted Intervention. Academic Press, Cambridge (2019)

A Neural Framework for Multi-variable Lesion Quantification Through B-Mode Style Transfer

SeokHwan Oh[1], Myeong-Gee Kim[1], Youngmin Kim[1], Hyuksool Kwon[2], and Hyeon-Min Bae[1(✉)]

[1] Electrical Engineering Department,
Korea Advanced Institute of Science and Technology, Daejeon 34141, South Korea
{joseph9337,myeonggee.kim,hmbae}@kaist.ac.kr
[2] Department of Emergency Medicine, Seoul National University Bundang Hospital,
Seong-nam 13620, South Korea

Abstract. In this paper, we present a scalable lesion-quantifying neural network based on b-mode-to-quantitative neural style transfer. Quantitative tissue characteristics have great potential in diagnostic ultrasound since pathological changes cause variations in biomechanical properties. The proposed system provides four clinically critical quantitative tissue images such as sound speed, attenuation coefficient, effective scatterer diameter, and effective scatterer concentration simultaneously by applying quantitative style information to structurally accurate b-mode images. The proposed system was evaluated through numerical simulation, and phantom and ex-vivo measurements. The numerical simulation shows that the proposed framework outperforms the baseline model as well as existing state-of-the-art methods while achieving significant parameter reduction per quantitative variables. In phantom and ex-vivo studies, the BQI-Net demonstrates that the proposed system achieves sufficient sensitivity and specificity in identifying and classifying cancerous lesions.

Keywords: Quantitative ultrasound imaging · Deep neural network · Neural style transfer

1 Introduction

Ultrasound (US) has become a standard medical imaging modality thanks to its cost-effective, non-invasive, and real-time nature. Although b-mode ultrasonography demonstrates high sensitivity at detecting abnormal tissues, the tissue properties cannot be retrieved from operator-dependent qualitative information. Therefore, there exist clear needs for quantitative US imaging techniques.

SH. Oh, and M.-G. Kim contributed equally.

Electronic supplementary material The online version of this chapter (https://doi.org/10.1007/978-3-030-87231-1_22) contains supplementary material, which is available to authorized users.

© Springer Nature Switzerland AG 2021
M. de Bruijne et al. (Eds.): MICCAI 2021, LNCS 12906, pp. 222–231, 2021.
https://doi.org/10.1007/978-3-030-87231-1_22

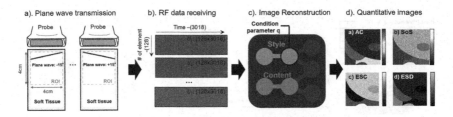

Fig. 1. The proposed quantitative US reconstruction workflow

Quantitative US imaging techniques have been developed to identify variations in tissue biomechanical properties caused by pathological changes. Elastography quantifies elastic properties of tissues by estimating shear wave velocity [1]. Ultrasound computed tomography offers a precise reconstruction of acoustic properties, such as the speed of sound. However, the tomographic system requires circular or rotating transducer arrangement which hinders widespread clinical use [2]. Recently, several attempts have been made to extract quantitative tissue parameters such as the speed of sound [3,4], attenuation coefficient [5], or backscatter coefficient [6] with a hand-held US probe [7]. Such solutions are based on the information contained in the pulse-echo radio frequency (RF) signals and show limitations in reconstructing complex tissue structures as compared to b-mode ultrasonography.

In this paper, a b-mode-to-quantitative imaging network referred to as BQI-Net performs multi-variable quantitative image reconstruction with enhanced specificity and sensitivity. The BQI-Net simultaneously synthesizes four clinically informative quantitative parameters: speed of sound (SOS), attenuation coefficient (AC), effective scatter diameter (ESD), and effective scatter concentration (ESC) [3–7]. Such diverse characteristics of tissues contribute to enhanced diagnostic specificity. The BQI-Net adopts a novel feed-forward neural style transfer framework to fully utilize the structurally faithful b-mode image as contents for accurate image reconstruction. Besides, the neural framework employs a conditional quantitative style encoding scheme for efficient quantitative image generation.

2 Method

Figure 1 illustrates the proposed quantitative US reconstruction system. A single 128-elements, 5 MHz US transducer is employed to measure RF signals obtained from plane waves with seven different incidence angles. The plane waves are emitted with a propagating angle ranging between −15° to 15° at 5° intervals and are received at a sampling rate of 62.5 MHz. Considering the width of the transducer, the reconstructing region of interest (ROI) is set to 40 mm × 40 mm. In order to exclude the transmission burst from the reflected signal of probe lens, the meaningful ROI requires minimum separation of 5 mm from the probe.

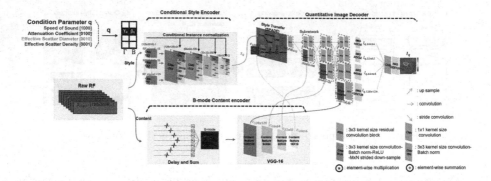

Fig. 2. BQI-Net structure

2.1 BQI-Net Architecture

In this section, the architecture of the proposed BQI-Net that generates multi-variable quantitative images from seven-angle plane wave RF signals ($\{u_{-15°}, u_{-10°}, \ldots, u_{angle}, \ldots, u_{+15°}\} \in U \sim \mathbb{R}^{128 \times 3018}$) is presented. The BQI-Net consists of three components: (1) b-mode contents encoder extracting geometrical information of lesions from B-mode ultrasonography, (2) conditional style encoder extracting designated quantitative information, and (3) quantitative image decoder synthesizing quantitative image from extracted content and style.

B-Mode Content Encoder. The b-mode content encoder generates semantic contents of spatial tissue geometry. Recently, a line of studies has demonstrated that progressive generation of an image from low resolution to higher resolution results in detailed image reconstruction [8]. Hence, the contents encoder extracts multi-resolution representation of the contents, $\{c_{16 \times 16}, c_{32 \times 32}, c_{64 \times 64}, c_{128 \times 128}\}$ $\in C$ and allows the quantitative image decoder to progressively synthesize the image. The encoder starts by reconstructing the b-mode image from U by using a time gain compensated (TGC) delay-and-sum (DAS) method. To generate C from b-mode image, convolutional and pooling layers of VGG-Net [9] are employed. The convolutional and pooling modules encode and compress the b-mode image sequentially into lower spatial feature space. The feature maps obtained by the convolutional layers are extracted prior to the pooling layers of the VGG-Net for the configuration of C.

Conditional Style Encoder. The conditional style encoder selectively extracts quantitative style information, $\{s_{AC}, s_{SOS}, s_{ESD}, s_{ESC}\} \in S$ embedded in U. Recent studies in neural style transfer have demonstrated that the neural network (NN) can be reformulated to perform multiple objective tasks just by adjusting the biases of a model [10,11]. Hence, the conditional instance normalization (CIN) [11] was performed on the style encoder to generate targeted quantitative profiles. The CIN is denoted as

$$CIN(x,q) = \gamma_q \frac{x - \mu(x)}{\sigma(x)} + \beta_q \tag{1}$$

where $\mu(x)$ and $\sigma(x)$ are the mean and standard deviation of the input x, respectively. The quantitative conditioning parameter, $\{AC, SOS, ESD, ESC\} \in q$, is encoded as a one-hot vector and is used to generate the corresponding scaling γ_q and shifting β_q factors through element-wise product with Γ, and B matrices, respectively. The conditional style encoder responds to each quantitative condition efficiently with only 10k parameters while demonstrating negligible accuracy degradation as compared to a simple approach of using $N = 4$ independent encoders requiring 11.9M additional parameters. The style encoder starts by individual processing of u_{angle} with 2D convolution filtering followed by rectified linear unit activation [12] and 1×2 stride down-sampling. Then, each filtered U is channel-wise concatenated. A series of conditional encoding processes parameterize the input into compressed style representation of $s_q \sim \mathbb{R}^{16 \times 16 \times 512}$.

Quantitative Image Decoder. The quantitative image decoder translates the b-mode contents C into the quantitative image, I_q, under the supervision of S. The b-mode to quantitative image translation is achieved using spatially adaptive demodulation (SPADE) followed by a series of residual convolution blocks [13,14]. The SPADE operation is formulated as

$$SPADE(c, s_q) = \gamma_{x,y,ch}(s_q) \frac{c - \mu_{ch}(c)}{\sigma_{ch}(c)} + \beta_{x,y,ch}(s_q) \tag{2}$$

where b-mode content c is channel-wise (ch) normalized, and $\gamma_{x,y,ch}(s_q)$ and $\beta_{x,y,ch}(s_q)$ are the learned modulation parameters generated by 2D convolution (Fig. 2). The modulation parameters are adaptively formulated depending on the spatial relationship of c and s_q. The SPADE operation offers the NN a precise interpretation of the correlation between quantitative values and the location of a lesion, where the pulse-echo RF are strongly influenced by the spatial distribution of the b-mode content. In addition, we empirically demonstrate that the SPADE outperforms other image translation methodologies including adaptive instance normalization [15] and channel-wise concatenation (Supplementary. A).

Inspired by HR-Net [16], the overall decoder architecture comprises parallel multi-resolution subnetworks to generate a detailed image unlike conventional up-sampling schemes [17,18]. Starting from translating $c_{16 \times 16}$ into quantitative image feature $I_{q,16 \times 16}$, the BQI-Net sequentially incorporates higher spatial resolution subnetworks supported by finer b-mode contents. Multi-resolution parallel convolution preserves low-resolution feature representation and multi-resolution fusion, implemented at every node of the subnetworks, enables detailed image reconstruction. Finally, the outputs subnetworks are concatenated into higher resolution representation and I_q is synthesized via 1×1 convolution.

2.2   Training Details

Dataset. For the training of the BQI-Net, diverse complex soft-tissue phantoms are created for numerical ultrasonic simulation using the k-wave simulation toolbox in MATLAB. The organs and lesions are modeled by placing 0–5 ellipses, with radii of 2–30, at random positions on a 50 × 50 mm background (Fig. 1.d). The biomechanical property of each lesion and background was set to cover general soft tissue characteristics (see Table 1) [19]. We have simulated 19.5k phantoms and randomly split them into three groups (17.5k for training, 1.5k for validation, and 0.5k for testing).

Table 1. Quantitative characteristic of simulated soft-tissue phantoms.

Feature	Property	Feature	Property
SOS	1400–1700 m/s	AC	0–1.5 dB/cm/MHz
ESD	25–150 μm	ESC	$0-10/wavelength(wav)^2$
Density	0.9–1.1 kg/m^3		

Implementation Details. The objective function of the BQI-Net is

$$G^* = \underset{G}{\arg\min} \; \mathbb{E}_{U,Y}[\|Y_q - G(U,q)\|^2] + L_{SUB} + \lambda L_2 \qquad (3)$$

where

$$L_{SUB} = \mathbb{E}_{U,Y_{q,R}} \sum_R \|Y_{q,R} - G_R(U,q)\|^2, \; L_2 = \sum_{i=1} w_i^2. \qquad (4)$$

Y_q is the ground truth quantitative image with given quantitative condition q, $Y_{q,R}$ is the down-sampled image of Y_q with spatial resolution $\{\mathbb{R}^{16\times16}, \mathbb{R}^{32\times32}, \mathbb{R}^{64\times64}, \mathbb{R}^{128\times128}\} \in \mathbb{R}$. The neural network G^* is trained to minimize the mean square difference between Y_q and inference output $G(U)$ while L_{SUB} regularizes each subnetwork to progressively synthesize corresponding resolution quantitative image $I_{q,R}$. L_2 ($\lambda = 10^{-6}$) is implemented to avoid overfitting by regularizing weights w_i of the NN. The Gaussian and quantization noise are added to U, and the b-mode image is generated applying TGC selected randomly in the range from 0 to 1.2 dB/μs. The BQI-Net alternates q at each iteration for balanced training of the NN. The BQI-Net was optimized by Adam [20] with a learning rate of 10^{-4} ($\beta_1 = 0.9$, $\beta_2 = 0.999$). To better generalize the NN, dropout [21], with a retentional probability of 0.5, was applied. The NN was trained up to 250 epochs, which is determined based on convergence of validation loss. The NN was implemented using TensorFlow, accelerated with RTX 2080Ti GPU.

3   Experiments

The performance of BQI-Net is assessed with 500 representative simulated test samples using root mean squared error (RMSE), mean normalized absolute error (MNAE), peak signal to noise ratio (PSNR), and structural similarity index (SSIM) metrics. A heterogeneous thyroid phantom, imitating anatomic structure and character of the thyroid, and a breast phantom, composed of breast-like background, cyst, and dense mass, is employed to assess preclinical efficacy in lesion identification. Ex-vivo bovine muscles, with insertions imitating cyst, benign, and malignant tissue, were fabricated to verify clinical possibility in cancer differentiation. The phantom and ex-vivo tests are measured using Vantage 64LE (Verasonics Inc.) with a 5 MHz linear array transducer (Humanscan Inc.)

3.1   Numerical Simulation

Table 2. Quantitative assessments of numerical phantoms.

	VAR	Structure	RMSE	MNAE	PSNR	SSIM	VAR/NN	PAR/VAR	Time/VAR
FCN	AC	Encoder-Decoder (ED)	0.152 [dB/cm/MHz]	0.075	20.21	0.812	1	186M	0.272 s
	SoS		48.09 [m/s]	0.131	16.66	0.753			
	ESD		29.66 [μm]	0.165	14.60	0.607			
	ESC		2.558 [wav^2]	0.233	12.48	0.661			
U-Net	AC	ED	0.136	0.065	21.13	0.825	1	107M	0.074
	SoS		43.40	0.114	17.32	0.751			
	ESD		21.76	0.111	17.07	0.654			
	ESC		2.168	0.188	13.68	0.676			
HR-Net	AC	ED with subnetworks	0.131	0.061	21.78	0.866	1	116M	0.068
	SoS		39.02	0.104	18.61	0.823			
	ESD		22.44	0.115	17.40	0.716			
	ESC		2.100	0.190	14.60	0.764			
BQI Net	AC	Feed-forward neural style transfer	**0.104**	**0.048**	**23.62**	**0.893**	4	86M	0.031
	SoS		**28.78**	**0.074**	**21.32**	**0.868**			
	ESD		**16.51**	**0.086**	**20.39**	**0.782**			
	ESC		**1.869**	**0.168**	**15.73**	**0.801**			

The performance of the BQI-Net and baseline models are reported in Table 2 (The standard deviation is presented in Supplementary. B). The fully convolutional network (FCN) [17] and U-Net [18] are employed as a baseline encoder-decoder framework. The HR-Net [16] is evaluated to validate the effectiveness of the multi-resolution subnetwork module in our application. The HR-Net achieves performance gain by 0.084, and 0.065 in SSIM compared to FCN (average SSIM is 0.7085) and U-Net (SSIM is 0.7270) respectively, demonstrating the subnetwork module provides a detailed image reconstruction. In a head-to-head comparison, the BQI-Net outperformed the baseline models for each quantitative variable reconstruction. The proposed BQI-Net demonstrates higher accuracy (PSNR is 20.27 dB) and structural similarity (SSIM is 0.8364) in image reconstruction as compared to HR-Net (PSNR and SSIM are 18.10 dB and 0.7928 respectively). The performance improvement of the BQI-Net is attributed to the fact that the NN can better recognize the quantitative features embedded in RF with the help of the geometric information of the b-mode image. Moreover, the NN achieves enhanced precision in the definition of lesion shape employing detailed description of the geometry of the lesion given through b-mode image (Supplementary. B). The proposed framework requires 36M parameter (PAR) per quantitative variable (VAR) which is a compelling enhancement on efficiency. In addition, the NN can generate N (≤ 4) quantitative images with a single inference with a batch size of N.

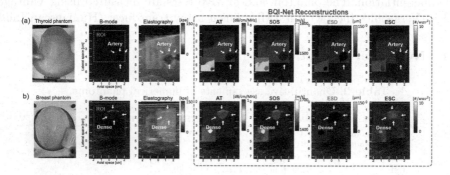

Fig. 3. Reconstruction image of the (a) thyroid and (b) breast phantoms.

Table 3. Reconstruction results of the thyroid and breast phantoms

			AC [dB/cm/MHz]	SoS [m/s]	ESD [μm]	ESC [$/wav^2$]
Thyroid phantom	Background	Ground truth	0.5 ± 0.1	1540 ± 10	60–100	N.A (high)
		Reconstruction	0.43	1551	62.0	3.83
	Artery	Ground truth	0.1 ± 0.1	1510 ± 10	0.00	0.00
		Reconstruction	0.12	1514	0.02	1.96
Breast phantom	Background	Ground truth	0.3 ± 0.05	1475 ± 10	N.A. (high)	N.A. (low)
		Reconstruction	0.23	1489	30.6	2.28
	Dense mass	Ground truth	0.54 ± 0.05	1610 ± 20	N.A. (low)	N.A. (high)
		Reconstruction	0.50	1607	2.76	5.72

3.2 Phantom, and Ex-Vivo Measurements

Phantom Measurements. Figure 3 presents quantitative images of biomimicking US phantoms generated by the proposed BQI-Net and compares them with conventional imaging methods. In a thyroid phantom, the BQI-Net clearly identifies the artery with the surrounding muscle background. The quantitative parameters of AT, SOS, and ESD extracted by the BQI-Net are within the range provided by the manufacturer as shown in Table 3. In the case of ESC reconstruction, the artery has less scatter concentration than the background which is in an agreement with the tissue characteristics. The breast phantom was measured to assess the efficacy of the system in differentiating cancerous lesions. The results show that the BQI-Net classifies the lesion as a dense mass correctly and offers high contrast AT and SOS images despite low acoustic impedance contrast. The BQI-Net reconstructs the AT, and SOS of the lesion with the errors less than $0.04\,\mathrm{dB/cm/MHz}$ and $3\,\mathrm{m/s}$, respectively. In addition, the reconstructed lesion shows lower ESD, and higher ESC than the background, which correlates well with the properties of a highly concentrated small dense mass.

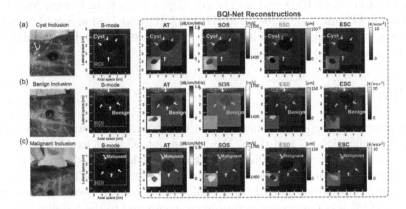

Fig. 4. Reconstruction images of the (a) Cyst, (b) Benign, and (c) Malignant objects.

Table 4. Reconstruction results and ground truth of inclusions in ex-vivo experiments.

			AC	SoS	ESD	ESC
Ex-vivo	Cyst tissue	Ground truth	0.17	1462	0.00	0.00
		Reconstruction	0.15	1487	0.32	1.14
	Benign tissue	Ground truth	0.38	1574	106	1.75
		Reconstruction	0.41	1550	114	2.91
	Malignant tissue	Ground truth	0.52	1609	37.5	8.75
		Reconstruction	0.48	1603	30.2	7.20

Ex-Vivo Measurements. In ex-vivo experiments, cyst [22], benign and malignant-like [23,24] objects are manufactured to represent general biomechanical properties of cancerous lesions (Table 4). The ground truth ESD and ESC of insertions are the actual diameter and concentration of the added scatter. The reference AC and SOS are acquired by computing attenuated amplitude and arrival time of traversed US waves measured from two US systems with ultrasound probes facing each other. The reference AC and SOS values are measured 5 times for each insertion and the standard deviations are $0.032\,\mathrm{dB/cm/MHz}$ and $2.75\,\mathrm{m/s}$, respectively. Figure 4 presents the quantitative reconstruction of the inclusion and bovine muscle background. The ex-vivo experiments demonstrate that the BQI-Net differentiates cancerous lesion with high specificity through a comprehensive analysis of diverse quantitative features. The benign object shows $0.26\,\mathrm{dB/cm/MHz}$, and $63\,\mathrm{m/s}$ higher AT and SOS than those of cystic object, with errors less than 2.0%, and 1.4%, respectively, which corresponds to the actual biomechanical difference of the cyst and benign tissue. Considering the malignant cell division and concentration, the malignant insertion is fabricated to have lower ESD and higher ESC. The reconstruction of the malignant lesion shows $83.8\,\mu\mathrm{m}$ lower ESD and $4.21/wav^2$ higher ESC than benign lesion. The background AC and SOS show $0.76\,\mathrm{dB/cm/MHz}$ and $1581\,\mathrm{m/s}$ which are in a good quantitative agreement with reported AC ($0.89 \pm 0.11\,\mathrm{dB/cm/MHz}$) and SOS ($1589 \pm 10\,\mathrm{m/s}$) [25], respectively.

4 Conclusion

In this paper, we proposed a single probe US system performing multi-variable lesion quantification in real-time. We implement a novel feed-forward neural style transfer framework to utilize a structurally faithful b-mode image for accurate image reconstruction. A scalable quantitative style encoding scheme is presented to simultaneously provide four clinically informative biomechanical properties with 68% less parameters requirement compared to the baseline model. In numerical simulation, the BQI-Net presents 21% and 24% enhancement in RMSE compared to existing state-of-the-art methods, HR-Net, and U-Net, respectively. Phantom and ex-vivo measurements demonstrate that the proposed NN achieves specified identification of cancerous lesions through diverse quantitative features analysis. The proposed system is readily available with the standard US system and shows high potential for clinical purpose, especially in early detection and differential diagnosis of cancer.

References

1. Greenleaf, J., Fatemi, M., Insana, M.: Selected methods for imaging elastic properties of biological tissues. Ann. Rev. Biomed. Eng. **5**, 57–78 (2003)
2. Zografos, G., et al.: Differentiation of birads-4 small breast lesions via multimodal ultrasound tomography. Eur. Radiol. **25**, 410–418 (2015)
3. Sanabria, S.J., Ozkan, E., Rominger, M., Goksel, O.: Spatial domain reconstruction for imaging speed-of-sound with pulse-echo ultrasound: simulation and in vivo study. Phys. Med. Biol. **63**(21), 215015 (2018)
4. Kim, M.-G., Oh, S., Kim, Y., Kwon, H., Bae, H.-M.: Robust single-probe quantitative ultrasonic imaging system with a target-aware deep neural network. IEEE Trans. Biomed. Eng. (2021). https://doi.org/10.1109/TBME.2021.3086856
5. Rau, R., Unal, O., Schweizer, D., Vishnevskiy, V., Goksel, O.: Attenuation imaging with pulse-echo ultrasound based on an acoustic reflector. In: Shen, D., et al. (eds.) MICCAI 2019. LNCS, vol. 11768, pp. 601–609. Springer, Cham (2019). https://doi.org/10.1007/978-3-030-32254-0_67

6. Oelze, M.L., Mamou, J.: Review of quantitative ultrasound: envelope statistics and backscatter coefficient imaging and contributions to diagnostic ultrasound. IEEE Trans. Ultrason. Ferroelectr. Freq. Control **63**(2), 336–351 (2016)
7. Oh, S., Kim, M. -G., Kim, Y., Bae, H.-M.: A learned representation for multi variable ultrasound lesion quantification. In: ISBI, pp. 1177–1181. IEEE (2021)
8. Chen, Q., Koltun, V.: Photographic image synthesis with cascaded refinement networks. In: ICCV, pp. 1520–1529. IEEE (2017)
9. Simonyan, K., Zisserman, A.: Very deep convolutional networks for large-scale image recognition. In: ICLR (2015)
10. Van den Oord, A., Kalchbrenner, N., Espeholt, L., Vinyals, O., Graves, A., Kavukcuoglu, K.: Conditional image generation with pixelcnn decoders. In: NIPS (2016)
11. Dumoulin, V., Shlens, J., Kudlur, M.: A learned representation for artistic style. arXiv preprint arXiv:1610.07629 (2016)
12. Xu, B., Wang, N., Chen, T., Li, M.: Empirical evaluation of rectified activations in convolutional network. arXiv preprint arXiv:1505.00853 (2015)
13. He, K., Zhang, X., Ren, S., Sun, J.: Deep residual learning for image recognition. In: CVPR, pp. 770–778 (2016)
14. Park, T., Liu, M.Y., Wang, T.C., Zhu, J.Y.: Semantic image synthesis with spatially-adaptive normalization. In: ICCV, pp. 2337–2346. IEEE (2019)
15. Huang, X., Belongie, S.: Arbitrary style transfer in real-time with adaptive instance normalization. In: ICCV, pp. 1501–1510. IEEE (2017)
16. Wang, J., et al.: Deep high-resolution representation learning for visual recognition. IEEE Trans. Pattern Anal. Mach. Intell. (2020)
17. Feigin, M., Freedman, D., Anthony, B.W.: A deep learning framework for single-sided sound speed inversion in medical ultrasound. IEEE Trans. Biomed. Eng. **67**(4), 1142–1151 (2019)
18. Ronneberger, O., Fischer, P., Brox, T.: U-Net: convolutional networks for biomedical image segmentation. In: Navab, N., Hornegger, J., Wells, W.M., Frangi, A.F. (eds.) MICCAI 2015. LNCS, vol. 9351, pp. 234–241. Springer, Cham (2015). https://doi.org/10.1007/978-3-319-24574-4_28
19. Mast, T.D.: Empirical relationships between acoustic parameters in human soft tissues. Acoust. Res. Lett. Online **1**(37), 37–43 (2000)
20. Kingma, D., Ba, J.: Adam: a method for stochastic optimization. In: ICLR (2014)
21. Srivastava, N., Hinton, G., Krizhevsky, A., Sutskever, I., Salakhutdinov, R.: Dropout: a simple way to prevent neural networks from overfitting. J. Mach. Learn. Res. **15**(1), 1929–1958 (2014)
22. Cabrelli, L.C., Grillo, F.W., Sampaio, D.R., Carneiro, A.A., Pavan, T.Z.: Acoustic and elastic properties of glycerol in oil-based gel phantoms. Ultrasound Med. Biol. **43**(9), 2086–2094 (2017)
23. Culjat, M.O., Goldenberg, D., Tewari, P., Singh, S.R.: A review of tissue substitutes for ultrasound imaging. Ultrasound Med. Biol. **36**(6), 861–873 (2010)
24. Anderson, J.J., et al.: Interlaboratory comparison of backscatter coefficient estimates for tissue-mimicking phantoms. Ultrason. Imaging **32**(1), 48–64 (2010)
25. Goss, S.A., Johnston, R.L., Dunn, F.: Comprehensive compilation of empirical ultrasonic properties of mammalian tissues. J. Acoust. Soc. Am. **64**(2), 423–457 (1978)

Temporal Feature Fusion with Sampling Pattern Optimization for Multi-echo Gradient Echo Acquisition and Image Reconstruction

Jinwei Zhang[1,3]($\boxtimes$), Hang Zhang[2,3], Chao Li[3,4], Pascal Spincemaille[3], Mert Sabuncu[1,2,3], Thanh D. Nguyen[3], and Yi Wang[1,2,3]

[1] Department of Biomedical Engineering, Cornell University, Ithaca, NY, USA
jz853@cornell.edu
[2] Department of Electrical and Computer Engineering, Cornell University, Ithaca, NY, USA
[3] Department of Radiology, Weill Medical College of Cornell University, New York, NY, USA
[4] Department of Applied Physics, Cornell University, Ithaca, NY, USA

Abstract. Quantitative imaging in MRI usually involves acquisition and reconstruction of a series of images at multi-echo time points, which possibly requires more scan time and specific reconstruction technique compared to conventional qualitative imaging. In this work, we focus on optimizing the acquisition and reconstruction process of multi-echo gradient echo pulse sequence for quantitative susceptibility mapping as one important quantitative imaging method in MRI. A multi-echo sampling pattern optimization block extended from LOUPE-ST is proposed to optimize the k-space sampling patterns along echoes. Besides, a recurrent temporal feature fusion block is proposed and inserted into a backbone deep ADMM network to capture the signal evolution along echo time during reconstruction. Experiments show that both blocks help improve multi-echo image reconstruction performance.

Keywords: Multi-echo images · Quantitative susceptibility mapping · Under-sampled k-p space · Deep ADMM

1 Introduction

Quantitative imaging (QI) is an emerging technology in magnetic resonance imaging (MRI) to quantify tissues' magnetic properties. Typical QI methods involve tissues' parameter mapping such as T1 and T2 relaxation time quantification [6,7], water/fat separation with R2* estimation [26], quantitative susceptibility

Electronic supplementary material The online version of this chapter (https://doi.org/10.1007/978-3-030-87231-1_23) contains supplementary material, which is available to authorized users.

mapping (QSM) [24] and etc. These parameters provide new biomarkers for the clinical assessment of diverse diseases. In QI, a multi-echo pulse sequence is used to acquire MRI signals at different echo times. After image reconstruction for all echo times, a temporal evolution model with respect to the parameters of interest is used to compute the parameters via a model-based nonlinear least square fitting. For example, QSM estimates the tissue induced local magnetic field by a nonlinear fitting of complex multi-echo signals [14] and solves tissue susceptibility using regularized dipole inversion [13].

Despite its merits for tissue parameter quantification, QI has a major drawback of extended scan time compared to conventional qualitative imaging, as the echo time needs to be long enough to cover the temporal evolution for all tissue types in order to obtain accurate parameter mapping. Classical acceleration techniques in MRI can be incorporated into QI. Parallel imaging (PI) [9,19], compressed sensing (CS) [15] and their combinations (PI-CS) [16,17] are widely used techniques for acquiring and reconstructing under-sampled k-space data to shorten scan time, which is referred to as under-sampled k-p space acquisition and reconstruction in QI [28,30]. To acquire k-p space data, a variable density under-sampling pattern at each echo is applied, where the design of the 'optimal' under-sampling patterns across echoes remains an open problem in QI. In fact, various methods have been proposed for single-echo sampling pattern optimization, which may inspire multi-echo optimal sampling pattern design for QI. Representative single-echo sampling pattern optimization methods include machine learning based optimization LOUPE [2] and its extension LOUPE-ST [27], experimental design with the constrained Cramer-Rao bound OEDIPUS [10] and greedy pattern selection [8].

After k-p space acquisition, the next step is to reconstruct multi-echo images from under-sampled data. Based on the observation that correlation exists among multi-echo image structures, simultaneous reconstruction of multi-echo images with joint sparsity regularization was introduced for QI [30]. Besides, globally low rank Casorati matrix was imposed assuming the profiles of signal decay curves at all spatial locations were highly redundant [18]. Accordingly, a locally low rank regularization was applied by restricting the Casorati matrix to different local image regions [28]. To reconstruct multi-echo images using the above regularizations, these methods minimized the corresponding objective functions via an iterative scheme. Recently, with the advance of convolutional neural network, unrolling the whole iterative optimization process with a learned regularizer has become popular in under-sampled k-space reconstruction. Pioneering methods include MoDL [1] and VarNet [11] for single-echo image reconstruction and cascaded [22] and recurrent network [20] for dynamic image sequence reconstruction, which also inspire the design of unrolled reconstruction network for multi-echo QI.

In this paper, we focus on quantitative susceptibility mapping (QSM) [24] acquired with Multi-Echo GRadient Echo (MEGRE) pulse sequence. In QSM, MEGRE signals are generated based on the following signal evolution model:

$$s_j = m_0 e^{-R_2^* t_j} e^{i(\phi_0 + f \cdot t_j)}, \tag{1}$$

where s_j is the signal at the j-th echo time t_j, m_0 is the proton density (water) in the tissue, R_2^* is the effective transverse magnetization relaxation factor, ϕ_0 is the initial phase at radio-frequency (RF) pulse excitation, and f is the total magnetic field data generated by the tissue and air susceptibility. After acquiring and reconstructing $\{s_j\}$ of all echoes, a nonlinear field estimation with Levenberg-Marquardt algorithm [14] is used to estimate f from Eq. 1, then a morphology enabled dipole inversion (MEDI) method [13] is deployed to compute the corresponding susceptibility map from f. In this work, we attempt to combining the optimization of acquisition and reconstruction process of MEGRE signals into one learning based approach. We extend LOUPE-ST [27] to the multi-echo scenario for temporal sampling pattern optimization and propose a novel temporal feature fusion block into a deep ADMM based unrolled reconstruction network to capture MEGRE signal evolution during reconstruction.

2   Method

The mulit-coil multi-echo k-space under-sampling process is as follows:

$$b_{jk} = U_j F E_k s_j + n_{jk}, \tag{2}$$

where b_{jk} is the measured under-sampled k-space data of the k-th receiver coil at the j-th echo time with N_C receiver coils and N_T echo times in total, U_j is the k-space under-sampling pattern at j-th echo time, F is the Fourier transform, E_k is the sensitivity map of k-th coil, s_j is the complex image of the j-th coil to be reconstructed, and n_{jk} is the voxel-by-voxel i.i.d. Gaussian noise when measuring b_{jk}. Notice that U_j may vary across echoes, which provides the flexibility of sampling pattern design for different echo time.

Having acquired $\{b_{jk}\}$ with fixed $\{U_j\}$, we aim at reconstructing $\{s_j\}$ of all echoes simultaneously with a cross-echo regularization $R(\{s_j\})$. Based on Eq. 2, the objective function to minimize is:

$$E(\{s_j\}; \{U_j\}) = R(\{s_j\}) + \sum_{j,k} \|U_j F E_k s_j - b_{jk}\|_2^2. \tag{3}$$

In the following sections, we will minimize Eq. 3 iteratively with a learned regularizer $R(\{s_j\})$ which captures the dynamic evolution along echoes. We will also design $\{U_j\}$ to boost the reconstruction performance under a fixed Cartesian under-sampling ratio.

2.1   Deep ADMM as Backbone

Assuming U_j's are pre-defined, we use alternating direction method of multiplier (ADMM) [4] to minimize Eq. 3. Introducing auxiliary variable $v_j = s_j$ for all j's, ADMM splits the original problem into a sequence of subproblems:

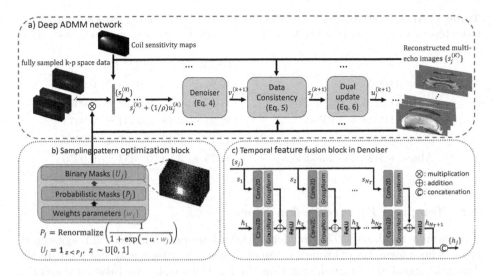

Fig. 1. Network architecture. Deep ADMM (a) was used as backbone for undersampled k-p space reconstruction. A sampling pattern optimization block (b) extended from LOUPE-ST was used to learn optimal multi-echo patterns. A temporal feature fusion block (c) was inserted into deep ADMM to capture signal evolution along echoes.

$$\{v_j^{(k+1)}\} = \underset{\{v_j\}}{\arg\min}\, R(\{v_j\}) + \frac{\rho}{2}\sum_j \|v_j - \tilde{v}_j^{(k)}\|_2^2, \tag{4}$$

$$\{s_j^{(k+1)}\} = \underset{\{s_j\}}{\arg\min}\, \sum_{j,k} \|U_j F E_k s_j - b_{jk}\|_2^2 + \frac{\rho}{2}\sum_j \|s_j - \tilde{s}_j^{(k)}\|_2^2, \tag{5}$$

$$\{u_j^{(k+1)}\} = \{u_j^{(k)}\} + \rho(\{s_j^{(k+1)}\} - \{v_j^{(k+1)}\}), \tag{6}$$

where $\tilde{v}_j^{(k)} = s_j^{(k)} + \frac{1}{\rho}u_j^{(k)}$, $\tilde{s}_j^{(k)} = v_j^{(k+1)} - \frac{1}{\rho}u_j^{(k)}$, $u_j^{(k)}$ is the dual variable and ρ is the penalty parameter in ADMM. To choose a regularizer $R(\{s_j\})$ in Eq. 4, a plug-and-play ADMM [5] strategy was proposed to replace the "denoising" step Eq. 4 by an off-the-shelf image denoising algorithm $\{v_j^{(k+1)}\} = \mathcal{D}(\{\tilde{v}_j^{(k)}\})$ for flexibility and efficiency. Inspired by a recent work MoDL [1] where an MR image reconstruction network was built by unrolling the quasi-Newton iterative scheme and learning the regularizer from fully-sampled data, we propose to unroll the iterative schemes in Eq. 4–6 to a data graph and design $\mathcal{D}(\{\tilde{v}_j^{(k)}\})$ as a convolutional neural network (CNN) to be trained from data. The whole reconstruction network architecture is shown in Fig. 1(a), which we call deep ADMM. Multi-echo real and imaginary parts of MR images are concatenated into the channel dimension, yielding $2N_T$ channels in $\mathcal{D}(\cdot)$. In the backbone deep ADMM, vanilla $\mathcal{D}(\cdot)$ consists of five convolutional layers with zero padding to preserve image spatial dimension. Data consistency subproblem Eq. 5 is solved by conjugate gradient (CG) descent algorithm. Both weights in $\mathcal{D}(\cdot)$ and penalty parameter ρ are learnable parameters in deep ADMM.

Table 1. Ablation study. Reconstruction performances were progressively improved as more blocks were added to deep ADMM (* denotes statistical significance; $p < 0.05$).

	PSNR (↑)	SSIM (↑)
Deep ADMM	40.95 ± 3.72	0.9820 ± 0.0257
Deep ADMM + single SPO	41.77 ± 3.24	0.9839 ± 0.0091
Deep ADMM + TFF	42.24 ± 3.28	0.9864 ± 0.0074
Deep ADMM + TFF + single SPO	42.77 ± 3.38	0.9867 ± 0.0076
Above + R2* and Field loss	41.75 ± 3.64	0.9853 ± 0.0085

2.2   Temporal Feature Fusion Block

The vanilla regularizer $\mathcal{D}(\cdot)$ of deep ADMM proposed in Sect. 2.1 aggregates cross-echo information through channel dimension, which may not capture the dynamic signal evolution along time in Eq. 1. In order to take into account the signal evolution based regularization, we propose a temporal feature fusion (TFF) block as shown in Fig. 1(c). In TFF, a recurrent module is repeated N_T times with shared weights, in which at the j-th repetition single echo image s_j and hidden state feature h_j are fed into the module to generate next hidden state feature h_{j+1}. Since h_j is meant to carry information from preceding echoes, the concatenated multi-echo hidden states $\{h_j\}$ is able to capture the echo evolution and fuse features across all echoes, which will then be fed into a denoising block to generate denoised multi-echo images. The signal evolution model along echo time is implicitly incorporated during the recurrent feed forward process due to the parameter sharing mechanism which effectively leverages the relationship between input of one echo and its temporal neighbors.

2.3   Sampling Pattern Optimization Block

Another goal is to design optimal k-p space sampling pattern $\{U_j\}$ for better image reconstruction performance. In this work, we focus on 2D variable density sampling pattern with a fixed under-sampling ratio. We accomplish this by extending single-echo LOUPE-ST [27] to the multi-echo scenario as our sampling pattern optiomization (SPO) block. The multi-echo LOUPE-ST block is shown in Fig. 1(b). Learnable weights $\{w_j\}$ with echo index j are used to generate multi-echo probabilistic patterns $\{P_j\}$ through sigmoid transformation and sampling ratio renormalization. Binary multi-echo under-sampling patterns $\{U_j\}$ are generated via stochastic sampling from $\{P_j\}$, i.e., $U_j = \mathbf{1}_{z<P_j}$ for each echo with $\mathbf{1}_x$ the indicator function on the truth value of x and z uniformly distributed between [0, 1]. Then $\{U_j\}$ are used to retrospectively acquire $\{b_{jk}\}$ from fully-sampled multi-echo k-p space dataset. LOUPE-ST applied a straight-through estimator [3] for back-propagation to the stochastic sampling layer where binary pattern was generated, which was reported to perform better than the vanilla LOUPE [2]. We refer readers to [27] for details.

Table 2. Ablation study. Reconstruction performances were progressively improved as more blocks were added to deep ADMM (* denotes statistical significance; $p < 0.05$).

	PSNR (↑)	SSIM (↑)
Deep ADMM	36.71 ± 3.10	0.9711 ± 0.0137
Deep ADMM + single SPO	37.39 ± 2.95	0.9734 ± 0.0124
Deep ADMM + TFF	40.12 ± 2.93	0.9835 ± 0.0086
Deep ADMM + TFF + single SPO	40.35 ± 2.74	0.9844 ± 0.0080

Table 3. Performance comparison. For each sampling pattern, Deep ADMM + TFF performed better than LLR and multi-echo MoDL (* denotes statistical significance; $p < 0.05$). For each reconstruction method, improvements were observed from manually designed to multi-echo SPO patterns.

	Variable density		Single SPO	
	PSNR (↑)	SSIM (↑)	PSNR (↑)	SSIM (↑)
LLR	36.68 ± 3.70*	0.9604 ± 0.0230*	36.72 ± 3.77*	0.9582 ± 0.0248*
Multi-echo MoDL	40.45 ± 3.45*	0.9809 ± 0.0101*	41.01 ± 3.30*	0.9817 ± 0.0099*
Deep ADMM + TFF	**42.24** ± 3.28	**0.9864** ± 0.0074	**42.77** ± 3.38	**0.9867** ± 0.0076

	Multi SPO	
	PSNR (↑)	SSIM (↑)
LLR	37.93 ± 3.65*	0.9699 ± 0.0179*
Multi-echo MoDL	42.28 ± 3.02*	0.9859 ± 0.0078*
Deep ADMM + TFF	**43.75** ± 3.02	**0.9894** ± 0.0058

3   Experiments

3.1   Data Acquisition and Preprocessing

Cartesian fully sampled k-space data of multi-echo images were acquired in 7 subjects using a 3D MEGRE sequence on a 3T GE scanner with a 32-channel head coil. Imaging parameters were: $256 \times 206 \times 80$ imaging matrix size with the corresponding readout × phase × phase encoding directions, $1 \times 1 \times 2$ mm^3 resolution, 10 echoes with 1.972 ms as the first TE and 3.384 ms echo spacing. 32-coil k-space data of each echo were compressed into 8 virtual coils using a geometric singular value decomposition based coil compression algorithm [29]. After compression, coil sensitivity maps of the first echo were estimated with a reconstruction null space eigenvector decomposition based algorithm ESPIRiT [23] using a centric $24 \times 24 \times 24$ auto-calibration k-space region for each compressed coil. The estimated coil sensitivity maps were also used for the remaining echoes during reconstruction. From the fully sampled k-space data, coil combined multi-echo images were computed to provide the ground truth labels for both network training and performance comparison. Central 200 slices along readout direction containing brain tissues of each fully sampled subject were extracted and 3/1/3 subjects (600/200/600 slices) were used as training, validation and test datasets.

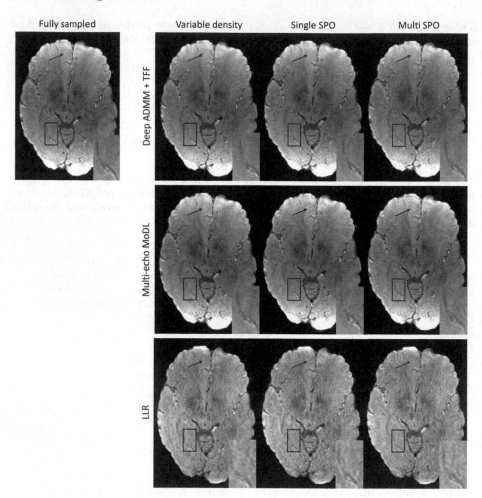

Fig. 2. Performance comparison of one echo-combined test slice. Heavy artifacts were present almost everywhere in LLR, while mild artifacts (red arrows) were shown in multi-echo MoDL. In contrast, deep ADMM with TFF produced clean images without visible artifacts. For each method, zoom-in veil structures (from red boxes) were progressively improved from manually designed to multi-echo SPO patterns. (Color figure online)

3.2   Implementation Details and Ablation Study

23% retrospective under-sampling on the coil-compressed fully sampled k-space data was applied in the experiment. The input 20-channel (real and imaginary parts with 10 echoes) zero-filled images were fed into deep ADMM with $K = 10$ unrolled iterations to generate reconstructed images (Fig. 1a). During training, both weights in deep ADMM network and SPO block were updated simultaneously by minimizing a channel-wise structural similarity index measure (SSIM)

[25] loss: $\sum_{k=1}^{K} \sum_{j=1}^{N_T} \text{SSIM}(s_j^{(k)}, s_j^*)$, where s_j^* denotes label and $s_j^{(k)}$ denotes k-th deep ADMM iteration at j-th echo. After training, a specific set of binary patterns $\{U_j\}$ were generated from $\{P_j\}$ and used to further train deep ADMM alone. We implemented in PyTorch using the Adam optimizer [12] (batch size 1, number of epochs 100 and initial learning rate 10^{-3}) on a RTX 2080Ti GPU. During test, the retrospectively under-sampled k-space data were fed into deep ADMM for inference. PSNR and SSIM on the echo-combined image $\sqrt{\sum_{j=1}^{N_T} |s_j|^2}$ were used to measure reconstruction quality.

An ablation study regarding TFF and SFO blocks were investigated and reconstruction performances on test dataset are shown in Table 1. A variable density sampling pattern was manually designed based on a multi-level sampling scheme [21] and used to train baseline deep ADMM without TFF and SFO (first row in Table 1). TFF, single-echo SPO and multi-echo SPO were progressively added to the baseline deep ADMM (2–5 rows in Table 1) to check the effectiveness of each block. For Deep ADMM without TFF, the number of convolutional kernels in the denoiser block were extended to match the memory consumption to deep ADMM with TFF. In Table 1, reconstruction performance was progressively improved as more blocks were added to the baseline deep ADMM, where Deep ADMM with both TFF and multi-echo SPO blocks achieved the best performance. All the sampling patterns designed/learned in the ablation study are shown in the Appendix.

3.3   Performance Comparison

The proposed deep ADMM + TFF network was compared with locally low rank (LLR) [28] and multi-echo MoDL [1], where the vanilla MoDL was modified to reconstruct multi-echo image simultaneously. Manually designed variable density, learned single-echo SPO and multi-echo SPO patterns in the ablation study were used for performance comparison. Reconstructed echo-combined images of one test slice using different sampling patterns and reconstruction methods are shown in Fig. 2. Heavy artifacts were present almost everywhere in LLR, while mild artifacts (red arrows) were shown in multi-echo MoDL. In contrast, deep ADMM with TFF produced clean images without visible artifacts. For each method, zoom-in veil structures (from red boxes) were progressively improved from manually designed pattern to multi-echo SPO patterns. Quantitative metrics are shown in Table 2. Deep ADMM with TFF performed the best for each type of sampling pattern. Besides, for each reconstruction method, improvements were observed from manually designed, single SPO to multi-echo SPO patterns. QSMs were estimated from deep ADMM + TFF reconstructed multi-echo images and shown in Fig. 3 with the same slice to Fig. 2. The same veil structures which appeared bright in QSMs were also progressively improved from manually designed pattern to multi-echo SPO patterns.

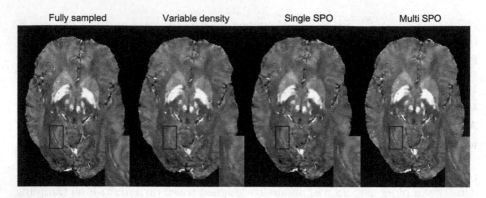

Fig. 3. QSMs estimated from "deep ADMM + TFF" reconstructions using different sampling patterns. The same veil structures from Fig. 2 appeared bright in QSMs and were progressively improved from manually designed variable density to multi-echo SPO patterns. (Color figure online)

4   Conclusion

We propose a unified method to optimize MEGRE signal acquisition and image reconstruction. The proposed reconstruction network inserted a recurrent TFF block into a deep ADMM to capture image evolution dynamics along echo time. The proposed SPO block extended the single-echo LOUPE-ST to multi-echo regime. Experimental results showed superior performance for both reconstruction and multi-echo sampling pattern compared to other methods and patterns.

References

1. Aggarwal, H.K., Mani, M.P., Jacob, M.: Modl: model-based deep learning architecture for inverse problems. IEEE Trans. Med. Imaging **38**(2), 394–405 (2018)
2. Bahadir, C.D., Wang, A.Q., Dalca, A.V., Sabuncu, M.R.: Deep-learning-based optimization of the under-sampling pattern in MRI. IEEE Trans. Comput. Imaging **6**, 1139–1152 (2020)
3. Bengio, Y., Léonard, N., Courville, A.: Estimating or propagating gradients through stochastic neurons for conditional computation. arXiv preprint arXiv:1308.3432 (2013)
4. Boyd, S., Parikh, N., Chu, E.: Distributed Optimization and Statistical Learning via the Alternating Direction Method of Multipliers. Now Publishers Inc. (2011)
5. Chan, S.H., Wang, X., Elgendy, O.A.: Plug-and-play ADMM for image restoration: fixed-point convergence and applications. IEEE Trans. Comput. Imaging **3**(1), 84–98 (2016)
6. Deichmann, R.: Fast high-resolution t1 mapping of the human brain. Magn. Reson. Med. Offic. J. Int. Soc. Magn. Reson. Med. **54**(1), 20–27 (2005)
7. Deoni, S.C., Peters, T.M., Rutt, B.K.: High-resolution t1 and t2 mapping of the brain in a clinically acceptable time with despot1 and despot2. Magn. Reson. Med. Offic. J. Int. Soc. Mag. Reson. Med. **53**(1), 237–241 (2005)

8. Gözcü, B., et al.: Learning-based compressive MRI. IEEE Trans. Med. Imaging **37**(6), 1394–1406 (2018)
9. Griswold, M.A., et al.: Generalized autocalibrating partially parallel acquisitions (grappa). Magn. Reson. Med. Offic. J. Int. Soc. Magn. Reson. Med. **47**(6), 1202–1210 (2002)
10. Haldar, J.P., Kim, D.: Oedipus: an experiment design framework for sparsity-constrained MRI. IEEE Trans. Med. imaging **38**(7), 1545–1558 (2019)
11. Hammernik, K., et al.: Learning a variational network for reconstruction of accelerated MRI data. Magn. Reson. Med. **79**(6), 3055–3071 (2018)
12. Kingma, D.P., Ba, J.: Adam: a method for stochastic optimization. arXiv preprint arXiv:1412.6980 (2014)
13. Liu, J., et al.: Morphology enabled dipole inversion for quantitative susceptibility mapping using structural consistency between the magnitude image and the susceptibility map. Neuroimage **59**(3), 2560–2568 (2012)
14. Liu, T., Wisnieff, C., Lou, M., Chen, W., Spincemaille, P., Wang, Y.: Nonlinear formulation of the magnetic field to source relationship for robust quantitative susceptibility mapping. Magn. Reson. Med. **69**(2), 467–476 (2013)
15. Lustig, M., Donoho, D., Pauly, J.M.: Sparse MRI: The application of compressed sensing for rapid MR imaging. Magn. Reson. Med. Offic. J. Int. Soc. Magn. Reson. Med. **58**(6), 1182–1195 (2007)
16. Murphy, M., Alley, M., Demmel, J., Keutzer, K., Vasanawala, S., Lustig, M.: Fast l_1-spirit compressed sensing parallel imaging MRI: scalable parallel implementation and clinically feasible runtime. IEEE Trans. Med. Imaging **31**(6), 1250–1262 (2012)
17. Otazo, R., Kim, D., Axel, L., Sodickson, D.K.: Combination of compressed sensing and parallel imaging for highly accelerated first-pass cardiac perfusion MRI. Magn. Reson. Med. **64**(3), 767–776 (2010)
18. Peng, X., Ying, L., Liu, Y., Yuan, J., Liu, X., Liang, D.: Accelerated exponential parameterization of t2 relaxation with model-driven low rank and sparsity priors (morasa). Magn. Reson. Med. **76**(6), 1865–1878 (2016)
19. Pruessmann, K.P., Weiger, M., Scheidegger, M.B., Boesiger, P.: Sense: sensitivity encoding for fast MRI. Magn. Reson. Med. Offic. J. Int. Soc. Magn. Reson. Med. **42**(5), 952–962 (1999)
20. Qin, C., Schlemper, J., Caballero, J., Price, A.N., Hajnal, J.V., Rueckert, D.: Convolutional recurrent neural networks for dynamic MR image reconstruction. IEEE Trans. Med. Imaging **38**(1), 280–290 (2018)
21. Roman, B., Hansen, A., Adcock, B.: On asymptotic structure in compressed sensing. arXiv preprint arXiv:1406.4178 (2014)
22. Schlemper, J., Caballero, J., Hajnal, J.V., Price, A.N., Rueckert, D.: A deep cascade of convolutional neural networks for dynamic MR image reconstruction. IEEE Trans. Med. Imaging **37**(2), 491–503 (2017)
23. Uecker, M., et al.: Espirit–an eigenvalue approach to autocalibrating parallel MRI: where sense meets grappa. Magn. Reson. Med. **71**(3), 990–1001 (2014)
24. Wang, Y., Liu, T.: Quantitative susceptibility mapping (QSM): decoding MRI data for a tissue magnetic biomarker. Magn. Reson. Med. **73**(1), 82–101 (2015)
25. Wang, Z., Bovik, A.C., Sheikh, H.R., Simoncelli, E.P.: Image quality assessment: from error visibility to structural similarity. IEEE Trans. Image Process. **13**(4), 600–612 (2004)
26. Yu, H., Shimakawa, A., McKenzie, C.A., Brodsky, E., Brittain, J.H., Reeder, S.B.: Multiecho water-fat separation and simultaneous r estimation with multifrequency fat spectrum modeling. Magn. Reson. Med. Offic. J. Int. Soc. Magn. Reson. Med. **60**(5), 1122–1134 (2008)

27. Zhang, J., et al.: Extending LOUPE for K-space under-sampling pattern optimization in Multi-coil MRI. In: Deeba, F., Johnson, P., Würfl, T., Ye, J.C. (eds.) MLMIR 2020. LNCS, vol. 12450, pp. 91–101. Springer, Cham (2020). https://doi.org/10.1007/978-3-030-61598-7_9
28. Zhang, T., Pauly, J.M., Levesque, I.R.: Accelerating parameter mapping with a locally low rank constraint. Magn. Reson. Med. **73**(2), 655–661 (2015)
29. Zhang, T., Pauly, J.M., Vasanawala, S.S., Lustig, M.: Coil compression for accelerated imaging with cartesian sampling. Magn. Reson. Med. **69**(2), 571–582 (2013)
30. Zhao, B., Lu, W., Hitchens, T.K., Lam, F., Ho, C., Liang, Z.P.: Accelerated MR parameter mapping with low-rank and sparsity constraints. Magn. Reson. Med. **74**(2), 489–498 (2015)

Dual-Domain Adaptive-Scaling Non-local Network for CT Metal Artifact Reduction

Tao Wang[1], Wenjun Xia[1], Yongqiang Huang[1], Huaiqiang Sun[2], Yan Liu[3], Hu Chen[1], Jiliu Zhou[3], and Yi Zhang[1(✉)]

[1] College of Computer Science, Sichuan University, Sichuan, China
`yzhang@scu.edu.cn`
[2] Department of Radiology, West China Hospital of Sichuan University, Sichuan, China
[3] School of Electrical Engineering Information, Sichuan University, Sichuan, China

Abstract. Metal implants can heavily attenuate X-rays in computed tomography (CT) scans, leading to severe artifacts in reconstructed images. Several network models have been proposed for metal artifact reduction (MAR) in CT. Despite the encouraging results were achieved, there is still much room to further improve performance. In this paper, a novel Dual-domain Adaptive-scaling Non-local Network (DAN-Net) is proposed for MAR. The corrupted sinogram was corrected using adaptive scaling first to preserve more tissue and bone details. Then, an end-to-end dual-domain network is adopted to successively process the sinogram and its corresponding reconstructed image is generated by the analytical reconstruction layer. In addition, to better suppress the existing artifacts and restrain the potential secondary artifacts caused by inaccurate results of the sinogram-domain network, a novel residual sinogram learning strategy and non-local module are leveraged in the proposed network model. Experiments demonstrate the performance of the proposed DAN-Net is competitive with several state-of-the-art MAR methods in both qualitative and quantitative aspects. The code is available online: https://github.com/zjk1988/DAN-Net.

Keywords: Computed tomography · Metal artifact reduction · Deep learning · Dual-domain network

1 Introduction

With the help of computed tomography (CT) images, medical diagnosis and treatments can be conducted effectively. Due to metallic implants, the reconstructed CT images are contaminated by heavy artifacts which degrade the imaging quality and severely compromise doctors' diagnoses. In particular, some

Electronic supplementary material The online version of this chapter (https://doi.org/10.1007/978-3-030-87231-1_24) contains supplementary material, which is available to authorized users.

© Springer Nature Switzerland AG 2021
M. de Bruijne et al. (Eds.): MICCAI 2021, LNCS 12906, pp. 243–253, 2021.
https://doi.org/10.1007/978-3-030-87231-1_24

artifacts and certain lesions have considerable commonalities, leading to misdiagnosis, and subsequent medical image analysis is difficult [1].

During the past several decades, numerous metal artifact reduction (MAR) methods have been proposed. Conventional MAR methods can be grouped into three categories [2]: projection completion methods, iterative reconstruction methods and image postprocessing methods. The projection completion methods regard projection data in the metal trace region as missing and fill in lost data with estimated values by different image inpainting methods [3–5] or interpolation strategies [6–10]. However, it is hard to guarantee smoothness at the metal trace boundaries. Iterative methods improve image quality gradually based on constrained optimization but are usually time-consuming and require manually well-designed regularizers [11,12]. Image postprocessing methods [13,14] aim to reduce metal artifacts in the image domain but usually cannot suppress the artifacts well and are apt to distort the anatomic structure [2,15].

Recently, several deep learning (DL)-based MAR methods have been proposed. [16–20] proposed to recover the missing data in the metal trace. Meanwhile, some DL-based studies [21,22] dedicated to reduce metal artifacts in image domain. Nowadays, end-to-end dual domain networks have become the mainstream for MAR. However, there are some critical limitations. [23,24] regarded projection data in the metal trace as missing, which resulted in the loss of details near the metal area in reconstructed CT images. [15,25] directly used metal-corrupted projection data and corresponding reconstructed CT images as inputs. Actually, the data in the metal trace have a much higher amplitude than the data outside the metal trace and they two can be regarded as obeying two different data distributions. It is difficult for neural networks to transform two different data distributions into a uniform distribution. Moreover, the change in the metal trace boundaries will cause weak continuity of the first derivative of projection data in a certain section, which will be further expanded by filtering and will generate extra artifacts [26]. Finally, metal artifacts are non-local in the image domain, which may fail to be reduced well by convolution operation.

To address the problems mentioned above, we propose a novel Dual-domain Adaptive-scaling Non-local Network (DAN-Net) for MAR. Our contributions is threefold: 1) Different from current dual-domain networks, the original sinogram is preprocessed using adaptive scaling and taking the scaled metal projection and corresponding preprocessed CT image reconstructed by filtered back-projection (FBP) as the inputs, which can preliminarily suppress metal artifacts and maintain tissue details. 2) A novel residual sinogram learning strategy is proposed to avoid transforming two different data distributions into a uniform one and to improve the smoothness of the corrected projection. 3) A non-local U-Net architecture is designed for image-domain enhancement, which can capture long-range dependencies of metal artifacts and further improve the image quality.

2   Method

2.1   Problem Formulation

In this work, we consider the case of a 2D attenuation distribution. The linear attenuation coefficients with metallic implants can be expressed as follows:

$$\mu(E) = \mu_t(E) \odot (1 - M) + \mu_m(E) \odot M \tag{1}$$

where $\mu_t(E)$ and $\mu_m(E)$ represent attenuation images of tissue and metal parts respectively; M denote the metal mask; and $\odot$ is the elementwise multiplication. Then, the projection data S_{ma} can be calculated as follows [25]:

$$
\begin{aligned}
S_{ma} &= -\ln \int \eta(E) exp(-\mathcal{P}(\mu(E)))dE \\
&= -\ln \int \eta(E) exp(-\mathcal{P}(\mu_t(E) \odot (1 - M) + \mu_m(E) \odot M))dE \\
&= -\ln \int \eta(E) exp(-\mathcal{P}(\mu_t(E) \odot (1 - M))dE+ \\
&\quad - \ln \int \eta(E) exp(-\mathcal{P}(\mu_m(E) \odot M))dE \\
&= S_{tissue} + S_{metal}
\end{aligned}
\tag{2}
$$

where $\mathcal{P}$ is the forward projection operation and $\eta(E)$ denotes the intensity distribution with spectral energy at E. Thus, S_{ma} are contributed by the attenuation of tissues and metal objects denoted as S_{tissue} and S_{metal}, respectively. As shown in Eq. 2, if we simply discard the projection data in the metal trace, both tissue and metal projections will be lost, and the reconstructed CT image has to take the risk of losing tissue details around the metallic implants. In this paper, we regard linear interpolation (LI) corrected sinogram S_{LI} as a rough estimation and attempt to restore the valuable residual information from $S_{sub} = S_{ma} - S_{LI}$.

2.2   The Proposed DAN-Net

Figure 1 depicts the overview of our proposed DAN-Net, which consists of three components: adaptive scaling, sinogram-domain network and image-domain network. More details are presented in subsequent sections.

Adaptive Scaling. To eliminate the abrupt change in projection data caused by the metal and maintain more useful information, Chen et al. [27] adopted a linear attenuation operation to restore the data in the metal trace as $S_{res} = \lambda \times S_{sub}$, $S_{pre} = S_{LI} + S_{res}$, where λ is the scaling parameter to control the trade-off between artifact reduction and detail preservation. S_{res} and S_{pre} represent the scaled metal projection and the corrected projection after adaptive scaling, respectively. In this paper, we chose $\lambda = 0.4$ experimentally. The corresponding adaptively scaled CT image is obtained as $X_{pre} = \mathcal{P}^{-1}(S_{pre})$, where $\mathcal{P}^{-1}$ denotes FBP operation.

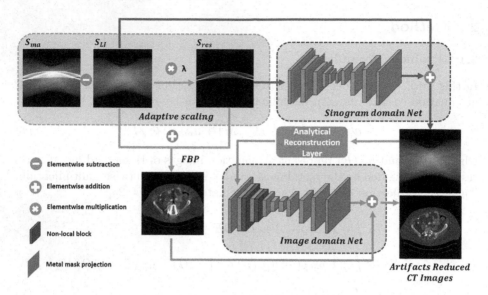

Fig. 1. Overview of the proposed DAN-Net.

Sinogram Domain Network. To complete the sinogram, we propose a neural network, G_{sino}, with a residual sinogram learning strategy, which takes S_{res} as input, and only modify the projection data within the metal trace. In this work, U-Net is utilized as the backbone of G_{sino} and the mask pyramid network (MPN) [21] is introduced to retain sufficient information of metal projection. And we have $M_t = \sigma(M_p > 0)$, where $\sigma(\bullet)$ is a binary indicator function. The detailed structure of G_{sino} is illustrated in the supplementary material.

Since our main goal is to retrieve information within the metal trace, we only refine the adaptively scaled residual sinogram in the metal trace. The corrected sinogram can be written as $S_{sino} = G_{sino}(S_{res}, M_p) \odot M_t + S_{LI}$. $\mathcal{L}_{sino}$ is adopted to measure the differences between S_{sino} and the ground truth S_{gt} as

$$\mathcal{L}_{sino} = ||(S_{sino} - S_{gt}) \odot M_t||_1 \tag{3}$$

Then, $X_{sino} = \mathcal{P}^{-1}(S_{sino})$ can be obtained using an analytical reconstruction layer. To alleviate the secondary artifacts in the reconstructed CT image, the reconstruction loss $\mathcal{L}_{FBP}$ between X_{sino} and the ground truth image X_{gt} is utilized as

$$\mathcal{L}_{FBP} = ||(X_{sino} - X_{gt}) \odot (1 - M)||_1 \tag{4}$$

Image Domain Net. We also utilize U-Net as the backbone to enhance the reconstructed CT images. For computational efficiency, we halve the channel numbers. It is well known that convolution is a local operator whose receptive field is limited by the size of filters. Metal artifacts are non-local, convolution-based post-processing methods may fail to remove the artifacts well. To tackle this problem, the non-local network modules [28] are embedded into G_{im} after the second and third down-sampling steps, as depicted in Fig. 1. To focus on the artifact-impacted regions, X_{sino} and X_{pre} are concatenated as inputs of G_{im}. A residual learning strategy is also adopted, which is written as: $X_{im} = X_{pre} + G_{im}(X_{sino}, X_{pre})$. The detailed structure of G_{im} is illustrated in the supplementary material. G_{im} is also optimized with L1 norm in the image domain:

$$\mathcal{L}_{im} = ||(X_{im} - X_{gt}) \odot (1 - M)||_1 \tag{5}$$

In summary, the total objective function is:

$$\mathcal{L} = \mathcal{L}_{sino} + \alpha \times \mathcal{L}_{FBP} + \beta \times \mathcal{L}_{im} \tag{6}$$

where α and β are the weighting parameters of different components. In our experiments, we empirically set $\alpha = \beta = 1$.

3   Experiments

3.1   Dataset

For data simulation, we followed the procedure of [23] and used the DeepLesion dataset [29]. For metal mask simulation, we employed the masks generated from [30], containing 100 manually segmented metal implants with all kinds of metal implants. Specifically, we randomly selected 1000 CT images from the DeepLesion dataset and 90 metal masks to synthesize 90,000 combinations in the training set. The remaining 200 CT images and 10 masks were adopted for evaluation. The original CT images were resized to 256×256 for computational efficiency. To simulate Poisson noise, a polychromatic X-ray source was employed, and the incident beam X-ray was set to 2×10^7 photons [31]. The partial volume effects and scatter were also taken into consideration. Without loss of generality, our experiments were restricted to 2D parallel-beam geometry. For the sampling condition, 367 detector bins and 361 sampling views uniformly distributed from $0°$ to $180°$ were assumed. Unlike [23], we truncated the CT values to $[0, 4095]$, which better conforms to the real situation.

Table 1. Quantitative comparison of different methods on the simulated dataset.

Methods	Uncorrected	LI	NMAR	CNNMAR	DuDoNet	ADN	DAN-Net
PSNR	15.33	30.74	30.83	32.15	36.82	33.60	**40.61**
SSIM	0.6673	0.9224	0.9270	0.9508	0.9777	0.9275	**0.9872**

3.2   Implementation Details

We trained the network in an end-to-end manner, and the model was implemented with the *PyTorch* framework [32]. The back-projection was implemented by the *numba* library in Python, which can improve the computational efficiency, aided by CUDA. The network was optimized by the Adam optimizer with the parameters $(\beta_1, \beta_2) = (0.5, 0.999)$. The learning rate was initialized to 0.0002 and halved every 20 epochs. The network was trained with 200 epochs on one NVIDIA 1080Ti GPU with 11 GB memory, and the batch size was 4.

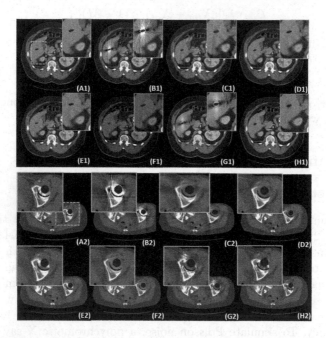

Fig. 2. Visual comparison on the simulated dataset. (A1)&(A2)- (H1)&(H2): Reference images, Uncorrected, LI, NMAR, CNNMAR, DuDoNet, ADN and DAN-Net. The display window is [−375, 560] HU.

3.3   Comparison with State-of-the-Art Methods

The proposed DAN-Net was compared with several state-of-the-art MAR methods: LI [6], NMAR [33], CNNMAR [30], DuDoNet [23] and ADN [21]. LI and NMAR are classic methods. CNNMAR demonstrates the effectiveness and potential of CNN-based methods. DuDoNet is a supervised dual-domain framework. ADN is a state-of-the-art unsupervised image-domain method. Structural similarity (SSIM) and peak signal-to-noise ratio (PSNR) are adopted as quantitative metrics. In Table 1, it is observed that the LI and NMAR significantly improve both SSIM and PSNR values compared with uncorrected CT images, and NMAR achieves better scores than LI, for NMAR takes advantages of both prior images and the LI method. However, these methods cannot guarantee the continuity of the matal trace boundries well. CNNMAR takes the outputs of different MAR methods as inputs and outperforms conventional methods. ADN achieves similar performance to CNNMAR without the need for paired training data. DuDoNet and our method attain remarkable improvements on both SSIM and PSNR since they simultaneously leverage the advantages of the sinogram domain and image domain. However, DuDoNet lost details around metals. Compared with DuDoNet, DAN-Net further raises the scores, supresses the metal artifacts and keeps the tissue details, which demonstrates the performance of our proposed method quantitatively.

For qualitative comparisons, Fig. 2 presents two representative metallic implants with different sizes. LI and NMAR contain some radial artifacts, while DL-based methods perform better. When metal objects get larger, LI and NMAR introduce obvious new artifacts in Fig. 2 (C2&D2). In the second case, other methods fail to preserve the details around metallic implants, while DAN-Net maintains these structural details more completely.

Table 2. Quantitative comparison of different variants of our method on the simulated dataset.

Methods	Sino-Net	Res-Sino-Net	IM-Net	Non-local-IM-Net	Ma-Dual-Net	DAN-Net
PSNR	31.43	31.71	33.79	34.75	34.15	**40.61**
SSIM	0.9232	0.9494	0.9520	0.9720	0.9597	**0.9872**

3.4 Clinical Study

To verify the performance of the proposed DAN-Net in clinical scenario, a clinical CT image with real metal artifacts was tested. Figure 3 presents the MAR results using different methods, the metal mask was empirically segmented using 2000 HU as the threshold. It is observed that DAN-Net suppresses most of the metal artifacts and preserves the fine-grained anatomical structures around the metals, which supplies coherent results to the simulated data and demonstrates the potential for real clinical application.

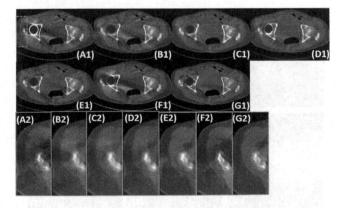

Fig. 3. Visual comparison with different MAR methods on a clinical CT image. A1–G1 and A2–G2 represent uncorrected CT images and corrected results using LI, NMAR, CNNMAR, DuDoNet, ADN and DAN-Net. The display window is [−375, 560] HU.

4   Ablation Study

In this section, we investigate the effectiveness of different modules of the proposed DAN-Net. The ablation study configurations are listed as follows:

(1) Sino-Net: the sinogram-domain network only without residual learning;
(2) Res-Sino-Net: the sinogram-domain network only with residual learning;
(3) IM-Net: the image-domain network only without a non-local module;
(4) Non-local-IM-Net: the image-domain network only with the non-local module;
(5) Ma-Dual-Net: a dual-domain network with sinogram-domain residual learning and image-domain non-local modules but without an adaptively scaled sinogram; and
(6) DAN-Net: the same architecture as Ma-Dual-Net with an adaptively scaled sinogram.

Effect of Sinogram-Domain Residual Learning. In Table 2, the Res-Sino-Net improves the SSIM and PSNR values. In Fig. 4 (B1), there are evident artifacts. In contrast, in Fig. 4 (C1), residual information is recovered from adaptively scaled projection data, thereby easing network learning. There are obvious differences between (B2) and (A2). However, (C2) and (A2) look more consistent.

Effect of Image-Domain Non-local Module. In Table 2, the Non-local-IM-Net has higher SSIM and PSNR values than IM-Net. For the qualitative comparison, in Fig. 5(A) and (B), artifacts are better suppressed in the results of Non-local-IM-Net than IM-Net.

Effect of Adaptive Scaling. Ma-Dual-Net takes the original sinogram and corresponding reconstructed image as the inputs. In Table 2, our approach outperforms Ma-Dual-Net in quantitative aspects. In Fig. 5 (C) and (D), we can

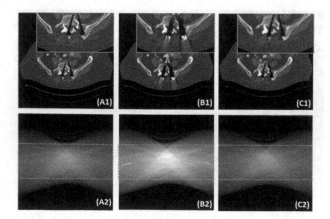

Fig. 4. Results of ablation study. (A): The ground truth, (B): Sino-Net and (C): Res-Sino-Net. The display window is [−375, 560] HU.

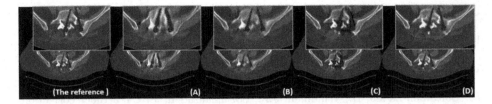

Fig. 5. Results of ablation study. (A): IM-Net, (B): Non-local-IM-Net, (C): Ma-Dual-Net and (D): Ma-Dual-Net. The display window is [−375, 560] HU.

observe that our method retrieves many more structural details around the metallic implants.

5   Conclusion

In this paper, we propose a novel dual-domain MAR method, dubbed as DAN-Net. Aided by adaptive scaling, artifacts can be reduced primarily and tissue details will be better preserved. A residual sinogram learning strategy and non-local module are leveraged to further improve the CT image quality. Experiments demonstrate the performance of DAN-Net is competitive with several state-of-the-art MAR methods in both qualitative and quantitative aspects.

References

1. Yazdi, M., Beaulieu, L.: A novel approach for reducing metal artifacts due to metallic dental implants. In: Nuclear Science Symposium Conference Record, pp. 2260–2263. IEEE (2007)
2. Gjesteby, L., et al.: Metal artifact reduction in CT: where are we after four decades? IEEE Access **4**, 5826–5849 (2016)
3. Zhang, Y., et al.: A new CT metal artifacts reduction algorithm based on fractional-order sinogram inpainting. J. X-ray Sci. Technol. **19**(3), 373–384 (2011)
4. Xue, H., et al.: Metal artifact reduction in dual energy CT by sinogram segmentation based on active contour model and TV inpainting. In: 2009 IEEE Nuclear Science Symposium Conference Record (NSS/MIC), pp. 904–908. IEEE (2009)
5. Duan, X., et al.: Metal artifact reduction in CT images by sinogram TV inpainting. In: 2008 IEEE Nuclear Science Symposium Conference Record, pp. 4175–4177. IEEE (2008)
6. Lewitt, R.M., Bates, R.H.T.: Image reconstruction from projections III: projection completion methods (theory). Optik **50**, 189–204 (1978)
7. Kalender, W.A., Hebel, R., Ebersberger, J.: Reduction of CT artifacts caused by metallic implants. Radiology **164**(2), 576–577 (1987)
8. Zhao, S., et al.: A wavelet method for metal artifact reduction with multiple metallic objects in the field of view. J. X-Ray Sci. Technol. **10**(1), 67–76 (2002)
9. Mehranian, A., et al.: X-ray CT metal artifact reduction using wavelet domain L-0 sparse regularization. IEEE Trans. Med. Imaging **32**(9), 1707–1722 (2013)

10. Lin, Z., Shi, Q.: Reduction of metal artifact in X-ray CT by quartic-polynomial interpolation. J. Image Graph. **6**(2), 142–147 (2001)
11. Yu, W., et al.: Low-dose computed tomography reconstruction regularized by structural group sparsity joined with gradient prior. Signal Process. **182**, 107945 (2021)
12. Gong, C., Zeng, L.: Adaptive iterative reconstruction based on relative total variation for low-intensity computed tomography. Signal Process. **165**, 149–162 (2019)
13. Soltanian-Zadeh, H., Windham, J.P., Soltanianzadeh, J.: CT artifact correction: an image-processing approach. In: Medical Imaging 1996: Image Processing. 1996: International Society for Optics and Photonics. https://doi.org/10.1117/12.237950
14. Ballhausen, H., et al.: Post-processing sets of tilted CT volumes as a method for metal artifact reduction. Radiat. Oncol. **9**(1), 114 (2014)
15. Yu, L., et al.: Deep sinogram completion with image prior for metal artifact reduction in CT images. IEEE Trans. Med. Imaging **40**(1), 228–238 (2020)
16. Park, H.S., et al.: CT sinogram-consistency learning for metal-induced beam hardening correction. Med. Phys. **45**(12), 5376–5384 (2018)
17. Ghani, M.U., Karl, W.C.: Deep learning based sinogram correction for metal artifact reduction. Electron. Imaging. **2018**(15), 472-1-4728 (2018)
18. Ghani, M.U., Karl, W.C.: Fast enhanced CT metal artifact reduction using data domain deep learning. IEEE Trans. Comput. Imaging **6**, 181–193 (2019)
19. Long, J., Shelhamer, E., Darrell, T.: Fully convolutional networks for semantic segmentation. IEEE Trans. Patt. Anal. Mach. Intell. **39**(4), 640–651 (2015)
20. Pimkin, A., et al.: Multidomain CT metal artifacts reduction using partial convolution based inpainting. In: 2020 International Joint Conference on Neural Networks (IJCNN). IEEE (2020). https://doi.org/10.1109/IJCNN48605.2020.9206625
21. Liao, H., et al.: ADN: artifact disentanglement network for unsupervised metal artifact reduction. IEEE Trans. Med. Imaging **39**(3), 634–643 (2019)
22. Philbin, J., Chum, O., Isard, M.: Lost in quantization: improving particular object retrieval in large scale image databases, Computer Vision and Pattern Recognition. In: IEEE Conference on CVPR 2008, vol. 2008, pp. 1–8 (2008)
23. Lin, W., et al.: DudoNet: dual domain network for CT metal artifact reduction. In: Proceedings of the IEEE Conference on Computer Vision and Pattern Recognition, pp. 10512–10521 (2019)
24. Peng, C., et al.: An irregular metal trace inpainting network for x-ray CT metal artifact reduction. Med. Phys. **47**(9), 4087–4100 (2020)
25. Lyu, Y., Lin, W.-A., Liao, H., Lu, J., Zhou, S.K.: Encoding metal mask projection for metal artifact reduction in computed tomography. In: Martel, A.L., et al. (eds.) MICCAI 2020. LNCS, vol. 12262, pp. 147–157. Springer, Cham (2020). https://doi.org/10.1007/978-3-030-59713-9_15
26. Pan, X.: Optimal noise control in and fast reconstruction of fan-beam computed tomography image. Med. Phys. **26**(5), 689–697 (1999)
27. Chen, L.M., et al.: Novel method for reducing high-attenuation object artifacts in CT reconstructions. In: Proceedings of SPIE - The International Society for Optical Engineering, vol. 4684, pp. 841–850 (2002)
28. Wang, X., et al.: Non-local neural networks. In: Proceedings of the IEEE Conference on Computer Vision and Pattern Recognition, pp. 7794–7803 (2018)
29. Yan, K., et al.: Deep lesion graphs in the wild: relationship learning and organization of significant radiology image findings in a diverse large-scale lesion database. In: Proceedings of the IEEE Conference on Computer Vision and Pattern Recognition, pp. 9261–9270 (2018)
30. Zhang, Y., Yu, H.: Convolutional neural network based metal artifact reduction in x-ray computed tomography. IEEE Trans. Med. Imaging **37**(6), 1370–1381 (2018)

31. Tang, S., et al.: Application of projection simulation based on physical imaging model to the evaluation of beam hardening corrections in X-ray transmission tomography. J. X-ray Sci. Technol. **16**(2), 95–117 (2008)
32. Paszke, A., et al.: Pytorch: an imperative style, high-performance deep learning library. arXiv preprint arXiv:1912.01703 (2019)
33. Meyer, E., et al.: Normalized metal artifact reduction (NMAR) in computed tomography. Med. Phys. **37**(10), 5482–5493 (2010)

Towards Ultrafast MRI via Extreme k-Space Undersampling and Superresolution

Aleksandr Belov[1,2], Joël Stadelmann[1], Sergey Kastryulin[1,2],
and Dmitry V. Dylov[2(✉)]

[1] Philips Innovation Labs RUS, Moscow, Russian Federation
{aleksandr.belov,joel.stadelmann,sergey.kastryulin}@philips.com
[2] Skolkovo Institute of Science and Technology, Moscow, Russian Federation
d.dylov@skoltech.ru

Abstract. We went below the MRI acceleration factors (*a.k.a.*, k-space undersampling) reported by all published papers that reference the original fastMRI challenge [29], and then used deep learning based image enhancement methods to compensate for the underresolved images. We thoroughly study the influence of the sampling patterns, the undersampling and the downscaling factors, as well as the recovery models on the final image quality for both the brain and the knee fastMRI benchmarks. The quality of the reconstructed images compares favorably against other methods, yielding an MSE of $11.4 \cdot 10^{-4}$, a PSNR of 29.6 dB, and an SSIM of 0.956 at $\times 16$ acceleration factor. More extreme undersampling factors of $\times 32$ and $\times 64$ are also investigated, holding promise for certain clinical applications such as computer-assisted surgery or radiation planning. We survey 5 expert radiologists to assess 100 pairs of images and show that the recovered undersampled images statistically preserve their diagnostic value.

Keywords: Fast MRI · Superresolution · Image-to-image translation

1 Introduction

Magnetic Resonance Imaging (MRI) is a non-invasive imaging modality that offers excellent soft tissue contrast and does not expose the patient to ionizing radiation. The raw MRI signal is recorded in the so-called k-space, with the digital images being then generated by Fourier transform. The filling of the k-space typically lasts 15–60 min, during which the patient must remain motionless (problematic for children, neurotic or claustrophobic patients). If a patient cannot stay still, a motion artifact will appear on the images, oftentimes demanding

Electronic supplementary material The online version of this chapter (https://doi.org/10.1007/978-3-030-87231-1_25) contains supplementary material, which is available to authorized users.

© Springer Nature Switzerland AG 2021
M. de Bruijne et al. (Eds.): MICCAI 2021, LNCS 12906, pp. 254–264, 2021.
https://doi.org/10.1007/978-3-030-87231-1_25

a complete re-scan [8]. Furthermore, the long acquisition time limits the applicability of MRI to dynamic imaging of the abdomen or the heart [11,25] and decreases the throughput of the scanner, leading to higher costs [3].

Contrasting chemicals [5] and physics-based acceleration approaches [20,24] alleviate the challenge and have been a subject of active research over the last two decades, solidifying the vision of the 'Ultrafast MRI' as the ultimate goal. A parallel pursuit towards the same vision is related to compressed sensing [5,11,25], where the methods to compensate for the *undersampled* k-space by proper image reconstruction have been proposed. This direction of research experienced a noticeable resurgence following the publication of the fastMRI benchmarks [29] and the proposal to remove the artifacts resulting from the gaps in the k-space by deep learning (DL) [14]. Given the original settings defined by the challenge, the majority of groups have been experimenting with the reconstruction for the acceleration factors of $\times 2$, $\times 4$, and $\times 8$, with only rare works considering stronger undersampling of $\times 16$ and $\times 32$.

In this article, we propose the use of powerful image-to-image translation models, originally developed for natural images, to tackle the poor quality of the underresolved MRI reconstructions. Specifically, we retrain Pix2Pix [12] and SRGAN [15] on the fastMRI data and study their performance alongside the reconstructing U-Net [22] at various downscaling and undersampling factors and sampling patterns. *Our best model compares favorably against other approaches at all conventional acceleration factors and allows to go beyond them to attempt extreme k-space undersampling, such as $\times 64$.*

2   Related Work

Several DL approaches have already been implemented to accelerate MRI. For instance, [27] proposed an algorithm that combines the optimal cartesian undersampling and MRI reconstruction using U-Net [22], which increased the fixed mask PSNR by 0.45–0.8 dB and SSIM by 0.012–0.017. The automated transform by manifold approximation (AUTOMAP) [30] learns a transform between k-space and the image domain using fully-connected layers and convolution layers in the image domain. The main disadvantage of AUTOMAP is a quadratic growth of parameters with the number of pixels in the image. Ref. [10] presented an approach, based on the interpolation of the missing k-space data using low-rank Hankel matrix completion, that consistently outperforms existing image-domain DL approaches on several masks. [19] used a cascade of U-Net modules to mimic iterative image reconstruction, resulting in an SSIM of 0.928 and a PSRN of 39.9 dB at $\times 8$ acceleration on multi-coil data.

The most recent works tackle the same challenge by applying parallel imaging [6], generative adversarial networks [2], ensemble learning with priors [16], trajectory optimization [26], and greedy policy search [1] (the only study that considered the factor $\times 32$). We summarize the metric values reported by these and other papers in the supplementary material. The issues of DL-based reconstruction robustness and the ultimate clinical value have been outlined in multiple works (*e.g.*, [4,26]), raising a reasonable concern of whether the images recovered by DL preserve their diagnostic value. Henceforth, we decided to engage a team of radiologists to perform such a user study to review our results.

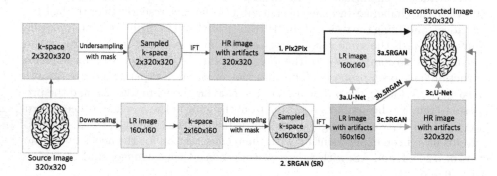

Fig. 1. Three possible approaches to downscale and undersample *k*-space. The sampling masks details are given in the supplementary material.

3   Method

3.1   Data Description

The Brain `fastMRI` dataset contains around 10,000 anonymized MRI volumes provided by the NYU School of Medicine [29]. Because of their greater number, we considered only T2 DICOM scans (as the largest class), padded, and downsized them to 320 × 320 pixels, discarding the smaller images. Finally, we selected 9 slices of each MRI volume at mid-brain height to obtain similar brain surface area in each image. The resulting 73,478 slices were split into train/validation/test subsets as 60%/20%/20%, making sure that slices from a given patient belong within the same subset. We applied min-max scaling of the intensities using the 2nd and the 98th percentiles.

We also validated our methods on the MRI scans of the knee. The same data preparation resulted in 167,530 T2 DICOM images, split into the subsets by the same proportion. Synovial liquid [5] noticeably shifted the upper percentiles in the knee data; thus, we normalized the intensities using the distance between the first and the second histogram extrema instead.

3.2   Models

An overview of the different models studied herein is shown in Fig. 1. Using the compressed sensing paradigm [5,14,25], we aim to accelerate MRI by sampling low resolution or/and partially filled *k*-space. We simulated the partial filling of the *k*-space by applying patterned masks on the Fourier transform of the original slices. We used the masks proposed by the `fastMRI` challenge (a central vertical band of 8% of the value is fixed and the remaining bands are sampled randomly). In addition, we used masks to approximate the radial and spiral sampling trajectories – after regridding – that are popular in the field of compressed sampling [28]. The three types of masks considered and the effect of them on the images are shown in the supplementary material.

We used Pix2Pix [12] as an image-to-image translation method to correct the artifacts introduced by the k-space undersampling and SRGAN [15] to upscale the low-resolution images to their input size of 320×320. Interested in preserving the high-frequency components, we used the MSE of the feature maps obtained with the VGG16 pre-trained on `ImageNet` 31th layer for the perceptual loss, relying on the reports that convolutional networks pre-trained on `ImageNet` are still relevant for the perceptual losses of MRI images [13,18,21].

We retrained Pix2Pix and SRGAN (SR, for superresolution alone) entirely on the datasets described above. We combined the U-Net generator from Pix2Pix with that from SRGAN to form a joined generator (paths 3a–c in Fig. 1), training it with the discriminator from SRGAN in a typical GAN manner. We also considered SRGAN for the direct reconstruction without the U-Net Pix2Pix generator (path 2 in Fig. 1). The quality of the produced images was assessed using the MSE, PSNR and SSIM [17] metrics, similarly to other studies [14,19].

4   Experiments

We ran extensive series of experiments outlined in Fig. 1. In all SRGAN variants, our models always outperformed bicubic interpolation, which we confirmed for the $\times 2$, $\times 4$, and $\times 8$ upscaling factors. We also observed that Pix2Pix models, following the undersampling with the radial and the spiral masks, always outperformed the `fastMRI` mask. The difference between the radial and the spiral masks, however, was less pronounced (see Tables 1 and 2). All experimental results are shown in the format of $\mu \pm \sigma$, where μ is the average metric value over the test set, and σ is its standard deviation.

Table 1. Pix2Pix model results.

Acc. / frac. of k-space	Mask	MSE, $\times 10^{-4}$	PSNR	SSIM, $\times 10^2$
$\times 2$ 50%	fastMRI	14.39 ± 3.33	28.53 ± 0.96	94.02 ± 3.19
	spiral	6.14 ± 2.13	32.37 ± 1.47	96.84 ± 2.22
	radial	5.08 ± 1.83	33.20 ± 1.47	97.70 ± 1.74
$\times 4$ 25%	fastMRI	27.93 ± 6.07	25.64 ± 0.93	90.65 ± 3.95
	spiral	13.96 ± 3.81	28.70 ± 1.15	93.56 ± 3.52
	radial	11.81 ± 4.62	29.51 ± 1.37	94.46 ± 3.93
$\times 8$ 12.5%	fastMRI	54.41 ± 12.13	22.75 ± 0.98	85.74 ± 4.21
	spiral	28.00 ± 7.67	25.68 ± 1.13	89.45 ± 4.38
	radial	21.12 ± 5.45	26.89 ± 1.08	91.23 ± 4.61
$\times 16$ 6.25%	spiral	45.04 ± 9.89	23.57 ± 1.00	85.93 ± 3.90
	radial	42.21 ± 9.78	23.87 ± 1.07	87.12 ± 4.53
$\times 32$ 3.125%	spiral	69.00 ± 17.48	21.76 ± 1.17	82.37 ± 4.84
	radial	65.62 ± 19.70	22.02 ± 1.31	83.06 ± 5.76

Table 2. SRGAN (SR) model results.

Acc. / frac. of k-space	Models	MSE, $\times 10^{-4}$	PSNR	SSIM, $\times 10^2$
$\times 4$ 25%	Bicubic	10.51 ± 4.29	30.16 ± 1.84	96.85 ± 1.51
	SRGAN	3.03 ± 1.61	35.72 ± 2.12	98.64 ± 0.93
$\times 16$ 6.25%	Bicubic	43.94 ± 14.66	23.84 ± 1.58	88.48 ± 3.17
	SRGAN	11.69 ± 4.21	29.58 ± 1.49	95.63 ± 2.35
$\times 64$ 1.5625%	Bicubic	121.59 ± 29.76	19.30 ± 1.21	73.04 ± 4.17
	SRGAN	28.76 ± 9.67	24.37 ± 1.13	88.49 ± 3.32

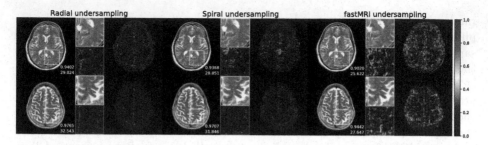

Fig. 2. Pix2Pix recovery results for ×4 *k*-space undersampling. The difference with the ground truth (multiplied by 3 for clarity) are shown on the right. Shown in the corner are SSIM and PSNR values. Note the inferior performance of original fastMRI masks.

Figure 2 compares the Pix2Pix reconstruction at ×4 undersampling for different sampling patterns. Visually, the reconstructions of all models look clear with sharp anatomic structures. However, the difference images reveal that the finest details are reconstructed best via radial sampling, which is probably a consequence of the more homogeneous sampling of the highest spatial frequencies.

For the combination of architectures, we downscaled the images to 160 × 160 pixels before applying *k*-space undersampling. We then trained SRGAN for both the upscaling and artifact removal or SRGAN+U-Net/U-Net+SRGAN, depending on the order of application of the corresponding generators. Visually, SRGAN alone performed worse than SRGAN+U-Net and U-Net+SRGAN combos, both of which seem to be statistically equivalent (see Tables 3, 4, and Fig. 3). A comprehensive presentation of the results obtained at different acceleration factors can be found in the supplementary material.

Table 3. Low-Resolution (160 × 160) model with ×2 undersampling, total acceleration: ×8, fraction of *k*-space: 12.5%

Mask	Model	MSE, ×10⁻⁴	PSNR	SSIM, ×10²
fastMRI	SRGAN	12.98 ± 3.79	29.04 ± 1.22	94.84 ± 2.55
	SRGAN+U-Net	8.81 ± 3.08	30.80 ± 1.44	95.97 ± 2.17
	U-Net+SRGAN	9.27 ± 3.28	30.56 ± 1.41	95.69 ± 2.69
spiral	SRGAN	7.66 ± 2.94	31.47 ± 1.66	96.28 ± 2.26
	SRGAN+U-Net	7.02 ± 2.84	31.88 ± 1.73	96.51 ± 2.06
	U-Net+SRGAN	7.09 ± 2.82	31.81 ± 1.65	96.39 ± 2.37
radial	SRGAN	6.25 ± 2.63	32.40 ± 1.78	96.95 ± 1.95
	SRGAN+U-Net	**6.05 ± 2.64**	**32.58 ± 1.87**	**97.01 ± 1.87**
	U-Net+SRGAN	6.13 ± 2.62	32.48 ± 1.75	96.92 ± 2.11

Table 4. Low-Resolution (160 × 160) model with ×4 undersampling, total acceleration: ×16, fraction of *k*-space: 6.25%

Mask	Model	MSE, ×10⁻⁴	PSNR	SSIM, ×10²
fastMRI	SRGAN	23.07 ± 6.30	26.53 ± 1.17	92.40 ± 2.93
	SRGAN+U-Net	15.97 ± 4.77	28.15 ± 1.26	94.00 ± 2.61
	U-Net+SRGAN	18.55 ± 5.88	27.51 ± 1.28	93.32 ± 3.27
spiral	SRGAN	17.29 ± 5.05	27.79 ± 1.20	93.32 ± 2.88
	SRGAN+U-Net	13.61 ± 4.11	28.85 ± 1.26	94.19 ± 2.58
	U-Net+SRGAN	13.49 ± 4.18	28.89 ± 1.27	94.08 ± 3.39
radial	SRGAN	12.71 ± 4.07	29.18 ± 1.38	94.54 ± 2.79
	SRGAN+U-Net	11.44 ± 3.90	29.66 ± 1.46	**94.92 ± 2.59**
	U-Net+SRGAN	**11.39 ± 3.92**	**29.67 ± 1.42**	94.79 ± 3.23

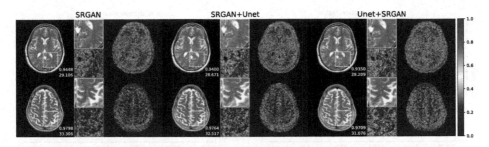

Fig. 3. Recovery of ×8 Low-Resolution k-space (undersampling with the radial mask). The subtraction from the ground truth (multiplied by 3 for clarity and shown on the right) suggests the SRGAN+U-Net combo to be the best model for MRI acceleration.

Our experiments show that undersampling with the radial mask, both for the Pix2Pix models and for the SRGAN+U-Net combo, always outperforms the other masks. Interestingly, SRGAN (SR) yields slightly better scores than the combo methods up to the ×16 acceleration (see Fig. 6) despite the visual analysis where the combos are somewhat more convincing to the eye of an expert (Fig. 8).

We applied the same models to the knee dataset, which, at first, had led to a low image reconstruction quality (both visually and quantitatively). We hypothesized that the low quality of reconstruction could be caused by the differences in the pixel intensities after the normalization, with the same tissue type having been mapped to a different intensity range on two neighboring slices in the volume. Indeed, swapping percentile normalization with the histogram-based normalization yielded more homogeneous results of pixel intensities across the knee slices, which, consequently, boosted the quality of image reconstruction to the level of the results on the brain data (see Fig. 4 and 5).

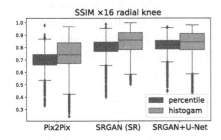

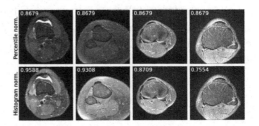

Fig. 4. Influence of the normalization method on the SSIM for the knee dataset with ×16 acceleration and radial sampling.

Fig. 5. Knee slice reconstructions with percentile-based (top) and histogram-based normalization (bottom).

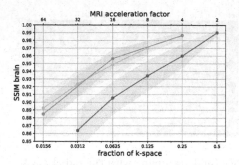
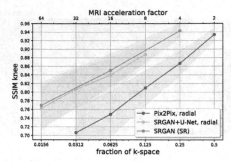

Fig. 6. SSIM dependence on the fraction of k-space for several best models, the brain dataset.

Fig. 7. SSIM dependence on the fraction of k-space for several best models, the knee dataset.

5   User Study

Image quality (IQ) assessment metrics such as SSIM help to estimate the reconstruction quality. However, they may not fully reflect how valuable these reconstructions are from the diagnostic perspective [7]. To solve this problem, we conducted a user study of the reconstruction quality with trained radiologists. We randomly selected 25 brain and 25 knee slices, downscaled (with a ×4 factor) and undersampled by factors of ×2 and ×4, using the radial masks. The resulting 100 slices were evaluated for their diagnostic content, compared to their corresponding fully-sampled full-resolution counterparts (the ground truth).

Five radiologists were involved in the study, using a simple survey tool [23]. The experts were shown the ground truth images and asked to rate the quality of our superresolved reconstructions based on three criteria vital for a typical diagnostic decision-making [5]: the presence of artifacts, the signal-to-noise ratio (SNR), and the contrast-to-noise ratio (CNR). Each criterion was scored on a 4-point scale as follows: fully acceptable (4), rather acceptable (3), weakly acceptable (2), not acceptable (1). Figure 8 shows high diagnostic value of the reconstructions, with around 80% of images getting the scores of (3) and (4) for both anatomies.

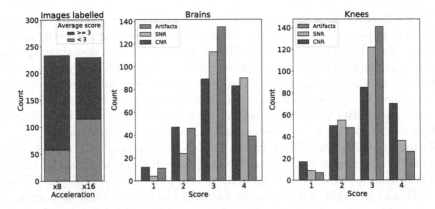

Fig. 8. Diagnostic quality scores from a survey of radiologists based on three IQ parameters for brain and knee anatomies. The majority of votes fall into rather acceptable (3) and fully acceptable (4) categories, showing little discrepancy between anatomies.

6   Discussion and Conclusion

We considered powerful image-to-image translation and superresolution models to compensate for the strong undersampling and downscaling of the simulated k-space. Our study comprises a comprehensive evaluation of the undersampling patterns and factors, downscaling, and reconstruction models.

Being more symmetrical and homogeneous at high frequencies, the radial and the spiral undersampling trajectories outperform corresponding original fastMRI sampling masks (for an equivalent fraction of k-space). Unlike them, the radial and the spiral sampling trajectories guarantee that the sampling is omnidirectional, without any given region of the k-space being omitted, including the high-frequency ones. Following the radial or spiral sampling, the acquired data can be translated to the Cartesian grid according to the established apparatus in the compressed sensing [25]. It is natural, thus, to extrapolate our results to the other currently proposed undersampling trajectories [3,25,26], which we anticipate to 'heal' similarly well by the Pix2Pix model. This extrapolation and the approbation of the proposed models on the raw k-space data (as acquired by the single and the multi-coil MRI machines) will be the subjects of future work.

We found that until the ×16 acceleration factor, SRGAN (SR) marginally outperforms the SRGAN+U-Net combo, and both of these approaches noticeably outperform Pix2Pix. The absence of significant differences between SRGAN (SR), SRGAN+U-Net, and U-Net+SRGAN at a given fraction of k-space suggests that the latter two can be adopted in general as the go-to acceleration method, allowing for more flexibility in terms of computation efficiency and image quality trade-off (downscaled images require fewer parameters for the final reconstructing generator, which may influence the selection of one model *vs.* the others for a particular application).

Different applications may indeed require a proper domain adaptation [9]. Herein, *e.g.*, the range of intensities in the knee data was noticeably skewed by the joint's liquids (the synovial fluid), requiring us to consider the histogram normalization approach in the data pre-processing stage (unlike percentile normalization sufficient for the brain data). More complex DL-based domain adaptation methods, translating the images between different MRI modalities and across different scanners [9,21], can also be conjoined to the proposed MRI acceleration pipeline and will be done elsewhere. Quantitatively, with the histogram-based normalization, SRGAN (SR) yielded an MSE of $8.8 \cdot 10^{-4}$, a PSNR of 32.8 dB, and an SSIM of 0.943 at $\times 4$ acceleration factor on the brain dataset.

Using the SRGAN+U-Net combo at $\times 8$ acceleration, the reconstruction achieved an MSE of $6.1 \cdot 10^{-4}$, a PSNR of 32.6 dB, and an SSIM of 0.97. At $\times 16$ acceleration, SRGAN+U-Net achieved an MSE of $11.4 \cdot 10^{-4}$, a PSNR of 29.6 dB, and an SSIM of 0.956 on the `fastMRI` brain data. These measurements correspond to a larger acceleration and compare favorably against other methods reported in literature, see, e.g., [1,2,4,10,16,30]. The quantitative outcome is concordant with the results of our user study, where the differences between the brain and the knee images reflected those in the measured metrics in the radiologists' evaluation.

In another ongoing survey, we are collecting the ranking votes for the pairs of images at the most extreme k-space undersampling factors generated in our work ($\times 32$ and $\times 64$). While we do not anticipate those images to be useful for typical clinical examinations, the radiologists still express their enthusiasm because these extreme MRI acceleration factors would be of interest in certain other (non-diagnostic) applications, where the coarse images oftentimes prove sufficient (e.g., surgical planning in the radiological suites or radiation therapy planning in oncological care). We believe these extremely undersampled k-space measurements are bound to improve with time due to the growth of the available data, which will gradually pave the way towards true ultrafast imaging of high diagnostic value.

References

1. Bakker, T., van Hoof, H., Welling, M.: Experimental design for MRI by greedy policy search. In: Larochelle, H., Ranzato, M., Hadsell, R., Balcan, M., Lin, H. (eds.) Advances in Neural Information Processing Systems 33: Annual Conference on Neural Information Processing Systems 2020 (2020)
2. Chen, Y., Christodoulou, A.G., Zhou, Z., Shi, F., Xie, Y., Li, D.: MRI superresolution with GAN and 3D multi-level DenseNet: smaller, faster, and better (2020)
3. Cohen, M.S., Weisskoff, R.M.: Ultra-fast imaging. Magn. Reson. Imaging 9(1), 1–37 (1991)
4. Darestani, M.Z., Chaudhari, A., Heckel, R.: Measuring robustness in deep learning based compressive sensing (2021)
5. Debatin, J.F., McKinnon, G.C.: Ultrafast MRI: Techniques and Applications. Springer, London (1998). https://doi.org/10.1007/978-3-642-80384-0
6. Dieckmeyer, M., et al.: Effect of MRI acquisition acceleration via compressed sensing and parallel imaging on brain volumetry. Magn. Reson. Mater. Phys., Biol. Med. 34(4), 487–497 (2021). https://doi.org/10.1007/s10334-020-00906-9

7. Esteban, O., et al.: Crowdsourced MRI quality metrics and expert quality annotations for training of humans and machines. Sci. Data **6**(1) (2019). Article number: 30. https://doi.org/10.1038/s41597-019-0035-4

8. Ferreira, P.F., Gatehouse, P.D., Mohiaddin, R.H., Firmin, D.N.: Cardiovascular magnetic resonance artefacts. J. Cardiovasc. Magn. Reson. **15**(2) (2013). Article number: 41. https://doi.org/10.1186/1532-429X-15-41

9. Ghafoorian, M., et al.: Transfer learning for domain adaptation in MRI: application in brain lesion segmentation. In: Descoteaux, M., Maier-Hein, L., Franz, A., Jannin, P., Collins, D.L., Duchesne, S. (eds.) MICCAI 2017. LNCS, vol. 10435, pp. 516–524. Springer, Cham (2017). https://doi.org/10.1007/978-3-319-66179-7_59

10. Han, Y., Sunwoo, L., Ye, J.C.: k-space deep learning for accelerated MRI. IEEE Trans. Med. Imaging **39**(2), 377–386 (2020)

11. Hsiao, A., et al.: Rapid pediatric cardiac assessment of flow and ventricular volume with compressed sensing parallel imaging volumetric cine phase-contrast MRI. Am. J. Roentgenol. **198**(3), 250–259 (2012)

12. Isola, P., Zhu, J.Y., Zhou, T., Efros, A.A.: Image-to-image translation with conditional adversarial networks. In: 2017 IEEE Conference on Computer Vision and Pattern Recognition (CVPR), pp. 5967–5976 (2017). https://doi.org/10.1109/CVPR.2017.632

13. Johnson, J., Alahi, A., Fei-Fei, L.: Perceptual losses for real-time style transfer and super-resolution. In: Leibe, B., Matas, J., Sebe, N., Welling, M. (eds.) ECCV 2016. LNCS, vol. 9906, pp. 694–711. Springer, Cham (2016). https://doi.org/10.1007/978-3-319-46475-6_43

14. Knoll, F., et al.: Advancing machine learning for MR image reconstruction with an open competition: overview of the 2019 fastMRI challenge. arXiv preprint arXiv:2001.02518 (2020)

15. Ledig, C., et al.: Photo-realistic single image super-resolution using a generative adversarial network. In: 2017 IEEE Conference on Computer Vision and Pattern Recognition (CVPR), pp. 4681–4690 (2017)

16. Lyu, Q., Shan, H., Wang, G.: MRI super-resolution with ensemble learning and complementary priors. IEEE Trans. Comput. Imaging **6**, 615–624 (2020). https://doi.org/10.1109/tci.2020.2964201

17. Nilsson, J., Akenine-Möller, T.: Understanding SSIM. arXiv preprint arXiv:2006.13846 (2020)

18. Panda, A., Naskar, R., Rajbans, S., Pal, S.: A 3D wide residual network with perceptual loss for brain MRI image denoising, pp. 1–7 (2019). https://doi.org/10.1109/ICCCNT45670.2019.8944535

19. Pezzotti, N., et al.: An adaptive intelligence algorithm for undersampled knee MRI reconstruction. IEEE Access **8**, 204825–204838 (2020)

20. Prakkamakul, S., et al.: Ultrafast brain MRI: clinical deployment and comparison to conventional brain MRI at 3T. J. Neuroimaging **26**(5), 503–510 (2016). https://doi.org/10.1111/jon.12365

21. Prokopenko, D., Stadelmann, J.V., Schulz, H., Renisch, S., Dylov, D.V.: Synthetic CT generation from MRI using improved dualgan. arXiv preprint arXiv:1909.08942 (2019)

22. Schonfeld, E., Schiele, B., Khoreva, A.: A U-Net based discriminator for generative adversarial networks. arXiv preprint arXiv:2002.12655v1 (2020)

23. Tkachenko, M., Malyuk, M., Shevchenko, N., Holmanyuk, A., Liubimov, N.: Label Studio: Data labeling software (2020–2021). https://github.com/heartexlabs/label-studio

24. Tsao, J.: Ultrafast imaging: principles, pitfalls, solutions, and applications. J. Magn. Reson. Imaging **32**(2), 252–266 (2010)
25. Vasanawala, S.S., Alley, M.T., Hargreaves, B.A., Barth, R.A., Pauly, J.M., Lustig, M.: Improved pediatric MR imaging with compressed sensing. Radiology **256**(2), 607–616 (2010)
26. Wang, G., Luo, T., Nielsen, J.F., Noll, D.C., Fessler, J.A.: B-spline parameterized joint optimization of reconstruction and K-space trajectories (BJORK) for accelerated 2D MRI (2021)
27. Weiss, T., Vedula, S., Senouf, O., Michailovich, O., Zibulevsky, M., Bronstein, A.: Joint learning of Cartesian under sampling andre construction for accelerated MRI. In: ICASSP 2020–2020 IEEE International Conference on Acoustics, Speech and Signal Processing (ICASSP), pp. 8653–8657 (2020). https://doi.org/10.1109/ICASSP40776.2020.9054542
28. Wright, K.L., Hamilton, J.I., Griswold, M.A., Gulani, V., Seiberlich, N.: Non-Cartesian parallel imaging reconstruction. J. Magn. Reson. Imaging **40**(5), 1022–1040 (2014). https://doi.org/10.1002/jmri.24521
29. Zbontar, J., et al.: An open dataset and benchmarks for accelerated MRI. arXiv preprint arXiv:1811.08839 (2019)
30. Zhu, B., Liu, J.Z., Rosen, B.R., Rosen, M.S.: Image reconstruction by domain transform manifold learning. Nature **555**, 487–492 (2017). https://doi.org/10.1038/nature25988

Adaptive Squeeze-and-Shrink Image Denoising for Improving Deep Detection of Cerebral Microbleeds

Hangfan Liu[1,2](✉), Tanweer Rashid[1], Jeffrey Ware[3], Paul Jensen[4], Thomas Austin[4], Ilya Nasrallah[2], Robert Bryan[5], Susan Heckbert[4], and Mohamad Habes[1,2]

[1] Neuroimage Analytics Laboratory and Biggs Institute Neuroimaging Core, Glenn Biggs Institute for Neurodegenerative Disorders, University of Texas Health Science Center at San Antonio, San Antonio, TX, USA
hfliu@upenn.edu, habes@uthscsa.edu
[2] Center for Biomedical Image Computing and Analytics, University of Pennsylvania, Philadelphia, USA
[3] Department of Radiation Oncology, University of Pennsylvania, Philadelphia, PA, USA
[4] University of Washington, Seattle, WA, USA
[5] The University of Texas at Austin, Austin, TX, USA

Abstract. Deep learning for medical image analysis requires large quantities of high-quality imaging data for training purposes, which could be often less available due to existence of heavy noise in particular imaging modalities. This issue is especially obvious in cerebral microbleed (CMB) detection, since CMBs are more discernable on long echo time (TE) susceptibility weighted imaging (SWI) data, which are unfortunately much noisier than those with shorter TE. In this paper we present an effective unsupervised image denoising scheme with application to boosting the performance of deep learning based CMB detection. The proposed content-adaptive denoising technique uses the log-determinant of covariance matrices formed by highly correlated image contents retrieved from the input itself to implicitly and efficiently exploit sparsity in PCA domain. The numerical solution to the corresponding optimization problem comes down to an adaptive squeeze-and-shrink (ASAS) operation on the underlying PCA coefficients. Obviously, the ASAS denoising does not rely on any external dataset and could be better fit the input image data. Experiments on medical image datasets with synthetic Gaussian white noise demonstrate that the proposed ASAS scheme is highly competitive among state-of-the-art sparsity based approaches as well as deep learning based method. When applied to the deep learning based CMB detection on the real-world TE3 SWI dataset, the proposed ASAS denoising could improve the precision by 18.03%, sensitivity by 7.64%, and increase the correlation between counts of ground truth and automated detection by 19.87%.

Keywords: Image denoising · Unsupervised learning · Deep learning · Cerebral microbleed detection

© Springer Nature Switzerland AG 2021
M. de Bruijne et al. (Eds.): MICCAI 2021, LNCS 12906, pp. 265–275, 2021.
https://doi.org/10.1007/978-3-030-87231-1_26

1   Introduction

Cerebral microbleeds (CMBs) are a type of small vessel ischemic disease associated with brain aging, vascular disease and neurodegenerative disorders [1]. Automated detection of CMBs is challenging because they are generally very small in size and may occur sporadically in the brain. They are more readily detected by magnetic resonance imaging (MRI) with susceptibility-weighted imaging (SWI) than modalities such as T2*-weighted images [2]. Since the blooming effect caused by the presence of iron in CMBs has a more pronounced effect on SWI for longer time of echo (TE), CMBs are more discernable on long TE SWIs, which unfortunately are much noisier than those with shorter TE and more susceptible to air and motion-related artifacts (see Fig. 1), and the additional noise would be more likely detected false positive CMBs.

A potentially effective and inexpensive solution is to enhance image quality via denoising before performing CMB detection. As a fundamental and important low-level computer vision task, image denoising aims to recover the high-quality clean images u from the noisy observations z, which is generally formulated as.

$$z = u + v, \tag{1}$$

where v is the additive noise. The desired denoising technique is expected to preserve the fine details in the original images while thoroughly removing noise. To solve such an ill-posed problem, we usually need to introduce prior knowledge to regularize the solution space, such as total variation [3–5], sparsity [6–10] or low rank [11–14]. Sparsity-based methods are generally based on the fact that image signals can be represented in certain low-dimensional spaces in which image contents of interest can be easily separated from noise. Total variation regularization can be seen as sparsity in the gradient domain [15], while low-rank regularization can be interpreted as sparsity of singular values [14]. Compared with old-fashioned approaches that use fixed bases to decompose the image signals [6], more advanced methods in this category generally train signal-adaptive dictionaries to achieve promoted sparsity [8–10]. In the past few years, with the success of deep learning, deep neural network based schemes [16, 17] were also proposed for image denoising and achieved promising performance, but usually they heavily rely on training data and the generalization of such models to all kinds of medical imaging data is not guaranteed.

In this paper we propose an effective unsupervised denoising technique combined with a deep convolution neural network (CNN) model for improved CMB detection. The proposed denoising approach optimizes sparse representation of images without explicitly training the signal-adaptive dictionaries. It uses the log-determinant of covariance matrices calculated from highly correlated image contents, which implicitly perform principal component analysis (PCA) and promotes sparsity by applying an adaptive squeeze-and-shrink (ASAS) operation to the latent PCA coefficients. When used as a pre-processing step for deep learning based CMB detection, ASAS boosted the performance by a large margin. To sum up, the theoretical and practical contributions of this work include:

1. We proposed an efficient denoising approach that achieves signal-adaptive sparsity and does not rely on any outer dataset. Empirical results on medical image datasets

demonstrate that the proposed denoising scheme is highly competitive among state-of-the-art denoising methods, in terms of both peak signal-to-noise ratio (PSNR) and structural similarity (SSIM) [18].

2. We investigated the physical meanings of the log-determinant based objective function and revealed the underlying connection between sparse representation. Our mathematical analysis revealed that, although there is no explicit transform applied to the images, the proposed approach essentially achieves near-optimal sparsity through an ASAS operation on the eigenvalues.

3. We demonstrated the practical value of the proposed approach in CMB detection. When applied to the real-world clinical data, the ASAS denoising significantly improved the performance of the deep neural network without to any additional burden of collecting training data.

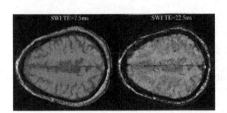

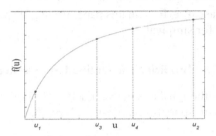

Fig. 1. SWI images with longer time of echo (TE) generally contain more information, but meanwhile are significantly noisier. Please enlarge the figure for better comparison.

Fig. 2. When $f(u)$ is upper convex, for any $0 < u_1 < u_3 < u_4 < u_2$ and $u_1 + u_2 = u_3 + u_4$, we have $f(u_1) + f(u_2) < f(u_3) + f(u_4)$.

2 Adaptive Squeeze-and-Shrink Denoising

In practice, medical imaging data are usually inevitable polluted by noise, which could severely influence the performance of various medical image analysis techniques. In this paper we aim to propose an effective denoising scheme with application to improving deep learning based cerebral microbleeds (CMB) segmentation. Since noise-free magnetic resonance imaging (MRI) data are usually insufficient to serve as ground truth for training deep neural networks, we try to learn the prior model from the input itself without relying on any outer dataset.

2.1 Overall Framework

The proposed approach evolves from rank minimization. Inspired by the observations that the ranks of matrices formed by nonlocal similar patches are low [14], and that compared with convex surrogates like nuclear norm [11], the log-determinant of the corresponding covariance matrices should be a better approximation for rank penalty [19], we estimate the noise-free image u from the noisy input z by solving.

$$\tilde{u} = \underset{u}{\operatorname{argmin}} \sum_i \log \left| \left(U_i - \mu_i \times 1^T \right) \left(U_i - \mu_i \times 1^T \right)^T \right| + \beta \| u - z \|_F^2, \qquad (2)$$

where $U_i \in \mathbb{R}^{s^2 \times n}$ is the matrix formed by the nonlocally retrieved n $(n > s^2)$ patches that are most similar with the i-th vectorized patch $p_i \in \mathbb{R}^{s^2}$, with the similarity measured by Euclidean distance of the voxel/pixel intensities; $\mu_i \in \mathbb{R}^{s^2}$ is the mean vector of the patches in U_i, $1 \in \mathbb{R}^{s^2}$ is a vector of ones. The regularization parameter β is inversely proportional to the noise variance σ_n^2, in that heavier noise (i.e., larger σ_n^2) indicates that the observation z is less reliable, hence the data fidelity term $\| u - z \|_F^2$ should be assigned with a smaller weight. Although in practice we can retrieve 3 dimensional (3D) patches from the input imaging data to form U_i for better denosing performance (in that case $p_i \in \mathbb{R}^{s^3}$), in this paper we use 2D patches of size $s \times s$ retrieved from each slice of z for simplicity.

2.2 Implicit Near-Optimal Sparse Representation

Although the objective function (1) looks relatively compact, it is interestingly connected with sparsity optimization. Note that.

$$\left| \overline{U} \overline{U}^T \right| = \prod_j \lambda_j \Rightarrow \log \left| \overline{U} \overline{U}^T \right| = \sum_j \log \lambda_j,$$

where $\overline{U} = U - \mu \times 1^T$ and λ_j is the j-th eigenvalue of the matrix $\overline{U} \overline{U}^T$, which is generated by eigenvalue decomposition $\overline{U} \overline{U}^T = \Phi^T \times diag[\dots, \lambda_j, \dots] \times \Phi$. Here $\Phi \in \mathbb{R}^{s^2 \times s^2}$ is the underlying signal-adaptive transform that decorrelate the patches to achieve near-optimal sparse representation, and λ_j is the variance of the coefficients in the j-th transform band.

Furthermore, solving problem (1) is implicitly promoting the sparsity of the PCA coefficients. At one extreme, minimizing $\sum_j \log \lambda_j$ alone would simply shrink the eigenvalues towards zero. At the other extreme, we consider the constraint case.

$$\min \sum_j \log \lambda_j \text{ s.t. } \sum_j \lambda_j' = \sum_j \lambda_j, \qquad (3)$$

where λ_j' is the eigenvalue of the corresponding noisy patch retrieved from z. Since $\log(\bullet)$ is an upper convex function, $\forall \lambda_p > \lambda_q$, when $\lambda_p + \lambda_q$ is constant, $\log(\lambda_p) + \log(\lambda_q)$ decreases when $\lambda_p - \lambda_q$ increases, which can be easily seen from Fig. 2. This conclusion can be extended to the s^2-element case, i.e., the more diversified these eigenvalues λ_j values are, the less $\sum_j \log \lambda_j$ is. Therefore, the solution to problem (3) basically squeezes the eigenvalues to diversify their distribution, thus encourages sparsity of PCA coefficients. Apparently, when combining the log-determinant term with the data fidelity term, the minimization of the two penalty terms is fundamentally an adaptive squeeze-and-shrink (ASAS) operation.

2.3 The Denoising Procedure

In the optimization problem (2), the log-determinant term is applied on patch groups while the data fidelity term is applied on the whole image, making it difficult to solve directly. Suppose Z_i is the noisy version of U_i that consists of corresponding patches from the input image z. Since the times of pixels/voxels being included in the patch groups U_i are roughly the same, we have

$$\|\boldsymbol{u} - z\|_F^2 \approx c \|U_i - Z_i\|_F^2, \tag{4}$$

where c is a constant. Based on (4), we can transform objective function (2) into a group level problem.

$$\tilde{\boldsymbol{u}} = \underset{\boldsymbol{u}}{\arg\min} \sum_i \log \left| \left(U_i - \boldsymbol{\mu}_i \times 1^T\right)\left(U_i - \boldsymbol{\mu}_i \times 1^T\right)^T \right| + \alpha \|U_i - Z_i\|_F^2, \tag{5}$$

with $\alpha = c\beta$. To solve (5), we introduce a positive semi-definite matrix $\Psi_i \in \mathbb{R}^{s^2 \times s^2}$ containing variational parameters and update Ψ_i, U_i iteratively [19, 20]:

$$\Psi_{(0)} = \overline{Z}\overline{Z}^T / N,$$

$$\Psi_{(k+1)} = (\overline{U}_{(k)}\overline{U}_{(k)}^T + (\Psi_{(k)} - \Psi_{(k)}(\Psi_{(k)} + \alpha I)^{-1}\Psi_{(k)}))/N, \tag{6}$$

$$U_{(k)} = \Psi_{(k)}(\Psi_{(k)} + \alpha I)^{-1}\overline{Z} + \boldsymbol{\mu}_i \times 1^T, \tag{7}$$

where $\overline{Z}$ is the noisy version of $\overline{U}$, $I \in \mathbb{R}^{s^2 \times s^2}$ is the identity matrix, k is the iteration number, and the subscripts i in (6) and (7) are omitted for conciseness. In this paper we consider the noise to be zero-mean, hence the mean vector $\boldsymbol{\mu}_i$ calculated from U_i and Z_i are equal. Since the derivation of (6) and (7) has been presented in existing publications, we refer interested readers to [19, 20] for further details.

Suppose R_i is the matrix to retrieve U_i from $\boldsymbol{u}$, after obtaining patch group level estimates $\tilde{U}_i$, we can calculate the whole denoised image as

$$\tilde{\boldsymbol{u}} = \left(\sum_i R_i R_i^T\right)^{-1} \sum_i R_i^T \tilde{U}_i, \tag{8}$$

which basically means putting back the patches and averaging overlaps.

3 The Deep Detection of Cerebral Microbleeds

Compared with conventional models [21], deep convolutional neural networks (CNN) have shown remarkable success in image classification. Most CNNs for classification tasks such as AlexNet [22], VGG [23], and ResNet [24] consist of, in general terms, a contracting stage, followed by one or more fully connected (or dense) layers. The contracting stage consists of a series of convolutional layers, activation layers and max-pooling layers which reduce the size of the layer's output, typically by half. For image segmentation, the U-Net [25] is a specific class of CNN that consists of a contracting stage similar to the classification CNN models, and followed by an expanding stage (also known as the upsampling stage). The purpose of the expanding stage is to reconstruct the image size with each progressive layer, and feature maps from contracting layers are typically concatenated to the respective expanding layers.

We used a modified version of U-Net [26] tailored for CMB detection. The model consists of a downsampling path with five layers, followed by an upsampling stage of five layers, plus a convolutional block between the downsampling stage and upsampling stage. Each downsampling layer consists of two layers of a 2D padded convolution layer with a 3×3 kernel and 1×1 stride, followed by batch normalization (BN) and rectified linear unit (ReLU) activation. The downsampling block ends with a 2×2 max pooling layer which reduces the resolution feature map by half in each spatial direction. The central block consists of two instances of padded 2D convolution with kernel size 3×3 and stride 1×1, followed by BN and ReLU activation. Each upsampling layer passes its input data through a 2D transpose convolution with kernel size of 2×2 and stride 2×2 in order to double the size of the feature map. The doubled feature map is then concatenated with the feature map (same size) of the corresponding downsampling stage (i.e., the feature map before max pooling layer), followed by two instances of a padded 2D convolution layer with kernel size 3×3 and stride 1×1, followed by BN and ReLU activation. Due to the use of padded convolutions throughout the model, the input and output image sizes are the same (256×256). The smallest downsampled image size is 8×8 in the central convolution block. Categorical cross-entropy is used as the loss function and intersection-over-union is adopted as the validation metric.

Table 1. Average PSNR (in dB)/SSIM scores of denoised results produced by BM3D, CSR, DnCNN and the proposed ASAS. The best scores are marked in bold.

	BM3D	CSR	DnCNN	ASAS
DX 1	42.16/0.9713	42.21/0.9710	39.79/0.9547	**42.38/0.9719**
DX 2	41.99/0.9707	42.09/0.9704	39.65/0.9538	**42.25/0.9712**
MMM	44.65/0.9798	44.72/0.9806	41.74/0.9574	**44.91/0.9818**

4   Gaussian White Noise Removal

Before examining the proposed adaptive squeeze-and-shrink (ASAS) approach on the real-world clinical dataset, we tested the ASAS method on datasets with known ground truth so as to measure the denoising performance. In this paper we use the two most widely used objective metrics peak signal-to-noise ratio (PSNR) and structural similarity (SSIM) [18]. Given the ground truth (i.e. noise-free) image y whose pixel/voxel intensities have been rescaled to [0, 255], the PSNR of the corresponding denoised version $\tilde{y}$ is calculated as.

$$PSNR(y, \tilde{y}) = \log(255^2 / \|y - \tilde{y}\|_F^2. \tag{9}$$

Apparently, the calculation of PSNR and SSIM requires noise-free images, while most of the brain imaging data in this study, are more or less polluted by noise. To this end, we used the three cleanest datasets available to us for evaluation of denoising performance, that are generally noise-free and are frequently used in the medical image processing literature for algorithms evaluation, including the two test sets that respectively include 150 and 100 images from dental X-ray (DX) dataset [27], as well as the 322 images from the mini mammography (MMM) database [28]. The noisy images were generated by adding white Gaussian noise with known variance to the ground truth. For comparison, we used the state-of-the-art sparsity based schemes BM3D [7] and CSR [10] as well as a typical deep learning based method DnCNN [16].

For the compared methods, we used the executables released by the authors without changing any parameter setting or model. All of them used noise variance as an input. In order to decide a proper set of hyperparameters for the proposed ASAS, we used the 12 natural images in [10] as validation data and tested a range of values for each parameter to find the one that leads to highest PSNR. In this way we set patch size $s = 5$, numbers of similar patches in a group $N = 60$, total number of iterations $K = 3$. It's worth noting that, although these parameters are selected using a handful of natural images, the proposed model is highly effective on medical imaging data (as shown below), demonstrating excellent generalization. We implemented ASAS in MATLAB.

The average PSNR and SSIM scores of the 4 competing methods tested on 3 datasets are shown in Table 1. The variance of added Gaussian noise was 25, which is commensurate with the noise level in real-world long echo time SWI data. Evidently, the proposed ASAS achieved the highest scores among the compared methods.

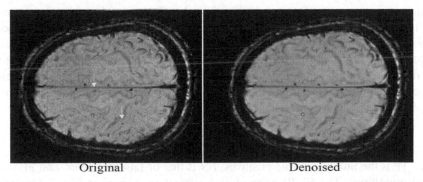

Original   Denoised

Fig. 4. Detection results on the original and denoised SWI image. The true positives are marked as red and false positives as yellow. Please enlarge the figure for better comparison. (Color figure online)

5 Experiments on Real-World Data

Since the noise level of the data is unknown, we first estimate the noise variance in the data by [29] and then use the data and the noise variance as input of the ASAS denoising algorithm. The denoised data are then fed in to the network described in Sect. 3.

5.1 Data

To explore the feasibly of the denoising approach proposed herein in detecting cerebral microbleeds in the brain, we used susceptibility weighted imaging (SWI) data [30] with long echo time of TE = 22.5 ms, which is a part of the multi-echo SWI sequence acquired for the Multi-Ethnic Study of Atherosclerosis (MESA) Atrial Fibrillation study [31–33]. We included MR data from 24 subjects in the MESA Exam 6 [33].

5.2 CMB Detection

To demonstrate the effectiveness of the proposed denoising method in deep neural network based CMB detection, we perform two experiments for detecting CMBs on SWI data from 24 subjects using the modified U-Net. In the first experiment, we used unaltered SWI data, and in the second experiment the SWI data were denoised by ASAS. For training and testing, in both experiments we used leave-one-out cross-validation on the 24 subjects to ensure generalization of the results. In each cross-validation fold, a single subject's data was left out for testing, and the remaining 23 subjects' data was split into training (17 subjects) and validation (6 subjects) datasets. Both training and validation datasets were augmented to improve the robustness of the training process.

Figure 4 shows the visual comparison between an original SWI image and its denoised version, as well as the corresponding detected CMBs. Obviously, the denoised image generated by ASAS is much cleaner, with most of the noise removed while textures well preserved, and there are much less false positives in the detection results produced by the denoised data. We Further examined the detection performance by sensitivity S and precision P, which are calculated as.

$$S = TP/(TP + FN), P = TP/(TP + FP), \tag{10}$$

where TP is the number of true positives, FN is that of false negatives, and FP stands for false positives. The (S, P) pairs of each subject are plotted in Fig. 5. One of the subjects has no CMB in ground truth, hence the corresponding $(S, P) = (0, 0)$. Besides, since for detection of small vessel diseases like CMB, the count of lesions is of interest clinically, we also measure the Pearson correlation C between the counts of CMBs in the ground truth and those detected by the neural network as in related works [34–36]. As can be seen from Table 2, the proposed ASAS denoising improved the average precision by 18.03%, average sensitivity by 7.64%, and increased the overall correlation between counts of ground truth and automated detection by 19.87%.

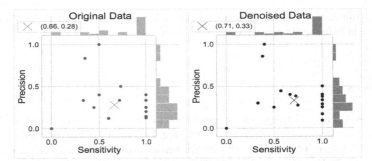

Fig. 5. Scatterplot of the results on original (left) and denoised (right) SWI data. Each dot represents the (S, P) pair of one subject, and $\times$ represents the average (S, P) of all subjects.

Table 2. Average sensitivity, precision and overall correlation.

	S	P	C
Denoised long TE	**0.7100**	**0.3286**	**0.7033**
Original long TE	0.6596	0.2784	0.5867

6 Discussion and Conclusions

This paper presents a combination of an unsupervised denoising technique with a deep learning model for improved cerebral microbleed (CMB) detection. By introducing a log-determinant term utilizing the correlation matrices of nonlocally correlated image data, the proposed denoising method learns the statistical characteristics of the image signal from the input itself, and implicitly achieves near-optimal sparsity in PCA domain. It does not rely on a training dataset thus will not add to the burden to search for a large-scale ground truth data, and does not have the issue of overfitting. It can be further improved by incorporating inter-slice correlations within the imaging data. Experimental results on medical image datasets with noise-free ground truth demonstrate that the proposed denoising scheme is highly competitive among the state-of-the-art methods in terms of PSNR and SSIM. Experiments on real-world clinical data show that the proposed denoising scheme can significantly improve the performance of the deep neural network for CMB detection.

Acknowledgement. This MESA research was supported by contracts 75N92020D00001, HHSN268201500 003I, N01-HC-95159, 75N92020D00005, N01-HC-95160, 75N92020D00002, N01-HC-95161, 75N92020D00003, N01-HC-95162, 75N92020D00006, N01-HC-95163, 75N92020D00004, N01-HC-95164, 75N92020D00007, N01-HC-95165, N01-HC-95166, N01-HC-95167, N01-HC-95168 and N01-HC-95169 and grant HL127659 from the National Heart, Lung, and Blood Institute, and by grants UL1-TR-000040, UL1-TR-001079, and UL1-TR-001420 from the National Center for Advancing Translational Sciences (NCATS). The content is solely the responsibility of the authors and does not necessarily represent the official views of the National Institutes of Health. The authors thank the other investigators, the staff, and the participants of the

MESA study for their valuable contributions. A full list of participating MESA investigators and institutions can be found at http://www.mesa-nhlbi.org.

References

1. Gouw, A.A., et al.: Heterogeneity of small vessel disease: a systematic review of MRI and histopathology correlations. J. Neurol. Neurosurg. Psychiatry **82**, 126–135 (2011)
2. Shi, Y., Wardlaw, J.M.: Update on cerebral small vessel disease: a dynamic whole-brain disease. Stroke Vasc. Neurol. **1**, 83–92 (2016)
3. Osher, S., Burger, M., Goldfarb, D., et al.: An iterative regularization method for total variation-based image restoration. Multiscale Model. Simul. **4**, 460–489 (2005)
4. Liu, H., Xiong, R., Ma, S., Fan, X., Gao, W.: Non-local extension of total variation regularization for image restoration. In: IEEE International Symposium on Circuits and Systems, pp. 1102–1105 (2014)
5. Vogel, C.R., Oman, M.E.: Iterative methods for total variation denoising. SIAM J. Sci. Comput. **17**, 227–238 (1996)
6. Sendur, L., Selesnick, I.W.: Bivariate shrinkage functions for wavelet-based denoising exploiting interscale dependency. IEEE Trans. Signal Process. **50**, 2744–2756 (2002)
7. Dabov, K., Foi, A., Katkovnik, V., et al.: Image denoising by sparse 3-D transform-domain collaborative filtering. IEEE Trans. Image Process. **16**, 2080–2095 (2007)
8. Liu, H., Zhang, J., Xiong, R.: CAS: correlation adaptive sparse modeling for image denoising. IEEE Trans. Comput. Imag. **7**, 638–647 (2021)
9. Zhang, L., Dong, W., Zhang, D., et al.: Two-stage image denoising by principal component analysis with local pixel grouping. Pattern Recogn. **43**, 1531–1549 (2010)
10. Dong, W., Li, X., Zhang, L., Shi, G.: Sparsity-based image denoising via dictionary learning and structural clustering. In: IEEE Conference on Computer Vision and Pattern Recognition, pp. 457–464 (2011)
11. Dong, W., Shi, G., Li, X.: Nonlocal image restoration with bilateral variance estimation: a low-rank approach. IEEE Trans. Image Process. **22**, 700–711 (2012)
12. Liu, H., Xiong, R., Liu, D., Wu, F., Gao, W.: Low rank regularization exploiting intra and inter patch correlation for image denoising. In: IEEE Visual Communications and Image Processing, St. Petersburg, USA (2017)
13. Zhang, Y., et al.: Image denoising via structure-constrained low-rank approximation. Neural Comput. Appl. **32**(16), 12575–12590 (2020). https://doi.org/10.1007/s00521-020-04717-w
14. Wang, S., Zhang, L., Liang, Y.: Nonlocal spectral prior model for low-level vision. In: Lee, K.M., Matsushita, Y., Rehg, J.M., Hu, Z. (eds.) ACCV 2012. LNCS, vol. 7726, pp. 231–244. Springer, Heidelberg (2013). https://doi.org/10.1007/978-3-642-37431-9_18
15. Liu, H., Xiong, R., Zhang, X., Zhang, Y., Ma, S., Gao, W.: Nonlocal gradient sparsity regularization for image restoration. IEEE Trans. Circuits Syst. Video Technol. **27**, 1909–1921 (2017)
16. Zhang, K., Zuo, W., Chen, Y., Meng, D., Zhang, L.: Beyond a Gaussian denoiser: residual learning of deep CNN for image denoising. IEEE Trans. Image Process. **26**, 3142–3155 (2017)
17. Mou, C., Zhang, J., Fan, X., Liu, H., Wang, R.: COLA-Net: collaborative attention network for image restoration. IEEE Trans. Multimedia (2021)
18. Wang, Z., Bovik, A.C., Sheikh, H.R., et al.: Image quality assessment: from error visibility to structural similarity. IEEE Trans. Image Process. **13**, 600–612 (2004)
19. Mohan, K., Fazel, M.: Iterative reweighted algorithms for matrix rank minimization. J. Mach. Learn. Res. **13**, 3441–3473 (2012)

20. Wipf, D.P., Rao, B.D., Nagarajan, S.: Latent variable Bayesian models for promoting sparsity. IEEE Trans. Inf. Theory **57**, 6236–6255 (2011)
21. Liu, H., Rashid, T., Habes, M.: Cerebral microbleed detection via Fourier descriptor with dual domain distribution modeling. In: IEEE International Symposium on Biomedical Imaging Workshops, pp. 1–4 (2020)
22. Krizhevsky, A., Sutskever, I., et al.: Imagenet classification with deep convolutional neural networks. Adv. Neural. Inf. Process. Syst. **25**, 1097–1105 (2012)
23. Simonyan, K., Zisserman, A.: Very deep convolutional networks for large-scale image recognition. arXiv preprint arXiv:1409.1556 (2014)
24. He, K., Zhang, X., Ren, S., Sun, J.: Deep residual learning for image recognition. In: IEEE Conference on Computer Vision and Pattern Recognition, pp. 770–778 (2016)
25. Ronneberger, O., Fischer, P., Brox, T.: U-net: convolutional networks for biomedical image segmentation. In: International Conference on Medical Image Computing and Computer-Assisted Intervention, pp. 234–241 (2015)
26. Rashid, T., Abdulkadir, A., et al.: DEEPMIR: a deep neural network for differential detection of cerebral Microbleeds and IRon deposits in MRI. Scientific Reports (2021)
27. Wang, C.-W., Huang, C.-T., Lee, J.-H., Li, C.-H., et al.: A benchmark for comparison of dental radiography analysis algorithms. Med. Image Anal. **31**, 63–76 (2016)
28. Suckling, J., et al.: The mammographic image analysis society digital mammogram database. Excerpta Med. Int. Congr. Ser. **1069**, 375–386 (1994)
29. Coupé, P., Manjón, J.V., Gedamu, E., Arnold, D., Robles, M., Collins, D.L.: An object-based method for Rician noise estimation in MR images. In: International Conference on Medical Image Computing and Computer-Assisted Intervention, pp. 601–608 (2009)
30. Haacke, E.M., Xu, Y., Cheng, Y.C.N., Reichenbach, J.R.: Susceptibility weighted imaging (SWI). Magn. Reson. Med.: Off. J. Int. Soc. Magn. Reson. Med. **52**, 612–618 (2004)
31. Bild, D.E., Bluemke, D.A., Burke, G.L., et al.: Multi-ethnic study of atherosclerosis: objectives and design. Am. J. Epidemiol. **156**, 871–881 (2002)
32. Olson, J.L., Bild, D.E., Kronmal, R.A., Burke, G.L.: Legacy of MESA. Glob. Heart **11**, 269–274 (2016)
33. Heckbert, S.R., Austin, T.R., Jensen, P.N., Floyd, J.S., Psaty, B.M., et al.: Yield and consistency of arrhythmia detection with patch electrocardiographic monitoring: the multi-ethnic study of atherosclerosis. J. Electrocardiol. **51**, 997–1002 (2018)
34. Boespflug, E.L., Schwartz, D.L., Lahna, D., et al.: MR imaging–based multimodal autoidentification of perivascular spaces (mMAPS): automated morphologic segmentation of enlarged perivascular spaces at clinical field strength. Radiology **286**, 632–642 (2018)
35. Dubost, F., Yilmaz, P., Adams, H., et al.: Enlarged perivascular spaces in brain MRI: automated quantification in four regions. Neuroimage **185**, 534–544 (2019)
36. Ballerini, L., et al.: Perivascular spaces segmentation in brain MRI using optimal 3D filtering. Sci. Rep. **8**, 1–11 (2018)

3D Transformer-GAN for High-Quality PET Reconstruction

Yanmei Luo[1], Yan Wang[1](✉), Chen Zu[2], Bo Zhan[1], Xi Wu[3], Jiliu Zhou[1,3], Dinggang Shen[4,5](✉), and Luping Zhou[6](✉)

[1] School of Computer Science, Sichuan University, Chengdu, China
[2] Department of Risk Controlling Research, JD.COM, Chengdu, China
[3] School of Computer Science, Chengdu University of Information Technology, Chengdu, China
[4] School of Biomedical Engineering, ShanghaiTech University, Shanghai, China
[5] Department of Research and Development, Shanghai United Imaging Intelligence Co., Ltd., Shanghai, China
[6] School of Electrical and Information Engineering, University of Sydney, Sydney, Australia
luping.zhou@sydney.edu.au

Abstract. To obtain high-quality positron emission tomography (PET) image at low dose, this study proposes an end-to-end 3D generative adversarial network embedded with transformer, namely Transformer-GAN, to reconstruct the standard-dose PET (SPET) image from the corresponding low-dose PET (LPET) image. Specifically, considering the convolutional neural network (CNN) can well describe the local spatial features, while the transformer is good at capturing the long-range semantic information due to its global information extraction ability, our generator network takes advantages of both CNN and transformer, and is designed as an architecture of EncoderCNN-Transformer-DecoderCNN. Particularly, the EncoderCNN aims to extract compact feature representations with rich spatial information by using CNN, while the Transformer targets at capturing the long-range dependencies between the features learned by the EncoderCNN. Finally, the DecoderCNN is responsible for restoring the reconstructed PET image. Moreover, to ensure the similarity of voxel-level intensities as well as the data distributions between the reconstructed image and the real image, we harness both the voxel-wise estimation error and the adversarial loss to train the generator network. Validations on the clinical PET data show that our proposed method outperforms the state-of-the-art methods in both qualitative and quantitative measures.

Keywords: Positron Emission Tomography (PET) · Generative Adversarial Network (GAN) · Transformer · Image reconstruction

1 Introduction

As an ultra-sensitive and non-invasive nuclear imaging technology, positron emission tomography (PET) enables the visualization of human metabolic processes and has been extensively used in clinics for early disease diagnosis and prevention [1]. In clinical practice, standard-dose PET imaging is preferred to provide more diagnostic information.

© Springer Nature Switzerland AG 2021
M. de Bruijne et al. (Eds.): MICCAI 2021, LNCS 12906, pp. 276–285, 2021.
https://doi.org/10.1007/978-3-030-87231-1_27

Yet, the inherent radiation of radioactive tracers is harmful to the human body. Especially for patients who require multiple PET scans, the radiation risk is cumulative. Nevertheless, reducing the tracer dose will inevitably involve unintended noise and artifacts, resulting in debased image quality. One solution for this clinical dilemma is to reconstruct standard-dose PET (SPET) from the corresponding low-dose PET (LPET), thereby obtaining clinically acceptable PET images while reducing the radiation exposure.

In the past decade, with the rise of deep learning, many research works have been developed for SPET reconstruction [2–10]. For example, Xiang et al. [2] combined multiple convolutional neural network (CNN) modules for PET synthesis following an auto-context strategy. Kim et al. [3] presented an iterative PET reconstruction framework using a deep learning-based prior. Wang et al. proposed a 3D conditional generative adversarial network (GAN) [4] and a locality adaptive GAN [5] for high-quality PET image estimation from LPET. Liu et al. [6] presented an end-to-end model which adopts two coupled networks to sequentially denoise low-dose sinogram and reconstruct activity map for PET reconstruction. Although these CNN-based methods have achieved relatively good reconstruction results, there are still some shortcomings. First, convolution operates in a fixed-sized window with a locally limited receptive field, and thus cannot capture long-range dependencies which are also crucial in the reconstruction task. Second, the convolution operation cannot flexibly adapt for different inputs owing to the fixed weights of the convolution filter after training [11].

Recently, transformer [12], which is originally proposed in natural language processing field, has also shown comparable or even superior performance in computer vision tasks, such as objective detection [13], panoptic segmentation [14] and image generation [15], compared with traditional convolutional networks. The basic architecture of a transformer is an encoder-decoder network, where the encoder is composed of several identical encoder layers, with each layer involving a multi-head self-attention sub-layer and a position-wise feed-forward neural network, and the decoder follows a similar structure, except for an additional encoder-decoder attention sub-layer embedded before the feed-forward network to perform multi-head attention on the outputs of the encoder. Thanks to the self-attention mechanism, compared with pure CNNs, the transformer can model the long-range interactions in the input domain and adapt its weight aggregation for different input dynamically, yielding stronger representation ability [11].

In this study, to maintain the high quality of PET images while reducing the radiation exposure damage, and also inspired by the great success of transformer, we innovatively propose a GAN based model embedded with transformer, namely Transformer-GAN, to directly reconstruct SPET images from the corresponding images at low dose. Our main contributions are summarized as follows. 1) Current applications using transformer mainly focus on 2D images. However, for the 3D PET data, the existing 2D-based methods can only estimate the 2D-slices of a volume independently, but neglect the discontinuity problem across slices and the spatial context loss along the other two directions. Therefore, we extend our Transformer-GAN to a 3D manner to preserve more spatial context information. 2) Since the original trans-former takes account of the pairwise relationships among all input sequences, it requires much higher computational cost than CNNs which only consider local correlations, raising the difficulty in processing 3D medical images. In fact, the low-level spatial structure of images can be well described by a convolutional architecture. In light of this, the proposed method effectively combines the locality prior of CNNs and the powerful expressivity of transformer, making both *the global semantic dependencies* and *the low-level spatial information of local connectivity* be well captured with less computational overhead. 3) Apart from the estimation error loss used to ensure similarity between voxel-level intensities of the reconstructed images and the real images, our Transformer-GAN model additionally leverages the adversarial loss of GAN to make the data distribution of the reconstructed image consistent with that of the real image, leading to more real-like reconstructed SPET images.

2   Methodology

The architecture of the proposed Transformer-GAN is depicted in Fig. 1, which consists of a generator network and a discriminator network. The generator network takes the LPET image as input, and outputs the corresponding SPET-like image. Specifically, the generator network comprises three components: (1) a CNN-based encoder (Encoder-CNN for simplicity) aiming to learn a compact feature representation with rich spatial information; (2) a transformer network used to model the long-range dependencies between the input sequences learned by EncoderCNN; and (3) a CNN-based decoder (DecoderCNN) applied to restore the feature representation modeled by the transformer to the final target image. The discriminator network inputs a pair of images, i.e., the source LPET image and its corresponding real/fake target SPET image, and outputs the probability that the input image pair is real.

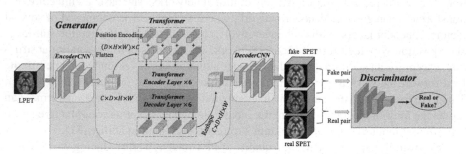

Fig. 1. Illustration of the proposed 3D Transformer-GAN architecture.

2.1 Architecture

EncoderCNN: Since a vanilla transformer takes account of the pairwise interactions among all input sequences, it requires enormous computational cost, limiting its applicability for 3D medical images. To address this, we first employ a CNN-based encoder (EncoderCNN) which effectively utilizes the inductive bias of CNNs on the locality to extract relatively small size features with low-level spatial information. Following U-Net [16], the EncoderCNN takes the source 3D LPET image as input, and outputs a learned compact 3D feature representation. Specifically, the EncoderCNN contains four convolution blocks, each involving two $3 \times 3 \times 3$ convolutional layers, followed by BatchNormalizaion and LeakyReLU activation function, and only the last three convolution blocks are used for down-sampling. The extracted feature maps are further processed by a convolutional layer to reduce the channel numbers from 512 to 192, for reducing the computational overhead and matching the channel dimension of the fixed position encoding. Finally, a learned feature map $f_{E\_C} \in R^{C \times D \times H \times W}$ is obtained and equipped to sequences which will be further input to the transformer.

Algorithm 1

---

1. **Encoder Input:** A set of input sequence $f_v = [f_{v1}, f_{v2}, f_{v3}, \cdots, f_{vN}]$, and a set of position encoding $P_v = [P_{v1}, P_{v2}, P_{v3}, \cdots, P_{vN}]$, where N is the total number of voxels of the feature map along the channel dimensions learned by EncoderCNN.
   Decoder Input: The Encoder output f_{E_T}, and the position encoding P_v.

2. **Parameters:** The number of layers of the transformer encoder and decoder L.
   Initialize: $m = n = 1$, $L=6$, f_{E_0} is initialized to f_v, and f_{D_0} is initialized to 0.

   Encoding stage:

3. **While** $m < L$
4. The MSA sub-layer takes P_v and $f_{E_{m-1}}$ as input, and processes them through dropout, residual connection and Layer Normalization (LN) in sequence, and outputs f'_{E_m} by

   $$f'_{E_m} = LN\left(f_{E_{m-1}} + dropout(MSA(Q_m, K_m, V_m))\right),$$ where Query Q_m equal to Key $K_m = f_{E_{m-1}} + P_v$, and Value $V_m = f_{E_{m-1}}$. The FFN sub-layer receives f'_{E_m} as input, then processes it through dropout, residual connection and LN, and outputs f_{E_m} by

   $$f_{E_m} = LN\left(f'_{E_m} + dropout\left(FFN(f'_{E_m})\right)\right).$$

5. $m = m + 1$
6. **End while**

   Decoding stage:

7. **While** $n < L$
8. The MSA sub-layer takes the output of the previous decoder layer $f_{D_{n-1}}$ and P_v as input, and outputs f'_{D_n}, whose process is the same as encoder. The MEDA sub-layer takes the Encoder output f_{E_T}, f'_{D_n} and P_v as input, and outputs $f''_{D_n} = LN\left(f_{E_{n-1}} + dropout(MSA(Q_n, K_n, V_n))\right)$, where $Q_n = f_{E\_T} + P_v$, $K_n = f'_{D_n} + P_v$, and $V_n = f_{E\_T}$. Similar to the encoding stage, the FFN sub-layer receives f''_{D_n} as input and outputs f_{D_n}.
9. $n = n + 1$
10. **End while**
11. **Output:** A set of output sequences $f_{D\_T} = [f_{D1}, f_{D2}, f_{D3}, \cdots, f_{DN}]$.

Transformer: The architecture of our transformer follows the standard encoder-decoder structure in [12]. However, different from the original transformer, our model applies parallel decoding to achieve parallel sequence prediction, since the original transformer predicting results sequence-by-sequence usually requires prohibitive inference cost [13].

Considering the transformer expects sequence-style as input, we treat each voxel of the EncoderCNN-learned feature map $f_{E\_C} \in R^{C \times D \times H \times W}$ along the channel dimension as a "sequence", and thus obtain N sequences $f_v \in R^{N \times C}$ after flattening, where $N = D \times H \times W$. Meanwhile, to preserve the spatial information, the position details

$P_v \in R^{N \times C}$ are obtained by position encoding according to [13]. Note that, since our proposed model is 3D-based, we encode the position information along three directions (x, y, z), respectively, and the embedding dimension of each direction is $C/3$. The final position encoding is obtained by concatenating the position encodings along these three directions.

The transformer encoder has six identical encoder layers, each with a multi-head self-attention (MSA) module cascading multiple self-attention modules, and a simple feed forward network (FFN) stacking two fully connected layers. Please refer to the original transformer [12] for more specific architectures regarding these two modules. For the m-th ($m = 1, 2,..., 6$) encoder layer, it receives the output of the previous layer $f_{E_{m-1}} \in R^{N \times C}$ and the position encoding P_v, and produces $f_{E_m} \in R^{N \times C}$. In particular, we set f_{E_0} as f_v. After repeating six times, we can obtain an encoded latent feature f_{E_T} $= f_{E_6} \in R^{N \times C}$.

The transformer decoder follows a similar architecture as the encoder, except for an additional multi-head encoder-decoder attention (MEDA) which is embedded before FFN. The main difference between MSA and MEDA is that MEDA models the correlations between the outputs of the transformer encoder and the previous MSA, while MSA calculates the attention map with its own input. Specifically, for the n-th ($n = 1, 2,..., 6$) decoder layer, it takes the output of encoder f_{E_T}, the output of the previous decoder layer $f_{D_{n-1}}$ (initially set to 0 when $n = 1$), and the positional encoding P_v as input, and models the long-range dependencies through multiple MSA sub-layers, MEDA sub-layer and FFN, sequentially. After six decoder layers, the output of the transformer, $f_{D\_T} \in R^{N \times C}$, can be generated, which will be further reshaped into the features $f_{D\_C} \in R^{C \times D \times H \times W}$, and then processed by a $1 \times 1 \times 1$ convolution with stride of 1 to increase the channel number to the same as that of the EncoderCNN output.

For more intuitive understanding, the procedure is summarized in Algorithm 1.

DecoderCNN: The CNN-based decoder (DecoderCNN) aims to restore the feature representations captured by the transformer to the reconstructed PET image. Its detailed architecture is similar to that of the EncoderCNN, except that the down-sampling operation is replaced with the up-sampling operation. At last, a $1 \times 1 \times 1$ convolution layer followed by a sigmoid activation function is applied to produce the final output. Please note that, the skip connections are constructed between the EncoderCNN layers and the corresponding DecoderCNN layers to combine hierarchical features, which can be beneficial for retaining the spatial details during the down-sampling.

Discriminator: The discriminator takes a pair of 3D PET images, i.e., source LPET image and the corresponding fake/real target SPET image as input, determining whether the input is real or not. Concretely, the discriminator consists of four convolution blocks with the same Convolution-BatchNormalizaion-LeakyRelu structure and a convolutional layer, followed by a sigmoid activation function to distinguish the pairs between fake and real.

2.2 Objective Functions

The objective function of the proposed model is mainly composed of two parts: 1) the voxel-wise estimation error between the real and reconstructed SPET images; and 2) the adversarial loss. Given an LPET image $x \sim P_{low}(x)$, and the corresponding SPET image $y \sim P_{std}(y)$, the reconstructed SPET image $G(x)$ is generated from x through the generator network G. In order to ensure that the reconstructed PET image approximates its corresponding real SPET image, we consider L1 loss as the voxel-wise estimation error to narrow the gap between them, as shown in Eq. (1).

$$V_{L1}(G) = E_{x \sim P_{low}(x), y \sim P_{std}(y)} \left[\| y - G(x)_1 \| \right] \tag{1}$$

Moreover, considering the consistency of the data distributions between the real and reconstructed SPET images, we further introduce the feedback of the discriminator D for training G. Different from the original GANs which takes the random noise vector as input, our Transformer-GAN conditions the model on an input LPET image. The adversarial loss is defined as:

$$\min_{G} \max_{D} V(G, D) = E_{x \sim P_{low}(x), y \sim P_{std}(y)} \left[\log D(x, y) \right] + E_{x \sim P_{low}(x)} \left[log(1 - D(x, G(x))) \right]$$

$$\tag{2}$$

In this way, the generator is required *not only* to produce an image similar to the real SPET image at voxel-level, *but also* to keep the data distribution of the reconstructed image consistent with that of the real image.

In summary, the final objective function of our Transformer-GAN is defined as:

$$V_{total} = \min_{G} \max_{D} V(G, D) + \lambda V_{L1}(G), \tag{3}$$

where $\lambda > 0$ is a hyperparameter used to control the balance between these two terms.

2.3 Training Details

The generator network G and the discriminator network D are trained alternately as the standard approach in [17]. Concretely, we first fix G to optimize D through the feedback of the objective functions, and then in turn fix D to optimize G. The whole training procedure is like playing a competitive max-min game, as shown in Eq. (2). In the testing stage, only the well-trained G is needed for reconstructing SPET image from the corresponding LPET. All networks are trained by Adam optimizer with batch size of 4. The training process lasts for 150 epochs, the learning rate is set to 2×10^{-4} for the first 100 epochs, and linearly decays to 0 in the last 50 epochs. λ in Eq. (3) is empirically set as 100.

3 Experiments and Results

Data Description: We conduct a set of experiments to verify our proposed method on a clinical dataset which includes eight normal control (NC) subjects and eight mild cognitive impairment (MCI) subjects. Each subject contains an LPET image and an SPET image, with the size of $128 \times 128 \times 128$. The PET data were acquired on a Siemens Biograph mMR PET-MR system, and the subjects were injected with an averaged

203 MBq of ^{18}F-Flurodeoxyglucose ([^{18}F] FDG). SPET and LPET images were acquired consecutively based on standard imaging protocols. Particularly, the SPET images were obtained in a 12-min period within one hour of tracer injection, and the LPET scans were acquired in a 3-min short period to simulate the acquisition at a quarter of the standard dose. The commonly used OSEM method was adopted for iterative PET reconstruction with 3 iterations, 21 subsets, and a 2 mm FWHM 3D Gaussian post-reconstruction filter. The SPET and LPET for the same subject used the same attenuation map computed by the Dixon fat-water method supplied by the manufacturer of the scanners.

Table 1. Quantitative comparison with four state-of-the-art SPET reconstruction methods in terms of PSNR, SSIM, and NMSE.

Method	NC subjects			MCI subjects			Parameters
	PSNR	SSIM	NMSE	PSNR	SSIM	NMSE	
LPET	20.684	0.979	LPET	21.541	0.976	0.058	
3D U-NET	23.744	0.981	0.0245	24.165	0.982	0.0256	24M
Auto-context CNN [2]	23.867	–	0.0235	24.435	–	0.0264	41M
3D-cGANS [4]	24.024	–	0.0231	24.617	–	0.0256	127M
LA-GANs [5]	24.296	0.982	–	24.729	0.983	–	127M
Proposed	**24.818**	**0.986**	**0.0212**	**25.249**	**0.987**	**0.0231**	**76M**

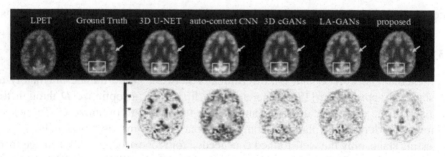

Fig. 2. Visual comparison with four state-of-the-art SPET reconstruction methods.

Experimental Settings: Considering the required GPU memory and the limited training samples, we extract 729 large patches of size $64 \times 64 \times 64$ from the whole image of size $128 \times 128 \times 128$ for each subject to train the proposed Transformer-GAN model. Also, to enhance the stability of the model with limited samples, we used the leave-one(subject)-out cross-validation (LOOCV) strategy, i.e., the training-test procedure is repeated for 16 times while each time one subject is used in turn for testing and the other 15 subjects are for training, and the averaged performance is reported to avoid potential bias. In this manner, the total training samples are increased from 15 to 10935 (15×729), which could provide a sufficient number of samples to train a good model. For quantitative evaluation, we use three indicators to study the performance of all method, including normalized mean squared error (NMSE), peak signal-to-noise (PSNR), and structural similarity index (SSIM).

Comparison with Other SPET Reconstruction Methods: We first compare the proposed method with four state-of-the-art SPET reconstruction methods: (1) 3D U-NET, (2) auto-context CNN [2], (3) 3D-cGANs [4], and (4) LA-GANs [5]. For a fair comparison, we had carefully tuned the hyper-parameters of the comparison methods and trained the networks with the same LOOCV strategy. The quantitative comparison results are reported in Table 1, from which we can see that our method attains the best performance. Compared with the LA-GANs which achieves the second-best result, our proposed method still boosts the PSNR and NMSE performance by 0.522 dB and 0.004 for NC subjects, and 0.52 dB and 0.004 for MCI subjects, respectively. Moreover, through paired t-test, the p-value between the proposed and other comparison methods are all less than 0.05, indicating the statistically significant improvement of the proposed Transformer-GAN.

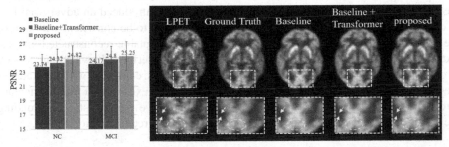

Fig. 3. Quantitative comparisons (left) and qualitative comparisons (right) of our proposed method with its two variants.

We also provide the visual comparison results in Fig. 2, where the first row presents the LPET, the real SPET (ground truth) as well as the reconstructed images by five different methods, and the second row displays the corresponding error maps. It can be clearly seen that the reconstructed SPET image generated by our proposed method has less artifacts and richer details with lighter error map than those by other comparison methods. Regions with obvious improvements are indicated by the red arrows and red boxes. Both qualitative and quantitative experimental results demonstrate that the proposed Transformer-GAN can reconstruct better SPET-like images, compared to the other state-of-the-art methods.

Ablation Study: To study the contributions of important components of our Transformer-GAN model, we respectively conduct ablation experiments by (1) removing the transformer from the proposed model (i.e., EncoderCNN + DecoderCNN, denoted as Baseline), (2) the proposed model without discriminator (i.e., EncoderCNN + Transformer + DecoderCNN, denoted as Baseline + Transformer), and (3) the proposed Transformer-GAN model. The quantitative comparison results are given in the left part of Fig. 3. We can clearly see that the performance for both NC and MCI subjects is progressively improved when the transformer module and discriminator are added to the Baseline framework. The qualitative comparisons are presented in the right part of Fig. 3. Consistent with the quantitative results, the proposed model achieves superior visual results with more content details, especially for the regions indicated by the red arrows. Please note that, compared with the case purely using Transformer, our model effectively reduces the number of learning parameters from 248M to 42M, demonstrating the lightweight design of our model.

4   Conclusion

In this paper, we proposed a novel 3D-based Transformer-GAN model to reconstruct high-quality SPET images from the LPET images. The proposed model effectively combines the locality prior of CNNs and the powerful expressivity of transformer, which can reduce computational overhead while capturing long-range dependencies between the input sequences. Moreover, besides the voxel-wise estimation error which encourages the reconstructed image similar to the real one, we also introduced an adversarial loss to ensure the data distribution of the reconstructed image to be consistent with that of the real image, thus obtaining reliable and clinically acceptable reconstructed images. Extensive experiments conducted on real human brain dataset have shown the superiority of our method in both qualitative and quantitative measures, compared with the state-of-the-art SPET reconstruction approaches. Considering the existence of complementary information between different modalities, in the future, we will investigate extending the single-modal reconstruction model to the multi-modal reconstruction model.

Acknowledgement. This work is supported by National Natural Science Foundation of China (NFSC 62071314) and Sichuan Science and Technology Program (2021YFG0326, 2020YFG0079).

References

1. Chen, W.: Clinical applications of PET in brain tumors. J. Nucl. Med. **48**(9), 1468–1481 (2007)
2. Xiang, L., Qiao, Y., Nie, D., et al.: Deep auto-context convolutional neural networks for standard-dose PET image estimation from low-dose PET/MRI. Neurocomputing **267**, 406–416 (2017)
3. Kim, K., Wu, D., Gong, K., et al.: Penalized PET reconstruction using deep learning prior and local linear fitting. IEEE Trans. Med. Imaging **37**(6), 1478–1487 (2018)
4. Wang, Y., Yu, B., Wang, L., et al.: 3D conditional generative adversarial networks for high-quality PET image estimation at low dose. Neuroimage **174**, 550–562 (2018)
5. Wang, Y., Zhou, L., Yu, B., et al.: 3D auto-context-based locality adaptive multi-modality GANs for PET synthesis. IEEE Trans. Med. Imaging **38**(6), 1328–1339 (2019)
6. Feng, Q., Liu, H.: Rethinking PET image reconstruction: ultra-low-dose, sinogram and deep learning. In: Martel, A.L., et al. (eds.) MICCAI 2020. LNCS, vol. 12267, pp. 783–792. Springer, Cham (2020). https://doi.org/10.1007/978-3-030-59728-3_76

7. Gong, K., Guan, J., Liu, C.C., et al.: PET image denoising using a deep neural network through fine tuning. IEEE Trans. Radiat. Plasma Med. Sci. **3**(2), 153–161 (2018)

8. Spuhler, K., Serrano-Sosa, M., Cattell, R., et al.: Full-count PET recovery from low-count image using a dilated convolutional neural network. Med. Phys. **47**(10), 4928–4938 (2020)

9. Xu, J., Gong, E., Pauly, J., et al.: 200x low-dose PET reconstruction using deep learning. arXiv preprint arXiv:1712.04119 (2017)

10. Xiang, L., Wang, L., Gong, E., Zaharchuk, G., Zhang, T.: Noise-aware standard-dose PET reconstruction using general and adaptive robust loss. In: Liu, M., Yan, P., Lian, C., Cao, X. (eds.) MLMI 2020. LNCS, vol. 12436, pp. 654–662. Springer, Cham (2020). https://doi.org/10.1007/978-3-030-59861-7_66

11. Khan, S., Naseer, M., Hayat, M., et al.: Transformers in vision: a Survey. arXiv preprint arXiv:2101.01169 (2021)

12. Vaswani, A., Shazeer, N., Parmar, N., et al.: Attention is all you need. arXiv preprint arXiv:1706.03762 (2017)

13. Carion, N., Massa, F., Synnaeve, G., et al.: End-to-end object detection with transformers. In: European Conference on Computer Vision, pp. 213–229. Springer, Cham (2020). https://doi.org/10.1007/978-3-030-58452-8_13

14. Wang, H., Zhu, Y., Adam, H., et al.: MaX-DeepLab: end-to-end panoptic segmentation with mask transformers. arXiv preprint arXiv:2012.00759 (2020)

15. Parmar, N., Vaswani, A., Uszkoreit, J., et al.: Image transformer. In: International Conference on Machine Learning. PMLR, pp. 4055–4064 (2018)

16. Ronneberger, O., Fischer, P., Brox, T.: U-net: convolutional networks for biomedical image segmentation. International Conference on Medical image Computing and Computer-Assisted Intervention, pp. 234–241. Springer, Cham (2015). https://doi.org/10.1007/978-3-319-24574-4_28

17. Goodfellow, I., Pouget-Abadie, J., Mirza, M.: Generative adversarial nets. In: Advances in Neural Information Processing Systems, pp. 2672–2680 (2014)

Learnable Multi-scale Fourier Interpolation for Sparse View CT Image Reconstruction

Qiaoqiao Ding[1], Hui Ji[1], Hao Gao[2], and Xiaoqun Zhang[3(✉)]

[1] Department of Mathematics, National University of Singapore,
119076 Singapore, Singapore
{matding,matjh}@nus.edu.sg
[2] Department of Radiation Oncology, University of Kansas Medical Center,
Kansas City 66160, KS, USA
Hgao2@kumc.edu
[3] Institute of Natural Sciences and School of Mathematical Sciences and MOE-LSC,
Shanghai Jiao Tong University, Shanghai 200240, China
xqzhang@sjtu.edu.cn

Abstract. Image reconstruction in sparse view CT is a challenging ill-posed inverse problem, which aims at reconstructing a high-quality image from few and noisy measurements. As a prominent tool in the recent development of CT reconstruction, deep neural network (DNN) is mostly used as a denoising post-process or a regularization sub-module in some optimization unrolling method. As the problem of CT reconstruction essentially is about how to convert discrete Fourier transform in polar coordinates to its counterpart in Cartesian coordinates, this paper proposed to directly learn an interpolation scheme, modeled by a multi-scale DNN, for predicting 2D Fourier coefficients in Cartesian coordinates from the available ones in polar coordinates. The experiments showed that, in comparison to existing DNN-based solutions, the proposed DNN-based Fourier interpolation method not only provided the state-of-the-art performance, but also is much more computationally efficient.

Keywords: CT reconstruction · Deep learning · Fourier interpolation

1 Introduction

Computed tomography (CT) is one imaging technique widely used in clinical and industry applications. Mathematically, in CT imaging, a projection data (or sinogram) p from a scanning angle $\theta \in [0, \pi)$ at a position $\xi \in \mathbb{R}$ is obtained via the following Radon transform [14]:

$$p_\theta(\xi) = \int_{-\infty}^{\infty} f(\xi\vec{s} + \eta\vec{s}^{\perp})d\eta, \tag{1}$$

where f is the target image to be reconstructed, $\vec{s} = (\cos\theta, \sin\theta)^{\top}$ and $\vec{s}^{\perp} = (-\sin\theta, \cos\theta)^{\top}$. Reconstructing the image f is then about inverting Radon

© Springer Nature Switzerland AG 2021
M. de Bruijne et al. (Eds.): MICCAI 2021, LNCS 12906, pp. 286–295, 2021.
https://doi.org/10.1007/978-3-030-87231-1_28

transform from a limited number of observations corrupted by noise. Fourier slice theorem [2,21] plays an important role in the inversion of Radon transform, which relates the 1D Fourier transform of projections with the 2D Fourier transform of the image in polar coordinates:

$$F(\omega, \theta) = F(\omega \cos \theta, \omega \sin \theta) = \widehat{p}_\theta(\omega),$$

where F denotes the 2D Fourier transform of the image f, and $\widehat{\cdot}$ denotes 1D Fourier transform. Recall that once $\{F(\omega_x, \omega_y)\}$ in 2D Cartesian coordinates are obtained, the image f can be reconstructed from $\{F(\omega_x, \omega_y)\}$ by inverse DFT. Thus, the CT image reconstruction can then be re-formulated as, given a set of 1D discrete Fourier coefficients $\{\widehat{p_{\theta_k}}(\omega)\}_k$ for a limited number of $\{\theta_k\}_k$ in polar coordinates, how to estimate $\{F(\omega_x, \omega_y)\}$ in 2D Cartesian coordinates.

The interpolation problem in 2D discrete Fourier domain, i.e. estimating all 2D Fourier coefficients on Cartesian coordinates from the available measurements on polar coordinates, is indeed a challenging problem. It cannot be simply solved by classic numerical interpolation schemes, e.g. linear interpolation. The main reason is that the 2D Fourier coefficients of an image are highly irregular, which contradicts the local smoothness assumption made by classic interpolation schemes. As a result, direct call of classic interpolation schemes will lead to very erroneous results with noticeable artifacts. In the case of sparse view CT where the available measurements are sparse along a limited number of angles, the results can be even worse. Zhang and Froment [23] proposed to tackle such an interpolation problem with a total variation (TV) regularization on the image. While many other regularization techniques introduced in CT reconstruction, e.g., [7,9,11,13,15,16,19,22], can also be used for regularizing the interpolation in discrete Fourier domain, these regularization strategies are based on some pre-defined image prior, which does not always hold true in practice.

In recent years, DNN-based deep learning has emerged as one powerful tool for CT image reconstruction. By treating the artifacts in the image reconstructed by some method, earlier works used the DNN as a powerful denoiser to remove the artifacts in the result [4,12,18,20]. More recently, the so-called optimization unrolling with DNN-based prior becomes a more preferred approach; see e.g. [1,3,5,6,8,10,17]. These methods take some iterative regularization method, and replace the regularization-relating modules by the DNNs with learnable parameters. Note that, in each iteration, these methods needs to perform the projection and back-projection operation, i.e. the Radon transform (1) and it adjoint operator. As these operations are computationally expensive, the inclusion of multiple such operations in the network not only makes the computation very time-consuming, but also increases the complexity of network training.

This paper aims at developing a deep learning method for image reconstruction, which not only provide better performance than existing solutions to sparse view CT, but also is much more computationally efficient. The basic idea is to interpret the problem of CT image reconstruction as an interpolation problem in discrete Fourier domain, not as a linear inverse problem in image domain as most methods do.In this paper, we proposed an multi-scale interpolation

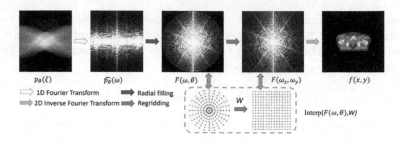

Fig. 1. The work flow of Fourier interpolation (regridding) for image reconstruction

scheme in discrete Fourier domain with adaptive interpolating weights, which is predicted by a learnable multi-scale DNN. Note that there is no projection or back-projection is involved in the proposed approach. For performance evaluation, the proposed method is compared with FBP, TV method [15], two deep CT reconstruction methods: LEARN [3] and Learned-PD [1]. The experiments showed that the proposed method outperformed these methods by a noticeable margin, while requiring much less training time.

2 Main Body

We first introduced some notations for facilitating the discussion. Let $f(x,y)$, $(x,y) \in \Omega \subset \mathbb{R}^2$ denote the image of spatial attenuation distribution, and $p_\theta(\xi)$ denote the Radon transform of the object with $\theta \in (0, \pi]$ and $\xi \in [-r, r]$, where r denotes the radius of Ω. For parallel scanning geometry, both $\theta = (\theta_1, \cdots, \theta_L)$ and $\xi = (\xi_1, \cdots, \xi_N)$ are uniformly distributed in the range $[0, \pi) \times [-r, r]$. The 1D Fourier transform of the projection at each view, denoted as $\widehat{q_\theta}$, are distributed on a discrete polar grid in Euclidean space. In other words, the input measurement is $\{F(\omega_k, \theta_\ell)\}$, the 2D Fourier coefficients of the image f on a polar coordinates $\{\omega_k, \theta_l\}_{k,l}$. In order to reconstruct the image f, one needs to estimate the full set of 2D discrete Fourier coefficients of the image f, denoted by $\{F(\omega_{k_1}, \omega_{k_2})\}_{k_1, k_2}$, on Cartesian coordinates. Once we have $\{F(\omega_{k_1}, \omega_{k_2})\}_{k_1, k_2}$, the image can be reconstructed by simply calling an inverse DFT. In short,

$$\{p_{\theta_k}\}_k \to \{F(\omega_k, \theta_l)\}_{k,l} := \{\widehat{p_{\theta_k}}\}_k \to \{F(\omega_{k_1}, \omega_{k_2})\}_{k_1, k_2} \to f := \mathcal{F}^{-1}F,$$

where $\mathcal{F}$ denotes 2D DFT and $\mathcal{F}^{-1}$ denotes its inversion. In the procedure above, the key step is to estimate $\{F(\omega_{k_1}, \omega_{k_2})\}_{k_1, k_2}$ from $\{F(\omega_k, \theta_l)\}_{k,l}$, which is often called the problem of *re-gridding*, an interpolation problem from one grid to a different type of grid. See Fig. 1 for an illustration of the work flow. As the density of measured spectral data decreases rapidly from low to high frequencies, it is a challenging task to interpolate the regions of high frequencies.

In this paper, we proposed a learnable method for re-gridding in discrete Fourier domain, which excludes the forward/backward projection in the network. The basic idea is to interpolate the missing Fourier coefficient using a weighted summation of its neighbors with adaptive weights. Specifically, for a frequency $(\omega_{k_1}, \omega_{k_2})$ on the Cartesian grid, the missing coefficient is predicted by a weighted average of its K nearest neighbors $\{F(\omega_{k(i)}, \theta_{l(i)})\}_{i=1}^{K}$ on the polar grid:

$$F(\omega_{k_1}, \omega_{k_2}) = \sum_{i=1}^{K} W_{\omega_{k_1}, \omega_{k_2}}(i) F(\omega_{k(i)}, \theta_{l(i)}) \qquad (2)$$

where $W_{\omega_{k_1}, \omega_{k_2}}(i)$ are learnable weights that are adaptive to different frequencies. Such an interpolation scheme is implemented in a multi-scale manner. For low frequencies with dense neighbors in polar coordinates, few neighbors are used for interpolation with smaller K. For high frequencies with sparse neighbors in polar coordinates, more neighbors are used for interpolation with larger K.

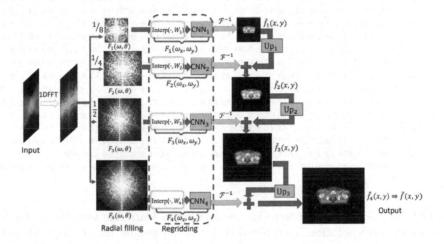

Fig. 2. Diagram of the proposed method.

After the interpolation is done, we have now an initial estimation of all Fourier coefficients on the Cartesian grid. Then, a learnable CNN is introduced to correct possible prediction errors arising in the interpolation. As prediction accuracy decreases from low to high frequencies, we propose a multi-scale CNN-based correction scheme. Briefly, we construct S sets of Fourier coefficients on the polar grid whose range increases from low to high frequencies, denoted by $\{F_s(\omega, \theta)\}_{s=1}^{S}$. Then, after the interpolation and CNN-based correction in discrete Fourier domain, the image can be reconstructed by inverse DFT with respect to different resolutions. Instead of using such a prediction for estimating the image f at the resolution s, the estimation is done by the concatenation of current estimation and the one from the coarser resolution $s - 1$ after upsampling.

In summary, for each resolution s, the resulting prediction of the image f, denoted by $\tilde{f}_s$, can be expressed as

$$\tilde{f}_s(x, y) = \mathcal{F}^{-1}(\text{CNN}_s(\text{Interp}(F_s(\omega, \theta), W_s), \theta_s)) + \text{Up}_{s-1}(\tilde{f}_{s-1}(x, y), \theta_{s-1}^{up}),$$

Table 1. Quantitative evaluation of the results (Mean±STD) from different methods.

Angle number	Noise level	Method	FBP	TV	FBPConvNet	Learned-PD	LEARN	Proposed
180 views	0%	RMSE	29.35 ± 3.20	25.08 ± 2.81	21.79 ± 2.73	16.21 ± 2.16	$\mathbf{10.79 \pm 1.70}$	13.14 ± 2.34
		PSNR	40.45 ± 0.97	41.35 ± 1.08	43.05 ± 1.04	45.56 ± 0.57	$\mathbf{49.19 \pm 1.29}$	47.50 ± 1.46
	10%	RMSE	36.72 ± 2.98	32.23 ± 2.42	26.91 ± 3.08	28.51 ± 2.39	19.90 ± 1.94	$\mathbf{16.61 \pm 2.57}$
		PSNR	38.48 ± 0.70	39.58 ± 1.05	41.20 ± 0.94	40.65 ± 0.61	43.81 ± 0.82	$\mathbf{45.43 \pm 1.27}$
	20%	RMSE	52.84 ± 4.72	35.64 ± 1.94	23.65 ± 2.53	32.78 ± 1.96	23.76 ± 2.03	$\mathbf{20.86 \pm 2.25}$
		PSNR	35.32 ± 0.75	38.71 ± 0.83	42.32 ± 0.90	39.45 ± 0.52	42.26 ± 0.72	$\mathbf{43.41 \pm 0.91}$
	30%	RMSE	72.05 ± 7.39	39.71 ± 2.07	27.85 ± 2.59	35.92 ± 1.86	34.65 ± 4.70	$\mathbf{22.31 \pm 2.59}$
		PSNR	32.64 ± 0.86	37.77 ± 1.03	40.88 ± 0.79	38.65 ± 0.45	39.03 ± 1.13	$\mathbf{42.83 \pm 1.00}$
45 views	0%	RMSE	82.85 ± 8.55	43.57 ± 2.31	29.95 ± 3.20	36.92 ± 2.62	32.81 ± 2.38	$\mathbf{20.80 \pm 2.57}$
		PSNR	32.42 ± 0.87	36.97 ± 0.86	40.27 ± 0.91	37.96 ± 0.54	39.45 ± 0.62	$\mathbf{43.45 \pm 1.05}$
	10%	RMSE	94.01 ± 7.90	44.87 ± 2.70	33.12 ± 3.34	39.09 ± 3.30	37.16 ± 2.40	$\mathbf{21.05 \pm 2.39}$
		PSNR	30.31 ± 0.71	36.84 ± 0.85	39.39 ± 0.87	37.93 ± 0.72	38.36 ± 0.55	$\mathbf{43.34 \pm 0.98}$
	20%	RMSE	121.28 ± 9.56	48.84 ± 2.23	34.54 ± 3.28	39.06 ± 2.24	39.91 ± 2.15	$\mathbf{25.55 \pm 2.78}$
		PSNR	28.10 ± 0.66	35.97 ± 0.72	39.39 ± 0.81	37.92 ± 0.50	37.86 ± 0.46	$\mathbf{41.65 \pm 0.93}$
	30%	RMSE	156.30 ± 14.11	55.12 ± 3.24	40.25 ± 3.29	48.02 ± 2.26	32.29 ± 2.93	$\mathbf{30.64 \pm 3.32}$
		PSNR	25.90 ± 0.76	34.92 ± 0.69	37.68 ± 0.69	36.13 ± 0.41	39.60 ± 0.79	$\mathbf{40.07 \pm 0.93}$
30 views	0%	RMSE	143.74 ± 11.18	47.21 ± 2.40	42.11 ± 3.91	48.13 ± 2.74	41.85 ± 2.38	$\mathbf{23.11 \pm 2.53}$
		PSNR	26.62 ± 0.68	36.27 ± 0.56	37.30 ± 0.79	36.11 ± 0.49	37.33 ± 0.50	$\mathbf{42.41 \pm 0.91}$
	10%	RMSE	154.24 ± 10.00	48.56 ± 2.41	39.82 ± 4.15	43.77 ± 2.99	45.07 ± 2.27	$\mathbf{26.16 \pm 2.58}$
		PSNR	26.00 ± 0.56	36.02 ± 0.85	37.79 ± 0.86	36.94 ± 0.59	36.68 ± 0.43	$\mathbf{41.43 \pm 0.84}$
	20%	RMSE	181.94 ± 10.17	52.48 ± 2.43	38.37 ± 3.34	53.41 ± 3.31	51.37 ± 2.57	$\mathbf{28.82 \pm 3.10}$
		PSNR	24.56 ± 0.48	35.35 ± 0.54	38.10 ± 0.74	35.21 ± 0.54	35.54 ± 0.43	$\mathbf{40.60 \pm 0.93}$
	30%	RMSE	220.13 ± 14.92	58.44 ± 3.52	43.90 ± 4.96	53.11 ± 3.71	56.23 ± 5.97	$\mathbf{31.82 \pm 3.38}$
		PSNR	22.91 ± 0.57	34.41 ± 0.52	36.94 ± 0.88	35.26 ± 0.61	34.79 ± 0.88	$\mathbf{39.74 \pm 0.90}$

where CNN_s denotes the correction network with learnable weights θ_s, Interp denotes the interpolation operator with learnable weight W_s, $F_s(\omega, \theta)$ denotes the input measurement of scale s, $\tilde{f}_s(x, y)$ denotes the output of scale s, and Up_s denotes the upsampling layers from lower to higher resolution image with learnable parameter θ_s^{up}. The image reconstruction $\tilde{f}(x, y)$ is defined as the estimation in the finest resolution $\tilde{f}_S$. In the implementation, we set $S = 4$ and $\tilde{f}_0(x, y) = 0$. See Fig. 2 for the diagram. Let $\{p^j, f^j\}_{j=1}^J$ denote the training set with J training samples, where each (p^j, f^j) denotes one pair of sinogram and true image. The network is trained by minimizing the following loss function

$$\mathcal{L}(\Theta) = \frac{1}{J} \sum_{j=1}^{J} \| \tilde{f}^j - f^j \|_2^2, \tag{3}$$

where $\Theta := \{\theta_s, W_s, \theta_s^{up}\}$ denotes the whole set of NN parameters.

3   Experiments

To evaluate the performance of the proposed method, we simulated projection data from CT images as follows. The dataset included 6400 prostate CT images of 256×256 pixels per image from 100 anonymized scans, where 80%, and 20% of the data is set for training and testing respectively.

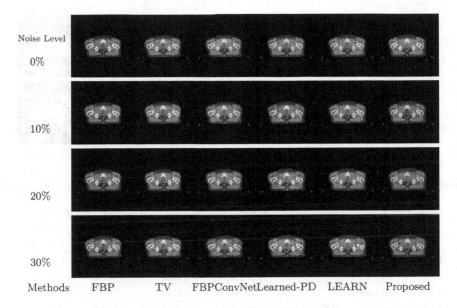

Noise Level						
0%						
10%						
20%						
30%						
Methods	FBP	TV	FBPConvNet	Learned-PD	LEARN	Proposed

Fig. 3. Parallel beam CT reconstruction with the projection data of 180 views.

Through the experiments, $K = 2, 4, 6, 8$ are used for the numbers of the nearest neighbors from low to high resolution. For an image of size $N \times N$, the learnable weights, $W_1^{N/8 \times N/8 \times 2}, W_2^{N/4 \times N/4 \times 4}$, $W_3^{N/2 \times N/2 \times 6}, W_4^{N \times N \times 8}$ at four levels, are initialized with the normalized distance for every point and learned along with the parameters in CNN and Upsampling layers. The standard CNN blocks are stacked with the structure Conv→BN→ReLU. For all the Conv layers in the CNN, the kernel size is set as 3×3, and the channel size is set to 64. The block numbers of CNN are increased over the four levels: starting with 10 blocks for the lowest frequencies and sequentially adding 4 blocks when the resolution level increases. Finally, a CNN with 10 blocks are used in the upsampling stage.

The NN was implemented with PyTorch on a NVIDIA Titan GPU, and trained by the Adam where momentum parameter $\beta = 0.9$, the mini-batch size is 8, and the learning rate is 10^{-4}. The model was trained for 50 epochs. The convolution weights are initialized with constant matrices and the biases are initialized with zeros. Both root mean square error (RMSE) and peak signal to noise ratio (PSNR) are used for quantitative assessment of image quality.

Parallel Beam CT Reconstruction. In the experiments for parallel beam CT reconstruction, the projection data was down-sampled from 180 to 45 and 30 views to simulate the sparse view geometry. For performance evaluation, the proposed methods is compared to FBP method [2], TV method [15], FBPConvNet [12] and two optimization-unrolling-based deep learning methods: Learned-PD [1] and LEARN [3].

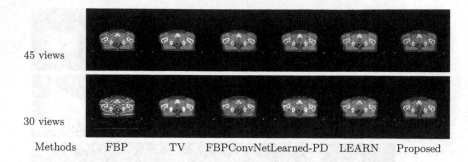

Fig. 4. Parallel beam CT reconstruction with noiseless data: 45 views and 30 views.

Table 2. Comparison of training/testing time of different methods.

		FBP	TV	FBPConvNet	Learned-PD	LEARN	Proposed
Training	180 views	–	–	40 min	6 day	6.2 day	6 h
	45 views	–	–	40 min	2 day	3 day	5 h
	30 view	–	–	40 min	1 day	1 day	5 h
Testing	180 views	0.20 s	42.67 s	0.003 s	2.16 s	2.27 s	0.05 s
	45 views	0.04 s	9.76 s	0.003 s	0.55 s	0.57 s	0.04 s
	30 view	0.02 s	5.76 s	0.003 s	0.44 s	0.39 s	0.04 s

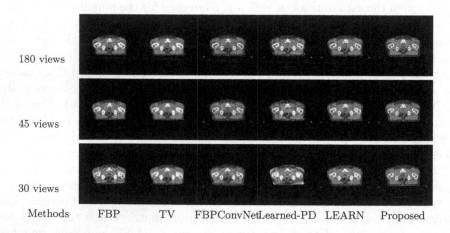

Fig. 5. Fan beam CT reconstruction with noiseless projection data of 45 and 30 views.

See Table 1 for quantitative comparison of the results from different methods, in terms of RMSE and PSNR. It can be seen that deep learning methods noticeably outperformed two non-learning methods, FBP and TV methods. For the data with 180 views, LEARN is the best performer when measurement is noise-free. In the presence of noise, the proposed method outperformed LEARN, Learned-PD and FBPConvNet by a noticeable margin for different noise levels.

See Fig. 3 for visual comparison of the results from different methods on one test data with 180 views under different noise levels, and Fig. 4 on the same noise-free test data with different views. Both quantitative and visual comparison showed the advantage of the proposed method over existing solutions to spare view CT reconstruction, especially in the presence of measurement noise.

Computational efficiency is another main motivation of our work. See Table 2 for the comparison of different methods on training/testing time. For the exclusion of the projection/back-projection operators, the proposed method is much more computational efficient than LEARN and Learned-PD, in both training and testing. While FBPConvNet is faster than the proposed method, its performance is significantly worse than the proposed one as shown in Table 1.

Fan Beam CT Reconstruction. The proposed method is also applied for fan beam CT reconstruction, after re-binning fan beam projection to parallel projection. The experiments are conducted on the noiseless projection data with X-ray source rotated with 180, 45 and 30 views spanned on half circle. See Table 3 for the quantitative comparison of the results from different methods and Fig. 5 for visual inspection of the results on sample data. The proposed method remains the top performer, in terms of both quantitative metric and visual quality.

Table 3. Quantitative evaluation of different methods for fan beam CT reconstruction.

		FBP	TV	FBPConvNet	Learned-PD	LEARN	Proposed
180 views	RMSE	85.87 ± 8.88	75.24 ± 2.80	43.14 ± 3.83	50.65 ± 3.04	45.27 ± 2.93	**28.16 ± 3.44**
	PSNR	31.11 ± 0.87	32.25 ± 0.50	37.08 ± 0.76	35.67 ± 0.52	36.65 ± 0.57	**40.91 ± 1.06**
45 views	RMSE	92.81 ± 9.07	78.10 ± 2.13	46.30 ± 5.52	62.54 ± 4.88	53.91 ± 3.04	**33.86 ± 3.97**
	PSNR	30.43 ± 0.83	31.90 ± 0.53	36.49 ± 0.99	33.85 ± 0.68	35.13 ± 0.50	**39.57 ± 1.01**
30 views	RMSE	103.44 ± 8.59	82.34 ± 2.04	52.84 ± 6.24	88.78 ± 10.07	63.36 ± 4.02	**38.95 ± 4.77**
	PSNR	29.48 ± 0.71	31.44 ± 0.51	35.34 ± 0.98	30.83 ± 0.96	33.73 ± 0.55	**38.07 ± 1.06**

Ablation Study. This ablation study is for evaluating the performance gain brought by three components in sparse view CT reconstruction: (1) CNN-based upsampling vs. bilinear, (2) multi-scale structure and (3) learnable interpolating weights $\{W_i\}$ vs. fixed weights. The study is conducted on noise-free sample data with 180 views. See Fig. 6 for the comparison of different network architecture with individual component being replaced. It can be seen all three components makes noticeable contributions toward the performance gain. In addition, we also evaluate the impact of the values of the hyper-parameter K to the performance. The experiments are conducted on the noisy data of 45 views with 10% noise level. It can be seen from Table 4 that the configuration with $K = 2, 4, 6, 8$ from low to high resolution achieves better performance than other configurations.

Methods	Fixed Weight	Single-Scale	Bilinear Upsample	Proposed	True
180 views RMSE	28.8885	27.3821	19.5619	12.6809	

Fig. 6. Reconstruction results of different NNs for ablation study.

Table 4. Quantitative evaluations of the results from different configurations on K.

K	2,2,2,2	4,4,4,4	6,6,6,6	8,8,8,8	2,4,6,8
RMSE	26.32 ± 2.81	25.53 ± 3.84	22.88 ± 2.57	21.34 ± 2.49	21.05 ± 2.39
PSNR	41.39 ± 0.94	41.70 ± 1.29	42.61 ± 0.96	43.22 ± 1.00	43.34 ± 0.98

4 Conclusion

This paper presented a multi-scale DNN for sparse view CT image reconstruction, whose key part is to learn an interpolation scheme that converts discrete Fourier transform on polar coordinates to its counterpart on Cartesian coordinates. The proposed method not only provide SOTA performance on CT image reconstruction, but also is very computationally efficient. In future, we would like to extend the proposed method to the case of 3D CT imaging geometry.

Acknowledgment. This work was supported by NSFC (No.11771288, No.12090024), Shanghai Municipal Science and Technology Major Project (2021SHZDZX0102) and Singapore MOE Academic Research Fund (MOE2017-T2-2-156, R-146-000-315-114).

References

1. Adler, J., Öktem, O.: Learned primal-dual reconstruction. IEEE Trans. Med. Imaging **37**(6), 1322–1332 (2018)
2. Buzug, T.M.: Computed tomography. In: Springer Handbook of Medical Technology, pp. 311–342. Springer, Berlin (2011)
3. Chen, H., et al.: LEARN: learned experts' assessment-based reconstruction network for sparse-data CT. IEEE Trans. Med. Imaging **37**(6), 1333–1347 (2018)
4. Chen, H., et al.: Low-dose CT with a residual encoder-decoder convolutional neural network. IEEE Trans. Med. Imaging **36**(12), 2524–2535 (2017)
5. Ding, Q., Chen, G., Zhang, X., Huang, Q., Ji, H., Gao, H.: Low-dose CT with deep learning regularization via proximal forward–backward splitting. Phys. Med. Biol. **65**(12), 125009 (2020)
6. Ding, Q., Nan, Y., Gao, H., Ji, H.: Deep learning with adaptive hyper-parameters for low-dose CT image reconstruction. IEEE Trans. Comput. Imaging, 1–1 (2021). https://doi.org/10.1109/TCI.2021.3093003

7. Dong, B., Shen, Z., et al.: MRA based wavelet frames and applications. IAS Lecture Notes Series, Summer Program on "The Mathematics of Image Processing", Park City Mathematics Institute. **19** (2010)
8. Gupta, H., Jin, K.H., Nguyen, H.Q., McCann, M.T., Unser, M.: CNN-based projected gradient descent for consistent CT image reconstruction. IEEE Trans. Med. Imaging **37**(6), 1440–1453 (2018)
9. Hara, A.K., Paden, R.G., Silva, A.C., Kujak, J.L., Lawder, H.J., Pavlicek, W.: Iterative reconstruction technique for reducing body radiation dose at CT: feasibility study. Am. J. Roentgenol. **193**(3), 764–771 (2009)
10. He, J., et al.: Optimizing a parameterized plug-and-play ADMM for iterative low-dose CT reconstruction. IEEE Trans. Med. Imaging **38**(2), 371–382 (2018)
11. Jia, X., Dong, B., Lou, Y., Jiang, S.B.: GPU-based iterative cone-beam CT reconstruction using tight frame regularization. Phys. Med. Biol. **56**(13), 3787 (2011)
12. Jin, K.H., McCann, M.T., Froustey, E., Unser, M.: Deep convolutional neural network for inverse problems in imaging. IEEE Trans. Med. Imaging **26**(9), 4509–4522 (2017)
13. Katsura, M., et al.: Model-based iterative reconstruction technique for radiation dose reduction in chest CT: comparison with the adaptive statistical iterative reconstruction technique. Eur. Radiol. **22**(8), 1613–1623 (2012)
14. Radon, J.: 1.1 Über die Bestimmung von Funktionen durch ihre Integralwerte längs gewisser Mannigfaltigkeiten. Ber. Verh. Sächs. Akad. Wiss., Math. -Nat. KI. 69, 262–277 (1917)
15. Sidky, E.Y., Pan, X.: Image reconstruction in circular cone-beam computed tomography by constrained, total-variation minimization. Phys. Med. Biol. **53**(17), 4777 (2008)
16. Silva, A.C., Lawder, H.J., Hara, A., Kujak, J., Pavlicek, W.: Innovations in CT dose reduction strategy: application of the adaptive statistical iterative reconstruction algorithm. Am. J. Roentgenol. **194**(1), 191–199 (2010)
17. Sun, J., Li, H., Xu, Z., et al.: Deep ADMM-Net for compressive sensing MRI. In: Advances in Neural Information Processing Systems, pp. 10–18 (2016)
18. Wang, G.: A perspective on deep imaging. IEEE Access **4**, 8914–8924 (2016)
19. Xu, Q., Yu, H., Mou, X., Zhang, L., Hsieh, J., Wang, G.: Low-dose X-ray CT reconstruction via dictionary learning. IEEE Trans. Med. Imaging **31**(9), 1682–1697 (2012)
20. Ye, J.C., Han, Y., Cha, E.: Deep convolutional framelets: a general deep learning framework for inverse problems. SIAM J. Imaging Sci. **11**(2), 991–1048 (2018)
21. Zeng, G.L.: Medical Image Reconstruction: A Conceptual Tutorial. Springer, New York (2010)
22. Zhang, X.-Q., Froment, J.: Constrained total variation minimization and application in computerized tomography. In: Rangarajan, A., Vemuri, B., Yuille, A.L. (eds.) EMMCVPR 2005. LNCS, vol. 3757, pp. 456–472. Springer, Heidelberg (2005). https://doi.org/10.1007/11585978_30
23. Zhang, X., Froment, J.: Total variation based fourier reconstruction and regularization for computer tomography. In: IEEE Nuclear Science Symposium Conference Record, 2005, vol. 4, pp. 2332–2336. IEEE (2005)

U-DuDoNet: Unpaired Dual-Domain Network for CT Metal Artifact Reduction

Yuanyuan Lyu[1], Jiajun Fu[2], Cheng Peng[3], and S. Kevin Zhou[4,5(✉)]

[1] Z2Sky Technologies Inc., Suzhou, China
[2] Beijing University of Posts and Telecommunications, Beijing, China
[3] Johns Hopkins University, Baltimore, MD, USA
[4] Medical Imaging, Robotics, and Analytic Computing Laboratory and Engineering (MIRACLE) Center, School of Biomedical Engineering and Suzhou Institute for Advanced Research, University of Science and Technology of China, Suzhou, China
[5] Key Lab of Intelligent Information Processing of Chinese Academy of Sciences (CAS), Institute of Computing Technology, CAS, Beijing, China

Abstract. Recently, both supervised and unsupervised deep learning methods have been widely applied on the CT metal artifact reduction (MAR) task. Supervised methods such as Dual Domain Network (Du-DoNet) work well on simulation data; however, their performance on clinical data is limited due to domain gap. Unsupervised methods are more generalized, but do not eliminate artifacts completely through the sole processing on the image domain. To combine the advantages of both MAR methods, we propose an unpaired dual-domain network (U-DuDoNet) trained using unpaired data. Unlike the artifact disentanglement network (ADN) that utilizes multiple encoders and decoders for disentangling content from artifact, our U-DuDoNet directly models the artifact generation process through additions in both sinogram and image domains, which is theoretically justified by an additive property associated with metal artifact. Our design includes a self-learned sinogram prior net, which provides guidance for restoring the information in the sinogram domain, and cyclic constraints for artifact reduction and addition on unpaired data. Extensive experiments on simulation data and clinical images demonstrate that our novel framework outperforms the state-of-the-art unpaired approaches.

Keywords: Metal artifact reduction · Dual-domain learning · Unpaired learning

1 Introduction

Computed tomography (CT) reveals the underlying anatomical structure within the human body. However, when a metallic object is present, metal artifacts

Electronic supplementary material The online version of this chapter (https://doi.org/10.1007/978-3-030-87231-1_29) contains supplementary material, which is available to authorized users.

© Springer Nature Switzerland AG 2021
M. de Bruijne et al. (Eds.): MICCAI 2021, LNCS 12906, pp. 296–306, 2021.
https://doi.org/10.1007/978-3-030-87231-1_29

appear in the image because of beam hardening, scatters, photon starvation, etc. [2,3,19], degrading the image quality and limiting its diagnostic value.

With the success of deep learning in medical image processing [27,28], deep learning has been used for metal artifact reduction (MAR). Single-domain networks [5,21,25] have been proposed to address MAR with success. Lin *et al.* are the first to introduce dual-domain network (DuDoNet) to reduce metal artifacts in the sinogram and image domain jointly and DuDoNet shows further advantages over single-domain networks and traditional approaches [4,10,12,13,19]. Following this work, variants of the dual-domain architecture [18,22,24] have been designed. However, all the above-mentioned networks are supervised and rely on paired clean and metal-affected images. Since such clinical data is hard to acquire, simulation data are widely used in practice. Thus, supervised models may over-fit to simulation and *do not generalize well to real clinical data.*

Learning from unpaired, real data is thus of interest. To this, Liao *et al.* propose ADN [14,16], which separates content and artifact in the latent spaces with multiple encoders and decoders and induces unsupervised learning via various forms of image generation and specialized loss functions. An artifact consistency loss is introduced to retain anatomical preciseness during MAR. The loss is based on the assumption that metal artifacts are additive. Later on, Zhao *et al.* [26] design a simple reused convolutional network (RCN) of encoders and decoders to recurrently generating both artifact and non-artifact images. RCN also adopts the additive metal artifacts assumption. However, *neither of the works has theoretically proved the property.* Moreover, without the aid of processing in sinogram domain, both methods have *limited effect on removing strong metal artifacts,* such as the dark and bright bands around metal objects.

In this work, we analytically derive the additive property associated with metal artifacts and propose an unpaired dual-domain MAR network (U-DuDoNet). Without using complicated encoders and decoders, our network *directly estimates* the additive component of metal artifacts, jointly using two U-Nets on two domains: a sinogram-based estimation net (S-Net) and an image-based estimation net (I-Net). S-Net first restores sinogram data and I-net removes additional streaky artifacts. Unpaired learning is achieved with cyclic artifact reduction and synthesis processes. Strong metal artifacts can be reduced in the sinogram domain with prior knowledge. Specifically, sinogram enhancement is guided by a self-learned sinogram completion network (P-Net) with clean images. Both simulation and clinical data show our method outperforms competing unsupervised approaches and has better generalizability than supervised approaches.

2 Additive Property for Metal Artifacts

Here, we prove metal artifacts are inherently additive up to mild assumptions. The CT image intensity represents the attenuation coefficient. Let $X^c(E)$ be a normal attenuation coefficient image at energy level E. In a polychromatic x-ray system, the ideal projection data (sinogram) S^c can be expressed as,

$$S^c = -ln \int \eta(E)e^{-\mathcal{P}(X^c(E))}dE, \tag{1}$$

where $\mathcal{P}$ and $\eta(E)$ denote forward projection (FP) operator and fractional energy at E. Comparing with metal, the attenuation coefficient of normal body tissue is almost constant with respect to E, thus we have $X^c = X^c(E)$ and $S^c = \mathcal{P}(X^c)$. Without metal, filtered back projection (FBP) operator $\mathcal{P}^*$ provides a clean CT image I^c as a good estimation of X^c, $I^c = \mathcal{P}^*(S^c) = \mathcal{P}^*(\mathcal{P}(X^c))$.

Metal artifacts appear mainly because of beam hardening. An attenuation coefficient image with metal $X^a(E)$ can be split into a relatively constant image without metal X^{ac} and a metal-only image $X^m(E)$ varies rapidly against E, $X^a(E) = X^{ac} + X^m(E)$. Often $X^m(E)$ is locally constrained. The contaminated sinogram S^a can be given as,

$$S^a = -ln \int \eta(E)e^{-\mathcal{P}(X^a(E))}dE = \mathcal{P}(X^{ac}) - ln \int \eta(E)e^{-\mathcal{P}(X^m(E))}dE. \tag{2}$$

And the reconstructed metal-affected CT image I^a is,

$$I^a = \mathcal{P}^*(\mathcal{P}(X^{ac})) - \mathcal{P}^*(ln \int \eta(E)e^{-\mathcal{P}(X^m(E))}dE) = I^{ac} + F(X^m(E)). \tag{3}$$

Here, $\mathcal{P}^*(\mathcal{P}(X^{ac}))$ is the MAR image I^{ac} and the second term introduces streaky and band artifacts, which is a function of $X^m(E)$. Since metal artifacts are caused only by $X^m(E)$, we can create a plausible artifact-affected CT I^{ca} by adding the artifact term to an arbitrary, clean CT image: $I^{ca} = I^c + F(X^m(E))$.

3   Methodology

Figure 1b shows the proposed cyclical MAR framework. In Phase I, our framework first estimates artifact components a_S and a_I through U-DuDoNet from I^a, see Fig. 1a. In Phase II, based on the additive metal artifact property (Sect. 2), plausible clean image I^{ac} and metal-affected image I^{ca} could be generated, $I^{ac} = I^a - a_S - a_I$, $I^{ca} = I^c + a_S + a_I$. Then, the artifact components should be removable from I^{ca} by U-DuDoNet, resulting in a'_S and a'_I. In the end, reconstructed images I^{aca}, I^{cac} can be obtained through subtracting or adding the artifact components, $I^{aca} = I^{ac} + a'_S + a'_I$, $I^{cac} = I^{ca} - a'_S - a'_I$.

3.1   Network Architecture

Artifact Component Estimation in Sinogram Domain. Strong metal artifacts like dark and bright bands can not be suppressed completely by image domain processing, while metal artifacts are inherently local in the sinogram domain. Thus, we aim to reduce metal shadows by sinogram enhancement.

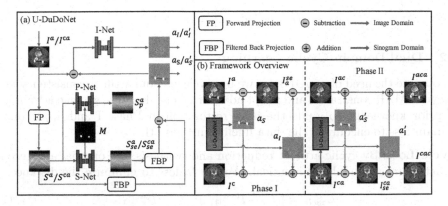

Fig. 1. (a) Unpaired dual-domain network (U-DuDoNet). The input is either a real artifact-affected image I^a, or a synthetic artifact-affected image I^{ca}. (b) The proposed cyclical MAR framework. The notations in Phase I and Phase II are marked with black and brown, respectively. (Color figure online)

First, we acquire metal corrupted sinograms (S^a and S^{ca}) by forward projecting I^a and I^{ca}: $S^a = \mathcal{P}(I^a)$, $S^{ca} = \mathcal{P}(I^{ca})$. Then, we use a pre-trained prior net (P-Net) to guide the sinogram restoration process. P-Net is an inpainting net ϕ_P that treats the metal-affected area in sinogram as missing and aims to complete it, i.e., $S_p^a = \phi_P(S^a \odot M_t, M_t)$. Here M_t denotes a binary metal trace, $M_t = \delta(\mathcal{P}(M))$, where M is a metal mask and $\delta(\cdot)$ is a binary indicator function. We adopt a mask pyramid U-Net as ϕ_P from [15,17]. To train ϕ_P, we artificially inject masks into clean sinograms S_c. We use $\{S^c \odot M_t, S_c\}$ as input and groundtruth pairs and apply L_1 loss.

Then, we use a sinogram network (S-Net) to predict enhanced sinogram S_{se}^a, S_{se}^{ca} from S^a, S^{ca}, respectively.

$$S_{se}^a = \phi_S(S^a, \mathcal{P}(M)) \odot M_t + S^a, S_{se}^{ca} = \phi_S(S^{ca}, \mathcal{P}(M)) \odot M_t + S^{ca}, \quad (4)$$

where ϕ_S represents a U-Net [20] of depth 2. Residual learning [8] is applied to ease the training process, and singoram prediction is limited to the M_t region. To prevent information loss from discrete operators, we obtain the sinogram artifact component as a difference image between reconstructed input image and reconstructed enhanced sinogram,

$$a_S = \mathcal{P}^*(S^a) - \mathcal{P}^*(S_{se}^a), a_S' = \mathcal{P}^*(S^{ca}) - \mathcal{P}^*(S_{se}^{ca}). \quad (5)$$

Artifact Component Estimation in Image Domain. As sinogram data inconsistency leads to secondary artifacts in the whole image, we further use an image domain network (I-Net) to reduce newly introduced and other streaky artifacts. Let ϕ_I denote I-Net, which is a 5-depth U-Net. First, sinogram enhanced images are obtained by subtracting sinogram artifact component from corrupted images, $I_{se}^a = I^a - a_S$, $I_{se}^{ca} = I^{ca} - a_S'$. Then, I-Net takes a sinogram enhanced image, and outputs an artifact component in image domain (a_I or a_I'),

$$a_I = \phi_I(I_{se}^a), a_I' = \phi_I(I_{se}^{ca}). \tag{6}$$

3.2 Dual-Domain Cyclic Learning

To obviate the need of paired data, we use cycle loss and artifact consistency loss as cyclic MAR constraints and adopt adversarial loss. Besides, we take advantage of prior knowledge to guide the data restoration in Phase I and apply dual-domain loss to encourage the data fidelity in Phase II.

Cycle Loss. By cyclic artifact reduction and synthesis, the original and reconstructed images should be identical. We use L_1 loss to minimize the distance,

$$\mathcal{L}_{cycle} = ||I^a - I^{aca}||_1 + ||I^c - I^{cac}||_1. \tag{7}$$

Artifact Consistency Loss. To ensure the artifacts components added to I^c could be removed completely when applying the same network on I^{ca}, the artifact components estimated from I^a and I^{ca} should be the same,

$$\mathcal{L}_{art} = ||a_S - a_S'||_1 + ||a_I - a_I'||_1. \tag{8}$$

Adversarial Loss. The synthetic images, I^{ca} and I^{ac}, should be indistinguishable to input images. Since paired groundtruth is not available, we adopt Patch-GAN [9] as discriminators D^a and D^c to apply adversarial learning. Since metal affected images always contain streaks, we add gradient image generated by Sobel operator $\bigtriangledown$ as an additional channel of the input of D^a and D^c to achieve better performance. The loss would be written as,

$$\mathcal{L}_{adv} = \mathbb{E}[\log D^a(I^a, \bigtriangledown I^a)] + \mathbb{E}[1 - \log D^a(I^{ca}, \bigtriangledown I^{ca})] \\ + \mathbb{E}[\log D^c(I^c, \bigtriangledown I^c)] + \mathbb{E}[1 - \log D^c(I^{ac}, \bigtriangledown I^{ac})]. \tag{9}$$

Fidelity Loss. To learning artifact reduction from generated I^{ca}, we minimize the distances between S_{se}^{ca} and S^c, I_{se}^{ca} and I^c,

$$\mathcal{L}_{fed} = ||S_{se}^{ca} - S^c||_1 + ||I_{se}^{ca} - I^c||_1. \tag{10}$$

Prior Loss. Inspired by DuDoNet [17], sinogram inpainting network provides smoothed estimation of sinogram data within M_t. Thus, we use a Gaussian blur operation $\mathcal{G}_{\sigma_s}$ with a scale of σ_s and L_2 loss to minimize the distance between blurred prior and enhanced sinogram. Meanwhile, inspired by [11], blurred sinogram enhanced image also serves as an good estimation of blurred MAR image. Also, we minimize the distance between low-pass versions of sinogram enhanced and MAR images with a Gaussian blur operation $\mathcal{G}_{\sigma_i}$ to stabilize the unsupervised training. The prior loss could be formulated as,

$$\mathcal{L}_{prior} = ||\mathcal{G}_{\sigma_s}(S_p^a) - \mathcal{G}_{\sigma_s}(S_{se}^a)||_2 + ||\mathcal{G}_{\sigma_i}(I_{se}^a) - \mathcal{G}_{\sigma_i}(I^{ac})||_2. \tag{11}$$

The overall objective function is the weighted sum of all the above losses, we empirically set the weight of $\mathcal{L}_{adv}$ to 1, and the weights of $\mathcal{L}_{prior}$, $\mathcal{L}_{art}$ to 10, and the weights of the other losses to 100. We set σ_s to 1 and σ_i to 3 in $\mathcal{L}_{prior}$.

4   Experiment

4.1   Experimental Setup

Datasets. Following [14], we evaluate our model on both simulation and clinical data. For simulation data, we generate images with metal artifacts using the method in [25]. From DeepLesion [23], we randomly choose 3,984 clean images combining with 90 metal masks for training and additional 200 clean images combining with 10 metal masks for testing. For unsupervised training, we spilt 3,984 images into two groups and randomly select one metal corrupted image and one clean image. For clinical data, we select 6,146 images with artifacts and 21,002 clean images for training from SpineWeb [6,7]. Additional 124 images with metal are used for testing.

Implementation and Metrics. We implement our model with the PyTorch framework and differential FP and FBP operators with ODL library [1]. We train the model for 50 epochs using an Adam optimizer with a learning rate of 1×10^{-4} and a batch size of 2. For clinical data, we train another model using unpaired images for 20 epochs and the metal mask is segmented with a threshold of 2,500 HU. We use peak signal-to-noise ratio (PSNR) and structural similarity index (SSIM) to evaluate the corrected image.

Baselines. We compare U-DuDoNet with multiple state-of-the-art (SOTA) MAR methods. DuDoNet [17], DuDoNet++ [18], DSCIP [24] and DAN-Net [22] are supervised methods which are trained with simulation data and tested on both simulation and clinical data. DuDoNet, DuDoNet++ and DAN-Net share the same SE-IE architecture with an image enhancement (IE) following sinogram enhancement (SE) network, while DSCIP adopts an IE-SE architecture that predicts a prior image first then outputs the sinogram enhanced image. RCN [26] and ADN [14] are unsupervised and can be trained and tested on each dataset.

Table 1. Quantitative comparison of different SOTA methods. Statistical significance tests are performed between ours and each method using paired t-test (*** indicates $p < 0.001$) and the effect size are measured by Cohen's d.

	Method	PSNR(dB)/SSIM	p value	Cohen's d
Metal	-	27.23/0.692	-/-	-/-
Supervised	DuDoNet [17]	36.95/0.927	***/***	1.33/0.31
	DuDoNet++ [18]	37.65/**0.953**	***/***	2.93/1.51
	DSCIP [24]	29.22/0.624	***/***	4.65/5.32
	DAN-Net [22]	**39.73**/0.944	***/***	3.83/1.58
Unsupervised	RCN [26]	32.98/0.918	***/***	1.26/1.19
	ADN [14]	33.81/0.926	***/***	0.54/0.51
	U-DuDoNet (ours)	**34.54/0.934**	-/-	-/-

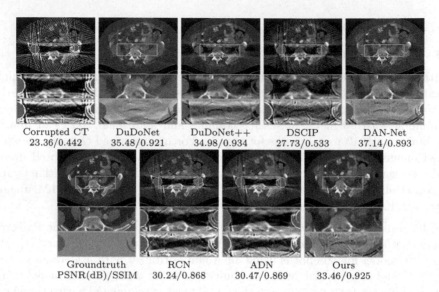

Fig. 2. Visual comparisons with the SOTA methods on simulation data. Zoomed images and difference images between groundtruth and MAR images are listed below CT images. The display window are [−200, 600] and [−300, 300] HU for original and difference images. Red pixels stand for metal implants.

4.2 Comparison on Simulated and Real Data

Simulated Data. From Table 1, we observe that DAN-Net achieves the highest PSNR and DuDoNet++ achieves the highest SSIM. All methods with SE-IE architecture outperform DSCIP. The reason is image enhancement network helps recover details and bridge the gap between real images and reconstructed images. Among all the unsupervised methods, our model attains the best performance, with an improvement of 0.73 dB in PSNR compared with ADN. Besides, our model runs as fast as the supervised dual-domain models but slower than image-domain unsupervised models. Figure 2 shows the visual comparisons of a case. The zoomed subfigure shows that metallic implants induce dark bands in the region between two implants or along the direction of dense metal pixels. Learning from linearly interpolated (LI) sinogram, DuDoNet removes the dark bands and streaky artifacts completely but smooths out the details around the metal. DuDoNet++ and DSCIP could not remove the dark bands completely as they learn from the corrupted images. DAN-Net contains fewer streaks than DuDoNet++ and DSCIP since it recovers from a blended sinogram of LI and metal-affected data. Among all the unsupervised methods, only our model recovers the bony structure in dark bands and contains least streaks.

Real Data. Figure 3 shows a clinical CT image with two rods on each side of the spinous process of a vertebra. The implants induce severe artifacts, which make some bone part invisible. DuDoNet recovers the bone but introduces strong secondary artifacts. DuDoNet++, DSCIP, and DAN-Net do not generalize to clinical data as the dark band remains in the MAR images. Besides, all the supervised methods output smoothed images as training images from DeepLesion might be reconstructed by a soft tissue kernel. The MAR images of the unsupervised method could retain the sharpness of the original image. But, RCN and ADN do not reduce the artifacts completely or retain the integrity of bone structures near the metal as these structures might be confused with artifacts. Our model removes the dark band while retaining the structures around the rods. More visual comparisons are in the supplemental material.

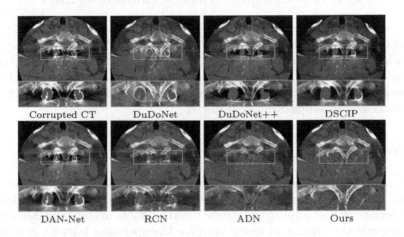

Fig. 3. Visual comparisons with the SOTA methods on clinical data.

Table 2. Quantitative comparison of different variants of our model. Statistical significance tests are performed between M3 and each model.

	PSNR(dB)/SSIM	p value	Cohen'd
M1 (Image Domain)	32.97/0.927	***/***	1.45/0.76
M2 (Dual-domain)	33.30/0.930	***/***	1.70/0.71
M3 (Dual-domain + prior)	**34.54/0.934**	-/-	-/-

4.3 Ablation Study

We evaluate the effectiveness of different components in our full model. Table 2 shows the configuration of our ablation models. Briefly, M1 refers to the model with I-Net and $\mathcal{L}_{cycle}$, $\mathcal{L}_{adv}$, M2 refers to M1 plus S-Net, $\mathcal{L}_{gt}$, $\mathcal{L}_{art}$, and M3 refers to M2 plus P-Net, $\mathcal{L}_{prior}$. As shown in Table 2 and Fig. 4, M1 has the capability of MAR in image domain, but strong artifacts like dark bands and streaks remain in the output image. Dual-domain learning increases the PSNR

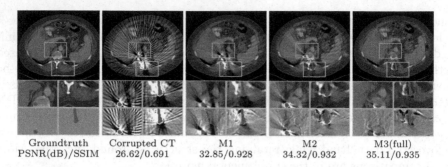

| Groundtruth | Corrupted CT | M1 | M2 | M3(full) |
| PSNR(dB)/SSIM | 26.62/0.691 | 32.85/0.928 | 34.32/0.932 | 35.11/0.935 |

Fig. 4. Visual comparisons of different variants of our model.

by 0.33 dB, and the dark bands are partial removed in the corrected image, but streaks show up as sinogram enhancement might be not perfect. With the aid of prior knowledge, M3 could remove the dark bands completely and further suppresses the secondary artifacts. Visualisations of the intermediate outputs of M3 can be seen in supplementary material.

5   Conclusion

In this paper, we present an unpaired dual-domain network (U-DuDoNet) that exploits the additive property of artifact modeling for metal artifact reduction. In particular, we first remove the strong metal artifacts in sinogram domain and then suppress the streaks in image domain. Unsupervised learning is achieved via cyclic additive artifact modeling, i.e. we try to remove the same artifact after inducing artifact in an unpaired clean image. We also apply prior knowledge to guide data restoration. Qualitative evaluations and visual comparisons demonstrate that our model yields better MAR performance than competing methods. Moreover, our model shows great potential when applied to clinical images.

References

1. Adler, J., Kohr, H., Oktem, O.: Operator discretization library (ODL) (2017). https://github.com/odlgroup/odl
2. Barrett, J.F., Keat, N.: Artifacts in CT: recognition and avoidance. Radiographics **24**(6), 1679–1691 (2004)
3. Boas, F.E., Fleischmann, D.: CT artifacts: causes and reduction techniques. Imag. Med. **4**(2), 229–240 (2012)
4. Chang, Z., Ye, D.H., Srivastava, S., Thibault, J.B., Sauer, K., Bouman, C.: Prior-guided metal artifact reduction for iterative x-ray computed tomography. IEEE Trans. Med. Imag. **38**(6), 1532–1542 (2018)
5. Ghani, M.U., Karl, W.C.: Fast enhanced CT metal artifact reduction using data domain deep learning. IEEE Trans. Comput. Imag. **6**, 181–193 (2019)

6. Glocker, B., Feulner, J., Criminisi, A., Haynor, D.R., Konukoglu, E.: Automatic localization and identification of vertebrae in arbitrary field-of-view CT scans. In: International Conference on Medical Image Computing and Computer-Assisted Intervention. pp. 590–598. Springer (2012)

7. Glocker, B., Zikic, D., Konukoglu, E., Haynor, D.R., Criminisi, A.: Vertebrae localization in pathological spine CT via dense classification from sparse annotations. In: International Conference on Medical Image Computing and Computer-Assisted Intervention. pp. 262–270. Springer (2013)

8. He, K., Zhang, X., Ren, S., Sun, J.: Deep residual learning for image recognition. In: Proceedings of the IEEE conference on computer vision and pattern recognition. pp. 770–778 (2016)

9. Isola, P., Zhu, J.Y., Zhou, T., Efros, A.A.: Image-to-image translation with conditional adversarial networks. In: Proceedings of the IEEE conference on computer vision and pattern recognition. pp. 1125–1134 (2017)

10. Jin, P., Bouman, C.A., Sauer, K.D.: A model-based image reconstruction algorithm with simultaneous beam hardening correction for X-ray CT. IEEE Trans. Comput. Imag. 1(3), 200–216 (2015)

11. Jin, X., Chen, Z., Lin, J., Chen, Z., Zhou, W.: Unsupervised single image deraining with self-supervised constraints. In: 2019 IEEE International Conference on Image Processing (ICIP), pp. 2761–2765. IEEE (2019)

12. Kalender, W.A., Hebel, R., Ebersberger, J.: Reduction of CT artifacts caused by metallic implants. Radiology 164(2), 576–577 (1987)

13. Karimi, S., Martz, H., Cosman, P.: Metal artifact reduction for CT-based luggage screening. J. X-ray Sci. Technol. 23(4), 435–451 (2015)

14. Liao, H., Lin, W., Zhou, S.K., Luo, J.: ADN: artifact disentanglement network for unsupervised metal artifact reduction. IEEE Trans. Med. Imag. (2019). https://doi.org/10.1109/TMI.2019.2933425

15. Liao, H., et al.: Generative mask pyramid network for CT/CBCT metal artifact reduction with joint projection-sinogram correction. In: International Conference on Medical Image Computing and Computer-Assisted Intervention. pp. 77–85. Springer (2019)

16. Liao, H., Lin, W.A., Yuan, J., Zhou, S.K.Z., Luo, J.: Artifact disentanglement network for unsupervised metal artifact reduction. In: International Conference on Medical Image Computing and Computer-Assisted Intervention (MICCAI) (2019)

17. Lin, W.A., et al.: DudoNet: dual domain network for CT metal artifact reduction. In: Proceedings of the IEEE Conference on Computer Vision and Pattern Recognition, pp. 10512–10521 (2019)

18. Lyu, Y., Lin, W.A., Liao, H., Lu, J., Zhou, S.K.: Encoding metal mask projection for metal artifact reduction in computed tomography. In: International Conference on Medical Image Computing and Computer-Assisted Intervention, pp. 147–157. Springer (2020)

19. Meyer, E., Raupach, R., Lell, M., Schmidt, B., Kachelrieß, M.: Normalized metal artifact reduction (RMAT) in computed tomography. Med. Phys. 37(10), 5482–5493 (2010)

20. Ronneberger, O., Fischer, P., Brox, T.: U-net: Convolutional networks for biomedical image segmentation. In: International Conference on Medical image computing and computer-assisted intervention, pp. 234–241. Springer (2015)

21. Wang, J., Zhao, Y., Noble, J.H., Dawant, B.M.: Conditional generative adversarial networks for metal artifact reduction in CT images of the ear. In: International Conference on Medical Image Computing and Computer-Assisted Intervention, pp. 3–11. Springer (2018)

22. Wang, T., et al.: Dan-net: dual-domain adaptive-scaling non-local network for CT metal artifact reduction. arXiv preprint arXiv:2102.08003 (2021)
23. Yan, K., et al..: Deep lesion graphs in the wild: relationship learning and organization of significant radiology image findings in a diverse large-scale lesion database. In: Proceedings of the IEEE Conference on Computer Vision and Pattern Recognition, pp. 9261–9270 (2018)
24. Yu, L., Zhang, Z., Li, X., Xing, L.: Deep sinogram completion with image prior for metal artifact reduction in CT images. IEEE Trans. Med. Imag. 40(1), 228–238 (2020)
25. Zhang, Y., Yu, H.: Convolutional neural network based metal artifact reduction in X-ray computed tomography. IEEE Trans. Med. Imag. 37(6), 1370–1381 (2018). https://doi.org/10.1109/TMI.2018.2823083
26. Zhao, B., Li, J., Ren, Q., Zhong, Y.: Unsupervised reused convolutional network for metal artifact reduction. In: International Conference on Neural Information Processing. pp. 589–596. Springer (2020)
27. Zhou, S.K., et al.: A review of deep learning in medical imaging: imaging traits, technology trends, case studies with progress highlights, and future promises. In: Proceedings of the IEEE (2021)
28. Zhou, S.K., Rueckert, D., Fichtinger, G.: Handbook of Medical Image Computing and Computer Assisted Intervention. Academic Press, London (2019)

Task Transformer Network for Joint MRI Reconstruction and Super-Resolution

Chun-Mei Feng[1,2], Yunlu Yan[1], Huazhu Fu[2], Li Chen[3], and Yong Xu[1(✉)]

[1] Shenzhen Key Laboratory of Visual Object Detection and Recognition, Harbin
Institute of Technology, Shenzhen, China
[2] Inception Institute of Artificial Intelligence, Abu Dhabi, UAE
[3] First Affiliated Hospital with Nanjing Medical University, Nanjing, China
https://github.com/chunmeifeng/T2Net

Abstract. The core problem of Magnetic Resonance Imaging (MRI) is
the trade off between acceleration and image quality. Image reconstruction
and super-resolution are two crucial techniques in Magnetic Resonance
Imaging (MRI). Current methods are designed to perform these tasks
separately, ignoring the correlations between them. In this work, we propose an end-to-end task transformer network (T^2Net) for joint MRI reconstruction and super-resolution, which allows representations and feature
transmission to be shared between multiple task to achieve higher-quality,
super-resolved and motion-artifacts-free images from highly undersampled and degenerated MRI data. Our framework combines both reconstruction and super-resolution, divided into two sub-branches, whose features are expressed as queries and keys. Specifically, we encourage joint
feature learning between the two tasks, thereby transferring accurate
task information. We first use two separate CNN branches to extract
task-specific features. Then, a task transformer module is designed to
embed and synthesize the relevance between the two tasks. Experimental
results show that our multi-task model significantly outperforms advanced
sequential methods, both quantitatively and qualitatively.

Keywords: Multi-task learning · MRI reconstruction ·
Super-resolution

1 Introduction

Magnetic resonance imaging (MRI) is a popular diagnostic modality. However,
the physics behind its data acquisition process makes it inherently slower than
other methods such as computed tomography (CT) or X-rays. Therefore, improving the acquisition speed of MRI has been an important research goal for decades.
MRI reconstruction and super-resolution (SR) are two main methods for this,
where the former accelerates MRI by reducing the k-space sampling rate, and the

C.-M. Feng and Y. Yan are contributed equally to this work. This work was done
during the internship of C.-M. Feng at Inception Institute of Artificial Intelligence.

© Springer Nature Switzerland AG 2021
M. de Bruijne et al. (Eds.): MICCAI 2021, LNCS 12906, pp. 307–317, 2021.
https://doi.org/10.1007/978-3-030-87231-1_30

latter achieves a high-resolution (HR) image by restoring a single degenerated low-resolution (LR) image [6].

Outstanding contributions have been made in both areas [5,7,9]. Specifically, compressed sensing (CS) [13], low-rank [25], dictionary learning [23,30], and manifold fitting [20] techniques utilize various priors to overcome aliasing artifacts caused by the violation of the Shannon-Nyquist sampling theorem for MRI reconstruction. With the renaissance of deep neural networks, different convolutional neural network (CNN) approaches have also been developed for fast MRI reconstruction [8,33]. Typical examples include model-based unrolling methods, e.g., VN-Net [10], which generalizes the CS formulation to a variational model, and ADMM-Net, which is derived from the iterative procedures [29]; end-to-end learning methods, e.g., using U-Net as the basic framework to solve the problem of MRI reconstruction [11,22]; and generative adversarial networks (GANs) [19,28]. In addition to various network structures, series of convolutions based on the characteristics of MRI data have also been designed to solve the problem of MRI reconstruction [8,26]. For MRI SR, iterative algorithms (e.g., low rank or sparse representation) take image priors into account as regularization items and try to obtain a higher-quality image from a single LR image [27,32]. Similarly, CNN approaches have achieved state-of-the-art performance in SR [2,17]. For example, residual learning can be used to extract multi-scale information and obtain higher-quality images [21,24]. GAN-based methods have also been used to recover HR details from an LR input [3,18].

However, these works are designed to perform one specific function, i.e., train a single model to carry out the desired task. While acceptable performance can be achieved in this way, information that might help the model perform better in certain metrics is often ignored, since too much focus is given to one single task. In the real world, a network that can perform multiple tasks simultaneously is far preferable to a set of independent networks, as it can provide a more complete visual system. Since related tasks often share features, real-world tasks tend to have strong dependencies. Recently, multi-task learning has been successfully applied to various fields, including natural language processing [4], speech recognition [12] and computer vision [15]. By sharing representations between related tasks, the model can better generalize to the original task. Compared with standard single-task learning, multi-task models can express both shared and task-specific characteristics. In natural images, multi-task learning has been widely used for image enhancement [1,31]. However, current models directly incorporate different tasks into the network in a sequential manner, without exploring the features shared across the tasks.

Inspired by the powerful visual capabilities of transformer and multi-task learning, we propose an end-to-end task transformer network, named T^2Net, for multi-task learning, which integrates both MRI reconstruction and SR. Our contributions are three-fold: **First**, to the best of our knowledge, we are the first to introduce the transformer framework into multi-task learning for MRI reconstruction and SR. Our network allows representations to be shared between the two tasks, leveraging knowledge from one task to speed up the learning process in the other and increase the flexibility for sharing complementary features. **Second**, we develop a framework with two branches for expressing task-specific features and a task transformer module for transferring shared features. More specifically, the task transformer module includes relevance embedding, transfer attention and soft attention, which enable related tasks to share visual features. **Third**, we demonstrate that our multi-task model generates superior results compared to various sequential combinations of state-of-the-art MRI reconstruction and super-resolution models.

2   Method

2.1   Task Transformer Network

Let $\mathbf{y}$ be the complex-valued, fully sampled k-space. The corresponding fully sampled HR image with a size of $h \times w$ can be obtained by $\mathbf{x} = \mathcal{F}^{-1}(\mathbf{y})$, where $\mathcal{F}^{-1}$ is the inverse 2D fast Fourier transform (FFT). To accelerate the MRI acquisition, a binary mask operator M defining the Cartesian acquisition trajectory is used to select a subset of the k-space points. Therefore, the undersampled k-space is obtained by $\hat{\mathbf{y}} = M \odot \mathbf{y}$, where $\odot$ denotes element-wise multiplication. Accordingly, the zero-filled image can be expressed as $\hat{\mathbf{x}} = \mathcal{F}^{-1}(\hat{\mathbf{y}})$. In MRI super-resolution, to obtain the LR image $\mathbf{x}_{LR}$ with a size of $\frac{h}{s} \times \frac{w}{s}$ (s is the scale factor), we follow [3], first downgrading the resolution by truncating the outer part of $\mathbf{y}$ with a desired factor to obtain $\mathbf{y}_{LR}$, and then applying $\mathcal{F}$ to it. Therefore, if we apply downgrading to $\hat{\mathbf{x}}$, we will obtain the undersampled, degenerated MRI data for our multi-task input $\hat{\mathbf{x}}_{LR}$.

To effectively achieve higher-quality, motion-artifact-free images from highly undersampled and degenerated MRI data $\hat{\mathbf{x}}_{LR}$, we propose a simple and effective end-to-end framework, named the Task Transformer Network (T²Net). As shown in Fig. 1, our multi-task framework consists of three parts: an SR branch, a reconstruction (Rec) branch and a task transformer module. The first two branches are used to extract task-specific features, providing our network the ability to learn features tailored to each task. The task transformer module is then used to learn shared features, encouraging the network to learn a generalizable representation. As can be seen, the input of the two branches is the undersampled, degenerated MRI data $\hat{\mathbf{x}}_{LR}$, which contains motion artifacts and blurring effects. The output of the Rec branch is the LR motion-free image $\mathbf{x}'_{LR}$, while the output of the SR branch is our final desired high-quality, super-resolved and motion-free image $\mathbf{x}'$. Our framework can be approximated using neural networks by minimizing an ℓ_1 loss function:

$$\hat{\theta} = \underset{\theta_1, \theta_2}{\arg\min} \sum_{j}^{N} \left(\alpha \left\| \mathbf{x}^j - f_{cnn}^{SR} \left(\hat{\mathbf{x}}_{LR}^j \mid \theta_1 \right) \right\|_1 + \beta \left\| \mathbf{x}_{LR}^j - f_{cnn}^{Rec} \left(\hat{\mathbf{x}}_{LR}^j \mid \theta_2 \right) \right\|_1 \right), \quad (1)$$

where f_{cnn}^{SR} and f_{cnn}^{Rec} represent the mapping functions of the SR and Rec branches with parameters θ_1 and θ_2, respectively, and α and β are used to balance the

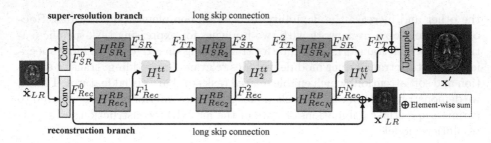

Fig. 1. Overview of the proposed multi-task framework, including an SR branch, a reconstruction (Rec) branch, and a task transformer module.

weights of the two branches. Note that with sufficient training data $\{\mathbf{x}^j, \hat{\mathbf{x}}_{LR}^j\}$ and the SGD algorithm, we can obtain well-trained weights $\hat{\theta}$.

SR Branch. Our SR branch is used to enlarge the image from an undersampled and degenerated input $\hat{\mathbf{x}}_{LR}$. As shown in Fig. 1, for an input image of size $\frac{h}{s} \times \frac{w}{s}$ with artifacts, a convolutional layer is used to extract the shallow feature F_{SR}^0 of the SR branch. Then we send it to the backbone of EDSR [14] to extract the SR features: $F_{SR}^1 = H_{SR_1}^{RB} \left(F_{SR}^0 \right)$, where $H_{SR_1}^{RB}$ represents the first Resblock in the SR branch. To enhance features from different tasks, we propose a task transformer module H^{tt} (Sect. 2.2), which transfers the motion-artifacts-free representation to the SR branch. Formally, we have

$$F_{TT}^i = H_i^{tt} \left(F_{SR}^i + F_{Rec}^i \right), \quad i = 1, 2, \ldots, N, \tag{2}$$

where N is the number of H^{tt}, F_{Rec}^i is the feature from the Rec branch (see Eq. (4)), and F_{SR}^i represents the i-th feature of the SR branch. The learned motion-artifacts-free representation F_{TT}^i is then sent to the following Resblock:

$$F_{SR}^{i+1} = H_{SR_{i+1}}^{RB} \left(F_{TT}^i \right). \tag{3}$$

Finally, a sub-pixel convolution $U_\uparrow$ is used as the upsampling module to generate the output $\mathbf{x}'$ of scale $h \times w$: $\mathbf{x}' = U_\uparrow \left(F_{SR}^N + F_{SR}^0 \right)$. The whole branch is trained under the supervision of the fully sampled HR image $\mathbf{x}$.

Reconstruction Branch. As discussed above, only relying on the SR module is not sufficient for recovering a high-resolution and motion-corrected image when starting from an LR image with artifacts as input. Reconstruction, on the other hand, can restore a clear image with correct anatomical structure from an input with motion artifacts $\hat{\mathbf{x}}_{LR}$, because it is trained under the supervision of $\mathbf{x}_{LR}$. This means that reconstruction can effectively remove the artifacts introduced by the undersampled k-space, which is helpful for our final multi-task goal. By comparing the input and output of the Rec branch in Fig. 1, we can easily see that the Rec branch is more powerful in eliminating artifacts. For this branch,

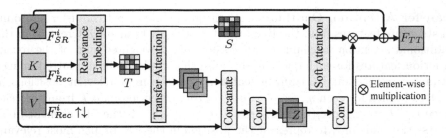

Fig. 2. Architecture of the proposed task transformer module. Q and K are the features inherited from the SR and Rec branches, respectively. V is the feature from the Rec branch with sequential upsampling and downsampling applied to it.

as shown in Fig. 1, we employ the same design as the SR branch to reduce the computational cost and generate high-quality results. We first use a convolutional layer to extract the shallow feature F_{Rec}^0 from the Rec branch. Then a series of $H_{Rec_i}^{RB}$ is used to extract the deep motion-corrected features

$$F_{Rec}^i = H_{Rec_i}^{RB}\left(F_{Rec}^{i-1}\right), \quad i = 1, 2, \ldots, N, \tag{4}$$

where $H_{Rec_i}^{RB}$ represents the i-th Resblocks, and F_{Rec}^i represents the i-th feature of the Rec branch. The Rec branch is trained under the supervision of the LR motion-artifacts-free image $\mathbf{x}_{LR}$, aiming to remove the artifacts from the input. In our multi-task framework, the output of this branch is fused to the SR branch to obtain the final super-resolved, motion-artifact-free image.

2.2 Task Transformer Module

Since the Rec branch contains a stronger artifact removal capacity than the SR branch, we introduce a task transformer module to guide the SR branch to learn SR motion-artifacts-free representation from the Rec branch. Our task transformer module consists of three parts: a relevance embedding, a transfer attention for feature transfer and a soft attention for feature synthesis. As shown in Fig. 2, the features F_{SR}^i and F_{Rec}^i inherited from the SR and Rec branches are expressed as the query (Q) and key (K). The value (V) is the feature $F_{Rec}^i \uparrow\downarrow$ obtained by sequentially applying upsampling $\uparrow$ and downsampling $\downarrow$ on F_{Rec}^i to make it domain-consistent with Q.

Relevance Embedding. Relevance embedding aims to embed the relevance information from the Rec branch by estimating the similarity between Q and K. To calculate the relevance $r_{i,j}$ between these two branches, we have

$$r_{i,j} = \left\langle \frac{q_i}{\|q_i\|}, \frac{k_j}{\|k_j\|} \right\rangle, \tag{5}$$

where $q_i \left(i \in \left[1, \frac{h}{s} \times \frac{w}{s}\right]\right)$ and $k_j \left(j \in \left[1, \frac{h}{s} \times \frac{w}{s}\right]\right)$ are the patches of Q and K, respectively.

Transfer Attention. Our transfer attention module aims to transfer anatomical structure features from the Rec branch to the SR branch. Different from the traditional attention mechanism, we do not take a weighted sum of the reconstruction features for each query q_i, because this would result in blurred images for image restoration. To transfer features from the most relevant positions in the Rec branch for each q_i, we obtain a transfer attention map T from the relevance $r_{i,j}$: $t_i = \arg\max_j(r_{i,j})$, where t_i $\left(i \in \left[1, \frac{h}{2} \times \frac{w}{2}\right]\right)$ is the i-th element in T. We use the value of t_i to represent the position in the Rec branch most relevant to the i-th position in the SR branch.

To obtain the anatomical structure features C without artifacts transferred from the Rec branch, an index selection operation is applied to the unfolded patches of V using t_i as the index: $c_i = v_{t_i}$, where c_i represents the value of C in the i-th position, which is equal to the t_i-th position of V.

Soft Attention. To synthesize the features from the two branches in our model, we first concatenate Q and C, and send them to a convolutional layer $Z = \mathrm{Conv}_z(\mathrm{Concat}(C, Q))$. Then, we use a soft attention module to aggregate the synthetic features Z and Q. To enhance the transferred anatomical structure information, we compute the soft attention map S from $r_{i,j}$ to represent the confidence of the transferred structure features for each position in C: $s_i = \max_j(r_{i,j})$, where s_i is the i-th position of the soft attention map S. To leverage more information from the SR branch, we first combine the synthetic feature Z with the original feature of the SR branch Q. Then, the final output of the task transformer module is obtained as follows:

$$F_{\mathrm{TT}} = Q \oplus \mathrm{Conv}_{out}(Z) \otimes S, \tag{6}$$

where $\oplus$ denotes the element-wise summation, $\otimes$ denotes the element-wise multiplication, and F_{TT} represents the final output of the task transformer module, which will be sent to the SR branch to restore a higher-quality, SR and motion-artifact-free image.

3 Experiments

Datasets. We employ the public IXI dataset and a clinical brain MRI dataset to evaluate our method. The clinical dataset is scanned with fully sampling using a clinical 3T Siemens Magnetom Skyra system on 155 patients. The imaging protocol is as follows: matrix size $320 \times 320 \times 20$, TR $= 4511$ ms, TE $= 112.86$ ms, field of view (FOV) $= 230 \times 200\,\mathrm{mm}^2$, turbo factor/echo train length TF $= 16$. For the IXI dataset, we exclude the first few slices of each volume since the frontal slices are much noisier than the others, making their distribution different. More details on the IXI dataset can be obtained from http://brain-development.org/ixi-dataset/. We split each dataset patient-wise into a ratio of 7:1:2 for training/validation/testing.

Table 1. Quantitative results on the two datasets under different enlargement scales.

Dataset	IXI dataset						Clinical dataset					
scale	2×			4×			2×			4×		
	PSNR	SSIM	NMSE	PSNR	SSIM	NMSE	PSNR	SSIM	NMSE	PSNR	SSIM	NMSE
Com-A	27.541	0.801	0.041	21.111	0.705	0.178	27.031	0.764	0.065	26.169	0.742	0.079
Com-B	28.439	0.847	0.033	21.323	0.687	0.170	28.750	0.816	0.044	27.539	0.803	0.058
Com-C	27.535	0.802	0.041	21.696	0.731	0.156	28.781	0.765	0.064	26.197	0.751	0.079
Com-D	28.426	0.847	0.033	21.895	0.710	0.149	28.839	0.817	0.043	27.700	0.815	0.056
w/o Rec	28.400	0.809	0.035	25.952	0.789	0.091	28.932	0.802	0.045	28.601	0.819	0.044
w/o H^{tt}	28.700	0.856	0.031	26.692	0.7730	0.089	29.510	0.817	0.037	29.528	0.821	0.037
T^2Net	**29.397**	**0.872**	**0.027**	**28.659**	**0.850**	**0.032**	**30.400**	**0.841**	**0.030**	**30.252**	**0.840**	**0.031**

Experimental Setup. For fair comparison, we implement four methods (two MRI reconstruction methods, ADMMNet [29] and MICCAN [11], and two MRI SR methods, MGLRL [24] and Lyu et al. [16]) with various sequential combinations, which we consider as baselines. These include: Com-A: ADMMNet-MGLRL, Com-B: ADMMNet-Lyu et al., Com-C: MICCAN-MGLRL, and Com-D: MICCAN-Lyu et al. The first model in each combination is used to remove artifacts, while the second is used to obtain higher-quality images. We implement our model in PyTorch using Adam with an initial learning rate of 5e-5, and train it on two NVIDIA Tesla V100 GPUs with 32 GB of memory per card, for 50 epochs. Parameters α and β are empirically set to 0.2 and 0.8, respectively. We use $N = 8$ residual groups in our network. All the compared methods are retrained using their default parameter settings.

Experimental Results. We evaluate our multi-task model under 6× Cartesian acceleration with 2× and 4× enlargement, respectively. In Table 1, we report the average PSNR, SSIM and NMSE scores with respect to the baselines on the two datasets, where w/o Rec and w/o H^{tt} will discussed in the ablation study. On the IXI dataset, our T^2Net achieves a PSNR of up to 29.397 dB under 2× enlargement. Further, compared to the best sequential combination, we improve the PSNR from 21.895 to 28.659 dB under 4× enlargement. Moreover, with higher enlargement, the sequential combinations obtain worse scores, while our T^2Net still preserves excellent results. On the clinical dataset, our T^2Net again achieves significantly better results than all combinations, under both enlargement scales. This suggests that our model can effectively transfer anatomical structure features to the SR branch, and that this is beneficial to multi-task learning.

We provide visual comparison results with corresponding error maps in Fig. 3. The first two rows show the restored images and error maps from the IXI dataset with 6× Cartesian acceleration and 2× enlargement, while the last two rows are the results for the clinical dataset with 6× Cartesian acceleration and 4× enlargement. As we can see, the input has significant aliasing artifacts and loss of anatomical details. The sequential combination methods can improve the image

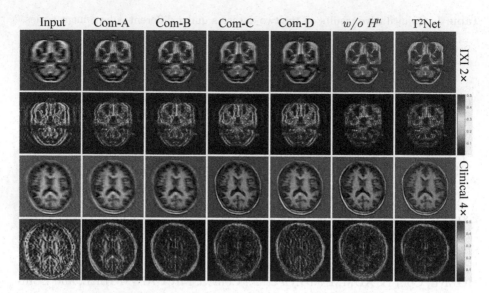

Fig. 3. Visual comparison with error maps of different methods on the two datasets.

quality, but are less effective than our multi-task methods. Our methods are obviously robust to aliasing artifacts and structural loss in the input. More importantly, at a high enlargement scale, our multi-task methods achieve much better results than the sequential combination methods.

Ablation Study. We evaluate the effectiveness of the two branches and the task transformer module in our multi-task network. Without loss of generality, the restoration results of two key components, including the Rec branch and task transformer module, are evaluated under 6× acceleration and 2× as well as 4× enlargement. We summarize the component analysis in Table 1, where w/o Rec indicates that only the SR branch is employed, while w/o H^{tt} indicates that both branches are used but the task transformer module H^{tt} is removed. As we can observe, w/o Rec obtains the worst results, which indicates the importance of both branches in our method, as each contains task-specific features for the target image restoration. Moreover, we can also see that w/o H^{tt} outperforms w/o Rec, demonstrating that transferring anatomical structure features to the target SR branch is necessary to achieve complementary representations. More importantly, our full T²Net further improves the results on both datasets and all settings. This demonstrates the powerful capability of our H^{tt} and two-branch structure in multi-task learning, which increases the model's flexibility to share complementary features for the restoration of higher-quality, super-resolved, and motion-artifacts-free images from highly undersampled and degenerated MRI data.

4   Conclusion

In this work, we focus on the multi-task learning of MRI reconstruction and super-resolution. For this purpose, we propose a novel end-to-end task transformer network (T^2Net) to transfer shared structure information to the task-specific branch for higher-quality and super-resolved reconstructions. Specifically, our model consists of two task-specific branches, *i.e.*, a target branch for SR and auxiliary branch for reconstruction, together with a task transformer module to transfer anatomical structure information to the target branch. The proposed task transformer consists of a feature embedding, hard attention and soft attention to transfer and synthesize the final reconstructions with correct anatomical structure, whilst maintaining fine details and producing less blurring and artifacts. In the future, we will design a network to automatically learn the loss weights.

References

1. Cai, J., Han, H., Shan, S., Chen, X.: FCSR-GAN?: joint face completion and super-resolution via multi-task learning. IEEE Trans. Biomet, Behav. Identity Sci. **2**(2), 109–121 (2019)
2. Chaudhari, A.S., et al.: Super-resolution musculoskeletal MRI using deep learning. Magn. Resonan. Med. **80**(5), 2139–2154 (2018)
3. Chen, Y., Shi, F., Christodoulou, A.G., Xie, Y., Zhou, Z., Li, D.: Efficient and accurate MRIi super-resolution using a generative adversarial network and 3D multi-level densely connected network. In: International Conference on Medical Image Computing and Computer-Assisted Intervention, pp. 91–99. Springer (2018)
4. Collobert, R., Weston, J.: A unified architecture for natural language processing: Deep neural networks with multitask learning. In: Proceedings of the 25th International Conference on Machine Learning, pp. 160–167 (2008)L
5. Feng, C.M., Fu, H., Yuan, S., Xu, Y.: Multi-contrast MRI super-resolution via a multi-stage integration network. In: International Conference on Medical Image Computing and Computer Assisted Intervention (MICCAI) (2021)
6. Feng, C.M., Wang, K., Lu, S., Xu, Y., Li, X.: Brain MRI super-resolution using coupled-projection residual network. Neurocomputing **456**, 190–199 (2021)
7. Feng, C.M., Yan, Y., Chen, G., Fu, H., Xu, Y., Shao, L.: Accelerated multi-modal MR imaging with transformers (2021)
8. Feng, C.M., Yang, Z., Chen, G., Xu, Y., Shao, L.: Dual-octave convolution for accelerated parallel MR image reconstruction. In: Proceedings of the 35th AAAI Conference on Artificial Intelligence (AAAI) (2021)
9. Feng, C.M., Yang, Z., Fu, H., Xu, Y., Yang, J., Shao, L.: DONet: dual-octave network for fast MR image reconstruction. IEEE Trans. Neural Netw. Learn. Syst. (2021)
10. Hammernik, K., et al.: Learning a variational network for reconstruction of accelerated MRI data. Magn, Resonan. Med. **79**(6), 3055–3071 (2018)
11. Huang, Q., Yang, D., Wu, P., Qu, H., Yi, J., Metaxas, D.: MRI reconstruction via cascaded channel-wise attention network. In: International Symposium on Biomedical Imaging (ISBI 2019), pp. 1622–1626. IEEE (2019)

12. Kim, S., Hori, T., Watanabe, S.: Joint CTC-attention based end-to-end speech recognition using multi-task learning. In: 2017 IEEE International Conference on Acoustics, Speech and Signal Processing (ICASSP), pp. 4835–4839. IEEE (2017)
13. Lai, Z., et al.: Image reconstruction of compressed sensing MRI using graph-based redundant wavelet transform. Med. Image Anal. **27**, 93–104 (2016)
14. Lim, B., Son, S., Kim, H., Nah, S., Mu Lee, K.: Enhanced deep residual networks for single image super-resolution. In: Proceedings of the IEEE Conference on Computer Vision and Pattern Recognition Workshops, pp. 136–144 (2017)
15. Liu, S., Johns, E., Davison, A.J.: End-to-end multi-task learning with attention. In: Proceedings of the IEEE/CVF Conference on Computer Vision and Pattern Recognition, pp. 1871–1880 (2019)
16. Lyu, Q., Shan, H., Wang, G.: MRI super-resolution with ensemble learning and complementary priors. IEEE Trans. Comput. Imag. **6**, 615–624 (2020)
17. Lyu, Q., You, C., Shan, H., Zhang, Y., Wang, G.: Super-resolution MRI and CT through gan-circle. In: Developments in X-Ray Tomography XII, vol. 11113, p. 111130X. International Society for Optics and Photonics (2019)
18. Mahapatra, D., Bozorgtabar, B., Garnavi, R.: Image super-resolution using progressive generative adversarial networks for medical image analysis. Comput. Med. Imag. Graph. **71**, 30–39 (2019)
19. Mardani, M., Gong, E., Cheng, J.Y., Vasanawala, S.S., Zaharchuk, G., Xing, L., Pauly, J.M.: Deep generative adversarial neural networks for compressive sensing MRI. IEEE Trans. Med. Imaging **38**(1), 167–179 (2018)
20. Nakarmi, U., Wang, Y., Lyu, J., Liang, D., Ying, L.: A kernel-based low-rank (KLR) model for low-dimensional manifold recovery in highly accelerated dynamic MRI. IEEE Trans. Medi. Imag. **36**(11), 2297–2307 (2017)
21. Oktay, O., Bai, W., Lee, M., Guerrero, R., Kamnitsas, K., Caballero, J., de Marvao, A., Cook, S., O'Regan, D., Rueckert, D.: Multi-input cardiac image superresolution using convolutional neural networks. In: International Conference on Medical Image Computing and Computer Assisted Intervention. pp. 246–254. Springer (2016). https://doi.org/10.1007/10704282
22. Qin, C., Schlemper, J., Caballero, J., Price, A.N., Hajnal, J.V., Rueckert, D.: Convolutional recurrent neural networks for dynamic MR image reconstruction. IEEE Trans. Medical Imag. **38**(1), 280–290 (2018)
23. Ravishankar, S., Bresler, Y.: MR image reconstruction from highly undersampled k-space data by dictionary learning. IEEE Trans. Med. Imag. **30**(5), 1028–1041 (2010)
24. Shi, J., Liu, Q., Wang, C., Zhang, Q., Ying, S., Xu, H.: Super-resolution reconstruction of MR image with a novel residual learning network algorithm. Phys. Med. Biol. **63**(8), 085011 (2018)
25. Shin, P.J., Larson, P.E., Ohliger, M.A., Elad, M., Pauly, J.M., Vigneron, D.B., Lustig, M.: Calibrationless parallel imaging reconstruction based on structured low-rank matrix completion. Magn. Resonan. Med. **72**(4), 959–970 (2014)
26. Wang, S., Cheng, H., Ying, L., Xiao, T., Ke, Z., Zheng, H., Liang, D.: Deepcomplexmri: exploiting deep residual network for fast parallel MR imaging with complex convolution. Magn. Resonan. Imag. **68**, 136–147 (2020)
27. Wang, Y.H., Qiao, J., Li, J.B., Fu, P., Chu, S.C., Roddick, J.F.: Sparse representation-based MRI super-resolution reconstruction. Measurement **47**, 946–953 (2014)

28. Yang, G., Yu, S., Dong, H., Slabaugh, G., Dragotti, P.L., Ye, X., Liu, F., Arridge, S., Keegan, J., Guo, Y., et al.: Dagan: Deep de-aliasing generative adversarial networks for fast compressed sensing MRI reconstruction. IEEE Trans. Med. Imag. **37**(6), 1310–1321 (2017)
29. Yang, Y., Sun, J., Li, H., Xu, Z.: Deep ADMM-net for compressive sensing MRI. In: Proceedings of the 30th International Conference on Neural Information Processing Systems, pp. 10–18 (2016)
30. Zhan, Z., Cai, J.F., Guo, D., Liu, Y., Chen, Z., Qu, X.: Fast multiclass dictionaries learning with geometrical directions in MRI reconstruction. IEEE Trans. Biomed. Eng. **63**(9), 1850–1861 (2015)
31. Zhang, M., Liu, W., Ma, H.: Joint license plate super-resolution and recognition in one multi-task gan framework. In: 2018 IEEE International Conference on Acoustics, Speech and Signal Processing (ICASSP), pp. 1443–1447. IEEE (2018)
32. Zhang, X., Lam, E.Y., Wu, E.X., Wong, K.K.Y.: Application of Tikhonov regularization to super-resolution reconstruction of brain MRI images. In: Gao, X., Müller, H., Loomes, M.J., Comley, R., Luo, S. (eds.) MIMI 2007. LNCS, vol. 4987, pp. 51–56. Springer, Heidelberg (2008). https://doi.org/10.1007/978-3-540-79490-5_8
33. Zhu, B., Liu, J.Z., Cauley, S.F., Rosen, B.R., Rosen, M.S.: Image reconstruction by domain-transform manifold learning. Nature **555**(7697), 487–492 (2018)

Conditional GAN with an Attention-Based Generator and a 3D Discriminator for 3D Medical Image Generation

Euijin Jung, Miguel Luna, and Sang Hyun Park[✉]

Department of Robotics Engineering, DGIST, Daegu, South Korea
{euijin,shpark13135}@dgist.ac.kr

Abstract. Conditional Generative Adversarial Networks (cGANs) are a set of methods able to synthesize images that match a given condition. However, existing models designed for natural images are impractical to generate high-quality 3D medical images due to enormous computation. To address this issue, most cGAN models used in the medical field process either 2D slices or small 3D crops and join them together in subsequent steps to reconstruct the full-size 3D image. However, these approaches often cause spatial inconsistencies in adjacent slices or crops, and the changes specified by the target condition may not consider the 3D image as a whole. To address these problems, we propose a novel cGAN that can synthesize high-quality 3D MR images at different stages of the Alzheimer's disease (AD). First, our method generates a sequence of 2D slices using an attention-based 2D generator with a disease condition for efficient transformations depending on brain regions. Then, consistency in 3D space is enforced by the use of a set of 2D and 3D discriminators. Moreover, we propose an adaptive identity loss based on the attention scores to properly transform features relevant to the target condition. Our experiments show that the proposed method can generate smooth and realistic 3D images at different stages of AD, and the image change with respect to the condition is better than the images generated by existing GAN-based methods.

Keywords: 3D image generation · Conditional GAN · Alzheimer's disease · 3D discriminator · Adaptive identity loss

1 Introduction

Developing an automatic tool to predict Alzheimer's disease (AD) progression is important to perform early detection and develop of personalized treatments that can potentially prevent the rapid deterioration caused by the disease [21]. However, modern learning based methods require a large number of longitudinal Magnetic Resonance images (MRI) to accurately train the model. Recently, many studies have proposed alternatives to generate synthetic medical images

© Springer Nature Switzerland AG 2021
M. de Bruijne et al. (Eds.): MICCAI 2021, LNCS 12906, pp. 318–328, 2021.
https://doi.org/10.1007/978-3-030-87231-1_31

to address the data limitation problem, e.g., cross modality image generation [1,15,17,18,27,28], contrast dependent image generation [16,23,26], age prediction [4,25,30], and progressive AD generation [3,20,22]. Also, several conditional Generative Adversarial Networks (cGANs) based methods are proposed to generate a transformed image depend on a source image and a target condition [5,8,19,29]. However, most existing cGANs based methods were designed in 2D space, thus the generated 3D image often includes discontinuity along a certain axis or suffers from partial or inconsistent feature transformation, i.e., the partial change of organ shape or severity of a lesion and the mixture of expansion and contraction within one organ, which can lead to incorrect diagnosis. For example, appropriate volume changes on gray matter and hippocampus are required to precisely match the AD stage shape defined by the target condition, but previous methods often generated partially shrunk gray matter or hippocampus. Thus, the distributions of synthesized AD, MCI, and NC images do not match the real image distributions and a classifier trained by synthetic data can not improve the prediction performance when applied on the real images. Meanwhile, training a 3D model is often unstable due to the large number of parameters and massive computation.

In this paper, we propose a novel cGAN that is able to generate a 3D MR image at a specified AD state by using a source MR image and a continuous target condition. We call our method Alzheimer's Disease Evolution Synthesizer (ADESyn). ADESyn generates realistic and high-quality 3D images with no spatial inconsistencies that accurately match the AD state image properties defined by the input condition. Our model is composed of an attention-based 2D generator to synthesize 2D MR images, a 2D discriminator to classify authenticity and predict similarity to target condition of 2D slices, and a 3D discriminator accounting for 3D structural information of independently generated 2D slices. The 3D discriminator utilizes sections of consecutive 2D slices synthesized by the 2D generator in the same mini-batch, which makes the 2D generator generate smooth transformations of consecutive 2D images. The idea of using a 3D discriminator was introduced in [12], but the method could not synthesize images at intermediate points of AD progress and often caused unstable deformations on several local regions since similar deformations occur in the entire brain area. In ADESyn, we use an attention module in the generator to emphasize transformations of the brain areas where major deformation occurs according to AD progression. Moreover, we propose an adaptive identity loss that makes use of the attention scores to improve conditional generation performance.

Key contributions of this work are as follows: (1) we propose a novel cGAN model that can generate a series of changing 3D brain MR images from Normal to MCI to AD according to a continuous stage condition. (2) Our attention-based 2D generator with a 3D discriminator can constrain the transformation to the areas that need to be deformed while considering the changes in 3D. (3) We show that an adaptive identity loss based on attention scores is effective to synthesize natural 3D images by suppressing excessive deformations. (4) Lastly, ADESyn is compared with several existing cGAN-based methods demonstrating better qualitative performance. For rigorous verification, we used a model pre-trained with publicly open brain MRI data to show accurate quantitative results. The code is available at https://github.com/EuijinMisp/ADESyn.

2   Method

We designed Alzheimer's Disease Evolution Synthesizer (ADESyn) following a cGAN 2D structure with the addition of a 3D discrimination module as shown in Fig. 1. Our network consists of three main components: an attention-based 2D generator, a 2D discriminator, and a 3D discriminator. The attention-based 2D generator G helps to generate 2D realistic synthetic slices s_{2D} to construct a full 3D brain MR scan s_{3D} that resembles the original patient MR image x_{3D} that also contains the changes specified by a disease condition c.

Normally, patient' data does not have the corresponding follow up MR scans which makes it unsuitable to train fully supervised networks. However, adversarial learning can help model the intrinsic features of different stages of Alzheimer's disease, even if the available MR images belong to different patient. Therefore, we use adversarial learning to teach the generator G to modify the brain areas that are usually affected as the disease progresses, while keeping track of the patient's identity by using an objective function that constrains the generated image. The 2D discriminator F determines the realism of the 2D synthetic slice s_{2D} compared to other real slices x_{2D} and verifies whether the disease target condition c has been matched. The 2D generator and discriminator designed in our method follow the structure proposed in GANimation [19]. As stated earlier, 2D generative models suffer from inconsistencies in 3D space and a 3D model increases the number of parameters and the required computation. Thus, we incorporate a 3D module H in ADESyn to aid with the discrimination task to ensure generation of better quality 3D MR images. To interconnect the 3D discriminator H and the 2D generator G, several 2D synthesized slices s_{2D} need to be concatenated to form a section of the 3D MR image s_{3D}. This is performed by generating sequences of consecutive input slices in a mini-batch, where the 3D discriminator H takes either a consecutive sequence of real slices x_{2D}^1, x_{2D}^2, ...,x_{2D}^k or synthetic ones s_{2D}^1, s_{2D}^2, ...,s_{2D}^k as input.

ADESyn Network Details: The generator G is composed of an encoder, a transition layer, and a decoder. It requires two inputs, the 2D slice x_{2D} and target condition c. c is an additional slice with the same dimensions as x_{2D} but it has only a single value for all pixels, i.e. 0, 0.5 or 1 for normal, MCI, or AD, respectively. Both x_{2D} and c are concatenated along the channel dimension, before passing them to the encoder. The encoder reduces the feature maps to $\frac{1}{4}$ of the original size of x_{2D} through three convolutional layers. Then, the transition layer processes the feature maps through six residual blocks, to finally be up-sampled by the decoder. The decoder has two outputs: one uses two transpose convolution layers in series while the second one uses a single transpose convolution layer in parallel. The purpose of the two outputs is to get a generated image g_{2D} and an attention mask a. g_{2D} is a reconstructed image with

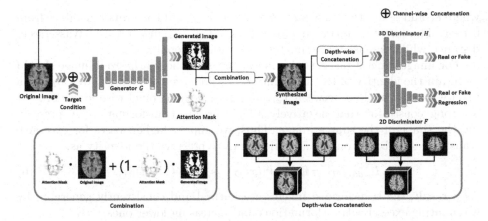

Fig. 1. ADESyn conditional generative adversarial network details.

the same dimensions of x_{2D} with its intensity output bounded by a *tanh* activation function. The attention mask a also has the same dimensions of x_{2D} and it is bounded by a sigmoid activation function. a determines what percentage of the input image x_{2D} pixels intensity pass to s_{2D} and the complement is taken from g_{2D}. In other words, the synthetic image s_{2D} is defined as the weighted combination of g_{2D}, x_{2D} by a, as follows:

$$s_{2D} = a \times x_{2D} + (1 - a) \times g_{2D}. \tag{1}$$

The 2D discriminator F is a convolutional neural network (CNN) with five convolutional layers and two fully connected layers that verify image realism with F_a (real or fake) and predict the condition c with F_b (intermediate stage from normal to AD). Similar to [12], the 3D discriminator H consists of five 3D convolutional layers and a single fully connected layer. The 3D discriminator determines whether its input corresponds to real 3D image x_{3D} showing a continuous 3D image x_{2D} or a synthesized one s_{3D} showing a concatenation of independently generated 2D slice s_{2D}. H maintains the spatial consistency of 3D images in all 3 axes by ensuring the model considers the consecutiveness of adjacent 2D slices. Depth-wise concatenation block in Fig. 1 visualizes how s_{3D} is reconstructed by concatenating k consecutive 2D slices s_{2D}.

Objective Function: We trained the network with GAN, adaptive identity, regression, attention, and variation regularization losses. As suggested in [12], the GAN loss is defined by Wasserstein distance with gradient penalty as:

$$L_{2D-adv} = \mathbb{E}_{x_{2D},c}[F_a(s_{2D})] - \mathbb{E}_{x_{2D}}[F_a(x_{2D})] + \lambda_{GP}(||\nabla_{z_{2D}} F_a(z_{2D})||_2 - 1)^2, \tag{2}$$

$$L_{3D-adv} = \mathbb{E}_{x_{3D},c}[H(s_{3D})] - \mathbb{E}_{x_{3D}}[H(x_{3D})] + \lambda_{GP}(||\nabla_{z_{3D}} H(z_{3D})||_2 - 1)^2, \tag{3}$$

$$z_{2D} = \epsilon_1 x_{2D} + (1 - \epsilon_1)s_{2D}, \epsilon_1 \sim U[0,1], \tag{4}$$

$$z_{3D} = \epsilon_2 x_{3D} + (1 - \epsilon_2)s_{3D}, \epsilon_2 \sim U[0,1], \tag{5}$$

where ϵ_1 and ϵ_2 are the same size as x_{2D} and x_{3D}, and have random values from 0 to 1. The first two terms in L_{2D-adv} and L_{3D-adv} measure the Wasserstein distance while the last term denotes the gradient penalty [7].

For image generation, it is also important that the synthesized image should maintain the identity of the input image. To preserve the identity, we propose a pixel-wise identity loss between x_{2D} and s_{2D}. Regular pixel-wise losses present a strong constraint that negatively affects feature transformation by preserving the shape of the source image. Thus, we impose higher losses for excessive transformation based on the attention map to preserve the identity as:

$$L_{AdaptID} = \mathbb{E}_{x_{2D},c}[|||(x_{2D} - s_{2D}) \times (1-a)||_1]. \tag{6}$$

As a result, the transformed images are natural and consistently generated by attenuating excessive transformations and increasing lower ones.

With the above criteria, s_{2D} should be similar to x_{2D} and meet the target condition c at the same time. To address this, F_b predicts the condition of a 2D input slice between 0 to 1. Since F_b can predict a continuous target condition c, we can generate continuous synthetic MRI from normal to AD. The predicted condition of s_{2D} is used to optimize G, while the predicted condition of x_{2D} is used to optimize F. As suggested in [19], the regression losses L_{reg}^r for x_{2D} and L_{reg}^s for s_{2D} are defined as:

$$L_{reg}^r = \mathbb{E}_{x_{2D}}[||F_b(x_{2D}) - c||_2^2], \tag{7}$$

$$L_{reg}^s = \mathbb{E}_{x_{2D},c}[||F_b(s_{2D}) - c||_2^2]. \tag{8}$$

The attention loss L_A and total variation regularization L_{TV} are also used to generate smooth attention maps as:

$$L_A = \mathbb{E}_{x_{2D},c}[||a||_1], \tag{9}$$

$$L_{TV} = \mathbb{E}_{x_{2D},c}[\sum_{i,j}^{H,W}[|a_{i+1,j} - a_{i,j}| + |a_{i,j+1} - a_{i,j}|]]. \tag{10}$$

Finally, the generator loss L_G is defined as:

$$L_G = (1 - \lambda_{adv})\mathbb{E}_{x_{2D},c}[F_a(s_{2D})] + \lambda_{adv}\mathbb{E}_{x_{3D},c}[H(s_{3D})] + \lambda_{reg}L_{reg}^s + \lambda_{ID}L_{AdaptID} + \lambda_A L_A + \lambda_{TV}L_{TV}, \tag{11}$$

where $\lambda_{adv}, \lambda_{reg}, \lambda_{ID}, \lambda_{GP}, \lambda_A, \lambda_{TV}$ are hyperparameters to balance the effect of each loss. In L_G, we determine the ratio λ_{adv} between L_{2D-adv} and L_{3D-adv} by considering k as: $\lambda_{adv} = \frac{1}{k-1}$ since the effect of H is affected by the number of concatenated images k. For example, λ_{adv} was set as $= 0.5$ when k is 3 as minimum. Other hyperparameters were heuristically set as $\lambda_{reg} = 30, \lambda_{ID} = 10, \lambda_{GP} = 10, \lambda_A = 0.1, \lambda_{TV} = 10^{-5}$. The discriminator losses L_F and L_H are defined as:

$$L_F = L_{2D-adv} + \lambda_{reg}L_{reg}^r, \tag{12}$$

$$L_H = L_{3D-adv}. \tag{13}$$

The generator and discriminators are trained iteratively to reduce the above losses.

Table 1. Results of proposed methods and the related methods for image quality measurements. FID and KID scores in axial(A), coronal(C), and sagittal(S) views measures the image quality, and GAN_{train} and GAN_{test} scores measures whether the images are created well with respect to the condition.

Models	FID (A)	FID (C)	FID (S)	KID (A)	KID (C)	KID (S)	GAN_{train}	GAN_{test}
Real images	0.65 ± 0.55	0.37 ± 0.36	0.59 ± 0.49	0.051 ± 0.05	0.028 ± 0.03	0.037 ± 0.03	75.95	76.88
CJAAE [29]	14.19 ± 4.22	13.55 ± 2.21	15.44 ± 3.35	1.486 ± 0.41	1.652 ± 0.25	1.608 ± 0.33	52.53	56.11
AttGAN [8]	2.14 ± 1.75	1.68 ± 0.93	4.53 ± 2.01	0.092 ± 0.09	0.071 ± 0.06	0.116 ± 0.11	62.03	66.71
StarGAN [5]	43.52 ± 8.94	51.10 ± 4.70	73.04 ± 8.72	3.313 ± 0.69	5.858 ± 0.47	7.062 ± 0.76	65.19	85.41
GANimation [19]	1.05 ± 1.05	0.54 ± 0.51	0.94 ± 0.91	0.060 ± 0.06	0.050 ± 0.04	0.050 ± 0.04	68.35	85.41
ADESyn(3DD)($k = 3$)	0.72 ± 0.69	0.33 ± 0.34	0.57 ± 0.57	0.054 ± 0.02	0.033 ± 0.01	0.042 ± 0.02	69.62	84.16
ADESyn(3DD)($k = 6$)	**0.61 ± 0.58**	0.28 ± 0.30	0.56 ± 0.58	**0.042 ± 0.02**	0.033 ± 0.01	0.042 ± 0.02	72.15	89.28
ADESyn(3DD)($k = 9$)	**0.61 ± 0.59**	0.28 ± 0.29	0.50 ± 0.50	0.043 ± 0.02	0.033 ± 0.01	0.040 ± 0.01	63.29	70.25
ADESyn(3DD+aID)($k = 3$)	0.69 ± 0.74	0.35 ± 0.29	0.70 ± 0.70	0.054 ± 0.02	**0.028 ± 0.01**	0.044 ± 0.03	71.52	87.41
ADESyn(3DD+aID)($k = 6$)	**0.61 ± 0.59**	**0.27 ± 0.20**	0.49 ± 0.52	0.045 ± 0.02	0.029 ± 0.01	0.042 ± 0.02	**75.32**	**89.53**
ADESyn(3DD+aID)($k = 9$)	**0.61 ± 0.59**	**0.27 ± 0.19**	**0.45 ± 0.44**	0.045 ± 0.02	0.029 ± 0.01	**0.038 ± 0.01**	74.68	82.04

3   Experimental Results

Implementation Details: We evaluated the efficacy of ADESyn using the Alzheimer's Disease Neuroimaging Initiative (ADNI) dataset [11]. We selected 200 MR images for AD patients, 400 MCI, and 200 images of normal control (NC) subjects. The experiments were performed on slices of size 192×192 in the axial view of the MR image. The model was trained for 200,000 iterations with Adam optimizer [13] and learning rate of 0.0001 with constant decay. The code was developed with the PyTorch deep learning framework run on a RTX 2080 Ti NVIDIA GPU.

To evaluate the model, we divided the dataset into two even subsets (Set_A and Set_B). Set_A has a total of 401 subjects including 99 subjects of AD, 202 MCI, and 100 subjects with NC. Set_B is composed of 400 subjects including 99 AD, 202 MCI, and 99 NC. From Set_A we generated 1203 synthetic subjects to form $Set_{A'}$ across all three target conditions (NC, MCI, and AD).

The quality of the generated 3D images was measured by applying the Frchet Inception Distance (FID) [9] and Kernel Inception Distance (KID) [2]. FID evaluates the similarity between real and generated images using mean and covariance on represented features encoded by pre-trained network. KID evaluates the difference between real and generated images based on Maximum mean discrepancy(MMD) [6] of embedded representation. In the testing stage, Set_B represents real images and $Set_{A'}$ synthetic. To measure similarity between $Set_{A'}$ and general real MR images, another MR image dataset that was not used for training is required to pre-train the models used to perform FID and KID. For these reasons, we used 410 subjects of Open Access Series of Imaging Studies-3 (OASIS3) data [14] to pre-train the model for embedded representation of brain MRI based on autoencoder [10]. To measure 3D image quality accurately, we measured FID and KID for each view (axial, coronal, and sagittal). We randomly sampled 5000 images out of total generated images to satisfy the IID condition, then performed evaluation. We obtained the mean and standard deviation of KID and FID by repeating the process 10 times.

Furthermore, to measure the correspondence between target condition and synthesized image, also called conditional generation performance, GAN_{train} and GAN_{test} scores [24] were used. To compute the GAN_{train} score, a classification model was trained on synthetic data $Set_{A'}$ and tested on real data Set_B. On the other hand, GAN_{test} is measured by training the model on real data Set_B and testing it on synthetic data $Set_{A'}$. A GAN model is considered to be effective if the average of both scores, GAN_{train} and GAN_{test}, is high and the standard deviation is low. We measured these scores on NC vs AD binary classification.

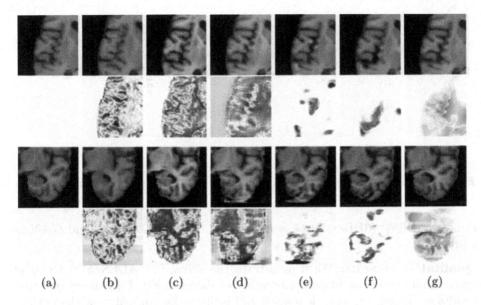

(a)   (b)   (c)   (d)   (e)   (f)   (g)

Fig. 2. Examples of generated AD images from normal images. (a) Normal images used as input, (b) CAAE, (c) AttGAN, (d) StarGAN (e) GANimation, (f) ADEsyn(3DD), (g) ADEsyn(3DD+aID). Transformed region can be shown in difference map (original input - fake image). Red and blue shows positive and negative, respectively. (Color figure online)

Quantitative Results: For comparison, we compared our method with an autoencoder based method (CAAE) [29], AttGAN [8], StarGAN [5] and GANimation [19] models. Table 1 shows the image quality scores for images generated by ADESyn and other related cGANs. Real image scores were measured between Set_A and Set_B to compare previous methods with our proposed method. The image quality scores of GANimation and ADESyn are close to those of real images as an effect of using an attention map to focus on the regions that need transformation. The use of the attention map allows the generation of better quality images compared to others. All ADESyn variants using 3D discriminator achieve better FID and KID scores than previous methods in all views. Also, FID and KID scores keep decreasing as k increases. This is mainly because our method uses a 3D discriminator on top of a 2D cGAN structure to generate smooth images not only in the axial view but also the coronal and sagittal views.

The conditional generation performance can be measured by GAN_{train} and GAN_{test} scores. The results of synthetic images are compared with the scores of real images. GAN_{train} and GAN_{test} scores of ADESyn(3DD) increase as the k increases until $k = 6$, but decrease after that. Conditional generation performance is increased as an effect of quality improvement, but the transformation ability is decreased since the 3D discriminator is largely influenced to follow the source distribution when k increases. However, by using the adaptive identity loss, we were able to mitigate large variations in the score changes based on k.

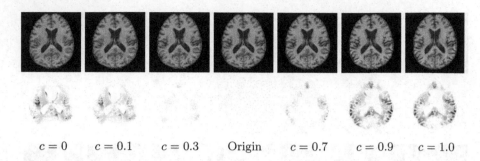

| $c = 0$ | $c = 0.1$ | $c = 0.3$ | Origin | $c = 0.7$ | $c = 0.9$ | $c = 1.0$ |

Fig. 3. Generated images based on a MCI image (middle) and continuous conditions

Finally, ADESyn(3DD+aID)($k = 6$) achieves the best GAN_{train} and GAN_{test} scores.

Qualitative Results: We demonstrate the efficacy of ADESyn in terms of image quality and conditional generation as shown in Fig. 2. In the second row, results of (f) and (g) show less noise and artifacts for coronal view than others thanks to the use of the 3D discriminator. We also evaluated image quality and conditional generation with a difference map. The difference map shows distinctive pixel-wise changes from a source image. The effect of adaptive identity loss can be confirmed by comparing the difference map of (f) and (g) in Fig. 2. In order for the generated AD image to be ideal, the areas of hippocampus and gray matter should be smoothly reduced while white matter intensity should smoothly increase. These features can be easily seen on the difference map of the (g)ADESyn (3DD+aID) through consistent red on hippocampus and gray matter, and blue on white matter region. In addition, continuous transformations guided by the target condition can be visualized in Fig. 3 where the changes on the difference map are consistently smooth. The attention mask allows the generator to modify the areas of the brain that need to change while keeping other areas intact. Thus, the generator does not need to learn how to generate the whole image, reducing its complexity and allowing it to focus on changes in the areas of interest.

4   Conclusion

In this work, we have presented a conditional GAN that is capable of synthesizing MR images at any point in time between normal status to AD from a MR image and a target condition. The proposed method uses a generator with attention masking to increase the quality of the generated images and improve conditional generation performance using a novel adaptive identity loss. We verified the effectiveness of the proposed method by comparison with several cGAN based methods in terms of image quality and conditional generation performances. Since our method can predict how the brain will be transformed for each patient according to the progression of AD, it is expected to be used for longitudinal studies and group analysis where data is difficult to construct.

Acknowledgement. This research was supported by the National Research Foundation of Korea (NRF) grant funded by the Korean Government (MSIT) (No. 2019R1C1C1008727)

References

1. Ben-Cohen, A., Klang, E., Raskin, S.P., Amitai, M.M., Greenspan, H.: Virtual PET images from CT data using deep convolutional networks: initial results. In: Simulation and Synthesis in Medical Imaging, pp. 49–57 (2017)
2. Bińkowski, M., Sutherland, D., Arbel, M., Gretton, A.: Demystifying MMD GANs. ICML (2018)
3. Bowles, C., Gunn, R., Hammers, A., Rueckert, D.: Modelling the progression of Alzheimer's disease in mri using generative adversarial networks. SPIE Medical Imaging, p. 55 (2018)
4. Choi, H., Kang, H., Lee, D.S., T.A.D.N.I.: Predicting aging of brain metabolic topography using variational autoencoder. Front. Aging Neurosc. **10**, 212 (2018)
5. Choi, Y., Choi, M.J., Kim, M., Ha, J.W., Kim, S., Choo, J.: StarGAN: unified generative adversarial networks for multi-domain image-to-image translation. In: Proceedings of the IEEE Conference on Computer Vision and Pattern Recognition (2018)
6. Gretton, A., Borgwardt, K.M., Rasch, M.J., Schölkopf, B., Smola, A.: A kernel two-sample test. JMLR (2012)
7. Gulrajani, I., Ahmed, F., Arjovsky, M., Dumoulin, V., Courville, A.C.: Improved Training of Wasserstein GANs. In: NIPS, pp. 5767–5777 (2017)
8. He, Z., Zuo, W., Kan, M., Shan, S., Chen, X.: Attgan: facial attribute editing by only changing what you want. IEEE Trans. Image Process. **28**(11), 5464–5478 (2019)
9. Heusel, M., Ramsauer, H., Unterthiner, T., Nessler, B., Hochreiter, S.: GANs trained by a two time-scale update rule converge to a local nash equilibrium. In: NIPS (2017)
10. Hinton, G., Salakhutdinov, R.: Reducing the dimensionality of data with neural networks. Science **313**, 504–507 (2006)
11. Jack, C., et al.: The Alzheimer's Disease Neuroimaging Initiative (ADNI): MRI methods. J. Magn. Reson. Imaging **27**(4), 685–691 (2008)
12. Jung, E., Luna, M., Park, S.H.: Conditional generative adversarial network for predicting 3d medical images affected by alzheimer's diseases. In: International Workshop on PRedictive Intelligence in MEdicine, pp. 79–90 (2020)
13. Kingma, D.P., Ba, J.: Adam: a method for stochastic optimization. In: 3rd International Conference for Learning Representations (2014)
14. LaMontagne, P.J., et al.: Marcus, D.: Oasis-3: Longitudinal neuroimaging, clinical, and cognitive dataset for normal aging and alzheimer disease. medRxiv (2019)
15. Lei, Y., et al.: MRI-only based synthetic CT generation using dense cycle consistent generative adversarial networks. Med. Phys. **46**(8), 3565–3581 (2019)
16. Muhammad, S., Muhammad, Naveed, R., Jing, W., Chengnian, L., Shaoyuan, L.: Unpaired multi-contrast MR image synthesis using generative adversarial networks. In: Simulation and Synthesis in Medical Imaging. pp. 22–31. Springer International Publishing (2019). https://doi.org/10.1007/978-3-030-32778-1_3

17. Pan, Y., Liu, M., Lian, C., Zhou, T., Xia, Y., Shen, D.: Synthesizing missing PET from MRI with cycle-consistent generative adversarial networks for Alzheimer's disease diagnosis. In: Medical Image Computing and Computer Assisted Intervention, pp. 455–463 (2018)
18. Prokopenko, D., Stadelmann, J., Schulz, H., Renisch, S., Dylov, D.: Synthetic CT generation from MRI using improved dualgan. arXiv:1909.08942 (2019)
19. Pumarola, A., Agudo, A., Martinez, A., Sanfeliu, A., Moreno-Noguer, F.: GANimation: one-shot anatomically consistent facial animation. In: International Journal of Computer Vision (IJCV) (2019)
20. Ravi, D., Alexander, D.C., Oxtoby, N.P.: Degenerative adversarial neuroimage nets: generating images that mimic disease progression. In: Medical Image Computing and Computer Assisted Intervention, pp. 164–172 (2019)
21. Reitz, C.: Toward precision medicine in Alzheimer's disease. Ann. Transl. Med. 4(6), 107 (2016)
22. Roychowdhury, S., Roychowdhury, S.: A modular framework to predict alzheimer's disease progression using conditional generative adversarial networks. In: 2020 International Joint Conference on Neural Networks (IJCNN), pp. 1–8 (2020)
23. Salman, Ul, H.D., Mahmut, Y., Levent, K., Aykut, E., Erkut, E., Tolga, C.: Image synthesis in multi-contrast mri with conditional generative adversarial networks. IEEE Trans. Med. Imaging 38(10), 2375–2388 (2019)
24. Shmelkov, K., Schmid, C., Alahari, K.: How good is my gan? The European Conference on Computer Vision (2018)
25. Wegmayr, V., Horold, M., Buhmann, J.: Generative aging of brain mri for early prediction of mci-ad conversion. In: International Symposium on Biomedical Imaging, pp. 1042–1046 (2019)
26. Welander, P., Karlsson, S., Eklund, A.: Generative adversarial networks for image-to-image translation on multi-contrast mr images - a comparison of cyclegan and unit. arXiv:1806.07777 (2018)
27. Wolterink, J.M., Dinkla, A.M., Savenije, M.H.F., Seevinck, P.R., van den Berg, C.A.T., Išgum, I.: Deep MR to CT synthesis using unpaired data. In: Simulation and Synthesis in Medical Imaging, pp. 14–23 (2017)
28. Zeng, G., Zheng, G.: Hybrid generative adversarial networks for deep MR to CT synthesis using unpaired data. Medical Image Computing and Computer Assisted Intervention, pp. 59–767 (2019)
29. Zhang, Z., Song, Y., Qi, H.: Age progression/regression by conditional adversarial autoencoder. CVPR (2017)
30. Zhao, Q., Adeli, E., Honnorat, N., Leng, T., Pohl, K.M.: Variational autoencoder for regression: Application to brain aging analysis. In: Medical Image Computing and Computer Assisted Intervention, pp. 823–831 (2019)

Multimodal MRI Acceleration via Deep Cascading Networks with Peer-Layer-Wise Dense Connections

Xiao-Xin Li[1,2], Zhijie Chen[1], Xin-Jie Lou[1], Junwei Yang[3], Yong Chen[4], and Dinggang Shen[2,5(✉)]

[1] College of Computer Science and Technology, Zhejiang University of Technology, Hangzhou, China
[2] School of Biomedical Engineering, ShanghaiTech University, Shanghai, China
[3] Department of Computer Science and Technology, University of Cambridge, Cambridge, UK
[4] Department of Radiology, Case Western Reserve University, Cleveland, OH, USA
[5] Shanghai United Imaging Intelligence Co., Ltd., Shanghai, China

Abstract. Medical diagnosis benefits from multimodal Magnetic Resonance Imaging (MRI). However, multimodal MRI has an inherently slow acquisition process. For acceleration, recent studies explored using a fully-sampled side modality (fSM) as a guidance to reconstruct the fully-sampled query modalities (fQMs) from their undersampled k-space data via convolutional neural networks. However, even aided by fSM, the reconstruction of fQMs from *highly* undersampled QM data (uQM) is still suffering from aliasing artifacts. To enhance reconstruction quality, we suggest to fully use both uQM and fSM via a deep cascading network, which adopts an iterative Reconstruction-And-Refinement (iRAR) structure. The main limitation of the iRAR structure is that its intermediate reconstruction operators impede the feature flow across subnets and thus leads to *short-term memory*. We therefore propose two typical Peer-layer-wise Dense Connections (PDC), namely, inner PDC (iPDC) and end PDC (ePDC), to achieve *long-term memory*. Extensive experiments on different query modalities under different acceleration rates demonstrate that the deep cascading network equipped with iPDC and ePDC consistently outperforms the state-of-the-art methods and can preserve anatomical structure faithfully up to 12-fold acceleration.

Keywords: MRI acceleration · Guidance-based reconstruction methods · Deep cascading networks · Peer-layer-wise dense connections

1 Introduction

Due to increasing the diversity of diagnostic information in a single examination, multimodal Magnetic Resonance Imaging (MRI) is commonly used in

Electronic supplementary material The online version of this chapter (https://doi.org/10.1007/978-3-030-87231-1_32) contains supplementary material, which is available to authorized users.

© Springer Nature Switzerland AG 2021
M. de Bruijne et al. (Eds.): MICCAI 2021, LNCS 12906, pp. 329–339, 2021.
https://doi.org/10.1007/978-3-030-87231-1_32

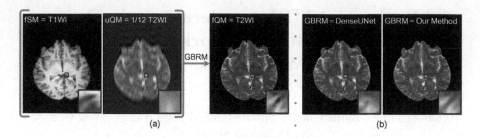

Fig. 1. (a) Framework of GBRMs. (b) Predictions of DenseUNet [22] and our method from 1/12 T2WI with the aid of T1WI. Aliasing artifacts can be seen clearly in the prediction of DenseUNet but can be well reduced by our method.

clinical application. However, all of these modalities have to be acquired in a sequential manner and data sampling of each modality in k-space is also slow due to MRI physics and physiological constraints. This makes the whole scanning procedure rather lengthy. To accelerate multimodal MRI acquisition, recent studies [2,4,5,11,20–22] explored the Guidance-Based Reconstruction Methods (GBRMs), as illustrated in in Fig. 1(a). That is, a particular modality requiring *shorter* scanning time, e.g., T1-weighted image (T1WI), is first fully sampled and used as a *side modality* (SM) [4,5], and the other modalities requiring *longer* scanning time, e.g., T2-weighted image (T2WI), referred to as *query modalities* (QMs), are then undersampled with a desired acceleration rate and subsequently reconstructed with the aid of the SM data. For simplicity, we denote by fSM, fQM, and uQM the fully-sampled SM data, fully-sampled QM data, and undersampled QM data acquired with the desired acceleration rate, respectively.

However, even aided by fSM and using the state-of-the-art Convolutional Neural Networks (CNNs) [2,11,21,22], the reconstruction result of the existing GBRMs from uQM with a high acceleration rate is still suffering from aliasing artifacts, as shown in the left subfigure of Fig. 1(b). One main reason is that both fSM and uQM have very low involvement in the forward mapping of the existing CNN models. The existing CNN models [2,11,21,22] were all based on the UNet architecture [15], which induces that fSM and uQM can be only directly used at the very beginning or very end of the network. As such, the sampled k-space data in uQM might be distorted heavily with the depth of the network increasing, and the guiding power of fSM will be weakened accordingly.

To fully use fSM and uQM, we suggest to adopt the Deep Cascading Networks [3,14,16], which employ an iterative Reconstruction-And-Refinement (iRAR) structure. In particular, the iRAR structure consists of several cascaded subnets and each subnet performs reconstruction by refining the reconstruction result of its previous subnet. Compared to UNet, the iRAR structure is much more flexible to repeatedly involve both uQM and fSM. Specifically, we can use the acquired k-space in uQM by adding a data consistency (DC) layer [16] at the end of each subnet, and reintegrate fSM into the network at the beginning of each subnet. However, the intermediate reconstructions of the iRAR structure impede the features flow from low-level subnets to high-level ones and thus lead to limited receptive field size and *short-term memory.*

To enhance information flow, we introduce two typical *peer-layer-wise dense connections* (PDC), namely inner PDC (iPDC) and end PDC (ePDC), to the iRAR structure. Here, peer layers mean the layers having the same index in different subnets. Image reconstruction benefits from dense connections [17,18,21–24], as dense connections are helpful for collecting rich knowledge and reducing feature redundancy [9]. For all we know, the existing dense connections are all designed in a *sequential-layer-wise* manner. That is, dense connections are only imposed on the layers/blocks which are cascaded sequentially. Clearly, using sequential-layer-wise dense connections (SDC) is not an efficient way to boost performance for the iRAR structure, as it cannot enlarge receptive field, and can only lead to *restricted long-term memory* [17] (i.e., the low-level features are still confined in the same subnet and cannot flow across subnets). By contrast, PDC can result in long-term memory. Please refer to Fig. 1 of the supplementary material for more clarity about the difference between PDC and SDC. In this work, we introduce iPDC and ePDC to enhance information and gradient flow between the hidden convolutional layers and the outputs of all subnets, respectively.

In summary, the study on the guidance-based multimodal MRI acceleration problem makes our work contribute in the following three aspects. *First*, to fully involve both fSM and uQM, we suggest to use the iRAR structure, instead of the UNet structure, to build GBRMs. *Second*, we create two typical *peer-layer-wise dense connections* (PDC), namely, iPDC and ePDC, to enhance information flow of the iRAR-base network. *Third*, based on iPDC and ePDC, we propose a novel iRAR-based network, namely, a Deep Cascading Network with PDC (DCNwPDC). As demonstrated in Fig. 1(b) and experiments in Sect. 4, our DCNwPDC outperforms the state-of-the-art methods for different query modalities under different acceleration rates.

2 Problem Formulation

Fig. 2. Horizontal Cartesian undersampling masks used in our experiments. Acceleration rates from left to right are 4, 8 and 12, respectively.

Let $y \in \mathbb{C}^{M \times N}/y' \in \mathbb{C}^{M \times N}$ represent fQM/fSM in k-space, and x/x' be the corresponding 2D MRI image reconstructed from y/y' via 2D Inverse Fourier

Transform (IFT), $\boldsymbol{F}^{-1}$, i.e., $x = \boldsymbol{F}^{-1}(y)$ and $x' = \boldsymbol{F}^{-1}(y')$. Our problem is to reconstruct x from uQM y_u with the aid of fSM x'/y', such that:

$$y_u = M \odot y = M \odot \boldsymbol{F}(x), \tag{1}$$

where $\odot$ is the element-wise multiplication operation, $\boldsymbol{F}$ denotes 2D Fourier Transform (FT), and $M \in \{0,1\}^{M \times N}$ represents a binary undersampling mask used for MRI acceleration. We define M by applying the widely-used horizontal Cartesian undersampling scheme [6,12,16,22], where k-space is fully-sampled in the frequency-encoding direction and undersampled in the phase-encoding direction. Figure 2 illustrates the undersampling masks used in this work. As y_u is highly undersampled, directly applying IFT on y_u will give rise to a highly aliased reconstruction of x: $x_u = \boldsymbol{F}^{-1}(y_u)$. Next, we will explore how to reconstruct x with the aid of x'/y' and the proposed DCNwPDC.

3   Proposed Method

The overall architecture of the proposed DCNwPDC is schematically illustrated in Fig. 3. It consists of T subnets. The whole architecture is well designed to adapt to our guidance-based multimodal MRI acceleration problem and to overcome the problem caused by short-term memory of the iRAR structure. Specially, it comprises the following three elements.

Long-term Involvement of uQM and fSM via Using the iRAR Structure. As shown in Fig. 3, our DCNwPDC comprises T subnets: $f_1, f_2, \cdots, f_T$. The long-term involvement of uQM and fSM is implemented by taking y_u, x' and y' as input of each subnet f_t. Let $x^{(t)}$ be the output of f_t, $x^{(0)} = x_u$, and $[\cdot, \cdot, \cdots]$ denote the concatenation operator. We define f_t in an iterative way:

$$x_{i\text{CNN}}^{(t)} = x^{(t-1)} + i\text{CNN}_t\left(\left[x^{(t-1)}, x'\right]; \hat{\mathbf{W}}_t\right) \tag{2}$$

$$x_{\text{DC}}^{(t)} = \text{DC}\left(y_u, x_{i\text{CNN}}^{(t)}; \lambda, M\right) \tag{3}$$

$$x^{(t)} = k\text{EL}_t\left(y_u, X^{(t)}, y'; \tilde{\mathbf{W}}_t\right), \tag{4}$$

where $X^{(t)} \triangleq \left\{x^{(1)}, \cdots, x^{(t-1)}, x_{\text{DC}}^{(t)}\right\}$, $x_{i\text{CNN}}^{(t)}$ and $x_{\text{DC}}^{(t)}$ are two intermediate outputs of f_t, and $\hat{\mathbf{W}}_t/\{\lambda, M\}/\tilde{\mathbf{W}}_t$ is the parameter set of $i\text{CNN}_t/\text{DC}/k\text{EL}_t$. Specially, $i\text{CNN}_t$ consists of D convolutional layers $\{\text{Conv}_{t,d}\}_{d=1}^{D}$ and goes in image domain; DC denotes the data consistence (DC) layer [16], and its parameter λ is decided by the noise level of uQM and set to be zero in this work; $k\text{EL}_t$ is the k-space ensemble-learning layer of f_t. We next focus on $i\text{CNN}_t$ and $k\text{EL}_t$.

Long-Term Memory of Internal Features via iPDC. The existing iRAR-based networks only have short-term memory [3,16] or restricted long-term memory [13,14]. To improve the information flow between subnets, we propose a

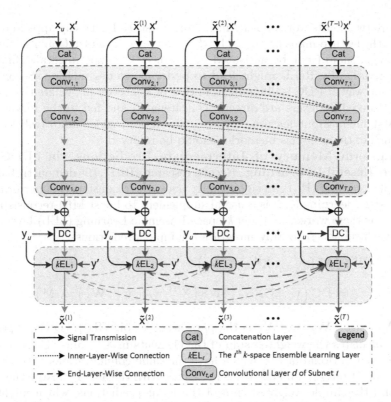

Fig. 3. The structure of the proposed Deep Cascading Network with two typical Peer-layer-wise Dense Connections: iPDC (marked by the arrows with dotted lines) and ePDC (marked by the arrows with dashed lines). The red and blue solid arrows indicate how uQM (x_u/y_u) and fSM (x') busily participate in the forward mapping. (Color figure online)

different connectivity pattern: inner Peer-layer-wise Dense Connections (iPDC). Here, peer layers mean the layers having the same index in different subnets. For instance, $\{\text{Conv}_{\iota,d}\}_{\iota=1}^{T}$ forms the peer-layer set of $\text{Conv}_{t,d}$ in Fig. 3. For any inner layer $\text{Conv}_{t,d}$ ($d \in \{2, 3, \cdots, D-1\}$), iPDC feeds its output forward to its child peer-layer set $\{\text{Conv}_{t+\iota,d+1}\}_{\iota=0}^{\tau}$. Here, $\tau = \min(T-t, \delta-1)$, and $\delta \leq T$ is the *maximum memory length* of iPDC, which is predefined as the maximum number of the forward connections between a layer and its child peer-layers. Consequently, $\text{Conv}_{t,d+1}$ receives the feature-maps from its parent peer-layer set $\{\text{Conv}_{t+\iota,d}\}_{\iota=-\ell}^{0}$ ($\ell = \min(t, \delta-1)$) as input:

$$H_{t,d+1} = \text{Conv}_{t,d+1}\left(H_{t-\ell,d}, \cdots, H_{t-1,d}, H_{t,d}; \hat{\mathbf{W}}_{t,d+1}\right)$$
$$= \hat{W}_{t,d+1}^{t} * \left[\hat{W}_{t,d+1}^{t-\ell} * H_{t-\ell,d}, \cdots, \hat{W}_{t,d+1}^{t-1} * H_{t-1,d}, \sigma\left(H_{t,d}\right)\right], \quad (5)$$

where $H_{t,d}$ is the output feature map of $\text{Conv}_{t,d}$, $*$ represents convolutional operator, σ denotes the rectifier linear unit (ReLU), and $\hat{\mathbf{W}}_{t,d+1} \triangleq \left\{\hat{W}_{t,d+1}^{\iota}\right\}_{\iota=1}^{t}$.

There are two points worth noting in Eq. (5). First, ReLU is only applied to $H_{t,d}$. That is, the feature-maps produced by the parent peer layers of $Conv_{t,d+1}$, except for $Conv_{t,d}$, should not be activated in advance. We experimentally find that early activation by ReLU might lead to performance degradation. Second, the connection way of PDC is different from that of DenseNet [9], where the feature-maps of preceding layers are directly concatenated and fed to the target layer. PDC first filters input feature-maps by convolution, which adaptively controls how much of the previous memories should be reserved.

Long-term Memory of Intermediate Predictions via ePDC and k-space Ensemble Learning. Recent research [10,17,19] demonstrated that Ensemble Learning (EL) can significantly boost the reconstruction performance. Also, k-space learning [6,7] is a useful supplement to boost MRI reconstruction. Inspired by these works, we use a k-space Ensemble Learning (kEL) layer at the end of each subnet f_t to fully integrate the input predictions,

$$\tilde{x}^{(t)} = kEL_t\left(y_u, X^{(t)}, y'; \tilde{W}_t\right),$$
$$= DC\left(y_u, kCNN_t\left(\left[WS\left(X^{(t)}; w_t\right), y'\right]; \tilde{W}_t^k\right); \lambda, M\right). \tag{6}$$

Here $\tilde{W}_t = w_t \cup \tilde{W}_t^k$, $w_t/\tilde{W}_t^k$ is the parameter set of $WS(\cdot)/kCNN_t$. Specially, $WS(\cdot)$ calculates the weighted sum of the elements in $X^{(t)}$, and $kCNN_t$ consists of three convolutional layers, which have the same configurations with $Conv_{t,1}$, $Conv_{t,2}$, and $Conv_{t,D}$, and performs in k-space with the aid of y'. Note that kEL_t is beyond the simple weighted sum of the input predictions, which might lead to unstable performance. The result of the weighted sum is further processed by $kCNN_t$, which also provides a chance to involve y'.

Network Training. Given a training dataset $\left\{x_u^{\langle\gamma\rangle}, y_u^{\langle\gamma\rangle}, x'^{\langle\gamma\rangle}, y'^{\langle\gamma\rangle}, x^{\langle\gamma\rangle}\right\}_{\gamma=1}^{\Gamma}$, our goal is to find the best model $f_{DCNwPDC}$ that accurately reconstructs x from (x_u, y_u, x', y'). Note that $f_{DCNwPDC}$ outputs multiple intermediate predictions. To boost performance, existing work [10,17] usually adopt the multi-supervision strategy. However, multi-supervision will introduce additional weight parameters to balance all predictions, and these weight parameters should be carefully chosen. For our network, multi-supervision is unnecessary, as all intermediate predictions are connected to the final output layer, which leads to implicit multi-supervision [9]. We therefore directly minimize $\frac{1}{2}\|x - f_{DCNwPDC}(x_u, y_u, x', y')\|_2^2$ averaged over the whole training set. We train our model using the Adam solver with a momentum of 0.9 and perform 100 epochs. The batch size is set to 6 and the initial learning rate is set to 10^{-4}, which is divided by 10 for every 50 epochs. The network is implemented in Python using PyTorch library. Our source code will be subsequently published online.

4   Experiments

We use the MSSEG dataset [1] to evaluate the efficacy of our DCNwPDC. MSSEG includes four MRI modalities: T1WI, T2WI, DP and FLAIR, which are

Table 1. Effect of iPDC and kEL for reconstruction of 1/8 T2WI assisted by T1WI. $\sim$ and $\sim\sim$ denote D5C5 and D5C5(Conv+Cat, 3), respectively.

	Subject 01		Subject 02		Subject 03		Average	
	PSNR	SSIM	PSNR	SSIM	PSNR	SSIM	PSNR	SSIM
$\sim$	37.54	0.9895	38.94	0.9922	38.67	0.9920	38.38	0.9912
$\sim$+iSDC(Cat, 4)	37.82	0.9903	38.74	0.9920	38.96	0.9925	38.51	0.9916
$\sim$+iPDC(Cat, 2)	37.67	0.9898	39.02	0.9925	38.93	0.9924	38.54	0.9916
$\sim$+iPDC(Res, 2)	37.56	0.9897	38.78	0.9921	38.74	0.9921	38.36	0.9913
$\sim$+iPDC(Conv+Cat, 2)	37.76	0.9899	39.00	0.9925	39.06	0.9927	38.61	0.9917
$\sim$+iPDC(ReLU+Conv+Cat, 2)	37.72	0.9900	38.91	0.9925	38.95	0.9926	38.53	0.9917
$\sim$+iPDC(Conv+Cat, 3)	37.84	0.9903	39.10	0.9926	38.95	0.9926	38.63	0.9918
$\sim$+iPDC(Conv+Cat, 4)	37.84	**0.9904**	**39.12**	**0.9928**	**39.10**	**0.9929**	**38.69**	**0.9920**
$\sim$+iPDC(Conv+Cat, 5)	**37.86**	**0.9904**	**39.12**	**0.9928**	39.09	0.9928	**38.69**	**0.9920**
$\sim\sim$+WS(X)	37.80	0.9903	39.12	0.9929	39.13	0.9929	38.68	0.9920
$\sim\sim$+kCNN(x_{DC})	37.90	0.9905	39.22	0.9930	39.12	0.9929	38.75	0.9921
$\sim\sim$+kCNN(x_{DC}, y')	37.87	0.9904	39.24	0.9930	39.22	0.9930	38.78	0.9921
$\sim\sim$+kCNN(WS(X))	37.84	0.9904	39.23	0.9930	39.13	0.9929	38.73	0.9921
$\sim\sim$+kCNN(WS(X), y')	**37.95**	**0.9906**	**39.26**	**0.9931**	**39.26**	**0.9933**	**38.82**	**0.9923**

Table 2. Quantitative evaluation of the compared methods in Regions of Interest (ROI) for the reconstruction performance of 1/4, 1/8, and 1/12 undersampled QM $\in$ {T2WI, DP} assisted by T1WI. Five-fold cross validations are performed. Numbers in red and blue indicate the first and second best performance, respectively.

	1/4 T2WI		1/8 T2WI		1/12 T2WI	
	PSNR	SSIM	PSNR	SSIM	PSNR	SSIM
Zero Filling	31.01±0.0671	0.9591±4.35e-06	29.47±0.61603	0.9038±3.94e-05	27.19±1.2760	0.8683±4.90e-05
UNet	37.48±0.0350	0.9784±8.27e-08	36.72±0.41053	0.9646±9.95e-07	35.69±0.6826	0.9632±1.19e-06
DenseUNet	37.83±0.0761	0.9785±2.45e-07	36.82±0.40868	0.9645±1.22e-06	35.76±0.6526	0.9631±1.15e-06
RefineGAN	37.93±0.0405	0.9776±0.83e-07	36.94±0.46610	0.9637±1.12e-06	36.16±0.5782	0.9624±1.21e-06
D5C5	39.04±0.0242	0.9785±4.57e-07	37.71±0.38307	0.9652±1.16e-06	36.73±0.5317	0.9637±1.26e-06
D5C5+iPDC	39.53±0.0420	0.9792±7.95e-08	38.25±0.38315	0.9663±9.03e-07	37.38±0.4986	0.9645±9.25e-07
D5C5+iPDC+kEL	**39.79±0.0338**	**0.9795±6.83e-08**	**38.39±0.33928**	**0.9668±8.13e-07**	**37.45±0.4882**	**0.9648±9.17e-07**

	1/4 DP		1/8 DP		1/12 DP	
	PSNR	SSIM	PSNR	SSIM	PSNR	SSIM
Zero Filling	33.17±0.0683	0.9692±3.98e-06	32.10±0.64329	0.9417±3.25e-05	29.21±0.6247	0.9281±5.42e-05
UNet	38.26±0.0889	0.9779±1.69e-07	37.72±0.44730	0.9655±1.29e-06	36.35±0.5251	0.9613±1.73e-06
DenseUNet	38.38±0.0791	0.9777±1.39e-07	38.33±0.45025	0.9653±1.12e-06	36.52±0.9043	0.9613±1.96e-06
RefineGAN	39.04±0.0354	0.9777±1.97e-07	39.93±0.49513	0.9668±1.63e-06	37.85±0.2215	0.9638±1.92e-06
D5C5	42.48±0.0586	0.9795±1.35e-07	40.92±0.35283	0.9680±1.47e-06	38.78±0.7644	0.9654±1.64e-06
D5C5+iPDC	42.83±0.0726	0.9803±8.59e-08	41.40±0.32191	0.9686±9.12e-07	39.27±0.9208	0.9659±7.17e-07
D5C5+iPDC+kEL	**42.87±0.0540**	**0.9806±8.31e-08**	**41.51±0.30215**	**0.9690±8.49e-07**	**39.41±0.9027**	**0.9661±7.69e-07**

scanned from three different scanners. We use T1WI as the side modality, and the rest T2WI, DP and FLAIR as the query modalities. As the three scanners generate images of different sizes, we select five subjects scanned by the same Philips Ingenia 3T scanner, which produces 3D volumes containing 336 × 336 × 261 voxels. By excluding the slices only containing the background, we obtain around 200 samples in axial direction for each subject. Training and test subjects are split to 4 : 1. We evaluate the reconstruction results using Peak Signal-to-Noise Ratio (PSNR) and Structural Similarity Index (SSIM). Throughout the

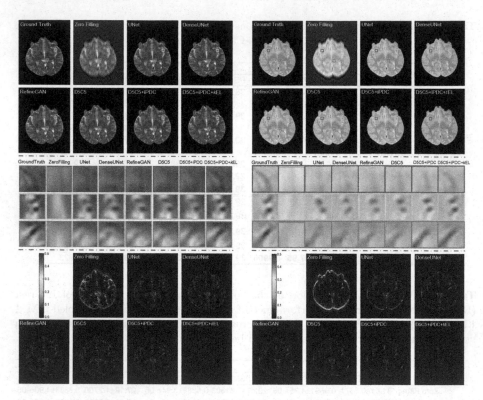

Fig. 4. Visual comparison of the reconstruction results (top row) and the error maps (bottom row) of the seven compared methods for 12-fold undersampled T2WI (left) and DP (right). Three subregions of interest for each reconstructed image are enlarged in the middle row for more clarity.

experiments, we use D5C5 [16] as a baseline network to implement our DCNwPDC. D5C5 is a classical iRAR-based network. By setting $D = 5$ and $T = 5$, we have DCNwPDC = D5C5 + iPDC + kEL.

We first evaluate the effects of two important elements of $f_{DCNwPDC}$: iPDC and kEL. We fix T1WI and 1/8 undersampled T2WI (shortly, 1/8 T2WI) as fSM and uQM, and sequentially select Subjects 01, 02 and 03 as the test subject.

Effect of iPDC. We compare iPDC defined in Eq. (5) against iSDC (inner SDC) and the variants of iPDC with other possible configurations. Equation (5) indicates that iPDC mainly has two configurations: the connection way between the input feature-maps and the maximum memory length δ. Equation (5) uses the Conv+Cat connections. The other possible operators that can replace Conv+Cat are are Cat, Res (Residual) [8], and ReLU+Conv+Cat. We represent a specific iPDC/iSDC by iPDC/iSDC(Connection-way, δ). The top sub-table of Table 1 reports how the connection way and δ influence the reconstruction performance. We can find that compared to $\sim$ ($\sim$ denotes D5C5), both $\sim$+iSDC(Cat, 4) and $\sim$+iPDC(Res, 2) lead to performance degradation on Subject 02. By comparing

$\sim$+iPDC(Conv+Cat, 2) and $\sim$+iPDC(ReLU+Conv+Cat, 2), we can find early ReLU might lead to performance dropping. δ usually boosts performance but a too big δ might lead to overfitting due to using more parameters. Table 1 also indicates our iPDC can effectively use the network capacity by comparing the performance of $\sim$, $\sim$+iSDC(Cat, 4) and $\sim$+iPDC(Cat, 2). Clearly, $\sim$+iSDC(Cat, 4) uses more parameters than $\sim$ but it might lead to performance degeneration against $\sim$. By contrast, $\sim$+iPDC(Cat, 2) leads to stable performance boost over $\sim$, although it has fewer parameters than $\sim$+iSDC(Cat, 4). This demonstrates the superiority of the network design of our iPDC against iSDC.

Effect of kEL. By observing Eq. (6), we can obtain several options to implement the main part (except DC) of kEL: e.g., WS(X), kCNN(x_{DC}), kCNN(x_{DC}, y') and etc. The bottom sub-table of Table 1 reports the related performance. Only kCNN (WS(X), y'), as defined in Eq. (6), stably outperforms the other options.

Comparisons with Existing Methods. We compare our DCNwPDC with the basic zero-filling method and four advanced CNN-based methods: UNet [15], DenseUNet [22], RefineGAN [14], and D5C5 [16]. Note that even if the GAN-based networks [2,11] are specifically designed for guided MRI reconstruction, we choose RefineGAN [14] for comparison due to its stronger reconstruction abilities. Specially, RefineGAN also adopts the iRAR structure by cascading two UNets. For fair comparison, we implement all iRAR-based methods by inputting fSM at the beginning of all subnets. Table 2 reports the quantitative evaluation results of the compared methods on two query modalities, T2WI and DP, in Regions of Interest (ROI). Note that we perform quantitative evaluations in ROI rather than in the whole image, as radiologists mainly concern the subtle abnormalities in ROI. Figure 3 in our supplementary material plots the ROI used for the query modalities. For each uQM acquired with three acceleration rates, our method (i.e., D5C5 + iPDC + kEL) and its simplified variant D5C5 + iPDC achieve the best and second best performance, respectively. In particular, our method gains better performance for higher acceleration rates. It indicates that enhancing information flow is particularly important for highly aggressive undersampling. Figure 4 further demonstrates the superiority of our method by visually comparing the reconstruction quality of the compared methods for 12-fold undersampled T2WI and DP. At this high acceleration rate, we can see clearly that our method better restores more anatomical details than the other compared methods. The quantity and quality evaluation results of the compared methods on FLAIR reconstruction are reported in our supplementary materials.

5   Conclusion

By integrating two typical peer-layer-wise dense connections (iPDC and ePDC) to the iRAR-based structure, we propose a novel deep cascading network, which outperforms the state-of-the-art methods. With the proposed network, the acquisition of query modalities can be accelerated up to 12 folds without evident sacrifice of image quality. Future studies will be focused on further improvement of DCNwPDC by optimizing the design of PDC, and extending to handle more sampling patterns beyond the Cartesian trajectory, such as radial, spiral, etc.

References

1. Commowick, O., Cervenansky, F., Ameli, R.: MSSEG challenge proceedings: multiple sclerosis lesions segmentation challenge using a data management and processing infrastructure. In: International Conference on Medical Image Computing and Computer-Assisted Intervention (2016)
2. Dar, S.U., Yurt, M., Shahdloo, M., Ildız, M.E., Tınaz, B., Çukur, T.: Prior-guided image reconstruction for accelerated multi-contrast MRI via generative adversarial networks. IEEE J. Select. Topics Signal Process. **14**(6), 1072–1087 (2020)
3. Denton, E.L., Chintala, S., Fergus, R., et al.: Deep generative image models using a Laplacian pyramid of adversarial networks. In: Advances in Neural Information Processing Systems, pp. 1486–1494 (2015)
4. Ehrhardt, M.J.: Multi-modality imaging with structure-promoting regularisers. arXiv:2007.11689 (2020)
5. Ehrhardt, M.J., Betcke, M.M.: Multicontrast MRI reconstruction with structure-guided total variation. SIAM J. Imaging Sci. **9**(3), 1084–1106 (2016)
6. Eo, T., Jun, Y., Kim, T., Jang, J., Lee, H.J., Hwang, D.: KIKI-net: cross-domain convolutional neural networks for reconstructing undersampled magnetic resonance images. Magn. Reson. Med. **80**(5), 2188–2201 (2018)
7. Han, Y., Sunwoo, L., Ye, J.C.: k-space deep learning for accelerated MRI. IEEE Trans. Med. Imaging **39**(2), 377–386 (2019)
8. He, K., Zhang, X., Ren, S., Sun, J.: Deep residual learning for image recognition. In: IEEE Conference on Computer Vision and Pattern Recognition, pp. 770–778. Las Vegas, USA (2016)
9. Huang, G., Liu, Z., van der Maaten, L., Weinberger, K.Q.: Densely connected convolutional networks. In: IEEE Conference on Computer Vision and Pattern Recognition, pp. 4700–4708 (2017)
10. Kim, J., Kwon Lee, J., Mu Lee, K.: Deeply-recursive convolutional network for image super-resolution. In: IEEE Conference on Computer Vision and Pattern Recognition, pp. 1637–1645 (2016)
11. Kim, K.H., Do, W.J., Park, S.H.: Improving resolution of MR images with an adversarial network incorporating images with different contrast. Med. Phys. **45**(7), 3120–3131 (2018)
12. Li, W., Feng, X., An, H., Ng, X.Y., Zhang, Y.J.: MRI reconstruction with interpretable pixel-wise operations using reinforcement learning. In: Proceedings of the AAAI Conference on Artificial Intelligence, vol. 34, pp. 792–799 (2020)
13. Qin, C., Schlemper, J., Caballero, J., Price, A.N., Hajnal, J.V., Rueckert, D.: Convolutional recurrent neural networks for dynamic MR image reconstruction. IEEE Trans. Med. Imaging **38**(1), 280–290 (2019)
14. Quan, T.M., Nguyen-Duc, T., Jeong, W.K.: Compressed sensing MRI reconstruction using a generative adversarial network with a cyclic loss. IEEE Trans. Med. Imaging **37**(6), 1488–1497 (2018)
15. Ronneberger, O., Fischer, P., Brox, T.: U-Net: Convolutional networks for biomedical image segmentation. In: International Conference on Medical Image Computing and Computer-Assisted Intervention, pp. 234–241. Springer (2015). https://doi.org/10.1007/978-3-319-24574-4_28

16. Schlemper, J., Caballero, J., Hajnal, J.V., Price, A.N., Rueckert, D.: A deep cascade of convolutional neural networks for dynamic MR image reconstruction. IEEE Trans. Med. Imaging **37**(2), 491–503 (2018)
17. Tai, Y., Yang, J., Liu, X., Xu, C.: Memnet: a persistent memory network for image restoration. In: IEEE International Conference on Computer Vision, pp. 4539–4547 (2017)
18. Tong, T., Li, G., Liu, X., Gao, Q.: Image super-resolution using dense skip connections. In: The IEEE International Conference on Computer Vision (2017)
19. Wang, L., Huang, Z., Gong, Y., Pan, C.: Ensemble based deep networks for image super-resolution. Pattern Recogn. **68**, 191–198 (2017)
20. Weizman, L., Eldar, Y.C., Ben Bashat, D.: Reference-based MRI. Med. Phys. **43**(10), 5357–5369 (2016)
21. Xiang, L., et al.: Ultra-fast T2-weighted MR reconstruction using complementary T1-weighted information. In: International Conference on Medical Image Computing and Computer-Assisted Intervention, pp. 215–223 (2018)
22. Xiang, L., et al.: Deep-learning-based multi-modal fusion for fast MR reconstruction. IEEE Trans. Biomed. Eng. **66**(7), 2105–2114 (2019)
23. Zhang, Y., Tian, Y., Kong, Y., Zhong, B., Fu, Y.: Residual dense network for image restoration. IEEE Trans. Pattern Anal. Mach. Intell. p. online (2020)
24. Zhang, Y., Tian, Y., Kong, Y., Zhong, B., Fu, Y.: Residual dense network for image super-resolution. In: IEEE Conference on Computer Vision and Pattern Recognition, pp. 2472–2481 (2018)

Rician Noise Estimation for 3D Magnetic Resonance Images Based on Benford's Law

Rosa Maza-Quiroga[1,3]([✉]) [iD], Karl Thurnhofer-Hemsi[1,3] [iD],
Domingo López-Rodríguez[2,3] [iD], and Ezequiel López-Rubio[1,3] [iD]

[1] Department of Computer Languages and Computer Science,
Universidad de Málaga, Málaga, Spain
{rosammq,karkhader,ezeqlr}@lcc.uma.es, dominlopez@uma.es
[2] Department of Applied Mathematics, Universidad de Málaga,
Bulevar Louis Pasteur, 35, 29071 Málaga, Spain
[3] Biomedic Research Institute of Málaga (IBIMA), Málaga, Spain

Abstract. In this paper, a novel method to estimate the level of Rician noise in magnetic resonance images is presented. We hypothesize that noiseless images follow Benford's law, that is, the probability distribution of the first digit of the image values is logarithmic. We show that this is true when we consider the raw acquired image in the frequency domain. Two measures are then used to quantify the (dis)similarity between the actual distribution of the first digits and the more theoretical Benford's law: the Bhattacharyya coefficient and the Kullback-Leibler divergence. By means of these measures, we show that the amount of noise directly affects the distribution of the first digits, thereby making it deviate from Benford's law. In addition, in this work, these findings are used to design a method to estimate the amount of Rician noise in an image. The utilization of supervised machine learning techniques (linear regression, polynomial regression, and random forest) allows predicting the parameters of the Rician noise distribution using the dissimilarity between the measured distribution and Benford's law as the input variable for the regression. In our experiments, testing over magnetic resonance images of 75 individuals from four different repositories, we empirically show that these techniques are able to precisely estimate the noise level present in the test T1 images.

Keywords: MRI · Rician noise · Benford's law · Noise estimation

1 Introduction

Real-world magnetic resonance imaging (MRI) data is very often corrupted by a noise component, generated in the acquisition process. The noise is a challenging problem since it degrades the reliability of both radiologists and automatic

M. de Bruijne et al. (Eds.): MICCAI 2021, LNCS 12906, pp. 340–349, 2021.
https://doi.org/10.1007/978-3-030-87231-1_33

computer-aided diagnosis. It also can affect the efficiency of automated quantitative post-processing methods, which are increasingly used nowadays, both in clinical practice and in research.

Denoising methods are an important part of the pre-processing of MRIs, and try to improve the image quality by increasing the Signal-to-Noise Ratio (SNR) while preserving the image features. Currently, many different denoising methods have appeared in the literature. We can find methods based, for example, on the wavelet transform [24], the anisotropic diffusion filter [12] or non-local filters [25], the linear minimum square error [6], a sparse representation learning [3], the singular value decomposition [27] or the maximum likelihood approach [19].

With the rise of deep learning, new denoising techniques have appeared, e.g., the stacked sparse auto-encoder [4], multi-layer perceptron [13] or convolutional neural networks [23]. Using residual learning, the authors in [26] developed a deep denoising conventional neural network for Gaussian denoising, achieving good performance. Also, in [8], a convolutional neural network is presented for medical image denoising. More recently, attention-guided models such as [22] have been presented, showing good performance.

It is already known that MRI noise follows a Rician distribution [7] and that there is around 60% underestimation of the true noise if the noise in MRI is assumed to be Gaussian. Note that the Rician distribution is signal-dependent, differently from the additive Gaussian noise. Thus, separating noise from the raw MRI without losing critical image features remains a challenging task.

Some of the different denoising algorithms assume that the deviation parameter σ of the Rician distribution which generates the noise is known. Noise parameters can be estimated by using methods based on principal components analysis [15] and on the wavelet transform [2]. The method based on PCA is best suited for weak texture images but not so good for Rician noise estimation. In the wavelet approach, the image is decomposed in sub-bands, of which the HH sub-band is composed of the wavelet noise coefficient. The median of these coefficients is used to compute the *median absolute deviation* estimator for σ. Although this wavelet model fits better Gaussian noise, it can be adequately modified [11] to estimate the σ parameter in Rician noise.

In this work, we propose an alternate technique to estimate the noise deviation parameter, based on Benford's law, that is, in the statistical distribution of the first significant digits in a dataset. Although it is well known that image histogram does not follow Benford's law, certain transformations in the image are consistent with such distribution. Particularly, in [9], it is shown that the gradient and Laplace transform magnitude follows Benford's law, even in medical images such as MRI [20]. Other transformations whose coefficients follow Benford's law are the discrete cosine and wavelet transforms [1].

Our proposal aims at demonstrating that the coefficients of the Fourier transform of an image follow closely Benford's law, and we hypothesize that larger amounts of noise in an image make these coefficients deviate from such distribution. Hence, the level of agreement between the expected distribution and the actual first digit distribution in the Fourier domain is an indicator of the noise parameter σ. As an application of these results, we show how regression techniques are able to accurately predict the noise level of an image.

2   Methodology

In this section, the proposed methodology to estimate Rician noise in 3D MRIs is presented. It is well known that the noise in magnitude MRIs can be modeled by Rician noise [7]. Let $\hat{x}$ be the original (noiseless) image pixel intensity, and x the measured pixel intensity in the presence of Rician noise of level σ, where σ is the standard deviation of the Gaussian noise affecting the real and imaginary MRIs, so that the Gaussian noise level σ in both real and imaginary images is assumed to be the same. Then the probability distribution for x is given by:

$$p\left(x\right) = \frac{x}{\sigma^2} \exp\left(-\frac{x^2 + \hat{x}^2}{2\sigma^2}\right) I_0 \left(\frac{x\hat{x}}{\sigma^2}\right) \tag{1}$$

where I_0 is the modified zeroth order Bessel function of the first kind.

In this work we propose to employ Benford's law to estimate the Rician noise level σ. Benford's law is an empirical law which states that the probability distribution $Q\left(n\right)$ of the first digit n of the decimal representation of a nonzero real number y is logarithmic [5]:

$$Q\left(n\right) = \log_{10}\left(1 + \frac{1}{n}\right), \quad \text{where} \quad n = \left\lfloor \frac{|y|}{10^{\lfloor \log_{10} |y| \rfloor}} \right\rfloor \tag{2}$$

where $n \in \{1, 2, ..., 9\}$, $|\cdot|$ stands for the absolute value of a real number, and $\lfloor \cdot \rfloor$ denotes rounding towards $-\infty$.

Despite its counter intuitive nature, Benford's law has been found to hold for many datasets coming from natural processes, such as natural images in a transformed domain [1]. Our hypothesis is that the higher the Rician noise corrupting a MRI, the farther that the image departs from Benford's law. This effect is better observed for the values of the 3D Fast Fourier Transform (FFT) of the MRI rather than the raw pixel intensity values because the distribution of the first digits of the latter is affected by the measurable pixel intensity range, while the former is relatively independent from the pixel intensity range. As seen in [21], distributions that have a large width, i.e. they spread their probability mass across several orders of magnitude, follow Benford's law more closely. The raw pixel intensity values have a limited range, so their distribution has a reduced width. On the contrary, the FFT of the raw values is not constrained by this limitation. In fact, the compression/expansion property of the FFT means that the narrower the range of the raw pixel values, the wider the range of their FFT.

Let us note y the 3D FFT of the measured pixel intensity values x, so that the first digit of the decimal representation of y is noted n, as given by (2). Then a measure $\mathcal{D}$ of the dissimilarity of the observed distribution $P\left(n\right)$ with respect to Benford's law distribution $Q\left(n\right)$ given by (2) must be selected. We

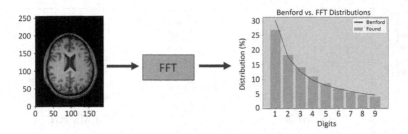

Fig. 1. First step: FFT histogram computation and comparison with Benford's law.

consider two such measures, namely the Bhattacharyya Coefficient (BC) and the Kullback-Leibler (KL) divergence:

$$\mathcal{D}_{BC} = \sum_{n=1}^{9} \sqrt{P(n)Q(n)}, \quad \mathcal{D}_{KL} = \sum_{n=1}^{9} P(n) \log \frac{P(n)}{Q(n)} \tag{3}$$

It must be noted that the lower $\mathcal{D}_{KL}$, the closer the observed distribution $P(n)$ to Benford's law distribution $Q(n)$. In contrast to this, the higher $\mathcal{D}_{BC}$, the closer the observed distribution $P(n)$ to Benford's law distribution $Q(n)$.

After this, the Rician noise level σ of the MRI is estimated as a function of the selected measure $\mathcal{D}$:

$$\sigma \approx f(\mathcal{D}) \tag{4}$$

where f is estimated by machine learning regression techniques.

3   Experimental Setting

Three types of experiments are presented. First of all, we verify the distribution of the first digit of voxels values of a 3D MRI in the frequency domain resembles Benford's law. Secondly, we check if the noise in an image disturbs Benford's law probability distribution. Finally, we propose three types of models to predict the quantity of noise in an MRI.

A total of 75 T1-weighted brain MRIs were selected, being publicly accessible with a non-restrictive license available in Mindboogle [10], from healthy participants, and of high quality to ensure right noise analysis. The images in NIfTI format come from 4 repositories: 12 from HLN [17] with dimension $256 \times 256 \times 170$, 21 from MMRR [14] with dimension $170 \times 256 \times 256$, 20 from NKI-RS [18] and 20 from NKI-TRT [18] with dimension $192 \times 256 \times 256$.

Using all the MRI voxels, the FFT is computed and the probability distribution of each voxel's frequency of the first digit is constructed. Finally, all the images in the Fourier domain are assessed to follow a distribution that matches Benford's law. Then, for each image, 20 new images with noised values were generated varying the distortion in the range $[0, 10)$ percent of Rician noise with respect to the amplitude of the signal, following a uniform continuous distribution. Thus, for each repository, we have a new image data set. As explained

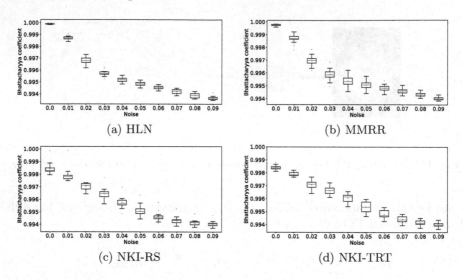

Fig. 2. Evolution of Bhattacharyya coefficient as noise increases in four image repositories. Average is in green diamond and median in red line. (Color figure online)

in Sect. 2, two methods are used to measure the similarity with Benford's law: BC, belonging to the range $[0, 1]$, where it will be 0 if there is no overlap and 1 if it matches perfectly; and KL divergence, where 0 indicates that the two distributions in question are identical. Since Benford's law appears to be a noise detector, three types of regressors were used to predict the noise in an image: Linear Regression (LR), Polynomial Regression of degree two (PR) and, Random Forest (RF). The data were split into training (80%) and testing (20%) samples selected from each dataset to validate the considered models.

In order to assess the accuracy of the estimation (4), the Mean Squared Error is employed. In addition, the regression score function, also known as the coefficient of determination, was used to measure the goodness of fit:

$$MSE = \frac{1}{M} \sum_{i=1}^{M} (\sigma_i - f(\mathcal{D}_i))^2, \quad R^2 = 1 - \frac{\sigma_r^2}{\sigma^2} \quad (5)$$

where M is the total number of test images, σ is the variance of the dependent variable, and the residual variance is σ_r.

In this work, a PC with Intel Core 7 CPU, 32 GB RAM, a NVIDIA TRX2080 Super Ventus, and 1 TB of SSD was used, running under Ubuntu 20.04, and using Python 3.8. The scientific libraries *matplotlib*, *nibabel*, *scipy*, *sklearn*, and *benford* [16] were used. All models were trained using default parameters[1].

---

[1] The source code with scripts and sample data is available in: https://github.com/icai-uma/RicianNoiseEst_3DMRI_BenfordsLaw.git.

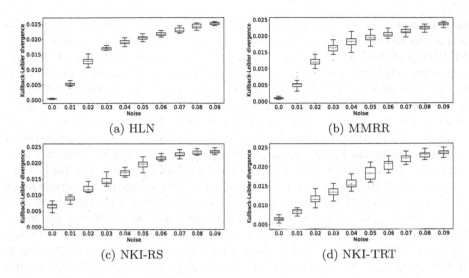

(a) HLN

(b) MMRR

(c) NKI-RS

(d) NKI-TRT

Fig. 3. Evolution of Kullback-Leibler divergence as noise increases in four image repositories. Average is in green diamond and median in red line. (Color figure online)

4 Results

The high quality of the original datasets ensures the right surface reconstruction without noise, and the evidence is shown in Figs. 2 and 3. When no noise is added, i.e. 0.0, BC takes values close to 1 and KL values close to 0, especially in HLN and MMRR. The values of BC and KL in NKI-RS and NKI-TRT might be due to the lack of enough quality in the acquisition protocol since BC is between 0.998 and 0.999, and KL is not so close to 0.005. Nevertheless, for all repositories, the overall trend remains presenting very good values of BC and KL without noise. Thus, we accept that a noiseless MRI in the Fourier frequency domain follows Benford's law.

Table 1. MSE and R^2 results of the prediction models by using the Bhattacharyya Distance and Kullback-Leibler divergence for all datasets. MSE is multiplied by 10^{-5}

Metric	Bhattacharyya coefficient						Kullback-Leibler					
Regressor	LR		PR		RF		LR		PR		RF	
Dataset	MSE	R^2	MSE	R^2	MSE	R^2	MSE	R^2	MSE	R^2	MSE	R^2
HLN	11.94	0.82	1.87	0.97	2.57	0.96	11.79	0.82	1.83	0.97	2.62	0.96
MMRR	16.93	0.81	5.37	0.94	6.98	0.92	16.58	0.82	5.039	0.94	5.39	0.94
NKI-RS	7.92	0.89	5.76	0.92	6.98	0.91	8.25	0.88	6.09	0.92	8.42	0.89
NKI-TRT	4.98	0.94	4.18	0.95	6.02	0.93	5.15	0.94	4.33	0.95	5.54	0.93

With the addition of noise in the image, the results shown in Figs. 2 and 3 indicate that the first digit distribution of an image changes. A dependency between noise and the parameter's value is clearly appreciated. The more noise is introduced into the image, the farther the distribution is from Benford's law. NKI-RS and NKI-TRT present a nearly linear decrease trend in BC and an approximately linear growth trend in KL. On the contrary, the HLN and MMRR repositories show a fast decrease in BC and a rapid KL increase. NKI-TRT boxes are the most extensive compared to other repositories showing a more considerable distance between quantiles, but NKI-TRT has no outliers.

Now, the results of estimating the noise level by means of the regression techniques mentioned above are presented. Note that BC and KL measures have a significancy in the order of 10^{-3}. The results of MSE presented in Table 1 are in the order of 10^{-5}, indicating good results with two orders of magnitude lower of the parameters. Besides, in the same table, the R^2 shows values close to 1 in general, so the models fit the data well globally. The best results are generated by PR and followed by RF, and finally LR. The worst results are obtained using LR for MMRR repository and the best are for the PR model with HLN. Models and training data are represented in Figs. 4 and 5. The models fit the training data well without losing test accuracy. As more noise is added, the BC and KL values have a wider range.

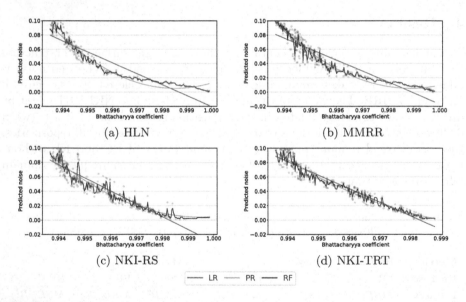

Fig. 4. Bhattacharyya Coefficient training data (black dots) are displayed along with the three models (LR in red, PR in green and RF in blue) in four repositories. (Color figure online)

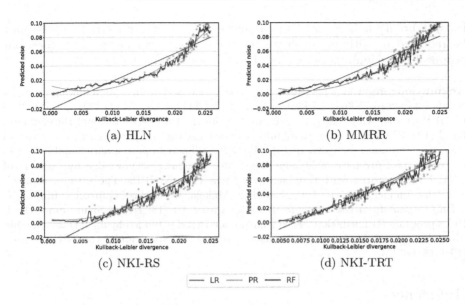

(a) HLN

(b) MMRR

(c) NKI-RS

(d) NKI-TRT

—— LR —— PR —— RF

Fig. 5. Kullback-Leibler divergence training data (blue dots) are displayed along with the three models (LR in red, PR in blue and RF in green) in four repositories. (Color figure online)

5   Conclusions

In this work, we have shown that the coefficients of the Fourier transform of a T1 MRI follow Benford's law, that is, their first digit distribution is logarithmic. We also demonstrate that the amount of Rician noise present in an image directly affects the first digit distribution of the Fourier transform of a T1 MRI, making it deviate from Benford's law. Hence, by measuring the level of agreement of the distribution of the first digits of the Fourier coefficients with Benford's law, using the the Bhattacharyya coefficient and the Kullback-Leibler divergence, one can estimate the noise level in a T1 MRI.

In addition, in this paper we show that supervised learning techniques allow estimating the noise level, using the distribution dissimilarity measures mentioned above as regressors. Although not all the datasets have a optimal quality, our experiments over MRIs of 75 individuals confirm that Benford's law is fulfilled by the Fourier coefficients of noiseless T1 MRIs and that it can be properly used to precisely estimate the noise level, since all the error measures followed a similar tendency. Therefore, further works on denoising algorithms could integrate this methodology to estimate the level of noise. This will make the algorithms work better and have a direct implication in improving radiologists' diagnoses.

To sum up, our work presents an empirical demonstration of the proposed hypothesis of Benfords law in the frequency domain and its novel use as a noise estimator with the help of machine learning, providing very promising results.

It must be highlighted that more than one image quality metrics (IQMs) may be required to faithfully evaluate the noise level of an image. Therefore, our proposal is a significant achievement in the search for reliable IQMs.

Acknowledgements. This work is partially supported by the following Spanish grants: TIN2016-75097-P, PIT.UMA.B1.2017, RTI2018-094645-B-I00 and UMA18-FEDERJA-084. All of them include funds from the European Regional Development Fund (ERDF). The authors thankfully acknowledge the computer resources, technical expertise and assistance provided by the SCBI (Supercomputing and Bioinformatics) center of the University of Málaga. They also gratefully acknowledge the support of NVIDIA Corporation with the donation of two Titan X GPUs. The authors acknowledge the funding from the Universidad de Málaga. Rosa Maza-Quiroga is funded by a Ph.D. grant from the Instituto de Salud Carlos III (ISCIII) of Spain under the i-PFIS program (IFI19/00009). Karl Thurnhofer-Hemsi is funded by a Ph.D. scholarship from the Spanish Ministry of Education, Culture and Sport under the FPU program (FPU15/06512).

References

1. Al-Bandawi, H., Deng, G.: Classification of image distortion based on the generalized Benford's law. Multimedia Tools Appl. **78**, 25611–25628 (2019)
2. Chang, S.G., Yu, B., Vetterli, M.: Spatially adaptive wavelet thresholding with context modeling for image denoising. IEEE Trans. Image Process. **9**(9), 1522–1531 (2000)
3. Chen, W., You, J., Chen, B., Pan, B., Li, L., Pomeroy, M., Liang, Z.: A sparse representation and dictionary learning based algorithm for image restoration in the presence of rician noise. Neurocomputing **286**, 130–140 (2018)
4. Dolz, J., et al.: Stacking denoising auto-encoders in a deep network to segment the brainstem on mri in brain cancer patients: a clinical study. Comput. Med. Imaging Graph. **52**, 8–18 (2016)
5. Fu, D., Shi, Y.Q., Su, W.: A generalized Benford's law for JPEG coefficients and its applications in image forensics. In: III, E.J.D., Wong, P.W. (eds.) Security, Steganography, and Watermarking of Multimedia Contents IX, vol. 6505, pp. 574–584. International Society for Optics and Photonics, SPIE (2007)
6. Golshan, H.M., Hasanzadeh, R.P., Yousefzadeh, S.C.: An mri denoising method using image data redundancy and local snr estimation. Magn. Reson. Imaging **31**(7), 1206–1217 (2013)
7. Gudbjartsson, H., Patz, S.: The Rician distribution of noisy MRI data. Magn. Reson. Med. **34**(6), 910–914 (1995)
8. Jifara, W., Jiang, F., Rho, S., Cheng, M., Liu, S.: Medical image denoising using convolutional neural network: a residual learning approach. J. Supercomput. **75**(2), 704–718 (2019)
9. Jolion, J.M.: Images and benford's law. J. Math. Imaging Vis. **14**(1), 73–81 (2001)
10. Klein, A., Tourville, J.: 101 labeled brain images and a consistent human cortical labeling protocol. Front. Neurosci. **6**, 171 (2012). https://doi.org/10.3389/fnins.2012.00171
11. Koay, C.G., Basser, P.J.: Analytically exact correction scheme for signal extraction from noisy magnitude MR signals. J. Magn. Reson. **179**(2), 317–322 (2006)

12. Krissian, K., Aja-Fernández, S.: Noise-driven anisotropic diffusion filtering of MRI. IEEE Trans. Image Process. **18**(10), 2265–2274 (2009)
13. Kwon, K., Kim, D., Park, H.: A parallel mr imaging method using multilayer perceptron. Med. Phys. **44**(12), 6209–6224 (2017)
14. Landman, B.A., et al.: Multi-parametric neuroimaging reproducibility: a 3-T resource study. Neuroimage **54**(4), 2854–2866 (2011). https://doi.org/10.1016/j.neuroimage.2010.11.047
15. Liu, X., Tanaka, M., Okutomi, M.: Noise level estimation using weak textured patches of a single noisy image. In: 2012 19th IEEE International Conference on Image Processing, pp. 665–668. IEEE (2012)
16. Marcel, M.: Benford_py: a Python Implementation of Benford's Law Tests (2017). https://github.com/milcent/benford_py
17. Morgan, V.L., Mishra, A., Newton, A.T., Gore, J.C., Ding, Z.: Integrating functional and diffusion magnetic resonance imaging for analysis of structure-function relationship in the human language network. PLOS ONE **4**(8), 1–8 (2009). https://doi.org/10.1371/journal.pone.0006660
18. Nooner, K.B., et al.: The NKI-Rockland sample: a model for accelerating the pace of discovery science in psychiatry. Front. Neurosci. **6**, 152 (2012). https://doi.org/10.3389/fnins.2012.00152
19. Rajan, J., Jeurissen, B., Verhoye, M., Van Audekerke, J., Sijbers, J.: Maximum likelihood estimation-based denoising of magnetic resonance images using restricted local neighborhoods. Phys. Med. Biol. **56**(16), 5221 (2011)
20. Sanches, J., Marques, J.S.: Image Reconstruction using the Benford Law. In: 2006 International Conference on Image Processing, pp. 2029–2032 (2006). https://doi.org/10.1109/ICIP.2006.312845
21. Smith, S.W.: The Scientist & Engineer's Guide to Digital Signal Processing. California Technical Publishing, San Diego, CA (1997)
22. Tian, C., Xu, Y., Li, Z., Zuo, W., Fei, L., Liu, H.: Attention-guided CNN for image denoising. Neural Netw. **124**, 117–129 (2020)
23. Tripathi, P.C., Bag, S.: Cnn-dmri: a convolutional neural network for denoising of magnetic resonance images. Pattern Recogn. Lett. **135**, 57–63 (2020)
24. Yang, X., Fei, B.: A wavelet multiscale denoising algorithm for magnetic resonance (MR) images. Measure. Sci. Technol. **22**(2), 025803 (2011)
25. Yu, H., Ding, M., Zhang, X.: Laplacian eigenmaps network-based nonlocal means method for MR image denoising. Sensors **19**(13), 2918 (2019)
26. Zhang, K., Zuo, W., Chen, Y., Meng, D., Zhang, L.: Beyond a gaussian denoiser: residual learning of deep CNN for image denoising. IEEE Trans. Image Process. **26**(7), 3142–3155 (2017)
27. Zhang, X., et al.: Denoising of 3d magnetic resonance images by using higher-order singular value decomposition. Med. Image Anal. **19**(1), 75–86 (2015)

Deep J-Sense: Accelerated MRI Reconstruction via Unrolled Alternating Optimization

Marius Arvinte$^{(\boxtimes)}$, Sriram Vishwanath, Ahmed H. Tewfik, and Jonathan I. Tamir

The University of Texas at Austin, Austin, TX 78705, USA
arvinte@utexas.edu

Abstract. Accelerated multi-coil magnetic resonance imaging reconstruction has seen a substantial recent improvement combining compressed sensing with deep learning. However, most of these methods rely on estimates of the coil sensitivity profiles, or on calibration data for estimating model parameters. Prior work has shown that these methods degrade in performance when the quality of these estimators are poor or when the scan parameters differ from the training conditions. Here we introduce Deep J-Sense as a deep learning approach that builds on unrolled alternating minimization and increases robustness: our algorithm refines both the magnetization (image) kernel and the coil sensitivity maps. Experimental results on a subset of the knee fastMRI dataset show that this increases reconstruction performance and provides a significant degree of robustness to varying acceleration factors and calibration region sizes.

Keywords: MRI acceleration · Deep learning · Unrolled optimization

1 Introduction

Parallel MRI is a multi-coil acceleration technique that is standard in nearly all clinical systems [5,15,21]. The technique uses multiple receive coils to measure the signal in parallel, and thus accelerate the overall acquisition. Compressed sensing-based methods with suitably chosen priors have constituted one of the main drivers of progress in parallel MRI reconstruction for the past two decades [4,10,16,25]. While parallel MRI provides additional degrees of freedom via simultaneous measurements, it brings its own set of challenges related to

Supported by ONR grant N00014-19-1-2590, NIH Grant U24EB029240, NSF IFML 2019844 Award, and an AWS Machine Learning Research Award.

Electronic supplementary material The online version of this chapter (https://doi.org/10.1007/978-3-030-87231-1_34) contains supplementary material, which is available to authorized users.

M. de Bruijne et al. (Eds.): MICCAI 2021, LNCS 12906, pp. 350–360, 2021.
https://doi.org/10.1007/978-3-030-87231-1_34

estimating the spatially varying *sensitivity maps* of the coils, either explicitly [15,25,26] or implicitly [5,21]. These algorithms typically use a fully sampled region of k-space or a low-resolution reference scan as an auto-calibration signal (ACS), either to estimate k-space kernels [5,11], or to estimate coil sensitivity profiles [15]. Calibration-free methods have been proposed that leverage structure in the parallel MRI model; namely, that sensitivity maps smoothly vary in space [16,26] and impose low-rank structure [6,20].

Deep learning has recently enabled significant improvement to image quality for accelerated MRI when combined with ideas from compressed sensing in the form of *unrolled iterative optimization* [1,7,18,22,23]. Our work falls in this category, where learnable models are interleaved with optimization steps and the entire system is trained end-to-end with a supervised loss. However, there are still major open questions concerning the robustness of these models, especially when faced with distributional shifts [2], i.e., when the scan parameters at test time do not match the ones at training time or the robustness of methods across different training conditions. This is especially prudent for models that use estimated sensitivity maps, and thus require reliable estimates as input.

Our contributions are the following: i) we introduce a novel deep learning-based parallel MRI reconstruction algorithm that unrolls an alternating optimization to jointly solve for the image and sensitivity map kernels directly in k-space; ii) we train and evaluate our model on a subset of the fastMRI knee dataset and show improvements in reconstruction fidelity; and iii) we evaluate the robustness of our proposed method on distributional shifts produced by different sampling parameters and obtain state-of-the-art performance. An open-source implementation of our method is publicly available[1].

2 System Model and Related Work

In parallel MRI, the signal is measured by an array of radio-frequency receive coils distributed around the body, each with a spatially-varying sensitivity profile. In the measurement model, the image is linearly mixed with each coil sensitivity profile and sampled in the Fourier domain (k-space). Scans can be accelerated by reducing the number of acquired k-space measurements, and solving the inverse problem by leveraging redundancy across the receive channels as well as redundancy in the image representation. We consider parallel MRI acquisition with C coils. Let $\mathbf{m} \in \mathbb{C}^n$ and $\mathbf{s} = [\mathbf{s}_1, \cdots, \mathbf{s}_C] \in \mathbb{C}^{k \times C}$ be the n-dimensional image (magnetization) and set of k-dimensional sensitivity map kernels, respectively, defined directly in k-space. We assume that $\mathbf{k}_i$, the k-space data of the i-th coil image, is given by the linear convolution between the two kernels as

$$\mathbf{k}_i = \mathbf{s}_i * \mathbf{m}. \tag{1}$$

Joint image and map reconstruction formulates (1) as a bilinear optimization problem where both variables are unknown. Given a sampling mask represented by the matrix $\mathbf{A}$ and letting $\mathbf{y} = \mathbf{Ak} + \mathbf{n}$ be the sampled noisy multicoil k-space

---

[1] https://github.com/utcsilab/deep-jsense.

data, where $\mathbf{k}$ is the ground-truth k-space data and $\mathbf{n}$ is the additive noise, the optimization problem is

$$\underset{\mathbf{s},\mathbf{m}}{\arg\min} \frac{1}{2} \|\mathbf{y} - \mathbf{A}(\mathbf{s} * \mathbf{m})\|_2^2 + \lambda_\mathbf{m} R_\mathbf{m}(\mathbf{m}) + \lambda_\mathbf{s} R_\mathbf{s}(\mathbf{s}). \tag{2}$$

$R_\mathbf{m}(\mathbf{m})$ and $R_\mathbf{s}(\mathbf{s})$ are regularization terms that enforce priors on the two variables (e.g., J-Sense [26] uses polynomial regularization for $\mathbf{s}$). We let $\mathbf{A}_\mathbf{m}$ and $\mathbf{A}_\mathbf{s}$ denote the linear operators composed by convolution with the fixed variable and $\mathbf{A}$. A solution of (2) by alternating minimization involves the steps

$$\mathbf{m} = \underset{\mathbf{m}}{\arg\min} \frac{1}{2} \|\mathbf{y} - \mathbf{A}_\mathbf{m}\mathbf{m}\|_2^2 + \lambda_\mathbf{m} R_\mathbf{m}(\mathbf{m}), \tag{3a}$$

$$\mathbf{s} = \underset{\mathbf{s}}{\arg\min} \frac{1}{2} \|\mathbf{y} - \mathbf{A}_\mathbf{s}\mathbf{s}\|_2^2 + \lambda_\mathbf{s} R_\mathbf{s}(\mathbf{s}). \tag{3b}$$

If ℓ_2-regularization is used for the R terms, each sub-problem is a linear least squares minimization, and the Conjugate Gradient (CG) algorithm [19] with a fixed number of steps can be applied to obtain an approximate solution.

2.1   Deep Learning for MRI Reconstruction

Model-based deep learning architectures for accelerated MRI reconstruction have recently demonstrated state-of-the-art performance [1,7,22]. The MoDL algorithm [1] in particular is used to solve (1) when only the image kernel variable is unknown and a deep neural network $\mathcal{D}$ is used in $R_\mathbf{m}(\mathbf{m})$ as

$$\underset{\mathbf{m}}{\arg\min} \frac{1}{2} \|\mathbf{y} - \mathbf{A}_\mathbf{m}\mathbf{m}\|_2^2 + \lambda \|\mathcal{D}(\mathbf{m}) - \mathbf{m}\|_2^2. \tag{4}$$

To unroll the optimization in (4), the authors split each step in two different sub-problems. The first sub-problem treats $\mathcal{D}(\mathbf{m})$ as a constant and uses the CG algorithm to update $\mathbf{m}$. The second sub-problem treats $\mathcal{D}$ as a proximal operator and is solved by direct assignment, i.e., $\mathbf{m}^+ = \mathcal{D}(\mathbf{m})$. In our work, we use the same approach for unrolling the optimization, but we use a pair of deep neural networks, one for each variable in (2). Unlike [1], our work does not rely on a pre-computed estimate of the sensitivity maps, but instead treats them as an optimization variable.

The idea of learning a sensitivity map estimator using neural networks was first described in [3]. Recently, the work in [22] introduced the E2E-VarNet architecture that addresses the issue of estimating the sensitivity maps by training a *sensitivity map estimation* module in the form of a deep neural network. Like the ESPiRiT algorithm [25], E2E-VarNet additionally enforces that the sensitivity maps are normalized per-pixel and uses the same sensitivity maps across all unrolls. This architecture – which uses gradient descent instead of CG – is then trained end-to-end, using the estimated sensitivity maps and the forward operator $\mathbf{A}_\mathbf{m}$. The major difference between our work and [22] is that we iteratively update the maps instead of using a single-shot data-based approach [22] or the ESPiRiT algorithm [1,7]. As our results show in the sequel, this has a significant impact on the out-of-distribution robustness of the approach on scans whose parameters differ from the training set.

Concurrent work also extends E2E-VarNet and proposes to jointly optimize the image and sensitivity maps via the Joint ICNet architecture [9]. Our main difference here is the usage of a forward model directly in k-space, and the fact that we do not impose any normalization constraints on the sensitivity maps. Finally, the work in [12] proposes a supervised approach for sensitivity map estimation, but this still requires an external algorithm and a target for map regression, which may not be easily obtained.

3   Deep J-Sense: Unrolled Alternating Optimization

We unroll (2) by alternating between optimizing the two variables as

$$\mathbf{s} = \arg\min_{\mathbf{s}} \frac{1}{2} \|\mathbf{y} - \mathbf{A_s}\mathbf{s}\|_2^2 + \lambda_\mathbf{s} R_\mathbf{s}(\mathbf{s}), \tag{5a}$$

$$\mathbf{s}^+ = \mathcal{D}_\mathbf{s}(\mathbf{s}), \tag{5b}$$

$$\mathbf{m} = \arg\min_{\mathbf{m}} \frac{1}{2} \|\mathbf{y} - \mathbf{A_m}\mathbf{m}\|_2^2 + \lambda_\mathbf{m} R_\mathbf{m}(\mathbf{m}), \tag{5c}$$

$$\mathbf{m}^+ = \mathcal{D}_\mathbf{m}(\mathbf{m}), \tag{5d}$$

where R is defined as

$$R(\mathbf{x}) = \left\| \mathcal{F}\{\mathcal{D}_\mathbf{x}(\mathcal{F}^{-1}\{\mathbf{x}\})\} - \mathbf{x} \right\|_2^2, \tag{6}$$

for both $\mathbf{m}$ and $\mathbf{s}$, $\mathcal{F}$ is the Fourier transform, and $\mathcal{D}$ is the deep neural network corresponding to each variable. Similar to MoDL, we set $\mathcal{D}_\mathbf{m}^{(j)} = \mathcal{D}_\mathbf{m}$ and $\mathcal{D}_\mathbf{s}^{(j)} = \mathcal{D}_\mathbf{s}$ across all unrolls, leading to the efficient use of learnable weights. The coefficients $\lambda_\mathbf{s}$ and $\lambda_\mathbf{m}$ are also learnable. The optimization is initialized with $\mathbf{m}^{(0)}$ and $\mathbf{s}^{(0)}$, obtained using a simple root sum-of-squares (RSS) estimate. Steps (5a) and (5c) are approximately solved with n_1 and n_2 steps of the CG algorithm, respectively, while steps (5b) and (5d) represent direct assignments. The two neural networks serve as generalized denoisers applied in the image domain and are trained in an end-to-end fashion after unrolling the alternating optimization for a number of N outer steps. A block diagram of one unroll is shown in Fig. 1.

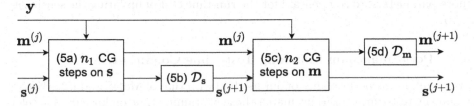

Fig. 1. A single unroll of the proposed scheme. The CG algorithm is executed on the loss given by the undersampled data $\mathbf{y}$ and the measurement matrices $\mathbf{A_s}$ and $\mathbf{A_m}$. Each block is matched to its corresponding equation.

Using the estimated image and map kernels after N outer steps, we train the end-to-end network using the estimated RSS image as

$$\hat{\mathbf{x}} = \sqrt{\sum_{i=1}^{C} |\mathcal{F}^{-1}\{\mathbf{s}_i^{(N)} * \mathbf{m}^{(N)}\}|^2}, \tag{7}$$

where the supervised loss is the structural similarity index (SSIM) between the estimated image and ground truth image as $L = -\text{SSIM}(\hat{x}, x)$.

Our model can be seen as a unification of MoDL and J-Sense. For example, by setting $n_1 = 0$ and $\mathcal{D}_\mathbf{s}(\mathbf{s}) = \mathbf{s}$, the sensitivity maps are never updated and the proposed approach becomes MoDL. At the same time, removing the deep neural networks by setting $\mathcal{D}(\cdot) = 0$ and removing steps (5b) and (5d) leads to L2-regularized J-Sense. One important difference is that, unlike the ESPiRiT algorithm or E2E-VarNet, our model does *not* perform pixel-wise normalization of the sensitivity maps, and thus does not impart a spatially-varying weighting across the image. We use a forward model in the k-space domain based on a linear convolution instead of prior work that uses the image domain. This allows us to use a small-sized kernel for the sensitivity maps as an implicit smoothness regularizer and reduction in memory.

4   Experimental Results

We compare the performance of the proposed approach against MoDL [1] and E2E-VarNet [22]. We train and evaluate all methods on a subset of the fastMRI knee dataset [27] to achieve reasonable computation times. For training, we use the five central slices from each scan in the training set, for a total of 4580 training slices. For evaluation, we use the five central slices from each scan in the validation set, for a total of 950 validation slices. All algorithms are implemented in PyTorch [14] and SigPy [13]. Detailed architectural and hyper-parameter choices are given in the material.

To evaluate the impact of optimizing the sensitivity map kernel, we compare the performance of the proposed approach with MoDL trained on the same data, that uses the same number of unrolls (both inner and outer) and the same architecture for the image denoising network $\mathcal{D}_\mathbf{m}$. We compare our robust performance with E2E-VarNet trained on the same data, and having four times more parameters, to compensate for the run-time cost of updating the sensitivity maps.

4.1   Performance on Matching Test-Time Conditions

We compare the performance of our method with that of MoDL and E2E-VarNet when the test-time conditions match those at training time on knee data accelerated by a factor of $R = 4$. For MoDL, we use a denoising network with the same number of parameters as our image denoising network and the same number of outer and inner CG steps. We use sensitivity maps estimated by the ESPiRiT algorithm via the BART toolbox [24], where a SURE-calibrated version [8] is used to select the first threshold, and we set the eigenvalue threshold to zero so as to not unfairly penalize MoDL, since both evaluation metrics are influenced by background noise. For E2E-VarNet, we use the same number of $N = 6$ unrolls (called *cascades* in [22]) and U-Nets for all refinement modules.

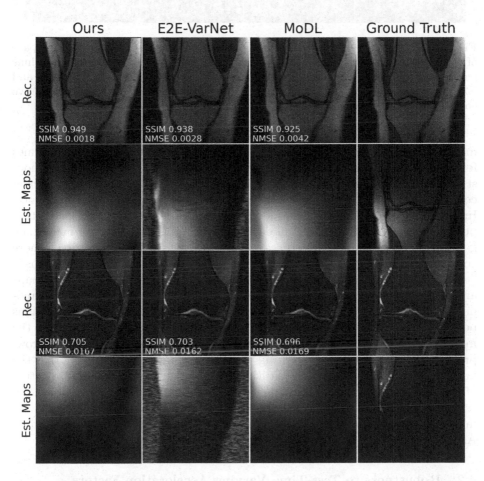

Fig. 2. Example reconstructions at $R = 4$ and matching train-test conditions. The first and third rows represent RSS images from two different scans. The second and fourth rows represents the magnitude of the estimated sensitivity map (no ground truth available, instead the final column shows the coil image) for a specific coil from each scan, respectively. The maps under the MoDL column are estimated with the ESPiRiT algorithm.

Table 1 shows statistical performance on the validation data. The comparison with MoDL allows us to evaluate the benefit of iteratively updating the sensitivity maps, which leads to a significant gain in both metrics. Furthermore, our

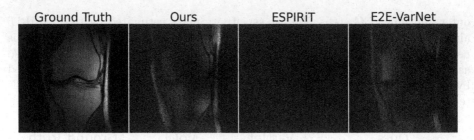

Fig. 3. Magnitude of the projection of the ground truth k-space data onto the null space of the normal operator given by the estimated sensitivity maps for various algorithms (residuals amplified 5×). The sample is chosen at random; both Deep J-Sense and E2E-VarNet produce map estimates that deviate from the linear model, but improve end-to-end reconstruction of the RSS image.

method obtains a superior performance to E2E-VarNet while using *four* times fewer trainable weights and the same number of outer unrolls. This demonstrates the benefit of trading off the number of parameters for computational complexity, since our model executes more CG iterations than both baselines. Importantly, Deep J-Sense shows a much lower variance of the reconstruction performance across the validation dataset, with nearly one order of magnitude gain against MoDL and three times lower than E2E-VarNet, delivering a more consistent reconstruction performance across a heterogeneity of patient scans.

Randomly chosen reconstructions of scans and the estimated sensitivity maps are shown in Fig. 2. We notice that our maps capture higher frequency components than those of MoDL (estimated via ESPiRiT), but do not contain spurious noise patterns outside the region spanned by the physical knee. In contrast, the maps from E2E-VarNet exhibit such patterns and, in the case of the second row, produce spurious patterns even *inside* the region of interest, suggesting that the knee anatomy "leaks" into the sensitivity maps. Figure 3 shows the projection of the fully-sampled k-space data on the null space of the pixel-wise normalized estimated sensitivity maps. Note that since we do not normalize the sensitivity maps this represents a different set of maps that those learned during training. As Fig. 3 shows, the residual obtained is similar to that of E2E-VarNet.

4.2   Robustness to Test-Time Varying Acceleration Factors

Figure 4 shows the performance obtained at acceleration factors between 2 and 6, with models trained only at $R = 4$. The modest performance gain for E2E-VarNet at $R < 4$ confirms the findings in [2]: certain models cannot efficiently use additional measurements if there is a train-test mismatch. At the same time, MoDL and the proposed method are able to overcome this effect, with our method significantly outperforming the baselines across all accelerations.

Table 1. Validation performance on a subset of the fastMRI knee dataset. Higher average/median SSIM (lower NMSE) indicates better performance. Lower standard deviations are an additional desired quality.

	Avg. SSIM	Med. SSIM	σ SSIM	Avg. NMSE	Med. NMSE	σ NMSE
MoDL	0.814	0.840	0.115	0.0164	0.0087	0.0724
E2E	0.824	0.851	0.107	0.0111	0.0068	0.0299
Ours	**0.832**	**0.857**	**0.104**	**0.0091**	**0.0064**	**0.0095**

Importantly, there is a significant decrease of the performance loss *slope* against MoDL, rather than an additive gain, showing the benefit of estimating sensitivity maps using all the acquired measurements.

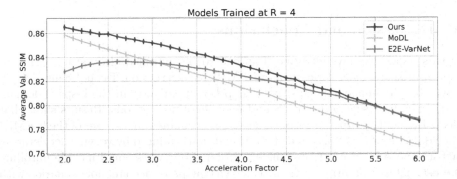

Fig. 4. Average SSIM on the fastMRI knee validation dataset evaluated at acceleration factors R between 2 and 6 (with granularity 0.1) using models trained at $R = 4$. The vertical lines are proportional to the SSIM standard deviation in each case, from which no noticeable difference can be seen.

4.3 Robustness to Train-Time Varying ACS Size

We investigate the performance of the proposed method and E2E-VarNet as a function of the ACS region size (expressed as number of acquired lines in the phase encode direction). All models are trained at $R = 4$ and ACS sizes $\{1, 6, 12, 26, 56\}$ and tested in the same conditions, giving a total of ten models. While there is no train-test mismatch in this experiment, Fig. 5 shows a significant performance gain of the proposed approach when the calibration region is small (below six lines) and shows that overall, end-to-end performance is robust to the ACS size, with the drops at 1 and 12 lines not being statistically significant. This is in contrast to E2E-VarNet, which explicitly uses the ACS in its map estimation module and suffers a loss when this region is small.

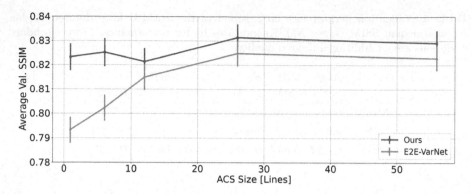

Fig. 5. Average SSIM on the fastMRI knee validation dataset evaluated at different sizes of the fully sampled auto-calibration region, at acceleration factor $R = 4$. The vertical lines are proportional to the SSIM standard deviation. Each model is trained and tested on the ACS size indicated by the x-axis.

5   Discussion and Conclusions

In this paper, we have introduced an end-to-end unrolled alternating optimization approach for accelerated parallel MRI reconstruction. Deep J-Sense jointly solves for the image and sensitivity map kernels directly in the k-space domain and generalizes several prior CS and deep learning methods. Results show that Deep J-Sense has superior reconstruction performance on a subset of the fastMRI knee dataset and is robust to distributional shifts induced by varying acceleration factors and ACS sizes. A possible extension of our work could include unrolling the estimates of multiple sets of sensitivity maps to account for scenarios with motion or a reduced field of view [17,25].

References

1. Aggarwal, H.K., Mani, M.P., Jacob, M.: Modl: model-based deep learning architecture for inverse problems. IEEE Trans. Med. Imaging **38**(2), 394–405 (2018)
2. Antun, V., Renna, F., Poon, C., Adcock, B., Hansen, A.C.: On instabilities of deep learning in image reconstruction and the potential costs of AI. Proc. Natl. Acad. Sci. **117**(48), 30088–30095 (2020)
3. Cheng, J.Y., Pauly, J.M., Vasanawala, S.S.: Multi-channel image reconstruction with latent coils and adversarial loss. ISMRM (2019)
4. Deshmane, A., Gulani, V., Griswold, M.A., Seiberlich, N.: Parallel mr imaging. J. Magn. Reson. Imaging **36**(1), 55–72 (2012)
5. Griswold, M.A., et al.: Generalized autocalibrating partially parallel acquisitions (grappa). Magn. Reson. Med.: Off. J. Int. Soc. Magn. Res. Med. **47**(6), 1202–1210 (2002)
6. Haldar, J.P.: Low-rank modeling of local k-space neighborhoods (loraks) for constrained MRI. IEEE Trans. Med. Imaging **33**(3), 668–681 (2013)
7. Hammernik, K., et al.: Learning a variational network for reconstruction of accelerated mri data. Magn. Reson. Med. **79**(6), 3055–3071 (2018)

8. Iyer, S., Ong, F., Setsompop, K., Doneva, M., Lustig, M.: Sure-based automatic parameter selection for espirit calibration. Magn. Reson. Med. **84**(6), 3423–3437 (2020)
9. Jun, Y., Shin, H., Eo, T., Hwang, D.: Joint deep model-based MR image and coil sensitivity reconstruction network (joint-ICNET) for fast MRI. In: Proceedings of the IEEE/CVF Conference on Computer Vision and Pattern Recognition (CVPR), pp. 5270–5279 (2021)
10. Lustig, M., Donoho, D., Pauly, J.M.: Sparse MRI: the application of compressed sensing for rapid mr imaging. Magn. Reson. Med.: Off. J. Int. Soc. Magn. Reson. Med. **58**(6), 1182–1195 (2007)
11. Lustig, M., Pauly, J.M.: Spirit: iterative self-consistent parallel imaging reconstruction from arbitrary k-space. Magn. Reson. Med. **64**(2), 457–471 (2010)
12. Meng, N., Yang, Y., Xu, Z., Sun, J.: A prior learning network for joint image and sensitivity estimation in parallel MR imaging. In: Shen, D., et al. (eds.) MICCAI 2019. LNCS. vol. 11767, pp. 732–740. Springer, Cham (2019). https://doi.org/10.1007/978-3-030-32251-9_80
13. Ong, F., Lustig, M.: Sigpy: a python package for high performance iterative reconstruction. In: Proceedings of the ISMRM 27th Annual Meeting, Montreal, Quebec, Canada, p. 4819 (2019)
14. Paszke, A., et al.: Pytorch: an imperative style, high-performance deep learning library. In: Wallach, H., Larochelle, H., Beygelzimer, A., d' Alché-Buc, F., Fox, E., Garnett, R. (eds.) Advances in Neural Information Processing Systems, vol. 32, pp. 8024–8035. Curran Associates, Inc. (2019). http://papers.neurips.cc/paper/9015-pytorch-an-imperative-style-high-performance-deep-learning-library.pdf
15. Pruessmann, K.P., Weiger, M., Scheidegger, M.B., Boesiger, P.: Sense: sensitivity encoding for fast MRI. Magn. Reson. Med.: Off. J. Int. Soc. Magn. Res. Med. **42**(5), 952–962 (1999)
16. Rosenzweig, S., Holme, H.C.M., Wilke, R.N., Voit, D., Frahm, J., Uecker, M.: Simultaneous multi-slice MRI using cartesian and radial flash and regularized nonlinear inversion: Sms-nlinv. Magn. Reson. Med. **79**(4), 2057–2066 (2018)
17. Sandino, C.M., Lai, P., Vasanawala, S.S., Cheng, J.Y.: Accelerating cardiac cine MRI using a deep learning-based espirit reconstruction. Magn. Reson. Med. **85**(1), 152–167 (2021)
18. Schlemper, J., Caballero, J., Hajnal, J.V., Price, A.N., Rueckert, D.: A deep cascade of convolutional neural networks for dynamic MR image reconstruction. IEEE Trans. Med. Imaging **37**(2), 491–503 (2017)
19. Shewchuk, J.R., et al.: An introduction to the conjugate gradient method without the agonizing pain (1994)
20. Shin, P.J., et al.: Calibrationless parallel imaging reconstruction based on structured low-rank matrix completion. Magn. Reson. Med. **72**(4), 959–970 (2014)
21. Sodickson, D.K., Manning, W.J.: Simultaneous acquisition of spatial harmonics (smash): fast imaging with radiofrequency coil arrays. Magn. Reson. Med. **38**(4), 591–603 (1997)
22. Sriram, A., et al.: End-to-end variational networks for accelerated MRI reconstruction. In: International Conference on Medical Image Computing and Computer-Assisted Intervention, pp. 64–73 (2020)
23. Sriram, A., Zbontar, J., Murrell, T., Zitnick, C.L., Defazio, A., Sodickson, D.K.: Grappanet: Combining parallel imaging with deep learning for multi-coil MRI reconstruction. In: Proceedings of the IEEE/CVF Conference on Computer Vision and Pattern Recognition (CVPR) (2020)

24. Tamir, J.I., Ong, F., Cheng, J.Y., Uecker, M., Lustig, M.: Generalized magnetic resonance image reconstruction using the berkeley advanced reconstruction toolbox. In: ISMRM Workshop on Data Sampling and Image Reconstruction, Sedona, AZ (2016)
25. Uecker, M., et al.: Espirit—an eigenvalue approach to autocalibrating parallel MRI: where sense meets grappa. Magn. Reson. Med. **71**(3), 990–1001 (2014)
26. Ying, L., Sheng, J.: Joint image reconstruction and sensitivity estimation in sense (jsense). Magn. Reson. Med.: Off. J. Int. Soc. Magn. Reson. Med. **57**(6), 1196–1202 (2007)
27. Zbontar, J., et al.: fastMRI: An open dataset and benchmarks for accelerated MRI. ArXiv e-prints (2018)

Label-Free Physics-Informed Image Sequence Reconstruction with Disentangled Spatial-Temporal Modeling

Xiajun Jiang[1(✉)], Ryan Missel[1], Maryam Toloubidokhti[1], Zhiyuan Li[1], Omar Gharbia[1], John L. Sapp[2], and Linwei Wang[1]

[1] Rochester Institute of Technology, Rochester, NY 14623, USA
{xj7056,rxm7244,mt6129,zl7904,oag1929,linwei.wang}@rit.edu
[2] Dalhousie University, Halifax, NS, Canada
john.sapp@nshealth.ca

Abstract. Traditional approaches to image reconstruction uses physics-based loss with data-efficient inference, although the difficulty to properly model the inverse solution precludes learning the reconstruction across a distribution of data. Modern deep learning approaches enable expressive modeling but rely on a large number of reconstructed images (labeled data) that are often not available in practice. To combine the best of the above two lines of works, we present a novel label-free image reconstruction network that is supervised by physics-based forward operators rather than labeled data. We further present an expressive yet disentangled spatial-temporal modeling of the inverse solution, where its latent dynamics is modeled by neural ordinary differential equations and its emission over non-Euclidean geometrical domains by graph convolutional neural networks. We applied the presented method to reconstruct electrical activity on the heart surface from body-surface potential. In simulation and real-data experiments in comparison to both traditional physics-based and modern data-driven reconstruction methods, we demonstrated the ability of the presented method to learn how to reconstruct using observational data without any corresponding labels.

Keywords: Image reconstruction · Neural ODE · Graph convolution

1 Introduction

Traditional approaches to medical image reconstruction are heavily physics-informed. They include typically a *forward operator* that relates the unknown inverse solution to the observed data based on the underlying imaging physics,

---

X. Jiang and R. Missel contributed equally to this work.

Electronic supplementary material The online version of this chapter (https://doi.org/10.1007/978-3-030-87231-1_35) contains supplementary material, which is available to authorized users.

© Springer Nature Switzerland AG 2021
M. de Bruijne et al. (Eds.): MICCAI 2021, LNCS 12906, pp. 361–371, 2021.
https://doi.org/10.1007/978-3-030-87231-1_35

and *a priori* constraints that regularize the ill-posedness of the inverse problem: a notable example being the temporal dynamics of the inverse solution observed over a sequence of images [26]. This generally gives rises to classic state space models (SSMs) with interpretable modeling of the dynamics of the inverse solution and its emission to observations, lending to data-efficient inference [9]. Traditional SSMs however rely on strong prior assumptions on model structures. Furthermore, the inverse solution is typically solved independently for any given observations, without any common knowledge learned or leveraged across data.

Alternatively, deep learning has shown great promise both in direct medical image reconstruction [8,13,27,28], and in combination with traditional physics-based approaches [4,19]. Till now, however, most existing approaches still require labeled data for supervision. In medical image reconstruction, *labeled data* are in the forms of the underlying tissue structure [27], property [25], and functioning [14] to be reconstructed, which are often difficult and sometimes impossible to obtain (*e.g.*, invasive catheter mapping of heart potential). In such scenario, training has to resort to *in-silico* simulation data or *in-vitro* phantom study, which unfortunately may be difficult to generalize to *in-vivo* test data [14,17]. Furthermore, to incorporate temporal dynamics into neural reconstruction of medical images, typical approaches utilize variants of interlaced recurrent and convolutional neural networks [14,17]. Compared to interpretable SSMs, these autoregressive models are difficult to interpret and data-hungry.

We present a novel label-free and physics-informed image reconstruction network to combine the best of the above two lines of works. Like traditional SSMs: 1) the network is supervised by physics-based forward operators and *a priori* constraints commonly used in classic inferences, removing the need of labeled data; and 2) we further disentangle the latent dynamics of the inverse solution and its emission to the image domain, to improve interpretibilty and learning from small unlabeled data. Unlike traditional SSMs, 1) we make minimal structural assumptions by modeling the spatial emission (and embedding) over non-Euclidean geometrical domains using graph convolutional neural networks (GCNNs), and describe the latent dynamics by a neural ordinary differential equation (ODE) [5]); and 2) we accumulate and improve our knowledge about the spatial-temporal model of the inverse solution by optimizing parameters of the presented ODE-GCNN supervised by the traditional physics-informed loss.

We tested the presented method in reconstructing electrical activity of the heart surface from body-surface potential sequences. In controlled simulation experiments, we demonstrated that the presented method is able to leverage larger unlabeled data to outperform supervised training with limited labeled data. We then demonstrated the advantage of the presented method in *in-vivo* experiments where data labels are not available, in comparison to both networks supervised by simulation data (typical data-driven approaches) and traditional physics-based inference (supervised with the physics loss without learning across data). Experimental results suggested the presented work as a promising intersection between physics-based and data-driven approaches for learning how to reconstruct using observational data without the corresponding labels.

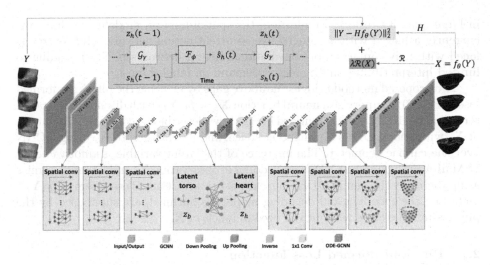

Fig. 1. Overview of the presented network: ODE-GCNN.

2 Methodology

Cardiac electrical excitation $\mathbf{X}_t$ generates time-varying voltage signals on the body surface $\mathbf{Y}_t$ following the quasi-static electromagnetic theory [21]. Solving the governing physics on any given pair of heart and torso geometry will give:

$$\mathbf{Y}_t = \mathbf{H}\mathbf{X}_t \quad \forall t \in \{1, ..., T\}. \tag{1}$$

Based on this physics model, traditional image reconstruction then seeks to find an inverse solution $\mathbf{X}_t$ that minimizes the fitting of the observational data $\mathbf{Y}_t$, while satisfying certain constraints $\mathcal{R}(\mathbf{X}_t)$ about $\mathbf{X}_t$ [10]:

$$\arg\min_{\mathbf{X}_t}||\mathbf{Y}_t - \mathbf{H}\mathbf{X}_t||_2^2 + \lambda\mathcal{R}(\mathbf{X}_t). \tag{2}$$

where the regularization parameter λ is often empirically tuned. In most existing works, $\mathbf{X}_t$ is solved independently for a given $\mathbf{Y}_t$. To be able to learn to reconstruct $\mathbf{X}_t$ across many observations of $\mathbf{Y}_t$'s requires constructing and estimating a parameterized model of $\mathbf{X}_t$, both tasks daunting in classic SSMs.

Alternatively, existing deep learning approaches learn the inverse mapping of $\mathbf{X}_t$ as a neural function of $\mathbf{Y}_t$: $\mathbf{X}_t = f_\theta(\mathbf{Y}_t)$. This neural function is typically supervised by signal pairs of $\{\mathbf{X}_t^i, \mathbf{Y}_t^i\}_{i=1}^N$ that can be loosely expressed as:

$$\arg\min_\theta \sum_i^N ||\mathbf{X}_t^i - f_\theta(\mathbf{Y}_t^i)||_2^2 \tag{3}$$

In comparison to Eq. (2), modeling $\mathbf{X}_i$ as a parameterized neural function of $\mathbf{Y}_i$ allows us to learn across many $\mathbf{Y}_i$'s. The learning, however, ignores the underlying imaging physics and requires labeled data $\mathbf{X}_i$ that is typically not available in large quantity *in-vivo*. In addition, the spatiotemporal modeling within f_θ is currently achieved with autoregressive recurrent nerual networks (RNNs) interlaced with fully-connected or convolutional neural networks (CNNs), leading to limited interpretibility and increased demand on labeled data.

The proposed method bridges the above gaps by utilizing the physics-informed loss in Eq. (2) to supervise a neural function $\mathbf{X}_t = f_\theta(\mathbf{Y}_t)$ with disentangled sptial-temporal modeling. As summarized in Fig. 1, we describe the continuous dynamics of $\mathbf{X}_t$ with neural ODEs in the latent space, and its emission to $\mathbf{X}_t$ using GCNN over the cardiac geometry. The inference of the latent variable, as motivated by SSM filtering, combines GCNN embedding of observation $\mathbf{Y}_t$ and latent dynamics as predicted by the neural ODE from the past time instant. Parameters of $f_\theta(\mathbf{Y}_t)$, including the encoding and decoding models as outlined, are supervised by the physics-informed loss without the need of labels of $\mathbf{X}_t$.

2.1   Physics-Informed Loss Function

The proposed physics-informed loss combines the strength of the two losses in Eq. (2) and Eq. (3): it utilizes the physics knowledge behind the forward operator $\mathbf{H}$ and the prior constraint $\mathcal{R}(\mathbf{X}_t)$ without asking for labels of $\mathbf{X}_t$ [24], while modeling $\mathbf{X}_t$ as a neural function of $f_\theta(\mathbf{Y})$ to allow learning the parameters θ of the spatia-temporal model of $\mathbf{X}_t$ across a distribution of unlabled data $\mathbf{Y}_t$:

$$\arg\min_\theta \sum_i^N ||\mathbf{Y}_t^i - \mathbf{H}f_\theta(\mathbf{Y}_t^i)||_2^2 + \lambda\mathcal{R}(\mathbf{X}_t), \tag{4}$$

The presented method can be applied to Eq. (4) with different choices of $\mathcal{R}(\mathbf{X}_t)$. As a first proof of concept, we adopt the popular Laplacian smoothing over the heart geometry, and λ is empirically tuned as in Eq. (2).

2.2   Spatial Modeling via GCNN

As $\mathbf{X}_t$ and $\mathbf{Y}_t$ live on 3D geometry, we describe their embedding or emission with encoding-decoding GCNNs with their relationship learned at the latent space.

Geometrical Representation in Graphs. We represent triangular meshes of the heart and torso as two separate undirected graphs, with edge attributes $\mathbf{u}(i,j)$ between vertex i and j as normalized differences in their 3D coordinates if an edge exists. Encoding and decoding are performed over hierarchical graph representations of the heart and torso geometry, which are obtained by a specialized mesh coarsening method [3,11,20] to preserve topology of the geometry.

Spatial Graph Convolution. A continuous spline kernel for spatial convolution is used such that it can be applied across graphs [12]. Given kernel $\mathbf{g} = (g_1, ..., g_M)$ and graph node features $\mathbf{f} \in \mathbb{R}^{V \times M}$ at each time instant, spatial convolution for vertex $i \in \mathcal{V}$ with its neighborhood $N(i)$ is then defined as:

$$(f_l * g_l)(i) = \sum_{j \in N(i), \mathbf{p} \in \mathcal{P}(\mathbf{u}(i,j))} f_l(j) \cdot g_l(\mathbf{u}(i,j)), \tag{5}$$

where $\mathcal{P}$ is the Cartesian product of the B-spline bases. To make the network deeper and more expressive, we introduce residual blocks here to pass the input of spatial convolution through a skip connection with 1D convolution before adding it to the output of the spatial convolution.

Latent Inverse Mapping. Following [17], we assume the linearity to hold between $\mathbf{X}_t$ and $\mathbf{Y}_t$ in the latent space during inverse imaging. We construct a bipartite graph between the graph embedding of the heart and torso geometry: the edge attribute $\mathbf{u}(i,j)$ between torso vertex i and heart vertex j describes their relative geometrical relationship. Using spline convolution, we model the latent representation $\mathbf{z}_h(i)$ on vertex i of the latent heart mesh as a linear combination of latent representation $\mathbf{z}_b(j)$ across all vertices j of the latent torso mesh as:

$$\mathbf{z}_h(i) = \sum_j \mathbf{z}_b(j) \cdot \hat{\mathbf{h}}(\mathbf{u}(i,j)), \tag{6}$$

where the linear coefficients $\hat{\mathbf{h}}$ are learned as the spline convolution kernel.

2.3 Temporal Modeling via Neural ODEs

For temporal modeling, we draw inspiration from the classic SSM Bayesian filtering setting that is composed of a series of prediction and update operations: during prediction, a temporal transition function $\mathcal{F}$ predicts the next state $\hat{\mathbf{s}}_h(t)$ based on the current state estimate $\mathbf{s}_h(t-1)$; during update, this prediction is combined with the state $\mathbf{z}_h(t)$ inferred from the observation $\mathbf{Y}_t$ to get the corrected state $\mathbf{s}_h(t)$ via a function $\mathcal{G}$:

$$\begin{aligned}\textbf{Prediction: } &\hat{\mathbf{s}}_h(t) = \mathcal{F}(\mathbf{s}_h(t-1)), \\ \textbf{Update: } &\mathbf{s}_h(t) = \mathcal{G}(\mathbf{z}_h(t), \hat{\mathbf{s}}_h(t)).\end{aligned} \tag{7}$$

Below we elaborate on each operation.

Latent Dynamics Prediction with Neural ODEs. As illustrated in Fig. 1, the dynamics of the unknown $\mathbf{X}_t$ in the latent space is modeled by an ODE $\mathcal{F}$, where the prediction of the latent state, *i.e.* $\mathbf{s}_h(t)$, is given by:

$$\hat{\mathbf{s}}_h(t) = \mathbf{s}_h(t-1) + \int_{t-1}^t \mathcal{F}_\phi(\mathbf{s}_h(\tau))d\tau, \tag{8}$$

where $\mathcal{F}_\phi$ is an ODE modeled by a neural network with learnable parameters ϕ, and $\mathbf{s}_h(t-1)$ is the output of the previous time instant $t-1$. Following [7], this latent-dynamics neural ODE is shared by both the encoder and the decoder.

Latent Dynamics Correction with GCN-GRU. The final estimate of the latent variable $\mathbf{s}_h(t)$ is given as a weighted combination of the latent representation $\mathbf{z}_h(t)$ of the measurement $\mathbf{Y}_t$ on the heart and the prediction of the latent dynamics $\hat{\mathbf{s}}_h(t)$ given in the previous section. This is achieved by a Gated Recurrent Unit (GRU) cell [6] whose underlying architecture contains GCN layers as both the hidden state and input to the cell are graphs. We call it a *GCN-GRU* cell and denote it as $\mathcal{G}_\gamma$:

$$\mathbf{s}_h(t) = \mathcal{G}_\gamma(\mathbf{z}_h(t), \hat{\mathbf{s}}_h(t)). \tag{9}$$

where $\mathbf{z}_h(t)$ is obtained from $\mathbf{Y}_t$ as described in Eq. (6).

This resembles a classic SSM that separately models the latent dynamics of a variable and its emission to the observation space, with the exception that both models as presented are expressive neural functions.

3 Experiments

In simulation experiments where both $\mathbf{X}_t$ and $\mathbf{Y}_t$ are available, we test the performance of the presented ODE-GCNN to learn from unlabeled data in comparison to two existing reconstruction networks supervised by small labeled data. We then test the presented ODE-GCNN in clinical data where labels of $\mathbf{X}_t$ are not available. In all experiments, the presented network consists of three GCNN blocks and two regular convolution layers in the encoder, one spline convolution layer for latent inverse mapping, one ODE-solver followed by a GCN-GRU block for latent dynamics, and four GCNN blocks and two regular convolution layers in the decoder. We used tanh activation for the final encoding layer and the ODE, and ELU for the rest. We used Adam optimizer [18] and a learning rate of 5×10^{-4}. All experiments were run on NVIDIA Tesla T4s with 16 GB memory. Our implementation is available here: https://github.com/john-x-jiang/phy_geo.

3.1 Synthetic Experiments

As described earlier, the presented temporal modeling consists of an integration of prediction (via neural ODE) and correction (via GCN-GRU) operations. When omitting the GRU-based correction module, the presented model became similar to latent neural ODE models [22] that are purely driven by ODE dynamics. This formulation unfortunately failed in our problem setting and was not included in the following experiments. When omitting the preditive neural ODE, we obtain a more traditional sequence-to-sequence model where temporal modeling is entangled with spatial modeling in both the encoder and the decoder. To compare with this type of formulations, we compared against an Euclidean [13] and a non-Euclidean sequence-to-sequence encoding-decoding model (ST-GCNN) [17], both supervised by data. Furthermore, to examine the benefit of disentangled spatial-temporal modeling by ODE-GCNN, we also compared it with the ST-GCNN supervised by the same physics-informed loss.

Data and Training. On one heart-torso mesh, we generated propagation sequences of action potential by the Aliev-Panfilov model [1], considering a combination of 186 different origins of activation and 17 different spatial segments of scar tissue. The extracellular potential on the heart ($\mathbf{X}_t$) and body surface ($\mathbf{Y}_t$) were then simulated and 30 dB Gaussian noises were added to $\mathbf{Y}_t$.

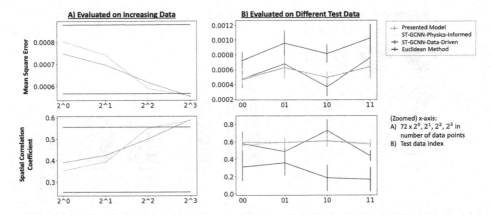

Fig. 2. Performance of different models under (A) increasing training data and (B) different testing conditions.

To generate disjoint training and testing sets, we partition the data based on the locations of activation origins and scar tissue, using 12 segments for training and five for testing. To train the physics-informed models, we randomly sample the unlabelled data at exponentially increasing rates, starting with 72 parameter sets (denoted as 2^0) to 569 (2^3). As the data-driven models require labelled data, we consider a setting where only a limited set of labelled data (2^0) is available.

We further group test data into four testing conditions based on how far the activation origin and scar tissue are located from those used in training.

Results. The reconstruction accuracy was quantitatively measured by mean square error (MSE) and spatial correlation coefficient (CC) between the reconstructed and actual heart surface potential. Figure 2A shows the performance change as the two physics-informed models (red and yellow) were trained with increasing unlabeled data, in comparison to the Euclidean (blue) and non-Euclidean ST-GCNN (green) models. The Euclidean baseline, similar to that reported in [17], had difficulty to generalize when using small labeled data. The data-driven ST-GCNN, as expected, outperformed the two physics-informed models when trained using the same number of data. Notably, as unlabeled data increased, the two physics-informed models showed a strong positive trend in performance increase and, at data size 2^3, was comparable to the data-driven ST-GCNN. Furthermore, as shown in Fig. 2B, the physics-informed models were more stable than data-driven models under different testing conditions. Between the two physics-informed models, as shown in Fig. 2A, the presented ODE-GCNN showed better performance when unlabelled data was small. Supplementary Material visually highlights the temporal reconstructions across the models on a single example.

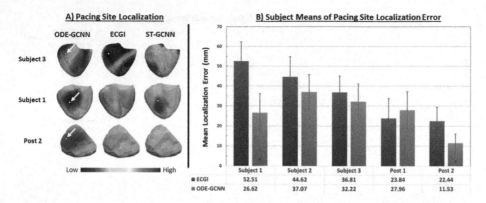

Fig. 3. A) Examples of reconstruction results on real data. The yellow arrows highlight actual pacing sites. Low potential in the reconstruction indicates regions of activation. B) Summary of localization errors for the origin of ventricular activation in each subject. Asterisks (*) mark significantly different results ($p < 0.05$, paired t-tests).

3.2   Clinical Data

Next, we evaluate the presented methods on *in-vivo* 120-lead ECG data from five subjects, including three healthy subjects [2,10] and two post-infarction subjects [2,23] undergoing ventricular pacing. As in most *in-vivo* settings, labeled data of heart-surface potential were not available. The physics-informed model as presented has a unique advantage here given its ability to learn from unlabeled data: here, we fine-tuned the presented ODE-GCNN (trained on 569 unlabeled simulation data in Sect. 3.1) on a subset of *in-vivo* ECG data. For comparison, we considered: 1) the data-driven ST-GCNN trained on simulation data in Sect. 3.1 (which cannot be fine-tuned due to the reliance on labeled data), and 2)standard physics-based reconstruction using the same optimization loss as the physics-informed network (but without learning across data), which we termed as electrocardiographic imaging (ECGI). All methods were tested on the same data not used in the fine-tuning of the physics-informed ODE-GCNN, totaling 36 cases across the five patients.

Due to the lack of *in-vivo* measurement of heart surface potential, quantitative accuracy was measured by the Euclidean distance between the reconstruction origin of activation and the known sites of pacing.

Results. Figure 3A highlights examples of the reconstructed propagation in two healthy subjects and one post-infarction subject. Across the runs, the data-driven ST-GCNN failed to produce stable temporal patterns. Both the physics-informed ODE-GCNN and ECGI were able to produce spatiotemporally smooth propagation patterns, highlighted in Supplementary Material. In some cases, as shown in the middle column in Fig. 3A, the ECGI method failed to capture the activation region.

Figure 3B summarizes the error for localizing the activation origin on each patient for the physics-informed ODE-GCNN against ECGI, summarized from in total 36 cases across the five patients. The unstable temporal patterns produced by the data-driven ST-GCNN precluded localization of the activation origin and was excluded from quantitative comparisons. The ODE-GCNN method was able to consistently outperform the ECGI method with the exception of patient Post 1, which was comparable in performance. In two cases, Subject 1 and Post 2, the difference in performance was statistically significant ($p < 0.05$, paired t-tests).

The results of the ODE-GCNN method were comparable to the method proposed in [10] despite their inclusion of a more advanced regularization term and activation localization method. We leave this up to future work to investigate the effect of stronger regularization terms for the proposed model.

4   Conclusion

In this work, we present a novel label-free and physics-informed image reconstruction network that is supervised by physics-based forward operators without the need of labeled data, with disentangled spatial-temporal modeling of the inverse solution by a novel ODE-GCNN architecture. This promising integration of physics-based and data-driven approach to learn reconstructions using observational data without corresponding labels was validated through the experimental results in both *in-silico* and *in-vivo* settings. While focusing on the problem of reconstructing electrical activity of the heart from surface recordings, the presented work is also related to the larger effort of using such surface data to personalize virtual heart models [15,16]. Future work will investigate the connection and comparison with these works. Other future works include to incorporate more advanced physics-informed objective functions and formulate the latent dynamics in a way that gives rise to stronger interpretability. To the best of our knowledge, this is the first label-free physics-informed image reconstruction network on geometrical space.

Acknowledgement. This work is supported by the National Institutes of Health (NIH) under Award Number R01HL145590 and R01NR018301.

References

1. Aliev, R.R., Panfilov, A.V.: A simple two-variable model of cardiac excitation. Chaos Solit. Fract. **7**(3), 293–301 (1996)
2. Aras, K., et al.: Experimental data and geometric analysis repository-EDGAR. J. Electrocardiol. **48**(6), 975–981 (2015)
3. Cacciola, F.: Triangulated surface mesh simplification. In: CGAL User and Reference Manual. CGAL Editorial Board, 5.0.2 edn. (2020). https://doc.cgal.org/5.0. 2/Manual/packages.html#PkgSurfaceMeshSimplification
4. Chen, E.Z., Chen, T., Sun, S.: MRI image reconstruction via learning optimization using neural odes. In: Medical Image Computing and Computer-Assisted Intervention (MICCAI) (2020)

5. Chen, R.T.Q., Rubanova, Y., Bettencourt, J., Duvenaud, D.: Neural ordinary differential equations (2019)
6. Cho, K., et al.: Learning phrase representations using RNN encoder-decoder for statistical machine translation (2014)
7. Chung, J., Kastner, K., Dinh, L., Goel, K., Courville, A., Bengio, Y.: A recurrent latent variable model for sequential data. Neural Inf. Proces. Syst. (2015)
8. Cohen, O., Zhu, B., Rosen, M.S.: MR fingerprinting deep reconstruction network (drone). Magn. Resonan. Med. **80**(3), 885–894 (2018)
9. Durbin, J., Koopman, S.J.: Time Series Analysis By State Space Methods. Oxford University Press, Oxford (2012)
10. Erem, B., Coll-Font, J., Orellana, R., Stovicek, P., Brooks, D.: Using transmural regularization and dynamic modeling for noninvasive cardiac potential imaging of endocardial pacing with imprecise thoracic geometry. IEEE Trans. Med. Imag. **33**, 726–738 (2014). https://doi.org/10.1109/TMI.2013.2295220
11. Fang, Q., Boas, D.A.: Tetrahedral mesh generation from volumetric binary and grayscale images. In: 2009 IEEE International Symposium on Biomedical Imaging: From Nano to Macro, pp. 1142–1145. Ieee (2009)
12. Fey, M., Eric Lenssen, J., Weichert, F., Müller, H.: Splinecnn: fast geometric deep learning with continuous b-spline kernels. In: The IEEE Conference on Computer Vision and Pattern Recognition (CVPR), pp. 869–877 (2018)
13. Ghimire, S., Dhamala, J., Gyawali, P.K., Sapp, J.L., Horacek, M., Wang, L.: Generative modeling and inverse imaging of cardiac transmembrane potential. In: Frangi, A.F., Schnabel, J.A., Davatzikos, C., Alberola-López, C., Fichtinger, G. (eds.) MICCAI 2018. LNCS, vol. 11071, pp. 508–516. Springer, Cham (2018). https://doi.org/10.1007/978-3-030-00934-2_57
14. Ghimire, S., Gyawali, P.K., Dhamala, J., Sapp, J.L., Horacek, M., Wang, L.: Improving generalization of deep networks for inverse reconstruction of image sequences. In: Chung, A.C.S., Gee, J.C., Yushkevich, P.A., Bao, S. (eds.) IPMI 2019. LNCS, vol. 11492, pp. 153–166. Springer, Cham (2019). https://doi.org/10.1007/978-3-030-20351-1_12
15. Giffard-Roisin, S., et al.: Transfer learning from simulations on a reference anatomy for ECGI personalized cardiac resynchronization therapy. IEEE Trans. Biomed. Eng. **66**(2), 343–353 (2018)
16. Giffard-Roisin, S., et al.: Noninvasive personalization of a cardiac electrophysiology model from body surface potential mapping. IEEE Trans. Biomed. Eng. **64**(9), 2206–2218 (2016)
17. Jiang, X., Ghimire, S., Dhamala, J., Li, Z., Gyawali, P.K., Wang, L.: Learning geometry-dependent and physics-based inverse image reconstruction. In: International Conference on Medical Image Computing and Computer-Assisted Intervention, pp. 487–496. Springer, Berlin (2020). https://doi.org/10.1007/10704282
18. Kingma, D.P., Ba, J.: Adam: A method for stochastic optimization. arXiv preprint arXiv:1412.6980 (2014)
19. Lai, K.W., Aggarwal, M., van Zijl, P., Li, X., Sulam, J.: Learned proximal networks for quantitative susceptibility mapping. In: Medical Image Computing and Computer-Assisted Intervention (MICCAI) (2020)
20. Lindstrom, P., Turk, G.: Fast and memory efficient polygonal simplification. In: Proceedings Visualization'98 (Cat. No. 98CB36276), pp. 279–286. IEEE (1998)
21. Plonsey, R.: Bioelectr. Phenomena. McGraw Hill, New York (1969)
22. Rubanova, Y., Chen, R.T., Duvenaud, D.: Latent odes for irregularly-sampled time series. arXiv preprint arXiv:1907.03907 (2019)

23. Sapp, J.L., Dawoud, F., Clements, J.C., Horáček, B.M.: Inverse solution mapping of epicardial potentials: quantitative comparison with epicardial contact mapping. Circul. Arrhyth. Electrophysiol. **5**(5), 1001–1009 (2012)
24. Stewart, R., Ermon, S.: Label-free supervision of neural networks with physics and domain knowledge. In: Proceedings of the AAAI Conference on Artificial Intelligence, vol. 31 (2017)
25. Ulas, C., et al.: Direct estimation of pharmacokinetic parameters from DCE-MRI using deep cnn with forward physical model loss. In: International Conference on Medical Image Computing and Computer-Assisted Intervention, pp. 39–47. Springer (2018)
26. Wang, L., Zhang, H., Wong, K.C., Liu, H., Shi, P.: Physiological-model-constrained noninvasive reconstruction of volumetric myocardial transmembrane potentials. IEEE Trans. Biomed. Eng. **57**(2), 296–315 (2009)
27. Yang, Y., Sun, J., Li, H., Xu, Z.: Deep admm-net for compressive sensing MRI. In: Proceedings of the 30th International Conference on Neural Information Processing Systems, pp. 10–18 (2016)
28. Zhu, B., Liu, J.Z., Cauley, S.F., Rosen, B.R., Rosen, M.S.: Image reconstruction by domain-transform manifold learning. Nature **555**(7697), 487–492 (2018)

High-Resolution Hierarchical Adversarial Learning for OCT Speckle Noise Reduction

Yi Zhou[1], Jiang Li[2], Meng Wang[1], Weifang Zhu[1], Yuanyuan Peng[1], Zhongyue Chen[1], Lianyu Wang[1], Tingting Wang[1], Chenpu Yao[1], Ting Wang[1], and Xinjian Chen[1,3(✉)]

[1] School of Electronics and Information Engineering, Soochow University, Suzhou, China
xjchen@suda.edu.cn
[2] Department of Electrical and Computer Engineering, Old Dominion University, Norfolk, VA, USA
[3] State Key Laboratory of Radiation Medicine and Protection, Soochow University, Suzhou, China

Abstract. Raw optical coherence tomography (OCT) images typically are of low quality because speckle noise blurs retinal structures, severely compromising visual quality and degrading performances of subsequent image analysis tasks. In this paper, we propose a novel end-to-end cross-domain denoising framework for speckle noise suppression. We utilize high quality ground truth datasets produced by several commercial OCT scanners for training, and apply the trained model to datasets collected by our in-house OCT scanner for denoising. Our model uses the high-resolution network (HRNet) as backbone, which maintains high-resolution representations during the entire learning process to restore high fidelity images. In addition, we develop a hierarchical adversarial learning strategy for domain adaption to align distribution shift among datasets collected by different scanners. Experimental results show that the proposed model outperformed all the competing state-of-the-art methods. As compared to the best of our previous method, the proposed model improved the signal to noise ratio (SNR) metric by a huge margin of 18.13 dB and only required 25 ms for denoising one image in testing phase, achieving the real-time processing capability for the in-house OCT scanner.

Keywords: OCT speckle noise reduction · HRNet · GAN

1 Introduction

OCT is a recent imaging modality for biological tissues [14]. OCT images usually suffer from speckle noises, which degrade quality of OCT images and make automated image analysis challenging. Traditional speckle noise suppression algorithms can be categorized into three groups: 1) Filter-based techniques

© Springer Nature Switzerland AG 2021
M. de Bruijne et al. (Eds.): MICCAI 2021, LNCS 12906, pp. 372–381, 2021.
https://doi.org/10.1007/978-3-030-87231-1_36

[1,4], 2) Sparse transform-based methods [6,8] and 3) Statistics and low-rank decomposition-based methods [2,3,9]. Though these traditional methods are effective for image denoising, there are remaining challenges such as insufficient image feature representation, and some of the methods are time-consuming.

In the past few years, there has been a boom of the application of deep learning to noise suppression. For example, Zhang et al. [16] used the residual learning strategy and batch normalization technique to improve the feedforward denoising convolutional neural network (DnCNN) for target blind Gaussian denoising. Dong et al. [5] proposed a denoising prior driven deep neural network (DPDNN) for image restoration, and both methods were designed based on the additive noise model. We proposed a model, Edge-cGAN [10] based on the conditional generative adversarial network (cGAN) [7] to remove speckle noise, and developed training dataset T by using commercial scanners for data collection. Later, we developed an in-house OCT scanner and improved Edge-cGAN to the second version (Mini-cGAN [17]) to suppress speckle noise for dataset B collected by the in-house scanner. While it achieved good performances on dataset B, Mini-cGAN is not an end-to-end learning model, requiring multiple steps of training. In addition, the testing time complexity is high, making it not suitable for real-time processing.

Different OCT scanners have different characteristics and datasets collected by different scanners may contain distribution shifts. In this paper, our goal is to compensate for the distribution shifts and leverage the high quality ground truth dataset T to achieve effective speckle noise suppression for dataset B collected by our in-house scanner. We propose a novel end-to-end learning framework to achieve our goal and the diagram is shown in Fig. 1. The proposed model 1) utilizes the HRNet backbone to maintain high-resolution representation for image restoration and 2) uses a dynamic hierarchical domain adaptation network to leverage the dataset T for training and apply the trained model to dataset B for noise removal. Our model not only significantly improves image quality as compared with Edge-cGAN and Mini-cGAN but also dramatically reduces testing time, satisfying the real-time requirement of our in-house scanner.

2   Method

2.1   Proposed Model

Overall Architecture: The proposed model (Fig. 1) consists of a generator G, a hierarchical discriminator D_H, and an output alignment discriminator D (omitted in the figure but its learning loss function shown as $\mathcal{L}_D$). Our objective is to leverage high quality dataset T to learn an end-to-end model for speckle noise removal and apply the model to dataset B collected by our in-house scanner. The distribution shift between the two datasets T and B is compensated by the loss function $\mathcal{L}_H$ through hierarchical adversarial learning. In addition, the denoised images $G(T)$ and $G(B)$ generated from the two data domains are aligned by the generative adversarial loss function $\mathcal{L}_D$ during training.

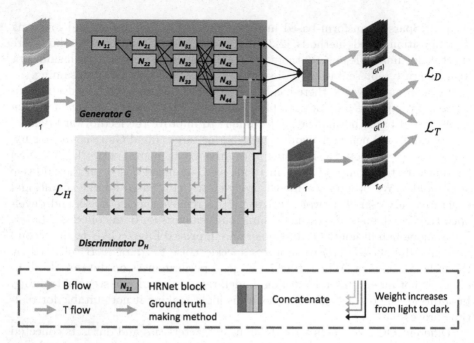

Fig. 1. Diagram of the proposed framework. During training, datasets B and T are alternatively input to the generator G (HRNet). Features of different resolutions are extracted by G and hierarchically combined in the discriminator D_H to perform adversarial learning such that features extracted from dataset B are indistinguishable from those from T. Features from T are then concatenated to generate denoised images as $G(T)$ and are compared to ground truth T_{GT} to minimize $\mathcal{L}_T$. In addition, $G(B)$, images generated from dataset B, are aligned to $G(T)$ by a discriminator D (omitted in this Figure) through the adversarial learning process governed by the loss $\mathcal{L}_D$. During testing, images from B are input to G to generate denoised images as $G(B)$.

HRNet Backbone for Image Restoration: HRNet [12] is an improved version of the U-Net structure that has been widely applied to image reconstruction and segmentation [11]. U-Net uses an encoder to compress input to a low-dimensional latent vector through convolution and pooling and utilizes a decoder to reconstruct input by expanding the compressed vector. There are also skip connections to copy information from encoder directly to decoder located at the same resolution level. Convolutional layers in deep learning models including U-Net extract and magnify useful information from input for classification or regression tasks. However, this convolutional processing only happens in the low resolution levels after pooling in U-Net and other levels copy information directly from encoder to decoder by the skip connections. In contrast, HRNet uses convolutional layers to learn useful information in all resolution levels as shown in Fig. 1. We select $HRNetV2$ layer [12] as output, which concatenates different resolution channels by bilinear interrelation to reconstruct denoised image.

We utilize the combination of mean square error (MSE) and L_1 losses for training: $\mathcal{L}_T = \mathcal{L}_{MSE} + \alpha \mathcal{L}_{L_1}$, as it has been shown in [7] that the L_1 loss led to sparse solutions, which are desired for the OCT speckle noise removal. The MSE loss and L_1 loss can be formulated as follows, respectively,

$$\mathcal{L}_{MSE} = \mathbb{E}_{t,t_{gt}\sim T}[\|t_{gt} - G(t)\|_2], \tag{1}$$

$$\mathcal{L}_{L_1} = \mathbb{E}_{t,t_{gt}\sim T}[\|t_{gt} - G(t)\|_1], \tag{2}$$

where t and t_{gt} represent a raw-clean image pair in training dataset T.

Hierarchical Adversarial Learning: Inspired by [15], we propose a hierarchical adversarial learning structure to compensate domain shift between datasets T and B. We combine outputs from different resolution levels in HRNet (G) using a hierarchial structure as D_H in Fig. 1. Low-level resolution features are domain informative while high-level resolution features are rich in semantic information. Direct stacking these features from different resolution levels for adversarial learning can impair adaptation performance because of the information conflicts among different resolution levels [15]. The hierarchical structure in D_H follows the resolution levels in the encoder part of G so that the conflicts can be mitigated. Each level has a separate objective function and G and D_H play a min-max game to minimize the overall loss, $\mathcal{L}_H = \sum_{k=1}^{K} \gamma_k \mathcal{L}_{h,k}$, where $\mathcal{L}_{h,k}$ is the loss function for level l_k, $k = 1, 2, 3..., K$ and

$$\mathcal{L}_{h,k} = \mathbb{E}_{l_k^t \in G(t)}[\log(D_H(l_k^t))] + \mathbb{E}_{l_k^b \in G(b)}[\log(1 - D_H(l_k^b))], \tag{3}$$

γ_k are mixing coefficients and it increases as k decreases, making the attention focus more on low-level domain information, and l_k^t and l_k^b represent features extracted at the kth layer from datasets T and B, respectively.

Output Alignment: At output of G, we utilize a discriminator D to align the reconstructed images $G(B)$ and $G(T)$ in image space by adversarial learning,

$$\mathcal{L}_D = \mathbb{E}_{t\sim T}[\log(D(G(t)))] + \mathbb{E}_{b\sim B}[\log(1 - D(G(b)))], \tag{4}$$

where b is one image in B. Finally, the total objective function of the framework is,

$$\min_{G} \max_{D_H, D} \mathcal{L}_T + \lambda_0 \mathcal{L}_H + \lambda_1 \mathcal{L}_D, \tag{5}$$

where λ_0, λ_1 are trade-off parameters between the two loss functions. With the above learning processes and loss functions, the generator G can be trained to restore high quality OCT images for our in-house scanner. The training and testing procedure descriptions are provided in the caption of Fig. 1.

2.2 Dataset

High quality dataset T was created in our previous study [10], which was approved by IRB of the University and informed consents were obtained from all subjects. The dataset contains 512 raw OCT images collected by four commercial scanners (Topcon DRI-1 Atlantis, Topcon 3D OCT 2000, Topcon 3D OCT 1000 and Zeiss Cirrus 4000, with their image sizes of 512×992, 512×885, 512×480 and 512×1024 respectively), and the training set is composed of 256 images from Topcon DRI-1 Atlantis and 256 images from Topcon 3D OCT 2000. For each raw image, a clean image was produced by registering and averaging raw images acquired repeatedly at the same location from the same subject. We used flipping along the transverse axis, scaling, rotation, and non-rigid transformations to augment T and increased the data size by 4× for effective training. Dataset B was collected by our recently developed non-commercial in-house OCT scanner and there were totally 1024 raw OCT images, of which 200 images had disease. We also set aside three disease raw images and six normal raw images from the in-house scanner for testing.

2.3 Evaluation Metrics

We utilize four performance metrics to evaluate the proposed model including signal-to-noise ratio (SNR), contrast-to-noise ratio (CNR), speckle suppression index (SSI) and edge preservation index (EPI). As Fig. 2a) and Fig. 3a) show, we manually selected regions of interest (ROIs) in background (green rectangles) and signal (red rectangles) areas for SNR and CNR calculation, and delimitated the blue boundaries for EPI computation. These four metrics are defined as,

$$SNR = 10\lg(\frac{\sigma_s^2}{\sigma_b^2}), CNR = 10\lg(\frac{|\mu_s - \mu_b|}{\sqrt{\sigma_s^2 + \sigma_b^2}}),$$

$$SSI = \frac{\sigma_r}{\mu_r} \times \frac{\mu_d}{\sigma_d}, EPI = \frac{\sum_i \sum_j |I_d(i+1,j) - I_d(i,j)|}{\sum_i \sum_j |I_r(i+1,j) - I_r(i,j)|}. \tag{6}$$

where μ_s, μ_b and σ_s, σ_b denote means and standard deviations of the defined signal and background regions in denoised image, respectively, for SNR and CNR computations. μ_r, σ_r and μ_d, σ_d denote means and standard deviations of raw and denoised images, respectively, for SSI computation. I_r and I_d represent raw and denoised images, and i and j represent longitudinal and lateral coordinates in image. SNR reflects noise level, CNR is the contrast between signal and background, SSI measures ratio between noise and denoised images, and EPI reflects the extent of details of edges preserved in denoised images. A small SSI value and large SNR, CNR and EPI values represent high quality images.

3 Experiments and Results

3.1 Implementation Details

Discriminators D_H and D followed the configurations in [7] consisting of six convolutional layers where the first three used instance normalization. For all the

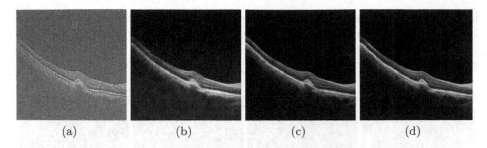

(a)   (b)   (c)   (d)

Fig. 2. Ablation study results on a disease OCT image. Red and green rectangles represent signal and background regions manually selected for SNR and CNR calculation, respectively. Blue curves are boundaries manually delimitated for EPI computation. (a) Raw OCT image (b) U-Net (c) HRNet (d) HRNet+Hier (Proposed). (Color figure online)

experiments, α was set to 100 in $\mathcal{L}_T$, hyper-parameters λ_0 and λ_1 were set to 10 in Eq. 5, and $\gamma_1, \gamma_2, \gamma_3, \gamma_4$ in $\mathcal{L}_H$ followed [15] to set to 4, 3, 2 and 1, respectively. We utilized the Adam solver with an initial learning rate of 2.0×10^{-4} and a momentum of 0.5 for optimization. The batch size was set to 2, and the number of training epochs was set to 400. Images were resized to 512×512 during training. The proposed method was implemented using Pytorch and was trained using one NVIDIA RTX 3080 GPU with 10 G memory.

3.2 Ablation Study

We conducted ablation study to investigate contribution of each of the three components in the proposed model, including 1) U-Net as backbone for the generator G with output alignment but no hierarchical adversarial learning (U-Net), 2) HRNet as backbone for G with output alignment but no hierarchical adversarial learning (HRNet) and 3) HRNet as backbone for G with both output alignment and hierarchical adversarial learning as the final proposed model (HRNet+Hier). Experimental results in Table 1 show that using HRNet to replace U-Net greatly improved performance of the model, increasing SNR from 22.69 dB to 35.41 dB. SSI and EPI were also improved with any exception of CNR that slightly degraded. HRNet maintains low- and high-resolution

Table 1. Ablation study on nine images from the in-house OCT scanner.

Method	SNR *(dB)*	CNR *(dB)*	SSI	EPI
Raw image	0.03 ± 0.54	3.80 ± 0.85	1.000 ± 0.00	1.00 ± 0.00
U-net	22.69 ± 7.45	$\mathbf{12.44 \pm 1.13}$	0.139 ± 0.01	0.83 ± 0.09
HRNet	35.41 ± 12.00	11.35 ± 1.34	$\mathbf{0.087 \pm 0.01}$	0.94 ± 0.07
HRNet+Hier	$\mathbf{40.41 \pm 7.69}$	11.15 ± 1.39	0.091 ± 0.01	$\mathbf{0.96 \pm 0.07}$

information throughout the entire process, making the denoised images much better quality. When the hierarchical adversarial loss was combined, SNR was improved further to 40.41 dB, EPI also slightly increased and the other two degraded slightly.

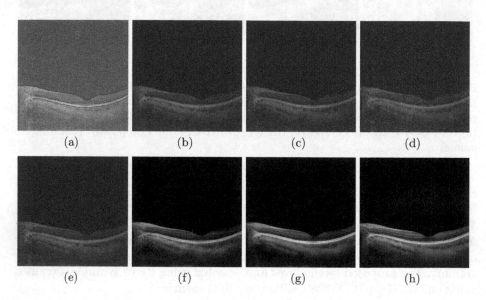

Fig. 3. Comparison study results. The red rectangles, green rectangles and blue curves are manually defined as in Fig. 2. (a) Raw image (b) NLM [1] (c) STROLLR [13] (d) DnCNN [16] (e) DPDNN [5] (f) Edge-cGAN [10] (g) Mini-cGAN [17] (h) Proposed (HRNet+Hier). (Color figure online)

3.3   Comparison Study

Table 2. Results by competing methods on nine images from the in-house scanner.

Method	SNR *(dB)*	CNR *(dB)*	SSI	EPI	Times(s)
Raw image	0.03 ± 0.54	3.80 ± 0.85	1.000 ± 0.00	1.00 ± 0.00	*None*
NLM [1]	19.50 ± 4.21	9.23 ± 1.98	0.697 ± 0.03	0.55 ± 0.06	0.089 ± 0.006
STROLLR [13]	18.01 ± 3.89	11.03 ± 2.01	0.707 ± 0.03	0.37 ± 0.02	182.203 ± 7.648
DnCNN [16]	14.99 ± 2.27	7.00 ± 1.10	0.670 ± 0.03	0.62±0.03	**0.022 ± 0.003**
DPDNN [5]	34.77 ± 8.40	8.40 ± 1.74	0.684 ± 0.03	0.52 ± 0.03	0.036 ± 0.001
Edge-cGAN [10]	24.35 ± 5.50	11.35 ± 0.97	0.105 ± 0.01	0.87 ± 0.08	0.929 ± 0.014
Mini-cGAN [17]	22.28 ± 4.65	**12.03 ± 1.61**	**0.087 ± 0.01**	0.93 ± 0.11	1.825 ± 0.022
Proposed	**40.41 ± 7.69**	11.15 ± 1.39	0.091 ± 0.01	**0.96 ± 0.07**	0.025 ± 0.008

We compared the proposed method with state-of-the-art methods, including non-local means (NLM) [1], sparsifying transform learning and low-rank method

(STROLLR) [13], deep CNN with residual learning (DnCNN) [16], denoising prior driven deep neural network for image restoration (DPDNN) [5], our previous method, Edge-cGAN [10] and its improved version, Mini-cGAN [17]. In these experiments, parameters for traditional methods were set to values so that the models can achieve best results for the application, and deep learning models followed their original configurations. The proposed method and Mini-cGAN compensated distribution shift existed between datasets T and B through adversarial learning. To have a fair comparison, we performed a separate histogram matching process for the testing images for all other competing methods before testing. In addition, we compared efficiency by recording testing time required by each method.

Visual inspection of Fig. 3 reveals that the proposed method achieved the best result (Fig. 3h). Our previous methods Mini-cGAN and Edge-cGAN ranked the second and third (Fig. 3g and Fig. 3f), respectively. Denoised images by all other methods have low visual qualities. Table 2 shows qualitative performance metrics indicating that the proposed method achieved the best SNR and EPI. Mini-cGAN obtained the best CNR and SSI, and DnCNN achieved the best computational efficiency, requiring only 22 ms to process one image. It is worth noting that the proposed model ranked the second and only needed 25 ms to denoise one image.

4   Discussion

The proposed method achieved the best visual quality as shown in Fig. 3. NLM (Fig. 3b) and STROLLR (Fig. 3c) suffered from excessive smoothing, leading to blurred regions at the boundaries between adjacent layers. The background regions were not very clean either. DnCNN (Fig. 3d) performed well in retina areas but left artifacts in background. DPDNN (Fig. 3e) obtained a very clean background, however, the interlayer details were not well maintained. Edge-cGAN (Fig. 3f) and Mini-cGAN (Fig. 3g) improved visual quality of the denoised image significantly. However, the signal was still weak in the top right retina area. The proposed model achieved the best image contrast, preserved the most details in the layers under retina, and resulted in a much sharper enhanced image (Fig. 3h).

The metrics of SNR and EPI represent signal to noise ratio and edge preservation performances in denoised images, respectively. The proposed model achieved the best SNR and EPI (Table 2), indicating that it restored the strongest signal by suppressing speckle noise and preserved the desired sharp detail information. In terms of testing time, DnCNN took 3 ms less than the proposed model. However, image quality by DnCNN was much worse. As compared to Mini-cGAN, the new model improved SNR by a huge margin of 18.13 dB to 40.41 dB and the testing time was accelerated by a factor of 73. The computational efficiency satisfied real-time requirement of our in-house scanner. DPDNN achieved the second best SNR of 34.77 dB, its other metrics including CNR, SSI and EPI were much worse than those by the proposed model, which can be confirmed in Fig. 3e.

CNR and SSI represent the contrast and speckle suppression performances in result images, respectively. Mini-cGAN uses U-Net as backbone and generates denoised images by averaging multiple overlapped patches outputted by the trained model during testing. In contrast, the proposed model utilizes the high-resolution HRNet to generate denoised images directly without averaging. The averaging processing in Mini-cGAN reduced variance and led to slightly better CNR and SSI with a cost of much longer testing time. We tested to conduct the same averaging process in the proposed model to generate denoised images and it did improve the CNR metric slightly but degraded SNR and required much longer testing time. We concluded that the HRNet was able to restore the strongest signal because of its unique structure and the repetition step in testing was not necessary.

5   Conclusion

In this paper, we proposed a novel end-to-end cross-domain denoising framework that significantly improved speckle noise suppression performance in OCT images. The proposed model can be trained and tested with OCT images collected by different scanners, achieving automatic domain adaptation. We utilized the HRNet backbone to carry high-resolution information and restored fidelity images. In addition, we developed a hierarchical adversarial learning module to achieve the domain adaptation. The novel model improved SNR by huge margins as compared to our previous models and all competing state of the arts, achieved a testing time of 0.025 s, and satisfied real-time process requirement.

References

1. Aum, J., Kim, J.h., Jeong, J.: Effective speckle noise suppression in optical coherence tomography images using nonlocal means denoising filter with double gaussian anisotropic kernels. Appl. Opt. **54**(13), D43–D50 (2015)
2. Cameron, A., Lui, D., Boroomand, A., Glaister, J., Wong, A., Bizheva, K.: Stochastic speckle noise compensation in optical coherence tomography using non-stationary spline-based speckle noise modelling. Biomed. Opt. Express **4**(9), 1769–1785 (2013)
3. Cheng, J., Tao, D., Quan, Y., Wong, D.W.K., Cheung, G.C.M., Akiba, M., Liu, J.: Speckle reduction in 3d optical coherence tomography of retina by a-scan reconstruction. IEEE Trans. Med. Imag. **35**(10), 2270–2279 (2016)
4. Chong, B., Zhu, Y.K.: Speckle reduction in optical coherence tomography images of human finger skin by wavelet modified BM 3D filter. Opt. Commun. **291**, 461–469 (2013)
5. Dong, W., Wang, P., Yin, W., Shi, G., Wu, F., Lu, X.: Denoising prior driven deep neural network for image restoration. IEEE Trans. Patt. Anal. Mach. Intel. **41**(10), 2305–2318 (2018)
6. Fang, L., Li, S., Cunefare, D., Farsiu, S.: Segmentation based sparse reconstruction of optical coherence tomography images. IEEE Trans. Med. Imag. **36**(2), 407–421 (2016)

7. Isola, P., Zhu, J.Y., Zhou, T., Efros, A.A.: Image-to-image translation with conditional adversarial networks. In: Proceedings of the IEEE Conference on Computer Vision and Pattern Recognition, pp. 1125–1134 (2017)
8. Kafieh, R., Rabbani, H., Selesnick, I.: Three dimensional data-driven multi scale atomic representation of optical coherence tomography. IEEE Trans. Med. Imag. **34**(5), 1042–1062 (2014)
9. Li, M., Idoughi, R., Choudhury, B., Heidrich, W.: Statistical model for oct image denoising. Biomed. Opt. Exp. **8**(9), 3903–3917 (2017)
10. Ma, Y., Chen, X., Zhu, W., Cheng, X., Xiang, D., Shi, F.: Speckle noise reduction in optical coherence tomography images based on edge-sensitive CCAN. Biomed. Opt. Exp. **9**(11), 5129–5146 (2018)
11. Ronneberger, O., Fischer, P., Brox, T.: U-net: Convolutional networks for biomedical image segmentation. U-Net: convolutional networks for biomedical image segmentation. In: Navab, N., Hornegger, J., Wells, W., Frangi, A. (eds.) Medical Image Computing and Computer-Assisted Intervention (MICCAI 2015) LNCS, vol . 9351, pp. 234–241. Springer, Cham (2015). https://doi.org/10.1007/978-3-319-24574-4_28
12. Wang, J., et al.: Deep high-resolution representation learning for visual recognition. In: IEEE Transactions on Pattern Analysis and Machine Intelligence, pp. 1–1 (2020). https://doi.org/10.1109/TPAMI.2020.2983686
13. Wen, B., Li, Y., Bresler, Y.: When sparsity meets low-rankness: transform learning with non-local low-rank constraint for image restoration. In: 2017 IEEE International Conference on Acoustics, Speech and Signal Processing (ICASSP), pp. 2297–2301. IEEE (2017)
14. Wojtkowski, M., et al.: In vivo human retinal imaging by Fourier domain optical coherence tomography. J. Biomed. Opt. **7**(3), 457–463 (2002)
15. Xue, Y., Feng, S., Zhang, Y., Zhang, X., Wang, Y.: Dual-task self-supervision for cross-modality domain adaptation. In: Martel, A.L., et al. (eds.) MICCAI 2020. LNCS, vol. 12261, pp. 408–417. Springer, Cham (2020). https://doi.org/10.1007/978-3-030-59710-8_40
16. Zhang, K., Zuo, W., Chen, Y., Meng, D., Zhang, L.: Beyond a gaussian denoiser: residual learning of deep CNN for image denoising. IEEE Trans. Image Process. **26**(7), 3142–3155 (2017)
17. Zhou, Y., et al.: Speckle noise reduction for oct images based on image style transfer and conditional CAN. IEEE J. Biomed. Health Inform. (2021). https://doi.org/10.1109/JBHI.2021.3074852

Self-supervised Learning for MRI Reconstruction with a Parallel Network Training Framework

Chen Hu[1,2], Cheng Li[1], Haifeng Wang[1], Qiegen Liu[5], Hairong Zheng[1], and Shanshan Wang[1,3,4(✉)]

[1] Paul C. Lauterbur Research Center for Biomedical Imaging, Shenzhen Institutes of Advanced Technology, Chinese Academy of Sciences, Shenzhen, Guangdong, China
ss.wang@siat.ac.cn
[2] University of Chinese Academy of Sciences, Beijing, China
[3] Peng Cheng Laboratory, Shenzhen, Guangdong, China
[4] Pazhou Lab, Guangzhou, Guangdong, China
[5] Department of Electronic Information Engineering, Nanchang University, Nanchang, Jiangxi, China

Abstract. Image reconstruction from undersampled k-space data plays an important role in accelerating the acquisition of MR data, and a lot of deep learning-based methods have been exploited recently. Despite the achieved inspiring results, the optimization of these methods commonly relies on the fully-sampled reference data, which are time-consuming and difficult to collect. To address this issue, we propose a novel self-supervised learning method. Specifically, during model optimization, two subsets are constructed by randomly selecting part of k-space data from the undersampled data and then fed into two parallel reconstruction networks to perform information recovery. Two reconstruction losses are defined on all the scanned data points to enhance the network's capability of recovering the frequency information. Meanwhile, to constrain the learned unscanned data points of the network, a difference loss is designed to enforce consistency between the two parallel networks. In this way, the reconstruction model can be properly trained with only the undersampled data. During the model evaluation, the undersampled data are treated as the inputs and either of the two trained networks is expected to reconstruct the high-quality results. The proposed method is flexible and can be employed in any existing deep learning-based method. The effectiveness of the method is evaluated on an open brain MRI dataset. Experimental results demonstrate that the proposed self-supervised method can achieve competitive reconstruction performance compared to the corresponding supervised learning method at high acceleration rates (4 and 8). The code is publicly available at https://github.com/chenhu96/Self-Supervised-MRI-Reconstruction.

Keywords: Image reconstruction · Deep learning · Self-supervised learning · Parallel network

© Springer Nature Switzerland AG 2021
M. de Bruijne et al. (Eds.): MICCAI 2021, LNCS 12906, pp. 382–391, 2021.
https://doi.org/10.1007/978-3-030-87231-1_37

1 Introduction

Magnetic resonance imaging (MRI) becomes an essential imaging modality in clinical practices thanks to its excellent soft-tissue contrast. Nevertheless, the inherently long scan time limits the wide employment of MRI in various situations. Acquiring MR data at sub-Nyquist rates followed by image reconstruction is one of the common approaches for MRI acceleration. However, image reconstruction is an ill-posed inverse problem, and high acceleration rates might lead to noise amplification and residual artifacts in the images [10,11]. Therefore, recover high-quality MR images from undersampled data is a meaningful but challenging task.

In the past few years, promising performance has been achieved in deploying deep learning-based methods for MRI reconstruction [2,9]. These methods can be broadly divided into two categories: data-driven networks and physics-based unrolled networks. The former can be described as training pure deep neural networks to learn the nonlinear mapping between undersampled data/corrupted images and fully-sampled data/uncorrupted images. Representative works include U-Net [7], GANCS [6], etc. [8]. Unrolled networks construct network architectures by unfolding Compressive Sensing (CS) algorithms. Examples of this category are ISTA-Net [14], ADMM-Net [13], MoDL [1], etc. [4]. Regardless of the approaches utilized, most existing deep learning-based methods rely on fully-sampled data to supervise the optimization procedure. However, it is difficult to obtain fully-sampled data in many scenarios due to physiological constraints or physical constraints. Recently, a self-supervised learning method (self-supervised learning via data undersampling, SSDU) was proposed specifically to solve the issue [12], where the undersampled data is split into two disjoint sets. One is treated as the input and the other is used to define the loss. Despite the impressive reconstruction performance achieved, there are two important issues. First, the two sets need to be split with caution. When the second set does not contain enough data, the training process becomes unstable. Second, since no constraint is imposed on the unscanned data points, there is no guarantee that the final outputs are the expected high-quality images and high uncertainties exist.

To address the above issues, we propose a novel self-supervised learning method with a parallel network training framework. Here, differently, we construct two subsets by randomly selecting part of k-space data from the undersampled data, and then feed them into two parallel networks. Accordingly, two reconstruction losses are defined using all of the undersampled data to facilitate the networks' capability of recovering the frequency information and ensure that stable model optimization can be achieved. In addition, a difference loss is introduced, which acts as an indirect and reasonable constraint on the unscanned data points, between the outputs of the two networks to better aid in the subsequent high-quality image reconstruction during model testing. In the test phase, the obtained undersampled data is fed to either of the two trained networks to generate the high-quality results. Our major contributions can be summarized as follows: 1) A parallel network training framework is constructed to accomplish self-supervised image reconstruction model development through recovering undersampled data. 2) A novel difference loss on the unscanned data points of the undersampled data is introduced with the parallel networks, which can effectively constrain the solution space and improve the reconstruction performance. 3) Our method outperforms the existing state-of-the-art self-supervised learning methods and achieves a reconstruction performance competitive to the corresponding supervised learning method at high acceleration rates on an open IXI brain scan MRI dataset.

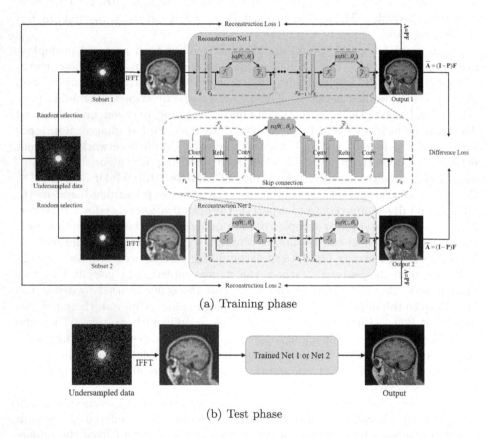

(a) Training phase

(b) Test phase

Fig. 1. The pipeline of our proposed framework for self-supervised MRI reconstruction.

2   Methods

Theoretically, the proposed method can be integrated with any existing deep learning-based method. In this work, an unrolled network, ISTA-Net [14], is utilized. The details are demonstrated as follows.

2.1   Mathematical Model of CS-MRI Reconstruction

Mathematically, the CS-MRI reconstruction problem can be written as:

$$\arg \min_{\mathbf{x}} \frac{1}{2} \|\mathbf{Ax} - \mathbf{y}\|_2^2 + \lambda R(\mathbf{x}) \tag{1}$$

where $\mathbf{x}$ is the desired image, $\mathbf{y}$ is the undersampled k-space measurement, $\mathbf{A}$ denotes the encoding matrix which include Fourier transform $\mathbf{F}$ and sampling matrix $\mathbf{P}$, $R(\mathbf{x})$ denotes the utilized regularization, and λ is the regularization parameter. The purpose of MRI reconstruction is to recover the desired image $\mathbf{x}$ from its measurement $\mathbf{y}$.

2.2   Brief Recap of ISTA-Net

ISTA-Net is an unrolled version of the Iterative Shrinkage Thresholding Algorithm (ISTA) [14], for which the regularization term in Eq. (1) is specified to be the L1 regularization. ISTA-Net solves the inverse image reconstruction problem in Eq. (1) by iterating the following two steps:

$$\mathbf{r}^{(k)} = \mathbf{x}^{(k-1)} - \rho \mathbf{A}^{\top} \left(\mathbf{Ax}^{(k-1)} - \mathbf{y} \right) \tag{2}$$

$$\mathbf{x}^{(k)} = \widetilde{\mathcal{F}}(soft(\mathcal{F}(\mathbf{r}^{(k)}), \theta)) \tag{3}$$

where k is the iteration index, ρ is the step size, $\mathcal{F}(\cdot)$ denotes a general form of image transform, $\widetilde{\mathcal{F}}(\cdot)$ denotes the corresponding left inverse, $soft(\cdot)$ is the soft thresholding operation, and θ is the shrinkage threshold. $\mathcal{F}(\cdot)$ and $\widetilde{\mathcal{F}}(\cdot)$ are realized through neural networks and $\widetilde{\mathcal{F}}(\cdot)$ has a structure symmetric to that of $\mathcal{F}(\cdot)$. All free parameters and functions can be learned by end-to-end network training. More details can be find in [14].

2.3   Proposed Self-supervised Learning Method

Figure 1 shows the overall pipeline of our proposed framework for self-supervised MRI reconstruction. It includes a training phase of network optimization with only undersampled data and a test phase of high-quality image reconstruction from undersampled data.

In the training phase, two subsets are constructed by randomly selecting part of k-space data from the undersampled data, and then fed into two parallel networks. In this work, the reconstruction network utilizes ISTA-Net$^+$, which is an enhanced version of ISTA-Net [14]. The architecture of ISTA-Net$^+$ is showed in Fig. 1, and the number of iterations is set to 9. As illustrated in Fig. 2, we construct the two subsets by taking the intersections of the undersampling mask and selection masks. The following strategies are adopted when choosing the selection masks: 1) The selection masks used by the two parallel networks should be different. 2) The input to the networks should include most of the low-frequency data points and part of the high-frequency data points. 3) We keep the numbers of the selected data points to be roughly half of the number of the undersampled data points. Two masks which are similar to the undersampling pattern are chosen as our selection masks. During network optimization, the undersampled data are used to calculate the reconstruction loss. Furthermore, a difference loss is defined to impose an indirect constraint to the unscanned data points, which can ensure that the learned unscanned data points of the two parallel networks are consistent. It is expected that the reconstructed images of the two networks are roughly the same since they are basically recovering the same thing. Overall, the network training process solves the following optimization problem:

$$\arg\min_{\mathbf{x}_1,\mathbf{x}_2} \frac{1}{2}\|\mathbf{A}_1\mathbf{x}_1 - \mathbf{y}_1\|_2^2 + \frac{1}{2}\|\mathbf{A}_2\mathbf{x}_2 - \mathbf{y}_2\|_2^2 + \lambda R(\mathbf{x}_1) + \mu R(\mathbf{x}_2) + \nu \mathcal{L}(\mathbf{x}_1, \mathbf{x}_2) \quad (4)$$

where $\mathbf{y}_1$ and $\mathbf{y}_2$ are the two subsets selected from the undersampled k-space data, $\mathbf{A}_1$ and $\mathbf{A}_2$ are the corresponding encoding matrices, and $\mathbf{x}_1$ and $\mathbf{x}_2$ are the two estimations of the ground-truth image $\mathbf{x}$ which should be theoretically consistent. $R(\cdot)$ is the regularization. $\mathcal{L}(\cdot)$ denotes some similarity metrics. λ, μ, and ν are the regularization parameters.

In the test phase, high-quality MR images are generated by inputting the raw undersampled data into either of the two trained networks. The network is expected to be generalizable enough and to be able to self-speculate the missing k-space data points.

Fig. 2. Construction of the two subsets. The autocalibrating signal (ACS) lines of the undersampling mask and the selection mask are 24 and 16, respectively. 2D random undersampling is utilized for both the undersampled data generation and the subsets construction. "up" and "down" refer to the two parallel networks.

2.4   Implementation Details

The loss function for model training is:

$$\mathcal{L}(\Theta) = \frac{1}{N} \left(\sum_{i=1}^{N} \mathcal{L}(\mathbf{y}^i, \mathbf{A}^i \mathbf{x}_1^i) + \sum_{i=1}^{N} \mathcal{L}(\mathbf{y}^i, \mathbf{A}^i \mathbf{x}_2^i) \right.$$
$$\left. + \alpha \sum_{i=1}^{N} \mathcal{L}(\overline{\mathbf{A}^i}\mathbf{x}_1^i, \overline{\mathbf{A}^i}\mathbf{x}_2^i) + \beta \mathcal{L}_{\text{cons1}} + \gamma \mathcal{L}_{\text{cons2}} \right) \tag{5}$$

where N is the number of the training cases, i denotes the ith training case. $\mathbf{x}_k = f(\mathbf{y}_k, \mathbf{A}_k; \theta_k), k = 1, 2$, where $f(\cdot)$ denotes the reconstruction network specified by the parameter set θ. $\mathbf{A} = \mathbf{PF}$ and $\overline{\mathbf{A}} = (\mathbf{I} - \mathbf{P})\mathbf{F}$ refer to the scanned k-space data points and the unknown/unscanned k-space data points which are utilized to calculate the reconstruction loss and the difference loss, respectively. $\mathcal{L}_{\text{cons}}$ denotes the constraint loss in [14], which is included to ensure the learned transform $\mathcal{F}(\cdot)$ satisfies the symmetry constraint $\widetilde{\mathcal{F}} \circ \mathcal{F} = \mathcal{I}$. α, β and γ are the regularization parameters. In our experiments, the loss metric $\mathcal{L}(\cdot)$ is set to the mean square error (MSE) loss, and $\alpha = \beta = \gamma = 0.01$.

The proposed method is implemented in PyTorch. We use Xavier [3] to initialize the network parameters with a gain of 1.0. To train the parallel networks, we use Adam optimization [5] with a learning rate warm up strategy (the learning rate is set to 0.0001 after 10 warm up epochs) and a batch size of 4. The learning rate is automatically reduced by a constant factor when the performance metric plateaus on the validation set. All experiments are conducted on an Ubuntu 18.04 LTS (64-bit) operating system utilizing two NVIDIA RTX 2080 Ti GPUs (each with a memory of 11 GB).

3   Experiments and Results

Dataset. The open-source dataset, Information eXtraction from Images (IXI), was collected from three hospitals in London. For each subject, T1, T2, PD-weighted, MRA, and diffusion-weighted images are provided. More details including scan parameters can be found on the official website[1]. Our experiments are conducted with 2D slices extracted from the brain T1 MR images, and the matrix size of each image is 256×256. The training set, validation set, and test set for our experiments contain 850 slices, 250 slices, and 250 slices, respectively. The image intensities are normalized to $[0, 1]$ before the retrospective undersampling process.

Comparison Methods. Our proposed method is compared to the following reconstruction methods to evaluate the effectiveness: 1) U-Net-256: A U-Net model trained in a supervised manner, where the number of channels of the

---

[1] http://brain-development.org/ixi-dataset/.

Table 1. Quantitative analysis of the different methods at two acceleration rates (4 and 8).

Methods	PSNR		SSIM	
	4×	8×	4×	8×
U-Net-256	30.833	29.252	0.89184	0.85748
SSDU	35.908	32.469	0.95130	0.91531
Ours	38.575	33.255	0.97177	0.92709
Supervised	39.471	33.928	0.97843	0.93919

last encoding layer is 256. 2) SSDU: An ISTA-Net$^+$ model trained in a self-supervised manner as in [12]. In this paper, a slight difference in the network loss calculation is made. To ensure stable network training, more k-space data points (including partial network input) are utilized to calculate the loss. 3) Supervised : An ISTA-Net$^+$ model trained in a supervised manner.

Table 1 lists the reconstruction results of the different methods. PSNR and SSIM represent the peak signal-to-noise ratio and the structural similarity index measurement values. Compared to U-Net-256 and SSDU, our method achieves

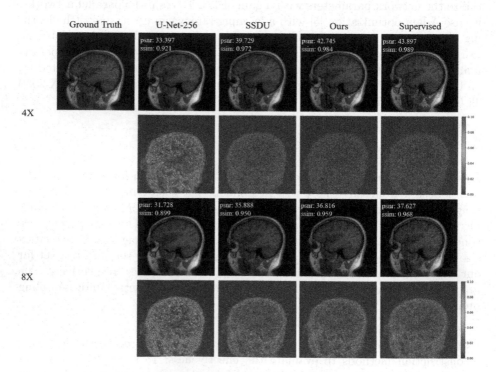

Fig. 3. Example reconstruction results of the different methods at two acceleration rates (4 and 8) along with their corresponding error maps.

significantly improved reconstruction performance at both acceleration rates. Moreover, our method generates very competitive results which are close to those generated by the corresponding supervised learning method that utilizes the exact same network architecture ("Supervised" in Table 1). Example reconstruction results are plotted in Fig. 3. It can be observed that our method recovers more detailed structural information compared to U-Net-256 and SSDU, which suggests that our method can reconstruct MR images with better visual qualities.

Ablation Analysis. To evaluate the effectiveness of the different components of the proposed method, ablation studies are performed and the results are reported in Table 2. It can be summarized that employing the defined reconstruction loss (utilizing all the available undersampled data to calculate the reconstruction loss) yields significantly better performance compared to SSDU, where only part of the undersampled data is used to calculate the loss. Besides, as illustrated in Table 2, with or without sharing the parameters of the parallel networks, the reconstruction performance is consistently improved when the difference loss is introduced, which suggests the effectiveness of the proposed difference loss for the high-quality image reconstruction task.

Table 2. Ablation study results at two acceleration rates (4 and 8). "wo DiffLoss" means without the difference loss. "share" represents sharing the parameters of the two parallel networks.

Methods	PSNR		SSIM	
	4×	8×	4×	8×
SSDU	35.908	32.469	0.95130	0.91531
Self-supervised/wo DiffLoss	38.054	33.077	0.97036	0.92439
Self-supervised/share	38.353	33.216	0.97086	0.92681
Self-supervised/wo share (Ours)	**38.575**	**33.255**	**0.97177**	**0.92709**

4   Conclusion

A novel self-supervised learning method for MRI reconstruction is proposed that can be employed by any existing deep learning-based reconstruction model. With our method, neural networks obtain the ability to infer unknown frequency information, and thus, high-quality MR images can be reconstructed without utilizing any fully sampled reference data. Extensive experimental results confirm that our method achieves competitive reconstruction performance when compared to the corresponding supervised learning method at high acceleration rates.

Acknowledgments. This research was partly supported by the National Natural Science Foundation of China (61871371, 81830056), Key-Area Research and Development Program of GuangDong Province (2018B010109009), Scientific and Technical

Innovation 2030-"New Generation Artificial Intelligence" Project (2020AAA0104100, 2020AAA0104105), Key Laboratory for Magnetic Resonance and Multimodality Imaging of Guangdong Province (2020B1212060051), the Basic Research Program of Shenzhen (JCYJ20180507182400762), Youth Innovation Promotion Association Program of Chinese Academy of Sciences (2019351).

References

1. Aggarwal, H.K., Mani, M.P., Jacob, M.: MoDL: model-based deep learning architecture for inverse problems. IEEE Trans. Med. Imag. **38**(2), 394–405 (2019). https://doi.org/10.1109/TMI.2018.2865356
2. Eo, T., Jun, Y., Kim, T., Jang, J., Lee, H.J., Hwang, D.: KIKI-net: cross-domain convolutional neural networks for reconstructing undersampled magnetic resonance images. Magn. Resonan. Med. **80**(5), 2188–2201 (2018). https://doi.org/10.1002/mrm.27201
3. Glorot, X., Bengio, Y.: Understanding the difficulty of training deep feedforward neural networks. In: Teh, Y.W., Titterington, M. (eds.) Proceedings of the Thirteenth International Conference on Artificial Intelligence and Statistics. Proceedings of Machine Learning Research, vol. 9, pp. 249–256. PMLR, Chia Laguna Resort, Sardinia, Italy, 13–15 May 2010. http://proceedings.mlr.press/v9/glorot10a.html
4. Hammernik, K., Klatzer, T., Kobler, E., Recht, M.P., Sodickson, D.K., Pock, T., Knoll, F.: Learning a variational network for reconstruction of accelerated MRI data. Magn. Resonan. Med. **79**(6), 3055–3071 (2018). https://doi.org/10.1002/mrm.26977
5. Kingma, D.P., Ba, J.: Adam: A Method for Stochastic Optimization (2017). https://arxiv.org/abs/1412.6980
6. Mardani, M., et al.: Deep generative adversarial neural networks for compressive sensing MRI. IEEE Trans. Med. Imag. **38**(1), 167–179 (2019). https://doi.org/10.1109/TMI.2018.2858752
7. Ronneberger, O., Fischer, P., Brox, T.: U-Net: convolutional networks for biomedical image segmentation. In: Navab, N., Hornegger, J., Wells, W.M., Frangi, A.F. (eds.) Medical Image Computing and Computer-Assisted Intervention (MICCAI 2015). pp. 234–241. Springer International Publishing, Cham (2015). https://doi.org/10.1007/978-3-319-24574-4_28
8. Wang, S., Cheng, H., Ying, L., Xiao, T., Ke, Z., Zheng, H., Liang, D.: Deep complex MRI: exploiting deep residual network for fast parallel MR imaging with complex convolution. Magn. Resonan. Imag. **68**, 136–147 (2020). https://doi.org/10.1016/j.mri.2020.02.002
9. Wang, S., et al.: Accelerating magnetic resonance imaging via deep learning. In: 2016 IEEE 13th International Symposium on Biomedical Imaging (ISBI), pp. 514–517 (2016). https://doi.org/10.1109/ISBI.2016.7493320
10. Wang, S., Tan, S., Gao, Y., Liu, Q., Ying, L., Xiao, T., Liu, Y., Liu, X., Zheng, H., Liang, D.: Learning joint-sparse codes for calibration-free parallel MR imaging. IEEE Trans. Med. Imag. **37**(1), 251–261 (2018). https://doi.org/10.1109/TMI.2017.2746086
11. Wang, S., Xiao, T., Liu, Q., Zheng, H.: Deep learning for fast MR imaging: a review for learning reconstruction from incomplete k-space data. Biomed. Sig. Proces. Control **68**, 102579 (2021). https://doi.org/10.1016/j.bspc.2021.102579

12. Yaman, B., Hosseini, S.A.H., Moeller, S., Ellermann, J., Uğurbil, K., Akçakaya, M.: Self-supervised learning of physics-guided reconstruction neural networks without fully sampled reference data. Magn. Resonan. Med. **84**(6), 3172–3191 (2020). https://doi.org/10.1002/mrm.28378
13. Yang, Y., Sun, J., Li, H., Xu, Z.: Deep ADMM-Net for compressive sensing MRI. In: Proceedings of the 30th International Conference on Neural Information Processing Systems (NIPS 2016), pp. 10–18. Curran Associates Inc., Red Hook (2016)
14. Zhang, J., Ghanem, B.: ISTA-Net: interpretable optimization-inspired deep network for image compressive sensing. In: Proceedings of the IEEE Conference on Computer Vision and Pattern Recognition (CVPR), June 2018

Acceleration by Deep-Learnt Sharing of Superfluous Information in Multi-contrast MRI

Sudhanya Chatterjee$^{(\boxtimes)}$, Suresh Emmanuel Joel, Ramesh Venkatesan, and Dattesh Dayanand Shanbhag

GE Healthcare, Bengaluru, India
sudhanya.chatterjee@ge.com

Abstract. The ability of magnetic resonance imaging (MRI) to obtain multiple imaging contrasts depicting various properties of tissue makes it an excellent tool for in-vivo differential diagnosis. Multi-contrast MRI (mcMRI) exams (e.g. T_2-w, FLAIR-T_2, FLAIR-T_1, etc. in a single examination) is a standard MRI protocol in a clinical setting. In this work we propose a method to accelerate mcMRI exams by acquiring only the essential contrast information for the slower contrast scans and sharing the un-acquired structure information from a fast fully sampled reference scan. The resulting artifact from the proposed structure-sharing method is then removed using a deep learning (DL) network. In addition, the proposed DL framework is driven by a smart loss function where weights of the loss function components are updated at end of each epoch using a transfer function based on its convergence to the ground truth. We show high quality accelerated reconstruction (up to 5x) at clinically accepted image resolution for (long) FLAIR-T_1 and FLAIR-T_2 scans using the proposed method with the faster T_2-w as the reference scan.

Keywords: MRI · Multi-contrast · FLAIR · Deep learning · Acceleration

1 Introduction

MRI provides excellent soft tissue contrast. Using appropriate acquisition techniques, a range of contrasts can be generated thereby providing information on specific tissue characteristics (e.g. transverse and longitudinal relaxation). In a typical clinical scenario, a set of contrasts (T_2-w, FLAIR-T_1, FLAIR-T_2, etc.) are acquired in an MRI examination for a single subject. MRI's comparatively longer acquisition time (and hence reduced throughput) as compared to other in-vivo imaging methods (such as CT) is a known limitation. Besides reducing throughput, longer scans are more vulnerable to motion artifacts which may compromise image quality and clinical diagnosis. Decades of research has been dedicated to image reconstruction techniques, pulse sequence development and hardware improvements to reduce MRI acquisition time [1,5]. Among the signal

© Springer Nature Switzerland AG 2021
M. de Bruijne et al. (Eds.): MICCAI 2021, LNCS 12906, pp. 392–401, 2021.
https://doi.org/10.1007/978-3-030-87231-1_38

processing based reconstruction techniques, random undersampling of MRI raw data (i.e. k-space) along with compressed sensing (CS) reconstruction has been the most accepted approach [11]. Recently, deep learning (DL) based methods have shown promise in ability to reconstruct high quality MRI data from highly undersampled acquisitions [4,7,8,13]. Most of the CS-MRI methods (DL and non-DL) for acceleration work on single MRI acquisition. However, information from all contrasts can be leveraged for better reconstruction of mcMRI exams. Sun et al. proposed a DL based CS-MRI to achieve mcMRI acceleration by first reconstructing each contrast (T_1, T_2, PD-w) independently and then using a feature sharing network which leverages inter-contrast information to further improve reconstruction quality [14]. Xiang et al. and Do et al. have accelerated T_2-w MRI by considering a fully sampled T_1-w MRI as reference [3,18].

We propose a method to accelerate mcMRI examinations by acquiring only the contrast information (low frequency (LF)) of lengthy scans and sharing structure information (high frequency (HF)) from a fully sampled reference scan. This is to ensure that all available phase encode (PE) lines are dedicated to acquire the critical contrast information of the accelerated scans. We leverage the k-space representation of MRI by grafting in the un-acquired HF structure information into the k-space of the accelerated scans from the reference scan k-space. While this method reduces the truncation artifacts of the accelerated scans, it also introduces systemic artifacts due to phase and/or tissue interface contrast mismatch. We then train a DL network to identify and remove these artifacts. In this framework, DL predicts the systematic artifacts introduced by k-space grafting rather than synthesize missing k-space as typically done in joint reconstruction of zero-filled random undersampling methods. Our DL network is trained on a smart loss function wherein weights of the loss function components (mean absolute error (MAE) and multi-scale structural similarity index metric [17]) are dynamically weighted depending upon the network convergence to ground truth. The scan with a shorter acquisition time and good signal-to-noise ratio (SNR) is chosen as the reference scan. In our work we choose T_2-w as the reference scan (short acquisition time and high SNR) to accelerate the typically lengthy but clinically significant FLAIR-T_1 and FLAIR-T_2 scans. Our method can be seen as an extension of keyhole imaging technique (which has been predominantly limited to dynamic MR imaging) [15] to cross-contrast anatomical imaging with deep learning enhancements to address artifacts arising from cross contrast structure sharing. To the best knowledge of the authors, this is the first such application of the keyhole imaging in structural mcMRI. We trained our model on over 200 unique MRI datasets from clinical sites which had data from field strength 1.5T and 3.0T. Similarly, our testing set had data from both field strengths. We demonstrate the superiority of our method over the popular approach of pesudo-random undersampling of MRI data acquisition while maintaining all other elements of the reconstruction workflow as-it-is. In addition to the popularly used metric of pSNR and SSIM for evaluation and comparison of the method(s), we also used Laplacian pyramidal Lap$_1$ loss (Laploss) [2] which enables us to quantify fine structure mismatch of the reconstructed image.

2   Method

In mcMRI, multiple MR image contrasts are obtained for the same subject, say $\{I_j\}_{j=1}^{N}$. Corresponding to each image, its k-space is obtained as $\{K_j\}_{j=1}^{N}$, which is a spatial frequency information of the image to be acquired. The relationship between image and the acquired k-space is hence $I_j = \mathcal{F}^{-1}(K_j)$, where $\mathcal{F}$ denotes Fourier transform. By the virtue of k-space representation, the LF image contrast information is contained at center of it and the HF structure information is present in the outer k-space regions.   When mcMRI is obtained

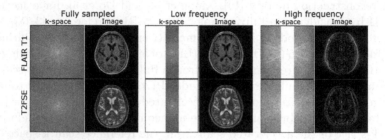

Fig. 1. The figure shows LF and HF content of FLAIR-T_1 and T_2-w for the same subject. The LF contain contrast information (PE lines around center of k-space) and the HF is structure information which is similar for the same subject. For this example, LF data was obtained by retaining 25% lines around center.

for a subject, the center k-space region (LF) contains exclusive contrast information that is significantly different for each scan. However, the outer k-space contains HF information of the image that is quite similar across contrasts. This has been illustrated in Fig. 1 (4x undersampling). Hence, it is possible to accelerate multi-contrast MR image examinations by exclusively acquiring the contrast for all scans but sharing the structure information from the reference scan. Since contrast is the most critical information of mcMRI data, we ensure that contrast information acquisition is not compromised for accelerated images. Among the multiple contrasts to be acquired, a fully sampled reference image is acquired (with k-space K_R) which has lowest acquisition time and high SNR. In our method, we use T_2-w as reference scan which is conventionally always acquired in a mcMRI exam. All other contrast acquisitions are accelerated by acquiring only the contrast information i.e. center of k-space (K_{acc} with sampling mask $\mathcal{M}_C$). For example, a 3x accelerated acquisition will acquire a third of the total PE lines around center of the k-space (LF content), thus ensuring full emphasis on contrast acquisition. Since structure is shared between contrasts of the same subject, un-acquired outer k-space data (predominantly structure information) is grafted in from k-space of fully sampled reference image (refer Fig. 2). Hence, the grafted k-space of the accelerated contrast is represented as

$$K_{graf} = K_{acc} \oplus ((1 - \mathcal{M}_C) \odot K_R) \tag{1}$$

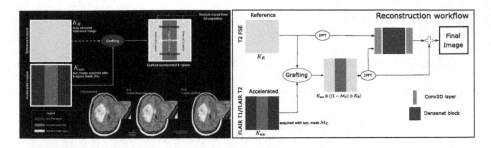

Fig. 2. *(Left)* The k-space grafting method to improve structure composition of the accelerated contrast $(\mathcal{F}^{-1}(K_{acc})$ obtained using k-space mask $\mathcal{M}_C)$ is shown here. The benefit of this operation is shown for an example. *(Right)* DL based reconstruction module is shown here. It is trained to predict grafting artifact (Δ), which is then added to the grafted accelerated contrast image to obtain a clean image $(\hat{I})$.

In Fig. 2 we can observe the benefit of image obtained from K_{graf} as compared to K_{acc} directly. Composure of the structures improve upon grafting structure information from K_R. A deep learning network is then trained on grafted image $(I_g = \mathcal{F}^{-1}(K_{graf}))$ and reference contrast data $(I_R = \mathcal{F}^{-1}(K_R))$ to predict the grafting artifact $(\Delta = I_{true} - I_g)$. Clean image is then obtained as $\hat{I} = I_g + \Delta$.

Objective. Accelerate a mcMRI examination which acquires T_2-w, FLAIR-T_1 and FLAIR-T_2 contrasts with T_2-w as reference scan. FLAIR sequences are clinically critical and typically lengthy acquisitions. Accelerating them in mcMRI will help in considerable reduction of exam time. T_2-w is treated as the reference scan since high SNR T_2-w scan can be acquired in short time.

Deep Learning Architecture. A residual learning based approach was adopted for this work to detect grafting artifact and obtain a clean image. The network architecture has been illustrated in Fig. 2. It is based on the residual dense block concept [20] which uses densely connected convolution layers [6]. A 10-layer dense block was used with filter size, filter count and growth factor of 3, 16 and 12 respectively (parameters as defined in [6]). All images were normalized using z-score normalization prior to feeding into the network. All experiments were run for 300 epochs. Separate networks are trained for FLAIR-T_1 and FLAIR-T_2. All networks are trained with 2D slices (256×256 matrix size) with accelerated and reference scans as two channels. All training was performed in Python using TensorFlow v1.14.0.

Loss Function. DL is driven using a dynamically weighted sum of mean absolute error (MAE) and multi-scale structural similarity image (msSSIM) [17] loss function. Initially while the predicted $(\hat{I})$ and ground truth (I_{true}) images are far from each other, MAE is weighed more than msSSIM. Eventually as $\hat{I}$ approaches

I_{true}, SSIM is given more weight. The dynamically weighted loss function formulation is realized as shown in Eq. 2.

$$\mathcal{L}(\hat{I}, I_{true}) = \alpha_{r_m} \times \text{MAE}(\hat{I}, I_{true}) + (1 - \alpha_{r_m})(1 - \text{msSSIM}(\hat{I}, I_{true})) \quad (2)$$

where α_{r_m} is updated at end of each epoch as per the formulation in Eq. 2.

$$\alpha_{r_m} = 1 - \frac{1}{1 + \exp(-0.1(r_m - R))} \quad (3)$$

where $r_m = \text{msSSIM/MAE}$. The choice of R in Eq. 2 determines the rate in which the weights are transferred from MAE to msSSIM as $\hat{I}$ approaches I_{true} (e.g. lower R implies faster decay in α_{r_m}, hence more weight for msSSIM, refer Fig. 3). For all experiments in this work, $R = 50$ was chosen.

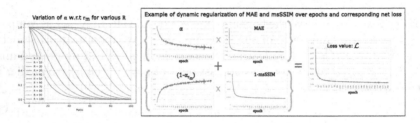

Fig. 3. Loss function. *(Left)* Dependence of α_{r_m} on choice of R (refer Eq. 2). *(Right)* Example (for 2x acceleration of proposed method) of dynamic regularization of MAE, msSSIM and final loss function (refer Eq. 2) for $R = 50$ over all epochs is shown here.

Data. DL network for FLAIR-T_1 was trained on 230 (5554 2D slices) unique MRI subjects (1.5T - 67 datasets, 3.0T - 163 datasets). The training dataset for FLAIR-T_2 contained MRI volumes from 217 (5557 2D slices) unique subjects (1.5T - 176 datasets, 3.0T - 41 datasets). The performance of proposed method was tested on 25 FLAIR-T_1 (1.5T - 5 datasets, 3.0T - 20 datasets) and 21 FLAIR-T_2 (1.5T - 18 datasets, 3.0T - 3 datasets) MRI volumes (unique subjects). All data was acquired on GE Healthcare MRI scanners. Three private datasets were used for the study. Data specifications: slice thickness - 5 mm for FLAIR-T_2, FLAIR-T_1 and T_2-w; spacing between slices - 6 mm for FLAIR-T_2, FLAIR-T_1 and T_2-w; in-plane resolution - in range of 0.43 mm–0.91 mm, 0.43 mm–0.56 mm and 0.43 mm–0.45 mm (all isotropic) for FLAIR-T_2, FLAIR-T_1 and T_2-w respectively. All images were resampled to matrix size of 256 × 256 for DL training. Only brain MRI data was used.

Comparison with Pseudo-random Undersampling. Pseudo-random undersampling of k-space along with compressed sensing is the most popular method to accelerate MRI acquisition by acquiring fewer samples [11].

Deep learning techniques have also been used to reconstruct clean images from randomly undersampled k-space data by removing zero-filled aliasing effects [9,10,16,19]. We compare our method with pseudo-randomly undersampled (1D) k-space sampling technique. Around 10% of the available PE lines for a given level of acceleration were reserved for center of k-space (in accordance with common practice of retaining auto-calibration signal lines for CS-MRI). Remaining all elements of the reconstruction workflow, including feeding both reference and accelerated images into DL is maintained. The comparison has been illustrated in Fig. 4. We show an example of accelerated image DL inputs for the proposed method and random undersampling approach for 3x acceleration level. We evaluated three metrics while comparing the methods on the test set: pSNR, SSIM and Laploss [2]. A perfect match will have Laploss and SSIM value of 0 and 1 respectively.

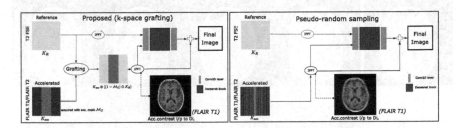

Fig. 4. Methods compared. *(Left)* Proposed method where k-space grafting is performed before feeding into DL *(Right)* DL reconstruction while using pseudo-random undersampling (used for CS based and DL approaches) to accelerate acquisition.

3   Results

Models were trained for 4 levels of acceleration: 2x, 3x, 4x and 5x. Performance of the methods were evaluated using three metrics: pSNR, SSIM and Laploss [2]. The performance metrics of the methods for all test cases (25 FLAIR-T_1 and 21 FLAIR-T_2 MRI volumes) are reported in Table 1. A test case performance for FLAIR-T_1 and FLAIR-T_2 are shown in Fig. 5 and 6 respectively. This case shows accelerated FLAIR-T_1 and FLAIR-T_2 scans of a patient from the same examination. The pathology appears as hyper-intense and hypo-intense on FLAIR-T_2 and FLAIR-T_1 scans respectively. Both FLAIR images were acquired at 0.9 mm isotropic in-plane resolution, slice thickness of 6 mm.

Table 1. Metrics reporting test cohort evaluation of the proposed method and pseudo-random undersampling (refer Fig. 4).

FLAIR T1	2x		3x		4x		5x	
	Proposed	Random	Proposed	Random	Proposed	Random	Proposed	Random
pSNR	**40.728**	36.665	**37.251**	35.144	**35.044**	33.878	**34.096**	33.681
SSIM	**0.990**	0.925	**0.980**	0.929	**0.967**	0.926	**0.955**	0.927
Laploss	**28.177**	61.524	**39.334**	71.552	**51.411**	76.994	**58.752**	78.705
FLAIR T2	2x		3x		4x		5x	
	Proposed	Random	Proposed	Random	Proposed	Random	Proposed	Random
pSNR	**39.675**	34.216	**35.749**	33.004	**33.294**	31.794	**32.162**	31.357
SSIM	**0.985**	0.934	**0.971**	0.928	**0.950**	0.918	**0.939**	0.912
Laploss	**6.235**	14.417	**8.660**	15.853	**11.514**	15.853	**11.514**	17.757

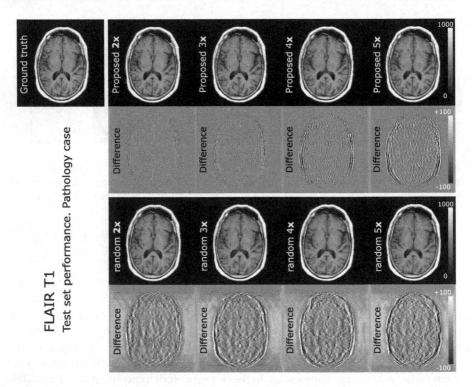

Fig. 5. FLAIR-T_1 test case results compared for two methods. Pathology case.

4   Discussion

Contrast is the most critical content of the mcMRI exam scans. The scan times differ for each contrast. While a high SNR T_2-w can be acquired in a short time, FLAIR sequences have long acquisition time. Hence accelerating FLAIR scans help in considerable reduction of the scan time. For the accelerated scans, the

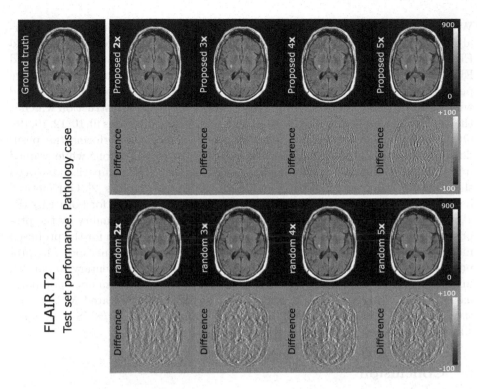

Fig. 6. FLAIR-T_2 test case results compared for two methods. Pathology case.

emphasis is to dedicate all available PE lines to acquire the contrast information (LF k-space region). This ensures that the critical contrast information of the mcMRI examination is not sacrificed in lieu of acceleration. By the virtue of k-space representation of MRI data, regions away from the center of k-space contain HF structure information. We utilize this representation of k-space along with the fact that all contrasts are acquired for the same subject (hence they share structure) to create grafted k-space representations for accelerated scans. This helps alleviate the zero-filling/truncation artifacts and intuitively provides DL module a good initial image for reconstruction. The resulting structure-sharing artifact is then removed using a DL network. We also proposed a framework to dynamically regularize components of loss function (MAE and msSSIM) as the training progresses. We realized from our experiments that when the accelerated data is heavily artifact ridden (i.e. at beginning of the training), MAE is better suited for robust reconstruction. As we approach the ground truth (later on in the training), msSSIM plays an important role in ensuring good quality reconstruction. The popular approach has been to fix regularization weights for the loss function components at beginning of the training. Our method allows user to set a function which decides the transfer of regularization weights from MAE to msSSIM as the network gets better at predicting the ground truth.

We demonstrate capability of our method to reconstruct high quality accelerated typically long sequences of a mcMRI examination (shown for FLAIR-T_1 and
FLAIR-T_2 contrasts) with SSIM greater than 0.93 and 0.95 for FLAIR-T_2 and
FLAIR-T_1 for highest level of acceleration tested (5x). All reconstructions were
performed at clinically used matrix size of 256×256. We compared our method
with pseudo-random undersampling of k-space with joint reconstruction which is
the one of the popular approaches to accelerate MRI acquisitions [9,10,12,16,19].
Our analysis demonstrates that the proposed method outperforms the pesu-
dorandom sampling approach to accelerate mcMRI examinations while trained
under similar setup with a fully sampled reference scan. The comparison between
the methods was performed based on pSNR, SSIM and Laploss [2]. Our method
has higher pSNR and SSIM values for all levels of acceleration for both FLAIR-
T_1 and FLAIR-T_2. This suggests better image reconstruction quality by the proposed method. Lower Laploss values for all levels of acceleration for the proposed
method indicates better retention of fine structures using our method. The proposed approach to accelerate mcMRI examinations is complementary to parallel
imaging (PI) and simultaneous multi-slice (SMS) readout techniques popularly
used to accelerate MRI scans. Joint optimization for accelerated MRI reconstruction using proposed mcMRI-keyholing approach and PI/SMS techniques
simultaneously can be a prospective study to this work.

5   Conclusion

We proposed a method to accelerate acquisitions of longer sequences in multi-
contrast MRI examinations by acquiring low frequency contrast data and sharing
structure information from a fully sampled reference scan. In future work we
intend to assess the impact of very high acceleration levels on small and low
contrast pathology structures.

References

1. Bernstein, M.A., King, K.F., Zhou, X.J.: Handbook of MRI Pulse Sequences. Elsevier, New York (2004)
2. Bojanowski, P., Joulin, A., Lopez-Paz, D., Szlam, A.: Optimizing the latent space
   of generative networks. arXiv preprint arXiv:1707.05776 (2017)
3. Do, W.J., Seo, S., Han, Y., Ye, J.C., Choi, S.H., Park, S.H.: Reconstruction of
   multicontrast MR images through deep learning. Med. Phys. 47(3), 983–997 (2020)
4. Duan, J., et al.: VS-Net: variable splitting network for accelerated parallel MRI
   reconstruction. In: Shen, D., et al. (eds.) MICCAI 2019. LNCS, vol. 11767, pp.
   713–722. Springer, Cham (2019). https://doi.org/10.1007/978-3-030-32251-9_78
5. Haacke, E.M., Brown, R.W., Thompson, M.R., Venkatesan, R., et al.: Magnetic
   Resonance Imaging: Physical Principles and Sequence Design, vol. 82. John Wiley
   & Sons, Hoboken
6. Huang, G., Liu, Z., Van Der Maaten, L., Weinberger, K.Q.: Densely connected
   convolutional networks. In: Proceedings of the IEEE Conference on Computer
   Vision and Pattern Recognition, pp. 4700–4708 (2017)

7. Hyun, C.M., Kim, H.P., Lee, S.M., Lee, S., Seo, J.K.: Deep learning for undersampled MRI reconstruction. Phys. Med. Biol. **63**(13), 135007 (2018)
8. Knoll, F., et al.: Advancing machine learning for MR image reconstruction with an open competition: Overview of the 2019 fast MRI challenge. Magn. Reson. Med. **84**(6), 3054–3070 (2020)
9. Lee, D., Yoo, J., Tak, S., Ye, J.C.: Deep residual learning for accelerated MRI using magnitude and phase networks. IEEE Trans. Biomed. Eng. **65**(9), 1985–1995 (2018)
10. Lee, D., Yoo, J., Ye, J.C.: Deep artifact learning for compressed sensing and parallel MRI. arXiv preprint arXiv:1703.01120 (2017)
11. Lustig, M., Donoho, D.L., Santos, J.M., Pauly, J.M.: Compressed sensing MRI. IEEE Sig. Process. Mag. **25**(2), 72–82 (2008)
12. Peng, C., Lin, W.A., Chellappa, R., Zhou, S.K.: Towards multi-sequence MR image recovery from undersampled k-space data. In: Medical Imaging with Deep Learning, pp. 614–623. PMLR (2020)
13. Schlemper, J., Caballero, J., Hajnal, J.V., Price, A., Rueckert, D.: A deep cascade of convolutional neural networks for MR image reconstruction. In: Niethammer, M., et al. (eds.) IPMI 2017. LNCS, vol. 10265, pp. 647–658. Springer, Cham (2017). https://doi.org/10.1007/978-3-319-59050-9_51
14. Sun, L., Fan, Z., Fu, X., Huang, Y., Ding, X., Paisley, J.: A deep information sharing network for multi-contrast compressed sensing MRI reconstruction. IEEE Trans. Image Process. **28**(12), 6141–6153 (2019)
15. Van Vaals, J.J., et al.: "Keyhole" method for accelerating imaging of contrast agent uptake. J. Magn. Reson. Imag. **3**(4), 671–675 (1993)
16. Wang, S., et al.: Accelerating magnetic resonance imaging via deep learning. In: 2016 IEEE 13th International Symposium on Biomedical Imaging (ISBI), pp. 514–517. IEEE (2016)
17. Wang, Z., Simoncelli, E.P., Bovik, A.C.: Multiscale structural similarity for image quality assessment. In: The Thrity-Seventh Asilomar Conference on Signals, Systems and Computers, vol. 2, pp. 1398–1402. IEEE (2003)
18. Xiang, L., et al.: Ultra-Fast T2-Weighted MR reconstruction using complementary T1-Weighted information. In: Frangi, A.F., Schnabel, J.A., Davatzikos, C., Alberola-López, C., Fichtinger, G. (eds.) MICCAI 2018. LNCS, vol. 11070, pp. 215–223. Springer, Cham (2018). https://doi.org/10.1007/978-3-030-00928-1_25
19. Yu, S., et al.: Deep de-aliasing for fast compressive sensing MRI. arXiv preprint arXiv:1705.07137 (2017)
20. Zhang, Y., Tian, Y., Kong, Y., Zhong, B., Fu, Y.: Residual dense network for image super-resolution. In: Proceedings of the IEEE Conference on Computer Vision and Pattern Recognition, pp. 2472–2481 (2018)

Sequential Lung Nodule Synthesis Using Attribute-Guided Generative Adversarial Networks

Sungho Suh[1], Sojeong Cheon[1,2,3], Dong-Jin Chang[4], Deukhee Lee[3],
and Yong Oh Lee[1(✉)]

[1] Smart Convergence Group, Korea Institute of Science and Technology Europe
Forschungsgesellschaft mbH, Saarbrücken, Germany
{s.suh,yongoh.lee}@kist-europe.de
[2] Imagoworks Inc., Seoul, Republic of Korea
sojeong@imagoworks.ai
[3] Center for Healthcare Robotics, Korea Institute of Science and Technology,
Seoul, Republic of Korea
dkylee@kist.re.kr
[4] Department of Ophthalmology, Yeouido St. Mary's Hospital, College of Medicine,
The Catholic University of Korea, Seoul, Republic of Korea
hpalways@catholic.ac.kr

Abstract. Synthetic CT images are used in data augmentation methods to tackle small and fragmented training datasets in medical imaging. Three-dimensional conditional generative adversarial networks generate lung nodule synthesis, controlling malignancy and benignancy. However, the synthesis still has limitations, such as spatial discontinuity, background changes, and vast computational cost. We propose a novel CT generation model using attribute-guided generative adversarial networks. The proposed model can generate 2D synthetic slices sequentially with U-Net architecture and bi-directional convolutional long short-term memory for nodule reconstruction and injection. Nodule feature information is considered as input in the latent space in U-Net to generate targeted synthetic nodules. The benchmark with LIDC-IDRI dataset showed that the lung nodule synthesis quality is comparable to 3D generative models in the Visual Turing test with lower computation costs.

Keywords: Conditional generative adversarial networks · Lung nodule synthesis · 2D sequential generation

---

S. Suh and S. Cheon—These authors contributed equally to this work.

Electronic supplementary material The online version of this chapter (https://doi.org/10.1007/978-3-030-87231-1_39) contains supplementary material, which is available to authorized users.

M. de Bruijne et al. (Eds.): MICCAI 2021, LNCS 12906, pp. 402–411, 2021.
https://doi.org/10.1007/978-3-030-87231-1_39

1   Introduction

Deep learning models improve object detection and segmentation of medical image analysis. However, the models rely on labeled large public datasets, which have limited availability due to sensitivity of private medical information, variability in visual appearance, and high labeling cost by a qualified expert level [11]. Additionally, medical image datasets with detailed annotation are limited due to the laborious labeling process by the trained expertise and legal issues of sharing publicly private medical information. Even among professional experts, the accuracy of the data annotation can have large inter-and intra- observer variability. Recently, data augmentation of synthetic data based on generative adversarial networks (GANs) [3] is a paradigm-shifting solution as large open datasets are scarce in supervised deep learning applications. Researchers have applied GAN-based data augmentation and generated synthetic lung nodule images improving object detection and segmentation [2,6,7,11,12]. However, various limitations arise, e.g., the approach has limited controllability of the synthetic nodules. 3D conditional GANs were employed to control malignancy [7,12], but they still showed unnaturally looking aspects due to background changes and blurring. Moreover, 3D GAN requires high computing operations for model training. Recently, in-painting approaches where the generated synthetic nodules are injected in the sub-area of the entire radiology images were proposed. Xu et al. controlled the synthetic lung nodule; nevertheless, the spatial continuity was lost due to the background image changes and remained boundaries [11].

To overcome the limitations, we propose a sequential lung nodule synthesis using attribute-guided GAN (SQ-AGGAN). Our model generates 2D synthetic slices sequentially, relying on U-Net [10]. As in [8], bi-directional convolutional long short-term memory (Bi-CLSTM) is employed at the encoder and decoder outputs in the U-Net architecture. Thus, the input of sub-sequential slices is computed as a single slice containing a lung nodule. The collection of sequential 2D synthetic slices represents the 3D synthetic images. In the latent space of U-Net, image and nodule information are added to control the synthetic nodules. The generator has two outputs: One is the reconstructed nodule image, and the other is the newly injected nodule image. The former generates nodule synthesis on the original image where the nodule is removed, while the latter generates nodule synthesis on the different original images where no nodule is shown. The two synthetic images enhance the generation capability of the synthetic nodules by verifying how well the removed nodule is reconstructed, and a new nodule is created in the background images. A discriminator is designed to distinguish the real from the fake image and to explore whether the feature input is conserved and the nodule is contained. Two outputs addition classifying real and fake images enhance the controllability of features of lung nodules and the generation of synthetic nodules.

The main contributions of this study are: A novel GAN model is proposed for (1) enhancing the feature control of synthetic nodules with lower computation; (2) improving the synthesis quality by minimizing the background changes, avoiding the boundary of the in-painting area and the effect of blurring; and (3) generating two types of synthesis: nodule reconstruction with control and nodule injection in the background image. The generation of sequential 2D slices requires less computation time compared to 3D GANs. The proposed GAN model can generate more realistic lung nodules controlled by the feature target. Finally, the evaluation of radiologists is conducted to measure the quality of synthesis.

2   Proposed Method

2.1   Model Architecture

The proposed SQ-AGGAN model architecture for lung nodule synthesis is shown in Fig. 1. The generator relies on a U-Net architecture with Bi-CLSTM blocks [8]. U-Net architecture comprises three sets of two convolutional layers and one max-pooling in the encoder and three sets of one up-sampling and two convolutional layers in the decoder with skip connections. Moreover, convolutional, max-pooling, and up-sampling layers are time-distributed layers passed to a time-distributed wrapper allowing the application of any layer to every temporal slice of the input independently. By employing two Bi-CLSTM blocks with a summation operator combining both directional outputs at the end of the encoder and decoder of U-Net, the spatio-temporal correlation of CT slices is preserved. The generation of sequential slices makes computational costs lower than the model for the entire 3D volume.

The feature vector is transformed by a fully connected layer and reshaped to concatenate with the encoder's output to create AGGAN by nodule features. Bi-CLSTM computes low dimensional high abstract features considering even nodule features using the concatenated data. The first Bi-CLSTM output undergoes the decoder and faces the last Bi-CLSTM that is computing high dimensional low abstract features. The sequential outputs from the last Bi-CLSTM are added. The following 1×1 convolutional layer generates a single slice with a synthesized nodule.

The generator creates two syntheses. The first is the nodule reconstruction image $G(x')$, where the lung nodule is regenerated after removing the nodule x' in the original nodule images x (nodule image). The second is the nodule injection image $G(y)$, where lung nodule is generated in nodule-free background images y

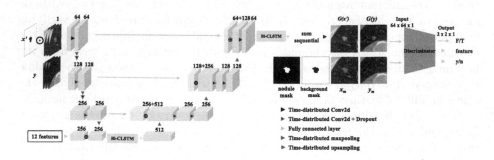

Fig. 1. The structure of the proposed model.

(background image). The nodule reconstruction can control nodule features in the nodule containing images without changing the nodule surroundings. The nodule injection can be used for data augmentation when an imbalance between no nodule images and nodule containing images exists in datasets.

The discriminator is a convolutional neural network with four convolutional layers, where the input is a pair of nodule reconstruction image $G(x')$ and its original image x, and a pair of nodule injection image $G(y)$ and its original nodule-free background image y. The additional input of the discriminator is the nodule and background masks of the nodule image x. The mask information is used in the loss functions to minimize the synthesis background change. The discriminator classifies into three categories: real and fake (F/T), features of the nodule $(feature)$, and the existence of nodule in the slice (y/n). The latter two classifiers compute the feature vector of the synthesized nodule and the label for the existence of the nodule, using a fully connected layer. The classification of nodule features enhances the controllability of the generator. The classification of nodule existence in the slice enhances the continuity of nodule synthesis in the sequential image generation.

2.2 Loss Functions

We adopt Wasserstein GAN with gradient penalty (WGAN-GP) [4] to generate realistic nodule in masked sequential slices x' and sequential slices without nodule y by training discriminator D and generator G stably. For the discriminator training, we design the WGAN-GP loss as

$$
\begin{aligned}
L_{adv,F} &= -E_{x_m}[D(x_m)] + E_{x',f}[D(G(x',f))] + \alpha E_{\hat{x}}[\|\nabla D(\hat{x})\|_2 - 1]^2 \\
L_{adv,B} &= -E_{y_m}[D(y_m)] + E_{y,f}[D(G(y,f))] + \alpha E_{\hat{y}}[\|\nabla D(\hat{y})\|_2 - 1]^2
\end{aligned}
\tag{1}
$$

where x_m and y_m are single slices extracted from the middle of the sequential slices with nodule x and without nodule y. $G(x',f)$ and $G(y,f)$ are the generated slices from the input set, including sequential slices and nodule features. f is a 12 nodule's feature vector, and α is the penalty coefficient $\hat{x}$ and $\hat{y}$ are random samples between real slice x_m, y_m and generated slice $G(x')$ and $G(y)$. In addition to WGAN-GP loss, we add L_f and L_e losses for feature and nodule existence classifier (C_F, C_{YN}) as

$$
\begin{aligned}
L_{D,f} &= E[\|C_F(x_m) - f\|_2] + E[\|C_F(G(x')) - f\|_2] + E[\|C_F(G(y)) - f\|_2] \\
L_{D,e} &= L_{BCE}(C_{YN}(x_m), 1) + L_{BCE}(C_{YN}(G(x')), 1) \\
&\quad + L_{BCE}(C_{YN}(G(y)), 1) + L_{BCE}(C_{YN}(y_m), 0) \\
L_D &= L_{adv,F} + L_{adv,B} + \lambda_f L_{D,f} + \lambda_{yn} L_{D,e}
\end{aligned}
\tag{2}
$$

where L_{BCE} is the binary cross-entropy (BCE) loss, and λ_f and λ_{yn} control the relative importance of different loss terms. L_f makes the model to synthesize the nodule with input feature f using L2 distance. We designate class 0 as "nodule existence = n" and class 1 as "nodule existence = y" to compute L_e. This loss ensures the existence of nodule in the synthesized slice, especially $G(y)$.

To enforce the generator to create the desired shape nodule while reducing the background change, we add $L_{G,F}$ and $L_{G,B}$ to the generator loss. The generator fools the discriminator and is near the ground truth nodule and background images. We use L1 distance, rather than L2 distance, between the ground truth mask and the generated image to encourage less blurring and ambiguity in the generation process. When $L_{G,F}$, we use nodule mask M_n and element-wise multiplication $\odot$ to extract the nodule part from the synthesized slice $G(x')$, $G(y)$, and from the nodule sequential slices x. Then, we multiply (the total number of pixels N_{total}/the number of nodule pixels N_n) to compensate for the small number of pixels with nodules. For the $L_{G,B}$, we use background mask M_{bg} opposing the nodule mask to extract the background part.

$$
\begin{aligned}
L_{G,F} &= \frac{N_{total}}{N_n}(E[\|M_n \odot G(x') - M_n \odot x\|_1] + E[\|M_n \odot G(y) - M_n \odot x\|_1]) \\
L_{G,B} &= E[\|M_{bg} \odot G(x') - M_{bg} \odot x\|_1] + E[\|M_{bg} \odot G(y) - M_{bg} \odot y\|_1] \\
L_{G,f} &= E[\|C_F(G(x')) - f\|_2] + E[\|C_F(G(y)) - f\|_2] \\
L_{G,e} &= L_{BCE}(C_{YN}(G(x')), 1) + L_{BCE}(C_{YN}(G(y)), 1) \\
L_G &= -L_{adv,F} - L_{adv,B} + \lambda_f L_{G,f} + \lambda_{yn} L_{G,e} + \lambda_{G,F} L_{G,F} + \lambda_{G,B} L_{G,B}
\end{aligned}
\tag{3}
$$

3   Experimental Results

3.1   Dataset and Implementation

The public Lung Image Database Consortium (LIDC-IDRI) dataset [1] provides 1,018 chest CT scans and notes from experienced thoracic radiologists. From CT scans, we crop 2,536 nodule containing volume of interests (VOIs) for x data, and randomly selected 1,462 background VOIs for y data. The size of VOI is $64 \times 64 \times d$ mm, where d being size of nodule in the case of x' and a random number between 3 and 30 mm in the case of y. The notes include the outline pixels of the nodules, the nodule diameter, and nine characteristics (subtlety, internal structure, calcification, sphericity, margin, spiculation, texture, lobulation, and malignancy) of the nodule with 5 or 6 levels. The feature vector considered the diameter and nine nodule characteristics of the LIDC-IDRI dataset. We added two more features: d from each VOI size and the slice number s between 0 and d, indicating the location of the corresponding sequential slices in VOI. This additional information compensated for the loss due to the sequential generation from 2D slices, not 3D volume generation.

The total number of sequential slices x' and y was 21,132 and 30,269. Moreover, we split x'/y data into training and test sets as 18,976/27,149 and 2,156/3,120. When training the model, y was randomly selected for each x'. Moreover, SQ-AGGAN was trained for 1000 epochs with a batch size of 32, using Adam optimizer with a fixed learning rate of 1e−4 and the standard values of $\beta_1 = 0.5$ and $\beta_2 = 0.999$ for the generator and discriminator, respectively. The model was implemented using PyTorch in Tesla V100 DGXS 32 GB. The codes are available at https://github.com/SojeongCheon/LSTM_GAN. For comparison purposes, we implemented MCGAN [5] and Tunable GAN [11] (denoted as T-GAN) as in [5,11], and the code from [9], which is the basis of T-GAN.

3.2 Analysis of Lung Nodule Synthesis and Computation Costs

Figure 2 shows the results of nodule synthesis generated by SQ-AGGAN using nodule reconstruction $(G(x'))$ and nodule injection $(G(y'))$, MCGAN, and T-GAN. The proposed SQ-AGGAN successfully generates realistic and feature-fitted nodules $G(x')$ and $G(y')$ from the masked nodule images x' and masked background image y'. The structural similarity index measure (SSIM) of nodule part of $G(x')$ and $G(y)$ are 0.9812 and 0.9152 respectively. MCGAN showed discontinuity on the boundary, which is a chronic problem of the in-painting method even though the training epoch increases. In case of T-GAN, the background image distortion occurs. Please refer video clip for more details.

The generated synthesis controllability was evaluated by generating different levels of malignancy lung nodule synthesis with control of malignancy value in the feature vector. Based on the original CT scans annotated as level five of malignancy, we generated lung nodule synthesis whose malignancy level was 1 (benign) and 5 (malignant). Figure 3 shows nodule synthesis results for different malignancy values in the input feature vector. The nodule synthesis controlled by SQ-AGGAN and T-GAN tends to be bright with calcification when malignancy = 1 (benign) and dark when malignancy = 5 (malignant).

Table 1 lists the computation cost for model training. We trained SQ-AGGAN and MCGAN for 32 batch sizes and T-GAN for 15 batch sizes due to memory constraints. SQ-AGGAN used the largest amount of memory, even though it had a small number of parameters. T-GAN had the longest, and MCGAN had

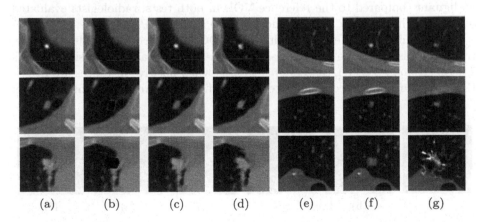

| (a)   (b)   (c)   (d)   (e)   (f)   (g) |

Fig. 2. Results of nodule synthesis: (a) LIDC-IDRI (original image), (b) masked nodule image x', (c) synthesis by SQ-AGGAN nodule reconstruction $G(x')$, (d) synthesis by MCGAN (e) background image y, (f) synthesis by SQ-AGGAN nodule injection $G(y')$, and (g) synthesis by T-GAN.

Table 1. Computation cost of SQ-AGGAN, MCGAN, T-GAN

	SQ-AGGAN	MCGAN	T-GAN
Batch size	32	32	15
Parameter amount	62,406,016	55,086,657	408,187,776
GPU training memory (MB)	32,155	18,525	31,483
Training time per epoch (s)	330	257	1,020
Total training time (s)	335,473	1,090,400	1,091,904

the shortest training time per epoch. Nevertheless, the proposed method spent the least total training time; 1000 epochs were sufficient for SQ-AGGAN, while MCGAN needed 4700 epochs for the bound box blended.

3.3 Visual Turing Test

Two experienced radiologists evaluated the synthesis quality and controllability of our method using the visual Turing test, which comprised two parts. **Test1** evaluated the realism of the synthesized VOI; 180 VOIs of lung nodule were provided (60 original LIDC-IDRI, 30 syntheses by MCGAN, 30 by T-GAN, 30 by SQ-AGGAN nodule reconstruction (SQ-AGGAN REC), 30 SQ-AGGAN nodule injection, (SQ-AGGAN INJ)). The radiologists determined whether the particular VOI was real or synthetic in 90 min. **Test2** evaluated the controllability of synthesis malignancy. We supplied five test sets, and each dataset had one reference VOI (malignancy = 3) and the other two VOIs, synthesized with malignancy = 1 (benign) and malignancy = 5 (malignant). The five test sets lasted 10 min. The radiologists judged whether the other two VOIs were benign or malignant compared to the reference VOI. In both tests, radiologists evaluated 3D VOI with $64 \times 64 \times 64$ sizes and used the ITK-SNAP viewer.

Table 2 lists the Test 1 results for LIDC-IDRI: true and false positive, true and false negative of synthesis, and F1-score of each synthesis, which is $2 \times (precision \times recall)/(precision + recall)$ when positive instances are LIDC-IDRI. The mean F1-score indicates that radiologists are more confused when

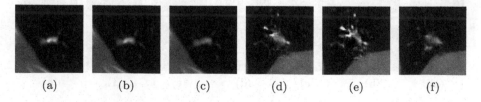

(a)	(b)	(c)	(d)	(e)	(f)

Fig. 3. Results of malignancy controllability test. Synthesis by SQ-AGGAN REC: malignancy = 1 for (a), 3 (b), 5 (c), and synthesis by T-GAN: malignancy = 1 for (c), 3 (d), 5 (f).

Table 2. Results of visual Turing test (classification of real and synthesis).

Visual Turing test		LIDC-IRDI	MCGAN	T-GAN	SQ-AGGAN REC	SQ-AGGAN INJ
Radiologist 1	Decision to real	35.00%	43.33%	13.33%	33.33%	10.00%
		(TP)	(FN)	(FN)	(FN)	(FN)
	Decision to synthesis	65.00%	56.67%	86.67%	66.67%	90.00%
		(FP)	(TN)	(TN)	(TN)	(TN)
Radiologist 2	Decision to real	52.63%	56.67%	74.07%	60.00%	48.28%
		(TP)	(FN)	(FN)	(FN)	(FN)
	Decision to synthesis	47.37%	43.33%	25.93%	40.00%	51.72%
		(FP)	(TN)	(TN)	(TN)	(TN)
Mean F1-Score			0.5038	0.5198	0.5086	0.5585

Note: Radiologist 2 did not determine real or synthesis for seven cases.

distinguishing between real and synthetic in the order of MCGAN, SQ-AGGAN REC, T-GAN, and SQ-AGGAN INJ.

MCGAN achieved the lowest mean F1-score, but the true negative rate is highly increased in the second half of the test. The radiologists learned the discontinuity caused by the in-painting method and noticed that the synthesis has unnatural lung parenchyma, which tends to be darker around the nodules. The radiologists revealed different opinions on synthesis using T-GAN. Radiologist 1 could easily distinguish T-GAN data as synthetic due to severe blurring while radiologist 2 was confused when focusing on the harmony between nodules and surroundings. For SQ-AGGAN, the true negative from radiologist 1 was relatively high due to the unnatural points in the coronal and sagittal views. Some nodules appeared with too much homogeneous intensity compared to the surroundings in SQ-AGGAN INJ, resulting in disharmony between the nodule and surroundings. After testing, we provided the entire CT for verifying the quality of lung nodule synthesis in familiar environment for radiologists. The radiologists commented on the unnatural border of in-painting-based MCGAN, and blurring on nodules and surrounding structures by T-GAN.

An 80% mean accuracy of SQ-AGGAN was obtained in Test 2 (radiologist 1 and 2 were 100% and 60% accurate, respectively), while T-GAN achieved 40% mean accuracy. Both radiologists judged nodule as benign when calcification was present and malignant otherwise. SQ-AGGAN could precisely control when combined with other feature values, such as size, calcification, sphericity, spiculation, because size, speculation, and growth rate are essential factors when determining malignancy.

3.4   Ablation Study

In order to verify the effectiveness of the Bi-CLSTM, we tested two different architectures, where some key features are excluded from SQ-AGGAN, under the same training conditions. For ablation study 1, the number of input slices

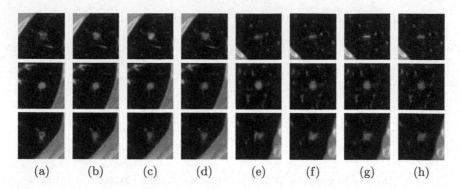

(a)   (b)   (c)   (d)   (e)   (f)   (g)   (h)

Fig. 4. Comparison of synthesis results in axial (a)–(d) and sagittal (e)–(h) views. (a), (e): LIDC-IDRI (original image), (b), (f): synthesis by SQ-AGGAN, (c), (g): ablation study 1, (d), (h): ablation study 2.

in the U-Net of SQ-AGGAN is changed from 3 to 1. This change can show the effectiveness of the sequence of slices with context because the Bi-CLSTM blocks do not preserve the spatio-temporal correlation of CT slices. For ablation study 2, the first Bi-CLSTM block of SQ-AGGAN is removed. If we remove both Bi-CLSTM blocks, the input and output of the proposed method should be changed and it is the same as 2D U-Net architecture. As shown in Fig. 4, we can find that the discontinuity of the generated nodule.

4   Conclusion

We proposed a novel sequential lung nodule synthesis method using AGGANs, which could generate realistic nodules from both masked nodule images and non-nodule images. The proposed SQ-AGGAN synthesized nodules using nodule masks and feature vectors while minimizing background changes, essential in the entire CT context. It is verified by the visual Turing test from trained radiologists that SQ-AGGAN improved the quality of synthesis by overcoming the limitations of the previous models, such as discontinuity problems around the bounding box and much-blurred synthesis, and controlled the malignancy of synthesis. For better synthesis generation, we are considering blend of multi-view in GAN training, and control of masked data margin as future works.

Acknowledgment. This work is supported by Korea Institute of Science and Technology Europe project 12120. KIST Europe collaborated with KIST Seoul (Hanyang University-KIST biomedical fellowship program) and Catholic Medical Center in Korea. The authors gave special thanks to Soyun Chang and Kyongmin Beck for activate participation to the visual turing test and valuable feedback.

References

1. The lung image database consortium (LIDC) and image database resource initiative (IDRI): a completed reference database of lung nodules on CT scans
2. Gao, C., Clark, S., Furst, J., Raicu, D.: Augmenting LIDC dataset using 3D generative adversarial networks to improve lung nodule detection. In: Medical Imaging 2019: Computer-Aided Diagnosis, vol. 10950, p. 109501K. International Society for Optics and Photonics (2019)
3. Goodfellow, I., et al.: Generative adversarial nets. In: Advances in Neural Information Processing Systems, pp. 2672–2680 (2014)
4. Gulrajani, I., Ahmed, F., Arjovsky, M., Dumoulin, V., Courville, A.C.: Improved training of Wasserstein GANs. In: Advances in Neural Information Processing Systems, pp. 5767–5777 (2017)
5. Han, C., et al.: Synthesizing diverse lung nodules wherever massively: 3D multi-conditional GAN-based CT image augmentation for object detection. In: 2019 International Conference on 3D Vision (3DV), pp. 729–737. IEEE (2019)
6. Jin, D., Xu, Z., Tang, Y., Harrison, A.P., Mollura, D.J.: CT-realistic lung nodule simulation from 3D conditional generative adversarial networks for robust lung segmentation. In: Frangi, A.F., Schnabel, J.A., Davatzikos, C., Alberola-López, C., Fichtinger, G. (eds.) MICCAI 2018. LNCS, vol. 11071, pp. 732–740. Springer, Cham (2018). https://doi.org/10.1007/978-3-030-00934-2_81
7. Liu, S., et al.: Decompose to manipulate: manipulable object synthesis in 3D medical images with structured image decomposition. arXiv preprint arXiv:1812.01737 (2018)
8. Novikov, A.A., Major, D., Wimmer, M., Lenis, D., Bühler, K.: Deep sequential segmentation of organs in volumetric medical scans. IEEE Trans. Med. Imag. **38**(5), 1207–1215 (2018)
9. Park, H., Yoo, Y., Kwak, N.: Mc-GAN: multi-conditional generative adversarial network for image synthesis. arXiv preprint arXiv:1805.01123 (2018)
10. Ronneberger, O., Fischer, P., Brox, T.: U-Net: convolutional networks for biomedical image segmentation. In: Navab, N., Hornegger, J., Wells, W.M., Frangi, A.F. (eds.) MICCAI 2015. LNCS, vol. 9351, pp. 234–241. Springer, Cham (2015). https://doi.org/10.1007/978-3-319-24574-4_28
11. Xu, Z., et al.: Tunable CT lung nodule synthesis conditioned on background image and semantic features. In: Burgos, N., Gooya, A., Svoboda, D. (eds.) SASHIMI 2019. LNCS, vol. 11827, pp. 62–70. Springer, Cham (2019). https://doi.org/10.1007/978-3-030-32778-1_7
12. Yang, J., et al.: Class-aware adversarial lung nodule synthesis in CT images. In: 2019 IEEE 16th International Symposium on Biomedical Imaging (ISBI 2019), pp. 1348–1352. IEEE (2019)

A Data-Driven Approach for High Frame Rate Synthetic Transmit Aperture Ultrasound Imaging

Yinran Chen[1(✉)], Jing Liu[2], Jianwen Luo[2], and Xiongbiao Luo[1]

[1] Department of Computer Science, Xiamen University, Xiamen 361005, China
{yinran_chen,xbluo}@xmu.edu.cn
[2] Department of Biomedical Engineering, Tsinghua University, Beijing 100084, China
luo_jianwen@tsinghua.edu.cn

Abstract. This paper proposes a deep learning approach for high frame rate synthetic transmit aperture ultrasound imaging. The complete dataset of synthetic transmit aperture imaging benefits image quality in terms of lateral resolution and contrast at the expense of a low frame rate. To achieve high-frame-rate synthetic transmit aperture imaging, we propose a self-supervised network, i.e., ApodNet, to complete two tasks. (i) The encoder of ApodNet guides the high-frame-rate plane wave transmissions to acquire channel data with a set of optimized binary apodization coefficients. (ii) The decoder of ApodNet recovers the complete dataset from the acquired channel data for the objective of two-way dynamic focusing. The image is finally reconstructed from the recovered dataset with conventional beamforming approach. We train the network with data from a standard tissue-mimicking phantom and validate the network with data from simulations and *in-vivo* experiments. Different loss functions are validated to determine the optimized ApodNet setup. The results of the simulations and the *in-vivo* experiments both demonstrate that, with a four-times higher frame rate, the proposed ApodNet setup achieves higher image contrast than other high-frame-rate methods. Furthermore, ApodNet has much shorter computational time for dataset recovery than the compared methods.

Keywords: Ultrasound imaging · Image reconstruction · Deep learning · Self-supervised · Synthetic transmit aperture · Beam focusing

1 Introduction

Ultrasound imaging is a commonly-used medical imaging modality for its real-time diagnosis procedure without adverse effects. The characteristics of beam focusing are crucial for imaging performance. In receive beamforming, dynamic focusing is achieved by effective delay-and-sum beamforming [1]. In transmit beamforming, various approaches have allowed different focusing characteristics. The conventional sequential scanning method only achieves single-spot focusing in transmission [1]. Alternatively, synthetic transmit aperture (STA) imaging employs sequential single-element transmission and

© Springer Nature Switzerland AG 2021
M. de Bruijne et al. (Eds.): MICCAI 2021, LNCS 12906, pp. 412–420, 2021.
https://doi.org/10.1007/978-3-030-87231-1_40

full-aperture reception [2]. Particularly, lateral resolution and contrast get benefits from the complete dataset of STA because of two-way dynamic focusing. However, STA suffers from a low frame rate due to a large number of transmissions, which equals the number of elements in the array.

Plane wave transmission is a popular approach for ultra-high frame rate imaging [3]. Plane wave imaging activates all the elements simultaneously in transmission. Transmit apodizations are commonly used to balance lateral resolution and contrast. However, plane wave imaging suffers from low image quality due to the lack of transmit focusing. Multiple plane waves with different steering angles can be coherently compounded to improve the image quality by sacrificing the frame rate [3].

The advantages of two-way dynamic focusing have motivated advanced imaging methods to recover the complete dataset of STA from other beamforming approaches. In [4], retrospective encoding for conventional ultrasound sequence (REFoCUS) recover the complete dataset from the channel data of focused beams. In [5], compressed sensing based STA (CS-STA) recovers the complete dataset from the channel data of high-frame-rate plane waves with random apodizations. In [6], CS-STA is improved with partial Hadamard matrix as the apodizations. However, CS-STA generally requires a long time to recover the dataset with sample-wise iterative algorithms.

In recent years, ultrasound imaging gets benefits from the data-driven approaches, e.g., deep learning. In [7], a learnable pipeline is proposed to improve the image quality of multi-line-transmit (MLT) and multi-line-acquisition (MLA) with single-line acquisition (SLA) as baseline. In [8], a stacked denoising autoencoder (SDA) is used to recover the sub-sampled image lines. In [9], adaptive beamforming by deep learning (ABLE) is proposed to calculate the optimized parameters for receive beamforming. In [10], CohereNet is proposed for spatial correlation calculation in coherence-based receive beamforming. In [11] and [12], CNN-based networks are trained to correct artifacts in multi-line-transmit and multi-line-acquisition, respectively. In [13], a modified sparse recovery CNN is used for ultrasound localization microscopy.

Motivated by CS-STA and deep learning, we propose a network for high-frame-rate STA. The technical novelty of this work is highlighted as follows. The network is composed of jointly-trained encoder and decoder modules. The encoding module trains a set of optimized coefficients for plane wave apodizations. Hence, we name our network as ApodNet, in which the term "Apod" is the abbreviation for "apodization". Furthermore, the decoding module can recover the complete dataset from the acquired channel data to achieve two-way dynamic focusing. We validate the effectiveness of ApodNet by comparing it with CS-STA in both computer simulations and *in-vivo* experiments.

2   Methods

2.1   Theory Basis and Network Architecture

Let the complete data set of STA with N times of transmissions be

$$\mathbf{E}(t) = \begin{bmatrix} \mathbf{e}_1(t) \dots \mathbf{e}_N(t) \end{bmatrix} = \begin{bmatrix} e_{1,1}(t) & \cdots & e_{1,N}(t) \\ \vdots & \ddots & \vdots \\ e_{N,1}(t) & \cdots & e_{N,N}(t) \end{bmatrix}, \tag{1}$$

where N is the number of elements and the number of STA transmissions. $e_{i,j}(t)$ is the backscattered echoes with the jth element as transmit and the ith element as receive.

According to the theory of linear acoustics, the channel data of a plane wave transmission is the superposition of the backscattered echoes of all the transmit-receive element pairs with apodizations [4, 5]. Let $\mathbf{W} \in \mathbb{R}^{N,K}$ be the apodization coefficients of K times of plane wave transmissions. Specifically, $w_{j,k}$ in $\mathbf{W}$ is the apodization coefficient for the jth element in the kth plane wave transmission. Hence, we formulate the channel data of the kth plane wave transmission, $\mathbf{p}_k(t)$, as

$$\mathbf{p}_k(t) = \mathbf{E}(t) \begin{bmatrix} w_{1,k} \\ \vdots \\ w_{N,k} \end{bmatrix}. \tag{2}$$

As a result, the channel data of K times of plane wave transmissions is expressed as a linear sampling of $\mathbf{E}(t)$ as

$$\mathbf{P}(t) = \begin{bmatrix} \mathbf{p}_1(t) \ldots \mathbf{p}_K(t) \end{bmatrix} = \mathbf{E}(t)\mathbf{W}. \tag{3}$$

Let the row vector in $\mathbf{P}(t)$ and $\mathbf{E}(t)$ be $\mathbf{y}^T \in \mathbb{R}^K$ and $\mathbf{x}^T \in \mathbb{R}^N$, respectively. The linear relationship is also formulated as

$$\mathbf{y} = \mathbf{W}^T\mathbf{x}. \tag{4}$$

Interestingly, the forward propagation in one layer of a fully-connected network is also modeled as matrix multiplication. Figures 1(a) and 1(b) illustrate the training and utilization of ApodNet. Specifically, ApodNet is built with a 4-layer stacked denoising autoencoder (SDA) [14]. The number of nodes in the input and output of the encoder equals N and K, respectively. The goal of the first layer, i.e., the encoder, is to train a set of optimized weights as the apodization coefficients for the plane wave transmissions. Note that the amplitude of transmit pulse in ultrasound imaging is preferred to be high under the safety restriction. The encoder is modified as a BinaryConnect [15], in which the weights are binarized to $+1$ or -1 with a Sign function in the forward propagation and kept as real numbers in the back propagation.

The well-trained ApodNet is utilized separately. The weights of the encoder are taken to be the apodizations for the K times of plane wave transmissions. The acquired channel data are then fed into the decoder to recover the complete dataset. The image is finally reconstructed from the recovered complete dataset with delay-and-sum beamforming.

2.2 Training Configurations

Figure 1(c) depicts the training data acquisition and pre-processing. STA sequence was run on an L11-4v linear array to acquire the complete datasets from a standard tissue-mimicking phantom (model 040GSE, CIRS, Norfolk, VA, USA). The first acquisition position contains a set of hyperechoic cylinders and wire-like targets. The second position contains a set of anechoic cylinders. The complete dataset is organized to a 3D matrix with dimensions of $M \times N \times N$, where $M = 2048$ is the number of samplings in each channel, ensuring a sampling depth over 50 mm. $N = 128$ is the number of channels, and also the number of STA transmissions. Hence, a total number of 524,288 samples of $\mathbf{x}^T \in \mathbb{R}^N$ are generated for training. The two complete datasets are then reshaped to a 2D matrix with dimensions of 524,288 $\times$ 128 and split into batches with dimension of 512 $\times$ 128. The batch is normalized to $[-1, +1]$ before being fed into the network. Because the ideal output of ApodNet is the complete dataset itself, the input and the output are identical, resulting in a self-supervised training scheme.

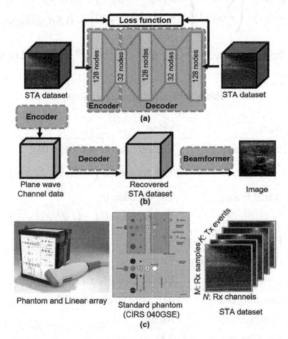

Fig. 1. Training and utilization of ApodNet. (a) Training of ApodNet. The core architecture is a 4-layer SDA with the encoder module modified to be a BinaryConnect. A self-supervised training is proposed by using the complete dataset of STA as the training data and labels. (b) Utilization of ApodNet. The binary weights of the encoder are used as the apodizations of the plane wave transmissions. The acquired channel data are then fed into the decoder to recover the complete dataset. The image is reconstructed with conventional delay-and-sum beamforming. (c) The training data are acquired by an L11-4v linear array from a standard phantom.

Table 1 lists the training configurations. We implemented ApodNet in Python 3.6.12 with PyTorch 1.7.0. The weights of the encoder are initialized with a uniform random distribution of $U(-0.01, 0.01)$ to minimize the impact of the initial signs. Tanh is taken as the non-linear activation considering the ultrasound radio-frequency (RF) data contain both positive and negative components with the same order of magnitude. To measure the similarity between the input and output data, which are the RF data with respect to the transmit events, different loss functions, i.e., the mean-square-error (MSE), mean-absolute-error (MAE), and Huber loss ($\beta = 1$) are validated, respectively. The loss functions are explicitly defined as follows:

$$\text{MSE} = \frac{1}{M_B} \sum_{m=1}^{M_B} \|\hat{\mathbf{x}}^{(m)} - \mathbf{x}^{(m)}\|_2^2, \tag{5}$$

$$\text{MAE} = \frac{1}{M_B} \sum_{m=1}^{M_B} \|\hat{\mathbf{x}}^{(m)} - \mathbf{x}^{(m)}\|_1, \tag{6}$$

$$\text{Huber} = \frac{1}{M_B} \sum_{m=1}^{M_B} \left(\sum_{n=1}^{N} \begin{cases} 0.5\left(\hat{x}_i^{(m)} - x_i^{(m)}\right)^2 / \beta, \ if \left|\hat{x}_i^{(m)} - x_i^{(m)}\right| < \beta \\ \left|\hat{x}_i^{(m)} - x_i^{(m)}\right| - 0.5\beta, \ otherwise \end{cases} \right). \tag{7}$$

where M_B is the batch size, $\mathbf{x}^{(m)}$ and $\hat{\mathbf{x}}^{(m)}$ are the mth input and output samples.

Table 1. Training configurations.

Item	Value
Initial weights	The encoder: $U(-0.01, 0.01)$
	The decoder: $U\left(-1/\sqrt{N}, 1/\sqrt{N}\right)$
Initial biases	The encoder: 0
	The decoder: $U\left(-1/\sqrt{N}, 1/\sqrt{N}\right)$
Batch normalization	Yes
Activation function	Tanh
Loss function	MSE, MAE, Huber ($\beta = 1$)
Learning rate	0.01 with an exponential decay
Optimizer	Adam
No. Epoch	200

2.3 Simulations and *In-Vivo* Experiments

We compare the proposed ApodNet with CS-STA by setting the number of plane wave transmissions to be $K = 32$. Under the premise of a 5 kHz pulse-repetition-frequency (PRF), the frame rate of ApodNet and CS-STA is ~156 Hz, which is four times higher than the that of conventional STA, i.e., ~39 Hz. CS-STA employs either partial Hadamard matrix [6] or Random matrix [5] as the apodizations (encoder). The decoder of CS-STA is a SPGL1 solver to iteratively solve a basis pursuit denoising (BPDN) problem to recover the complete dataset [16].

The simulations were implemented in Field II [17] by generating a numerical phantom including an anechoic cylinder with 3 mm radius and a wire-like target. The density of the scatterers are 5/mm^3 to generate fully developed speckles. White Gaussian noise with 1.21 dB SNR was added to the simulated channel data. Such noise level is pre-determined by calculating the SNR of the *in-vivo* channel data. The *in-vivo* experiments were performed by imaging the common carotid artery (CCA) of a healthy volunteer (male, 28 y/o.). The experimental protocol was approved prior to use by Beijing Friendship Hospital, Capital Medical University.

2.4 Metrics

In the simulations, we calculated the lateral resolution of the wire-like target, the contrast ratio (CR) and contrast-to-noise ratio (CNR) of the anechoic cylinder. The region-of-interest (ROI) of the interior of the anechoic cylinder is a disc with 2 mm radius. The ROI of the background is a disc with the same radius on the left side of the anechoic cylinder. In the *in-vivo* experiments, we calculated the CR and CNR of the carotid artery by setting the ROI as a disc with 2 mm radius in the interior of the artery and the ROI of the background as a disc on the left side of the artery. Furthermore, we measured the time for recovering the complete dataset in ApodNet and CS-STA, respectively. Specifically, CS-STA was implemented in Matlab 2015b (The MathWorks, Inc., Natick, MA, USA) on a desktop computer (Intel ® Core (TM) i7-10750H CPU @ 2.60 GHz, 16 GB RAM). For a fair comparison, we transformed the parameters of ApodNet, including the weights, biases, and coefficients of the batch normalization layers, from Python to Matlab and run the dataset recovery with forward propagations.

3 Results

Figure 2 presents the B-mode images of ApodNet with different loss functions in the simulations and the *in-vivo* experiments. The wire-like target is zoomed in for better observation. Qualitatively, the arrows in Fig. 2(b) indicate lower gray-scale levels in both sides of the field-of-view (FOV), showing background inconsistency for the MAE loss. The arrow in Fig. 2(c) points out shadowing artifacts below the anechoic cylinder. Moreover, the arrow in Fig. 2(e) indicates smoke-like artifacts and partially-missing fibrous structures for the MAE loss.

Figure 3 shows the B-mode images for the comparison between ApodNet and CS-STA. For ApodNet, the images with MSE loss are presented. The arrow in Fig. 3(e) points out the smoke-like artifacts near the hyperechoic region in the image of CS-STA with Hadamard apodization. For CS-STA with Random apodization, global noise in the entire FOV can be observed.

Tables 1 and 2 summarize the quantitative assessments for the simulations and the *in-vivo* experiments. Although ApodNet with MAE loss has higher CR and CNR than the setups with MSE and Huber losses, the existing artifacts deteriorate the overall imaging performance. ApodNet with MSE loss achieves moderate effectiveness in terms of CR and CNR without artifacts compared with ApodNet with other loss functions. Particularly, ApodNet with MSE loss achieve 8.91 dB higher CR and 6.91 dB higher CNR than CS-STA with Hadamard apodization, and 8.92 dB higher CR and 7.45 dB higher CNR than CS-STA with Random apodization in *in-vivo* imaging of the CCA. Furthermore, ApodNet can achieve a much faster dataset recovery (~2 s) than the CS-STA setups, for which the time is over 1,800 s for the Hadamard apodization and even over 35,000 s for the Random apodization.

By comparing the apodization coefficients, i.e., the binary weights of ApodNet encoder, in the initial state and after training, we found that, starting from random initialization, the signs of binary weights become the same with each other locally, showing block-like distributed patterns. Also note that the patterns stayed almost unchanged when the training procedure converged.

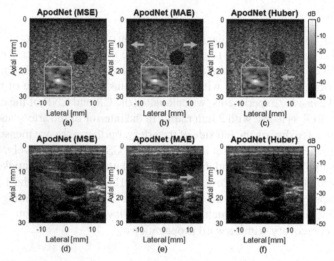

Fig. 2. B-mode images of ApodNet. (a)–(c) Images of the numerical phantom in the simulations with MSE, MAE, and Huber ($\beta = 1$) losses. (d)–(f) Images of the CCA in the *in-vivo* experiments.

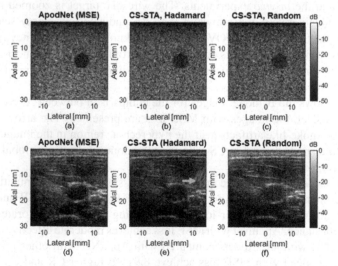

Fig. 3. B-mode images of ApodNet and CS-STA. (a)–c) Images of the numerical phantom in the simulations. (d)–(f) Images of the CCA in the *in-vivo* experiments.

Table 2. Quantitative assessments of the simulations.

Metrics	ApodNet (MSE loss)	ApodNet (MAE loss)	ApodNet (Huber loss)	CS-STA (Hadamard)	CS-STA (Random)
CR [dB]	11.66	14.52	11.23	10.00	8.26
CNR [dB]	4.24	6.32	3.95	3.27	1.37
Resolution [mm]	0.61	0.61	0.64	0.61	0.62
Artifacts	✗	○	○	○	✗

✗: without artifacts. ○: with artifacts

Table 3. Quantitative assessments of the *in-vivo* experiments.

Metrics	ApodNet (MSE loss)	ApodNet (MAE loss)	ApodNet (Huber loss)	CS-STA (Hadamard)	CS-STA (Random)
CR [dB]	15.08	16.98	16.52	6.17	6.16
CNR [dB]	5.89	6.34	6.64	−1.02	−1.56
Running time [s]	1.95	2.24	2.33	1855.95	37598.95
Artifacts	✗	○	○	○	✗

✗: without artifacts. ○: with artifacts

4 Conclusion

This paper proposes a self-supervised network, i.e., ApodNet, as a data-driven approach for high-frame-rate STA ultrasound imaging. The well-trained network can provide plane wave transmissions with optimized binary apodizations, and recover the complete dataset from the channel data of apodized plane waves for two-way dynamic focusing. Different loss functions for ApodNet are investigated to determine the MSE loss as the recommended similarity measurement. In the simulations and the *in-vivo* experiments, the proposed ApodNet setup achieves higher CR and CNR with much shorter computational time when compared with CS-STA, and obtains a four-times higher frame rate than the conventional STA imaging.

Acknowledgements. This work was supported in part by the National Natural Science Foundation of China (No. 62001403, No. 61871251, and No. 61971367), Natural Science Foundation of Fujian Province of China (No. 2020J05003), and Fundamental Research Funds for the Central Universities China (No. 20720200093, No. 20720210097).

References

1. Cikes, M., Tong, L., Sutherland, G.R., D'hooge, J.: Ultrafast cardiac ultrasound imaging: technical principles, applications, and clinical benefits. JACC: Cardiovascular Imaging 7(8), 812–823 (2014)

2. Jensen, J.A., Nikolov, S.I., Gammelmark, K.L., Pedersen, M.H.: Synthetic aperture ultrasound imaging. Ultrasonics **44** e5–e15 (2006)
3. Montaldo, G., Tanter, M., Bercoff, J., Benech, N., Fink, M.: Coherent plane-wave compounding for very high frame rate ultrasonography and transient elastography. IEEE Trans. Ultrason. Ferroelectr. Freq. Control **56**(3), 489–506 (2009)
4. Bottenus, N.: Recovery of the complete data set from focused transmit beams. IEEE Trans. Ultrason. Ferroelectr. Freq. Control **65**(1), 30–38 (2018)
5. Liu, J., He, Q., Luo, J.: A compressed sensing strategy for synthetic transmit aperture ultrasound imaging. IEEE Trans. Med. Imaging **36**(4), 878–891 (2016)
6. Liu, J., Luo, J.: Compressed sensing based synthetic transmit aperture for phased array using Hadamard encoded diverging wave transmissions. IEEE Trans. Ultrason. Ferroelectr. Freq. Control **65**(7), 1141–1152 (2018)
7. Vedula, S., Senouf, O., Zurakhov, G., Bronstein, A., Michailovich, O., Zibulevsky, M.: Learning beamforming in ultrasound imaging. In: The 2nd International Conference on Medical Imaging with Deep Learning, PMLR, vol. 102, pp. 126–134 (2018)
8. Perdios, D., Besson, A., Arditi, M., Thiran, J.: A deep learning approach to ultrasound image recovery. In: Proceedings on IEEE International Ultrasonics Symposium, pp. 1–4. IEEE (2017)
9. Luijten, B., et al.: Adaptive ultrasound beamforming using deep learning. IEEE Trans. Med. Imaging **39**(12), 3967–3978 (2020)
10. Wiacek, A., González, E., Bell, M.A.L.: CohereNet: a deep learning architecture for ultrasound spatial correlation estimation and coherence-based beamforming. IEEE Trans. Ultrasonics Ferroelectr. Freq Control **67**(12), 2574–2583 (2020)
11. Vedula, S., et al.: High quality ultrasonic multi-line transmission through deep learning. In: Knoll, F., Maier, A., Rueckert, D. (eds.) MLMIR 2018. LNCS, vol. 11074, pp. 147–155. Springer, Cham (2018). https://doi.org/10.1007/978-3-030-00129-2_17
12. Senouf, O., et al.: High frame-rate cardiac ultrasound imaging with deep learning. In: Frangi, A.F., Schnabel, J.A., Davatzikos, C., Alberola-López, C., Fichtinger, G. (eds.) MICCAI 2018. LNCS, vol. 11070, pp. 126–134. Springer, Cham (2018). https://doi.org/10.1007/978-3-030-00928-1_15
13. Liu, X., Zhou, T., Lu, M., Yang, Y., He, Q., Luo, J.: Deep learning for ultrasound localization microscopy. IEEE Trans. Med. Imaging **39**(10), 3064–3078 (2020)
14. Vincent, P., Larochelle, H., Lajoie, I., Bengio, Y., Manzagol, P.: Stacked denoising autoencoders: learning useful representations in a deep network with a local denoising criterion. J. Mach. Learn. Res. **11**, 3371–3408 (2010)
15. Courbariaux, M., Bengio, Y., David, J.: BinaryConnect: training deep neural networks with binary weights during propagations. In: Neural Information Processing Systems, pp. 3123–3131 (2015)
16. den Berg, E.V., Friedlander, M.P.: Probing the pareto frontier for basis pursuit solutions. SIAM J. Sci. Comput. **31**(2), 890–912 (2008)
17. Jensen, J.A., Svendsen, N.B.: Calculation of pressure fields from arbitrarily shaped, apodized, and excited ultrasound transducers. IEEE Trans. Ultrasonics Ferroelectrics Freq. Control **39**(2), 262–267 (1992)

Interpretable Deep Learning for Multimodal Super-Resolution of Medical Images

Evaggelia Tsiligianni[1]([✉])[iD], Matina Zerva[1][iD], Iman Marivani[2,3][iD],
Nikos Deligiannis[2,3][iD], and Lisimachos Kondi[1][iD]

[1] Department of Computer Science and Engineering, University of Ioannina,
Ioannina, Greece
{etsiligia,szerva,lkon}@cse.uoi.gr
[2] Department of Electronics and Informatics, Vrije Universiteit Brussel,
Brussels, Belgium
{imarivan,ndeligia}@etrovub.be
[3] imec, Kapeldreef 75, 3001 Leuven, Belgium

Abstract. In medical image acquisition, hardware limitations and scanning time constraints result in degraded images. Super-resolution (SR) is a post-processing approach aiming to reconstruct a high-resolution image from its low-resolution counterpart. Recent advances in medical image SR include the application of deep neural networks, which can improve image quality at a low computational cost. When dealing with medical data, accuracy is important for discovery and diagnosis, therefore, interpretable neural network models are of significant interest as they enable a theoretical study and increase trustworthiness needed in clinical practice. While several interpretable deep learning designs have been proposed to treat unimodal images, to the best of our knowledge, there is no multimodal SR approach applied for medical images. In this paper, we present an interpretable neural network model that exploits information from multiple modalities to super-resolve an image of a target modality. Experiments with simulated and real MRI data show the performance of the proposed approach in terms of numerical and visual results.

Keywords: Medical image super-resolution · Interpretable neural networks · Deep unfolding · Coupled sparse representations

1 Introduction

Image super-resolution (SR) is a well-known inverse problem in imaging applications. Depending on the number of the employed imaging modalities, SR techniques can be divided into single modal and multimodal. Single modal SR aims to

This work has been co-funded by the European Union and Greek national funds through the Operational Program Competitiveness, Entrepreneurship and Innovation, under the call RESEARCH-CREATE-INNOVATE (project code: T1EDK03895).

© Springer Nature Switzerland AG 2021
M. de Bruijne et al. (Eds.): MICCAI 2021, LNCS 12906, pp. 421–429, 2021.
https://doi.org/10.1007/978-3-030-87231-1_41

reconstruct a high-resolution (HR) counterpart of a given low-resolution (LR) image of the same modality. Multimodal SR uses complementary information from multiple modalities to recover a target modality. In medical imaging, multiple modalities are coming from different scanning devices or different hardware configurations. Acquisition time constraints, hardware limitations, human body motion etc., result in low-resolution images; therefore, applying a post-processing SR technique to improve the quality of the modality of interest is a considered approach in medical applications [6].

Existing image reconstruction approaches include conventional methods and data-driven techniques. Conventional methods model the physical processes underlying the problem and incorporate domain knowledge; however, the associated iterative optimization algorithms, typically, have a high computational complexity. Among data-driven techniques, deep learning (DL) is popular [12,17,21,24,29,30] as it can dramatically reduce the computational cost at the inference step [11]. Nevertheless, neural networks have generic architectures and it is unclear how to incorporate domain knowledge. As a result, one can hardly say what a model has learned. When dealing with medical data, the accuracy and trustworthiness of reconstruction is critical for discovery and diagnosis. Therefore, finding a balance between accuracy and latency raises a significant challenge [27].

Bridging the gap between conventional methods and DL has motivated the design of interpretable neural networks [15,18]. Recently, a principle referred to as deep unfolding has received a lot of attention [3,7,22]. The idea is to unfold the iterations of an inference algorithm into a deep neural network, offering interpretability of the learning process. The model parameters across layers are learned from data and the inference is performed at a fixed computational cost. The approach has been applied to medical imaging in [1,19,26]; however, existing works deal with unimodal data. To the best of our knowledge, no interpretable DL design has been reported for multimodal medical image reconstruction.

In this paper, we assume that the similarity between different imaging modalities can be captured by coupled sparse representations that are similar by means of the ℓ_1-norm. We formulate a coupled sparse representation problem which can be solved with an iterative thresholding algorithm. The algorithm is unfolded into a neural network form, resulting in a learned multimodal convolutional sparse coding model (LMCSC). We incorporate LMCSC into a network that can reconstruct an HR image of a target modality from an LR input with the aid of another guidance modality.

We apply our model to multi-contrast Medical Resonance Imaging (MRI). MRI images with different contrast mechanisms (T1-weighted, T2-weighted, FLAIR) provide different structural information about body tissues. However, the long acquisition process can result in motion-related artifacts. To reduce the acquisition time, a compromise is to generate an LR T2W image and a corresponding HR T1W (or FLAIR) image with a short acquisition time and then obtain an HR T2W image by using multimodal SR methods. Our experiments are conducted on two benchmark datasets, showing that the proposed model achieves state-of-the-art performance.

The paper is organized as follows. Section 2 provides the necessary background on sparse modelling, and Sect. 3 reports related work on deep unfolding. The proposed model is presented in Sect. 4, while experiments are included in Sect. 5. Finally, conclusions are drawn in Sect. 6.

2 Sparse Modelling for Image Reconstruction

Linear inverse problems in imaging are typically formulated as follows [20]:

$$y = Lx + \eta, \tag{1}$$

where $x \in \mathbb{R}^k$ is a vectorized form of the unknown source image, $y \in \mathbb{R}^n$ denotes the degraded observations and $\eta \in \mathbb{R}^n$ is the noise[1]. The linear operator $L \in \mathbb{R}^{n \times k}$, $n < k$, describes the observation mechanism. In image SR, L can be expressed as the product of a downsampling operator E and a blurring filter H [25].

Even when the linear observation operator L is given, problem (1) is ill-posed and needs regularization. Following a sparse modelling approach, we assume that $x = D_x u$, with $D_x \in \mathbb{R}^{k \times m}$, $k \leq m$, denoting a representation dictionary, and $u \in \mathbb{R}^m$ being a sparse vector. Then, (1) can be written as $y = A_x u + \eta$, with $A_x = L D_x$, $A_x \in \mathbb{R}^{n \times m}$, and finding x reduces to the sparse approximation problem

$$\min_u \frac{1}{2} \|y - A_x u\|_2^2 + \lambda \|u\|_1, \tag{2}$$

where λ is a regularization parameter, and $\|u\|_1 = \sum_{i=1}^m |u_i|$ is the ℓ_1-norm, which promotes sparsity. Sparse approximation was first used for single image SR in [25].

According to recent studies [16], the accuracy of sparse approximation problems can be improved if a signal ω correlated with the target signal x is available; we refer to ω as side information (SI). Let $\omega \in \mathbb{R}^d$ have a sparse representation $z \in \mathbb{R}^m$ under a dictionary $D_\omega \in \mathbb{R}^{d \times m}$, $d \leq m$; assume that z is similar to u by means of the ℓ_1-norm. Then, given the observations y, we can obtain u as the solution of the ℓ_1-ℓ_1 minimization problem

$$\min_u \frac{1}{2} \|y - A_x u\|_2^2 + \lambda(\|u\|_1 + \|u - z\|_1). \tag{3}$$

Similarity in terms of the ℓ_1-norm holds for representations with partially common support and a number of similar nonzero coefficients; we refer to them as *coupled sparse representations*.

---

[1] Notation: Lower case letters are used for scalars, boldface lower case letters for vectors, boldface upper case letters for matrices and boldface upper case letters in math calligraphy for tensors.

3 Deep Unfolding

Deep unfolding was first proposed in [7] where a sparse coding algorithm was unfolded into a neural network form. The resulting model, coined LISTA, is a learned version of the iterative soft thresholding algorithm (ISTA) [4]. Each layer of LISTA computes:

$$u^t = \phi_\gamma\big(S^t u^{t-1} + W y\big), \tag{4}$$

where ϕ_γ denotes the soft-thresholding operator $\phi_\gamma(\alpha_i) = \text{sign}(\alpha_i)(|\alpha_i| - \gamma)$, $i = 1, \ldots, k$; the parameters S^t, W, γ are learned from data. The authors of [10] integrated LISTA into a neural network design for image SR, obtaining an end-to-end reconstruction architecture that incorporates a sparse prior.

Multimodal image SR via deep unfolding was first addressed in [13], where the authors introduced the assumption that correlated images of multiple modalities can have coupled sparse representations. According to this assumption, given an HR image ω of a guidance modality, we can compute a sparse representation u from the observations of the target modality y by solving a problem of the form (3). The multimodal network presented in [13] incorporates LeSITA [23], a deep unfolding design that learns coupled sparse representations. Implementing iterations of a side-information-driven thresholding algorithm that solves (3), each layer of LeSITA computes:

$$u^t = \xi_\mu\big(S^t u^{t-1} + W y; z\big), \tag{5}$$

where ξ_μ is a proximal operator [23] that integrates the information coming from another modality (in the form of z) into the reconstruction process.

4 A Multimodal Convolutional Deep Unfolding Design for Medical Image Super-Resolution

Due to the large size of images, sparse modelling techniques are typically applied to image patches. Alternatively, Convolutional Sparse Coding (CSC) [28] can be directly applied to the entire image. Let $X \in \mathbb{R}^{n_1 \times n_2}$ be the image of interest. A sparse modelling approach with respect to a convolutional dictionary $\mathcal{D}^X \in \mathbb{R}^{p_1 \times p_2 \times k}$ has the form $X = \sum_{i=1}^{k} D_i^X * U_i$, where $D_i^X \in \mathbb{R}^{p_1 \times p_2}$, $i = 1, ..., k$, are the atoms of $\mathcal{D}^X$, and $U_i \in \mathbb{R}^{n_1 \times n_2}$, $i = 1, ..., k$, are the corresponding sparse feature maps; the symbol $*$ denotes a convolution operation. Then, an observation of X can be written as $Y = \sum_{i=1}^{k} A_i^X * U_i$, with $A_i^X = L D_i^X$.

When, besides the observation of the target image modality, another image modality Ω, correlated with X is available, we can reconstruct X by solving a convolutional form of (3), that is,

$$\min_{U_i} \frac{1}{2} \Big\| Y - \sum_{i=1}^{k} A_i^X * U_i \Big\|_F^2 + \lambda\Big(\sum_{i=1}^{k} \|U_i\|_1 + \sum_{i=1}^{k} \|U_i - Z_i\|_1 \Big), \tag{6}$$

where $Z_i \in \mathbb{R}^{n_1 \times n_2}$, $i = 1, ..., k$, are the sparse feature maps of the modality Ω with respect to a convolutional dictionary $\mathcal{D}^\Omega \in \mathbb{R}^{p_1 \times p_2 \times k}$.

The linear properties of convolution allow to write (6) in the form of (3). Then, LeSITA (5) can be used for the computation of the convolutional sparse codes. However, it is computationally more efficient to write (5) in a convolutional form [14], obtaining a learned multimodal convolutional sparse coding (LMCSC) model. LMCSC includes the following stages:

$$\mathcal{U}^t = \xi_\mu(\mathcal{U}^{t-1} - \mathcal{Q} * \mathcal{R} * \mathcal{U}^{t-1} + \mathcal{P} * Y; \mathcal{Z}), \tag{7}$$

with ξ_μ the proximal operator defined in [23]. The parameters $\mathcal{Q} \in \mathbb{R}^{p_1 \times p_2 \times c \times k}$, $\mathcal{R} \in \mathbb{R}^{p_1 \times p_2 \times k \times c}$, $\mathcal{P} \in \mathbb{R}^{p_1 \times p_2 \times c \times k}$ correspond to learnable convolutional layers; c is the number of channels of the employed images; $\mu > 0$ is also learnable.

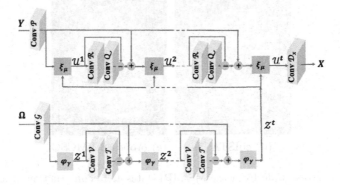

Fig. 1. The proposed multimodal SR model. The lower branch computes the sparse codes $\mathcal{Z}$ of the guidance modality, while the upper (main) branch computes the sparse codes $\mathcal{U}$ of the target modality with the aid of $\mathcal{Z}$. The target HR image X is the result of a convolution operation between $\mathcal{U}$ and a learned dictionary $\mathcal{D}^X$.

We will apply this model for the super-resolution of LR T2W images with the aid of HR T1W (or FLAIR) images. We assume that images of both modalities have coupled sparse representations under different convolutional dictionaries. We also assume that the LR and HR T2W images can have the same sparse representation under different convolutional dictionaries. Therefore, the reconstruction of the HR T2W image reduces to the computation of the convolutional coefficients of the corresponding LR image. The final HR T2W image can be obtained by a convolutional operation between the sparse coefficients and a convolutional dictionary. The model is depicted in Fig. 1. The training process results in learning the convolutional dictionary $\mathcal{D}^X$ as well as the parameters of the unfolded algorithm (7). The sparse codes $\mathcal{Z}$ of the guidance modality are obtained using a convolutional LISTA model [22], computing at the t-th layer:

$$\mathcal{Z}^t = \phi_\gamma(\mathcal{Z}^{t-1} - \mathcal{T} * \mathcal{V} * \mathcal{Z}^{t-1} + \mathcal{G} * \Omega), \tag{8}$$

with $\mathcal{T} \in \mathbb{R}^{p_1 \times p_2 \times c \times k}$, $\mathcal{G} \in \mathbb{R}^{p_1 \times p_2 \times c \times k}$, $\mathcal{V} \in \mathbb{R}^{p_1 \times p_2 \times k \times c}$ and γ learnable parameters.

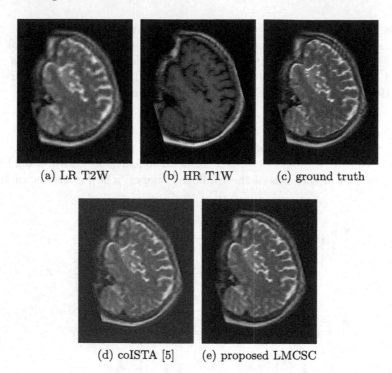

| (a) LR T2W | (b) HR T1W | (c) ground truth |

| (d) coISTA [5] | (e) proposed LMCSC |

Fig. 2. A ×6 SR example from the MS-MRI dataset. Reconstruction of an HR T2W image from an LR T2W image (PSNR = 17.51 dB) with the aid of an HR T1W image. Reconstruction PSNR values are 37.53 dB for coISTA and 38.45 dB for LMCSC.

5 Experiments

We have used LMCSC with two multimodal MRI databases from the Laboratory of Imaging Technologies[2], namely, a brain MR database [2], which contains simulated data from 20 patients, and an MR Multiple Sclerosis (MS) database [9], which contains real data from 30 patients. Both databases include co-registered T1W, T2W and FLAIR 3D images. From each database, we reserve data from five patients for testing. We create the training dataset by selecting cropped image slices of size 44×44, and apply data augmentation by flipping and rotating images, obtaining 22K training samples for the MS-MRI dataset and 25K samples for the brain-MRI dataset. We use whole image slices for testing. Each T2W image is blurred with a 3×3 Gaussian filter and downsampled. We obtain an input LR image of the desired dimensions after bicubic interpolation.

We implement the proposed model with three unfolding stages for each network branch. The size of the learned parameters is set to $7 \times 7 \times 1 \times 85$ for $\mathcal{P}$, $\mathcal{Q}$, $\mathcal{T}$, $\mathcal{G}$, and $7 \times 7 \times 85 \times 1$ for $\mathcal{R}$, $\mathcal{V}$, $\mathcal{D}^X$; a random gaussian distribution with standard deviation 0.01 is used for initialization. The initial value of the

---

[2] http://lit.fe.uni-lj.si/tools.php?lang=eng.

parameters γ, μ is set to 0.1. We set the learning rate equal to 0.0001 and use the Adam optimizer [8] with the mean square error loss function to train the network end-to-end for 100 epochs. As a baseline method, we use a convolutional form of the coISTA model proposed in [5]. We follow the same initialization and training procedure for coISTA. All experiments have been performed on a desktop with AMD Ryzen 5 1600 Six-Core 3.7 GHz CPU, 16 GB RAM, and an NVIDIA GeForce GTX 1070 GPU.

Numerical results, in terms of Peak Signal-to-Noise Ratio (PSNR), presented in Table 1 include ×4 and ×6 SR. Besides LMCSC and coISTA, we also report results for bicubic interpolation. The results show the superior performance of the proposed approach. A visual example presented in Fig. 2, shows that reconstruction with the proposed LMCSC results in a high-contrast and more clear image compared to coISTA [5].

We also report results for different realizations of the proposed model with varying number t of unfolding stages as described by (7). We only vary the number of stages of the main network branch computing the representation of the target modality. The number of ACSC unfoldings is kept fixed, i.e., equal to three. Experiments for this study have been conducted on the MS-MRI dataset for ×4 SR of T2W with the aid of T1W. As can be seen in Table 2, the best performance is achieved with three unfolding stages.

Table 1. Super-resolution of T2W with the aid of T1W or FLAIR images (SI). Results are presented in terms of PSNR (in dB) for two multimodal datasets.

SI	T1W				FLAIR			
Dataset	Brain-MRI		MS-MRI		Brain-MRI		MS-MRI	
SR-scale	×4	×6	×4	×6	×4	×6	×4	×6
Bicubic interpolation	28.56	25.30	17.16	17.15	28.55	25.30	17.18	17.14
coISTA [5]	32.57	28.34	40.54	36.56	**32.39**	28.10	40.28	36.66
LMCSC	**34.86**	**31.97**	**40.94**	**37.28**	32.11	**28.44**	**40.66**	**36.80**

Table 2. Performance of LMCSC [in terms of PSNR (in dB)] for varying number of unfolding stages. Experiments are conducted for ×4 SR of T2W with the aid of T1W images on the MS-MRI dataset.

Model configuration	#stages = 2	#stages = 3	#stages = 4
LMCSC	40.85	**40.94**	40.92

6   Conclusion

Interpretable deep learning is a promising approach for the recovery of medical images as it combines trustworthiness and fast inference. Following the principle of deep unfolding, we have presented LMCSC, an interpretable multimodal deep learning architecture that computes coupled convolutional sparse codes. LMCSC was applied to super-resolve multi-contrast MRI images. The model is designed to address linear inverse problems with side information, therefore, it can be applied for other multimodal recovery tasks such as denoising or compressive sensing reconstruction, while it can also include other medical imaging modalities. We will investigate these applications in our future work.

References

1. Adler, J., Öktem, O.: Learned primal-dual reconstruction. IEEE Trans. Med. Imag. **37**(6), 1322–1332 (2018)
2. Aubert-Broche, B., Griffin, M., Pike, G.B., Evans, A.C., Collins, D.L.: Twenty new digital brain phantoms for creation of validation image data bases. IEEE Trans. Med. Imag. **25**(11), 1410–1416 (2006)
3. Borgerding, M., Schniter, P., Rangan, S.: AMP-inspired deep networks for sparse linear inverse problems. IEEE Trans. Sig. Process. **65**(16), 4293–4308 (2017)
4. Daubechies, I., Defrise, M., Mol, C.D.: An iterative thresholding algorithm for linear inverse problems with a sparsity constrain. Commun. Pure Appl. Math. **57**, 1413 (2004)
5. Deng, X., Dragotti, P.L.: Deep coupled ISTA network for multi-modal image super-resolution. IEEE Trans. Image Process. **29**, 1683–1698 (2020)
6. Greenspan, H.: Super-resolution in medical imaging. Comput. J. **52**(1), 43–63 (2009)
7. Gregor, K., LeCun, Y.: Learning fast approximations of sparse coding. In: Proceedings of the 27th International Conference on Machine Learning, pp. 399–406. ICML 2010, Omnipress, USA (2010)
8. Kingma, D.P., Ba, J.: Adam: a method for stochastic optimization. arXiv preprint arXiv:1412.6980 (2014)
9. Lesjak, Ž, et al.: A novel public MR image dataset of multiple sclerosis patients with lesion segmentations based on multi-rater consensus. Neuroinformatics **16**(1), 51–63 (2018)
10. Liu, D., Wang, Z., Wen, B., Yang, J., Han, W., Huang, T.S.: Robust single image super-resolution via deep networks with sparse prior. IEEE Trans. Image Process. **25**(7), 3194–3207 (2016)
11. Lucas, A., Iliadis, M., Molina, R., Katsaggelos, A.K.: Using deep neural networks for inverse problems in imaging: beyond analytical methods. IEEE Sig. Process. Mag. **35**(1), 20–36 (2018)
12. Mansoor, A., Vongkovit, T., Linguraru, M.G.: Adversarial approach to diagnostic quality volumetric image enhancement. In: 2018 IEEE 15th International Symposium on Biomedical Imaging (ISBI 2018), pp. 353–356. IEEE (2018)
13. Marivani, I., Tsiligianni, E., Cornelis, B., Deligiannis, N.: Multimodal image super-resolution via Deep Unfolding with Side Information. In: European Signal Processing Conference (EUSIPCO) (2019)
14. Marivani, I., Tsiligianni, E., Cornelis, B., Deligiannis, N.: Multimodal deep unfolding for guided image super-resolution. IEEE Trans. Image Process. **29**, 8443–8456 (2020)
15. Monga, V., Li, Y., Eldar, Y.C.: Algorithm unrolling: Interpretable, efficient deep learning for signal and image processing. arXiv preprint arXiv:1912.10557 (2019)

16. Mota, J.F.C., Deligiannis, N., Rodrigues, M.R.D.: Compressed sensing with prior information: strategies, geometry, and bounds. IEEE Trans. Inf. Theory **63**(7), 4472–4496 (2017)
17. Nehme, E., Weiss, L.E., Michaeli, T., Shechtman, Y.: Deep-STORM: super-resolution single-molecule microscopy by deep learning. Optica **5**(4), 458–464 (2018)
18. Papyan, V., Romano, Y., Elad, M.: Convolutional neural networks analyzed via convolutional sparse coding. J. Mach. Learn. Res. **18**(1), 2887–2938 (2017)
19. Qin, C., Schlemper, J., Caballero, J., Price, A.N., Hajnal, J.V., Rueckert, D.: Convolutional recurrent neural networks for dynamic MR image reconstruction. IEEE Trans. Med. Imag. **38**(1), 280–290 (2018)
20. Ribes, A., Schmitt, F.: Linear inverse problems in imaging. IEEE Sig. Process. Mag. **25**(4), 84–99 (2008)
21. Schlemper, J., Caballero, J., Hajnal, J.V., Price, A.N., Rueckert, D.: A deep cascade of convolutional neural networks for dynamic MR image reconstruction. IEEE Trans. Med. Imag. **37**(2), 491–503 (2017)
22. Sreter, H., Giryes, R.: Learned convolutional sparse coding. In: 2018 IEEE International Conference on Acoustics, Speech and Signal Processing (ICASSP), pp. 2191–2195. IEEE (2018)
23. Tsiligianni, E., Deligiannis, N.: Deep coupled-representation learning for sparse linear inverse problems with side information. IEEE Sig. Process. Lett. **26**, 1768 (2019)
24. Xiang, L., et al.: Deep-learning-based multi-modal fusion for fast MR reconstruction. IEEE Trans. Biomed. Eng. **66**(7), 2105–2114 (2018)
25. Yang, J., Wright, J., Huang, T.S., Ma, Y.: Image super-resolution via sparse representation. IEEE Trans. Image Process. **19**, 2861–2873 (2010)
26. Yang, Y., Sun, J., Li, H., Xu, Z.: Deep ADMM-Net for compressive sensing MRI. In: Advances in Neural Information Processing Systems (NIPS), pp. 10–18 (2016)
27. Yedder, H.B., Cardoen, B., Hamarneh, G.: Deep learning for biomedical image reconstruction: a survey. Artif. Intell. Rev. **54**, 1–37 (2020)
28. Zeiler, M.D., Krishnan, D., Taylor, G.W., Fergus, R.: Deconvolutional networks. In: IEEE Conference on Computer Vision and Pattern Recognition (CVPR) (2010)
29. Zeng, K., Zheng, H., Cai, C., Yang, Y., Zhang, K., Chen, Z.: Simultaneous single- and multi-contrast super-resolution for brain MRI images based on a convolutional neural network. Comput. Biol. Med. **99**, 133–141 (2018)
30. Zhou, B., Zhou, S.K.: DuDoRNet: learning a dual-domain recurrent network for fast MRI reconstruction with deep T1 prior. In: Proceedings of the IEEE/CVF Conference on Computer Vision and Pattern Recognition, pp. 4273–4282 (2020)

MRI Super-Resolution Through Generative Degradation Learning

Yao Sui[1,2(✉)], Onur Afacan[1,2], Ali Gholipour[1,2], and Simon K. Warfield[1,2]

[1] Harvard Medical School, Boston, MA, USA
[2] Boston Children's Hospital, Boston, MA, USA
{yao.sui,onur.afacan,ali.gholipour,simon.warfield}@childrens.harvard.edu

Abstract. Spatial resolution plays a critically important role in MRI for the precise delineation of the imaged tissues. Unfortunately, acquisitions with high spatial resolution require increased imaging time, which increases the potential of subject motion, and suffers from reduced signal-to-noise ratio (SNR). Super-resolution reconstruction (SRR) has recently emerged as a technique that allows for a trade-off between high spatial resolution, high SNR, and short scan duration. Deconvolution-based SRR has recently received significant interest due to the convenience of using the image space. The most critical factor to succeed in deconvolution is the accuracy of the estimated blur kernels that characterize how the image was degraded in the acquisition process. Current methods use handcrafted filters, such as Gaussian filters, to approximate the blur kernels, and have achieved promising SRR results. As the image degradation is complex and varies with different sequences and scanners, handcrafted filters, unfortunately, do not necessarily ensure the success of the deconvolution. We sought to develop a technique that enables accurately estimating blur kernels from the image data itself. We designed a deep architecture that utilizes an adversarial scheme with a generative neural network against its degradation counterparts. This design allows for the SRR tailored to an individual subject, as the training requires the scan-specific data only, i.e., it does not require auxiliary datasets of high-quality images, which are practically challenging to obtain. With this technique, we achieved high-quality brain MRI at an isotropic resolution of 0.125 cubic mm with six minutes of imaging time. Extensive experiments on both simulated low-resolution data and

---

This work was supported in part by the National Institutes of Health (NIH) under grants R01 NS079788, R01 EB019483, R01 EB018988, R01 NS106030, R01 EB031849, IDDRC U54 HD090255, S10OD025111; a research grant from the Boston Children's Hospital Translational Research Program; a Technological Innovations in Neuroscience Award from the McKnight Foundation; a research grant from the Thrasher Research Fund; and a pilot grant from National Multiple Sclerosis Society under Award Number PP-1905-34002.

Electronic supplementary material The online version of this chapter (https://doi.org/10.1007/978-3-030-87231-1_42) contains supplementary material, which is available to authorized users.

M. de Bruijne et al. (Eds.): MICCAI 2021, LNCS 12906, pp. 430–440, 2021.
https://doi.org/10.1007/978-3-030-87231-1_42

clinical data acquired from ten pediatric patients demonstrated that our approach achieved superior SRR results as compared to state-of-the-art deconvolution-based methods, while in parallel, at substantially reduced imaging time in comparison to direct high-resolution acquisitions.

Keywords: MRI · Super-resolution · Deep learning

1 Introduction

MRI is critically important in clinical and scientific research studies. High spatial resolution in MRI allows for the precise delineation of the imaged tissues. However, the high spatial resolution requires a long scan time, and in turn unfortunately increases the potential of subject motion [1,22] and reduces the signal-to-noise ratio [2,16]. Super-resolution reconstruction (SRR) has recently emerged as a technique of post-acquisition processing, which allows for obtaining MRI images at a high spatial resolution, high SNR, and short scan time [7–9,19,26]. Among these SRR techniques, deconvolution-based methods [16,20,21,23,24] have recently received significant interest due to the convenience of using the image space only. However, the deconvolution requires an accurate estimate in the blur kernel that characterizes the degradation process of the acquired images, which are usually difficult to obtain. Current methods take handcrafted filters as an approximation of the blur kernel, such as Gaussian filters [7,20]. As the image degradation is complex and varies with different sequences and scanners, however, handcrafted approximations do not necessarily ensure the success of the deconvolution, and even possibly lead to unusable reconstructions.

Blur kernel estimation is often used in natural image deblurring, also known as blind deconvolution [3], in particular equipped with deep neural networks [6]. Unfortunately, no techniques focus on blur kernel estimation for deconvolution-based SRR in MRI. Although deep learning-based techniques have been widely used in natural image super-resolution, the majority of deep SRR methods for MRI are performed with 2D slices [4,28,30] as the large-scale, auxiliary datasets of high-quality volumetric MRI images are practically challenging to obtain at a high resolution and suitable SNR. It has been shown that the deep SRR model learned on volumetric data achieved better results than on 2D slice data [15].

We sought to develop a methodology that allows for accurately estimating blur kernels from the image data itself. We designed a deep architecture that utilizes an adversarial scheme with a generative neural network against its degradation counterparts. Our design enables the SRR tailored to an individual subject, as the training of our deep SRR model requires the scan-specific image data only. We achieved high-quality brain MRI images through our SRR method at an isotropic resolution of 0.125 cubic mm with six minutes of imaging time. We assessed our method on simulated low-resolution (LR) image data and applied it to ten pediatric patients. Experimental results show that our approach achieved superior SRR as compared to state-of-the-art deconvolution-based methods, while in parallel, at substantially reduced imaging time in comparison to direct high-resolution (HR) acquisitions.

2   Methods

2.1   Theory

It is difficult for SRR to enhance 2D in-plane or true 3D MRI resolution due to the frequency encoding scheme [14,18], but effective to improve through-plane resolution for 2D slice stacks [10]. Therefore, we focus on reducing the slice thickness of LR scans[1] with large matrix size and thick slices. The thick slices lead to reduced scan time and increased SNR. Thus, we can acquire a set of LR images with different orientations to facilitate the SRR to capture the HR signals distributed in different directions, while keeping the total scan time short.

Forward Model. Let $\mathbf{x}$ denote the HR reconstruction and $\{\mathbf{y}_j\}_{j=1}^{n}$ denote the n acquired LR images. The forward model that describes the acquisition process is formulated as

$$\mathbf{y}_j = \mathbf{D}_j \mathbf{B}_j \mathbf{T}_j \mathbf{x} + \boldsymbol{\nu}, \quad j = 1, 2, \ldots, n. \tag{1}$$

The transform $\mathbf{T}_j$ compensates for subject motion and is obtained by aligning the n LR images together. $\mathbf{B}_j$ describes the image degradation defined by a blur kernel. $\mathbf{D}_j$ is the downsampling operator that discards a certain amount of data from the HR signal. The noise $\boldsymbol{\nu}$ is additive and Gaussian when $\text{SNR} > 3$ [11].

Generative Degradation Learning. Our SRR targets a joint estimate in both the HR reconstruction and blur kernel. We use a generative neural network to generate an HR estimate by $\mathbf{x} = f_\theta(\mathbf{z}_x)$, and n generative networks to generate n blur kernels by $\mathbf{B}_j = g_{\omega_j}(\mathbf{z}_{b_j})$. These generations are constrained by the image degradation process as described in the forward model. Therefore, the generative degradation learning is found by

$$\min_{\mathbf{x}, \theta, \mathbf{B}_j, \omega_j} \ell\left(\mathbf{x} - f_\theta(\mathbf{z}_x)\right) + \sum_{j=1}^{n} \ell_{b_j}\left(\mathbf{B}_j - g_{\omega_j}(\mathbf{z}_{b_j})\right), \tag{2}$$

$$s.t. \quad \mathbf{y}_j = \mathbf{D}_j \mathbf{B}_j \mathbf{T}_j \mathbf{x} + \boldsymbol{\nu}, \quad j = 1, 2, \ldots, n,$$

where f_θ and g_{ω_j} are nonlinear functions that generate data, parameterized by θ and ω_j, respectively, ℓ and ℓ_{b_j} are the loss functions in the optimization, and $\mathbf{z}_x$ and $\mathbf{z}_{b_j}$ are the initial guesses for $\mathbf{x}$ and $\mathbf{B}_j$, respectively. The function f_θ here is also known as the deep image prior [25]. When the loss functions are ℓ_2 loss, we can substitute $\mathbf{x}$ and $\mathbf{B}_j$ with the generative functions, and the optimization is then re-formulated as

$$\min_{\theta, \omega_j} \sum_{j=1}^{n} \left\| \mathbf{y}_j - \mathbf{D}_j g_{\omega_j}(\mathbf{z}_{b_j}) \mathbf{T}_j f_\theta(\mathbf{z}_x) \right\|_2^2 + \lambda \mathcal{R}_{TV}\left(f_\theta(\mathbf{z}_x)\right), \tag{3}$$

with a total variation (TV) [17] term imposed to regularize the HR reconstruction for edge sharpness, and a weight parameter $\lambda > 0$.

---

[1] We refer to a 2D slice stack as a scan or an image hereafter.

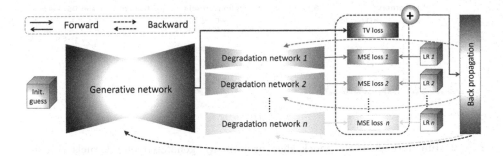

Fig. 1. Architecture of our proposed approach to generative degradation learning.

We implement these generative functions by deep neural networks. f_θ is realized by a 3D UNet-like encoder-decoder network [31]. g_{ω_j} is implemented by a fully connected network containing four hidden linear layers and one Tanh layers. Each layer is followed by a dropout regularization and a ReLU activation. The architecture of our generative degradation networks (GDN) is shown in Fig. 1. A degradation network comprises a generative function for the blur kernel in combination with the constraint delivered by the forward model. We solve the optimization by an Adam algorithm [12]. The training for the GDN is an optimization problem. It is thus carried out on the scan-specific LR images $\mathbf{y}_j$ themselves only, and in turn, allows for the SRR tailored to an individual subject.

The HR reconstruction is obtained directly from $\mathbf{x} = f_\theta(\mathbf{z}_x)$ once the GDN model has been trained. Also, to ensure the appropriate scale of the voxel intensity, a standard TV-based SRR with the learned blur kernels $\mathbf{B}_j$ can be optionally applied to obtain the HR reconstruction.

2.2 GDN-Based SRR

Since we focus on enhancing the through-plane resolution, the degradation process is assumed to be associated with the blurs from the slice excitation and downsampling. The downsampling is carried out by truncating the high frequency in the Fourier domain, and it is thus determined by a sinc low-pass filter. The slice excitation is characterized by the slice profile that is generated by the radio frequency (RF) pulse during scans. Therefore, the blur kernel we estimate is the convolution between the slice profile and the sinc filter.

The motion compensation is implemented by a rigid body image registration as we focus on brain MRI in this work. We interpolate the LR images to those with the resolution and size the same as the HR reconstruction, and align them together. The obtained transformations are used as $\mathbf{T}_j$ in the forward model.

We initialize the blur kernels as a sinc filter convolved with a Gaussian low-pass filter whose full width at half maximum (FWHM) is equal to the slice thickness of the LR image over that of the HR reconstruction. We compute an initial guess of the HR reconstruction by a standard TV-based SRR [16] with the above initialized blur kernels. We set $\lambda = 0.01$ in Eq. (3) according to our

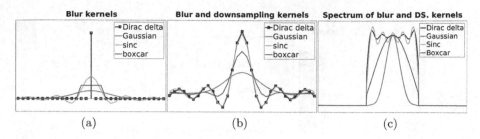

Fig. 2. Illustrations of the mock slice profiles (a), generated blur kernels (b), and spectrum magnitudes of these kernels (c).

empirical results. We run 4k iterations with a learning rate of 0.01 to minimize Eq. (3). It took about 2.5 h to reconstruct an HR image of size $384 \times 384 \times 384$ voxels from three LR scans on an NVIDIA Titan RTX GPU with PyTorch [13].

2.3   Materials

We simulated an LR dataset based on the Human Connectome Project (HCP) database [5]. We randomly selected forty subjects from the HCP database, including five T1w and five T2w HR images with an isotropic resolution of 0.7 mm as the ground truths. We generated four blur kernels, as shown in Figs. 2(b) and 2(c), based on a Dirac pulse, a Gaussian, a sinc, and a boxcar functions depicted in Fig. 2. We simulated four groups of LR images according to the four types of blur kernels. We downsampled each ground truth image to three LR images in the complementary planes with a slice thickness of 2 mm after convolving it with the blur kernel in its group. Gaussian noise was added in each LR image with a standard deviation of 5% of maximum voxel intensity.

We acquired thirty LR T2-FSE scans from ten pediatric patients - three images in complementary planes per patient - on a 3T scanner. The field of view (FOV) was 192 mm × 192 mm, matrix size was 384 × 384, slice thickness was 2 mm, TE/TR = 93/12890 ms with an echo train length of 16 and a flip angle of 160°. It took about two minutes in acquiring an LR image.

2.4   Experimental Design

As our SRR is an unsupervised deconvolution-based approach, we compared our approach to state-of-the-art methods in the same category, including a TV-based method [16], a gradient guidance regularized (GGR) method [20], and a TV-based method with joint blur kernel learning (TV-BKL) implemented by ourselves. The parameters of the baselines were set according to their best PSNR.

Experiment 1: Assessments on Simulated Data. The goal of this experiment is two-fold: to assess the accuracy of the blur kernel estimates and to

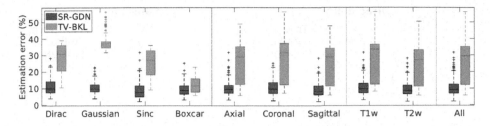

Fig. 3. Estimation errors in the blur kernels obtained from the two methods. Our methods, SR-GDN, considerably outperformed TV-BKL on the HCP dataset.

demonstrate that the estimated blur kernels lead to superior HR reconstructions. We reconstructed each HR image at an isotropic resolution of 0.7 mm from three LR images on the HCP dataset. We investigated the estimation error defined by the ℓ_2 norm of the difference between the estimated kernel and ground truth over the ℓ_2 norm of ground truth $\|k_{est} - k_{gt}\|_2 / \|k_{gt}\|_2$. To evaluate the reconstruction quality, we investigated the PSNR and SSIM [27] of the HR reconstructions.

Experiment 2: Assessments on Clinical Data. This experiment aimed at assessing the applicability of our SRR for clinical use. We reconstructed each HR image at an isotropic resolution of 0.5 mm on the clinical dataset. We evaluated the sharpness of these reconstructions by average edge strength [29]. We checked the estimated blur kernels and qualitatively assessed these reconstructions.

Experiment 3: Impact on Deconvolution-Based Algorithms. This experiment aimed at evaluating the impact of our generative degradation learning on deconvolution-based algorithms. We expected our estimated blur kernels can improve the TV and GGR algorithms in their SRR quality on the HCP dataset.

3 Results

Experiment 1: Assessments on Simulated Data. Figure 3 shows the estimation errors in the blur kernels obtained from our method and TV-BKL. Our method, SR-GDN, considerably outperformed TV-BKL on the HCP dataset. The average errors of SR-GND and TV-BKL were respectively 10.1% ± 4.6% and 25.8% ± 12.0%. Two-sample t-test showed that SR-GND offered significantly lower errors on the simulated data than TV-BKL at the 5% significance level ($p = 2.16e^{-118}$).

Figure 4 shows the spectrum of true and estimated blur kernels from SR-GDN and TV-BKL on four representative simulations. Our estimates (SR-GDN) closely followed the true kernels and offered higher accuracy than TV-BKL with all types of kernels on the HCP dataset.

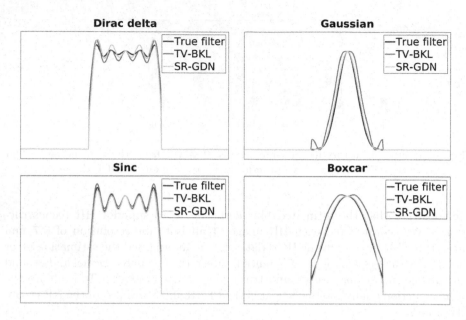

Fig. 4. Spectrum of true and estimated blur kernels from SR-GDN and TV-BKL on four representation simulations. Our estimates (SR-GDN) closely followed the true kernels and offered higher accuracy than TV-BKL with all types of kernels.

Figure 5 shows the quantitative assessment of our approach (SR-GDN) and the three baselines on the HCP dataset in terms of PSNR and SSIM. Our method achieved an average PSNR at 41.2 dB $\pm$ 3.74 dB and SSIM at 0.98 ± 0.01, which were considerably higher than obtained from the three baselines. Two-sample t-tests showed that SR-GDN significantly outperformed the three baselines in terms of SSIM at the 5% significance level with $p = 2.36e^{-18}$ against TV, $4.37e^{-18}$ against GGR, and $1.81e^{-7}$ against TV-BKL.

Experiment 2: Assessments on Clinical Data. Figures 6(a) and 6(b) show the mean and standard deviation of the blur kernels estimated by TV-BKL and SR-GDN from ten sets of clinical scans acquired on a 3T scanner. The results show that both methods led to small standard deviations of estimations. Figure 6(c) shows the spectrum magnitudes of the handcraft Gaussian filter and the mean blur kernels estimated by TV-BKL and SR-GND. In comparison to the simulation results shown in Fig. 4, the results show that our approach (SR-GND) estimated the slice profile as a close approximation to a boxcar function.

We evaluated the average edge strength (AES) of the HR reconstructions obtained from our approach and the three baselines on the ten sets of clinical scans. The results, as shown in Table 1, suggests that our approach, SR-GDN achieved the highest AES. TV-BKL yielded the lowest AES as the noise in the reconstructions led to blurry edges.

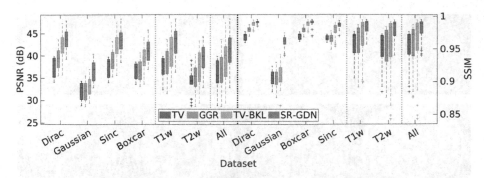

Fig. 5. Quantitative assessment of the four SRR methods in terms of PSNR and SSIM.

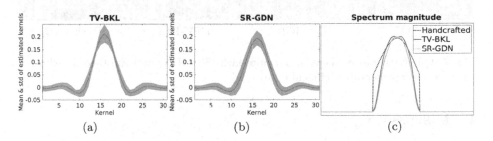

Fig. 6. Mean and standard deviation of the blur kernels estimated by (a) TV-BKL and (b) SR-GDN from ten sets of clinical scans. (c) Spectrum magnitudes of the handcraft Gaussian filter and the mean blur kernels estimated by TV-BKL and SR-GDN.

Figure 7 shows the qualitative assessment of our approach and the three baselines in the slices of representative clinical data. The results show that SR-GDN offered high-quality brain MRI with the clinical LR data. GGR smoothed the images excessively so missed some anatomical details, while TV and TV-BKL generated noisy images and caused artifacts as shown in the coronal image.

Experiment 3: Impact on Deconvolution-Based Algorithms. We replaced the handcrafted Gaussian filter used in TV and GRR with our estimated blur kernels and ran the two methods on the HCP dataset. The increased PSNR/SSIM were respectively: TV = 13.90%/3.15% and GGR = 3.35%/2.44%. The results show that our estimated blur kernels led to improved deconvolution-based algorithms.

Table 1. Average edge strength obtained from our approach and the three baselines on the ten sets of clinical scans.

TV	GGR	TV-BKL	SR-GDN
2.73 ± 0.72	3.12 ± 0.90	1.78 ± 0.50	3.49 ± 0.97

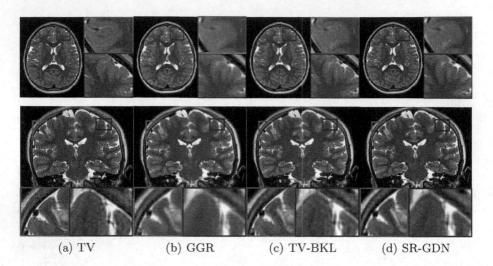

(a) TV   (b) GGR   (c) TV-BKL   (d) SR-GDN

Fig. 7. Qualitative assessment of our approach (SR-GDN) and the three baselines in the slices of representative clinical data.

4   Discussion

We have developed a deconvolutional technique that enabled accurately estimating blur kernels from the image data itself. We have designed a deep architecture that utilizes an adversarial scheme with a generative neural network against its degradation counterparts. This design has been demonstrated to allow for the SRR tailored to an individual subject. We have thoroughly assessed the accuracy of our approach on a simulated dataset. We have successfully applied our approach to ten pediatric patients, and have achieved high-quality brain MRI at an isotropic resolution of 0.125 cubic mm with six minutes of imaging time. Experimental results have shown that our approach achieved superior SRR results as compared to state-of-the-art deconvolution-based methods, while in parallel, at substantially reduced imaging time in comparison to direct HR acquisitions.

References

1. Afacan, O., et al.: Evaluation of motion and its effect on brain magnetic resonance image quality in children. Pediatr. Radiol. **46**(12), 1728–1735 (2016). https://doi.org/10.1007/s00247-016-3677-9
2. Brown, R.W., Cheng, Y.C.N., Haacke, E.M., Thompson, M.R., Venkatesan, R.: Magnetic Resonance Imaging: Physical Principles and Sequence Design, 2nd edn. Wiley, New York (2014)
3. Chan, T., Wong, C.K.: Total variation blind deconvolution. IEEE Trans. Image Process. **7**(3), 370–5 (1998)
4. Cherukuri, V., Guo, T., Schiff, S.J., Monga, V.: Deep MR brain image super-resolution using spatio-structural priors. IEEE Trans. Image Process. **29**, 1368–1383 (2020)

5. Essen, D., Smith, S., Barch, D., Behrens, T.E.J., Yacoub, E., Ugurbil, K.: The Wu-Minn human connectome project: an overview. NeuroImage **80**, 62–79 (2013)
6. Gao, H., Tao, X., Shen, X., Jia, J.: Dynamic scene deblurring with parameter selective sharing and nested skip connections. In: 2019 IEEE/CVF Conference on Computer Vision and Pattern Recognition (CVPR), pp. 3843–3851 (2019)
7. Gholipour, A., Estroff, J.A., Warfield, S.K.: Robust super-resolution volume reconstruction from slice acquisitions: application to fetal brain MRI. IEEE Trans. Med. Imag. **29**(10), 1739–1758 (2010)
8. Gholipour, A., et al.: Super-resolution reconstruction in frequency, image, and wavelet domains to reduce through-plane partial voluming in MRI. Med. Phys. **42**(12), 6919–6932 (2015)
9. Greenspan, H.: Super-resolution in medical imaging. Comput. J. **52**(1), 43–63 (2009)
10. Greenspan, H., Oz, G., Kiryati, N., Peled, S.: MRI inter-slice reconstruction using super-resolution. Magn. Reson. Imag. **20**(5), 437–446 (2002)
11. Gudbjartsson, H., Patz, S.: The Rician distribution of noisy MRI data. Magn. Reson. Med. **34**, 910–914 (1995)
12. Kingma, D.P., Ba, J.: Adam: a method for stochastic optimization. In: International Conference on Learning Representations (ICLR) (2015)
13. Paszke, A., et al.: Automatic differentiation in pytorch. In: Neural Information Processing Systems (NIPS) Workshop (2017)
14. Peled, S., Yeshurun, Y.: Super-resolution in MRI - Perhaps sometimes. Magn. Reson. Med. **48**, 409 (2002)
15. Pham, C., Ducournau, A., Fablet, R., Rousseau, F.: Brain MRI super-resolution using deep 3D convolutional networks. In: International Symposium on Biomedical Imaging, pp. 197–200 (2017)
16. Plenge, E., et al.: Super-resolution methods in MRI: can they improve the trade-off between resolution, signal-to-noise ratio, and acquisition time? Magn. Reson. Med. **68**, 1983–1993 (2012)
17. Rudin, L., Osher, S., Fatemi, E.: Nonlinear total variation based noise removal algorithms. Physica D Nonlinear Phenom. **60**, 259–268 (1992)
18. Scheffler, K.: Super resolution in MRI? Magn. Reson. Med. **48**, 408 (2002)
19. Scherrer, B., Gholipour, A., Warfield, S.K.: Super-resolution reconstruction to increase the spatial resolution of diffusion weighted images from orthogonal anisotropic acquisitions. Med. Image Anal. **16**(7), 1465–1476 (2012)
20. Sui, Y., Afacan, O., Gholipour, A., Warfield, S.K.: Isotropic MRI super-resolution reconstruction with multi-scale gradient field prior. In: Shen, D., et al. (eds.) MICCAI 2019. LNCS, vol. 11766, pp. 3–11. Springer, Cham (2019). https://doi.org/10.1007/978-3-030-32248-9_1
21. Sui, Y., Afacan, O., Gholipour, A., Warfield, S.K.: Learning a gradient guidance for spatially isotropic MRI super-resolution reconstruction. In: Martel, A.L., et al. (eds.) MICCAI 2020. LNCS, vol. 12262, pp. 136–146. Springer, Cham (2020). https://doi.org/10.1007/978-3-030-59713-9_14
22. Sui, Y., Afacan, O., Gholipour, A., Warfield, S.K.: SLIMM: slice localization integrated MRI monitoring. NeuroImage **223**(117280), 1–16 (2020)
23. Sui, Y., Afacan, O., Gholipour, A., Warfield, S.K.: Fast and high-resolution neonatal brain MRI through super-resolution reconstruction from acquisitions with variable slice selection direction. Front. Neurosci. **15**(636268), 1–15 (2021)
24. Tourbier, S., Bresson, X., Hagmann, P., Thiran, J., Meuli, R., Cuadra, M.: An efficient total variation algorithm for super-resolution in fetal brain MRI with adaptive regularization. Neuroimage **118**, 584–597 (2015)

25. Ulyanov, D., Vedaldi, A., Lempitsky, V.: Deep image prior. In: 2018 IEEE/CVF Conference on Computer Vision and Pattern Recognition (CVPR), pp. 9446–9454 (2018)
26. Van Reeth, E., Tham, I.W., Tan, C.H., Poh, C.L.: Super-resolution in magnetic resonance imaging: a review. Concepts Magn. Reson. **40**(6), 306–325 (2012)
27. Wang, Z., Bovik, A.C., Sheikh, H.R., Simoncelli, E.P.: Image quality assessment: from error measurement to structural similarity. IEEE Trans. Image Process. **13**(1), 600–612 (2004)
28. Xue, X., Wang, Y., Li, J., Jiao, Z., Ren, Z., Gao, X.: Progressive sub-band residual-learning network for MR image super-resolution. IEEE J. Biomed. Health Inform. **24**(2), 377–386 (2020)
29. Zaca, D., Hasson, U., Minati, L., Jovicich, J.: A method for retrospective estimation of natural head movement during structural MRI. J. Magn. Reson. Imag. **48**(4), 927–937 (2018)
30. Zhao, X., Zhang, Y., Zhang, T., Zou, X.: Channel splitting network for single MR image super-resolution. IEEE Trans. Image Process. **28**(11), 5649–5662 (2019)
31. Çiçek, Ö., Abdulkadir, A., Lienkamp, S.S., Brox, T., Ronneberger, O.: 3D U-Net: learning dense volumetric segmentation from sparse annotation. In: Ourselin, S., Joskowicz, L., Sabuncu, M.R., Unal, G., Wells, W. (eds.) MICCAI 2016. LNCS, vol. 9901, pp. 424–432. Springer, Cham (2016). https://doi.org/10.1007/978-3-319-46723-8_49

Task-Oriented Low-Dose CT Image Denoising

Jiajin Zhang, Hanqing Chao, Xuanang Xu, Chuang Niu, Ge Wang,
and Pingkun Yan[✉]

Department of Biomedical Engineering and Center for Biotechnology and
Interdisciplinary Studies, Rensselaer Polytechnic Institute, Troy, NY 12180, USA
{zhangj41,chaoh,xux12,niuc,wangg6,yanp2}@rpi.edu

Abstract. The extensive use of medical CT has raised a public concern over the radiation dose to the patient. Reducing the radiation dose
leads to increased CT image noise and artifacts, which can adversely
affect not only the radiologists judgement but also the performance of
downstream medical image analysis tasks. Various low-dose CT denoising methods, especially the recent deep learning based approaches, have
produced impressive results. However, the existing denoising methods
are all downstream-task-agnostic and neglect the diverse needs of the
downstream applications. In this paper, we introduce a novel Task-
Oriented Denoising Network (TOD-Net) with a task-oriented loss leveraging knowledge from the downstream tasks. Comprehensive empirical analysis shows that the task-oriented loss complements other task-
agnostic losses by steering the denoiser to enhance the image quality in
the task related regions of interest. Such enhancement in turn brings general boosts on the performance of various methods for the downstream
task. The presented work may shed light on the future development of
context-aware image denoising methods. Code is available at https://
github.com/DIAL-RPI/Task-Oriented-CT-Denoising_TOD-Net.

Keywords: Low-dose CT · Image denoising · Task-oriented loss ·
Downstream task

1 Introduction

Computed tomography (CT) is one of the most important breakthroughs in
modern medicine, which is routinely used in hospitals with millions of scans
performed per year world wide. However, the ionizing radiation associated with
CT puts patients at the risk of genetic damage and cancer induction. Low-dose
CT (LDCT) uses lower radiation dose for imaging, which helps reduce the risks
but increases noise and artifacts in reconstructed images. These compromised
CT images with artifacts can affect not only the diagnosis by physicians but also
the downstream medical image processing tasks.

J. Zhang and H. Chao are co-first authors.

© Springer Nature Switzerland AG 2021
M. de Bruijne et al. (Eds.): MICCAI 2021, LNCS 12906, pp. 441–450, 2021.
https://doi.org/10.1007/978-3-030-87231-1_43

In order to tackle this problem, the research community has developed various LDCT denoising methods. They can be grouped into the three categories, *i.e.*, sinogram filtration [15,18,19], iterative reconstruction [2,10], and image post-processing [6,9,20]. Classic post-processing methods include K-SVD [6], non-local means (NLM) [14] and block-matching 3D (BM3D) [9,12]. With the renaissance of artificial intelligence in the past decade, various deep neural networks were proposed to denoise LDCT images, which became the main stream methods. Chen et al. [5] first presented a 3-layer convolution neural network (CNN) trained to minimize the mean square error (MSE) for LDCT post denoising. Yang et al. [20] introduced a Wasserstein Generative Adversarial Network (WGAN) with a perceptual loss to keep more detailed information in denoising. Most recently, the self-attention mechanisms and self-supervised learning methods have been introduced in the field to further improve the performance [13]. Aiming at alleviating the difficulties brought by the noise to downstream tasks, this work also focuses on deep learning based post-processing methods for both image denoising and downstream task performance improvement.

Although many efforts have been made in LDCT denoising, existing methods are all downstream-task-agnostic. Specifically, deep learning based denoising methods all intend to reduce a distance between denoised LDCT images and the corresponding normal-dose CT (NDCT) counterparts. The diverse needs from the downstream tasks have been largely overlooked. In this paper, we argue that the denoising module should be trained with the awareness of downstream applications. This will bring two major benefits. One is that using the downstream task requirements can enhance the denoising performance on those task-related regions. The other is that this image quality enhancement can in turn boost the performance of the downstream tasks. To achieve this goal, here we propose a novel Task-Oriented Denoising Network (TOD-Net) with a task-oriented loss in the WGAN framework. Specifically, in training the TOD-Net, we incorporate a fixed representative network of the downstream task and use its task specific loss to guide the optimization of the TOD-Net's parameters. We demonstrate that in the whole workflow, the TOD-Net can not only significantly improve the performance of the specific network in the training phase but also generally boost the performance of various other networks for the same downstream task. Experiments on two datasets show that the image quality of the TOD-Net on the task-related regions is clearly superior to that of the other denoising methods.

The contributions of this paper can be summarized in two aspects: 1) This is the first work leveraging the knowledge of downstream tasks in LDCT denoising. Compared with existing task-agnostic denoising methods, the proposed Task-Oriented Denoising Network (TOD-Net) can significantly improve the denoised CT image quality on the task-related regions. 2) With the targeted image quality improvement, integrating TOD-Net with a downstream model can lead to a significant performance boost on the downstream task. In addition, TOD-Net can significantly improve the performance of various other networks for the same downstream task.

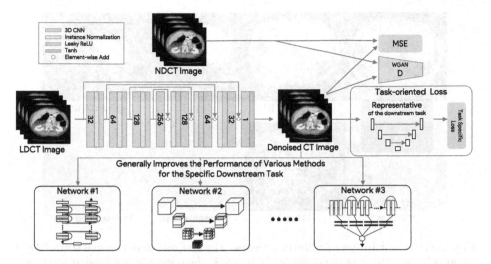

Fig. 1. Illustration of the proposed TOD-Net. The proposed TOD-Net consists of a denoiser and 3 loss estimators. The size of the 3D kernels in the denoiser is $3 \times 3 \times 3$. The numbers in the 3D CNN block are the channel number.

2 Method

The proposed TOD-Net generates denoised image from LDCT guided by a unique combination of three losses. These losses include an mean square error (MSE) loss to control the global Euclidean distance between a generated image and its NDCT counterpart, a WGAN discriminator to shrink the distance bewteen their distributions, and a Task-oriented loss to reflect downstream task-related requirements. Figure 1 shows an overview of our TOD-Net.

2.1 WGAN for LDCT Denoising

The TOD-Net is based on the well-known WGAN [1]. Let $\mathbf{x}$ denote a LDCT image and $\mathbf{x}^*$ be its counterpart NDCT image. WGAN consists of a denoiser G and a discriminator D to generate an image $\hat{\mathbf{x}} = G(\mathbf{x})$ with a distribution of $\mathbf{x}^*$.

Denoting the distribution of $\mathbf{x}$ as Q and the distribution of $\mathbf{x}^*$ as P, the optimization of loss functions of the WGAN can be formulated as:

$$L_D(\theta_D) = \mathbb{E}_{\mathbf{x}^* \sim P}[D(\mathbf{x}^*; \theta_D)] - \mathbb{E}_{\mathbf{x} \sim Q}[D(G(\mathbf{x}; \theta_G); \theta_D)]; \text{subject to } ||\theta_D||_1 \leq \epsilon, \quad (1)$$

$$L_{GAN}(\theta_G) = \mathbb{E}_{\mathbf{x} \sim Q}[D(G(\mathbf{x}; \theta_G); \theta_D)], \quad (2)$$

where θ_D and θ_G are the parameters of D and G respectively. Instead of using the Jensen–Shannon (JS) divergence or Kullback-Leibler (KL) divergence to discriminate P and Q, the discriminator of WGAN applies a more smooth measurement, Wasserstein distance, to evaluate the discrepancy between P and Q which can effectively stabilize the training of the network.

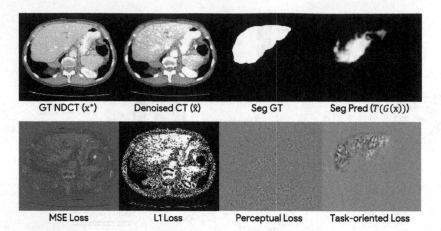

Fig. 2. Gradient maps of different losses. The first row shows ground truth NDCT image $\mathbf{x}^*$, the output of the denoising network $\hat{\mathbf{x}}$, the segmentation ground truth, and the out put of the representative model incorporated in the task-oriented loss. The second row shows the map of $\frac{\partial L}{\partial \hat{\mathbf{x}}}$ of 4 different losses.

2.2   Analysis of Task-Oriented Loss

The proposed task-oriented loss, our main innovation, is to equip the denoiser with an explicit awareness of the downstream task in the training phase. Under this task-oriented guidance, the denoising network will enhance the specific features essential to the downstream task and boost the whole medical image processing pipeline's performance. Specifically, a representative model T of a downstream task is incorporated into the training process. As shown in Fig. 1, the denoised images are fed to T to compute the task-oriented loss. In this paper, we demonstrate our TOD-Net based on the medical image segmentation task. Therefore, we minimize the prevalent Dice loss to compute the task-oriented loss

$$L_t(\theta_G) = \mathbb{E}_{\mathbf{x} \sim Q}[1 - Dice(T(G(\mathbf{x}; \theta_G)))]. \tag{3}$$

To analyze how the task-oriented loss complements other commonly used denoising losses, we examine the similarities and differences between their partial derivatives with respect to the denoiser's parameters θ_G. Specifically, we compare the task-oriented loss to MSE loss L_{MSE}, L_1 loss, and perceptual loss (L_p). For brevity, we use $\hat{\mathbf{x}}$ to represent $G(\mathbf{x})$. Then we have

$$\frac{\partial L_t}{\partial \theta_G} = \frac{\partial L_t}{\partial \hat{\mathbf{x}}} \frac{\partial \hat{\mathbf{x}}}{\partial \theta_G}, \tag{4}$$

$$L_{MSE} = \frac{1}{2}||\hat{\mathbf{x}} - \mathbf{x}^*||_2^2; \frac{\partial L_{MSE}}{\partial \theta_G} = ||\hat{\mathbf{x}} - \mathbf{x}^*||_2 \frac{\partial \hat{\mathbf{x}}}{\partial \theta_G}, \tag{5}$$

$$L_1 = ||\hat{\mathbf{x}} - \mathbf{x}^*||_1; \frac{\partial L_1}{\partial \theta_G} = \mathbb{1}[\hat{\mathbf{x}} - \mathbf{x}^*] \frac{\partial \hat{\mathbf{x}}}{\partial \theta_G}, \tag{6}$$

$$L_p = \frac{1}{2}||f(\hat{\mathbf{x}}) - f(\mathbf{x}^*)||_2^2; \frac{\partial L_p}{\partial \theta_G} = ||f(\hat{\mathbf{x}}) - f(\mathbf{x}^*)||_2 \frac{\partial f}{\partial \hat{\mathbf{x}}} \frac{\partial \hat{\mathbf{x}}}{\partial \theta_G}, \tag{7}$$

where $f(\cdot)$ denotes the network used in the perceptual loss. In Eq. 6, $\mathbb{1}[\cdot] = 1$ when $\cdot \geq 0$ and $\mathbb{1}[\cdot] = -1$ when $\cdot < 0$. From Eqs. 4–7, we can see that the partial derivatives of all these four losses share the same term of $\partial\hat{\mathbf{x}}/\partial\theta_G$. The differences between these loss functions lie only in the other part of the derivatives, denoted by $\partial L/\partial\hat{\mathbf{x}}$. For intuitive understanding, Fig. 2 visualizes the values of $\partial L/\partial\hat{\mathbf{x}}$ calculated on an output $\hat{\mathbf{x}}$ generated by a half-trained denoiser, $i.e.$, TOD-Net before converging. Liver segmentation is the downstream task in this demonstration. Since the decoder has not been fully trained, the "denoised" image $\hat{\mathbf{x}}$ includes many artifacts, which significantly degrade the performance of the downstream segmentation network. The second row shows the differences between the task-driven loss and the other three losses, with the partial derivatives of the task-oriented loss being focused on the liver. Such an attention to the task-related regions steers the denoiser to improve the image quality in these regions. Since the improvements are not for a specific representative network the improvements can enhance various other methods for this task.

2.3 Training Strategy

The total loss used for optimizing the denoiser G is given by

$$L_G = L_{GAN} + L_t + \lambda L_{MSE}, \tag{8}$$

where λ is a hyper-parameter to weigh the MSE loss. The task-oriented loss and the MSE loss work together to enhance the local features required by the downstream task, while maintaining a high global quality. The loss for training the discriminator D is denoted as L_D as defined in Eq. 1. In each training iteration, the two losses L_D and L_G are updated alternately.

3   Experiments

The proposed TOD-Net was evaluated on two datasets from two aspects: 1) image quality of the denoised image, and 2) performance of the downstream task taking the denoised image as the input. The image quality is measured by root-mean-square error (RMSE), structural similarity index (SSIM), and peak signal-to-noise ratio (PSNR). The performance of the downstream task, medical image segmentation, is evaluated by the Dice score. Our model was compared with three baselines, VGG-WGAN [20] SA-WGAN [13] and MSE-WGAN. We trained the VGG-WGAN and SA-WGAN on the two datasets with the exact procedure and hyper-parameters used in their original papers. The MSE-WGAN has the same structure as the TOD-Net but without the task-oriented loss. It was trained with the same protocol as the TOD-Net.

3.1   Datasets

We evaluated our model on two publicly available datasets for single-organ segmentation, $i.e.$, LiTS (Liver tumor segmentation challenge) [3], and KiTS (Kidney tumor segmentation challenge) [11]. LiTS consists of 131 abdominal NDCT images collected from different hospitals and the in-plane resolution varies from 0.45mm to 6mm and the slice spacing is between 0.6 mm and 1.0 mm. We split the dataset into training set, validation set and test set with a ratio of 70% (91 images), 10% (13 images), and 20% (27 images). KiTS includes abdominal NDCT images from 300 patient, 210 of which are publicly available. Since the number of samples is relatively large, we split them into training, validation and test set with a ratio of 60% (126 images), 20% (42 images), and 20% (42 images).

To train and test the proposed TOD-Net, we synthesized the low-dose counterparts of all NDCT images in the two datasets using an industrial CT simulator, CatSim by GE Global Research [7], which incorporates finite focal spot size, realistic x-ray spectrum, finite detector pitch and other data imperfections, scatter with and without an antiscatter grid, and so on. We denote these two LDCT datasets as LD-LiTS and LD-KiTS in the following sections.

3.2 Segmentation Networks

In this paper, four different segmentation networks were used as our downstream models, including U-Net [17], V-Net [16], Res-U-Net [8] and Vox-ResNet [4]. Each segmentation network was pretrained independently on the LiTS and KiTS training sets. On the LiTS test set, the four segmentation networks in the above order achieved Dice of 94.31%, 92.80%, 93.51%, and 92.08%, respectively. On the KiTS test set, these four segmentation networks achieved Dice of 90.87%, 89.61%, 89.78%, and 89.35%, respectively. All the four networks did NOT achieved a SoTA Dice score (>95%), because our data split is different from the original challenge. We split the challenge training set into training, validation, and test sets. The segmentation models are trained with fewer data and evaluated on a different test set than the original challenge.

3.3 Implementation Details

For all the datasets, the pixel intensity is clamped with a window of [−200,200]HU and normalized to [0, 1]. In Sect. 3.1, to train a denoising network, we first resized the images to have an in-plane resolution of 1.5 mm × 1.5 mm and a slice spacing of 1 mm. Then, the axial slices are center cropped/zero padded to a size of 256 × 256 pixels. To generate training batches, the 3D LDCT images and segmentation ground truth are split into multiple overlapping 256 × 256 × 32 sub-volumes. In the validation and test phase, the TOD-Net works directly on the original volume with size of 512 × 512 × #slices.

When training our TOD-Net, only the above trained U-Net [17] was used as the representative downstream model in the task-oriented loss. Then, the trained TOD-Net was directly applied to all the four segmentation networks to verify the generalizability. The WGAN discriminator used in the training is a 3 layers 3D CNN followed by a fully connected layer with a single output. Each 3D convolutional layer is followed by a batch normalization layer and a leaky ReLU layer. The kernel size of all convolutional layers is 3 × 3 × 3. The channel sizes of the layers are 32, 64, and 128, respectively. Due to the limitation of GPU

Table 1. Image quality analysis of denoised LDCT images on LiTS and KiTS datasets.

Denoiser	LD-LiTS						LD-KiTS					
	ROIs			Whole Image			ROIs			Whole image		
	SSIM	RMSE	PSNR	SSIM	RMSE	PSNR	SSIM	RMSE	PSNR	SSIM	RMSE	PSNR
VGG-WGAN	29.5	53.0	17.7	72.3	32.9	22.3	46.7	58.3	16.9	67.4	37.8	21.1
MSE-WGAN	35.1	41.8	19.8	74.7	31.8	22.3	50.0	53.5	17.6	65.9	37.0	21.9
SA-WGAN	37.5	40.1	20.1	73.1	31.1	22.7	54.5	53.7	18.3	67.2	34.9	**22.0**
TOD-Net	**40.7**	**35.4**	**21.3**	**76.7**	**28.5**	**23.3**	**61.1**	**45.5**	**19.0**	69.3	**32.9**	**22.0**

memory, in the training phase of TOD-Net, we further cropped 3D LDCT image into $256 \times 256 \times 32$ patches with a sliding window. On LD-LiTS and LD-KiTS, the TOD-Net was separately trained by the RMSprop optimizer with a learning rate of 0.0005 and a batch size of 4 for 50 epochs. The hyperparameter λ in L_G was set to be 0.5 and the gradient clipping value ϵ in Eq. 1 was set to be 0.01. One checkpoint was saved at the end of each epoch and the checkpoint with the best performance on the validation set was used as the final model.

3.4   Enhancement on Task-Related Regions

We first quantitatively evaluated the denoising quality of the proposed TOD-Net. We computed SSIM, RMSE and PSNR on both whole image and regions of interest (ROIs) of the downstream task, i.e., liver for LiTS and kidney for KiTS. Results are shown in Table 1. Compared with the other two WGAN denoisers, TOD-Net ranked the top in all evaluation metrics at both whole image and ROI levels.

It is worth noting that image quality enhancement of TOD-Net on ROIs is even more significant than on whole images. This verifies that with information from the pretrained downstream model, the task-oriented loss locally promotes image quality of task-related regions. Combining it with the MSE loss for global regularization, TOD-Net achieved the best performance both locally and globally.

3.5   Boosting Downstream Task Performance

In this experiment, we cascaded the TOD-Net with each of the four segmentation networks to evaluate its influence on the downstream task. The results on LD-LiTS and LD-KiTS are shown in Table 2. Since LDCT images contains more noise and artifacts than NDCT images, directly applying segmentation models trained on NDCT to LDCT images would cause a significant performance degradation, as shown by the rows of "No denoiser" in Table 2.

In contrast, using the denoised images improved the segmentation performance. Measured by Dice scores, TOD-Net significantly outperformed all other denoising models ($p < 0.05$) not only on the U-Net, which was used in the training of TOD-Net, but also on all three other segmentation networks. Such a generalizability can be explained by the significant image quality enhancement on the task-related regions mentioned earlier in Sect. 3.4.

Table 2. Comparison of TOD-Net and other denoising models on the downstream task Dice score (%) and generalizability on **(top)** LD-LiTS and **(bottom)** LD-KiTS.

Denoiser	U-Net	Generalize to other Seg. models		
		Vox-ResNet	V-Net	Res-U-Net
No denoiser	88.75	79.82	89.75	90.36
VGG-WGAN	92.34 ($p = 0.006$)	90.44 ($p = 0.002$)	91.21 ($p = 0.004$)	91.82 ($p = 0.004$)
MSE-WGAN	92.45 ($p = 0.005$)	89.49 ($p < 0.001$)	91.76 ($p = 0.005$)	91.95 ($p = 0.005$)
SA-WGAN	92.99 ($p = 0.007$)	89.26 ($p < 0.001$)	91.37 ($p = 0.007$)	91.46 ($p = 0.005$)
TOD-Net	**93.91**	**91.86**	**92.44**	**92.77**
No denoiser	75.30	43.09	80.75	80.57
VGG-WGAN	89.43 ($p = 0.044$)	86.88 ($p < 1e - 3$)	88.06 ($p = 0.005$)	89.67 ($p = 0.004$)
MSE-WGAN	88.82 ($p = 0.004$)	86.58 ($p < 0.001$)	88.46 ($p = 0.004$)	88.66 ($p = 0.004$)
SA-WGAN	88.87 ($p = 0.005$)	86.39 ($p < 0.001$)	88.66 ($p = 0.006$)	88.75 ($p = 0.005$)
TOD-Net	**90.21**	**89.81**	**89.83**	**90.03**

Table 3. TOD-Net performance on real CT images on the downstream task Dice score (%) and generalizability on **(top)** LiTS and **(bottom)** KiTS.

Denoiser	U-Net	Generalize to other Seg. models		
		Vox-ResNet	V-Net	Res-U-Net
No denoiser	94.33	92.19	92.82	**93.51**
TOD-Net	**95.01** ($p = 0.04$)	**92.43** ($p = 0.06$)	**93.44** ($p = 0.04$)	93.21 ($p = 0.08$)
No denoiser	90.93	89.39	89.58	89.81
TOD-Net	**91.52** ($p = 0.03$)	**90.41** ($p = 0.01$)	**91.37** ($p = 0.01$)	**91.03** ($p = 0.03$)

Since real clinical datasets including paired LDCT and NDCT with segmentation annotations are hard to obtain, we evaluate the performance of the TOD-Net on real CT images by applying TOD-Net on the original LiTS and KiTS datasets. The results in Table 3 show that TOD-Net brings significant improvement for all segmentation networks on KiTS. On LiTS, all the downstream networks perform better except the performance of Res-U-Net decrease a little with non-significant difference ($p > 0.05$).

In addition to the quantitative analysis to demonstrate the high performance and great generalizability of TOD-Net, we further visualized the differences brought by the task-oriented loss. Figure 3 shows two example cases. It can be seen that, in columns (e) and (f), the errors of TOD-Net are significantly less than MSE-WGAN, especially on the ROIs, which leads to better segmentation results in Fig. 3(h) compared to Fig. 3(g).

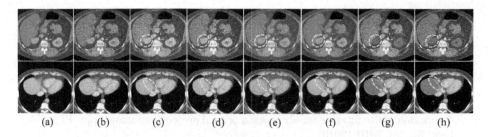

| (a) | (b) | (c) | (d) | (e) | (f) | (g) | (h) |

Fig. 3. Case study. Slices selected from **(top row)** KiTS and **(bottom row)** LiTS. From left to right, each column shows **(a)** NDCT, **(b)** LDCT, denoised image with **(c)** MSE-WGAN,**(d)** TOD-Net, **(e)** difference map between (a) and (c), **(f)** difference map between (a) and (d) (Red and green represent error above zero and below zero, respectively), **(g)** segmentation results of (c) with U-Net, **(h)** segmentation results of (d) with U-Net. The red, blue and green depict the true positive, false negative, and false positive regions, respectively. (Color figure online)

4   Conclusion

In conclusion, we proposed a TOD-Net with a task-oriented loss for LDCT image denoising. Quantitative evaluations and case studies show that introducing the knowledge of downstream applications into the training of the denoising model can significantly enhance the denoised image quality especially in the downstream task-related regions. Such a targeted improvement of image quality in turn boosts the performance of not only the downstream model used for training the denoising model but also various other independent models for the same task. The presented work may potentially enable a new group of image denoising methods as well as performance enhancement of existing tasks. In our future work, we will investigate the performance our task-oriented loss on other downstream tasks in medical image analysis.

References

1. Arjovsky, M., Chintala, S., Bottou, L.: Wasserstein generative adversarial networks. In: International Conference on Machine Learning, pp. 214–223. PMLR (2017)
2. Beister, M., Kolditz, D., Kalender, W.A.: Iterative reconstruction methods in x-ray CT. Physica Medica **28**(2), 94–108 (2012)
3. Bilic, P., Christ, P.F., Vorontsov, E., Chlebus, G., Chen, H., Dou, Q., et al.: The liver tumor segmentation benchmark (LiTS). arXiv preprint arXiv:1901.04056 (2019)
4. Chen, H., Dou, Q., Yu, L., Qin, J., Heng, P.A.: Voxresnet: deep voxelwise residual networks for brain segmentation from 3D MR images. NeuroImage **170**, 446–455 (2018)
5. Chen, H., et al.: Low-dose CT via convolutional neural network. Biomed. Opt. Express **8**(2), 679–694 (2017)
6. Chen, Y., et al.: Improving abdomen tumor low-dose CT images using a fast dictionary learning based processing. Phys. Med. Biol. **58**(16), 5803 (2013)

7. De Man, B., et al.: CatSim: a new computer assisted tomography simulation environment. In: Medical Imaging 2007: Physics of Medical Imaging, vol. 6510, p. 65102G. International Society for Optics and Photonics (2007)

8. Diakogiannis, F.I., Waldner, F., Caccetta, P., Wu, C.: ResUNet-a: a deep learning framework for semantic segmentation of remotely sensed data. ISPRS J. Photogrammetry Remote Sens. **162**, 94–114 (2020)

9. Feruglio, P.F., Vinegoni, C., Gros, J., Sbarbati, A., Weissleder, R.: Block matching 3D random noise filtering for absorption optical projection tomography. Phys. Med. Biol. **55**(18), 5401 (2010)

10. Hara, A.K., Paden, R.G., Silva, A.C., Kujak, J.L., Lawder, H.J., Pavlicek, W.: Iterative reconstruction technique for reducing body radiation dose at CT: feasibility study. Am. J. Roentgenol. **193**(3), 764–771 (2009)

11. Heller, N., Sathianathen, N., Kalapara, A., Walczak, E., Moore, K., Kaluzniak, H., et al.: The KiTS19 challenge data: 300 kidney tumor cases with clinical context, CT semantic segmentations, and surgical outcomes. arXiv preprint arXiv:1904.00445 (2019)

12. Kang, D., et al.: Image denoising of low-radiation dose coronary CT angiography by an adaptive block-matching 3D algorithm. In: Medical Imaging 2013: Image Processing, vol. 8669, p. 86692G. International Society for Optics and Photonics (2013)

13. Li, M., Hsu, W., Xie, X., Cong, J., Gao, W.: SaCNN: self-attention convolutional neural network for low-dose CT denoising with self-supervised perceptual loss network. IEEE Trans. Med. Imag. **39**(7), 2289–2301 (2020)

14. Ma, J., et al.: Low-dose computed tomography image restoration using previous normal-dose scan. Med. Phys. **38**(10), 5713–5731 (2011)

15. Manduca, A., et al.: Projection space denoising with bilateral filtering and CT noise modeling for dose reduction in CT. Med. Phys. **36**(11), 4911–4919 (2009)

16. Milletari, F., Navab, N., Ahmadi, S.A.: V-net: Fully convolutional neural networks for volumetric medical image segmentation. In: 2016 Fourth International Conference on 3D Vision (3DV), pp. 565–571. IEEE (2016)

17. Ronneberger, O., Fischer, P., Brox, T.: U-Net: convolutional networks for biomedical image segmentation. In: Navab, N., Hornegger, J., Wells, W.M., Frangi, A.F. (eds.) MICCAI 2015. LNCS, vol. 9351, pp. 234–241. Springer, Cham (2015). https://doi.org/10.1007/978-3-319-24574-4_28

18. Wang, J., Lu, H., Li, T., Liang, Z.: Sinogram noise reduction for low-dose CT by statistics-based nonlinear filters. In: Medical Imaging 2005: Image Processing, vol. 5747, pp. 2058–2066. International Society for Optics and Photonics (2005)

19. Wang, R., et al.: Image quality and radiation dose of low dose coronary CT angiography in obese patients: sinogram affirmed iterative reconstruction versus filtered back projection. Eur. J. Radiol. **81**(11), 3141–3145 (2012)

20. Yang, Q., et al.: Low-dose CT image denoising using a generative adversarial network with Wasserstein distance and perceptual loss. IEEE Trans. Med. Imag. **37**(6), 1348–1357 (2018)

Revisiting Contour-Driven and Knowledge-Based Deformable Models: Application to 2D-3D Proximal Femur Reconstruction from X-ray Images

Christophe Chênes[1,2](✉) ⓘ and Jérôme Schmid[1] ⓘ

[1] Department of Medical Radiology Technology, Geneva School of Health Sciences, HES-SO University of Applied Sciences and Arts of Western Switzerland, Delémont, Switzerland
{christophe.chenes,jerome.schmid}@hesge.ch
[2] Medical Image Processing Lab, Institute of Bioengineering, School of Engineering, Ecole Polytechnique Fédérale de Lausanne, Lausanne, Switzerland

Abstract. In many clinical applications, 3D reconstruction of patient-specific structures is of major interest. Despite great effort put in 2D-3D reconstruction, gold standard bone reconstruction obtained by segmentation on CT images is still mostly used – at the expense of exposing patients to significant ionizing radiation and increased health costs. State-of-the-art 2D-3D reconstruction methods are based on non-rigid registration of digitally reconstructed radiographs (DRR) – aiming at full automation – but with varying accuracy often exceeding clinical requirements. Conversely, contour-based approaches can lead to accurate results but strongly depend on the quality of extracted contours and have been left aside in recent years. In this study, we revisit a patient-specific 2D-3D reconstruction method for the proximal femur based on contours, image cues, and knowledge-based deformable models. 3D statistical shape models were built using 199 CT scans from THA patients that were used to generate pairs of high fidelity DRRs. Convolutional neural networks were trained using the DRRs to investigate automatic contouring. Experiments were conducted on the DRRs, and calibrated radiographs of a pelvis phantom and volunteers – with an analysis of the quality of contouring and its automatization. Using manual contours and DRR, the best reconstruction error was 1.02 mm. With state-of-the-art results for 2D-3D reconstruction of the proximal femur, we highlighted the relevance and challenges of using contour-driven reconstruction to yield patient-specific models.

Keywords: 2D-3D X-ray reconstruction · Shape priors · Deformable model

1 Introduction and Related Work

3D reconstruction of patient-specific bone anatomy is of growing interest in many clinical applications, such as surgery planning, personalized implant design, and postoperative analysis [1, 2]. Focusing on total hip arthroplasty (THA), 3D reconstruction of

© Springer Nature Switzerland AG 2021
M. de Bruijne et al. (Eds.): MICCAI 2021, LNCS 12906, pp. 451–460, 2021.
https://doi.org/10.1007/978-3-030-87231-1_44

the proximal femur enables better surgical planning and design of patient-specific surgical instruments – ultimately leading to more positive outcome for the patient. Today, gold standard reconstruction of bones is mainly obtained by segmentation on computed tomography (CT) images [3] for planification purpose. However, this comes at the expense of exposing patients to significant ionizing radiation and increased health costs [1]. Alternatively, 2D-3D reconstruction offers a cost-efficient mean of obtaining patient 3D models from a few calibrated 2D X-ray images, while reducing ionizing dose.

2D-3D reconstruction tries to solve the challenging problem of recovering 3D information from partial 2D information. In orthopedic surgery, the shape and pose of 3D bony structures need to be recovered from single or multiple X-ray images [2]. If the 3D shape is known, 2D-3D reconstruction is equivalent to a rigid registration problem solved by optimization techniques [4]. Otherwise, shape and pose are estimated by non-rigid registration techniques often coupled with statistical models of shape and intensity appearance of the structures [5]. Most of these methods assume that X-ray imaging is calibrated[1] or implement a calibration procedure. Though it is the case in our study, we will not cover the calibration process – being out of the scope of this publication.

Non-rigid 2D-3D reconstruction approaches encode prior information about the 3D shape and appearance of the targeted structure in deformable models. These knowledge-based models can be manipulated to best fit the patient's target structure in the 2D images. Most approaches use a reference template of the structure features, whose variations are modeled by analytical [2, 6] but mostly statistical [5, 7–15] representations. To obtain patient-specific 3D reconstructions, deformable models are transformed using information inferred from X-ray images using two possible strategies: (a) correspondence of features or (b) simulated X-ray images. (a) tries to minimize the distance between features (e.g., contours) extracted on both the deformable model and the X-ray images – making the strategy particularly sensitive to the quality of extracted features [16]. (b) generates DRR from the deformable model and optimizes the image similarity between the DRR and X-ray images [5]. This strategy bypasses explicit feature extraction but is computationally time-consuming and usually requires close initialization.

Historically, 2D-3D reconstruction methods only took shape into account, using statistical shape models (SSMs). Since then, models including intensity or appearance – referred to as statistical shape and intensity models (SSIMs) – have been studied to include bone density. Lately, deep learning (DL) algorithms have also been applied to 2D-3D reconstruction of bones [17, 18].

In terms of reconstruction accuracy, state-of-the-art SSIMs methods yield reconstruction errors ranging from 1.18 to 4.24 mm [5]. A bone reconstruction error of 1 to 1.5 mm with successful registration in 95% of the time has been suggested as a good performance indicator [5, 19]. However, the maximum tolerated error depends on the target application, as e.g., this suggested maximum error can be excessive to design patient-specific implants and surgical instruments in case of THA. Lastly, comparison of bone reconstruction errors on different bony areas should be considered with care – the degree of superimposition of confounding structures in X-ray images being different (e.g., more challenging for the hip compared to the knee).

---

[1] Calibration parameters include e.g., source-detector distance, pixel size, central point, relative transformations between multiple images, etc.

These observations motivated us to revisit the development of a patient-specific 2D-3D reconstruction method for the proximal femur based on a novel combination of contours, image cues, and knowledge-based deformable models, with a focus on the detailed analysis of the quality of the contours, their impact on the reconstruction and their possible automation. Another contribution also lies in the preparation and evaluation of a high-quality synthetic dataset derived from a large number of real pathological cases of the hip joints prior to THA surgeries.

2 Method and Material

In the following, we present our methodology with the processed datasets to train and evaluate our approach.

Statistical Shape Models. We collected 199 anonymized CT images of the hips from THA patients (aged 16 to 94, median 71 years) including gold-standard bone reconstructions performed semi-automatically by expert radiographers – resulting in 115 right and 84 left proximal femurs. Using R3DS Wrap software, we fit, for left or right, a same triangular mesh (left: 4585 vertices, right: 4626 vertices) of the proximal femur to the gold-standard reconstructions – yielding point correspondence across wrapped shapes. Using Generalized Procrustes similarity alignment, all shapes were co-aligned, and a subsampled version of each aligned shape was produced (1153 and 1156 vertices for left and right sides). Using Principal Components Analysis, left and right multiresolution SSMs were finally built using the fine and coarse resolutions [20]. In the following, the mean shapes of the SSMs will be denoted as the template shapes.

Contour Definition. Image contours can be manually delineated, estimated by an automatic approach, or derived from a ground-truth. In any case, we consider contours as a collection of 2D pixel points. However, to inject knowledge and robustify the segmentation process, we also produce *mapped* contours. The idea is to associate groups of 3D vertices M_j of the template shape to specific subsets of contour points m_j – resulting in a 2D-3D mapping. For the proximal femur, we defined 6 groups corresponding to the femoral head and neck, the greater and lesser trochanters, and the intertrochanteric line and crest. To automatically derive contours, we used a state-of-the-art DL architecture, the U-Net++ model [21] that improves the established U-Net approach and that we coupled with a RegNetY encoder [22]. Contours were extracted as the boundaries of inferred segmentations.

Knowledge-Based Deformable Model. Model reconstruction aims at deforming the template shapes using the calibrated X-ray images to recover the unknown femoral shape and pose with respect to the imaging system. Our approach follows the Newtonian dynamics-driven model deformation described in [20], in which a shape is deformed over time under the effect of forces applied to its vertices. Internal forces enforce model smoothness, local compliance to the initial shape ("shape memory") and similarity to the closest shape generated from the corresponding SSM.

Image forces ensure meaningful deformation using image gradient and contours. Based on the calibrated geometry information, we can identify which shape vertex P_i

contributes to salient image edges, known as the silhouette [23], and project them on the X-ray images as image positions p_i. Then, a sampling strategy in a neighborhood centered on p_i and directed by the projected mesh normal at P_i is used to probe image candidate positions r_i that maximize, gradient direction and magnitude [20] for the image force, and the proximity to contours for the contour force. If the contours are mapped, we only consider candidate positions mapped with the group of the corresponding 3D vertex P_i – preventing the incorrect attraction to confounding edges not easily distinguishable by gradient information. Image candidates in X-ray image I_k for gradient r_i^g and contour r_i^c forces are then back-projected as 3D lines $l_i^g = (O_k, r_i^g)$ and $l_i^c = (O_k, r_i^g)$, where O_k denotes the X-ray source position. Finally, gradient and contour forces are defined as spring forces attracting the vertex P_i to the closest points on lines l_i^g and l_i^c [20] with a parameter α weighting the force intensity.

3D landmarks are computed as the closest points to back-projected lines of 2D landmarks manually clicked in the images. By defining once the same landmarks on the template shapes, we can initialize the pose and global scale of the coarse shape in the calibrated geometry coordinate system by similarity alignment of corresponding 3D landmarks. In our case, landmarks were placed on the greater and lesser trochanters, and the center and fovea capitis of the femoral head.

Synthetic X-ray Dataset. Since CT images were acquired and used for patient-specific THA planning, X-ray images were not included in the dataset. We thus generated for each patient a pair of high fidelity DRRs from their CT scan – using a DL approach able to model polychromatic X-ray spectra as well as stochastic noise and X-ray scattering [24]. We extended the approach to better model detector response and the effect of anti-scatter grids. All acquisition parameters were controlled, leading to a perfect calibration for the 398 synthetic images produced. AP and lateral (LAT) views at a relative $75°$ angle were computed with a focal length of 120 cm, a resolution of 3000×3000 px and a square pixel spacing of 0.15 mm. The simulated X-ray beam characteristics were: 2.5 mm Al filtration, 80 kV and 25 mAs (AP), and 90 kV and 40 mAs (LAT). An anti-scatter grid with a 16:1 ratio was modelled. Though DRRs are not perfect X-ray images, the adopted DL method produces more realistic results than simple ray-casting approaches. An example of data for one patient is shown in Fig. 1a&b. By projecting the silhouette of the reconstructed models onto the DRRs, high quality 2D contours were created for both views. Since we wrapped a template model for building the SSM, we could identify by closest distance the mapped regions on reconstructed models – leading to projected mapped contours. Initialization landmarks were also available in 3D segmented models and were projected similarly.

Real X-ray Dataset. Real X-ray images were acquired for a pelvis phantom and three volunteers (3 healthy males, aged 32, 37 and 44). After duly filled informed consent, a pair of calibrated X-ray images were shot for each volunteer in AP and LAT views ($75°$ angle) in a clinical setup – using a designed protocol aiming at not disrupting the clinical routine, while including a calibration device, and a rotative platform for ensuring angle accuracy and minimal leg movement between the shots. Acquisition parameters were set to 120 cm, 70 kV (AP) and 75 kV (LAT) with automatic exposure control. Femoral bone models were derived from previous MRI acquisitions.

For the pelvis phantom – containing human bones – a CT scan and manual reconstruction of the bones were produced. Three AP/LAT acquisitions were performed in different setups with varying calibration device position and parameters settings. An example of AP X-ray image for the phantom is shown in Fig. 1c.

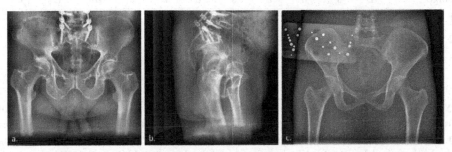

Fig. 1. Example of synthetic X-ray in AP (a.) and LAT (b.) views, and of real X-ray in AP view for the pelvis phantom (c.). The calibration device composed of metallic spheres can be spotted.

3   Results and Discussion

Experiments for 2D-3D reconstruction of the proximal femur were conducted on both synthetic (DRRs) and real X-ray datasets. To thoroughly assess our method accuracy with different contouring strategies, we compared the use of reference contours vs automatically estimated contours, in presence or absence of mapped contours.

Evaluation Metrics and Implementation Details. For the reconstruction metric, average absolute surface distance (ASD) was used and statistical significance between different strategy results were assessed using the two-sided Wilcoxon signed-rank test with a significance level of 0.01 – results not being normally distributed. Dice similarity coefficient (DSC) was used to assess the 2D image segmentation accuracy of the DL approach. In synthetic experiments, we used a leave-one-out strategy to train the SSMs, while a 10-fold cross-validation was chosen to evaluate the DL segmentation. An automatic image augmentation [25] was applied to enlarge the training dataset, and a 20–80% validation-training split was used for each fold.

Following [20], we empirically set the force parameters based on a fine-tuning with 3 randomly chosen DRR cases. Deformation took place during 3 stages with 500, 400 and 300 iterations respectively, during which the coarse shape resolution was used at the first stage, and the fine resolution for the remaining stages. Internal forces were weighted at each stage according to a specific schedule [20]: shape memory force $\alpha = 0.5; 0.1; 0.3$, SSM force $\alpha = 0.1; 0.4; 0.1$ and smoothing force $\alpha = 0.1$. For image forces, parameters were dependent on the use or not of DL-derived contours (Table 1).

Tests were performed on a computer running under Windows 10 with an AMD Ryzen 5 3600 6-Core 3.6 GHz processor and 32 GB RAM. The deformable model was implemented in C++ while the U-Net++ approach used Python 3.9 and PyTorch 1.7.1.

Table 1. Parameters of image forces in experiments, s being in mm [20].

| Stage | Without DL-based contours | | | | | | With DL-based contours | | | | | |
| | Gradient force | | | Contour force | | | Gradient force | | | Contour force | | |
	α	L	s	α	L	s	α	L	s	α	L	s
1	–	–	–	0.1	201	0.2	–	–	–	0.1	201	0.2
2	–	–	–	0.2	141	0.2	–	–	–	0.2	141	0.2
3	0.1	21	0.1	0.2	141	0.2	0.2	21	0.1	0.05	141	0.2

Synthetic Dataset Results. On the large DRR dataset our 2D-3D reconstruction method was tested in optimal (perfect calibration and reference contours, as exemplified in Fig. 2a. in green) and automatic (DL method for extraction of contours) conditions. The higher complexity of LAT view prevented for this view the use of the DL approach. Hence, for experiments on automatic contours, AP contours were extracted with DL (e.g., Fig. 2a. in red) while reference LAT contours obtained by projection were used. In this case, the contours were not mapped.

Table 2. Mean ASD errors with standard deviations for the different contouring strategies tested on the synthetic dataset.

| Reference contours | | Automatic contours | |
Not mapped	Mapped	Not mapped	Not mapped, gradient boost
1.16 ± 1.01 mm	1.02 ± 0.89 mm	1.98 ± 1.89 mm	1.68 ± 1.57 mm

Table 2 presents the results of our method applied on the synthetic dataset with reference and automatic contours. We obtained the best 3D reconstructions (ASD 1.02 $\pm$ 0.89 mm) when using the mapped reference contours (e.g., Fig. 2b.&d.) – the most complete approach. The use of mapped contours improved reconstruction accuracy (p-value $< 1e^{-12}$). Despite a satisfactory 2D DSC error of 0.957 (CI at 95%:[0.95,0.965]), automatic contours were less accurate and led to greater 3D ASD error, but still within the range of results of published works. To counterbalance the quality of the contours, our approach also takes advantage of image gradients. In the case of less accurate contours - as here for automatic contours - the gradient force should make it possible to catch up to a certain precision. To verify this, we repeated an experiment using the automatic contours with a "boosted" gradient by adding a gradient force at stage 2 ($\alpha = 0.1, L = 31, s = 0.5$) and by increasing the gradient force coverage at stage 3 ($s = 0.2$). These results (ASD 1.68 $\pm$ 1.57 mm vs 1.98 $\pm$ 1.89 mm, p-value $< 1e^{-11}$) highlight the relevance of using image gradients to improve the robustness of the overall approach (cf., Fig. 2c.&e.). From the clinical THA point of view, the highest reconstruction errors were located on the femoral head, which is of a lesser interest since it will be resected. The pathological "quality" of a random sample of 58 femurs (e.g., presence of several or severe osteophytes) was visually checked by 3 radiographers and classified into 3 balanced categories: "good" (19), "fair" (15) and "poor" (24). Our manual (mean ASD of 0.98, 1.03 and 0.96) and automatic (DSC of 0.96, 0.94 and 0.97) approaches appear to be robust to pathological variations.

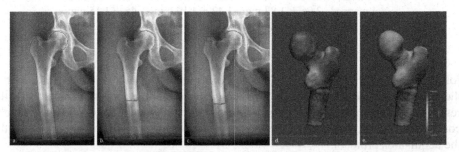

Fig. 2. Example of (a.) reference (in green) and DL-based (in red) input contours and the projection of their 3D reconstruction result silhouette in AP for (b.) reference contours and (c.) automatic contours. 3D models with mean ASD error for 115 right femurs are shown for (d.) mapped reference contours (ASD 1.06 ± 0.97 mm) and (e.) automatic contours (ASD 1.63 ± 1.53 mm). (Color figure online)

Real X-ray Dataset Results. Our method has also been tested on six pairs of real X-ray images. First on the phantom data, then on the less controlled volunteers' data.

For the pelvis phantom, acquisitions were repeated three times with some variations of the attached calibration device (visible in Fig. 1c). Contours and landmarks were drawn manually. Results on the pelvis phantom for the different setups and with mapped and unmapped contours strategies are reported in Table 3. The mapping improved the results for setups 1 and 3 (p-value $< 1e\text{-}^{16}$), while the difference was not statistically significant for setup 2 (p-value $= 0.013$).

Table 3. Mean ASD errors with standard deviations for the different setups and with and without mapped contours for the pelvis phantom real X-ray dataset

	Setup 1	Setup 2	Setup 3
Not mapped	1.85 ± 1.65 mm	1.2 ± 1.03 mm	1.60 ± 1.39 mm
Mapped	1.13 ± 1.18 mm	1.21 ± 1.21 mm	1.21 ± 1.23 mm

As for the phantom, the contours and landmarks were manually annotated for volunteers' data, and the impact of the contour mapping was analyzed. The results for mapped and not mapped contour strategies are displayed in Table 4 for all three volunteers. Similarly, the mapping improved the results for volunteers 1 and 3 (p-value $< 1e\text{-}^{15}$), but for volunteer 2 the difference was not significant (p-value $= 0.013$).

We noticed that acquisitions on volunteers were made by radiographers not sufficiently informed on the procedure – resulting in X-ray images acquired without strictly

Table 4. Mean ASD errors with standard deviations for the different contouring strategies tested on the volunteers' real X-ray dataset.

	Volunteer 1	Volunteer 2	Volunteer 3
Not mapped	2.06 ± 1.87 mm	1.42 ± 1.19 mm	1.45 ± 1.15 mm
Mapped	1.84 ± 1.99 mm	1.39 ± 1.19 mm	1.29 ± 1.19 mm

following the provided guidelines, mainly concerning the importance of not moving the subject between the two shots. This may explain the inferior results obtained with respect to the phantom. Indeed, phantom bones are fully static between shots, which convinces us of the potential of the method on real patients by better complying with the acquisition protocol.

4   Conclusion

Under optimal calibration conditions and with precise contours, our method has demonstrated a reconstruction accuracy (ASD 1.02 ± 0.89 mm) comparable or even superior to state-of-the-art methods. Methods reporting accurate results were either tested on synthetic results [9] or dry bones [7, 9, 12, 26], in perfect calibration setup [7, 9], or for the anatomically simpler distal femur [8, 15]. Yu et al. 2016 [9] reported an ASD error of 0.9 mm on synthetic proximal femurs using manual contouring on two fluoroscopy images, but for a database of 40 patients. When switching to real X-ray, in the case of simpler dry femur bone acquired with a C-arm greatly easing the calibration, their error increased to 1.2 mm. Our experiments carried out on a large database of pathological cases support the robustness of the method and its ability to successfully reconstruct complex cases.

To achieve accurate results on clinically interesting cases – in opposition to dry bones for example – the manual segmentation of a pair of 2D mapped contours do not seem a price too high to pay. With an appropriate ergonomic tool, such a task would be less tedious and could replace semi-automatic segmentation on CT scans used today for patient-specific THA planning – benefiting the patient's health and the health costs. However, aware of the importance of automatic tools in the clinical context, we have investigated a DL method for the automatic extraction of AP contours in DRRs with encouraging results (1.68 ± 1.57 mm). The extensive contour analysis has proven their usefulness in achieving accurate results, and their automation has been demonstrated on DRRs with results, albeit not as good, but already usable. The reconstruction is also fast – with a runtime below 30 s (including 1.8 s for DL segmentation) – and produces ready-to-use 3D models (e.g., anatomical markers for surgical planning can be predefined on the template models and be immediately available on the reconstructed structures). We particularly envision future work in the automatic and accurate extraction of mapped contours and landmarks from real radiographs albeit the collection of lateral views, less common in clinical practice, may be an interesting yet challenging problem in the context of data hungry algorithms such as deep learning approaches.

Acknowledgement. This work was supported by the Swiss CTI project MyPlanner (no. 25258.1). Authors would like to thank E. Ambrosetti, A. Al-Musibli, L. Assassi and B. Lokaj at the Geneva School of Health Sciences and M. Bernardoni, M. Parravicini and D. Ascani at Medacta International SA.

References

1. Sarkalkan, N., Weinans, H., Zadpoor, A.A.: Statistical shape and appearance models of bones. Bone **60**, 129–140 (2014). https://doi.org/10.1016/j.bone.2013.12.006
2. Gamage, P., Xie, S.Q., Delmas, P., Xu, W.L.: Diagnostic radiograph based 3D bone reconstruction framework: application to the femur. Comput. Med. Imaging Graph. **35**(6), 427–437 (2011). https://doi.org/10.1016/j.compmedimag.2010.09.008
3. Galibarov, P.E., Prendergast, P.J., Lennon, A.B.: A method to reconstruct patient-specific proximal femur surface models from planar pre-operative radiographs. Med. Eng. Phys. **32**(10), 1180–1188 (2010). https://doi.org/10.1016/j.medengphy.2010.08.009
4. Markelj, P., Tomaževič, D., Likar, B., Pernuš, F.: A review of 3D/2D registration methods for image-guided interventions. Med. Image Anal. **16**(3), 642–661 (2012). https://doi.org/10.1016/j.media.2010.03.005
5. Reyneke, C.J.F., Lüthi, M., Burdin, V., Douglas, T.S., Vetter, T., Mutsvangwa, T.E.M.: Review of 2-D/3-D reconstruction using statistical shape and intensity models and X-ray image synthesis: toward a unified framework. IEEE Rev. Biomed. Eng. **12**, 269–286 (2019). https://doi.org/10.1109/RBME.2018.2876450
6. Karade, V., Ravi, B.: 3D femur model reconstruction from biplane X-ray images: a novel method based on Laplacian surface deformation. Int. J. Comput. Assist. Radiol. Surg. **10**(4), 473–485 (2014). https://doi.org/10.1007/s11548-014-1097-6
7. Zheng, G., Yu, W.: Chapter 12 - Statistical shape and deformation models based 2D–3D reconstruction. In: Zheng, G., Li, S., Székely, G. (eds.) Statistical Shape and Deformation Analysis, pp. 329–349. Academic Press (2017)
8. Klima, O., Kleparnik, P., Spanel, M., Zemcik, P.: Intensity-based femoral atlas 2D/3D registration using Levenberg-Marquardt optimization. In: Medical Imaging 2016: Biomedical Applications in Molecular Structural, and Functional Imaging, vol. 9788, p. 97880F (2016)https://doi.org/10.1117/12.2216529
9. Yu, W., Chu, C., Tannast, M., Zheng, G.: Fully automatic reconstruction of personalized 3D volumes of the proximal femur from 2D X-ray images. Int. J. Comput. Assist. Radiol. Surg. **11**(9), 1673–1685 (2016). https://doi.org/10.1007/s11548-016-1400-9
10. Baka, N., et al.: 2D–3D shape reconstruction of the distal femur from stereo X-ray imaging using statistical shape models. Med. Image Anal. **15**(6), 840–850 (2011). https://doi.org/10.1016/j.media.2011.04.001
11. Zheng, G.: Personalized X-Ray reconstruction of the proximal femur via intensity-based non-rigid 2D-3D registration. In: Fichtinger, G., Martel, A., Peters, T. (eds.) MICCAI 2011. LNCS, vol. 6892, pp. 598–606. Springer, Heidelberg (2011). https://doi.org/10.1007/978-3-642-23629-7_73
12. Boussaid, H., Kadoury, S., Kokkinos, I., Lazennec, J.-Y., Zheng, G., Paragios, N.: 3D model-based reconstruction of the proximal femur from low-dose Biplanar X-Ray images. In: The 22nd British Machine Vision Conference - BMVC 2011, Dundee, United Kingdom, pp. 1–10, August 2011. https://doi.org/10.5244/C.25.35
13. Cerveri, P., Sacco, C., Olgiati, G., Manzotti, A., Baroni, G.: 2D/3D reconstruction of the distal femur using statistical shape models addressing personalized surgical instruments in knee arthroplasty: A feasibility analysis. Int. J. Med. Robot. Comput. Assisted Surgery **13**(4), e1823 (2017). https://doi.org/10.1002/rcs.1823

14. Youn, K., Park, M.S., Lee, J.: Iterative approach for 3D reconstruction of the femur from un-calibrated 2D radiographic images. Med. Eng. Phys. **50**, 89–95 (2017). https://doi.org/10.1016/j.medengphy.2017.08.016

15. Wu, J., Mahfouz, M.R.: Reconstruction of knee anatomy from single-plane fluoroscopic x-ray based on a nonlinear statistical shape model. JMI **8**(1), 016001 (2021). https://doi.org/10.1117/1.JMI.8.1.016001

16. Mahfouz, M.R., Hoff, W.A., Komistek, R.D., Dennis, D.A.: Effect of segmentation errors on 3D-to-2D registration of implant models in X-ray images. J. Biomech. **38**(2), 229–239 (2005). https://doi.org/10.1016/j.jbiomech.2004.02.025

17. Grupp, R.B., et al.: Automatic annotation of hip anatomy in fluoroscopy for robust and efficient 2D/3D registration. Int. J. Comput. Assist. Radiol. Surg. **15**(5), 759–769 (2020). https://doi.org/10.1007/s11548-020-02162-7

18. Kasten, Y., Doktofsky, D., Kovler, I.: End-To-end convolutional neural network for 3D reconstruction of knee bones from bi-planar X-ray images. In: Deeba, F., Johnson, P., Würfl, T., Ye, J.C. (eds.) MLMIR 2020. LNCS, vol. 12450, pp. 123–133. Springer, Cham (2020). https://doi.org/10.1007/978-3-030-61598-7_12

19. Livyatan, H., Yaniv, Z., Joskowicz, L.: Gradient-based 2-D/3-D rigid registration of fluoroscopic X-ray to CT. IEEE Trans. Med. Imaging **22**(11), 1395–1406 (2003). https://doi.org/10.1109/TMI.2003.819288

20. Damopoulos, D., et al.: Segmentation of the proximal femur in radial MR scans using a random forest classifier and deformable model registration. Int. J. Comput. Assist. Radiol. Surg. **14**(3), 545–561 (2019). https://doi.org/10.1007/s11548-018-1899-z

21. Zhou, Z., Siddiquee, M.M.R., Tajbakhsh, N., Liang, J.: UNet++: redesigning skip connections to exploit multiscale features in image segmentation. IEEE Trans. Med. Imaging **39**(6), 1856–1867 (2020). https://doi.org/10.1109/TMI.2019.2959609

22. Radosavovic, I., Kosaraju, R.P., Girshick, R., He, K., Dollar, P.: Designing network design spaces. In: 2020 IEEE/CVF Conference on Computer Vision and Pattern Recognition (CVPR), Seattle, WA, USA, pp. 10425–10433, June. 2020. https://doi.org/10.1109/CVPR42600.2020.01044

23. Benameur, S., Mignotte, M., Parent, S., Labelle, H., Skalli, W., de Guise, J.: 3D/2D registration and segmentation of scoliotic vertebrae using statistical models. Comput. Med. Imaging Graph. **27**(5), 321–337 (2003). https://doi.org/10.1016/S0895-6111(03)00019-3

24. Unberath, M., et al.: DeepDRR – a catalyst for machine learning in fluoroscopy-guided procedures. In: Frangi, A.F., Schnabel, J.A., Davatzikos, C., Alberola-López, C., Fichtinger, G. (eds.) MICCAI 2018. LNCS, vol. 11073, pp. 98–106. Springer, Cham (2018). https://doi.org/10.1007/978-3-030-00937-3_12

25. Hataya, R., Zdenek, J., Yoshizoe, K., Nakayama, H.: Faster AutoAugment: learning augmentation strategies using Backpropagation. In: Vedaldi, A., Bischof, H., Brox, T., Frahm, J.-M. (eds.) ECCV 2020. LNCS, vol. 12370, pp. 1–16. Springer, Cham (2020). https://doi.org/10.1007/978-3-030-58595-2_1

26. Kurazume, R., et al.: 3D reconstruction of a femoral shape using a parametric model and two 2D fluoroscopic images. Comput. Vis. Image Underst. **113**(2), 202–211 (2009). https://doi.org/10.1016/j.cviu.2008.08.012

Memory-Efficient Learning for High-Dimensional MRI Reconstruction

Ke Wang[1,6(✉)], Michael Kellman[2], Christopher M. Sandino[3], Kevin Zhang[1], Shreyas S. Vasanawala[4], Jonathan I. Tamir[5], Stella X. Yu[1,6], and Michael Lustig[1]

[1] Electrical Engineering and Computer Sciences, UC Berkeley, Berkeley, CA, USA
kewang@berkeley.edu
[2] Pharmaceutical Chemistry, UCSF, San Francisco, CA, USA
[3] Electrical Engineering, Stanford University, Stanford, CA, USA
[4] Radiology, Stanford University, Stanford, CA, USA
[5] Electrical and Computer Engineering, UT Austin, Austin, TX, USA
[6] International Computer Science Institute, UC Berkeley, Berkeley, CA, USA

Abstract. Deep learning (DL) based unrolled reconstructions have shown state-of-the-art performance for under-sampled magnetic resonance imaging (MRI). Similar to compressed sensing, DL can leverage high-dimensional data (e.g. 3D, 2D+time, 3D+time) to further improve performance. However, network size and depth are currently limited by the GPU memory required for backpropagation. Here we use a memory-efficient learning (MEL) framework which favorably trades off storage with a manageable increase in computation during training. Using MEL with multi-dimensional data, we demonstrate improved image reconstruction performance for in-vivo 3D MRI and 2D+time cardiac cine MRI. MEL uses far less GPU memory while marginally increasing the training time, which enables new applications of DL to high-dimensional MRI. Our code is available at https://github.com/mikgroup/MEL_MRI.

Keywords: Magnetic resonance imaging (MRI) · Unrolled reconstruction · Memory-efficient learning

1 Introduction

Deep learning-based unrolled reconstructions (Unrolled DL recons) [1,3,6,12,20, 22] have shown great success at under-sampled MRI reconstruction, well beyond the capabilities of parallel imaging and compressed sensing (PICS) [5,13,17]. These methods are often formulated by unrolling the iterations of an image reconstruction optimization [1,6,22] and use a training set to learn an implicit regularization term represented by a deep neural network. It has been shown that

---

Electronic supplementary material The online version of this chapter (https://doi.org/10.1007/978-3-030-87231-1_45) contains supplementary material, which is available to authorized users.

M. de Bruijne et al. (Eds.): MICCAI 2021, LNCS 12906, pp. 461–470, 2021.
https://doi.org/10.1007/978-3-030-87231-1_45

increasing the number of unrolls improves upon finer spatial and temporal textures in the reconstruction [1,6,18]. Similar to compressed sensing and other low-dimensional representations, DL recons can take advantage of additional structure in very high-dimensional data (e.g. 3D, 2D+time, 3D+time) to further improve image quality. However, these large-scale DL recons are currently limited by GPU memory required for gradient-based optimization using backpropagation. Therefore, most Unrolled DL recons focus on 2D applications or are limited to a small number of unrolls. In this work, we use our recently proposed memory-efficient learning (MEL) framework [10,24] to reduce the memory needed for backpropagation, which enables the training of Unrolled DL recons for 1) larger-scale 3D MRI; and 2) 2D+time cardiac cine MRI with a large number of unrolls (Fig. 1). We evaluate the spatio-temporal complexity of our proposed method on the Model-based Deep Learning (MoDL) architecture [1] and train these high-dimensional DL recons on a single 12 GB GPU. Our training uses far less memory while only marginally increasing the computation time. To demonstrate the advantages of high-dimensional reconstructions to image quality, we performed experiments on both retrospectively and prospectively under-sampled data for 3D MRI and cardiac cine MRI. Our in-vivo experiments indicate that by exploiting high-dimensional data redundancy, we can achieve better quantitative metrics and improved image quality with sharper edges for both 3D MRI and cardiac cine MRI.

2   Methods

2.1   Memory-Efficient Learning

As shown in Fig. 2 a), unrolled DL recons are often formulated by unrolling the iterations of an image reconstruction optimization [1,6]. Each unroll consists of two submodules: CNN based regularization layer and data consistency (DC) layer. In conventional backpropagation, the gradient must be computed for the entire computational graph, and intermediate variables from all N unrolls need to be stored at a significant memory cost. By leveraging MEL, we can process the full graph as a series of smaller sequential graphs. As shown in Fig. 2 b), first, we forward propagate the network to get the output $\mathbf{x}^{(N)}$ without computing the gradients. Then, we rely on the invertibility of each layer (required) to recompute each smaller auto-differentiation (AD) graph from the network's output in reverse order. MEL only requires a single layer to be stored in memory at a time, which reduces the required memory by a factor of N. Notably, the required additional computation to invert each layer only marginally increases the backpropagation runtime.

2.2   Memory-Efficient Learning for MoDL

Here, we use a widely used Unrolled DL Recon framework: MoDL [1]. We formulate the reconstruction of $\hat{\mathbf{x}}$ as an optimization problem and solve it as below:

$$\hat{\mathbf{x}} = \arg\min_{\mathbf{x}} \|\mathbf{A}\mathbf{x} - \mathbf{y}\|_2^2 + \mu\|\mathbf{x} - R_w(\mathbf{x})\|_2^2, \tag{1}$$

a) High-dimensional unrolled DL recon: from 2D to 3D

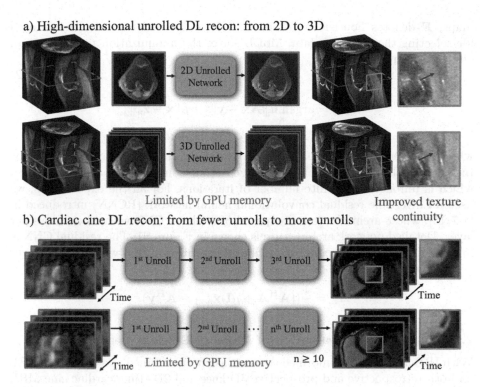

b) Cardiac cine DL recon: from fewer unrolls to more unrolls

Fig. 1. GPU memory limitations for high-dimensional unrolled DL recons: a) Compared to a 2D unrolled network, the 3D unrolled network uses a 3D slab during training to leverage more redundancy, but is limited by GPU memory. b) Cardiac cine DL recons are often performed with a small number of unrolls due to memory limitations.

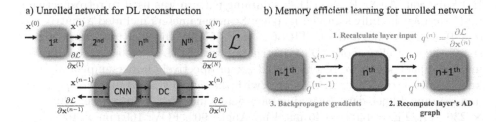

Fig. 2. a) In conventional DL recon training, gradients of all layers are evaluated as a single computational graph, requiring significant GPU memory. b) In MEL, we sequentially evaluate each layer by: i) Recalculate the layer's input $\mathbf{x}^{(n-1)}$, from the known output $\mathbf{x}^{(n)}$. ii) Reform the AD graph for that layer. iii) Backpropagate gradients $q^{(n-1)}$ through the layer's AD graph.

where $\mathbf{A}$ is the system encoding matrix, $\mathbf{y}$ denotes the k-space measurements and R_w is a learned CNN-based denoiser. For multi-channel MRI reconstruction, $\mathbf{A}$ can be formulated as $\mathbf{A} = \mathbf{PFS}$, where $\mathbf{S}$ represent the multi-channel sensitivity maps, $\mathbf{F}$ denotes Fourier Transform and $\mathbf{P}$ is the undersampling mask used for selecting the acquired data. MoDL solves the minimization problem by an alternating procedure:

$$\mathbf{z}_n = R_w(\mathbf{x}_n) \tag{2}$$

$$\mathbf{x}_{n+1} = \arg\min_{\mathbf{x}} \|\mathbf{Ax} - \mathbf{y}\|_2^2 + \mu\|\mathbf{x} - \mathbf{z}_n\|_2^2,$$
$$= (\mathbf{A}^H\mathbf{A} + \mu\mathbf{I})^{-1}(\mathbf{A}^H\mathbf{y} + \mu\mathbf{z}_n) \tag{3}$$

which represents the CNN-based regularization layer and DC layer respectively. In this formulation, the DC layer is solved using Conjugate Gradient (CG) [21], which is unrolled for a finite number of iterations. For all the experiments, we used an invertible residual convolutional neural network (RCNN) introduced in [4,7,15], whose architecture is composed of a 5-layer CNN with 64 channels per layer. Detailed network architecture is shown in Figure S1. The residual CNN is inverted using the fixed-point algorithm as described in [10], while the DC layer is inverted through:

$$\mathbf{z}_n = \frac{1}{\mu}((\mathbf{A}^H\mathbf{A} + \mu\mathbf{I})\mathbf{x}_{n+1} - \mathbf{A}^H\mathbf{y}). \tag{4}$$

2.3   Training and Evaluation of Memory-Efficient Learning

With IRB approval and informed consent/assent, we trained and evaluated MEL on both retrospective and prospective 3D knee and 2D+time cardiac cine MRI. We conducted 3D MoDL experiments with and without MEL on 20 fully-sampled 3D knee datasets (320 slices each) from mridata.org [19]. 16 cases were used for training, 2 cases were used for validation and other 2 for testing. Around 5000 3D slabs with size 21 × 256 × 320 were used for training the reconstruction networks. All data were acquired on a 3T GE Discovery MR 750 with an 8-channel HD knee coil. An 8x Poisson Disk sampling pattern was used to retrospectively undersample the fully sampled k-space. Scan parameters included a matrix size of 320 × 256 × 320, and TE/TR of 25 ms/1550 ms. In order to further demonstrate the feasibility of our 3D reconstruction with MEL on realistic prospectively undersampled scans, we reconstructed 8× prospectively undersampled 3D FSE knee scans (available at mridata.org) with the model trained on retrospectively undersampled knee data. Scanning parameters includes: Volume size: 320 × 288 × 236, TR/TE = 1400/20.46 ms, Flip Angle: 90°, FOV: 160 mm × 160 mm × 141.6 mm.

For the cardiac cine MRI, fully-sampled bSSFP cardiac cine datasets were acquired from 15 volunteers at different cardiac views and slice locations on 1.5T and 3.0T GE scanners using a 32-channel cardiac coil. All data were coil compressed [25] to 8 virtual coils. Twelve of the datasets (around 190 slices) were used for training, 2 for validation, and one for testing. k-Space data were retrospectively under-sampled using a variable-density k-t sampling pattern to simulate 14-fold acceleration with 25% partial echo. We also conducted experiments on a prospectively under-sampled scan (R = 12) which was acquired from a pediatric patient within a single breath-hold on a 1.5T scanner.

We compared the spatio-temporal complexity (GPU memory, training time) with and without MEL. In order to show the benefits of high-dimensional DL recons, we compared the reconstruction results of PICS, 2D and 3D MoDL with MEL for 3D MRI, and 2D+time MoDL with 4 unrolls and 10 unrolls for cardiac cine MRI. For both 2D MoDL and 3D MoDL with MEL, we used 5 unrolls, 10 CG steps and Residual CNN as the regularization layer. A baseline PICS reconstruction was performed using BART [23]. Sensitivity maps were computed using BART [23] and SigPy [14]. Common image quality metrics such as Peak Signal to Noise Ratio (pSNR), Structual Similarity (SSIM) [9] and Fréchet Inception Distance (FID) [8] were reported. FID is a widely used measure of perceptual similarity between two sets of images. All the experiments were implemented in Pytorch [16] and used Nvidia Titan XP (12 GB) and Titan V CEO (32 GB) GPUs. Networks were trained end-to-end using a per-pixel l_1 loss and optimized using Adam [11] with a learning rate of 1×10^{-4}.

3   Results

We first evaluate the spatio-temporal complexity of MoDL with and without MEL (Fig. 3). Without MEL, for a 12 GB GPU memory limit, the maximum slab size decreases rapidly as the number of unrolls increases, which limits the performance of a 3D reconstruction. In contrast, using MEL, the maximum slab size is roughly constant. Figure 3b) and c) show the comparisons from two different perspectives: 1)GPU memory usage; 2)Training time per epoch. Results indicate that for both 3D and 2D+time MoDL, MEL uses significantly less GPU memory than conventional backpropagation while marginally increasing training time. Notably, both MoDL with and without MEL have the same inference time.

Figure 4 shows a comparison of different methods for 3D reconstruction. Instead of learning from only 2D axial view slices (Fig. 1a), 3D MoDL with MEL captures the image features from all three dimensions. Zoomed-in details indicate that 3D MoDL with MEL is able to provide more faithful contrast with more continuous and realistic textures as well as higher pSNR over other methods. Figure 5 demonstrates that MEL enables the training of 2D+time MoDL with a large number of unrolls (10 unrolls), which outperforms MoDL with 4 unrolls with respect to image quality and y-t motion profile. With MEL, MoDL with 10 unrolls resolves the papillary muscles (yellow arrows) better than MoDL with 4 unrolls. Also, the y-t profile of MoDL with 10 unrolls depicts motion in a more natural way while MoDL with 4 unrolls suffers from blurring. Meanwhile, using 10 unrolls over 4 unrolls yields an improvement of 0.6dB in validation pSNR.

Table 1 shows the quantitative metric comparisons (pSNR, SSIM and FID) between different methods on both 3D MRI and cardiac cine MRI reconstructions. Here, we also included feed-forward U-Net [2] as a baseline. The results indicate that both 3D MoDL with MEL and 2D+time MoDL with MEL outperform other methods with respect to pSNR, SSIM and FID.

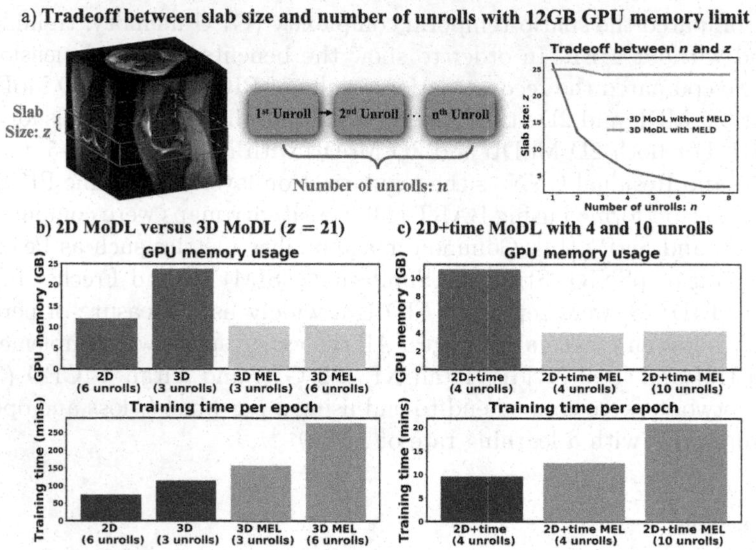

a) **Tradeoff between slab size and number of unrolls with 12GB GPU memory limit**

b) **2D MoDL versus 3D MoDL ($z = 21$)**

c) **2D+time MoDL with 4 and 10 unrolls**

Fig. 3. Spatio-temporal complexity of MoDL with and without MEL. a) Tradeoff between 3D slab size z and a number of unrolls n with a 12 GB GPU memory limitation. b) and c) show the memory and time comparisons for MoDL with and without MEL.

Table 1. Quantitative metrics (pSNR, SSIM and FID) of different methods on 3D MRI and cardiac cine MRI reconstructions (mean ± standard deviation of pSNR and SSIM).

Metric	Method	3D MRI	2D cardiac cine MRI
pSNR (dB)	PICS	31.01 ± 1.97	24.69 ± 2.74
	2D MoDL	31.44 ± 2.07	–
	3D U-Net	29.55 ± 1.86	–
	3D MoDL with MEL	$\mathbf{32.11 \pm 2.05}$	–
	2D+time MoDL: 4 unrolls	–	26.87 ± 2.98
	2D+time MoDL with MEL: 10 unrolls	–	$\mathbf{27.42 \pm 3.21}$
SSIM	PICS	0.816 ± 0.046	0.824 ± 0.071
	2D MoDL	0.821 ± 0.044	–
	3D U-Net	0.781 ± 0.039	–
	3D MoDL with MEL	$\mathbf{0.830 \pm 0.038}$	–
	2D+time MoDL: 4 unrolls	–	0.870 ± 0.042
	2D+time MoDL with MEL: 10 unrolls	–	$\mathbf{0.888 \pm 0.042}$
FID	PICS	46.71	39.40
	2D MoDL	43.58	–
	3D U-Net	60.10	–
	3D MoDL with MEL	**41.48**	–
	2D+time MoDL: 4 unrolls	–	36.93
	2D+time MoDL with MEL: 10 unrolls	–	**31.64**

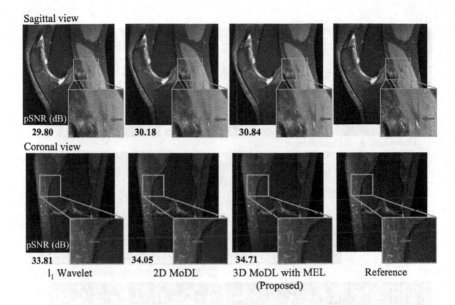

Fig. 4. A representative comparison of different methods (PICS, 2D MoDL, 3D MoDL with MEL) on 3D knee reconstruction (Sagittal view and Coronal view are shown). pSNRs are shown under each reconstructed image.

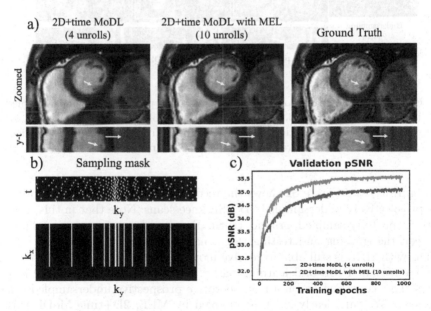

Fig. 5. a) Short-axis view cardiac cine reconstruction of a healthy volunteer on a 1.5T scanner. k-Space data was retrospectively undersampled to simulate 14-fold acceleration with 25% partial echo (shown in b) and reconstructed by: 2D+time MoDL with 4 unrolls, 2D+time MoDL with MEL and 10 unrolls. c) Validation pSNR of MoDL with 4 unrolls and MoDL with 10 unrolls.

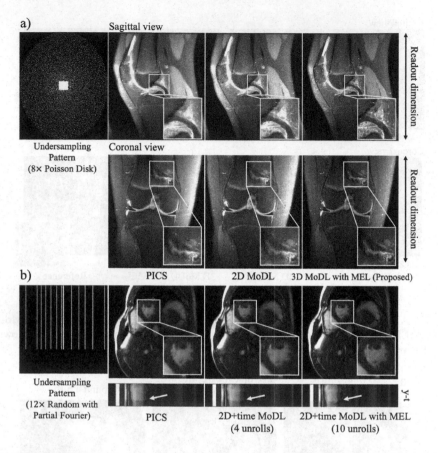

Fig. 6. a) Representative reconstruction results on a prospectively undersampled 3D FSE knee scan using different methods (PICS, 2D MoDL and 3D MoDL with MEL). b) Representative reconstruction results on a prospectively undersampled cardiac cine dataset. y-t motion profiles are shown along with the reconstructed images.

Figure 6a) and Figure S2 shows the reconstruction results on two representative prospectively undersampled 3D FSE knee scans. Note that in this scenario, there is no fully-sampled ground truth. Despite there exists some difference between the training and testing (e.g., matrix size, scanning parameters), 3D MoDL with MEL is still able to resolve more detailed texture and sharper edges over traditional PICS and learning-based 2D MoDL. Figure 6b) and Video S1 shows the reconstruction on a representative prospective undersampled cardiac cine scan. We can clearly see that enabled by MEL, 2D+time MoDL with 10 unrolls can better depicts the finer details as well as more natural motion profile.

4   Conclusions

In this work, we show that MEL enables learning for high-dimensional MR reconstructions on a single 12 GB GPU, which is not possible with standard backpropagation methods. We demonstrate MEL on two representative large-scale MR reconstruction problems: 3D volumetric MRI, 2D cardiac cine MRI with a relatively large number of unrolls. By leveraging the high-dimensional image redundancy and a large number of unrolls, we were able to get improved quantitative metrics and reconstruct finer details, sharper edges, and more continuous textures with higher overall image quality for both 3D and 2D cardiac cine MRI. Furthermore, 3D MoDL reconstruction results from prospectively undersampled k-space show that the proposed method is robust to the scanning parameters and could be potentially deployed in clinical systems. Overall, MEL brings a practical tool for training the large-scale high-dimensional MRI reconstructions with much less GPU memory and is able to achieve improved reconstructed image quality.

Acknowledgements. The authors would like to thank Dr. Gopal Nataraj for his helpful discusses and paper editing. We also acknowledge support from NIH R01EB009690, NIH R01HL136965, NIH R01EB026136 and GE Healthcare.

References

1. Aggarwal, H.K., Mani, M.P., Jacob, M.: Modl: model-based deep learning architecture for inverse problems. IEEE Trans. Med. Imaging **38**(2), 394–405 (2018)
2. Çiçek, Ö., Abdulkadir, A., Lienkamp, S.S., Brox, T., Ronneberger, O.: 3D U-Net: learning dense volumetric segmentation from sparse annotation. In: Ourselin, S., Joskowicz, L., Sabuncu, M.R., Unal, G., Wells, W. (eds.) MICCAI 2016. LNCS, vol. 9901, pp. 424–432. Springer, Cham (2016). https://doi.org/10.1007/978-3-319-46723-8_49
3. Diamond, S., Sitzmann, V., Heide, F., Wetzstein, G.: Unrolled optimization with deep priors. arXiv preprint arXiv:1705.08041 (2017)
4. Gomez, A.N., Ren, M., Urtasun, R., Grosse, R.B.: The reversible residual network: backpropagation without storing activations. arXiv preprint arXiv:1707.04585 (2017)
5. Griswold, M.A., et al.: Generalized autocalibrating partially parallel acquisitions (GRAPPA). Magn. Reson. Med. Official J. Int. Soc. Magn. Reson. Med. **47**(6), 1202–1210 (2002)
6. Hammernik, K., et al.: Learning a variational network for reconstruction of accelerated MRI data. Magn. Reson. Med. **79**(6), 3055–3071 (2018)
7. He, K., Zhang, X., Ren, S., Sun, J.: Deep residual learning for image recognition. In: Proceedings of the IEEE Conference on Computer Vision and Pattern Recognition, pp. 770–778 (2016)
8. Heusel, M., Ramsauer, H., Unterthiner, T., Nessler, B., Hochreiter, S.: Gans trained by a two time-scale update rule converge to a local nash equilibrium. arXiv preprint arXiv:1706.08500 (2017)

9. Isola, P., Zhu, J.Y., Zhou, T., Efros, A.A.: Image-to-image translation with conditional adversarial networks. In: Proceedings of the IEEE Conference on Computer Vision and Pattern Recognition, pp. 1125–1134 (2017)

10. Kellman, M., et al.: Memory-efficient learning for large-scale computational imaging. IEEE Trans. Comput. Imaging **6**, 1403–1414 (2020)

11. Kingma, D.P., Ba, J.: Adam: a method for stochastic optimization. arXiv preprint arXiv:1412.6980 (2014)

12. Küstner, T., et al.: CINENet: deep learning-based 3D cardiac cine MRI reconstruction with multi-coil complex-valued 4D spatio-temporal convolutions. Sci. Rep. **10**(1), 1–13 (2020)

13. Lustig, M., Donoho, D., Pauly, J.M.: Sparse MRI: the application of compressed sensing for rapid MR imaging. Magn. Reson. Med. Official J. Int. Soc. Magn. Reson. Med. **58**(6), 1182–1195 (2007)

14. Ong, F., Lustig, M.: SigPy: a python package for high performance iterative reconstruction. In: Proceedings of ISMRM (2019)

15. Ovadia, Y., et al.: Can you trust your model's uncertainty? Evaluating predictive uncertainty under dataset shift. arXiv preprint arXiv:1906.02530 (2019)

16. Paszke, A., et al..: PyTorch: an imperative style, high-performance deep learning library. In: Wallach, H., Larochelle, H., Beygelzimer, A., d' Alché-Buc, F., Fox, E., Garnett, R. (eds.) Advances in Neural Information Processing Systems, vol. 32, pp. 8024–8035. Curran Associates, Inc. (2019). http://papers.neurips.cc/paper/9015-pytorch-an-imperative-style-high-performance-deep-learning-library.pdf

17. Pruessmann, K.P., Weiger, M., Scheidegger, M.B., Boesiger, P.: Sense: sensitivity encoding for fast MRI. Magn. Reson. Med. Official J. Int. Soc. Magn. Reson. Med. **42**(5), 952–962 (1999)

18. Sandino, C.M., Lai, P., Vasanawala, S.S., Cheng, J.Y.: Accelerating cardiac cine MRI using a deep learning-based ESPiRiT reconstruction. Magn. Reson. Med. **85**(1), 152–167 (2021)

19. Sawyer, A.M., et al.: Creation of fully sampled MR data repository for compressed sensing of the knee. Citeseer (2013)

20. Schlemper, J., Caballero, J., Hajnal, J.V., Price, A.N., Rueckert, D.: A deep cascade of convolutional neural networks for dynamic MR image reconstruction. IEEE Trans. Med. Imaging **37**(2), 491–503 (2017)

21. Shewchuk, J.R., et al.: An introduction to the conjugate gradient method without the agonizing pain (1994)

22. Tamir, J.I., Yu, S.X., Lustig, M.: Unsupervised deep basis pursuit: learning inverse problems without ground-truth data. arXiv preprint arXiv:1910.13110 (2019)

23. Uecker, M., et al.: Berkeley advanced reconstruction toolbox. In: Proceedings of International Society for Magnetic Resonance in Medicine, No. 2486 in 23 (2015)

24. Zhang, K., Kellman, M., Tamir, J.I., Lustig, M., Waller, L.: Memory-efficient learning for unrolled 3D MRI reconstructions. In: ISMRM Workshop on Data Sampling and Image Reconstruction (2020)

25. Zhang, T., Pauly, J.M., Vasanawala, S.S., Lustig, M.: Coil compression for accelerated imaging with cartesian sampling. Magn. Reson. Med. **69**(2), 571–582 (2013)

SA-GAN: Structure-Aware GAN for Organ-Preserving Synthetic CT Generation

Hajar Emami[1], Ming Dong[1(✉)], Siamak P. Nejad-Davarani[2],
and Carri K. Glide-Hurst[3]

[1] Department of Computer Science, Wayne State University, Detroit, MI 48202, USA
mdong@wayne.edu
[2] Department of Radiation Oncology, University of Michigan, Ann Arbor,
MI 48109, USA
[3] Department of Human Oncology, University of Wisconsin Madison, Madison,
WI 53792, USA

Abstract. In medical image synthesis, model training could be challenging due to the inconsistencies between images of different modalities even with the same patient, typically caused by internal status/tissue changes as different modalities are usually obtained at a different time. This paper proposes a novel deep learning method, Structure-aware Generative Adversarial Network (SA-GAN), that preserves the shapes and locations of in-consistent structures when generating medical images. SA-GAN is employed to generate synthetic computed tomography (synCT) images from magnetic resonance imaging (MRI) with two parallel streams: the global stream translates the input from the MRI to the CT domain while the local stream automatically segments the inconsistent organs, maintains their locations and shapes in MRI, and translates the organ intensities to CT. Through extensive experiments on a pelvic dataset, we demonstrate that SA-GAN provides clinically acceptable accuracy on both synCTs and organ segmentation and supports MR-only treatment planning in disease sites with internal organ status changes.

Keywords: Structure-Aware GAN · Synthetic CT · Radiation therapy

1 Introduction

Multimodal medical imaging is crucial in clinical practice such as disease diagnosis and treatment planning. For example, Computed tomography (CT) imaging is an essential modality in treatment planning. Compared with CT, Magnetic Resonance Imaging (MRI) is a safer modality that does not involve patient's

---

Electronic supplementary material The online version of this chapter (https://doi.org/10.1007/978-3-030-87231-1_46) contains supplementary material, which is available to authorized users.

© Springer Nature Switzerland AG 2021
M. de Bruijne et al. (Eds.): MICCAI 2021, LNCS 12906, pp. 471–481, 2021.
https://doi.org/10.1007/978-3-030-87231-1_46

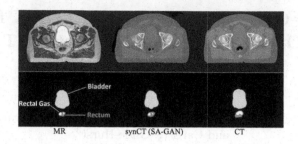

Fig. 1. Example of pelvic MRI with the corresponding CT, highlighting significant inconsistencies between the MRI and CT in bladder, rectum and rectal gas. The synCT generated by SA-GAN preserves the location and shape of organs as in MRI while changing their intensity to the CT, leading to more accurate dose calculation.

exposure to radiation. Due to the excellent soft tissue contrast in MRI, its integration into CT-based radiation therapy is expanding at a rapid pace. However, separate acquisition of multiple modalities is time-consuming, costly and increases unnecessary irradiation to patients. Thus, a strong clinical need exists to synthesize CT images from MRI [6,15,17].

Recently, deep learning methods have been employed in medical image synthesis. Initially, Fully Convolutional Networks (FCNs) were used in medical image synthesis [5,13,23]. The adversarial training in conditional generative adversarial networks (cGANs) [11,16,19] can retain fine details in medical image synthesis and has further improved the performance over CNN models [1,3,4,25]. Cycle generative adversarial network (CycleGAN) [30] has also been used for unsupervised medical image translation [12,18,26,28] when paired images are not available. While deep learning techniques have made enormous achievement in medical image synthesis, model training could be challenging due to the inconsistencies between images of different modalities even with the same patient, typically caused by internal status/tissue changes as different modalities are usually obtained at a different time. For instance, in prostate cancer treatment, MRI and the corresponding CT slices of pelvis are often not consistent due to variations in bladder filling/emptying, irregular rectal movement, and local deformations of soft tissue organs [22]. Thus, it is of paramount importance to generate synCTs from MRI that can accurately depict these variations to spare critical organs while ensuring accurate dose calculation in MR-only treatment planning.

In Fig. 1, an example of pelvic MRI is shown with the corresponding CT with a clear inconsistency in the rectal gas regions. Also note that the variations in bladder filling result in significant organ shape changes between MRI and the corresponding CT. This would adversely impact the dose delivered to the prostate as accurately preserving bladder status is important for radiation therapy. The desired synCT from the MRI example is shown in the second column, which should preserve the location and shapes of the inconsistent organs as in MRI while accurately changing the corresponding image intensities to the real CT to ensure high fidelity dose calculation in radiation therapy.

To improve MRI-CT consistency, pre-processing is typically employed. Chen et al. [2] utilized deformable registration to handle the MRI-CT inconsistencies. However, the accuracy of synCTs will largely depend upon the performance of the deformable registration, which is limited in multi-modality workflows with large volume changes and introduces geometrical uncertainties in the pelvic region [29]. Maspero et al. [21] assigned gas regions from MRI to CT as a pre-processing step to ensure the consistency of gas regions. This manual intervention is simple but time consuming. More importantly, it is restricted to a particular type of inconsistency (rectal gas in a pelvic), lacking the generality in the applicability to other types of inconsistencies or other disease sites. More recently, Ge et al. [9] used shape consistency loss in training to align the skin surface and overall shape of the synCT images with the original input image.

This work proposes a novel deep learning method, a structure-aware GAN (SA-GAN) model, that preserves the shapes and locations of inconsistent structures when generating synCTs. Our SA-GAN model is developed on the basis of GAN with two parallel streams in the generator. The global stream translates the input from the source domain (MRI) to the target domain (CT) while the local stream automatically segments the inconsistent organs between the two domains using a structure segmentation network, maintains their locations and shapes in MRI, and translates the organ intensities to CT using a novel adaptive organ normalization (AdaON) module. Our major contributions are: **1)** SA-GAN is the first automated framework that systematically addresses the MR-CT inconsistent issues in synCT generation in an end-to-end fashion. **2)** By fusing outputs from both the global and local streams, SA-GAN jointly minimizes the reconstruction and the structure segmentation losses together with the GAN loss through adversarial training. **3)** Without the time-consuming and error-prone pre-processing, SA-GAN offers strong potential for near real-time MR-only treatment planning while preserving features necessary for high precision dose calculation.

2   Method

The goal of the proposed SA-GAN model is to estimate a mapping $F_{MR \to CT}$ from the source domain (MRI) to the target domain (CT). The mapping F is learned from paired training data $S = \{(mr_i, ct_i) | mr_i \in MR, ct_i \in CT, i = 1, 2, ..., N\}$, where N is the number of MRI-CT pairs. This translation is a challenging task as some regions are not completely matched between two domains, e.g., bladder, rectum and rectal gas regions in a pelvic dataset. A conventional synCT model typically defines the same loss function across all regions in an input MRI. This would produce erroneous tissue assignments in the inconsistent regions between two domains. To better preserve organ features and shapes in synCT generation, we need to solve two important tasks simultaneously: (1) segmenting the inconsistent areas and translating only the image intensities in these regions (2) translating the remaining regions paired between MRI and CT.

The architecture of our proposed solution, SA-GAN, is shown in Fig. 2. The main components include the generator (G), the discriminator (D), the structure

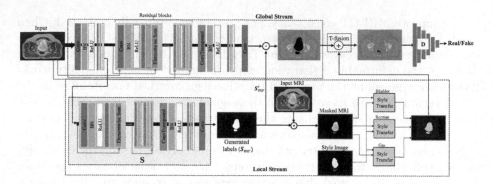

Fig. 2. SA-GAN architecture with two parallel streams.

segmentation network (S), the style transfer modules with adaptive organ normalization (AdaON) layers, and the fusion layer (T_{fusion}). SA-GAN contains two parallel streams in the generator. The global stream translates the input from MRI to CT domain while the local stream automatically segments the inconsistent regions using network S, keeps their locations and shapes in the MRI space, and translates the organ intensities to CT intensities using AdaON. The outputs of two streams are combined in the fusion layer to generate the final synCT. SA-GAN's global stream is used for global translation of consistent regions between the two domains and produces the reconstruction loss, and the local stream segments the inconsistent organs and produces the structure segmentation loss. These losses are jointly minimized together with the GAN loss through adversarial training in an end-to-end fashion.

2.1   Global Stream in SA-GAN

The generator G includes an encoder-decoder network with three convolution layers followed by nine residual blocks. Each residual block consist of two convolutional layers followed by batch normalization and a nonlinear activation function: Rectified Linear Unit (ReLU). In addition to the skip connections in the residual blocks, we also add connections from the input layer to the layers of the encoder. The input is down-sampled and is concatenated with the feature maps of the encoder layers. Adding these connections can help the network to preserve the low level features of the input MRI such as edges and boundaries. Finally, we have three transposed convolutional layers after the residual blocks to generate synCTs of the input size. These transposed convolution layers are usually used in the decoder part of an encoder-decoder architecture to project feature maps to a higher dimensional space. The dropout units are also used in the layers of the generator for regularization and to help prevent overfitting.

The discriminator (D) is a CNN with six convolutional layers followed by batch normalization and ReLU. G tries to generate synCTs as close as possible to real CTs while D tries to distinguish between the generated synCTs and real

CTs. Following [20], we replace the negative log likelihood objective by a least square loss to achieve more stable training:

$$\mathcal{L}_{GAN}(G, D) = \mathbb{E}_{ct\sim P(ct)}[(D(ct) - 1)^2] + \mathbb{E}_{mr\sim P(mr)}[D(G(mr))^2] \qquad (1)$$

where P is the data probability distribution. Moreover, L_1 norm is also used as the regularization in the reconstruction error:

$$\mathcal{L}_{L_1}(G) = \mathbb{E}_{mr\sim P(mr),ct\sim P(ct)}[||G(mr) - ct||_1] \qquad (2)$$

Using L_1 distance encourages less blurred results when compared with using L_2 to compute the reconstruction error [16]. Because of the potential MRI-CT inconsistencies, the reconstruction loss should not be calculated for all regions of the input. If inconsistent regions are included when calculating L_1 reconstruction loss, the error could be agnostic and will possibly mislead the generator. Thus, we need to exclude these regions in our calculations. To do so, we first define a binary mask u for the excluded regions as the union of inconsistent regions within MRI ($label_{mr}$) and CT ($label_{ct}$) provided in the ground truth:

$$u = label_{mr} \cup label_{ct} \qquad (3)$$

where u is 0 for regions to be excluded and 1 for all other regions. Then, the reconstruction loss is modified by performing element-wise multiplication on the real and the generated CT with u:

$$\mathcal{L}_{exc}(G) = \mathbb{E}_{mr\sim P(mr),ct\sim P(ct)}[||u \odot G(mr) - u \odot ct||_1] \qquad (4)$$

where $\odot$ denotes the element-wise multiplication operator. Note that MRI and CT labels of inconsistent regions are used here to calculate the modified reconstruction loss. After training, these labels are not required anymore for generating synCT. Finally, the objective is the minimax optimization defined as:

$$G^*_{MR\rightarrow CT}, D^* = \underset{G}{arg min}\underset{D}{max}\mathcal{L}_{GAN}(G_{MR\rightarrow CT}, D) + \lambda\mathcal{L}_{exc}(G) \qquad (5)$$

where we set the loss hyper-parameter $\lambda = 10$ throughout our experiments.

2.2 Segmentation Network in the Local Stream

The segmentation network in SA-GAN takes an input MRI image and segments out the pre-defined inconsistent structures. The network S has three convolution layers at the beginning, followed by six residual blocks and three transposed convolutional layers. The loss function to update S is a multi-class cross-entropy:

$$\mathcal{L}_S = \mathbb{E}_{mr\sim P(mr)}[-\frac{1}{N}\sum_i w_i^{-1}y^i log(S(mr)_i)] \qquad (6)$$

where y is the ground truth labels, and N is the total number of pixels. In order to balance the size differences of structures and their contribution to the loss, a coefficient w_i is adopted to weight each structure i with the invert sampling count: $w_i = \frac{n_i}{N}$ where n_i is the number of pixels belonging to structure i.

2.3 Organ Style Transfer with AdaON

Research shows that the image style and content are separable, and it is possible to change the style of an image while preserving its content [8,10]. In adaptive instance normalization (AdaIN) [14], an encoder-decoder is used to take a content image and a style image, and synthesizes an output image. After encoding the content and style images in feature space using encoder f, the mean and the standard deviation of the content input feature maps are aligned with those of the style input in an AdaIN layer. Then a randomly initialized decoder g is trained to generate the stylized image from the output of the AdaIN layer.

In SA-GAN, we use adaptive organ normalization (AdaON) layers to transfer the style of inconsistent organs to CT domain. To generate accurate results, we define a style for each inconsistent structure in the pelvic. Then, $AdaON_B$, $AdaON_R$ and $AdaON_G$ layers are trained separately for bladder, rectum and rectal gas regions by taking a masked CT style image and a masked MRI from an inconsistent region as inputs and minimizing style and content losses. After training, they are used to generate the inconsistent organs in the CT domain, which are then combined through element-wise addition as the output of the local stream. Finally, the outputs of the global and local streams are combined in the fusion layer T, also through element-wise addition.

3 Experiments and Results

Dataset. In our experiments, 3,375 2D images from 25 subjects with prostate cancer (median age = 71 years, Range: 53–96 years) were retrospectively analyzed. CT images were acquired on a Brilliance Big Bore scanner with 512 × 512 in-plane image dimensions, 1.28 × 1.28 mm^2 in-plane spatial resolution, and 3 mm slice thickness. Corresponding MR images were acquired in the radiation therapy treatment position on a 1.0 T Panorama High Field Open MR simulator (576 × 576 × 250 mm^3, Voxel size = 0.65 × 0.65 × 2.5 mm^3). A single physician delineated bladder and rectum volumes on CT and MRI images in the Eclipse Treatment Planning System. Rectal gas was identified by thresholding and applying morphological operators. Next, using Statistical Parametric Mapping (SPM) 12, MRIs and corresponding binary masks of each subject were rigidly co-registered to the CT images of the same subject.

The size of our dataset is consistent with the ones used for synCT in the literature [4,6,12,13,18,23–25]. Additionally, we used data augmentation (e.g., flipping) to increase the number of training images four times. To evaluate our model performance and avoid overfitting, a five-fold cross-validation was used in our model training and testing.

Implementation Details. The weights in SA-GAN were all initialized from a Gaussian distribution with parameter values of 0 and 0.02 for mean and standard deviation, respectively. The model is trained with ADAM optimizer with an initial learning rate of 0.0002 and with a batch size of 1. We trained the model for 200 epochs on a NVIDIA GTX 1080 Ti GPU. SA-GAN training took approximately 13 h. The model was implemented using PyTorch 0.2.

Evaluation Metrics. Three commonly-used quantitative measures are adopted for evaluation: Mean Absolute Error (MAE), Peak Signal-to-Noise Ratio (PSNR), and Structure Similarity Index (SSIM). For PSNR and SSIM, higher values indicate better results. For MAE, the lower the value, the better is the generation. Consideration was given to agreement using the full field of view, bone, rectal gas, bladder and rectum regions. Since input MRIs and their corresponding CTs include inconsistent regions in bladder, rectum and rectal gas, we calculate the MAE between the intersected regions in real CTs and synCTs. To segment the bone regions in the images, a threshold of +150 HU [7] was set in both real and generated CTs. We used, Dice similarity coefficient (DSC), the well-established metric to evaluate the organ segmentation performance.

Ablation Study. We performed the following ablation study. 1) We removed the local stream in SA-GAN. In this case, our model is reduced to cGAN. 2) We removed the local stream but added style loss and content loss calculated from the whole image to the model's objective (SA-GAN-wo-*seg*). 3) We had the segmentation network in the local stream but removed the AdaON layers (SA-GAN-wo-*AdaON*). 4) We had the whole local stream but the reconstruction loss is calculated from all regions in an MRI input image (SA-GAN-wo-$\mathcal{L}_{exc}$). MAE, PSNR and SSIM for different configuration of our model are reported in the supplementary materials. In summary, SA-GAN with all its components achieved the best results in our model ablation.

Comparison to State-of-the-Arts. Extensive experiments were performed to compare SA-GAN with current state-of-the-art models for synCT generation. Since SA-GAN's generator and discriminator share a similar architecture to cGAN [16,27] and CycleGAN [30], we included these methods in our comparison. In particular, CycleGAN is widely used for medical image synthesis and provides a potential solution to handle the organ inconsistencies in the pelvic dataset as it does not require paired input images. It is worth mentioning that because of the flexible architecture of SA-GAN, other variants of cGAN and CycleGAN, e.g., [1,12], can easily be incorporated as the generator in the global stream of SA-GAN. All the comparisons are done using the same training and testing splits.

SynCT Generation Results. Figure 3 shows synCT results for three different test cases that have inconsistencies in the bladder, rectum, and rectal gas between MRI and CT. The CNN model with no discriminator block generated blurry results. Since the discriminator in cGAN, CycleGAN, and SA-GAN can easily classify blurry images as fake, they do not generate blurry results. However, cGAN and CycleGAN are not able to preserve the inconsistent regions, particularly for the transient rectal gas regions which has been shown to cause dose calculation discrepancies [17]. As shown in Fig. 3, for Patient 3, substantial rectal gas appeared in the MRI that did not correspond to the CT. Consequently, CNN, cGAN, and CycleGAN yielded inaccurate intensities (similar to tissues) in this region, whereas rectal gas was accurately depicted in the results of SA-GAN.

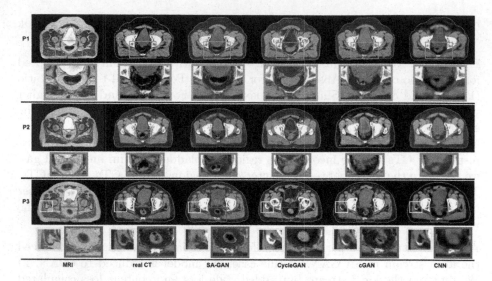

Fig. 3. Qualitative comparison of synCTs generated by SA-GAN and state-of-the-art models for three different patients (P1–P3). Regions of interest are drawn on a slice and enlarged in the row below for a more detailed comparison with other models.

Similarly, Patients 1 and 2 had less air in their rectal regions in MRI than their corresponding CTs, and these were apparent in the SA-GAN results whereas all other methods failed to reproduce the appropriate tissue type and shape. SynCTs generated by the CNN, cGAN, and CycleGAN models showed larger errors in the bone regions. Bones were generated in non-physical regions, and their intensities were underestimated in the spinal region. Clearly, bone anatomy was better represented by SA-GAN. In general, our qualitative comparison shows that SA-GAN produces more realstic synCTs with a higher spatial resolution and preserved organ shapes in MRI when compared to other methods. Please see the supplementary material for more synCT examples with a bigger display.

The average MAE, PSNR, and SSIM computed based on the real and synthetic CTs for all test cases are listed in Table 1. To better evaluate the performance, the MAEs for different regions are also reported. SA-GAN achieved better performance in generating both consistent (e.g., bone) and inconsistent (e.g., rectal gas) regions when compared to other models over all test cases. Clearly, the improvement in the inconsistent regions is due to the structure segmentation and AdaON in the local stream of SA-GAN. Additionally, in the global stream, we compute the reconstruction error between synCTs and real CTs after excluding the inconsistent regions. This allows our global translation model to focus on the consistent regions when calculating the loss, leading to more accurate results than other models. Finally, it is worth mentioning that **the combination of consistent and inconsistent regions in MRI-CT pairs presents a significant challenge** for synCT methods based on pixel correspondences, e.g., cGAN and CNN-based methods. Although CycleGAN was first introduced

Table 1. Performance comparison of SA-GAN for synCT generation. The average MAEs are computed from entire pelvis, bone, rectal gas, rectum and bladder. Since MRIs and CTs include inconsistent regions, the MAE is calculated between the intersected masked regions of MRIs and real CTs in bladder, rectum and rectal gas.

Method	MAE (HU)					PSNR	SSIM
	Entire pelvis	Bone	Rectal Gas	Rectum	Bladder		
CNN	42.7 ± 4.9	239.4 ± 35.6	676.4 ± 85.7	28.3 ± 8.4	13.9 ± 2.1	29.9 ± 1.2	0.88 ± 0.05
cGAN	54.6 ± 6.8	267.2 ± 43.6	652.6 ± 97.4	35.7 ± 11.0	20.2 ± 7.8	28.0 ± 0.9	0.85 ± 0.04
CycleGAN	59.8 ± 6.1	290.3 ± 43.2	678.8 ± 75.6	35.9 ± 7.1	41.1 ± 16.6	27.2 ± 1.0	0.83 ± 0.05
SA-GAN	**38.5 ± 4.9**	**210.6 ± 34.0**	**279.9 ± 64.1**	**28.2 ± 7.2**	**13.5 ± 2.6**	**30.5 ± 1.3**	**0.90 ± 0.04**

Table 2. Dice similarity coefficient (higher is better) computed between the ground truth MRI labels and the segmented organs using SA-GAN's segmentation network.

	Bladder	Rectal gas	Rectum
DSC	0.93 ± 0.03	0.90 ± 0.05	0.86 ± 0.03

to handle inconsistencies between two domains in the absence of paired data, as shown in our experimental results, it was **unable** to maintain the appropriate consistency between the bladder and rectal intensity values.

Organ Segmentation Results. The DSCs between the ground truth MRI labels and the segmented organs using SA-GAN are summarized in Table 2. High DSC suggests excellent ability of SA-GAN to localize inconsistent organs. Some segmentation examples are provided in the supplementary material.

Once trained, SA-GAN generates both the synCTs and segmentation labels of the inconsistent regions in the input MRIs in a short time of ~12 s. Thus, SA-GAN provides a practical tool that can be integrated into near real-time applications for producing synCTs in datasets with inconsistent structures.

4   Conclusion

In this paper, we proposed a novel SA-GAN to generate synCTs from MRI for disease sites where internal anatomy may change between different image acquisitions. Our experiments show that SA-GAN achieves clinically acceptable accuracy on both synCTs and organ-at-risk segmentation, and thus supports MR-only treatment planning (e.g., high fidelity dose calculation in radiation therapy) in disease sites with internal organ status changes.

Acknowledgments. This work was partially supported by the National Cancer Institute of the National Institutes of Health under Award Number R01CA204189.

References

1. Armanious, K., et al.: MedGAN: medical image translation using GANs. Computerized Med. Imag. Graph. **79**, 101684 (2020)
2. Chen, S., Qin, A., Zhou, D., Yan, D.: U-net-generated synthetic CT images for magnetic resonance imaging-only prostate intensity-modulated radiation therapy treatment planning. Med. Phys. **45**(12), 5659–5665 (2018)
3. Emami, H., Dong, M., Glide-Hurst, C.K.: Attention-guided generative adversarial network to address atypical anatomy in synthetic CT generation. In: 2020 IEEE 21st International Conference on Information Reuse and Integration for Data Science (IRI), pp. 188–193. IEEE (2020)
4. Emami, H., Dong, M., Nejad-Davarani, S.P., Glide-Hurst, C.K.: Generating synthetic CTs from magnetic resonance images using generative adversarial networks. Med. Phys. **45**(8), 3627–3636 (2018)
5. Emami, H., Liu, Q., Dong, M.: FREA-UNet: frequency-aware U-Net for modality transfer. arXiv preprint arXiv:2012.15397 (2020)
6. Fu, J., et al.: Deep learning approaches using 2D and 3D convolutional neural networks for generating male pelvic synthetic computed tomography from magnetic resonance imaging. Med. Phys. **46**(9), 3788–3798 (2019)
7. Fu, J., Yang, Y., Singhrao, K., Ruan, D., Low, D.A., Lewis, J.H.: Male pelvic synthetic CT generation from t1-weighted MRI using 2D and 3D convolutional neural networks. arXiv preprint arXiv:1803.00131 (2018)
8. Gatys, L.A., Ecker, A.S., Bethge, M.: Image style transfer using convolutional neural networks. In: Proceedings of the IEEE Conference on Computer Vision and Pattern Recognition, pp. 2414–2423 (2016)
9. Ge, Y., et al.: Unpaired MR to CT synthesis with explicit structural constrained adversarial learning. In: 2019 IEEE 16th International Symposium on Biomedical Imaging (ISBI 2019), pp. 1096–1099. IEEE (2019)
10. Ghiasi, G., Lee, H., Kudlur, M., Dumoulin, V., Shlens, J.: Exploring the structure of a real-time, arbitrary neural artistic stylization network. arXiv preprint arXiv:1705.06830 (2017)
11. Goodfellow, I., et al.: Generative adversarial nets. In: Advances in Neural Information Processing Systems, pp. 2672–2680 (2014)
12. Hamghalam, M., Lei, B., Wang, T.: High tissue contrast MRI synthesis using multistage attention-GAN for glioma segmentation. arXiv preprint arXiv:2006.05030 (2020)
13. Han, X.: MR-based synthetic CT generation using a deep convolutional neural network method. Med. Phys. **44**(4), 1408–1419 (2017)
14. Huang, X., Belongie, S.: Arbitrary style transfer in real-time with adaptive instance normalization. In: Proceedings of the IEEE International Conference on Computer Vision, pp. 1501–1510 (2017)
15. Huynh, T., et al.: Estimating CT image from MRI data using structured random forest and auto-context model. IEEE Trans. Med. Imag. **35**(1), 174–183 (2015)
16. Isola, P., Zhu, J.Y., Zhou, T., Efros, A.A.: Image-to-image translation with conditional adversarial networks. In: Proceedings of the IEEE Conference on Computer Vision and Pattern Recognition, pp. 1125–1134 (2017)
17. Kim, J., et al.: Dosimetric evaluation of synthetic CT relative to bulk density assignment-based magnetic resonance-only approaches for prostate radiotherapy. Radiat. Oncol. **10**(1), 239 (2015)

18. Lei, Y., et al.: MRI-only based synthetic CT generation using dense cycle consistent generative adversarial networks. Med. Phys. **46**(8), 3565–3581 (2019)
19. Lin, J., Xia, Y., Qin, T., Chen, Z., Liu, T.Y.: Conditional image-to-image translation. In: Proceedings of the IEEE Conference on Computer Vision and Pattern Recognition, pp. 5524–5532 (2018)
20. Mao, X., Li, Q., Xie, H., Lau, R.Y., Wang, Z., Paul Smolley, S.: Least squares generative adversarial networks. In: Proceedings of the IEEE International Conference on Computer Vision, pp. 2794–2802 (2017)
21. Maspero, M., et al.: Dose evaluation of fast synthetic-CT generation using a generative adversarial network for general pelvis MR-only radiotherapy. Phys. Med. Biol. **63**(18), 185001 (2018)
22. Nakamura, N., et al.: Variability in bladder volumes of full bladders in definitive radiotherapy for cases of localized prostate cancer. Strahlentherapie und Onkologie **186**(11), 637–642 (2010)
23. Nie, D., Cao, X., Gao, Y., Wang, L., Shen, D.: Estimating CT image from MRI data using 3D fully convolutional networks. In: Carneiro, G., et al. (eds.) LABELS/DLMIA -2016. LNCS, vol. 10008, pp. 170–178. Springer, Cham (2016). https://doi.org/10.1007/978-3-319-46976-8_18
24. Nie, D., et al.: Medical image synthesis with deep convolutional adversarial networks. IEEE Trans. Biomed. Eng. **65**(12), 2720–2730 (2018)
25. Nie, D., et al.: Medical image synthesis with context-aware generative adversarial networks. In: Descoteaux, M., Maier-Hein, L., Franz, A., Jannin, P., Collins, D.L., Duchesne, S. (eds.) MICCAI 2017. LNCS, vol. 10435, pp. 417–425. Springer, Cham (2017). https://doi.org/10.1007/978-3-319-66179-7_48
26. Pan, Y., Liu, M., Lian, C., Zhou, T., Xia, Y., Shen, D.: Synthesizing missing PET from MRI with cycle-consistent generative adversarial networks for Alzheimer's disease diagnosis. In: Frangi, A.F., Schnabel, J.A., Davatzikos, C., Alberola-López, C., Fichtinger, G. (eds.) MICCAI 2018. LNCS, vol. 11072, pp. 455–463. Springer, Cham (2018). https://doi.org/10.1007/978-3-030-00931-1_52
27. Tie, X., Lam, S.K., Zhang, Y., Lee, K.H., Au, K.H., Cai, J.: Pseudo-CT generation from multi-parametric MRI using a novel multi-channel multi-path conditional generative adversarial network for nasopharyngeal carcinoma patients. Med. Phys. **47**(4), 1750–1762 (2020)
28. Wolterink, J.M., Dinkla, A.M., Savenije, M.H.F., Seevinck, P.R., van den Berg, C.A.T., Išgum, I.: Deep MR to CT synthesis using unpaired data. In: Tsaftaris, S.A., Gooya, A., Frangi, A.F., Prince, J.L. (eds.) SASHIMI 2017. LNCS, vol. 10557, pp. 14–23. Springer, Cham (2017). https://doi.org/10.1007/978-3-319-68127-6_2
29. Zhong, H., Wen, N., Gordon, J.J., Elshaikh, M.A., Movsas, B., Chetty, I.J.: An adaptive MR-CT registration method for MRI-guided prostate cancer radiotherapy. Phys. Med. Biol. **60**(7), 2837 (2015)
30. Zhu, J.Y., Park, T., Isola, P., Efros, A.A.: Unpaired image-to-image translation using cycle-consistent adversarial networks. In: Proceedings of the IEEE International Conference on Computer Vision, pp. 2223–2232 (2017)

Clinical Applications - Cardiac

Distortion Energy for Deep Learning-Based Volumetric Finite Element Mesh Generation for Aortic Valves

Daniel H. Pak[1]($\boxtimes$), Minliang Liu[2], Theodore Kim[3], Liang Liang[4],
Raymond McKay[5], Wei Sun[2], and James S. Duncan[1]

[1] Biomedical Engineering, Yale University, New Haven, CT, USA
daniel.pak@yale.edu
[2] Biomedical Engineering, Georgia Institute of Technology, Atlanta, GA, USA
[3] Computer Science, Yale University, New Haven, CT, USA
[4] Computer Science, University of Miami, Miami-Dade County, FL, USA
[5] Division of Cardiology, The Hartford Hospital, Hartford, CT, USA

Abstract. Volumetric meshes with hexahedral elements are generally best for stress analysis using finite element (FE) methods. With recent interests in finite element analysis (FEA) for Transcatheter Aortic Valve Replacement (TAVR) simulations, fast and accurate generation of patient-specific volumetric meshes of the aortic valve is highly desired. Yet, most existing automated image-to-mesh valve modeling strategies have either only produced surface meshes or relied on simple offset operations to obtain volumetric meshes, which can lead to undesirable artifacts. Furthermore, most recent advances in deep learning-based meshing techniques have focused on watertight surface meshes, not volumetric meshes. To fill this gap, we propose a novel volumetric mesh generation technique using template-preserving distortion energies under the deep learning-based deformation framework. Our model is trained end-to-end for image-to-mesh prediction, and our mesh outputs have good spatial accuracy and element quality. We check the FEA-suitability of our model-predicted meshes using a valve closure simulation. Our code is available at https://github.com/danpak94/Deep-Cardiac-Volumetric-Mesh.

Keywords: 3D image to volumetric mesh · ARAP energy · CNN-GCN · Finite element analysis · Aortic valve modeling

1 Introduction

For stress analyses using finite element (FE) methods, volumetric meshes with hexahedral elements lead to most accurate results and better convergence [25]. With recent interests in finite element analysis (FEA) for Transcatheter Aortic Valve Replacement (TAVR) simulations [2,28], fast and accurate generation

© Springer Nature Switzerland AG 2021
M. de Bruijne et al. (Eds.): MICCAI 2021, LNCS 12906, pp. 485–494, 2021.
https://doi.org/10.1007/978-3-030-87231-1_47

of patient-specific volumetric meshes of the aortic valve is highly desired. Yet, most existing automated valve modeling strategies have only focused on voxel-wise segmentation [15], surface meshes, [7,18] or volumetric meshes generated by simple offset operations [6,12]. Although offsetting is viable for certain initial surface meshes, it quickly becomes ill-defined when combining multiple components or modeling structures with high curvature. Post-processing can mitigate some of these problems, but it limits usability by non-experts during test time. In this work, we aim to address this limitation by learning an image-to-mesh model that directly outputs optimized volumetric meshes.

Most existing valve modeling approaches have used template deformation strategies [4,7,12,16]. We adopt a similar approach, as it ensures mesh correspondence between model predictions for easy application to downstream tasks (e.g. shape analysis or batch-wise FEA). Some previous works have used sequential localization + deformation-along-surface-normals and/or hand-crafted image features [7,12], both of which limit the methods' adaptability to image and template mesh variations. Instead, we focus on deep learning-based deformation methods [3,16,27,30], which addresses both of these limitations by (1) not limiting the deformation to the surface normal directions and (2) learning image features via end-to-end training. Deep learning also has additional benefits such as fast inference and the ability to generate diffeomorphic deformation field [3].

Mesh deformation in computer graphics aims to match the user-defined locations of handle points while preserving the mesh's geometric detail [1,24]. Unfortunately, the same formulation is not ideal for valve modeling because it is difficult to define proper handle points and their desired locations on 3D images for the flexible valve components. Instead, we apply the idea of minimizing the mesh distortion energy into our deep learning pipeline, while enforcing spatial accuracy through surface distance metrics.

In summary, we propose a novel deep learning image-to-mesh model for volumetric aortic valve meshes. Our contributions include: (1) identifying two effective deformation strategies for this task, (2) incorporating distortion energy into both strategies for end-to-end learning, and (3) generating volumetric meshes from just the base surface training labels (i.e. surface before adding thickness).

2   Methods

2.1   Template Deformation-Based Mesh Generation

Template deformation strategies aim to find the optimal displacement vectors δ for every vertex $v_i \in V$ of a mesh M, where $M = (V, \mathcal{E})$ is a graph with nodes V and edges $\mathcal{E}$. Then, the optimization over a loss $\mathcal{L}$ is:

$$\delta^* = \arg\min_{\delta} \mathcal{L}(M, M_0, \delta) \tag{1}$$

where M and M_0 are target and template meshes, respectively. We used deep learning models as our function approximator $h_\theta(I; M_0) = \delta$, where I is the image and θ is the network parameters. Thus, we ultimately solved for θ:

$$\theta^* = \arg\min_{\theta} \left[\mathbb{E}_{(I,M)\sim\Omega}[\mathcal{L}(M, M_0, h_\theta(I; M_0))] \right] \tag{2}$$

where Ω is the training set distribution. We experimented with two variations of h_θ, as detailed below. Both models are shown schematically in Fig. 1.

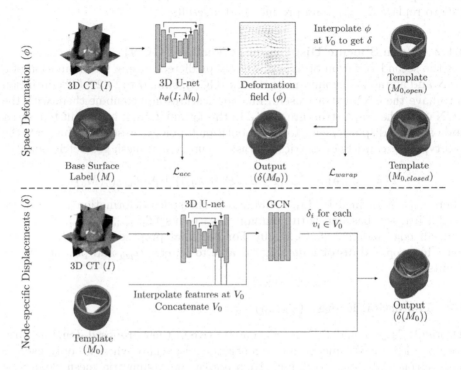

Fig. 1. (Top) Training steps using space deformation. (Bottom) Inference steps using node-specific displacements; training is performed with the same losses as space deformation using $\delta(M_0)$, M, $M_{0,open}$ and $M_{0,closed}$.

2.1.1 Space Deformation Field (U-Net)

For the first variation, we designed h_θ to be a convolutional neural network (CNN) that predicts a space deforming field $\phi \in \mathbb{R}^{H \times W \times D \times 3}$ for each $I \in \mathbb{R}^{H \times W \times D}$. From ϕ, we trilinearly interpolated at V_0 to obtain δ. To obtain a dense topology-preserving smooth field, we used the diffeomorphic B-spline transformation implemented by the Airlab library [22]. In this formulation, the loss typically consists of terms for task accuracy and field regularization [3,21]:

$$\mathcal{L}(M, M_0, \phi) = \mathcal{L}_{acc}(P(M), P(\phi(M_0))) + \lambda\mathcal{L}_{smooth}(\phi) \tag{3}$$

where $\mathcal{L}$ from Eq. 1 is modified to include ϕ, which fully defines δ. P is point sampling on the mesh surface, where for a volumetric mesh such as $\phi(M_0)$, points are sampled on the extracted base surface. In this work, L_{acc} is fixed for all methods to be the symmetric Chamfer distance:

$$\mathcal{L}_{acc}(A, B) = \frac{1}{|A|} \sum_{a \in A} \min_{b \in B} \|\mathbf{a} - \mathbf{b}\|_2^2 + \frac{1}{|B|} \sum_{b \in B} \min_{a \in A} \|\mathbf{b} - \mathbf{a}\|_2^2 \qquad (4)$$

For baseline comparison, we used the bending energy for $\mathcal{L}_{smooth}$ [11,21]. For our final proposed method, however, we show that the proposed distortion energy is able to replace $\mathcal{L}_{smooth}$ and produce better results.

2.1.2 Node-Specific Displacement Vectors (GCN)

For the second variation of h_θ, we directly predicted δ using a combination of a CNN and a graph convolutional network (GCN), similar to [5,27]. The intuition is to have the CNN extract useful imaging features and combine them with the GCN using the graph structure of M_0. In this formulation, it is difficult to restrict node-specific displacements to be smooth or topology-preserving. Instead, the loss typically consists of metrics for task accuracy and mesh geometric quality:

$$\mathcal{L}(M, M_0, \delta) = \mathcal{L}_{acc}(P(M), P(\delta(M_0))) + \boldsymbol{\lambda}^T \mathcal{L}_{geo}(\delta(M_0)) \qquad (5)$$

where $\mathcal{L}_{acc}$ is defined by Eq. 4. Similar to the space deformation method, we established the baseline with common $\mathcal{L}_{geo}$ terms ($\mathcal{L}_{normal}$, $\mathcal{L}_{edge}$, $\mathcal{L}_{lap}$ with uniform edge weights) [5,10,27,30]. For our final proposed method, we show that the proposed distortion energy is able to replace $\mathcal{L}_{geo}$ and produce better results.

2.2 Distortion Energy ($\mathcal{L}_{arap}$)

Although $\mathcal{L}_{smooth}$ and $\mathcal{L}_{geo}$ have been effective in their proposed domains, they are not ideal for volumetric mesh generation, especially when we only use the base surface labels for training. To preserve the volumetric mesh quality of $\delta(M_0)$, we used the deformation gradient $\mathbf{F}$ to allow for the calculation of various distortion energies [8]. For each tetrahedral element with original nodes $\bar{\mathbf{x}}_i$ and transformed nodes $\mathbf{x}_i = \delta_i(\bar{\mathbf{x}}_i)$:

$$\mathbf{F} = \left[\mathbf{x}_1 - \mathbf{x}_0 \middle| \mathbf{x}_2 - \mathbf{x}_0 \middle| \mathbf{x}_3 - \mathbf{x}_0 \right] \left[\bar{\mathbf{x}}_1 - \bar{\mathbf{x}}_0 \middle| \bar{\mathbf{x}}_2 - \bar{\mathbf{x}}_0 \middle| \bar{\mathbf{x}}_3 - \bar{\mathbf{x}}_0 \right]^{-1} \qquad (6)$$

which can be broken down into rotation and stretch components using the polar decomposition: $\mathbf{F} = \mathbf{RS}$. More specifically, we can use singular value decomposition (SVD) to obtain $\mathbf{F} = \mathbf{U\Sigma V}^T$, from which we can calculate $\mathbf{R} = \mathbf{UV}^T$ and $\mathbf{S} = \mathbf{V\Sigma V}^T$. Using these components, we can derive various task-related distortion energies [8,24]. We used the as-rigid-as-possible (ARAP) energy, a widely used energy for geometry processing. The ARAP energy density for each ith element can be expressed as:

$$\Psi_{arap}(i) = \|\mathbf{F} - \mathbf{R}\|_F^2 = \|\mathbf{R}(\mathbf{S} - \mathbf{I})\|_F^2 = \|(\mathbf{S} - \mathbf{I})\|_F^2 \qquad (7)$$

where $\mathbf{I}$ is the identity matrix and $\|\cdot\|_F$ is the Frobenius norm. Assuming equal weighting, $\mathcal{L}_{arap} = \frac{1}{N}\sum_{i=1}^{N}\Psi_{arap}(i)$ for N elements. Note that all operations are fully differentiable and therefore suitable for end-to-end learning, as long as $\mathbf{F}$ is full rank (i.e. no degenerate elements) and Σ has distinct singular values (i.e. $\mathbf{F} \neq \mathbf{I}$). In our experiments, both conditions were satisfied as long as we initialized δ with randomization. Computing $\mathbf{F}$ for hexahedral elements involves using quadrature points, but we were able to obtain just as accurate results in less training time by simply splitting each hexahedron into 6 tetrahedra and using the above formulation.

2.3 Weighted $\mathcal{L}_{arap}$ ($\mathcal{L}_{warap}$)

Due to the large structural differences in the leaflets during valve opening and closing, imposing $\mathcal{L}_{arap}$ with one template leads to suboptimal results. We addressed this with a simple weighting strategy:

$$\mathcal{L}_{warap} = \alpha_{closed}\mathcal{L}_{arap,closed} + \alpha_{open}\mathcal{L}_{arap,open} \tag{8}$$

where α is the softmax of distances from the output to the closed and open templates: $\alpha(i) = 1 - \exp(\mathcal{L}_{acc}(M_{0,i}, \delta(M_0)))/\sum_i \exp(\mathcal{L}_{acc}(M_{0,i}, \delta(M_0)))$. The final loss of our proposed method is then:

$$\mathcal{L}(M, M_0, \delta) = \mathcal{L}_{acc}(P(M), P(\delta(M_0))) + \lambda\mathcal{L}_{warap}(\delta(M_0)) \tag{9}$$

3 Experiments and Results

3.1 Data Acquisition and Preprocessing

We used a dataset of 88 CT scans from 74 different patients, all with tricuspid aortic valves. Of the 88 total scans, 73 were collected from IRB-approved TAVR patients at the Hartford hospital, all patients being 65–100 years old. The remaining 15 were from the training set of the MM-WHS public dataset [31]. For some Hartford scans, we included more than one time point. The splits for training, validation, and testing were 40, 10, 38, respectively, with no patient overlap between the training/validation and testing sets. We pre-processed all scans by thresholding the Hounsfield Units and renormalizing to [0, 1]. We resampled all images to a spatial resolution of $1 \times 1 \times 1\,\text{mm}^3$, and cropped and rigidly aligned them using three manually annotated landmarks, resulting in final images with [64, 64, 64] voxels.

We focused on 4 aortic valve components: the aortic wall and the 3 leaflets. The ground truth mesh labels were obtained via a semi-automated process [12], which included manually annotating the component boundaries and points on the surface. Commissures and hinges were separately labeled to assess correspondence accuracy. Two mesh templates for open and closed valves were created using Solidworks and Hypermesh, with the representative anatomical parameters in [26]. Each template has 19086 nodes and 9792 linear hexahedral elements.

Table 1. All evaluation metrics for baseline ($\mathcal{L}_{smooth}/\mathcal{L}_{geo}$), weighting ablation, ($\mathcal{L}_{arap}$), and proposed ($\mathcal{L}_{warap}$) methods. Values are combined across all patients and valve components (mean(std)). U-net: space deformation, GCN: node-specific displacements, CD: Chamfer Distance, HD: Hausdorff Distance, Corr: Correspondence error, Jac: scaled Jacobian determinant, (1): unitless, *: $p < 0.01$ between baseline and $\mathcal{L}_{warap}$, †: $p < 0.01$ between $\mathcal{L}_{arap}$ and $\mathcal{L}_{warap}$. Lower is better for all metrics.

	CD (mm)	HD (mm)	Corr (mm)	1 - Jac (1)	Skew (1)
U-net ($\mathcal{L}_{smooth}$)	0.63(19)	**3.60(128)**	1.72(117)	0.18(14)	0.40(19)
U-net ($\mathcal{L}_{arap}$)	0.61(20)	3.65(140)	1.61(101)	0.14(13)	0.32(20)
U-net ($\mathcal{L}_{warap}$)	**0.60(19)***	3.71(139)	**1.60(109)**	**0.12(12)***†	**0.30(18)***†
GCN ($\mathcal{L}_{geo}$)	0.70(23)	3.91(174)	**1.68(116)**	0.48(32)	0.62(21)
GCN ($\mathcal{L}_{arap}$)	0.72(23)	3.77(140)	1.81(110)	0.14(13)	0.33(20)
GCN ($\mathcal{L}_{warap}$)	**0.69(20)**†	**3.58(122)***†	1.74(108)	**0.11(11)***†	**0.28(18)***†

$$\text{U-net } (\mathcal{L}_{smooth}) \qquad \text{U-net } (\mathcal{L}_{warap}) \qquad \text{GCN } (\mathcal{L}_{geo}) \qquad \text{GCN } (\mathcal{L}_{warap})$$

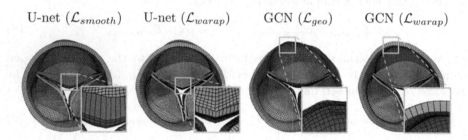

Fig. 2. Mesh predictions using space deformation (U-net) and node-specific displacements (GCN), with baseline regularization terms vs. $\mathcal{L}_{warap}$. The zoomed-in parts demonstrate the main advantage of our approach - good volumetric mesh quality throughout the entire mesh (shape closer to cube is better).

3.2   Implementation Details

We used Pytorch ver. 1.4.0 [17] to implement a variation of a 3D U-net for our CNN [20], and Pytorch3d ver. 0.2.0 [19] to implement the GCN. The basic CNN Conv unit was Conv3D-InstanceNorm-LeakyReLu, and the network had 4 encoding layers of ConvStride2-Conv with residual connections and dropout, and 4 decoding layers of Concatenation-Conv-Conv-Upsampling-Conv. The base number of filters was 16, and was doubled at each encoding layer and halved at each decoding layer. The GCN had 3 layers of graph convolution operations defined as $ReLU(\mathbf{w}_0^T \mathbf{f}_i + \sum_{j \in \mathcal{N}(i)} \mathbf{w}_1^T \mathbf{f}_j)$ and a last layer without $ReLU$. The input to the initial GCN layer was concatenation of vertex positions and point-sampled features from the last 3 U-net decoding layers. The GCN feature sizes were 227 for input, 128 for hidden, and 3 for output layers. We found λ for every experiment with a grid search based on validation error, ranging 5 orders of magnitude. $\lambda = 5$ for $\mathcal{L}_{warap}$. The value of λ is crucial for all experiments, but results were generally not too sensitive within one order of magnitude.

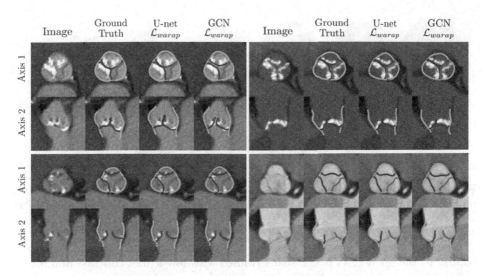

Fig. 3. CT images and predicted meshes at 2 orthogonal viewing planes. Each block of 8 images is a different test set patient. Y: aortic wall and R, G, B: valve leaflets

We used the Adam optimizer [9] with a fixed learning rate of 1e−4, batch size of 1, and 2000 training epochs. The models were trained with a B-spline deformation augmentation step, resulting in 80k training samples. All operations were performed on a single NVIDIA GTX 1080 Ti, with around ∼24 h of training time and maximum GPU memory usage of ∼1.2 GB. Inference takes ∼20 ms per image.

3.3   Spatial Accuracy and Volumetric Mesh Quality

We evaluated the mean and worst-case surface accuracy of our predicted meshes using the symmetric Chamfer distance (divided by 2 for scale) and Hausdorff distance, respectively. Note that the ground truth meshes are surface meshes, so we extracted the base surface of our predicted volumetric meshes for these calculations. For correspondence error, we measured the distance between hand-labeled landmarks (3 commissures and 3 hinges) and specific node positions on the predicted meshes. We also checked the predicted meshes' geometric quality using the scaled Jacobian determinant (−1 to 1; 1 being an optimal cube) and skew metrics (0 to 1; 1 being a degenerate element) [23]. Statistical significance was evaluated with a paired Student's t-test between our proposed method vs. the baseline/ablation experimental groups. The baseline was established with U-net + $\mathcal{L}_{smooth}$ and GCN + $\mathcal{L}_{geo}$, and the ablation study was for comparing against non-weighted $\mathcal{L}_{arap}$.

For both deformation strategies, our proposed method with $\mathcal{L}_{warap}$ holistically outperformed the baseline and non-weighted $\mathcal{L}_{arap}$ (Table 1, Fig. 2, 3). As expected, the most significant improvement was in element quality, and our method also showed slight improvements in spatial accuracy. Our model was

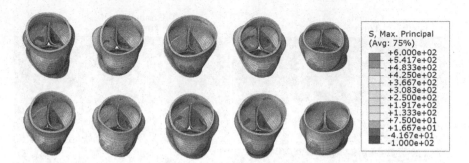

Fig. 4. FEA results using U-net $+ \mathcal{L}_{warap}$ meshes for 10 test set patients. Values indicate maximum principal stress in the aortic wall and leaflets during diastole (kPa).

robust to the noisy TAVR CT scans riddled with low leaflet contrast and lots of calcification, and was applicable to various phases of the cardiac cycle.

3.4   FE Stress Analysis During Valve Closure

Figure 4 shows the results of FEA performed with volumetric meshes generated directly from our method (i.e. no post-processing). We used an established protocol [12,29] with the static and nonlinear analysis type on Abaqus/Standard. Briefly, we simulated valve closure during diastole by applying an intraluminal pressure (P = 16 kPa) to the upper surface of the leaflets and coronary sinuses and a diastolic pressure (P = 10 kPa) to the lower portion of the leaflets and intervalvular fibrosa. The resulting maximum principal stresses in the aortic wall and leaflets were approximately 100–500 kPa (Fig. 4), consistent with previous studies [12,29]. This demonstrates the predicted meshes' viability for FEA, and thus potential clinical relevance in the form of biomechanics studies and TAVR planning.

Note that we can easily extend the analysis using the same predicted meshes, such as by using a material model that incorporates the strain energy function of fibrous and anisotropic structures. In this work, we evaluated the stresses based on the fact that the aortic valve is approximately statically determinate [13,14].

3.5   Limitations and Future Works

There were no hard failure cases of our model, but in future works, we hope to enable expert-guided online updates for more rigorous quality control during test time. We will also aim to address our main limitation of requiring two well-defined volumetric mesh templates. Lastly, we will expand our framework to other important structures for TAVR simulations, such as calcification, myocardium, and ascending aorta.

4   Conclusion

We presented a novel approach for predicting aortic valve volumetric FE meshes from 3D patient images. Our method provides a principled end-to-end learnable way to optimize the volumetric element quality within a deep learning template deformation framework. Our model can predict meshes with good spatial accuracy, element quality, and FEA viability.

Acknowledgments and Conflict of Interest. This work was supported by the NIH R01HL142036 grant. Dr. Wei Sun is a co-founder and serves as the Chief Scientific Advisor of Dura Biotech. He has received compensation and owns equity in the company.

References

1. Botsch, M., Kobbelt, L., Pauly, M., Alliez, P., Lévy, B.: Polygon Mesh Processing. CRC Press, Boca Raton (2010)
2. Caballero, A., Mao, W., McKay, R., Sun, W.: The impact of balloon-expandable transcatheter aortic valve replacement on concomitant mitral regurgitation: a comprehensive computational analysis. J. Roy. Soc. Interface **16**(157), 20190355 (2019)
3. Dalca, A.V., Balakrishnan, G., Guttag, J., Sabuncu, M.R.: Unsupervised learning of probabilistic diffeomorphic registration for images and surfaces. Med. Image Anal. **57**, 226–236 (2019)
4. Ghesu, F.C., et al.: Marginal space deep learning: efficient architecture for volumetric image parsing. IEEE Trans. Med. Imaging **35**(5), 1217–1228 (2016)
5. Gkioxari, G., Malik, J., Johnson, J.: Mesh R-CNN. In: Proceedings of the IEEE International Conference on Computer Vision, pp. 9785–9795 (2019)
6. Grbic, S., et al.: Image-based computational models for TAVI planning: from CT images to implant deployment. In: Mori, K., Sakuma, I., Sato, Y., Barillot, C., Navab, N. (eds.) MICCAI 2013. LNCS, vol. 8150, pp. 395–402. Springer, Heidelberg (2013). https://doi.org/10.1007/978-3-642-40763-5_49
7. Ionasec, R.I., et al.: Patient-specific modeling and quantification of the aortic and mitral valves from 4-D cardiac CT and tee. IEEE Trans. Med. Imaging **29**(9), 1636–1651 (2010)
8. Kim, T., Eberle, D.: Dynamic deformables: implementation and production practicalities. In: ACM SIGGRAPH 2020 Courses, pp. 1–182 (2020)
9. Kingma, D.P., Ba, J.: Adam: a method for stochastic optimization. arXiv preprint arXiv:1412.6980 (2014)
10. Kong, F., Wilson, N., Shadden, S.C.: A deep-learning approach for direct whole-heart mesh reconstruction. arXiv preprint arXiv:2102.07899 (2021)
11. Lee, M.C.H., Petersen, K., Pawlowski, N., Glocker, B., Schaap, M.: TeTrIS: template transformer networks for image segmentation with shape priors. IEEE Trans. Med. Imaging **38**(11), 2596–2606 (2019)
12. Liang, L., et al.: Machine learning-based 3-D geometry reconstruction and modeling of aortic valve deformation using 3-D computed tomography images. Int. J. Numer. Methods Biomed. Eng. **33**(5), e2827 (2017)
13. Liu, M., Liang, L., Liu, H., Zhang, M., Martin, C., Sun, W.: On the computation of in vivo transmural mean stress of patient-specific aortic wall. Biomech. Model. Mechanobiol. **18**(2), 387–398 (2019). https://doi.org/10.1007/s10237-018-1089-5

14. Miller, K., Lu, J.: On the prospect of patient-specific biomechanics without patient-specific properties of tissues. J. Mech. Behav. Biomed. Mater. **27**, 154–166 (2013)
15. Pak, D.H., Caballero, A., Sun, W., Duncan, J.S.: Efficient aortic valve multilabel segmentation using a spatial transformer network. In: 2020 IEEE 17th International Symposium on Biomedical Imaging (ISBI), pp. 1738–1742. IEEE (2020)
16. Pak, D.H., et al.: Weakly supervised deep learning for aortic valve finite element mesh generation from 3D CT images. In: Feragen, A., Sommer, S., Schnabel, J., Nielsen, M. (eds.) IPMI 2021. LNCS, vol. 12729, pp. 637–648. Springer, Cham (2021). https://doi.org/10.1007/978-3-030-78191-0_49
17. Paszke, A., et al.: Automatic differentiation in PyTorch (2017)
18. Pouch, A.M., et al.: Medially constrained deformable modeling for segmentation of branching medial structures: application to aortic valve segmentation and morphometry. Med. Image Anal. **26**(1), 217–231 (2015)
19. Ravi, N., et al.: Accelerating 3D deep learning with PyTorch3D. arXiv:2007.08501 (2020)
20. Ronneberger, O., Fischer, P., Brox, T.: U-Net: convolutional networks for biomedical image segmentation. In: Navab, N., Hornegger, J., Wells, W.M., Frangi, A.F. (eds.) MICCAI 2015. LNCS, vol. 9351, pp. 234–241. Springer, Cham (2015). https://doi.org/10.1007/978-3-319-24574-4_28
21. Rueckert, D., Sonoda, L.I., Hayes, C., Hill, D.L., Leach, M.O., Hawkes, D.J.: Non-rigid registration using free-form deformations: application to breast MR images. IEEE Trans. Med. Imaging **18**(8), 712–721 (1999)
22. Sandkühler, R., Jud, C., Andermatt, S., Cattin, P.C.: AirLab: autograd image registration laboratory. arXiv preprint arXiv:1806.09907 (2018)
23. Schroeder, W.J., Lorensen, B., Martin, K.: The visualization toolkit: an object-oriented approach to 3D graphics. Kitware (2004)
24. Smith, B., Goes, F.D., Kim, T.: Analytic eigensystems for isotropic distortion energies. ACM Trans. Graph. (TOG) **38**(1), 1–15 (2019)
25. Smith, M.: ABAQUS/Standard User's Manual, Version 6.9. Dassault Systèmes Simulia Corp, United States (2009)
26. Sun, W., Martin, C., Pham, T.: Computational modeling of cardiac valve function and intervention. Ann. Rev. Biomed. Eng. **16**, 53–76 (2014)
27. Wang, N., Zhang, Y., Li, Z., Fu, Y., Liu, W., Jiang, Y.G.: Pixel2Mesh: generating 3D mesh models from single RGB images. In: Proceedings of the European Conference on Computer Vision (ECCV), pp. 52–67 (2018)
28. Wang, Q., Kodali, S., Primiano, C., Sun, W.: Simulations of transcatheter aortic valve implantation: implications for aortic root rupture. Biomech. Model. Mechanobiol. **14**(1), 29–38 (2015). https://doi.org/10.1007/s10237-014-0583-7
29. Wang, Q., Primiano, C., McKay, R., Kodali, S., Sun, W.: CT image-based engineering analysis of transcatheter aortic valve replacement. JACC Cardiovasc. Imaging **7**(5), 526–528 (2014)
30. Wickramasinghe, U., Remelli, E., Knott, G., Fua, P.: Voxel2Mesh: 3D mesh model generation from volumetric data. In: Martel, A.L., et al. (eds.) MICCAI 2020. LNCS, vol. 12264, pp. 299–308. Springer, Cham (2020). https://doi.org/10.1007/978-3-030-59719-1_30
31. Zhuang, X., Shen, J.: Multi-scale patch and multi-modality atlases for whole heart segmentation of MRI. Med. Image Anal. **31**, 77–87 (2016)

Ultrasound Video Transformers
for Cardiac Ejection Fraction Estimation

Hadrien Reynaud[1(✉)], Athanasios Vlontzos[1], Benjamin Hou[1], Arian Beqiri[2],
Paul Leeson[2,4], and Bernhard Kainz[1,3]

[1] Department of Computing, Imperial College London, London, UK
hadrien.reynaud19@imperial.ac.uk
[2] Ultromics Ltd., Oxford, UK
[3] Friedrich–Alexander University Erlangen–Nürnberg, DE, Erlangen, Germany
[4] John Radcliffe Hospital, Cardiovascular Clinical Research Facility, Oxford, UK

Abstract. Cardiac ultrasound imaging is used to diagnose various heart
diseases. Common analysis pipelines involve manual processing of the
video frames by expert clinicians. This suffers from intra- and inter-
observer variability. We propose a novel approach to ultrasound video
analysis using a transformer architecture based on a Residual Auto-
Encoder Network and a BERT model adapted for token classification.
This enables videos of any length to be processed. We apply our model
to the task of End-Systolic (ES) and End-Diastolic (ED) frame detection
and the automated computation of the left ventricular ejection fraction.
We achieve an average frame distance of 3.36 frames for the ES and 7.17
frames for the ED on videos of arbitrary length. Our end-to-end learnable approach can estimate the ejection fraction with a MAE of 5.95
and R^2 of 0.52 in 0.15 s per video, showing that segmentation is not the
only way to predict ejection fraction. Code and models are available at
https://github.com/HReynaud/UVT.

Keywords: Transformers · Cardiac · Ultrasound

1 Introduction

Measurement of Left Ventricular Ejection Fraction (LVEF) is a commonly used
tool in clinical practice to aid diagnosis of patients with heart disease and to
assess options for life-prolonging therapies. The LVEF is the ratio between the
stroke volume, which is the difference between End-Diastolic (ED) and End-
Systolic (ES) volumes, and the ED volume of the left ventricle.

In primary and secondary care, 2D Ultrasound (US) video acquisition of the
standard apical four-chamber view is used to approximate LVEF from manually
delineated luminal areas in the left ventricle in one chosen ED and one ES frame.

Electronic supplementary material The online version of this chapter (https://
doi.org/10.1007/978-3-030-87231-1_48) contains supplementary material, which is
available to authorized users.

The biplane method of disks, which requires both 2-chamber and 4-chamber views, is the currently recommended two-dimensional method to assess LVEF [6], with its limitations known in literature [20]. The laborious nature of the data processing and substantial inter- and intra-operator variability makes this approach infeasible for high throughput studies, e.g., population health screening applications. Current clinical practice already neglects the recommendation to repeat this process on at least five heartbeats [15] and commonly only a single measurement is acquired to mitigate clinicians' workload. Our objective is to estimate LVEF accurately from US video sequences of arbitrary length, containing arbitrarily many heart-beats, and to localize relevant ES and ED frames.

This problem has been recognized in the medical image analysis community. Initially, techniques to automatically segment the left ventricle have been proposed [21,22] to support the manual LVEF estimation process. Recently, robust step-by-step processing pipelines have been proposed to identify relevant frames, segment them, calculate the LVEF and predict the risk for cardiac disease [17].

To the best of our knowledge, all existing techniques for automatically processing US video sequences follow the paradigm of discrete frame processing with limited temporal support. US videos, however, can be of arbitrary length and the cardiac cycle varies in length. Frame-by-frame [1,2] processing neglects the information encoded in the change over time, or requires heuristic frame sampling methods to form a stack of selected frames to enable spatio-temporal support in deep convolutional networks [17].

In this paper, we postulate that processing US videos should be considered more similar to processing language. Thus, we seek a model that can interpret sequences across their entire temporal length, being able to reason through comparison between heartbeats and to solve tasks on the entirety of the acquired data without introducing too many difficult to generalize heuristics. Following this idea, we propose a new US video ES/ED recognition and LVEF prediction network. We use an extended transformer architecture to regress simultaneously the LVEF and the indices of the ES and ED frames.

Our **contribution** is two-fold. (a) We evaluate a new paradigm for US video analysis based on sequence-to-sequence transformer models on a large database of more than 10,000 scans [17] for the clinically relevant task of LVEF estimation. (b) We introduce a modified transformer architecture that is able to process image sequences of variable length.

Related Works: Early automatic detection algorithms embed videos on a manifold and perform low-dimensional clustering [7]. Other methods use convolutional neural networks (CNN) to locate and monitor the center of the left ventricle (LV) of the heart to extract the necessary information to categorize frames as ES or ED [26]. In recent work, [11] introduces a CNN followed by an RNN that utilizes Doppler US videos to detect ED.

For the estimation of both the ES/ED frame indices and the LVEF, [14] assess three algorithms performing LVEF estimation in a multi-domain imaging scenario. Deep convolutional networks have been extensively used in various steps. In [21], the LV is segmented from standard plane US images with the help of a

U-Net [19]. ES, ED frames are heuristically identified. In [5] the authors extract spatio-temporal features directly from the input video, classifying whether the frames belong to a systole or a diastole, identifying ES and ED as the switching points between the two states. The authors of [22] leverage deep learning techniques throughout their proposed method and directly learn to identify the ES and ED frames which they subsequently feed into a U-Net to segment the LV followed by a metrics-based estimation of the LVEF. Segmentation has been the most explored method for the analysis of cardiac functions with neural networks [18,24]. [17] propose an end-to-end method that leverages spatio-temporal convolutions to estimate the LVEF. The extracted features in conjunction with semantic segmentation of the LV enable a beat-by-beat assessment of the input. Direct estimation of the Systolic Volume (SV) has been explored on cine MRI [13] and cine US [4]. Trained to leverage temporal information, their network directly estimates the SV, and the ES/ED frames from single-beat cardiac videos. However, these methods require a fixed number of input frames containing a single cardiac cycle and use LSTMs for temporal feature extraction, which are well known for forgetting initial elements when the sequences become longer [23]. Our approach is more closely related to the latter three approaches, which we will be comparing against. We explore the strength of transformers applied to video data like [8,12,16] have done recently, and build our own architecture.

2   Method

Just as a piece of text in natural language can have variable length and is required in its entirety to be understood, a US video can be of arbitrary duration and is needed in full for accurate interpretation and reasoning. To this end, we propose to use a new transformer model to interpret US videos. Our end-to-end trainable method comprises three distinct modules: (a) An encoder tasked with dimensionality reduction, (b) a Bidirectional Encoder Representations from Transformers (BERT)-based [3] module providing spatio-temporal reasoning capabilities, and (c) two regressors, one labeling ES and ED frames and the other estimating the LVEF. An overview of this model is shown in Fig. 1.

Dimensionality Reduction: Using full size image frames directly as inputs to a transformer is infeasible, as it would require an excessive amount of computational power. To make the problem computationally tractable and to allow BERT to understand the spatial structure of the images, we make use of the encoding part of a ResNetAE [10] to distil the US frames into a smaller dimensional embedding.

The network uses multi-scale residual blocks [9] in both the encoder and the decoder in order to incorporate information across dimensions. We optimize the hyper-parameters associated with the AE architecture, such as the depth of the encoder, the size of the latent space, by first performing a reconstruction task on the US dataset. The encoder setup of the optimal architecture is then used as the encoding module of our method. Each frame from the clip is distilled by the encoder into a $1024D$ vector. The resulting embeddings are stacked together

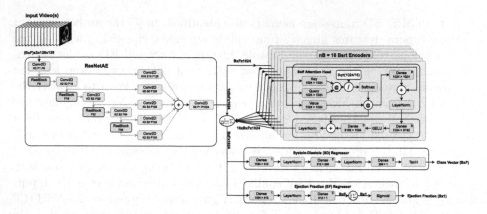

Fig. 1. Overview of the proposed architecture; Left to right: Clips are reduced in dimensions through the ResAE, spatio-temporal information is extracted through the BERT and then passed to each regression branch. The Systole-Diastole (SD) Regressor predicts the ES/ED indices while the EF Regressor predicts the LVEF. The @ operation is the dot product.

to produce the initial embedding of the clip, characterized by a shape $Batch \times N_{frames} \times 1024$. We use these latent embeddings as the input of the transformer and the combined architecture is trained as a whole end-to-end. Cascading the encoder with the transformer model presents an important benefit as the weights of the encoder are learned to optimize the output from the transformer. This is in contrast to using a pretrained dimensionality reduction network where the encoder would be task-agnostic.

Spatio-Temporal Reasoning: In order to analyze videos of arbitrary length we use a BERT encoder [25] to which we attach a regression network to build a Named Entity Recognition (NER) model for video. This acts as a spatio-temporal information extractor. The extracted embeddings E from the ResNetAE encoding step, are used as inputs for the BERT encoder. As in [3], the kth encoder can be characterized as $B_k(E) = \text{LayerNorm}(D_{k,c}(\text{GELU}(D_{k,b}(A_k(E)))) + S_k(E)$, where,

$$S_k(E) = \text{Softmax}\left(\frac{Q_k(E)K_k^T(E)}{\sqrt{\frac{nD}{nB}}}\right)V_k(E),$$

$$A_k(E) = \text{LayerNorm}(D_{k,a}(S_k(E)) + E).$$

$S_k(E)$ and $A_k(E)$ describe the Self-Attention block and Attention block respectively. The Query, Key and Value are parameterized as linear layers Q_k, K_k, V_k. While, $D_{k,\{a,b,c\}}$ are the intermediate linear (dense) layers in the BERT Encoder. We keep the key parameters nB, the number of BERT encoders, and nD the dimensionality of the embeddings, similar to [3], setting them to $nB = 16$

and $nD = 1024$ respectively. During training, dropout layers act as regularizers with a drop probability of 0.1.

Regressing the Output: Following the spatio-temporal information extraction, the resulting features from the k BERT encoders $B_k(E)$ are averaged together with the ResNetAE output to

$$M(E) = \frac{1}{nB + 1} \left(E + \sum_{k}^{nB} B_k(E) \right), \tag{1}$$

and passed through two regressors tasked with predicting ES, ED frame indices, $R_{SD}(M(E))$, and the LVEF, $R_{EF}(M(E))$, respectively. We define the output of $R_{SD}(M(E))$ as the output of three linear layers interleaved with layer-normalization, with a *tanh* activation at the end. LVEF is characterized by:

$$R_{EF}(M(E)) = \text{Sigmoid} \left(\frac{1}{nF} \sum_{f}^{nF} (\text{D}_{\text{EF},2} (\text{LayerNorm} (\text{D}_{\text{EF},1} (M(E))))) \right), \tag{2}$$

with nF the number of input frames. Thus, LVEF is estimated through a regression network which reduces the embedding dimension to 1 for each input frame. We then take the average of the predictions over the frames to output a single LVEF prediction per video. The LVEF prediction training is done with a combination of losses and regularization to address imbalance in the distribution of LVEF in the training set. We use both Mean Squared Error $\mathcal{L}_{MSE}(\hat{y}, y) = \frac{1}{nF} \sum_{f}^{nF} (\hat{y_f}, y_f)^2$ and Mean Average Error $\mathcal{L}_{MAE} = \frac{1}{nF} \sum_{f}^{nF} ||\hat{y_f}, y_f||$ to ensure that the network will be penalized exponentially when the error is large, but that the loss will not decrease too much when reaching small errors. Thus, the network will continue to learn even if its predictions are already close to the ground truth. A regularization term $\mathcal{R}(y) = (1 - \alpha) + \left(\alpha \cdot \frac{||y - \gamma||}{\gamma} \right)$ helps the network to emphasize training on the LVEF objective, weighing down LVEF estimates which are away from γ, where γ is chosen close to the mean of all the LVEF on the training set. The α parameter is a scalar which adjusts the maximum amount of regularization applied to LVEF close to γ. Thus, our overall objective loss can be written as $\mathcal{L}_{EF} = (\mathcal{L}_{MSE} + \mathcal{L}_{MAE}) \cdot \mathcal{R}(y)$.

3   Experimentation

Dataset: For all of our experiments, we use the Echonet-Dynamic dataset [17] that consists of a variety of pathologies and healthy hearts. It contains 10,030 echocardiogram videos of varied length, frame rate, and image quality, all containing 4-chamber views. We pad all the frames with zeros to go from 112×112 to 128×128 pixel size frames. The videos represent at least one full heart cycle, but most of them contain three or more cycles. In each clip, however, only one cycle has the ES and ED frames labelled. For each labelled frame, the index in the video and the corresponding ground truth segmentation of the left ventricle is available. We have access to the frame rate, left ventricular volumes in milliliters for the ES and ED frames and the ejection fraction, computed from these heart volumes via LVEF $= \frac{EDV-ESV}{ESV} * 100$. Following Echonet [17], we split the dataset in training, validation and testing subsets; dedicating 75% to our training subset and equally splitting validation and testing with 12.5% each.

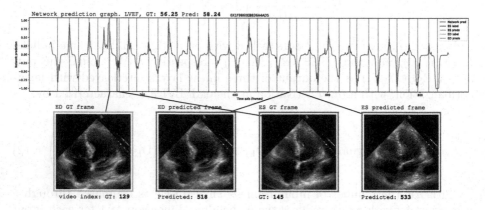

Fig. 2. Example network response (top), labelled pair and one predicted pair of US frames (bottom) for a very long video. The frames are shown for the labelled ED and ES indices and one of the multiple sets of the ES/ED frames produced by the network from a different heartbeat than those of the labelled indices.

Video Sampling Process: For training, we need videos of fixed size length to enable batch training and we need to know where the ES and ED frames are. In our data, only single ES and ED frames are labelled for each video, leaving many of the other true ES and ED frames unlabeled. We choose to create 128 frames long sequences, based on the distribution of distances between consecutive ES and ED frames in the training set. Any sequence longer than 128 frames is sub-sampled by a factor of 2. As training the transformer with unlabeled ES and ED frames would considerably harm performance, we tried two approaches to mitigate the lack of labels: (1) Guided Random Sampling: We sample the labeled frames and all frames in-between. We then add 10% to 70% of the distance between the two labelled frames before and after the sampled frames. To match the desired 128 length clip, where appropriate we pad it with black frames. The resulting clip is masked such that the attention heads ignore the empty frames. (2) Mirroring: We augment our clips by mirroring the transition frames between the two labelled frames, and placing them after the last labelled frame. Given the sequence $S = [f_{ES}, f_{T_1}, ..., f_{T_N}, f_{ED}]$, where f_{T_N} stands for the Nth transition frame, we augment the sequence such that it reaches a length superior to 128 frames by mirroring the frames around f_{ED} creating the new sequence $S' = [f_{ES}, f_{T_1}, ..., f_{T_N}, f_{ED}, f_{T_N}, ..., f_{T_1}, f_{ES}...]$. Once the sequences exceed 128 indices, we randomly crop them to 128 frames. Doing

Table 1. *Top:* Results for ES and ED detection on the test set (1024 videos) stated in **average Frame Distance (aFD) (standard deviation)** over all frame distances. Video sampling methods are equivalent for training and testing in each experiment. The rejected column indicates the number of videos for which the network did not find clear index positions, *i.e.*, not enough zero crossings in the output sequence. Our clinical expert confirmed not ideal image quality for these. US [4] and MRI [13] both train and test their approaches on single beat videos of fixed length from private datasets. *Bottom:* Results for LVEF prediction, compared to ground truth, labeled LVEF. Echonet (1 & 2), R3D and MC3 all come from [17] and use combinations of neural networks and heuristics. "R." and "M." refer to the random and mirror video sampling methods; "/" means not available, *e.g.*, from literature. Echonet (1) is restrictive through the use of segmentations and a fixed number of frames per beat. It aggregates per-beat predictions into a final estimate for entire sequences. Processing time is 1.6 s per beat and a minimum number of frames, *e.g.*, 32, have to be available for each beat during retrospective analysis. Echonet (2) uses segmentation and all the available frames over a single, selected beat, with no further processing. Our average processing time is 0.15 s for entire videos and our method runs in real-time during examination. $\uparrow/\downarrow$ mean higher/lower values are better.

Video sampling	Single heartbeat		Full video		
	ED and ES index detection				
	ES $\downarrow$	ED $\downarrow$	ES $\downarrow$	ED $\downarrow$	Rejected $\downarrow$
US [4]	4.1	3.7	/	/	/
MRI [13]	0.44 (0.46)	0.38 (0.39)	/	/	/
R./Cla. (ours)	1.84 (2.73)	2.34 (3.15)	**2.86 (6.43)**	7.88 (11.03)	3
R./Reg. (ours)	2.14 (2.46)	3.05 (3.74)	3.66 (7.96)	9.60 (30.71)	4
M./Cla. (ours)	0.09 (1.25)	**0.14 (1.49)**	5.49 (12.94)	9.21 (14.15)	31
M./Reg. (ours)	**0.08 (1.53)**	0.15 (1.89)	3.35 (6.79)	**7.17 (12.92)**	6
	LVEF prediction				
	MAE $\downarrow$	RMSE $\downarrow$	R^2 $\uparrow$ MAE $\downarrow$	RMSE $\downarrow$	R^2 $\uparrow$
EchoNet (1) [17]	4.22	5.56	0.79 4.05	5.32	0.81
EchoNet (2) [17]	7.35	9.53	0.40 /	/	/
R3D [17]	7.63	9.75	0.37 /	/	/
MC3 [17]	6.59	9.39	0.42 /	/	/
R. (ours)	5.54 (5.17)	7.57	0.61 6.77 (5.47)	8.70	0.48
M. (ours)	5.32 (4.90)	7.23	0.64 5.95 (5.90)	8.38	0.52

so ensures that the first frame is not always labelled. This augmentation follows common practices when using a transformer model in order to provide seamless computational implementation. In addition, it ensures that all the ES and ED frames that our transformer sees, are labelled as such, while having no empty frames and retaining spatio-temporal coherence.

The labels for these frames are defined depending on the method we use to predict the frame type. When regressing, we set the ES to -1 and the ED to 1. We adapt the heuristics used in [4,13] to smooth the transition between the ES

Table 2. Ablation study. ES|ED columns show the aFD and MAE|RMSE|R^2 shows LVEF scores. (Em. = Embeddings; Seq. len. = Sequence length)

Exp	BERTs	Em.	Linear	Seq. len.	ES|ED	MAE|RMSE|R^2	Inference	Params.
Ours	16	1024	8192	128	3.35| 7.17	5.95|8.38|0.52	0.15 s	346.8M
Reduced 1	4	256	1024	64	4.28|10.12	6.06|8.37|0.51	0.13 s	6.8M
Reduced 2	1	128	512	64	8.18|35.42	7.34|9.96|0.30	0.13 s	2.7M
No EF	16	1024	8192	128	3.71| 7.24	/	0.15 s	346.3M
No SD	16	1024	8192	128	/	5.95|8.38|0.52	0.15 s	346.2M

and ED volumes. When using a classification objective, we define three classes: transition (0), ED (1), ES (2), with no form of smoothing. In the mirror-sampling method we cannot use the same heuristics to smooth the volume transition when regressing. Instead, we apply an x^3 function scaled between the ES and ED to soften the loss around the class peaks $(-1, 1)$.

ES and ED Frame Detection: We train our end-to-end model to predict the ES, ED indices and LVEF using the two video sampling methods and two architecture variations for the SD branch. The architecture variation consists of outputting a $3 \times nF$ matrix instead of a $1 \times nF$ vector for the ES/ED prediction and replacing the activation function by *Softmax*. Instead of approximating the volume, like [4,13], of the left ventricle chamber, our architecture classifies each frame in either *transition*, ES or ED. For the regression (Reg.) model, we use the MSE loss. For the classification (Cla.) model, we use the weighted Cross-Entropy Loss, with weights $(1, 5, 5)$, corresponding to our classes [transition, ES, ED]. The results presented in the top half of Table 1 are obtained while training simultaneously the SD and EF branches of the network. As common in literature [4,13] we state average Frame Distance (aFD) as $aFD = \frac{1}{N} \sum_{n=1}^{N} |\hat{i}_n - i_n|$ and $std = sqrt\left(\frac{1}{N} \sum_{n=1}^{N} |\hat{i}_n - \bar{i}|^2\right)$, where i_n is the ES or ED label index for the nth test video and $\hat{i}_n$ is the corresponding predicted index.

LVEF Prediction: The prediction of the LVEF is done in parallel with the ES and ED detection. We apply a single pre-processing step which is the scaling of the value from 0–100 to 0–1. The results are presented in the 0–100 range. For the regularization term $\mathcal{R}(y)$ we empirically chose $\alpha = 0.7$ and set $\gamma = 0.65$ to match the over-represented LVEF values from the LVEF distribution in the training set. Results for LVEF are summarized in the bottom half of Table 1 and compared to results from literature.

Ablation Study: We tested the capacity of our M. Reg. model. From Table 2 we observe that using only four BERT encoders achieves results similar to the original architecture. Removing the EF or SD branch has little to no impact on the other branch. Most of the computing time is used in the ResNetAE encoder while most of the parameters are in the BERT encoders.

Discussion: Manual measurement of LVEF exhibits an inter-observer variability of 7.6%–13.9% [17]. From Table 1, we can speculate that transformer models are en par with human experts for LVEF estimation and ED/ES identification.

We observe that our model predicts on average a 3% higher LVEF than measured by the manual ground truth observers, especially on long videos. We hypothesize that the model learned to correlate spatio-temporal image features to ventricular blood volume, therefore sometimes selecting different heartbeats than labelled in the ground truth, *e.g.*, Fig. 2. While guidelines state that operators should select the cycle with the largest change between the ED area and the ES area, in practice, operators will usually just pick a "good" heart cycle, which might also be the case for our ground truth. This hypothesis requires clinical studies and cross-modal patient studies in future work.

A limitation of our study is that aFD is not an ideal metric, since over-prediction would lead to inflated scores. Our model predicts distinct ES/ED frames as shown in Fig. 2. This can also be shown by calculating the average heart rate from peak predictions. When accounting for frame rate differences this results in a reasonable 53 beats per minutes.

Implementation: PyTorch 1.7.1+cu110 with two Nvidia Titan RTX GPUs.

4   Conclusion

We have discussed probably the first transformer architecture that is able to analyze US videos of arbitrary length. Our model outperforms all existing techniques when evaluated on adult cardiac 4-chamber view sequences, where the task is to accurately find ES and ED frame indices. We also outperform heuristic-free methods on LVEF prediction. In the future we would expect transformers to play a more prominent role for temporally resolved medical imaging data.

Acknowledgements. This work was supported by Ultromics Ltd., the UKRI Centre for Doctoral Training in Artificial Intelligence for Healthcare (EP/S023283/1), and the UK Research and Innovation London Medical Imaging and Artificial Intelligence Centre for Value Based Healthcare.

References

1. Baumgartner, C.F., et al.: SonoNet: real-time detection and localisation of fetal standard scan planes in freehand ultrasound. IEEE Trans. Med. Imaging **36**(11), 2204–2215 (2017)
2. Carneiro, G., Nascimento, J.C., Freitas, A.: The segmentation of the left ventricle of the heart from ultrasound data using deep learning architectures and derivative-based search methods. IEEE Trans. Image Process. **21**(3), 968–982 (2011)
3. Devlin, J., Chang, M.W., Lee, K., Toutanova, K.: BERT: pre-training of deep bidirectional transformers for language understanding. arXiv:1810.04805 (2018)

4. Dezaki, F.T., et al.: Deep residual recurrent neural networks for characterisation of cardiac cycle phase from echocardiograms. In: Cardoso, M.J., et al. (eds.) DLMIA/ML-CDS-2017. LNCS, vol. 10553, pp. 100–108. Springer, Cham (2017). https://doi.org/10.1007/978-3-319-67558-9_12

5. Fiorito, A.M., Østvik, A., Smistad, E., Leclerc, S., Bernard, O., Lovstakken, L.: Detection of cardiac events in echocardiography using 3D convolutional recurrent neural networks. In: 2018 IEEE IUS, pp. 1–4 (2018)

6. Folland, E., Parisi, A., Moynihan, P., Jones, D.R., Feldman, C.L., Tow, D.: Assessment of left ventricular ejection fraction and volumes by real-time, two-dimensional echocardiography. A comparison of cineangiographic and radionuclide techniques. Circulation 60(4), 760–766 (1979)

7. Gifani, P., Behnam, H., Shalbaf, A., Sani, Z.A.: Automatic detection of end-diastole and end-systole from echocardiography images using manifold learning. Physiol. Measur. 31(9), 1091–1103 (2010)

8. Girdhar, R., Carreira, J., Doersch, C., Zisserman, A.: Video action transformer network. In: Proceedings of the IEEE/CVF Conference on Computer Vision and Pattern Recognition, pp. 244–253 (2019)

9. He, K., Zhang, X., Ren, S., Sun, J.: Identity mappings in deep residual networks. In: Leibe, B., Matas, J., Sebe, N., Welling, M. (eds.) ECCV 2016. LNCS, vol. 9908, pp. 630–645. Springer, Cham (2016). https://doi.org/10.1007/978-3-319-46493-0_38

10. Hou, B.: ResNetAE (2019). https://github.com/farrell236/ResNetAE. Accessed 22 June 2021

11. Jahren, T.S., Steen, E.N., Aase, S.A., Solberg, A.H.S.: Estimation of end-diastole in cardiac spectral doppler using deep learning. IEEE Trans. Ultrason. Ferroelectr. Freq. Control 67(12), 2605–2614 (2020)

12. Kalfaoglu, M.E., Kalkan, S., Alatan, A.A.: Late temporal modeling in 3D CNN architectures with BERT for action recognition. In: Bartoli, A., Fusiello, A. (eds.) ECCV 2020. LNCS, vol. 12539, pp. 731–747. Springer, Cham (2020). https://doi.org/10.1007/978-3-030-68238-5_48

13. Kong, B., Zhan, Y., Shin, M., Denny, T., Zhang, S.: Recognizing end-diastole and end-systole frames via deep temporal regression network. In: Ourselin, S., Joskowicz, L., Sabuncu, M.R., Unal, G., Wells, W. (eds.) MICCAI 2016. LNCS, vol. 9902, pp. 264–272. Springer, Cham (2016). https://doi.org/10.1007/978-3-319-46726-9_31

14. Kupinski, M.A., et al.: Comparing cardiac ejection fraction estimation algorithms without a gold standard. Acad. Radiol. 13(3), 329–337 (2006)

15. Lang, R.M., et al.: Recommendations for cardiac chamber quantification by echocardiography in adults: an update from the American society of echocardiography and the European association of cardiovascular imaging. Eur. Heart J. Cardiovasc. Imaging 16(3), 233–271 (2015)

16. Måløy, H.: EchoBERT: a transformer-based approach for behavior detection in echograms. IEEE Access 8, 218372–218385 (2020)

17. Ouyang, D., et al.: Video-based AI for beat-to-beat assessment of cardiac function. Nature 580, 252–256 (2020)

18. Qin, C., et al.: Joint learning of motion estimation and segmentation for cardiac MR image sequences. In: Frangi, A.F., Schnabel, J.A., Davatzikos, C., Alberola-López, C., Fichtinger, G. (eds.) MICCAI 2018. LNCS, vol. 11071, pp. 472–480. Springer, Cham (2018). https://doi.org/10.1007/978-3-030-00934-2_53

19. Ronneberger, O., Fischer, P., Brox, T.: U-Net: convolutional networks for biomedical image segmentation. In: Navab, N., Hornegger, J., Wells, W.M., Frangi, A.F. (eds.) MICCAI 2015. LNCS, vol. 9351, pp. 234–241. Springer, Cham (2015). https://doi.org/10.1007/978-3-319-24574-4_28
20. Russo, C., Hahn, R.T., Jin, Z., Homma, S., Sacco, R.L., Di Tullio, M.R.: Comparison of echocardiographic single-plane versus biplane method in the assessment of left atrial volume and validation by real time three-dimensional echocardiography. J. Am. Soc. Echocardiogr. **23**(9), 954–960 (2010)
21. Smistad, E., Østvik, A., Salte, I.M., Leclerc, S., Bernard, O., Lovstakken, L.: Fully automatic real-time ejection fraction and MAPSE measurements in 2D echocardiography using deep neural networks. In: 2018 IEEE IUS, pp. 1–4 (2018)
22. Smistad, E., et al.: Real-time automatic ejection fraction and foreshortening detection using deep learning. IEEE Trans. Ultrason. Ferroelectr. Freq. Control **67**(12), 2595–2604 (2020)
23. Vaswani, A., et al.: Attention is all you need. In: Advances in Neural Information Processing Systems, pp. 5998–6008 (2017)
24. Wei, H., et al.: Temporal-consistent segmentation of echocardiography with colearning from appearance and shape. In: Martel, A.L., et al. (eds.) MICCAI 2020. LNCS, vol. 12262, pp. 623–632. Springer, Cham (2020). https://doi.org/10.1007/978-3-030-59713-9_60
25. Wolf, T., et al.: Huggingface's transformers: state-of-the-art natural language processing. arXiv preprint arXiv:1910.03771 (2019)
26. Zolgharni, M., et al.: Automatic detection of end-diastolic and end-systolic frames in 2d echocardiography. Echocardiography **34**(7), 956–967 (2017)

EchoCP: An Echocardiography Dataset in Contrast Transthoracic Echocardiography for Patent Foramen Ovale Diagnosis

Tianchen Wang[1], Zhihe Li[2], Meiping Huang[2], Jian Zhuang[2], Shanshan Bi[3], Jiawei Zhang[4], Yiyu Shi[1], Hongwen Fei[2], and Xiaowei Xu[2(✉)]

[1] University of Notre Dame, Notre Dame, USA
{twang9,yshi4}@nd.edu
[2] Guangdong Provincial People's Hospital, Guangdong, China
zhuangjian5413@tom.com, xuxiaowei@gdph.org.cn
[3] Missouri University of Science and Technology, Rolla, USA
[4] Fudan University, Shanghai, China
17110240008@fudan.edu.cn

Abstract. Patent foramen ovale (PFO) is a potential separation between the septum, primum and septum secundum located in the anterosuperior portion of the atrial septum. PFO is one of the main factors causing cryptogenic stroke which is the fifth leading cause of death in the United States. For PFO diagnosis, contrast transthoracic echocardiography (cTTE) is preferred as being a more robust method compared with others. However, the current PFO diagnosis through cTTE is extremely slow as it is proceeded manually by sonographers on echocardiography videos. Currently there is no publicly available dataset for this important topic in the community. In this paper, we present EchoCP, as the first echocardiography dataset in cTTE targeting PFO diagnosis. EchoCP consists of 30 patients with both rest and Valsalva maneuver videos which covers various PFO grades. We further establish an automated baseline method for PFO diagnosis based on the state-of-the-art cardiac chamber segmentation technique, which achieves 0.89 average mean Dice score, but only 0.70/0.67 mean accuracies for PFO diagnosis, leaving large room for improvement. We hope that the challenging EchoCP dataset can stimulate further research and lead to innovative and generic solutions that would have an impact in multiple domains. Our dataset is released [1].

1 Introduction

The foramen ovale is a physiological channel of the atrial septum and is usually functionally closed at birth. Patent foramen ovale (PFO) arises when children

Electronic supplementary material The online version of this chapter (https://doi.org/10.1007/978-3-030-87231-1_49) contains supplementary material, which is available to authorized users.

M. de Bruijne et al. (Eds.): MICCAI 2021, LNCS 12906, pp. 506–515, 2021.
https://doi.org/10.1007/978-3-030-87231-1_49

older than 3 years do not have their foramen ovale closed, while approximately 20%–25% of adults have a PFO in the general population [6]. PFO is one of the main factors causing cryptogenic stroke [8,17], which is the fifth leading cause of death in the United States. Recent studies show that the transcatheter closure of PFO reduces the recurrence of stroke at higher rates compared with medical therapy [12,16]. Thus, it is essential to diagnose PFO in a fast and accurate way.

Various clinical methods have been used to diagnose PFO, such as contrast transcranial doppler echocardiography (cTCD), transesophageal echocardiography (TEE), and contrast transthoracic echocardiography (cTTE). Although TEE is considered as the silver bullet for PFO diagnosis, patients find it difficult to successfully complete the Valsalva maneuver (VM) during TEE's invasive examination, which leads to a lower detection rate of right-to-left shunt (RLS) [4,11]. A large number of false negatives due to invalid VM during TEE examination further raise the need for other inspection methods. As noninvasive methods are more acceptable for patients, both cTCD and cTTE are used to predict RLS by observing the number of microbubbles in the cranial circulation at the resting state and after VM. However cTCD is not preferred as about 5% of detected shunts do not correspond with PFO, which leads to lower sensitivity (68%) and specificity (65%) [3,5]. On the other hand, cTTE can isolate the source of RLS with a specificity of 97% albeit with a slightly lower sensitivity of about 63% [18], which makes cTTE a simple, safe and reliable base as either the main PFO diagnosis method or a supplement for other methods.

When using cTTE for PFO diagnosis, the apical four-chamber view is generally selected which includes right/left atrium (RA/LA), and right/left ventricle (RV/LV) [19]. The presence of RLS is confirmed when microbubbles are observed in LV/LA within the first three cardiac cycles after contrast appearance in the right atrium during normal respiration or the VM. To perform an accurate and fast PFO diagnosis, there are several challenges that need to be addressed. First, the existing approach is extremely time-consuming because the sonographers need to manually perform microbubble estimation as well as the PFO grade diagnosis on echocardiography videos. Second, the existing approach is expensive. The representation of PFO in cTTE is easily confused with other cardiac diseases such as pulmonary arteriovenous fistula (PAVF), which requires experienced sonographers to diagnose accurately. Third, the data quality of cTTE is limited. In addition to the low resolution of echocardiography videos (e.g. 112×112 [14]), the heavy noise in images, the ambiguity at the components' boundaries, and the expensive data annotation further limit the quality of available data. Meanwhile, the microbubbles to be expected in the cardiac chambers complicate the scenes. The automatic cTTE based PFO diagnosis still remains a missing piece due to the complexity of diagnosis and the lack of a dataset.

To the best of our knowledge, there are only three related echocardiographic datasets, however, all focus on other tasks with different annotation protocols, which is not applicable to PFO diagnosis through cTTE. **CETUS** [2] is the first echocardiographic dataset introduced in MICCAI 2014. The main challenge is the LV endocardium identification, which contains 45 patients sequences with the annotations only contouring the LV chamber on end-diastolic (ED)

Table 1. PFO diagnosis grade based on RLS through cTTE [22].

Grade	RLS state	Description
0	None	No microbubble observed in LV per frame
1	Minor	1 to 10 microbubble observed in LV per frame
2	Moderate	11 to 30 microbubble observed in LV per frame
3	Severe	More than 30 microbubble observed in LV per frame

Table 2. Characteristics of the EchoCP dataset.

Disease (#PFO (%)/#Normal (%))	20 (66.7%)/10 (33.3%)
Sex (#Male (%)/#Female (%))	17 (57%)/13 (43%)
Age (Mean ± SD)	35.0 ± 17.6
Manufacturer	Philip EPIQ 7C (Probe S5-1, 1–5 MHz)
Heart beat rate (Mean ± SD)	73.9 ± 11.6
Systolic/Diastolic blood pressure (mmHg)	120 ± 11.6/74.3 ± 10.8
Patient's height (cm)	160.1 ± 8.5
Patient's weight (cm)	59.2 ± 13.3
Spatial size of 2D images (pixels)	640 × 480
Video length (frames)	954 ± 232

and end-systolic (ES) instants. **CAMUS** [10] was introduced to enhance the delineation of the cardiac structures from 2D echocardiographic images. The datasets contains 500 patients with manual expert annotations for the LV, LA, and the myocardium regions. **EchoNet-Dynamic** [14] was introduced as a large echocardiography video dataset to study cardiac function evaluation and medical decision making. The dataset contains 10,271 apical four-chamber echocardiography videos with tracings and labels of LV chambers on ED and ES instants.

In this paper we present EchoCP, the first dataset for cTTE based PFO diagnosis. EchoCP contains both VM and rest echocardiography videos captured from 30 patients. Data annotation including diagnosis annotation and segmentation annotation are performed by four experienced cardiovascular sonographers. As there are more than a thousand images in each patient's video, sparse labeling (only select representative frames) of the segmentation is adopted [20]. Other than existing echocardiography datasets, EchoCP is the first dataset targeting cTTE based PFO diagnosis. In addition to the dataset, we also introduce a multi-step PFO diagnosis method based on a state-of-the-art chamber segmentation algorithm. Results show that the baseline method can achieve a 0.89 mean Dice score in segmentation and mean PFO diagnosis accuracies of 0.70/0.67 for VM and rest videos. As such, this dataset still calls for breakthroughs in segmentation/diagnosis algorithms for further improvement.

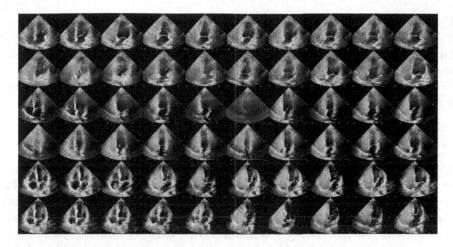

Fig. 1. Representative frames from EchoCP. The ECG data, text labels, and ultrasound acquisition information are removed in pre-processing. The rest and VM states from three patients (top three rows and bottom three rows, respectively) are sampled.

2   The EchoCP Dataset

2.1   Data Characteristics

Data Acquisition. Our EchoCP is made of echocardiography videos in cTTE for PFO diagnosis. EchoCP contains cTTE videos from 30 patients. For each patient, two videos corresponding to the rest and VM state of the patients are captured. Note that in the rest state, patients just relax and breathe normally. While in the VM, patients need to close their mouths, pinch their noses shut while expelling air out as if blowing up a balloon. The video is captured in the apical-4-chamber view and contains at least ten cardiac cycles. For the VM state, the action is performed three to five times during acquisition, and we selected the most representative one. More characteristics of our EchoCP dataset are shown in Table 2 and the representative frames are shown in Fig. 1.

Data Annotation. The data annotation of EchoCP includes two aspects: diagnosis annotation and segmentation annotation. For diagnosis annotation, patients' reports are extracted from our medical system, and the diagnosis is further evaluated and confirmed by two sonographers. For segmentation annotation, two sonographers are involved, and each annotation is annotated by one and evaluated by the other [21]. In segmentation annotation, the sonographers manually draw the regions of RV/LV/RA/LA, which do not include myocardium and valve regions. Due to the long echocardiography videos and the similarities among the frames, we select only representative frames for annotation. Frames in several representative situations where: (a) no microbubbles in all chambers, (b) many microbubbles in only RA, (c) many microbubbles in only RA and RV,

(d) many microbubbles in all chambers, (e) moderate microbubbles in LA and LV, and (f) a few microbubbles in RA and RV are selected. For each selected situation, four frames are annotated (two in the diastole and two in the systole). Due to the variety of patients, the videos from several patients may not cover all the above representative situations. Thus, usually 20 to 24 frames are annotated for each patient.

2.2   PFO Diagnosis and Evaluation Protocol

As mentioned above, PFO diagnosis through cTTE relies on the states of microbubble in LV after the microbubble solution injection. Specifically, during the test, a mixture of saline and room air is agitated to present as a good contrast agent, which is then vertically injected into the left cubital vein as a bolus both at rest and during VM. After a few cardiac cycles, the microbubbles would enter RV and then RA regions, and the two chambers would be full of microbubbles. We then observed the number of microbubbles shown in the LV region within three cardiac cycles. If the microbubbles are not shown in LA and LV chambers, the patient may have a negative PFO (grade 0). Otherwise, we proceed with the diagnosis by classifying RLS states based on the number of observed microbubbles, as shown in Table 1.

It is noted that for PFO diagnosis through cTTE, the decision on VM videos plays the dominant role. However, we still proceed with diagnosis method on both VM and rest videos because the result from the rest videos can act as a reference. Meanwhile for cardiac function diagnosis on other diseases such as PAVF, the RLS states classification in the rest videos would carry more weight in the diagnosis [15]. Therefore in the following proposed method, we process on both VM and rest videos for a more complete study.

3   Experiments of Baseline Method

In this section, we introduce a baseline method for PFO diagnosis using EchoCP. The method is based on the criterion in Sect. 2.2 and a state-of-the-art segmentation framework. Note that although our goal is to perform video-based PFO diagnosis which is basically a video classification task, we cannot apply an end-to-end video classification method to it. The first reason is that the lengths of captured videos are long with a large variety, ranging from 537 to 1386. It is not computationally feasible to propagate all frames at the same time for classification. The second reason is that the PFO diagnosis labels are video-level rather than frame-level, which means that we do not know the PFO grade of videos until the whole videos are seen and analyzed. We cannot classify or label a random small clip. Therefore, a multi-step method based on diagnosis criterion and a state-of-the-art framework is an applicable approach. As shown in Fig. 2, the proposed method has four steps as illustrated below.

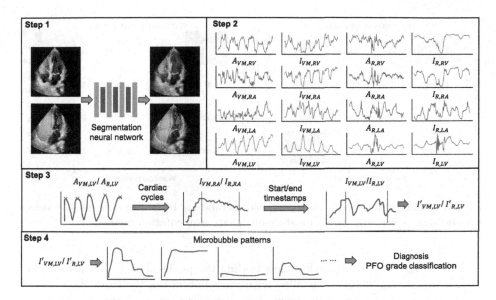

Fig. 2. The proposed baseline method has four steps: (1) segment to get the regions of four chambers through videos; (2) calculate the area (A) and the mean intensity (I) in rest (R) and VM (VM) states of four chambers (RV, RA, LA, LV); (3) get the start timestamp (RA full of microbubbles) and the end timestamp (three cardiac cycles), cropping $I_{VM,LV}$ and $I_{R,LV}$ to get $I'_{VM,LV}$ and $I'_{R,LV}$; (4) Fit $I'_{VM,LV}$ and $I'_{R,LV}$ to microbubble patterns for diagnosis.

- **Step 1.** We use an encoder-decoder style neural network to segment all the frames in both VM and rest videos, and obtain the regions of four chambers (RV, RA, LV, LA) as the output. As our task is to segment four chambers that are not available in previous echocardiography datasets [2,10,14], we trained the model only with the labeled frames in EchoCP. We use a representative state-of-the-art 2D Dynamic U-net (DynUnet) [7] to proceed segmentation task with the resized frames (384 × 384). For data augmentation, since the number of available annotations is limited, we do not proceed with a heavy augmentation such as distortion and affine transformation due to the relatively fixed positions of target chambers in the echocardiography videos. Thus, we choose a light augmentation that only consists of a random Gaussian noise added to images. The learning rate is set to 1e−3, the optimizer is set to Adam [9], and the loss is Dice loss [13].
- **Step 2.** We further process the obtained segmentation maps of four chambers in VM and rest videos. For each chamber (RV, RA, LA, LV) in each video (VM, R), two values through all the frames are calculated: area (A), and the mean intensity (I). Correspondingly, a total of 16 values are obtained. Note that the curves shown in Fig. 2 are cropped for visualization.
- **Step 3.** We first obtain the cardiac cycles, which can be estimated by the ED/ES cycles of LV. This can be achieved by locating the maximum area of

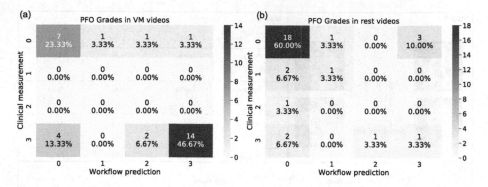

Fig. 3. Confusion matrices of the proposed method for PFO diagnosis on VM videos (a) and rest videos (b), built on a state-of-the-art segmentation method.

LV during the cardiac cycles, thus finding the peaks of A_{LV}. Then we locate the timestamp when RA is filled with microbubbles as the start. This can be estimated as the first time when I_{RA} is larger than an intensity threshold. Then we crop the I_{LV} between the start timestamp and the three cardiac cycles after that as I'_{LV} for the next step.

– **Step 4.** We apply the I'_{LV} to fit the predefined microbubble patterns. Along with the maximum intensity in I'_{LV}, we proceed with the PFO diagnosis by classifying the grade.

4   Results and Analysis

Following the proposed baseline method, we obtained $A_{VM,LV}$, $I_{VM,RA}$, $I_{VM,LV}$ and $A_{R,LV}$, $I_{R,RA}$, $I_{R,LV}$ for each patient. The four chambers segmentation task in Step (1) has achieved a mean Dice score of 0.89. The confusion matrices between the PFO prediction results and the human clinical measurements (as ground truth) on both VM and rest videos are shown in Fig. 3. The labels align the PFO grades in Table 1.

In PFO diagnosis with VM videos, the method achieves an average accuracy of 0.70. with precision/recall of 0.93/0.70, 0.0/0.0, 0.0/0.0, and 0.64/0.70 on PFO grade 3/2/1/0. For rest videos, the method achieves an average accuracy of 0.67, with precision/recall of 0.93/0.70, 0.0/0.0, 0.50/0.33, and 0.78/0.82 on PFO grade 3/2/1/0. The first major misdiagnosis is in VM videos where the grade-3 PFO cases are predicted as negative (indicated by a number of 4 in Fig. 3(a)). The second major misdiagnosis is in rest videos where the negative cases are predicted as grade-3 PFO (indicated by a number of 3 in Fig. 3(b)).

We further show the success and failure examples in Fig. 4. For segmentation, the good prediction examples would have the complete chamber regions and the correct contouring around and between the chambers' boundaries, which are not shown in bad prediction examples. For PFO diagnosis, the good examples would

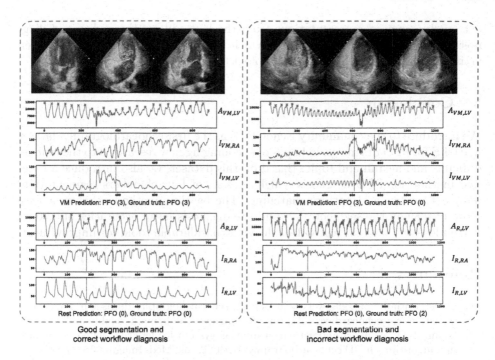

Fig. 4. Examples of good/bad segmentation results, and correct/incorrect diagnosis.

have the accurate estimation of cardiac cycles, the starting timestamp as observation, and number of microbubbles for fitting patterns. In the bad diagnosis examples of VM videos, the cardiac cycles are not accurately detected. Meanwhile due to the inaccurate segmentation, the chambers' regions are not correctly detected, which results in an intensity peak in LV, and the corresponding false negative. For the bad diagnosis examples in rest videos, although the cardiac cycles are well detected and the segmentation performs well, the microbubble estimation through intensity in LV does not show the pattern, leading to a false positive. This shows that the current proposed microbubble estimation approach by chamber intensity may not apply to all scenarios in echocardiography videos, where the periodical appearance of the myocardium background in target chambers accounts for more intensity changes compared with the filled microbubbles.

There are several workable approaches that could improve the diagnostic accuracy. First, more accurate segmentation is appreciated, especially when microbubbles flow in the chambers. The current segmentation method we used is image-based that ignores the temporal correlation between the frames in videos. A segmentation method with tracking function may work better. Second, more accurate microbubble estimation is required. Our current approach estimates the number of microbubbles by chamber's intensity. A more precise approach is to localize and count the microbubbles. Third, few shot end-to-end video classification is an attractive overall solution. By learning the limited data and extracting extremely compact representations from the long videos, the end-to-end diagnosis approaches would be possible, which requires deep collaboration between artificial intelligence experts and radiologists.

5 Conclusion

We introduce to the community EchoCP [1] as the first echocardiography dataset in cTTE for PFO diagnosis, in hopes of encouraging new research into the unique, difficult, and meaningful topic. Based on this dataset, we also introduce a baseline method for PFO diagnosis which consists of four chambers segmentation, data extraction, and pattern matching. The baseline method achieves a Dice score of 0.89 on the segmentation and the diagnosis mean accuracies of 0.70 on VM videos and 0.67 on rest videos, leaving considerable room for improvement.

References

1. https://github.com/xiaoweixu/echocp-an-echocardiography-dataset-in-contrast-transthoracic-echocardiography-for-pfo-diagnosis/blob/main/readme.md
2. Bernard, O., et al.: Standardized evaluation system for left ventricular segmentation algorithms in 3D echocardiography. IEEE Trans. Med. Imaging **35**(4), 967–977 (2015)
3. Dao, C.N., Tobis, J.M.: PFO and paradoxical embolism producing events other than stroke. Cathet. Cardiovasc. Interv. **77**(6), 903–909 (2011)
4. Del Sette, M., et al.: Migraine with aura and right-to-left shunt on transcranial doppler: a case-control study. Cerebrovasc. Dis. **8**(6), 327–330 (1998)
5. Faggiano, P., et al.: Low cerebrovascular event rate in subjects with patent foramen ovale and different clinical presentations: results from a prospective non randomized study on a population including patients with and without patent foramen ovale closure. Int. J. Cardiol. **156**(1), 47–52 (2012)
6. Hagen, P.T., Scholz, D.G., Edwards, W.D.: Incidence and size of patent foramen ovale during the first 10 decades of life: an autopsy study of 965 normal hearts. In: Mayo Clinic Proceedings, vol. 59, pp. 17–20. Elsevier (1984)
7. Isensee, F., Jäger, P.F., Kohl, S.A., Petersen, J., Maier-Hein, K.H.: Automated design of deep learning methods for biomedical image segmentation. arXiv preprint arXiv:1904.08128 (2019)
8. Khessali, H., Mojadidi, M.K., Gevorgyan, R., Levinson, R., Tobis, J.: The effect of patent foramen ovale closure on visual aura without headache or typical aura with migraine headache. JACC Cardiovasc. Interv. **5**(6), 682–687 (2012)
9. Kingma, D.P., Ba, J.: Adam: a method for stochastic optimization. arXiv preprint arXiv:1412.6980 (2014)
10. Leclerc, S., T., et al.: Deep learning for segmentation using an open large-scale dataset in 2D echocardiography. IEEE Trans. Med. Imaging **38**(9), 2198–2210 (2019)
11. Lip, P.Z., Lip, G.Y.: Patent foramen ovale and migraine attacks: a systematic review. Am. J. Med. **127**(5), 411–420 (2014)
12. Mas, J.L., et al.: Patent foramen ovale closure or anticoagulation vs. antiplatelets after stroke. N. Engl. J. Med. **377**(11), 1011–1021 (2017)

13. Milletari, F., Navab, N., Ahmadi, S.A.: V-Net: fully convolutional neural networks for volumetric medical image segmentation. In: 2016 Fourth International Conference on 3D Vision (3DV), pp. 565–571. IEEE (2016)
14. Ouyang, D., et al.: Video-based AI for beat-to-beat assessment of cardiac function. Nature **580**(7802), 252–256 (2020)
15. Sagin-Saylam, G., Somerville, J.: Contrast echocardiography for diagnosis of pulmonary arteriovenous fistulas late after construction of a Glenn anastomosis. Cardiol. Young **8**(2), 228–236 (1998)
16. Saver, J.L., et al.: Long-term outcomes of patent foramen ovale closure or medical therapy after stroke. N. Engl. J. Med. **377**(11), 1022–1032 (2017)
17. Torti, S.R., et al.: Risk of decompression illness among 230 divers in relation to the presence and size of patent foramen ovale. Eur. Heart J. **25**(12), 1014–1020 (2004)
18. Truong, T., et al.: Prevalence of migraine headaches in patients with congenital heart disease. Am. J. Cardiol. **101**(3), 396–400 (2008)
19. Wang, T., et al.: MSU-Net: multiscale statistical U-Net for real-time 3D cardiac MRI video segmentation. In: Shen, D., et al. (eds.) MICCAI 2019. LNCS, vol. 11765, pp. 614–622. Springer, Cham (2019). https://doi.org/10.1007/978-3-030-32245-8_68
20. Wang, T., et al.: ICA-UNet: ICA inspired statistical UNet for real-time 3D cardiac cine MRI segmentation. In: Martel, A.L., et al. (eds.) MICCAI 2020. LNCS, vol. 12266, pp. 447–457. Springer, Cham (2020). https://doi.org/10.1007/978-3-030-59725-2_43
21. Xu, X., et al.: ImageCHD: a 3D computed tomography image dataset for classification of congenital heart disease. In: Martel, A.L., et al. (eds.) MICCAI 2020. LNCS, vol. 12264, pp. 77–87. Springer, Cham (2020). https://doi.org/10.1007/978-3-030-59719-1_8
22. Zhao, E., Du, Y., Xie, H., Zhang, Y.: Modified method of contrast transthoracic echocardiography for the diagnosis of patent foramen ovale. BioMed Res. Int. 2019 (2019)

Transformer Network for Significant Stenosis Detection in CCTA of Coronary Arteries

Xinghua Ma, Gongning Luo[✉], Wei Wang[✉], and Kuanquan Wang[✉]

Harbin Institute of Technology, Harbin, China
{luogongning,wangwei2019,wangkq}@hit.edu.cn

Abstract. Coronary artery disease (CAD) has posed a leading threat to the lives of cardiovascular disease patients worldwide for a long time. Therefore, automated diagnosis of CAD has indispensable significance in clinical medicine. However, the complexity of coronary artery plaques that cause CAD makes the automatic detection of coronary artery stenosis in Coronary CT angiography (CCTA) a difficult task. In this paper, we propose a Transformer network (TR-Net) for the automatic detection of significant stenosis (i.e. luminal narrowing > 50%) while practically completing the computer-assisted diagnosis of CAD. The proposed TR-Net introduces a novel Transformer, and tightly combines convolutional layers and Transformer encoders, allowing their advantages to be demonstrated in the task. By analyzing semantic information sequences, TR-Net can fully understand the relationship between image information in each position of a multiplanar reformatted (MPR) image, and accurately detect significant stenosis based on both local and global information. We evaluate our TR-Net on a dataset of 76 patients from different patients annotated by experienced radiologists. Experimental results illustrate that our TR-Net has achieved better results in ACC (0.92), Spec (0.96), PPV (0.84), F1 (0.79) and MCC (0.74) indicators compared with the state-of-the-art methods. The source code is publicly available from the link (https://github.com/XinghuaMa/TR-Net).

Keywords: Coronary artery stenosis · Transformer · Coronary CT angiograph · Automatic detection

1 Introduction

Coronary artery disease (CAD), as a common cardiovascular disease, has been a leading threat to human health around the world [1,2]. It is caused by atherosclerotic plaques in the main blood supply branches of coronary artery trees and induces the stenosis or blockage of blood vessels, resulting in the symptoms of heart disease, such as myocardial ischemia, angina pectoris and heart failure [3]. Coronary CT angiography (CCTA) is a practical non-invasive vascular imaging technique, which plays an important role in the perioperative period of interventional treatment of CAD. Analyzing grades of stenosis accurately through

© Springer Nature Switzerland AG 2021
M. de Bruijne et al. (Eds.): MICCAI 2021, LNCS 12906, pp. 516–525, 2021.
https://doi.org/10.1007/978-3-030-87231-1_50

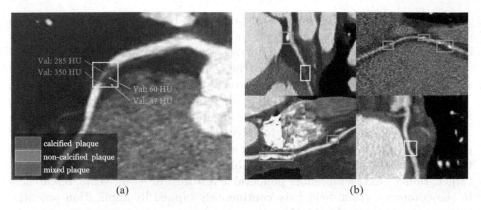

Fig. 1. The HU values of non-calcified plaques are similar to adjacent tissues (a), and types of plaques are complicated and shapes vary from different plaques (b).

the pathological information in CCTA scans is essential for clinical applications related to CAD [4].

In recent years, computer vision technology has been used to detect coronary artery stenosis through CCTA scans of patients to assist clinicians in the diagnosis of CAD. However, it is challenging to realize the automatic diagnosis of CAD, because the contrast of the HU values among adjacent tissues and structures is low. Besides, types of plaques that cause coronary artery stenosis are complicated and there is no shape feature that can be used to strictly describe plaques, as shown in Fig. 1.

Originally, most of the proposed methods were semi-automatic detection methods that require a lot of manual interaction [5]. And then, several machine learning-based methods have been proposed to describe changes in coronary lumens to quantify stenosis [6]. To a certain extent, these methods demonstrated that the geometric information of coronary lumen is considerable in clinical medicine. Several deep learning-based methods have been reported for the automatic detection of coronary artery stenosis in recent related literature [7,8]. These works mainly employed the model combining convolutional neural network (CNN) and recurrent neural network (RNN) to complete the task. Zreik et al. achieved the detection of coronary artery plaque and stenosis by a recurrent convolutional neural network (RCNN). Particularly, they firstly reconstructed multiplanar reformatted (MPR) images based on the centerlines of coronary arteries. Next, they employed a 3D-CNN to extract features from small volumes and achieved the classification of two tasks using an RNN [9]. Denzinger et al. improved the network structure of RCNN and predicted significant stenosis (i.e. luminal narrowing > 50%) with the combination of deep learning approach and radiomic features [10]. Also, Tejero-de-Pablos et al. extracted features from five views of coronary arteries and employed a Fisher vector to predict the classification probability of significant stenosis according to the features of varied views [11].

Although RNN can capture the dependencies between semantic features in a single direction to a certain extent, the global intervention of coronary artery branches to detect coronary artery stenosis is hardly considered in related work. To ensure that the model can learn the semantic features of entire coronary artery branches before local coronary artery stenoses are detected, we introduce Transformer into our method to analyze MPR images from a different perspective from others. Transformer is a type of deep neural network based on self-attention module [12], which was invented to solve related tasks in the natural language processing (NLP) field [13]. Transformer employs an attention mechanism to capture global context information to establish a long-distance dependence on the target, thereby extracting more ponderable features. In recent years, researchers in the computer vision field have continuously tapped its application potential in computer vision tasks [14,15].

In this work, we propose a novel Transformer Network (TR-Net) to detect significant stenosis in MPR images. The proposed TR-Net combines CNN and Transformer. As shown in Fig. 2, the former has a relatively large advantage in extracting local semantic information, while the latter can more naturally associate global semantic information. We employ a shallow 3D-CNN to extract local semantic features of coronary arteries. The shallow CNN enables the model to obtain the semantic information of each position in an MPR image while ensuring the efficiency of our model. Then, Transformer encoders are used to analyze feature sequences, which can mine the underlying dependence of local stenosis on each position of a coronary artery.

Our main contributions can be summarized as follows: (1) To achieve a more accurate diagnosis of coronary artery stenosis, we introduce Transformer to solve the challenging problem. To the best of our knowledge, this is the first attempt employing Transformer structure to complete the task of detecting coronary artery stenosis. (2) The proposed TR-Net can effectively integrate local and global information to detect significant stenosis. Experimental results illustrate that the proposed method has higher accuracy than state-of-the-art methods.

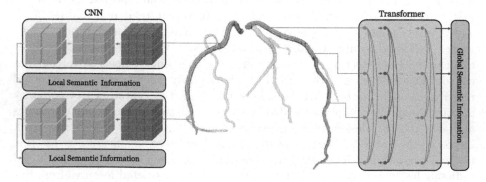

Fig. 2. CNN for obtaining local semantic information and transformer for obtaining global semantic information.

2   Method

In this section, we detail the proposed TR-Net for significant stenosis detection. Figure 3 illustrates the architecture of TR-Net. TR-Net mainly consists of two components. One part is the 3D-CNN used to extract local semantic features at different positions of a coronary artery. The other part is Transformer structure used to associate the local feature maps of each position, analyzing the dependence of different positions, and classifying the significant stenosis at each position, as shown in Fig. 2.

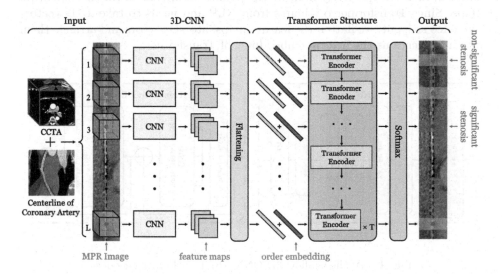

Fig. 3. Transformer network (TR-Net).

2.1   Semantic Feature Extraction for Local Cubic Volumes

For a certain locality of a coronary artery, the detail of the partial image is an indispensable reference basis for doctors when diagnosing CAD. This is also a prerequisite for our method to detect coronary artery stenosis. To efficiently extract local semantic features of coronary arteries, we design a shallow 3D-CNN as the first part of our method. The shallow CNN can not only prevent the overfitting of semantic information but also improve the efficiency of the model.

The input of our method is a coronary artery MPR image. We employ the voxels on the centerline of the coronary artery as center points to select cubic volumes from the MPR image, and the side length of cubic volumes is N voxels. Then, we arrange these cubic volumes into a sequence of length L according to the topological relationship of the coronary artery centerline. The semantic features of cubic volumes extracted by 3D-CNN are treated as the input of Transformer structure.

The structure of the 3D-CNN to extract semantic features of cubic volumes in volume sequences is inspired by [9], as shown in Fig. 4. The 3D-CNN consists of four sequentially connected substructures. Each substructure includes a convolutional layer with a convolution kernel size of $3 \times 3 \times 3$, a rectified linear unit (ReLU) and a $2 \times 2 \times 2$ max-pooling layer. The number of filters of the convolutional layer is 16 in the first part. In the remaining part, the number of filters of the convolutional layer is the number of the previous part multiplied by 2.

The feature maps obtained by the 3D-CNN are defined as $x \in \mathbb{R}^{C \times H \times H \times H}$, where C and H respectively indicate the number of filters and the size of feature maps. Since Transformer originates from NLP and needs to take a 1D vectors sequence as input, we flatten the feature maps into 1D vectors and arrange them into a sequence as the feature embeddings of Transformer.

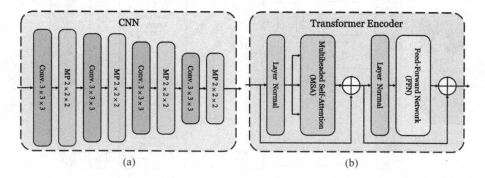

Fig. 4. (a) The shallow 3D-CNN. (b) Transformer encoder.

2.2   Transformer Structure for Global Sequence Analysis

According to clinical experience, a coronary artery branch may have multiple plaques, and each plaque affects the blood flow velocity in the patient's coronary lumen. Therefore, analyzing the potential relationship between plaques at different locations is valuable for clinical diagnosis. In this work, for a voxel on the centerline, there is image information in both two directions (the direction of ascending aorta and the direction of the coronary end) that can affect the detection result. To treat the feature maps of each coronary artery segment as the basis for judgment when detecting local coronary artery stenosis, we design Transformer structure to analyze feature sequence bidirectionally.

To introduce the order information of each cubic volume into our model, we add the learnable order embeddings [16] of the same dimension to feature embeddings before inputting embeddings into Transformer structure. The input embedding for Transformer structure Z_0 can be obtained by adding feature embeddings and order embeddings, expressed as follows:

$$Z_0 = [x_1 + o_1, x_2 + o_2, ..., x_L + o_L] \in \mathbb{R}^{L \times (C \cdot H^3)} \tag{1}$$

where x_i and o_i respectively indicate the feature embedding and order embedding for the ith cubic volume.

The Transformer structure of TR-Net contains T Transformer encoders, and the number T of Transformer encoders is 12 in this work. Each Transformer encoder consists of two sub-blocks connected in sequence, multiheaded self-attention (MSA) and feed-forward network (FFN), where FFN consists of two linear layers with a ReLU activation. Layer normal (LN) and residual connections are respectively employed before and after both two sub-blocks [17, 18], as shown in Fig. 4. For each Transformer encoder, the size of the input is the same as the output to ensure the consistency of Transformer encoders. The output of the previous Transformer encoder is treated as the input of the next Transformer encoder, the output of the tth Transformer encoder Z_t can be defined as:

$$\begin{aligned} Z'_t &= \text{MSA}(\text{LN}(Z_{t-1})) \in \mathbb{R}^{L \times (C \cdot H^3)} \\ Z_t &= \text{FFN}(\text{LN}(Z'_t) + Z_{t-1})) + Z'_t + Z_{t-1} \in \mathbb{R}^{L \times (C \cdot H^3)} \end{aligned} \tag{2}$$

where Z_{t-1} indicates the output of the $t - 1^{th}$ Transformer encoder. For the output of the last Transformer encoder $Z_T \in \mathbb{R}^{L \times (C \cdot H^3)}$, we split it into L embeddings, where the ith embedding is denoted as $Z_T^i \in \mathbb{R}^{1 \times (C \cdot H^3)}$. The order of these embeddings corresponds to the order of the cubic volumes which input model. These embeddings are fed into softmax classifiers to detect significant stenosis of the corresponding cubic volume.

3 Experiment

3.1 Dataset

We conducted experiments on a dataset consisting of 76 CCTA scans from different patients and evaluated our method. These scan data have a total of 158 significant stenoses. We extracted the MPR image of the main coronary artery branches in each CCTA scan. For the entire dataset, we extracted a total of the MPR images of 609 coronary artery branches. For these MPR images, 42425 voxels belonging to the centerline of coronary artery branches could be selected as the volume center points. The dataset was annotated by experienced radiologists, and each voxel on the centerline was marked with non-significant stenosis (i.e. luminal narrowing $\leq 50\%$) or significant stenosis (i.e. luminal narrowing $> 50\%$).

We selected voxels in the MPR image at intervals of 5 voxels along centerlines of coronary arteries and employed these voxels as volume center points to construct volume sequences. To extract local information properly, the side length

N of cubic volumes was set to 29 and the length L of volume sequences was 30 at most. Considering that in most coronary artery branches, the proportion of significant stenosis in the entire branch is low, we cut the non-significant stenosis part of MPR images appropriately to make training samples as balanced as possible when constructing volume sequences. To ensure that the model has stronger robustness, we made the volume center points randomly move up to three voxels in any direction along 6 neighborhoods and rotated the cubic volumes at random angles perpendicular to the centerline. Finally, we obtained 849 volume sequences, of which 3326 center points corresponded to significant stenosis.

Table 1. Evaluation results for significant stenosis detection.

Method	Metric	ACC	Sens	Spec	PPV	NPV	F1	MCC
Texture CLS [11]	Orig data	0.81	0.90	0.80	–	–	–	–
3D-RCNN [9]	Orig data	0.94	0.63	0.97	0.65	0.97	0.64	0.60
2D-RCNN+PT [10]	Orig data	0.87	0.60	0.93	0.68	0.91	0.64	0.56
3D-RCNN [9]	Our data	0.87	0.66	0.91	0.56	0.93	0.60	0.53
2D-RCNN+PT [10]	Our data	0.89	**0.82**	0.89	0.50	**0.97**	0.62	0.58
TR-Net	Our data	**0.92**	0.74	**0.96**	**0.84**	0.93	**0.79**	**0.74**

3.2   Experimental Results

Quantitative evaluation of experimental results can demonstrate the reliability of methods in clinical application, so we compared our proposed TR-Net with several state-of-the-art methods. To demonstrate the effectiveness of our TR-Net scientifically, we evaluated accuracy (ACC), sensitivity (Sens), specificity (Spec), predictive value (PPV), negative predictive value (NPV), F1-score, the Matthews correlation coefficient (MCC) based on the classification results. For all model experiments, we performed ten-fold cross-validation on centerline-level, where the validation set accounted for 10% of the training data. Models were trained for 200 epochs, and the models were saved with the best performance on the validation set to predict the data on the test set. As shown in Table 1, compared with state-of-the-art methods, TR-Net achieved the best performance on ACC, Spec, PPV, F1 and MCC indicators on our dataset.

The dataset we employed was obtained by marking whether each voxel was significant stenosis on the centerlines of coronary arteries. However, the model selected one for every 5 voxels along centerlines for significant stenosis detection. Therefore, there was a tolerable error between correct detection results and annotations at significant stenosis boundaries. If the error was less than 5 voxels, the detection results obtained by the model were considered correct. According to several representative examples of significant stenosis detection in Fig. 5, annotations and our detection results had high consistency. Experimental results demonstrated that TR-Net could effectively detect significant stenosis caused by various types of plaques, including non-calcified plaques that are difficult to detect. The detection results of our method had outstanding continuity, and there was almost no interruption when dealing with long-length significant stenosis.

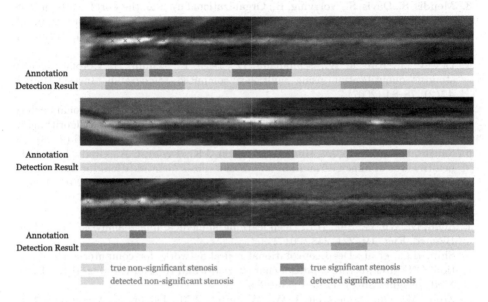

Fig. 5. Examples of significant stenosis detection. Volume center points are denoted by orange ×. (Color figure online)

4 Conclusion

In this work, we have proposed TR-Net to solve the challenging task of automatically detecting significant stenosis in MPR images. Our TR-Net can well combine the information of local areas adjacent to stenoses and the global information of coronary artery branches when detecting significant stenosis. Experimental results have demonstrated that TR-Net has better performance on multiple indicators compared with state-of-the-art methods. Through more comprehensive information analysis of coronary arteries in MPR images, our method can achieve the purpose of the computer-assisted diagnosis of CAD.

Acknowledgment. This work was supported by the National Natural Science Foundation of China under Grant 62001144 and Grant 62001141, and by China Postdoctoral Science Foundation under Grant 2021T140162 and Grant 2020M670911, and by Heilongjiang Postdoctoral Fund under Grant LBH-Z20066, and by Shandong Provincial Natural Science Foundation (ZR2020MF050).

References

1. Luo, G., et al.: Dynamically constructed network with error correction for accurate ventricle volume estimation. Med. Image Anal. **64**, 101723 (2020)
2. Luo, G., et al.: Commensal correlation network between segmentation and direct area estimation for bi-ventricle quantification. Med. Image Anal. **59**, 101591 (2020)
3. Mendis, S., Davis, S., Norrving, B.: Organizational update: the world health organization global status report on noncommunicable diseases 2014; one more landmark step in the combat against stroke and vascular disease. Stroke **46**(5), e121–e122 (2015)
4. Dewey, M., Rutsch, W., Schnapauff, D., Teige, F., Hamm, B.: Coronary artery stenosis quantification using multislice computed tomography. Invest. Radiol. **42**(2), 78–84 (2007)
5. Kirişli, H., et al.: Standardized evaluation framework for evaluating coronary artery stenosis detection, stenosis quantification and lumen segmentation algorithms in computed tomography angiography. Med. Image Anal. **17**(8), 859–876 (2013)
6. Sankaran, S., Schaap, M., Hunley, S.C., Min, J.K., Taylor, C.A., Grady, L.: HALE: healthy area of lumen estimation for vessel stenosis quantification. In: Ourselin, S., Joskowicz, L., Sabuncu, M.R., Unal, G., Wells, W. (eds.) MICCAI 2016. LNCS, vol. 9902, pp. 380–387. Springer, Cham (2016). https://doi.org/10.1007/978-3-319-46726-9_44
7. Shen, D., Wu, G., Suk, H.I.: Deep learning in medical image analysis. Annu. Rev. Biomed. Eng. **19**, 221–248 (2017)
8. Shin, H.C., et al.: Deep convolutional neural networks for computer-aided detection: CNN architectures, dataset characteristics and transfer learning. IEEE Trans. Med. Imaging **35**(5), 1285–1298 (2016)
9. Zreik, M., Van Hamersvelt, R.W., Wolterink, J.M., Leiner, T., Viergever, M.A., Išgum, I.: A recurrent CNN for automatic detection and classification of coronary artery plaque and stenosis in coronary CT angiography. IEEE Trans. Med. Imaging **38**(7), 1588–1598 (2018)
10. Denzinger, F., et al.: Coronary artery plaque characterization from CCTA scans using deep learning and radiomics. In: Shen, D., et al. (eds.) MICCAI 2019. LNCS, vol. 11767, pp. 593–601. Springer, Cham (2019). https://doi.org/10.1007/978-3-030-32251-9_65
11. Tejero-de-Pablos, A., et al.: Texture-based classification of significant stenosis in CCTA multi-view images of coronary arteries. In: Shen, D., et al. (eds.) MICCAI 2019. LNCS, vol. 11765, pp. 732–740. Springer, Cham (2019). https://doi.org/10.1007/978-3-030-32245-8_81
12. Vaswani, A., et al.: Attention is all you need. In: NIPS (2017)
13. Han, K., et al.: A survey on visual transformer. arXiv preprint arXiv:2012.12556 (2020)
14. Carion, N., Massa, F., Synnaeve, G., Usunier, N., Kirillov, A., Zagoruyko, S.: End-to-end object detection with transformers. In: Vedaldi, A., Bischof, H., Brox, T., Frahm, J.-M. (eds.) ECCV 2020. LNCS, vol. 12346, pp. 213–229. Springer, Cham (2020). https://doi.org/10.1007/978-3-030-58452-8_13
15. Wang, X., Girshick, R., Gupta, A., He, K.: Non-local neural networks. In: Proceedings of the IEEE Conference on Computer Vision and Pattern Recognition, pp. 7794–7803 (2018)
16. Dosovitskiy, A., et al.: An image is worth 16x16 words: transformers for image recognition at scale. arXiv preprint arXiv:2010.11929 (2020)

17. Baevski, A., Auli, M.: Adaptive input representations for neural language modeling. In: International Conference on Learning Representations (2018)
18. Wang, Q., et al.: Learning deep transformer models for machine translation. In: Proceedings of the 57th Annual Meeting of the Association for Computational Linguistics, pp. 1810–1822 (2019)

Training Automatic View Planner for Cardiac MR Imaging via Self-supervision by Spatial Relationship Between Views

Dong Wei[✉], Kai Ma, and Yefeng Zheng

Tencent Jarvis Lab, Shenzhen, China
{donwei,kylekma,yefengzheng}@tencent.com

Abstract. View planning for the acquisition of cardiac magnetic resonance imaging (CMR) requires acquaintance with the cardiac anatomy and remains a challenging task in clinical practice. Existing approaches to its automation relied either on an additional volumetric image not typically acquired in clinic routine, or on laborious manual annotations of cardiac structural landmarks. This work presents a clinic-compatible and annotation-free system for automatic CMR view planning. The system mines the spatial relationship—more specifically, locates and exploits the intersecting lines—between the source and target views, and trains deep networks to regress heatmaps defined by these intersecting lines. As the spatial relationship is self-contained in properly stored data, *e.g.*, in the DICOM format, the need for manual annotation is eliminated. Then, a multi-view planning strategy is proposed to aggregate information from the predicted heatmaps for all the source views of a target view, for a globally optimal prescription. The multi-view aggregation mimics the similar strategy practiced by skilled human prescribers. Experimental results on 181 clinical CMR exams show that our system achieves superior accuracy to existing approaches including conventional atlas-based and newer deep learning based ones, in prescribing four standard CMR views. The mean angle difference and point-to-plane distance evaluated against the ground truth planes are 5.98° and 3.48 mm, respectively.

Keywords: Cardiac magnetic resonance imaging · Automatic imaging view planning · Self-supervised learning

1 Introduction

Cardiac magnetic resonance imaging (CMR) is the gold standard for the quantification of volumetry, function, and blood flow of the heart [11]. A great deal

Electronic supplementary material The online version of this chapter (https://doi.org/10.1007/978-3-030-87231-1_51) contains supplementary material, which is available to authorized users.

of effort is put into the development of algorithms for accurate, robust, and automated analysis of CMR images, with a central focus on the segmentation of cardiac structures [2, 14, 16, 20, 21, 25]. However, much less attention has been paid to automatic view planning for the acquisition of CMR, which still remains challenging in clinical practice. First, imaging planes of CMR are customized for each individual based on specific cardiac structural landmarks [10], and the planning process demands specialist expertise [19]. Second, the planning process adopts a multi-step approach involving several localizers defined by the cardiac anatomy, which is complex, time-consuming, and subject to operator-induced variations [13]. These factors may constrain the use of CMR in clinical practice. Hence, automatic planing system is expected to increase the impact of the specific imaging technology on care of patients suffering from cardiovascular diseases.

A few works attempted automatic view planning for CMR [1, 3, 5, 13]. For example, both Lu *et al.* [13] and Frick *et al.* [5] tackled this challenging task from the perspective of classical atlas-based methods by fitting triangular mesh-based heart models into a 3D volume. Later, Alansary *et al.* [1] proposed to employ reinforcement learning to prescribe the standard four-chamber long-axis CMR view from a thoracic volume. A common foundation of these works was the use of a 3D MRI volume from which the standard CMR views were prescribed. However, such a 3D volume is not typically acquired in current clinic routine, where the standard CMR views are sequentially optimized based on a set of 2D localizers. In an effort to develop a clinic-compatible system, Blansit *et al.* [3] proposed to sequentially prescribe standard CMR views given a vertical long-axis localizer (also known as the pseudo two-chamber (p2C) localizer), driven by deep learning based localization of key landmarks. However, this method relied on extensive manual annotations to train the deep convolutional neural networks (CNNs). Besides, the p2C localizer—which was the starting point of the entire system—was assumed given. Yet in practice it requires expertise in cardiac anatomy to prescribe this localizer from scout images in normal body planes (such as the axial view), which is an obstacle to a fully automatic workflow.

In this work, we propose a clinic-compatible automatic view planning system for CMR. Similar to [3], our system takes advantage of the power of deep CNNs, but eliminates the need for annotation via self-supervised learning [2, 8, 24]. Above all, we make a critical observation that the way how the existing CMR data have been prescribed is self-contained in correctly recorded data, *e.g.*, in the Digital Imaging and Communications in Medicine (DICOM) format, in the form of the spatial relationship between views. Then, inspired by the recent progress in keypoint-based object detection [4, 12, 22, 23], we propose to regress the intersecting lines between the views, which can be readily computed using the spatial relationship. Training the networks to predict these intersecting lines actually teaches them to reason about the key cardiac landmarks that defined these lines when the operators produced the CMR data, while at the same time eliminates the need for manual annotation. After the intersecting lines are predicted in the localizers, the standard CMR views are eventually planned by aggregating the predictions in multiple localizers to search for a globally optimal prescription.

In summary, our contributions are three folds. First, we propose a CNN-based, clinical compatible system for automatic CMR view planning, which eliminates the need for manual annotation via self-supervised learning of the spatial relationship between views. Second, we propose to aggregate multi-view information for globally optimal plane prescription, to mimic the clinical practice of multi-view planning. Third, we conduct extensive experiments to study the proposed system and demonstrate its competence/superiorty to existing methods. In addition, we demonstrate prescription of the cardiac anatomy defined localizers (including the p2C) given the axial localizer, bridging the gap between normal body planes and CMR views for a more automated workflow.

2   Methods

Preliminary. Different from the commonly used axial, sagittal, or coronal plane oriented with respect to the long axis of the body (the body planes), CMR adopts a set of double-oblique, cardiac anatomy defined views customized for each individual. Often these views are prescribed along the long axis (LAX) or short axis (SAX) of the left ventricle (LV) and with respect to specific cardiac structural landmarks (*e.g.*, the apex and mitral valve), for optimal visualization of structures of interest and evaluation of cardiac functions. The mostly used standard CMR views include the LV two-chamber (2C) LAX, three-chamber (3C) LAX, four-chamber (4C) LAX, and SAX views, which provide complementary information for comprehensive evaluation of the heart. In clinical practice, the 2C and 4C LAX views are also called the vertical long-axis (VLA) and horizontal long-axis (HLA) views, whereas the 3C LAX view is called the left ventricular outflow tract (LVOT) view.

Our imaging protocol generally follows [10] for LV structure and function. After adjusting the heart to the isocenter of the bore, a multi-slice axial dark-blood localizer is firstly prescribed through the chest from sagittal and coronal scouts, serving as the basis for the following cardiac anatomy oriented views. Next, a single-slice pseudo 2C (p2C) localizer—the first cardiac anatomy oriented view—is prescribed from the axial localizer. Then, a single-slice pseudo 4C localizer (p4C) and a multi-slice pseudo SAX (pSA) localizer are prescribed based on the p2C localizer. Although similarly defined by the cardiac anatomy, these localizers usually cannot provide accurate anatomic and functional characterization of the heart like the standard views due to the obliquity of the heart walls to the body planes [6], hence are often referred to as the "pseudo" LAX (pLA) and SAX views [7]. Eventually, the p2C, p4C, and pSA localizers are used to sequentially optimize the imaging planes for the standard 2C, 3C, 4C, and SAX views. As in the literature [1,3,5,13], we aim at planning the standard LAX and SAX views (only the most basal SAX plane is considered) in this work. In addition, we investigate the prescription of the pseudo-view localizers from the axial localizer. As the first step of deriving cardiac anatomy oriented imaging planes from the body planes, it is indispensable for a clinic-compatible system yet remains uninvestigated in existing literature [3].

For clarity, we refer to a view plane to prescribe as the *target* plane, and the view(s) leading to the specific target plane as its *source* view(s). The source to target mappings for the standard LAX and SAX views are shown in Fig. 1 (see supplement Fig. S1 and Table S1 for more details).

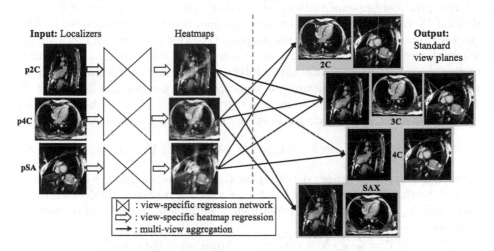

Fig. 1. Overview of the proposed approach to automatic CMR view planning (illustrated with the task of planning the standard CMR view planes from the localizers). Left: prediction of standard view planes in localizers via heatmap regression. Right: prescription of standard view planes by aggregating heatmaps predicted in multiple localizers; the prescriptions are presented as intersecting lines (red) with the localizers. (Color figure online)

Method Overview. The overview of our proposed system is shown in Fig. 1, including two main steps. First, given a set of localizers as input, heatmaps for the target planes are predicted by the view-specific regression networks. Second, multiple heatmaps for a specific target plane are aggregated to prescribe the target plane as output. Below we elaborate the two steps.

Target Plane Regression with Self-supervised Learning. A previous work on CNN-based automatic CMR view planning regressed locations of the cardiac structural landmarks, then prescribed the view planes by connecting the landmarks [3]. A drawback was that the landmark regression networks needed extensive annotations for training. In this work, we propose to mine the spatial relationship among the CMR data, and directly regress the intersecting lines of the target plane with the source views. As the spatial information of each CMR slice is recorded in its data header (*e.g.*, of the DICOM format), we can readily compute the intersecting lines. Therefore, the supervising ground truth of our proposed regression task is self-contained in the data, requiring no manual annotation at all. In addition, the intersecting lines are actually the prescription lines defined by the operator using cardiac landmarks during imaging. Hence, our networks are still trained to learn the cardiac anatomy by the proposed regression task.

For practical implementation, we train the networks to regress a heatmap defined by the distance to the intersecting line, instead of the line itself. This strategy is commonly adopted in the keypoint detection literature [4,12,15,22,23], where the regression target is the heatmaps defined by the keypoints. The benefits of regressing heatmaps include better interpretability and ease of learning for the network [15]. Formally, denoting the equation of the intersecting line in the *2D image coordinate system of a source view* by $Ax + By + C = 0$, where (x, y) are coordinates and (A, B, C) are parameters, the heatmap is computed as

$$H(x,y) = \exp\left[-(Ax + By + C)^2/(2\sigma^2(A^2 + B^2))\right], \tag{1}$$

where σ is a hyperparameter denoting the Gaussian kernel size. We define σ with respect to the slice thickness of the target view for concrete physical meaning, and study its impact later with experiments. An L2 loss is employed to train the network:

$$\mathcal{L} = \frac{1}{T}\frac{1}{|\Omega|}\sum_{t=1}^{T}\sum_{(x,y)\in\Omega}\left\|H_t(x,y) - \hat{H}_t(x,y)\right\|^2, \tag{2}$$

where T is the total number of target planes that are prescribed from a source view (also the number of channels of the network's output), (x, y) iterates over all pixels in the source view image domain Ω, and $\hat{H}$ is the heatmap predicted by the network (see Fig. 1 for example predictions by trained networks).

Plane Prescription with Multi-view Aggregation. After the heatmaps are predicted for all the source views of a specific target plane, we prescribe the target plane by aggregating information from all these heatmaps. Specifically, by prescribing the plane with the greatest collective response aggregated over its intersecting lines with all the source heatmaps, we mimic the clinical practice where skilled operators plan the CMR views by checking the locations of a candidate plane within multiple source views. Let us denote a candidate plane by $P : (p, \theta, \phi)$, where p is a point on the plane, θ and ϕ are the polar and azimuthal angles of its normal in the spherical coordinate system; and denote the intersecting line segment between P and the vth source view (of V total source views) by the exhaustive set of pixels lying on the line within the source view image I_v: $l_v = \{(x, y)|(x, y) \in P \cap I_v\}$, where $P \cap I_v$ denotes the intersection. Then, the optimal target plane can be formally expressed as

$$\hat{P} = \text{argmax}_{(p,\theta,\phi)}\sum_{v=1}^{V}\sum_{(x,y)\in l_v}\hat{H}_v(x, y). \tag{3}$$

We propose to grid search for the optimal triplet $(\hat{p}, \hat{\theta}, \hat{\phi})$ with two strategies to reduce the search space. First, we constrain the search for $\hat{p}$ along a line segment with which the target plane should intersect. In practice, the intersecting line segment of a source view with another source view is used. Second, a three-level coarse-to-fine pyramidal search is employed, with the steps set to 15, 5, and 1 pixel(s) (for p) and degree(s) (for θ and ϕ). Note that our multi-view aggregation scheme naturally takes into account the networks' confidence in the predictions, where higher values in the regressed heatmaps indicate higher confidence.

3   Experiments

Dataset and Evaluation Metrics. With institutional approval, we retrospectively collected 181 CMR exams from 99 infarction patients. Of these patients, 82 had two exams and the rest had only one. The exams were performed on a 1.5T MRI system; details about the image acquisition parameters are provided in supplement Table S3. The dataset is randomly split into training, validation, and test sets in the ratio of 64:16:20; both exams of the same patient are in the same set. The validation set is used for empirical model optimization such as hyperparameter tuning. Then, the training and validation sets are mingled to train a final model with optimized settings for evaluation on the held-out test set. Following existing literature [1,3,5,13], we use the angular deviation and point-to-plane distance as evaluation metrics. Specifically, the absolute angular difference between the normals to the automatic and ground truth planes is computed, and the distance from the center of the ground truth view (image) to the automatic plane is measured. For both metrics, smaller is better.

Implementation. The PyTorch [18] framework (1.4.0) is used for all experiments. We use the U-Net [17] as our backbone network. A model is trained for each of the four localizers (axial, p2C, p4C, and pSA). For stacked localizers (the axial and pSA), all slices of a patient are treated as a mini batch with the original image size as input size, thus both the mini batch size and input size vary with individual. For the others, we use a mini batch size of eight images, whose sizes are unified according to data statistics by cropping or padding, where appropriate, for training; specifically, the p2C and p4C localizers are unified to 192×176 (rows by columns) pixels and 160×192 pixels, respectively. The Adam [9] optimizer is used with a weight decay of 0.0005. The learning rate is initialized to 0.001 and halved when the validation loss does not decrease for 10 consecutive epochs. The exact numbers of training epochs vary with the specific localizer and value of σ, and are determined with the validation set (range 75–250). Online data augmentation including random scaling ($[0.9, 1.1]$), rotation ($[-10°, 10°]$), cropping, and flipping (left-right for the p2C, upside-down for the p4C, and none for stacked localizers) is conducted during training to alleviate overfitting. For simple preprocessing, the z-score standardization (mean subtraction followed by division by standard deviation) is performed per localizer per exam. A Tesla P40 GPU is used for model training and testing. The source code is available at: https://github.com/wd111624/CMR_plan.

Table 1. Impact of multi-view planning on prescription accuracy on the validation set (with $\sigma = 0.5t$). '–' indicates that a specific plane cannot be prescribed solely from the corresponding subset of localizers (loc.). Data format: mean ± standard deviation.

Target plane	Normal deviation (°)			Point-to-plane distance (mm)		
Loc.=	pLA	pSA	pLA & pSA	pLA	pSA	pLA & pSA
2C	–	4.96 ± 2.77	4.86 ± 2.95	–	3.33 ± 2.33	2.78 ± 1.92
3C	41.88 ± 29.98	7.68 ± 4.67	7.18 ± 4.62	7.83 ± 7.02	3.11 ± 2.59	2.76 ± 2.70
4C	–	6.80 ± 4.16	6.85 ± 4.54	–	3.92 ± 3.29	3.76 ± 3.09
Mean	–	6.48 ± 1.39	**6.30 ± 1.26**	–	3.45 ± 0.42	**3.10 ± 0.57**

Table 2. Impact of the Gaussian kernel size (σ) on plane prescription accuracy on the validation set. σ is defined as a ratio/multiple of the slice thickness t: $\sigma = \alpha \cdot t$ with α being a factor. Data format: mean ± standard deviation.

Target plane	Normal deviation (°)				Point-to-plane distance (mm)			
$\sigma =$	0.25t	0.5t	1.0t	2.0t	0.25t	0.5t	1.0t	2.0t
2C	4.71 ± 2.44	4.86 ± 2.95	5.82 ± 3.46	6.29 ± 4.14	2.66 ± 2.02	2.78 ± 1.92	3.47 ± 2.51	4.88 ± 3.23
3C	7.14 ± 4.18	7.18 ± 4.62	7.42 ± 4.90	7.97 ± 5.29	2.44 ± 2.47	2.76 ± 2.70	3.32 ± 3.35	3.87 ± 3.45
4C	7.33 ± 5.16	6.85 ± 4.54	7.09 ± 3.96	7.50 ± 4.98	3.76 ± 3.03	3.76 ± 3.09	3.63 ± 2.83	4.59 ± 3.56
SAX	8.40 ± 5.10	7.97 ± 4.61	8.04 ± 5.91	8.21 ± 4.53	3.11 ± 2.37	3.07 ± 2.45	3.77 ± 2.87	4.38 ± 2.57
Mean	6.89 ± 1.56	**6.71 ± 1.32**	7.09 ± 0.94	7.49 ± 0.85	**2.99 ± 0.58**	3.09 ± 0.47	3.55 ± 0.20	4.43 ± 0.43

Impact of Multi-view Aggregation. We first investigate the impact of the proposed multi-view planning on plane prescription accuracy. In theory, it is possible to define a target plane with only a subset of available source views, as the intersecting lines within two non-coinciding planes are sufficient to define any plane. We examine such cases in our scenario and present the results in Table 1. The results show that despite being theoretically feasible, prescribing the target planes using only a subset of localizers leads to inferior performance in general, suggesting the importance of multi-view planning. This is consistent with the clinical practice, where the operators consider multiple localizers to prescribe a standard view. Notably, when using only the pLA (p2C and p4C) localizers to plan the 3C view, the performance collapses with a mean normal deviation of 41.88°. We speculate this is because the key landmark (i.e., the aortic valve) that defines the 3C view is only visible in the stack of pSA localizer.

Impact of Gaussian Kernel Size. Next, we study the impact of the Gaussian kernel size σ in Eq. (1) by setting it to different values and comparing the validation performance. Specifically, σ is defined as a ratio/multiple of the slice thickness t of the target view: $\sigma = \alpha \cdot t$, with $\alpha = 0.25, 0.5, 1,$ and 2. The results are shown in Table 2. Based on a comprehensive evaluation of both metrics on all the four target planes, we choose $\sigma = 0.5t$ for subsequent experiments.

Evaluation on Test Set. We now evaluate our proposed system on the held-out test set, and compare the performance with existing approaches [1,3,5,13]. The results are charted in Table 3. Above all, our method achieves the best mean performance averaged over the four standard CMR views in both evaluation metrics. The second best method in terms of the mean normal deviation is [3] ($6.56 \pm 1.24°$ versus $5.98 \pm 0.79°$). This method is similar to ours in that it also exploited the power of deep CNNs. However, it required extensive annotations for training, whereas our method completely eliminates this demanding requirement and directly learns from data. Further analysis of each specific target plane indicates that our method is also competent for each view: it yields the best point-to-plane distances for all the four standard views, the lowest normal deviations for the 2C and 3C views, and the second lowest normal deviations for the 4C and SAX views. Lastly, our method can also successfully prescribe the pseudo-view localizers from the axial localizer, with a mean performance comparable to that on prescribing the standard CMR views. Prescribing these localizers is the first step from body-oriented imaging planes to the cardiac anatomy oriented imaging planes of CMR. So far as the authors are aware of, it has not been previously demonstrated for any clinic-compatible system. These results demonstrate the competence and effectiveness of our proposed system in automatic CMR view planning. Examples of the automatic prescription results are shown in Fig. 2 (more examples in supplement Fig. S2).

Table 3. Test set evaluation results and comparison to previous works. Bold and italic fonts indicate best and second best results, respectively. NA: not applicable. −: not reported. *: significance at 0.05 level for pairwise comparison to the proposed method (Bonferroni correction applied where appropriate); since different works used different data, independent samples t-test is employed. Data format: mean ± standard deviation.

Methods	2C	3C	4C	SAX	Mean	Localizers[a]
Normal deviation (°)						
Lu et al. [13]	18.9 ± 21.0*	12.3 ± 11.0*	17.6 ± 19.2*	8.6 ± 9.7	14.35 ± 4.78	NA
Alansary et al. [1]	−	−	8.72 ± 7.44	−	−	NA
Frick et al. [5]	7.1 ± 3.6*	9.1 ± 6.3	7.7 ± 6.1	6.7 ± 3.6	7.65 ± 1.05	NA
Blansit et al. [3]	8.00 ± 6.03*	7.19 ± 4.97	**5.49 ± 5.06**	**5.56 ± 4.60**	*6.56 ± 1.24*	−
Proposed	**4.97 ± 4.00**	**6.84 ± 4.16**	*5.84 ± 3.19*	*6.28 ± 3.48*	**5.98 ± 0.79**	5.88 ± 0.53
Point-to-plane distance (mm)						
Lu et al. [13]	6.6 ± 8.8*	4.6 ± 7.7	5.7 ± 8.5	13.3 ± 16.7*	7.55 ± 3.92	NA
Alansary et al. [1]	−	−	5.07 ± 3.33*	−	−	NA
Proposed	**2.68 ± 2.34**	**3.44 ± 2.37**	**3.61 ± 2.63**	**4.18 ± 3.15**	**3.48 ± 0.62**	4.18 ± 0.50

[a] Mean across the localizers; separate results for each localizer are presented in supplement Table S2.

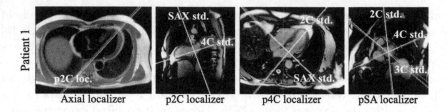

Fig. 2. Example CMR view plane prescription results by the proposed system. Green: ground truth; red: automatic prescription; yellow: overlap. Due to the closeness (sometimes coincidence) of the standard (std.) 3C view with std. 2C and 4C views in the p4C and p2C localizers (loc.), the std. 3C view is visualized only in the pSA loc. (Color figure online)

4   Conclusion

In this work, we proposed a CNN-based, clinic-compatible system for automatic CMR view planning. Our system was distinctive from a closely related work in that it eliminated the burdensome need for manual annotations, by mining the spatial relationship between CMR views. Also, the importance of the proposed multi-view aggregation—an analogue to the behaviour of human prescribers—was empirically validated. Experimental results showed that our system was superior to existing approaches including both conventional atlas-based and newer CNN-based ones, in prescription accuracy of four standard CMR views. In addition, we demonstrated accurate prescription of the pseudo-view localizers from the axial localizer and filled the gap in existing literature. Lastly, based on the encouraging results, we believe our work has opened up a new direction for automatic view planning of anatomy-oriented medical imaging beyond CMR.

Acknowledgment. This work was supported by the Key-Area Research and Development Program of Guangdong Province, China (No. 2018B010111001), and Scientific and Technical Innovation 2030 - "New Generation Artificial Intelligence" Project (No. 2020AAA0104100).

References

1. Alansary, A., et al.: Automatic view planning with multi-scale deep reinforcement learning agents. In: Frangi, A.F., Schnabel, J.A., Davatzikos, C., Alberola-López, C., Fichtinger, G. (eds.) MICCAI 2018. LNCS, vol. 11070, pp. 277–285. Springer, Cham (2018). https://doi.org/10.1007/978-3-030-00928-1_32
2. Bai, W., et al.: Self-supervised learning for cardiac MR image segmentation by anatomical position prediction. In: Shen, D., et al. (eds.) MICCAI 2019. LNCS, vol. 11765, pp. 541–549. Springer, Cham (2019). https://doi.org/10.1007/978-3-030-32245-8_60
3. Blansit, K., Retson, T., Masutani, E., Bahrami, N., Hsiao, A.: Deep learning-based prescription of cardiac MRI planes. Radiol. Artif. Intell. **1**(6), e180069 (2019)

4. Duan, K., Bai, S., Xie, L., Qi, H., Huang, Q., Tian, Q.: CenterNet: keypoint triplets for object detection. In: Proceedings of the IEEE/CVF International Conference on Computer Vision, pp. 6569–6578 (2019)
5. Frick, M., et al.: Fully automatic geometry planning for cardiac MR imaging and reproducibility of functional cardiac parameters. J. Magn. Reson. Imaging **34**(2), 457–467 (2011)
6. Ginat, D.T., Fong, M.W., Tuttle, D.J., Hobbs, S.K., Vyas, R.C.: Cardiac imaging: part 1, MR pulse sequences, imaging planes, and basic anatomy. Am. J. Roentgenol. **197**(4), 808–815 (2011)
7. Herzog, B.: The CMR pocket guide App. Eur. Heart J. **38**(6), 386–387 (2017)
8. Jing, L., Tian, Y.: Self-supervised visual feature learning with deep neural networks: a survey. IEEE Trans. Pattern Anal. Mach. Intell. (2020)
9. Kingma, D.P., Ba, J.: Adam: a method for stochastic optimization. arXiv preprint arXiv:1412.6980 (2014)
10. Kramer, C.M., Barkhausen, J., Bucciarelli-Ducci, C., Flamm, S.D., Kim, R.J., Nagel, E.: Standardized cardiovascular magnetic resonance imaging (CMR) protocols: 2020 update. J. Cardiovasc. Magn. Reson. **22**(1), 1–18 (2020)
11. La Gerche, A., et al.: Cardiac MRI: a new gold standard for ventricular volume quantification during high-intensity exercise. Circ. Cardiovasc. Imag. **6**(2), 329–338 (2012)
12. Law, H., Deng, J.: CornerNet: detecting objects as paired keypoints. In: Ferrari, V., Hebert, M., Sminchisescu, C., Weiss, Y. (eds.) Computer Vision – ECCV 2018. LNCS, vol. 11218, pp. 765–781. Springer, Cham (2018). https://doi.org/10.1007/978-3-030-01264-9_45
13. Lu, X., et al.: Automatic view planning for cardiac MRI acquisition. In: Fichtinger, G., Martel, A., Peters, T. (eds.) MICCAI 2011. LNCS, vol. 6893, pp. 479–486. Springer, Heidelberg (2011). https://doi.org/10.1007/978-3-642-23626-6_59
14. Painchaud, N., Skandarani, Y., Judge, T., Bernard, O., Lalande, A., Jodoin, P.-M.: Cardiac MRI segmentation with strong anatomical guarantees. In: Shen, D., et al. (eds.) MICCAI 2019. LNCS, vol. 11765, pp. 632–640. Springer, Cham (2019). https://doi.org/10.1007/978-3-030-32245-8_70
15. Pfister, T., Charles, J., Zisserman, A.: Flowing ConvNets for human pose estimation. In: Proceedings of the IEEE International Conference on Computer Vision, pp. 1913–1921 (2015)
16. Robinson, R., et al.: Automatic quality control of cardiac MRI segmentation in large-scale population imaging. In: Descoteaux, M., Maier-Hein, L., Franz, A., Jannin, P., Collins, D.L., Duchesne, S. (eds.) MICCAI 2017. LNCS, vol. 10433, pp. 720–727. Springer, Cham (2017). https://doi.org/10.1007/978-3-319-66182-7_82
17. Ronneberger, O., Fischer, P., Brox, T.: U-Net: convolutional networks for biomedical image segmentation. In: Navab, N., Hornegger, J., Wells, W.M., Frangi, A.F. (eds.) MICCAI 2015. LNCS, vol. 9351, pp. 234–241. Springer, Cham (2015). https://doi.org/10.1007/978-3-319-24574-4_28
18. Steiner, B., et al.: PyTorch: an imperative style, high-performance deep learning library. Adv. Neural. Inf. Process. Syst. **32**, 8026–8037 (2019)
19. Suinesiaputra, A., et al.: Quantification of LV function and mass by cardiovascular magnetic resonance: multi-center variability and consensus contours. J. Cardiovasc. Magn. Reson. **17**(1), 1–8 (2015)
20. Wei, D., Li, C., Sun, Y.: Medical image segmentation and its application in cardiac MRI. In: Biomedical Image Understanding, Methods and Applications, pp. 47–89 (2015)

21. Wei, D., Sun, Y., Ong, S.H., Chai, P., Teo, L.L., Low, A.F.: A comprehensive 3-D framework for automatic quantification of late gadolinium enhanced cardiac magnetic resonance images. IEEE Trans. Biomed. Eng. **60**(6), 1499–1508 (2013)
22. Zhou, X., Wang, D., Krähenbühl, P.: Objects as points. arXiv preprint arXiv:1904.07850 (2019)
23. Zhou, X., Zhuo, J., Krahenbuhl, P.: Bottom-up object detection by grouping extreme and center points. In: Proceedings of the IEEE/CVF Conference on Computer Vision and Pattern Recognition, pp. 850–859 (2019)
24. Zhou, Z., et al.: Models genesis: generic autodidactic models for 3D medical image analysis. In: Shen, D., et al. (eds.) MICCAI 2019. LNCS, vol. 11767, pp. 384–393. Springer, Cham (2019). https://doi.org/10.1007/978-3-030-32251-9_42
25. Zotti, C., Luo, Z., Lalande, A., Jodoin, P.M.: Convolutional neural network with shape prior applied to cardiac MRI segmentation. IEEE J. Biomed. Health Inform. **23**(3), 1119–1128 (2018)

Phase-Independent Latent Representation for Cardiac Shape Analysis

Josquin Harrison[1], Marco Lorenzi[1], Benoit Legghe[2,3], Xavier Iriart[2,3], Hubert Cochet[2,3], and Maxime Sermesant[1,2(✉)]

[1] Inria, Université Côte d'Azur, Nice, France
maxime.sermesant@inria.fr
[2] IHU Liryc, Université de Bordeaux, Bordeaux, France
[3] Bordeaux University Hospital, Bordeaux, France

Abstract. Atrial fibrillation (AF) is a complex cardiac disease impacting an ever-growing population and increases 6-fold the risk of thrombus formation. However, image based bio-markers to predict thrombosis in presence of AF are not well known. This lack of knowledge comes from the difficulty to analyse and compare the shape of the Left Atrium (LA) as well as the insufficiency of data that limits the complexity of models we can use. Conducting data analysis in cardiology exacerbates the small dataset problem because the heart cycle renders impossible to compare images taken at systole and diastole time. To address these issues, we first propose a graph representation of the LA, to focus on the impact of pulmonary veins (PV) and LA Appendage (LAA) positions, giving a simple object easy to analyse. Secondly, we propose a meta-learning framework for heterogeneous datasets based on the consistent representation of each dataset in a common latent space. We show that such a model is analogous to a meta-classifier, where each dataset is characterised by specific projection in a common latent space, while sharing the same separating boundary. We apply this model to the graph representation of the LA and interpret the model to give novel time-dependant bio-markers related to PV and LAA configurations for the prediction of thrombosis.

Keywords: Shape analysis · Atrial fibrillation · Thrombosis · Graph representation · Latent space model · Multi-task learning · Meta-learning

1 Introduction

Atrial fibrillation (AF) is a complex cardiac disease impacting an ever-growing population, creating a hemodynamic environment prone to clot formation and ischemic stroke. The stratification of stroke risk in AF has significant clinical implications for the management of anticoagulation, which was shown to effectively limit the occurrence of strokes but at the cost of increased risk of bleeding. The net benefit of introducing preventive anticoagulation is currently estimated

© Springer Nature Switzerland AG 2021
M. de Bruijne et al. (Eds.): MICCAI 2021, LNCS 12906, pp. 537–546, 2021.
https://doi.org/10.1007/978-3-030-87231-1_52

by computing a score based on patient demographics, clinical condition and past history, but the ability of this score to accurately assess stroke risk in AF patients is largely suboptimal [8]. Thus, to better understand AF, important studies were conducted on LA hemodynamics [12] or LAA morphology [13]. Indeed, as the immense majority of clots occur in the LAA which is known to show high inter-individual variability, series of studies have focused on characterizing the LAA shape, demonstrating moderate association with stroke risk [3]. However, those studies are mostly qualitative, tools available to clinician for decision making are still limited and while imaging data availability is growing, it is under-utilised to quantitatively explore novel image-based bio-markers. An important hurdle making it difficult to use is its heterogeneity. In cardiology, images are taken at different times in the heart cycle, using different imaging systems, and for different tasks. Developing analysis methods allowing to integrate information across heterogeneous datasets to enable statistical studies is therefore a necessity.

In this work we follow this idea. Inspired by [1], we propose a multi-channel formulation to merge multiple heterogeneous datasets into a common latent representation. This framework realises a combination of Multi-task Learning (MTL) [9] and Meta Learning (ML) [4]: the common representation across datasets imposed in the latent space induces homogeneity across latent projections while enriching the amount of data provided to train a classifier for automated image-based diagnosis. In addition, the learned latent distribution of the joint representation is a meta parameter that constitutes an excellent prior for future datasets as it is robust to multiple datasets sharing a common task.

We propose to use such a model to explore underlying links between Thrombosis and Pulmonary Veins (PV), Appendage (LAA) positions and orientations in the Left Atrium. In the following sections we propose a lightweight graph representation of the LA to focus on PV and LAA positions and formulate the classification methodology within a supervised framework where the common representation is improved in terms of Kullback-Leibler divergence. We finally apply the model to the joint analysis of LA Graphs where the data is split in systole and diastole subsets. This constitutes a multi-label classification problem across datasets where the labels are consistent but the dataset is heterogeneous.

2   Methodology

2.1   Pre-processing Pipeline

Our study was performed on 3D Computational Tomography scans (CT-Scans) along with clinical data from 107 patients suffering atrial fibrillation, of which 64 are labelled Thrombus positive, a composite criterium composed of a detection of LAA thrombus on CT scan and/or past history embolism. In particular, our database is composed of 50 patients in systole (of which 27 are Thrombus positive) and 57 patients in diastole (of which 37 are Thrombus positive). Cardiac segmentation was first conducted automatically with a 3D U-Net neural network as proposed by [5], and then hand-corrected by experts. All the data was acquired by the Bordeaux University Hospital.

From the available segmentation masks we use the open-source package MMG[1] for meshing the shapes. First we apply a marching cube algorithm, giving a rough meshing of the surface, the mmg3d algorithm is then applied with specific parameters to keep the number of triangles under 2500.

2.2   Graph Representation of the LA

To study the impact of PVs and LAA geometry in clot formation, we propose to represent the LA as a graph similar to its centre-line. To do so, we first label automatically the PVs and LAA of every mesh with the help of the LDDMM framework and the varifold representation of shapes (see [7]). We compute the *Atlas*, or mean shape of the population, giving us a diffeomorphic registration from the Atlas to every shape. The population can then be fully represented by the Atlas T and a set of deformations $\{\varphi_i\}_{i \leqslant n}$. After hand-labeling of the Atlas, we warped the labels through the deformations in order to label every patient.

In practice, due to different types of vein anatomy in the population, the atria were separated in three classes to prevent big deformations from moving the labels too far from the roots of the veins. The analysis was performed with the deformetrica software [2] and the deformation and varifold kernels widths were both set to 10 mm.

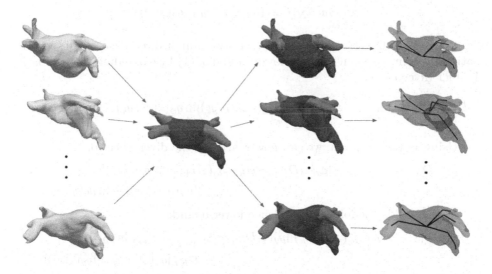

Fig. 1. Pipeline for the graph representation of the LA.

As a result, the variations of each mesh were captured well up to the residual noise on the surface. Therefore, atlas labels were warped to each subject faithfully to the anatomy of the atrium, as shown in Fig. 1.

---

[1] https://www.mmgtools.org/.

To achieve the graph representation, we extract the centre of mass of each label (i.e. PVs, LAA and body of LA) as well as the centre of each junction between labels, representing the ostium[2]. These points are the graph nodes, each branch is connected to the centre of the body of the LA.

Finally, to have a unified coordinate system between graphs, we set the body centre as the origin, fix the x-axis as the direction from the centre to the LAA ostium, and chose the left anterior PV ostium as the second direction.

2.3   Design of Fusion and Classification Loss Function

Let's denote $D = \{D_k\}_{k=1}^N$ N datasets with respective dimension d_k, and $(x, y) = \{x_k, y_k\}_{k=1}^N$ a set of pairs of observation x_k and label y_k from every dataset. Let $z \in \mathbb{R}^d$ be the latent variable shared by all elements of (x, y), with d the dimension of the latent space such that $d \ll \inf\{d_k | 1 \leqslant k \leqslant N\}$. We aim at having a common representation across datasets D_k, thus for every k a common distribution $p(z|x_k, y_k)$.

To do so, we use variational inference by introducing $\phi = \{\phi_k\}_{k=1}^N$ the inference parameters for each datasets, θ the common generative parameters and density functions $q_\phi(z|x_k, y_k) \in \mathcal{Q}$, which we want, on average, to be as close as possible to the common posterior $p_\theta(z|D)$. By using the Kullback-Leibler divergence, this problem translates to:

$$\underset{q \in \mathcal{Q}}{\operatorname{argmin}} \, \mathbb{E}_N[\mathcal{D}_{KL}(q_\phi(z|x_k, y_k)||p_\theta(z|D))] \tag{1}$$

Because of the intractability of $p_\theta(z|D)$ we cannot directly solve this optimisation problem. We aim to find a lower bound of (1) by expanding the Kullback-Leibler divergence:

$$\mathcal{D}_{KL}[q_\phi(z|x_k, y_k)||p_\theta(z|D))] = \int_{\mathbb{R}^d} q_\phi(z|x_k, y_k)[\ln q_\phi(z|x_k, y_k) - \ln p_\theta(z|D)]dz \tag{2}$$

Using Bayes' theorem, we can now rearrange the divergence to:

$$\begin{aligned} \mathcal{D}_{KL}[q_\phi(z|x_k, y_k)||p_\theta(z|D))] = \mathcal{D}_{KL}[q_\phi(z|x_k, y_k)||p_\theta(z|x))] \\ - \mathbb{E}_{z \sim q_{\phi_k}}[\ln p_\theta(y|z, x)] + \ln p_\theta(y|x) \end{aligned} \tag{3}$$

Which yields the following evidence lower bound:

$$\begin{aligned} \ln p_\theta(y|x) - \mathcal{D}_{KL}[q_\phi(z|x_k, y_k)||p_\theta(z|D))] = \mathbb{E}_{z \sim q_{\phi_k}(z|x_k, y_k)}[\ln p_\theta(y|z, x)] \\ - \mathcal{D}_{KL}[q_\phi(z|x_k, y_k)||p_\theta(z|x))] \end{aligned} \tag{4}$$

We impose this constraint over all datasets, by supposing that every dataset is conditionally independent, we have the following evidence lower bound:

$$\begin{aligned} \mathcal{L}(x, y, \theta, \phi) = \frac{1}{N} \sum_{k=1}^N \mathbb{E}_{z \sim q_{\phi_k}(z|x_k, y_k)} \left[\sum_{k=1}^N \ln p_\theta(y_k|z, x_k) \right] \\ - \mathcal{D}_{KL}[q_{\phi_k}(z|x_k, y_k)||p_\theta(z|x))] \end{aligned} \tag{5}$$

---

[2] The ostium is the centre of the root of the vein.

Maximising the lower bound $\mathcal{L}$ is therefore equivalent to optimising the initial problem (1). The distribution $p_\theta(y_k|z, x_k)$ of shared parameters θ is learned by a common decoder from the latent space, and acts on the labels y_k, in this sense the decoder is a classifier on the set of all labels in D. In addition the learned distribution p_θ is a meta-parameter that contains information from every datasets in D.

Unlike variational auto-encoders, the reconstruction objective in (5) is over the labels y, which transforms the traditional decoder into a classifier. Moreover having more than one encoder impacts the reconstruction loss which becomes a cross reconstruction of the labels from every dataset. This constraint forces the encoders to identify a common latent representation across all the datasets.

In practice, we assume $\mathcal{Q}$ is the Gaussian family, parameters θ and ϕ are initialised randomly, the optimisation is done by stochastic gradient descent, using an adaptive learning rate with an Adam optimiser and back-propagation.

3   Synthetic Experiments

The aim of our synthetic experiments[3] is to highlight the possibilities offered by our method, for the sake of interpretability, we chose to use a relatively simple parameterisation for our model, consisting of a neural network with three fully connected layers as our encoder and 2 fully connected layers for the classifier. All activation functions are ReLU, and a Softmax function is used for classification.

The synthetic data was generated with the `make-classification` function from the `scikit-learn`[4] library, by generating clusters of points for multi-label classification by sampling from a normal distribution. We then generate similar datasets by applying transformation and adding noise to the initial problem. Thanks to this method, we can generate a high variety of independent datasets where the features have different distributions, but sharing the same target space.

We generate a collection of 5 independent datasets of dimensions varying between 10 and 20, sharing three labels. The latent dimension is set to 4 and we train our model jointly across datasets, as well as on each dataset separately. On 100 experiments, results show a generally increased overall accuracy score on test data when the datasets are trained jointly. With a 90% confidence level, the accuracy for joint training lies in $[0.83, 0.99]$ with a median of 0.92, while the accuracy for separate training lies in $[0.73, 0.94]$ with a median of 0.86, and about the same amount of epochs are required for convergence. Moreover, we can see in Fig. 2 that training multiple datasets jointly leads to a coherent latent space shared between labels (Fig. 2b), while trying to independently represent the latent spaces of each dataset in a common space leads to poor consistency of the latent space representation across datasets (Fig. 2a).

---

[3] The code is available at https://github.com/Inria-Asclepios/mcvc.
[4] https://scikit-learn.org/.

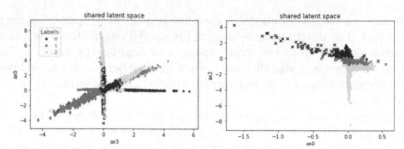

(a) The 5 independent latent spaces (b) Our model's Joint latent space
in a common space

Fig. 2. Latent space after convergence on 5 datasets with 3 labels. Our model provides a unified representation (Fig. 2b), not achievable when datasets are trained independently (Fig. 2a)

3.1 Noisy Labels

Because of the shared latent space between the collection of datasets D, some subspace is allocated for labels that may not belong to a given dataset D_i. This brings robustness to the classification when there is an uncertainty on the ground truth labels. In fact, if we assign a wrong label to a given dataset D_i the model is capable of assigning the observation to another label due to the constraint of obtaining a coherent latent space. To illustrate this point we created four datasets sharing three common possible labels; for one dataset we modified the third label into a fourth label. In Fig. 3 we show the evolution of the joint latent space throughout training, which highlights the preference for obtaining a coherent latent space rather than high-accuracy, the test accuracy dropped from 0.88 to 0.75 while the fourth label completely disappeared in the predictions.

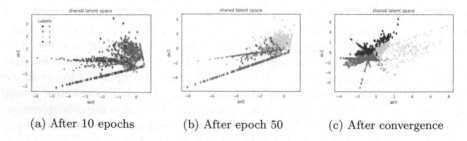

(a) After 10 epochs   (b) After epoch 50   (c) After convergence

Fig. 3. Evolution of joint latent space whith noisy labels. The wrong label (red) disappears completely from predictions to become coherent label (green) (Color figure online)

4   Application to LAA Graphs

Our clinical database is composed of 50 patients in systole (of which 27 are Thrombus positive) and 57 patients in diastole (of which 37 are Thrombus positive), thus we use two encoders in which we feed the nodes of the graphs to enable a joint analysis of the dataset. The labels are 1 if the patient is Thrombus positive and 0 otherwise. As such, the classes are well balanced for both datasets. We train the model with the same architecture and hyper-parameters as for synthetic data.

After 10-fold cross-validation, the model yields an average test accuracy of 0.89, for the diastolic set 0.92, and for the systolic set 0.86. In contrast, attempts to classify the subsets independently suffer from mode collapse; even a careful hyper-parameters fine-tuning results in very poor accuracy scores, reaching at best 0.65 in both cases. This highlights the robustness of the model and the clear advantage of joint analysis. Figure 4b shows the shared latent space with the systole subset as ($\times$) and diastole subset as ($\bullet$). We see a clear separation of classes as well as a common separator for the subsets, while systole and diastole are well grouped together for each class.

When we attempt to classify the complete dataset without any splitting, disregarding the considerable changes in shape during the cardiac cycle, we obtain slightly worse accuracy results of 0.84. In addition, such a model is less interpretable as important features for a given class can be contradictory in between subsets. To highlight this we investigate important features and show possible bio-markers.

As an additional baseline we performed PCA followed by logistic regression with cross validation and grid search on the number of principal components. We observed a much lower accuracy both on the complete dataset or individual subsets (0.65 at diastole, 0.71 at systole and 0.66 on the complete dataset).

We compare results from three interpretation algorithm (Integrated Gradient [11], DeepLIFT [10], and KernelSHAP [6]) implemented in `captum`[5] python library. We first feed the tests 100 times during cross validation and compute the feature importance algorithms on samples that are predicted right with more than a 95% certainty. Mean values over all samples are the final feature attribution scores. Figure 4a highlights the necessity to split the datasets to keep clinical coherence; It shows the Integrated gradient score on the x value of the Right Inferior PV (RIPV) point when the dataset is split (systole and diastole) and when it is trained commonly (i.e. not split). We see the model without splitting the data disregards this feature when in fact it is important for the diastole subset.

Figure 5 shows the possible bio-markers, in black are atlases of the population at systole and diastole, in blue are important feature values for control patients, in red are the ones for Thrombus positive patients. At diastole, the model seems to focus on the left PVs; their ostia being closer to and rotated towards the interior compared to the rest of the frame in Thrombus positive

---

[5]   https://captum.ai/.

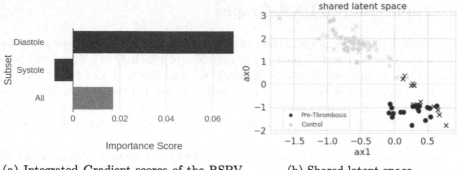

(a) Integrated Gradient scores of the RSPV center's x-axis for joint (purple) and without split (gray) training.

(b) Shared latent space.

Fig. 4. Results of the method. Joint analysis enables better interpretation of biomarkers (4a) as well as successful representation in the latent space (4b).

cases; In addition, for those cases, all veins intersection with the LA body are more horizontal, this 'fold' could impact the hemodynamic environment of the LA, which plays an important part in clot formations. At systole, the left PV and the angulation of the appendage are the focus. For Thrombus positive cases, PVs are rotated towards the interior with the left interior PV being on top and aligned with the LAA; The LAA ostium tends to be closer to the center and the LAA lower. Finally we see again the PVs tendency to being more 'folded'. While being interpretable, this also highlights the importance of analysing systole and diastole images separately.

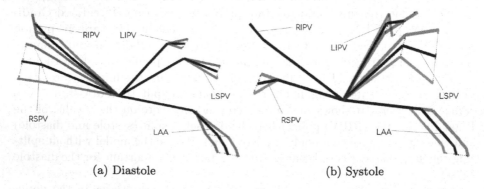

(a) Diastole

(b) Systole

Fig. 5. Visualisation of important features for predicting presence (red) and absence (blue) of thrombosis, both at diastole and systole. (Color figure online)

5   Conclusion

In this work, we provided a graph representation of the LA to analyse possible image-based bio-markers. In order to enable joint analysis of systole and diastole graphs we presented a new method at the crossroad between multi-task learning and meta learning to tackle the joint analysis of multiple heterogeneous datasets. By leveraging on the idea that the whole is better than its parts, we proposed a classification scheme with good interpretation properties of the latent space highlighted in the study of LA Graphs. We believe that the coherent latent space inherited from our model makes it possible to have deep neural network as encoders while conserving the interpretability of simpler models. We aim at further exploiting this property by applying the method to joint analysis of datasets containing much more heterogeneity. Finally, we believe the lightweight graph representation can be added in a more complete and multi-disciplinary study of the LA.

Acknowledgment. This work is funded by ERACoSysMed PARIS Project, and is in collaboration with the IHU Liryc of Bordeaux and the Pompeu Fabra University in Barcelona. This work has been supported by the French government, through the 3IA Côte d'Azur Investments in the Future project managed by the National Research Agency (ANR) with the reference number ANR-19-P3IA-0002, as well as the IHU LIRYC with the reference number ANR-10-IAHU-04.

References

1. Antelmi, L., Ayache, N., Robert, P., Lorenzi, M.: Sparse multi-channel variational autoencoder for the joint analysis of heterogeneous data. In: International Conference on Machine Learning, pp. 302–311. PMLR (2019)
2. Bône, A., Louis, M., Martin, B., Durrleman, S.: Deformetrica 4: an open-source software for statistical shape analysis. In: Reuter, M., Wachinger, C., Lombaert, H., Paniagua, B., Lüthi, M., Egger, B. (eds.) ShapeMI 2018. LNCS, vol. 11167, pp. 3–13. Springer, Cham (2018). https://doi.org/10.1007/978-3-030-04747-4_1
3. Di Biase, L., et al.: Does the left atrial appendage morphology correlate with the risk of stroke in patients with atrial fibrillation? Results from a multicenter study. J. Am. Coll. Cardiol. **60**(6), 531–538 (2012)
4. Hospedales, T., Antoniou, A., Micaelli, P., Storkey, A.: Meta-learning in neural networks: a survey. arXiv e-prints pp. arXiv-2004 (2020)
5. Jia, S., et al.: Automatically segmenting the left atrium from cardiac images using successive 3D U-Nets and a contour loss. In: Pop, M., et al. (eds.) STACOM 2018. LNCS, vol. 11395, pp. 221–229. Springer, Cham (2019). https://doi.org/10.1007/978-3-030-12029-0_24
6. Lundberg, S., Lee, S.I.: A unified approach to interpreting model predictions. Adv. Neural. Inf. Process. Syst. **30**, 4765–4774 (2017)
7. Pennec, X., Sommer, S., Fletcher, T.: Riemannian Geometric Statistics in Medical Image Analysis. Academic Press, Amsterdam (2019)
8. Quinn, G.R., Severdija, O.N., Chang, Y., Singer, D.E.: Wide variation in reported rates of stroke across cohorts of patients with atrial fibrillation. Circulation **135**(3), 208–219 (2017)

9. Ruder, S.: An overview of multi-task learning in deep neural networks. arXiv e-prints pp. arXiv-1706 (2017)
10. Shrikumar, A., Greenside, P., Kundaje, A.: Learning important features through propagating activation differences. In: International Conference on Machine Learning, pp. 3145–3153. PMLR (2017)
11. Sundararajan, M., Taly, A., Yan, Q.: Axiomatic attribution for deep networks. In: International Conference on Machine Learning, pp. 3319–3328. PMLR (2017)
12. Watson, T., Shantsila, E., Lip, G.Y.: Mechanisms of thrombogenesis in atrial fibrillation: Virchow's triad revisited. Lancet **373**(9658), 155–166 (2009)
13. Yaghi, S., et al.: The left atrial appendage morphology is associated with embolic stroke subtypes using a simple classification system: a proof of concept study. J. Cardiovasc. Comput. Tomogr. **14**(1), 27–33 (2020)

Cardiac Transmembrane Potential Imaging with GCN Based Iterative Soft Threshold Network

Lide Mu and Huafeng Liu[✉]

State Key Laboratory of Modern Optical Instrumentation, Zhejiang University,
Hangzhou 310027, China
liuhf@zju.edu.cn

Abstract. Accurate reconstruction and imaging of cardiac transmembrane potential through body surface ECG signals can provide great help for the diagnosis of heart disease. In this paper, a cardiac transmembrane potential reconstruction method (GISTA-Net) based on graph convolutional neural network and iterative soft threshold algorithm is proposed. It fully combines the rigor of mathematical derivation of traditional iterative threshold shrinkage algorithm and the powerful expression ability of deep learning method, as well as the characterization ability of graph convolutional neural network for non-Euclidean space data. We used this algorithm to simulate ectopic pacing data and simulated myocardial infarction data. The experimental results show that this algorithm can not only accurately locate the ectopic pacing point, but also accurately reconstruct the edge details of myocardial infarction scar while graph convolution makes full use of the connection information between the nodes on the heart surface.

Keywords: Cardiac electrophysiological reconstruction · ISTA · Graph convolution neural network

1 Introduction

The noninvasive imaging of the electrical activity and its propagation on the heart based on the measurement of body surface potential (BSP) has significant values for better understanding and diagnosis of cardiac diseases [1, 2]. Particularly, the cardiac transmembrane potential (TMP) can not only locate the source of premature ventricular beats and tachycardia, but also accurately reconstruct the scar boundaries of ischemic diseases such as myocardial ischemia and myocardial infarction, and many methods for the inverse problem in transmembrane potential reconstruction have been proposed which can be divided into two categories: 1) Traditional regularization methods [3–5]: In this kind of methods, a regularization matrix is usually defined first, such as wavelet transform [6], gradient transform [7], etc., then the regularization matrix and its solution are constrained, and finally solved by iteration. Although these methods have strong mathematical interpretation, the artificially defined regularization matrix is not completely suitable for the reconstruction of cardiac transmembrane potential, resulting

© Springer Nature Switzerland AG 2021
M. de Bruijne et al. (Eds.): MICCAI 2021, LNCS 12906, pp. 547–556, 2021.
https://doi.org/10.1007/978-3-030-87231-1_53

in poor reconstruction accuracy and relatively little information for the diagnosis of heart disease; 2) Deep-learning based methods: most of these methods often learns the mapping relation between TMP and BSP from a large number of data by CNN with strong expressive ability. However, this kind of methods has poor mathematical interpretation and requires a lot of data for CNN training, at the same time, CNN is more suitable for data in Euclidean space, but not very good for data in non-Euclidean space such as transmembrane potential and body surface potential. Recently, some researchers have proposed to reconstruct cardiac transmembrane potential through spatiotemporal graph convolutional neural network combined with encoding and decoding module [8]. Although processing data in non-Euclidean space has improved the accuracy of the reconstruction results, this method is still a black box and does not make full use of the forward propagation relationship between TMP and BSP.

In recent years, many researchers have proposed some approaches combining traditional regularization-based algorithms with deep learning to solve the inverse problem in the similar image compressive sensing [9–11], such as ISTA-Net, ADMM-Net, variational network. These methods add convolutional neural network to the iterative steps of solving the traditional regularization algorithm, which not only retains the mathematical interpretability of the traditional algorithm, but also greatly reduces the number of iterations and calculation time of the traditional algorithm. However, image data belongs to Euclidean space data and CNN is used, these methods are not completely suitable for the reconstruction of cardiac transmembrane potential which are non-Euclidean space data, while the geometric information of the heart surface is also not utilized.

In this paper, we proposed a graph convolution neural network and iterative soft threshold algorithm (GISTA-Net) for Cardiac Transmembrane potential imaging. The innovation of this paper is twofold: 1) Due to the powerful expression ability of graph convolution neural network, the regularization matrix used in the traditional regularization method based on L1 norm can be learned from the data; 2) We not only make use of the forward propagation relationship between TMP and BSP through the traditional regularization method, but also take advantage of the geometric structure of the heart surface which is the connection of each node on the heart surface, through the graph convolutional neural network. A large number of experiments have proved that the method proposed in this paper is very effective for heart transmembrane potential reconstruction, especially for the reconstruction of the edge of infarction scar.

2   Methodology

2.1   GISTA-Net Architecture

The body surface potential and cardiac transmembrane potential satisfies a linear forward relationship,

$$\Phi = Hu \tag{1}$$

where Φ is the BSP, H denotes the transfer matrix which can be obtained from the geometric relationship between heart and body surface by BEM and FEM, and u is the TMP.

It is a rank-deficient and ill-posed problem to reconstruct TMP from known BSP and transfer matrix as the nodes on the heart to be reconstructed is much more than the nodes measured on body surface. Traditional methods based on L1 norm often transform the inverse problem into a constraint minimization problem which is often convex by adding a L1 norm constraint term,

$$\min_u \frac{1}{2}\|Hu - \Phi\|_2^2 + \lambda\|Tu\|_1 \tag{2}$$

where T is some kind of prior transform operator such as gradient operator or wavelet operator based on different prior information and λ is the regularization parameter. This kind of L1-norm based optimization problem can be solved by iterative soft threshold algorithm (ISTA) [16] through proximal operator,

$$u^{(i)} = \text{prox}_{\lambda,F}\left(u^{(i-1)} - t\nabla G\left(u^{(i)}\right)\right) \quad \text{where } G(u) = \left\|Hu^{(i)} - \Phi\right\|_2^2 \tag{3}$$

where t is the step length in the gradient descent, $prox_{\lambda,F}$ denotes the proximal operator and $prox_{\lambda,T}(x) = T'\text{soft}(Tx, \lambda)$ for gradient operator and wavelet operator, $soft(\cdot)$ denotes the soft threshold shrinkage and T' is the transpose matrix of T.

Nevertheless, the prior transform operator is hard to find as well as the regularization parameter. Therefore, we define another nonlinear transformation $\Gamma(x) = \mathbf{D}\cdot ReLU(\mathbf{C}\cdot x)$ to replace the traditional prior operator instead of prior matrix so that the prior of TMP can be learned by a graph convolution network, where $\mathbf{C}$ and $\mathbf{D}$ are graph convolution blocks which can take advantage of the geometry of the heart which is further discussed in Sect. 2.2, and $ReLU(\cdot)$ is the Rectified Linear Units activation function. Then we can use the ISTA algorithm to solve the L1-norm constraint problem which can be expressed as two steps, the gradient descent for the rough solution and solving the proximal operator as shown in Eqs. (4) and (5),

$$\tilde{u}^{(i)} = u^{i-1} - t\nabla G\left(u^{(i-1)}\right) \tag{4}$$

$$u^{(i)} = \text{prox}\left(\tilde{u}^{(i)}\right) = \arg\min_u \frac{1}{2}\|\Gamma(u) - \Gamma(\tilde{u})\|_2^2 + \beta\|\Gamma(u)\|_1 \tag{5}$$

where $\tilde{u}^{(i)}$ denotes the rough solution of i-th iteration and $\beta = \lambda\alpha$ is the regularization parameter. It has been proved in [9] that $\left\|\Gamma(u) - \Gamma(\tilde{u}^i)\right\|_2^2 \approx \alpha\left\|u - \tilde{u}^i\right\|_2^2$. Then, the solution of the i-th iteration applying soft threshold shrinkage can be expressed as,

$$u^{(i)} = \Gamma'^{(i)}\left(\text{soft}\left(\Gamma^{(i)}\left(\tilde{u}^i\right), \beta^i\right)\right) \tag{6}$$

where $\Gamma'(x) = \mathbf{F}\cdot ReLU(\mathbf{E}\cdot x)$, $\mathbf{E}$, $\mathbf{F}$ are the graph convolution blocks and $\Gamma'(x)$ is also the transpose form of $\Gamma(x)$ in a sense that $\Gamma \circ \Gamma = I$ which can be constrained by the loss function defined later, I is the identity matrix.

In this way, the traditional ISTA algorithm for solving the traditional L1 norm constraint problem can be improved by a deep graph convolution neural network (GISTA-Net), which not only retains the proximal operator in the traditional ISTA algorithm, but also obtains the hyperparameters and the TMP prior information through network learning, while the geometric information of the heart is also used. And instead of learning the mapping between BSP and TMP directly, the residuals [12] of the TMP are learned by GISTA-net. The overall architecture of GISTA-net is shown in Fig. 1 and the loss function is defined with an error term and a symmetry regularization term,

$$L = \frac{1}{N} \sum_{k=1}^{N} \| \hat{u}_k - u_k \|^2 + \frac{\eta}{N} \sum_{k=1}^{N} \sum_{i=1}^{N_i} \left\| \tilde{\Gamma}^{(i)} \left(\Gamma^{(i)} (\tilde{u}_k) \right) - u_k \right\|_2^2 \tag{7}$$

where N is the number of training sets in a training batch, N_i is the iteration number of the GISTA-blocks and η is the regularization parameter which is set to 0.005 in our experiments.

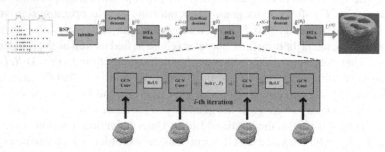

Fig. 1. GCN based iterative soft threshold network for cardiac transmembrane potential imaging (the body surface potential and the geometry of the heart surface are input into the network, and then the reconstructed transmembrane potential is output)

2.2 Implementation of Graph Convolution Network

Graph Convolution Network (GCN) was used in classification task on graph-structured data. The cardiac transmembrane potential data is also graph-structured data, while connection between each node is very close, therefore we apply the GCN to TMP reconstruction, which meet the calculation rules shown in the following equation,

$$H^{(l+1)} = \sigma \left(\tilde{D}^{-\frac{1}{2}} \tilde{A} \tilde{D}^{-\frac{1}{2}} H^{(l)} W^{(l)} \right) \tag{8}$$

where $\tilde{A} = A + I_N$ is adjacency matrix of self-connected undirected graph, I_N is identity matrix, $\tilde{D}_{ii} = \sum_j \tilde{A}_{ij}$ and $W^{(l)}$ are the trainable weight matrix of each layer, $\sigma(\cdot)$ is the activation function.

3   Experiments

3.1   Ectopic Pacing Experiment

We use ECGSIM software [13] to simulate 90000 pairs of data for GISTA-Net training and 1000 pairs of data for testing. The simulated data included heart transmembrane potential (low-potential region is un-activated tissue with a potential of about −85 mV, and high-potential region is activated tissue with a potential of about +15 mV), body surface potential and transport matrix. In order to verify the performance of GISTA-Net, we compare it with traditional total variation (TV) algorithm and ISTA net using 1D convolution and 2D convolution, respectively.

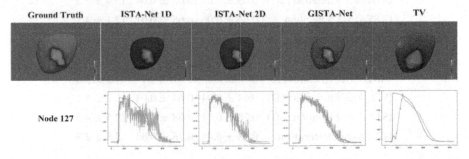

Fig. 2. The results of transmembrane potential reconstruction in an example of ectopic pacing experiment (The first line is the spatial presentation of TMP of a moment while activation starts, the second line is the temporal waveform of the pacing point, the blue line is the ground Truth and the orange line is the reconstruction results.) (Color figure online)

The results of transmembrane potential reconstruction in an example of ectopic pacing experiment are shown in Fig. 2, although the traditional TV algorithm can roughly reconstruct the location of ectopic pacing, there is still a large gap between the reconstruction result and the ground truth which also takes a long time. For ISTA-Net with one-dimensional convolution, because the number of learnable parameters of one-dimensional convolution is small, although the pace position can be roughly reconstructed, the boundary between the activated region and the inactive region is too smooth and fuzzy, and the reconstructed results can not provide enough detailed information. With the increase of parameters, the performance of two-dimensional convolution on ISTA-Net is better than that of one-dimensional convolution, but some boundary regions are still quite different from ground truth. In contrast, GISTA-Net using graph convolution network makes full use of the relationship between the adjacent nodes on the heart surface, and the reconstruction results of details and boundary area are better, we can obtain more accurate results in the case of relatively few iterations, which greatly reduces the amount of parameters and calculation of the algorithm. We also calculated the correlation coefficient (CC) [14], structural similarity (SSIM) [14] and the location error of the pacing point between the reconstruction results and the true value in the ectopic pacing experiment. In the case of the same number of iterations, the GISTA-Net proposed in this chapter has the highest CC and SSIM, and the smallest location error, the average location error of GISTA-Net of 9 iterations is only 9.8 mm as shown in Table 1.

Table 1. Quantitative analysis results of ectopic pacing experiment

	CC	SSIM	LE (mm)
TV	0.56 ± 0.04	0.58 ± 0.07	18.3 ± 4.5
ISTA-Net 5 iteration (1D)	0.60 ± 0.05	0.59 ± 0.03	17.6 ± 3.6
ISTA-Net 5 iteration (2D)	0.63 ± 0.08	0.63 ± 0.04	15.4 ± 5.2
GISTA-Net 5 iteration	0.66 ± 0.06	0.65 ± 0.01	12.1 ± 4.1
ISTA-Net 7 iteration (1D)	0.61 ± 0.05	0.62 ± 0.05	17.1 ± 4.2
ISTA-Net 7 iteration (2D)	0.65 ± 0.02	0.66 ± 0.03	13.2 ± 4.6
GISTA-Net 7 iteration	0.72 ± 0.05	0.69 ± 0.04	11.5 ± 3.2
ISTA-Net 9 iteration (1D)	0.64 ± 0.04	0.64 ± 0.08	16.2 ± 5.1
ISTA-Net 9 iteration (2D)	0.70 ± 0.01	0.68 ± 0.02	12.5 ± 4.5
GISTA-Net 9 iteration	$\mathbf{0.78 \pm 0.08}$	$\mathbf{0.75 \pm 0.04}$	$\mathbf{9.8 \pm 2.7}$

3.2 Myocardial Infarction Experiment

Fig. 3. Comparison of TMP reconstruction results between GISTA-NET and other algorithms in myocardial infarction experiment (the blue area is the infarct area, and the red area is the normal tissue) (Color figure online)

TMP is of great clinical significance in the diagnosis of myocardial ischemia and myocardial infarction. Therefore, we also reconstructed the data of simulated myocardial infarction in this section. Figure 3 shows the TMP reconstruction results of traditional TV algorithm, one-dimensional convolutional ISTA-Net, two-dimensional convolutional ISTA-Net and GISTA-Net while the whole heart is activated. Among them, the red area is the area of high potential that has been activated, and the blue area is the area of low potential that has not been activated, that is, the infarct area.

Unlike the diagnosis of ectopic pacing, which requires accurate positioning, the diagnosis of myocardial infarction also requires rich detailed information about the

Table 2. Quantitative analysis results of myocardial infarction experiment

	CC	SSIM
TV	0.58 ± 0.09	0.58 ± 0.04
ISTA-Net 5 iteration (1D)	0.61 ± 0.04	0.60 ± 0.02
ISTA-Net 5 iteration (2D)	0.65 ± 0.09	0.65 ± 0.04
GISTA-Net 5 iteration	0.68 ± 0.04	0.69 ± 0.08
ISTA-Net 7 iteration (1D)	0.62 ± 0.06	0.61 ± 0.04
ISTA-Net 7 iteration (2D)	0.68 ± 0.01	0.67 ± 0.05
GISTA-Net 7 iteration	0.72 ± 0.04	0.73 ± 0.08
ISTA-Net 9 iteration (1D)	0.65 ± 0.05	0.66 ± 0.07
ISTA-Net 9 iteration (2D)	0.72 ± 0.03	0.71 ± 0.06
GISTA-Net 9 iteration	**0.79 ± 0.09**	**0.76 ± 0.04**

infarction scar boundary. Although TV algorithm found the location of infarction in the right ventricular outflow tract, its reconstruction results completely lost the boundary information between infarction and normal tissue. Because one-dimensional convolution and two-dimensional convolution are directly calculated after randomly reshape the dimensions of TMP data, and graph convolutional neural network makes full use of the connection information between nodes on the heart surface, GISTA-Net has the best detail and edge reconstruction performance. The quantitative analysis of the experimental results is shown in Table 2, which is completely consistent with the results in Fig. 3. We also calculated the relative normalized mean squared error (MSE) of the results and the truth in order to evaluate the reconstruction quantitatively as illustrated in Table 2, the maximum MSE of proposed GISTA-Net was 0.005.

3.3 Cardiac Activation Sequence Reconstruction

The activation sequence of cardiomyocytes on the surface of the heart may provide crucial information about the origin and evolution of heart disease, therefore we also reconstructed the activation sequence of the heart based on the simulated data of ectopic pacing induced by myocardial infarction. Figure 4 shows an example of myocardial activation timing reconstruction. As shown in the figure, at 46 ms, the earliest activation point appears at the apex of the left ventricle, and then the activation slowly spreads to the distal right ventricle, and the infarct location is in the right ventricular outflow tract. Although the TV algorithm reconstructed the earliest activated location in the apex of the left ventricle, the reconstruction of activated propagation is not so satisfactory, especially the infarct location is not particularly prominent. Compared with TV algorithm, ISTA-Net using 1D convolution and 2D convolution can relatively accurately reconstruct the active propagation, but in the details of the propagation process, the edge reconstruction effect of the active propagation around the infarct scar is not as good as GISTA-Net using graph convolution neural network.

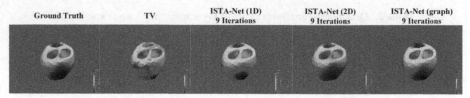

Fig. 4. Reconstruction of myocardial activation time sequence by simulating ectopic pacing data induced by myocardial infarction (the blue area is the earliest activated position, and the red area is the latest activated position) (Color figure online)

4 Discussion

In the actual clinical measurement process, it is inevitable to introduce a variety of different types and sizes of noise, so this section will discuss the transmembrane potential reconstruction in the presence of noise. We use paired BSP and TMP without noise to train GISTA-Net. During the test, 5 dB, 15 dB and 25 dB Gaussian white noise, shot noise and multiplicative noise are added to the body surface potential of the test data respectively. At the same time, we also carry out experiments in the case of multiple noises, and the results are shown in Table 3, GISTA-Net has good robustness to different levels and types of noise.

Table 3. Quantitative analysis of TMP reconstruction in the presence of noise

Method	Noise	CC	SSIM
GISTA-Net (9 iterations)	Without noise	0.79 ± 0.09	0.76 ± 0.04
	5 dB Gaussian noise	0.78 ± 0.05	0.75 ± 0.02
	15 dB Gaussian noise	0.76 ± 0.04	0.73 ± 0.04
	25 dB Gaussian noise	0.73 ± 0.08	0.70 ± 0.08
	25 dB Shot noise	0.72 ± 0.05	0.70 ± 0.05
	25 dB multiplicative noise	0.71 ± 0.04	0.71 ± 0.06
	15 dB mixed noise	0.75 ± 0.05	0.74 ± 0.05

In GISTA-Net proposed in this paper, most of the parameters can be learned from paired data, but the regularization weight parameter λ in the algorithm still needs to be artificially defined, which determines the influence of the regularization term on the final solution. Therefore, determining the appropriate regularization weight parameter λ will also have a decisive impact on the accuracy of the final solution. In this section, the sensitivity of the regularization weight parameter λ is analyzed, as shown in Fig. 5. When $\lambda = 0.5$, the mean value of CC and SSIM between the reconstructed result and the true value is the maximum. However, the increase or decrease of λ will result in the decrease of CC and SSIM, so $\lambda = 0.5$ in the experiment in this paper.

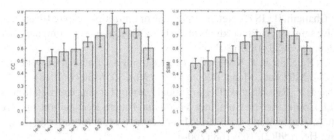

Fig. 5. Sensitivity analysis of regularization parameter λ

5 Conclusion

In this paper, GISTA-Net is proposed for cardiac transmembrane potential reconstruction, which GCN is used to make full use of the geometric structure information of the heart to adapt to the transmembrane potential data of the heart in non-Euclidean space. GISTA-NET performed very well in terms of ectopic pacing, myocardial infarction data for reconstruction of TMP and activation compared to existing methods.

Acknowledgement. This work is supported in part by the National Key Technology Research and Development Program of China (No: 2017YFE0104000, 2016YFC1300302), and by the National Natural Science Foundation of China (No: U1809204, 61525106, 61701436).

References

1. Paul, T., Windhagen-Mahnert, B., Kriebel, T., et al.: Atrial reentrant tachycardia after surgery for congenital heart disease endocardial mapping and radiofrequency catheter ablation using a novel, noncontact mapping system. Circulation **103**(18), 2266 (2001)
2. Ramanathan, C., Ghanem, R.N., Jia, P., et al.: Noninvasive electrocardiographic imaging for cardiac electrophysiology and arrhythmia. Nat. Med. **10**, 422–428 (2004)
3. Ghosh, S., Rudy, Y.: Application of L1-norm regularization to epicardial potential solution of the inverse electrocardiography problem. Ann. Biomed. Eng. **37**(5), 902–912 (2009)
4. Fang, L., Xu, J., Hu, H., et al.: Noninvasive imaging of epicardial and endocardial potentials with low rank and sparsity constraints. IEEE Trans. Biomed. Eng. **66**(9), 2651–2662 (2019)
5. Mu, L., Liu, H.: Noninvasive electrocardiographic imaging with low-rank and non-local total variation regularization. Pattern Recogn. Lett. **138**, 106–114 (2020)
6. Langer, M., Cloetens, P., Peyrin, F.: Fourier-wavelet regularization of phase retrieval in x-ray in-line phase tomography. J. Opt. Soc. Am. A Opt. Image Sci. Vis. **26**(8), 1876–1881 (2009)
7. Shou, G., Xia, L., Jiang, M.: Total variation regularization in electrocardiographic mapping. In: Li, K., Jia, L., Sun, X., Fei, M., Irwin, G.W. (eds.) ICSEE/LSMS - 2010. LNCS, vol. 6330, pp. 51–59. Springer, Heidelberg (2010). https://doi.org/10.1007/978-3-642-15615-1_7
8. Bacoyannis, T., Krebs, J., Cedilnik, N., et al.: Deep learning formulation of ECGI for data-driven integration of spatiotemporal correlations and imaging information. In: Coudière, Y., Ozenne, V., Vigmond, E., Zemzemi, N. (eds.) FIMH 2019. LNCS, vol. 11504, pp. 20–28. Springer, Cham (2019). https://doi.org/10.1007/978-3-030-21949-9_3
9. Jiang, X., Ghimire, S., Dhamala, J., et al.: Learning geometry-dependent and physics-based inverse image reconstruction. In: Martel, A.L., et al. (eds.) MICCAI 2020. LNCS, vol. 12266, pp. 487–496. Springer, Cham (2020). https://doi.org/10.1007/978-3-030-59725-2_47

10. Zhang, J., Ghanem, B.: ISTA-Net: interpretable optimization-inspired deep network for image compressive sensing. In: Proceedings of the IEEE Conference on Computer Vision and Pattern Recognition, pp. 1828–1837 (2018)
11. Hammernik, K., Klatzer, T., Kobler, E., et al.: Learning a variational network for reconstruction of accelerated MRI data. Magn. Reson. Med. **79**(6), 3055–3071 (2018)
12. He, K., Zhang, X., Ren, S., et al.: Deep residual learning for image recognition. In: Proceedings of the IEEE Conference on Computer Vision and Pattern Recognition, pp. 770–778 (2016)
13. Oosterom, A., Oostendorp, T.: ECGSIM: an interactive tool for studying the genesis of QRST waveforms. Heart **90**(2), 165–168 (2004)
14. Wang, L., Zhang, H., Wong, K.C.L., et al.: Physiological-model-constrained noninvasive reconstruction of volumetric myocardial transmembrane potentials. IEEE Trans. Biomed. Eng. **57**(2), 296–315 (2009)

AtrialGeneral: Domain Generalization for Left Atrial Segmentation of Multi-center LGE MRIs

Lei Li[1,2,3], Veronika A. Zimmer[3], Julia A. Schnabel[3], and Xiahai Zhuang[1(✉)]

[1] School of Data Science, Fudan University, Shanghai, China
zxh@fudan.edu.cn
[2] School of Biomedical Engineering, Shanghai Jiao Tong University, Shanghai, China
[3] School of Biomedical Engineering and Imaging Sciences, King's College London, London, UK

Abstract. Left atrial (LA) segmentation from late gadolinium enhanced magnetic resonance imaging (LGE MRI) is a crucial step needed for planning the treatment of atrial fibrillation. However, automatic LA segmentation from LGE MRI is still challenging, due to the poor image quality, high variability in LA shapes, and unclear LA boundary. Though deep learning-based methods can provide promising LA segmentation results, they often generalize poorly to unseen domains, such as data from different scanners and/or sites. In this work, we collect 140 LGE MRIs from different centers with different levels of image quality. To evaluate the domain generalization ability of models on the LA segmentation task, we employ four commonly used semantic segmentation networks for the LA segmentation from multi-center LGE MRIs. Besides, we investigate three domain generalization strategies, i.e., histogram matching, mutual information based disentangled representation, and random style transfer, where a simple histogram matching is proved to be most effective.

Keywords: Atrial fibrillation · LGE MRI · Left atrial segmentation · Domain generalization

1 Introduction

Radiofrequency (RF) ablation is a common technique in clinical routine for the atrial fibrillation (AF) treatment via electrical isolation. However, the success rate of some ablation procedures is low due to the existence of incomplete ablation pattern (gaps) on the left atrium (LA). Late gadolinium enhanced magnetic resonance imaging (LGE MRI) has been an important tool to detect gaps in ablation lesions, which are located on the LA wall and pulmonary vein (PV).

Electronic supplementary material The online version of this chapter (https://doi.org/10.1007/978-3-030-87231-1_54) contains supplementary material, which is available to authorized users.

© Springer Nature Switzerland AG 2021
M. de Bruijne et al. (Eds.): MICCAI 2021, LNCS 12906, pp. 557–566, 2021.
https://doi.org/10.1007/978-3-030-87231-1_54

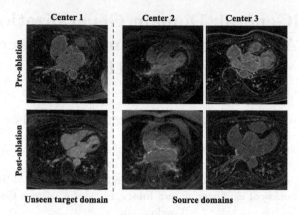

Fig. 1. Multi-center pre- and post-ablation LGE MRIs. The images differ in contrast, enhancement and background.

Thus, it is important to segment LA from LGE MRI for the AF treatment. Manual delineations of the LA from LGE MRI can be subjective and labor-intensive, and automating this segmentation remains challenging.

In recent years, many algorithms have been proposed to perform automatic LA segmentation from medical images, but mostly for non-enhanced imaging modalities. Conversely, LGE MRI has received less attention with respect to developed methods of LA segmentation to assist the ablation procedures. Most of the current studies on LA segmentation from LGE MRI are still based on time-consuming and error-prone manual segmentation methods [5,12]. This is mainly because LA segmentation methods in non-enhanced imaging modalities are difficult to directly apply to LGE MRI, due to the existence of the contrast agent and its low-contrast boundaries. Therefore, existing conventional automated LA segmentation of LGE MRI approaches generally require hard available supporting information, such as shape priors [20] or additional MRI sequences [8]. Recently, with the development of deep learning (DL) in medical image computing, some DL-based algorithms have been proposed for automatic LA segmentation directly from LGE MRI [7,15].

However, the generalization ability of the DL-based models is limited, i.e., the performance of a trained model on the known domain (source domain) will be degraded drastically on an unseen domain (target domain). This is mainly due to the existence of a *domain shift* or *distribution shift*, which is common among the data collected from different centers and vendors, as shown in Fig. 1. In the clinic, it is impractical to retrain a model each time for the data collected from new vendors or centers. Therefore, improving the model generalization ability is important to avoid the need of retraining. Current domain generalization (DG) methods can be categorized into three types: (1) domain-invariant feature learning approaches, such as disentangled representation [11]; (2) model-agnostic meta-learning algorithms, which optimize on the meta-train and meta-test domain split from the available source domain [4]; (3) data augmentation strategies, which increase the diversity of available data [1].

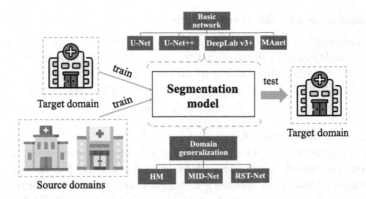

Fig. 2. Illustration of the LA segmentation models for multi-center LGE MRIs.

In this work, we investigate the generalization abilities of four commonly used segmentation models, i.e., U-Net [14], UNet++ [19], DeepLab v3+ [3] and multi-scale attention network (MAnet) [2]. As Fig. 2 shows, we select two different sources of training data, i.e., target domain (TD) and source domains (SD) to evaluate the model generalization ability. Besides, we compare three different DG schemes for LA segmentation of multi-center LGE MRIs. The schemes include histogram matching (HM) [10], mutual information based disentangled (MID) representation [11], and random style transfer (RST) [9,18].

2 Methodology

In this section, we describe the segmentation models we employ, formulate the DG problem (illustrated in Fig. 2) and describe the investigated three DG strategies.

2.1 Image Segmentation Models

All our segmentation models are supervised approaches based on convolutional neural networks. Typically, the models are trained using a training database $\mathcal{T}_\mathcal{D} = \{(X_m, Y_m), m = 1, \ldots, M\}$ with images $X \in \mathcal{D}$ from a single domain $\mathcal{D}$ and corresponding labels Y. The segmentation model $f(X)$ can be defined as,

$$f(X) \rightarrow Y, X \in \mathcal{D}, \tag{1}$$

where $X, Y \in \mathbb{R}^{1 \times H \times W}$ denote the image set and corresponding LA segmentation set.

We consider four commonly used segmentation models, all with an encoder-decoder architecture. The first model is a vanilla U-Net. U-Net++ is a modified version of the U-Net with a more complex decoder. DeepLab v3+ employs atrous spatial convolutions, and MAnet introduce multi-scale attention blocks.

2.2   Domain Generalization Models

The generalization ability of such models is limited, i.e., a model trained on a source domain $\mathcal{D}$ might perform poorly for images $X \notin \mathcal{D}$. DG strategies are therefore proposed to generalize models to unseen (target) domains. Given N source domains $\mathcal{D}_s = \{\mathcal{D}_1, \mathcal{D}_2, \cdots, \mathcal{D}_N\}$, we aim to construct a DG model $f^{DG}(X)$,

$$f^{DG}(X) \rightarrow Y, X \in \mathcal{D}_s \cup \mathcal{D}_t, \tag{2}$$

where $\mathcal{D}_t$ are unknown target domains.

We investigate three DG strategies for LA segmentation from LGE MRI. In the first and simplest approach, HM is performed on the images from the target domain to match its intensity histogram onto that of the source domains. The model and training process do not change. The second (MID-Net [11]) and third method (RST-Net [9,18]) are state-of-the-art methods employing different approaches to achieve DG. In MID-Net, domain-invariant features are extracted by mutual information based disentanglement in the latent space, while in RST-Net available domains are augmented via pseudo-novel domains.

3   Materials

3.1   Data Acquisition and Pre-processing

LGE MRIs with various image qualities, types and imaging parameters were collected from three centers, as Table 1 shows. The centers consist of Utah School of Medicine (Center 1), Beth Israel Deaconess Medical Center (Center 2), and Imaging Sciences at King's College London (Center 3). The dataset were selected from two public challenge, i.e., *MICCAI 2018 Atrial Segmentation Challenge* [17] and *ISBI 2012 Left Atrium Fibrosis and Scar Segmentation Challenge* [13]. A total of 140 images were collected and acquired either pre- or post-ablation. The acquisition time of pre-ablation scans varied slightly among 1 to 7 days, but that of post-ablation had a range from 1 to 27 months depending on the imaging center.

The LGE MRIs from center 1, 2, 3 and 4 were reconstructed to $0.625 \times 0.625 \times 1.25\,\mathrm{mm}$, $(0.7\text{--}0.75) \times (0.7\text{-}0.75) \times 2\,\mathrm{mm}$, $0.625 \times 0.625 \times 2\,\mathrm{mm}$, and $0.625 \times 0.625 \times 1.25\,\mathrm{mm}$, respectively. All 3D images were divided into 2D slices as network inputs and then were cropped into a unified size of 192×192 centering at the heart region, with a intensity normalization via Z-score. Random rotation, random flip and Gaussian noise augmentation were applied during training. The data distribution in the subsequent experiments is presented in Table 2.

3.2   Gold Standard and Evaluation

All the LGE MRIs were manually delineated by the experts from the corresponding centers. The manual LA segmentation were regarded as the gold standard. For LA segmentation evaluation, Dice score, average surface distance (ASD) and Hausdorff distance (HD) were applied. Each image from the three centers were assigned an image quality score by averaging the scores from two experts, mainly based on the visibility of enhancements and the existence of image artefacts (please see the Supplementary Material file).

Table 1. Image acquisition parameters of the multi-center LGE MRIs.

Parameters	Center 1	Center 2	Center 3
No. subject	100	20	20
Scanner	1.5T Siemens Avanto; 3T Siemens Vario	1.5T Philips Achieva	1.5T Philips Achieva
Resolution	1.25 × 1.25 × 2.5 mm	1.4 × 1.4 × 1.4 mm	1.3 × 1.3 × 4 mm
TI, TE/TR	270–310 ms, 2.3/5.4 ms	280 ms, 2.1/5.3 ms	280 ms, 2.1/5.3 ms
Pre-scan	N/A	<7 days	<2 days
Post-scan	3–27 months	30 days	3–6 months

Table 2. The distribution of training dataset and test dataset of LGE MRI from the three centers (C-i: center i).

Source of training data	Post/Pre	No. training data	No. test data
Target domain (TD)	Post	20 C-1	40 C-1
	Pre	20 C-1	20 C-1
Source domains (SD)	Post	10 C-2 + 10 C-3	40 C-1
	Pre	10 C-2 + 10 C-3	20 C-1

3.3 Implementation

The proposed framework was implemented in PyTorch, running on a computer with 2.20 GHz Intel(R) Xeon(R) E5-2630 v4 CPU and a GeForce GTX 1080 Ti GPU. We employed the released *Segmentation Models* [16] for experiments. All the backbones of the four semantic segmentation models are the *efficientnet-b6*. We used the Adam optimizer to update the network parameters. The initial learning rate was set to 5e−5 and multiplied by 0.95 every 10 epochs.

4 Experiment

4.1 Comparisons of Different Semantic Segmentation Networks

Table 3 summarizes the LA segmentation results in terms of Dice, ASD and HD based on the four semantic segmentation models. One can see that all the segmentation models had a performance decrease when the target domain was not included in the training data. It proves that the generalization capabilities of currently commonly used DL-based segmentation models are still very limited. When we observe the Dice value of the LA segmentation, the obtained performances of the four models training on the TD are very close. However, DeepLab v3+ achieved significantly better ASD and HD than the other three models. It may be attributed to its atrous convolution and spatial pyramid pooling module, which promote the network to learn more spatial information. When training on the SD, the performance decrease of was DeepLab v3+ was smaller than other three models. Therefore, in this work DeepLab v3+ is regarded as the baseline model, and we will improve its generalization ability using the proposed DG schemes.

Table 3. Performance of the four segmentation models on the multi-center LGE MRI for LA segmentation. The training and test data distribution refer to Table 2, i.e., test data is from Center 1 dataset, while training data is from the TD (C-1 dataset) and SD (C-2, 3 dataset), respectively.

Source	Metrics	U-Net [14]	U-Net++ [19]	DeepLab v3+ [3]	MAnet [2]
TD	Dice	0.908 ± 0.039	0.908 ± 0.037	0.900 ± 0.041	0.904 ± 0.057
	ASD (mm)	1.35 ± 0.722	1.26 ± 0.577	1.31 ± 0.557	1.37 ± 0.849
	HD (mm)	36.6 ± 12.3	28.5 ± 14.9	15.3 ± 5.78	32.7 ± 12.6
SD	Dice	0.652 ± 0.107	0.633 ± 0.104	0.678 ± 0.089	0.610 ± 0.116
	ASD (mm)	6.86 ± 1.51	7.12 ± 1.60	6.19 ± 1.65	8.26 ± 1.98
	HD (mm)	48.2 ± 7.93	51.1 ± 7.62	42.7 ± 8.62	53.3 ± 7.45

4.2   Comparisons of Post- and Pre-ablation LGE MRI

As Fig. 1 shows, the pre- and post-ablation LGE MRI can have high variability of tissue appearance. There are already several studies that have shown the performance of LA scar segmentation and quantification varied among pre- and post-ablation LGE MRI [6]. This is mainly because that when comparing to post-ablation images, the scars on pre-ablation LGE MRIs are hard to distinguish even for experts. In contrast, as far as we know, there are to this date no studies comparing the LA segmentation performance for pre- and post-ablation LGE MRI. Here, we compared and analyzed the LA segmentation performance on pre- and post-ablation LGE MRI on the four basic segmentation models.

Figure 3 presents the Dice and HD value obtained by the four models on the pre- and post-ablation images, separately. One can see that, the four models all suffered from an accuracy deterioration caused by the domain shift on both pre- and post-ablation LGE MRIs, which is consistent with the results in Table 3. Besides, the Dice obtained by the four models is similar on both pre- and post-ablation LGE MRIs, but DeepLab v3+ performed better in terms of HD, especially on pre-ablation data.

In summary, there is no evident performance difference between pre- and post-ablation data for the four models. However, the standard deviations of the Dice and HD values of the LA segmentation on the pre-ablation data are generally lower than those of post-ablation images. It may indicate that the segmentation model is more robust for the pre-ablation data of the multi-center LGE MRIs.

4.3   Comparisons of Different Generalization Models

Table 4 summarized three DG schemes to compare with the baseline DeepLab v3+ model training on multi-source domains. One can see that three tested generalization strategies worked when comparing with baseline results. Among the three methods, the conventional histogram matching algorithm performed best. The MID-Net and RST-Net obtained similar results in terms of Dice, but the ASD and HD of MID-Net were worse.

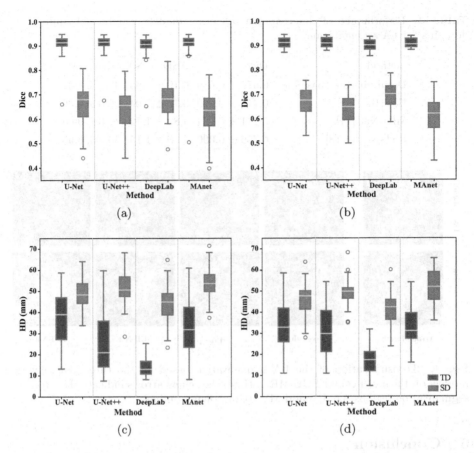

Fig. 3. LA segmentation results of four segmentation models with different source of training datasets: (a) Dice of post-ablation cases; (b) Dice of pre-ablation cases; (c) HD of post-ablation cases; (d) HD of pre-ablation cases.

Figure 4 presents the 2D visualization results of the four methods on post-/pre-ablation LGE MRI. In the post-ablation case, three DG schemes could identify some missing PV regions by the DeepLab v3+. Similarly, in the pre-ablation subject, MID-Net and RST-Net both mitigated the segmentation errors in the mitral valve (MV) area. It proved that for the both post- and pre-ablation cases, the employed DG methods worked.

Table 4. Performance of different generalization models training on multi-source domains for LA segmentation.

Method	Dice	ASD (mm)	HD (mm)
DeepLab v3+ (baseline)	0.678 ± 0.089	6.19 ± 1.65	42.7 ± 8.62
HM [10]	0.772 ± 0.089	3.76 ± 1.04	26.5 ± 5.30
MID-Net [11]	0.741 ± 0.064	4.83 ± 1.12	42.6 ± 9.66
RST-Net [9,18]	0.756 ± 0.090	4.20 ± 1.15	30.3 ± 6.90

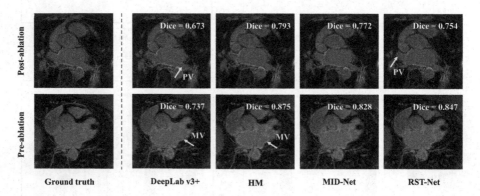

Fig. 4. 2D visualization of the LA segmentation based on the three generalization models on the multi-center LGE MRIs. Here, the yellow arrows indicate the wrong LA segmentation regions, i,e. PV and MV. (Color figure online)

5   Conclusion

In this work, we first investigated the generalization abilities of different semantic segmentation models for LA segmentation from multi-center LGE MRIs. The results showed that all the performance of the commonly used segmentation models degraded dramatically on the unknown domain. It emphasized the importance of promoting deep models with efficient inherent generalization abilities for LGE MRI data processing from different centers. We then introduced three DG strategies, which were all able to alleviate the performance decrease. Our study found that, quite surprisingly, the simple histogram matching strategy is the most effective method for DG on the LA segmentation of multi-center LGE MRI data. It may indicate that there is still large scope for further algorithmic developments in DG. In future, we will find the inherent differences of multi-center LGE MRIs, and develop a targeted and effective DG strategy to solve this problem. Moreover, we will further study the domain shift between post- and pre-ablation LGE MRI from the same center, and the label variations of LGE MRIs from different centers.

Acknowledgement. This work was funded by the National Natural Science Foundation of China (grant no. 61971142, 62111530195 and 62011540404) and the development fund for Shanghai talents (no. 2020015). L. Li was partially supported by the CSC Scholarship. JA Schnabel and VA Zimmer would like to acknowledge funding from a Wellcome Trust IEH Award (WT 102431), an EPSRC programme grant (EP/P001009/1), and the Wellcome/EPSRC Center for Medical Engineering (WT 203148/Z/16/Z).

References

1. Chen, C., et al.: Improving the generalizability of convolutional neural network-based segmentation on CMR images. Front. Cardiovasc. Med. **7**, 105 (2020)
2. Chen, L.C., Yang, Y., Wang, J., Xu, W., Yuille, A.L.: Attention to scale: scale-aware semantic image segmentation. In: Proceedings of the IEEE Conference on Computer Vision and Pattern Recognition, pp. 3640–3649 (2016)
3. Chen, L.C., Zhu, Y., Papandreou, G., Schroff, F., Adam, H.: Encoder-decoder with atrous separable convolution for semantic image segmentation. In: Proceedings of the European Conference on Computer Vision (ECCV), pp. 801–818 (2018)
4. Dou, Q., de Castro, D.C., Kamnitsas, K., Glocker, B.: Domain generalization via model-agnostic learning of semantic features. In: Advances in Neural Information Processing Systems, pp. 6450–6461 (2019)
5. Higuchi, K., et al.: The spatial distribution of late gadolinium enhancement of left atrial magnetic resonance imaging in patients with atrial fibrillation. JACC: Clin. Electrophysiol. **4**(1), 49–58 (2018)
6. Karim, R., et al.: Evaluation of current algorithms for segmentation of scar tissue from late gadolinium enhancement cardiovascular magnetic resonance of the left atrium: an open-access grand challenge. J. Cardiovasc. Magn. Reson. **15**(1), 1–17 (2013). Article number: 105
7. Li, L., Weng, X., Schnabel, J.A., Zhuang, X.: Joint left atrial segmentation and scar quantification based on a DNN with spatial encoding and shape attention. In: Martel, A.L., et al. (eds.) MICCAI 2020. LNCS, vol. 12264, pp. 118–127. Springer, Cham (2020). https://doi.org/10.1007/978-3-030-59719-1_12
8. Li, L., et al.: Atrial scar quantification via multi-scale CNN in the graph-cuts framework. Med. Image Anal. **60**, 101595 (2020)
9. Li, L., et al.: Random style transfer based domain generalization networks integrating shape and spatial information. In: Puyol Anton, E., et al. (eds.) STACOM 2020. LNCS, vol. 12592, pp. 208–218. Springer, Cham (2021). https://doi.org/10.1007/978-3-030-68107-4_21
10. Ma, J.: Histogram matching augmentation for domain adaptation with application to multi-centre, multi-vendor and multi-disease cardiac image segmentation. In: Puyol Anton, E., et al. (eds.) STACOM 2020. LNCS, vol. 12592, pp. 177–186. Springer, Cham (2021). https://doi.org/10.1007/978-3-030-68107-4_18
11. Meng, Q., et al.: Mutual information-based disentangled neural networks for classifying unseen categories in different domains: application to fetal ultrasound imaging. IEEE Trans. Med. Imaging **40**(2), 722–734 (2020)
12. Njoku, A., et al.: Left atrial volume predicts atrial fibrillation recurrence after radiofrequency ablation: a meta-analysis. Ep Europace **20**(1), 33–42 (2018)
13. Rhode, K., Karim, R.: ISBI 2012: left atrium fibrosis and scar segmentation challenge (2012). http://www.cardiacatlas.org/challenges/left-atrium-fibrosis-and-scar-segmentation-challenge/

14. Ronneberger, O., Fischer, P., Brox, T.: U-Net: convolutional networks for biomedical image segmentation. In: Navab, N., Hornegger, J., Wells, W.M., Frangi, A.F. (eds.) MICCAI 2015. LNCS, vol. 9351, pp. 234–241. Springer, Cham (2015). https://doi.org/10.1007/978-3-319-24574-4_28

15. Xiong, Z., et al.: A global benchmark of algorithms for segmenting the left atrium from late gadolinium-enhanced cardiac magnetic resonance imaging. Med. Image Anal. **67**, 101832 (2020)

16. Yakubovskiy, P.: Segmentation models (2019). https://github.com/qubvel/segmentation_models

17. Zhao, J., Xiong, Z.: MICCAI 2018: Atrial segmentation challenge (2018). http://atriaseg2018.cardiacatlas.org/

18. Zhou, K., Yang, Y., Hospedales, T., Xiang, T.: Learning to generate novel domains for domain generalization. In: Vedaldi, A., Bischof, H., Brox, T., Frahm, J.-M. (eds.) ECCV 2020. LNCS, vol. 12361, pp. 561–578. Springer, Cham (2020). https://doi.org/10.1007/978-3-030-58517-4_33

19. Zhou, Z., Rahman Siddiquee, M.M., Tajbakhsh, N., Liang, J.: UNet++: a nested U-Net architecture for medical image segmentation. In: Stoyanov, D., et al. (eds.) DLMIA/ML-CDS -2018. LNCS, vol. 11045, pp. 3–11. Springer, Cham (2018). https://doi.org/10.1007/978-3-030-00889-5_1

20. Zhu, L., Gao, Y., Yezzi, A., Tannenbaum, A.: Automatic segmentation of the left atrium from MR images via variational region growing with a moments-based shape prior. IEEE Trans. Image Process. **22**(12), 5111–5122 (2013)

TVnet: Automated Time-Resolved Tracking of the Tricuspid Valve Plane in MRI Long-Axis Cine Images with a Dual-Stage Deep Learning Pipeline

Ricardo A. Gonzales[1,2,3](✉) [iD], Jérôme Lamy[1] [iD], Felicia Seemann[1,3,4] [iD],
Einar Heiberg[3,4,5] [iD], John A. Onofrey[1] [iD], and Dana C. Peters[1] [iD]

[1] Department of Radiology and Biomedical Imaging, Yale School of Medicine,
Yale University, New Haven, CT, USA
[2] Department of Electrical Engineering, Universidad de Ingeniería y Tecnología,
Lima, Peru
ricardo.gonzales@utec.edu.pe
[3] Department of Clinical Physiology, Lund University, Skåne University Hospital,
Lund, Sweden
[4] Department of Biomedical Engineering, Lund University, Lund, Sweden
[5] Wallenberg Center for Molecular Medicine, Lund University, Lund, Sweden

Abstract. Tracking the tricuspid valve (TV) in magnetic resonance imaging (MRI) long-axis cine images has the potential to aid in the evaluation of right ventricular dysfunction, which is common in congenital heart disease and pulmonary hypertension. However, this annotation task remains difficult and time-demanding as the TV moves rapidly and is barely distinguishable from the myocardium. This study presents TVnet, a novel dual-stage deep learning pipeline based on ResNet-50 and automated image linear transformation, able to automatically derive tricuspid annular plane systolic excursion. Stage 1 uses a trained network for a coarse detection of the TV points, which are used by stage 2 to reorient the cine into a standardized size, cropping, resolution, and heart orientation and to accurately locate the TV points with another trained network. The model was trained and evaluated on 4170 images from 140 patients with diverse cardiovascular pathologies. A baseline model without standardization achieved a Euclidean distance error of 4.0 ± 3.1 mm and a clinical-metric agreement of ICC = 0.87, whereas a standardized model improved the agreement to 2.4 ± 1.7 mm and an ICC = 0.94, on par with an evaluated inter-observer variability of 2.9 ± 2.9 mm and an ICC = 0.92, respectively. This novel dual-stage deep learning pipeline substantially improved the annotation accuracy compared to a

Electronic supplementary material The online version of this chapter (https:// doi.org/10.1007/978-3-030-87231-1_55) contains supplementary material, which is available to authorized users.

The original version of this chapter was revised: an error in Reference 12 has been corrected. The correction to this chapter is available at
https://doi.org/10.1007/978-3-030-87231-1_60

baseline model, paving the way towards reliable right ventricular dysfunction assessment with MRI.

Keywords: Right ventricular dysfunction · Cine MRI · Annotation · Residual neural networks

1   Introduction

The analysis of the tricuspid valve (TV) motion provides anatomic and functional evaluation of the right ventricle (RV) [1]. Compared to the left ventricle, whose functional evaluation is paramount in the diagnosis of cardiovascular disease, the RV involvement in conditions such as pulmonary hypertension and congenital heart disease, and the high prognostic value of RV function evaluation are often overlooked [2]. The maximal longitudinal displacement, known as tricuspid annular plane systolic excursion (TAPSE), and the annular early diastolic velocity (RV e') are commonly evaluated in echocardiography to assess right ventricular dysfunction [3]. However, this imaging modality can be limited by the dependency of image quality on beam angle [4] and influenced by translational motion [5].

Magnetic resonance imaging (MRI) is a more reproducible imaging modality and is considered the reference standard for cardiac volume assessment. Its accuracy and reliability serve as a promising tool for serial examinations of RV function [6]. Although MRI would never be performed just to measure TAPSE or RV e', developing its capability to do so is important as these parameters can be assessed retrospectively from standard cardiovascular cine images acquired systematically in clinical practice. Furthermore, this analysis of both mitral and tricuspid valve motion by MRI has been shown to be reliable and to provide an evaluation of both systolic and diastolic function [7,8]. However, their measurement requires a precise annotation in the valve insertion points in every temporal frame of a long-axis cine, typically consisting of 30 temporal frames, which is time-demanding and prone to errors [9]. Although some progress has been made to semi-automatically track the valve insertion points in these images [10,11], a fully automated, fast, and accurate method for tracking the TV points using standard clinical MRI images is lacking.

The variability of long-axis views of the heart in orientation and location within the cine image is one obstacle towards precise automated identification of the TV points. We addressed this using a novel two-stage framework with an intermediate automated linear transformation task. A first trained neural network produces coarse annotation, which in turn is used to fully standardize the image with respect to size, centering, and rotation. A second network is then used to precisely annotate in these preprocessed images, to automatically track the tricuspid valve insertion points and derive TAPSE and RV e' on par with expert human-level performance. This method, trained and validated in a diverse population of 140 patients, showed an excellent agreement with the inter-observer variability and with a smoother trajectory annotation.

2   Methods

2.1   Imaging Data and Manual Annotation

One hundred forty patients were retrospectively included in this study (59 females, 53 ± 14 years old) as part of an IRB-approved chart-review study. The reported pathologies included pulmonary hypertension (n = 25), atrial fibrillation (n = 23), ventricular tachycardia or paroxysmal fibrillation (n = 16), hypertrophic cardiomyopathy (n = 17), heart failure (n = 9), right or left bundle branch block (n = 6), coronary artery disease (n = 6), sarcoid (n = 6), obstructive sleep apnea (n = 5), other cardiac diseases (n = 5), Lyme disease (n = 4), and healthy volunteers (n = 18). All subjects were scanned on a 1.5T (n = 115) or 3T (n = 25) conventional clinical MRI scanner (Siemens Healthcare, Erlangen) including a standard four-chamber long-axis cine exam, systematically acquired in a clinical setting. Such images, mostly retrospectively ECG-gated, were acquired during breath-hold with the following parameters: repetition time of 3 ms, echo time of 1.5 ms, and flip angle of $60°$. Four chamber data had an average spatial resolution of 2×2 mm^2, slice thickness of 8 mm, and mostly 30 temporal frames per cardiac cycle were reconstructed. The final dataset comprised a total of 4170 images (140 sets of time-resolved images) analyzed in previous studies [12–14].

RV lateral and septal points were manually placed at the intersection between the tricuspid valve and the right ventricular myocardium by observer 1, a researcher with 4 years of cardiac MRI experience. The landmark annotations were performed in every image, using the software Segment [15]. The landmarks, annotated by observer 1, were considered as the ground truth.

2.2   Dual-Stage Residual Neural Network

The proposed framework, illustrated in Fig. 1, involves two trained networks, using an established convolutional neural network with a residual framework of 50 layers, ResNet-50 [16]. The first stage, with a network, provides valve points sufficiently accurate to define the TV plane. The second stage uses these points to perform a linear transformation to the image and, with another network, predicts highly accurate points. For both networks, the input and output layers were adapted to input a grayscale image with a fixed size, and to output four numbers (i.e. two pairs of coordinates) with a regression layer. For both networks, the training and testing sets were partitioned into 3330 images (112 subjects) and 840 images (28 subjects), respectively. This distribution was performed homogeneously along the reported cardiovascular diseases in a random manner. To further mitigate selection bias, a 5-fold cross-validation with homogeneous distribution was additionally performed.

Initial Coarse Annotation. The first ResNet-50 network was trained with the ground-truth manual annotations and images "as is" with only an image resizing to 160×160 with cubic interpolation. Every input image was processed independently. The resulting coordinates were sufficiently accurate to localize both TV points and to roughly follow the valve motion.

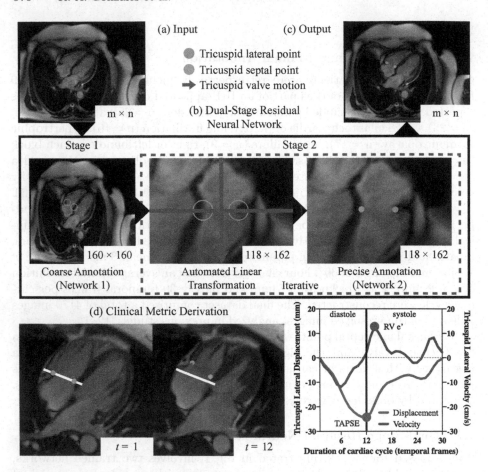

Fig. 1. TVnet pipeline. (a) The input cine images with an inherent clinical variability (size $(m \times n)$, resolution, orientation, and cropping) are fed to (b) the proposed dual-stage residual neural network. Stage 1 produces coarse annotations, marked in circumferences representing uncertainty, in every cine image in a fixed image size of 160×160. Stage 2 uses these points to apply a linear transformation to a standard spatial resolution of 0.75 mm, orientation and cropping around the TV center for a size of 118×162, and predicts precise annotations, marked in circles representing higher accuracy, which are adjusted again to the original input image. Stage 2, dashed red box, can be done iteratively as indicated. (c) The output time-resolved coordinates are used to (d) derive the tricuspid lateral displacement curve by measuring the perpendicular distance from the initial TV plane, marked in white at end-diastole $(t = 1)$, to the lateral point at every temporal frame, and the tricuspid lateral velocity, as its time-derivative. TAPSE is calculated as the maximum displacement, and RV e' is calculated as the second global velocity peak. (Color figure online)

Automated Linear Transformation. The output coordinates of the first network were used to standardize the input of the second network: (i) the image was interpolated to a standard spatial resolution of 0.75 mm, (ii) the TV was oriented horizontally (rotated) with the apex pointing down and with the left ventricle placed to the left, and (iii) the image was finally cropped around the TV center for a size of 118×162. The orientation and centering tasks were based on the first temporal frame and applied to the rest of the cine, accordingly. The first network processed the cine images independently and its output trajectory allowed to determine the valve plane for each image. The direction of the RV apex was determined by the valve movement direction along the cardiac cycle.

Final Precise Annotation. The second ResNet-50 network was trained with the ground-truth manual annotations and the linearly transformed images. These standardized images were used as input to the second network which generated the TV points with high accuracy, processing every image independently. These new predicted points were readjusted, with an inverse standardization transformation, to match the original input cine images.

Network Training. Both networks were trained in the same manner. A pre-trained version of the ResNet-50, from the ImageNet database [17], was used for weights initialization as part of a transfer learning process to reduce convergence time. Image pixel distribution was centered and reduced using the median and the interquartile range. Training data augmentation was performed 10-fold by scaling $\pm 10\%$, rotating $\pm 10°$ and translating ± 3 pixels, to add more inherent variability in stage 1, and to compensate for any error from stage 1 in stage 2, i.e., a slight misalignment of the valve plane center from the ground truth. The Adam method [18] was used for optimizing the mean square error loss function with a learning rate of 1×10^{-4}. Training consisted of 20 epochs with a mini-batch size of 8. The networks were trained and tested on MATLAB R2019b (Mathworks, Natick, MA) with a NVIDIA Titan RTX GPU.

2.3   Evaluation and Statistical Analysis

Clinical-Metric Derivation. Clinically relevant RV parameters extracted were the TAPSE and the RV e', which were calculated from the lateral displacement curve, as illustrated in Fig. 1(d). The perpendicular distance from the lateral point to the initial TV plane, defined at the beginning of the cardiac cycle (end-diastole, usually $t = 1$) with both lateral and septal points, was measured in every temporal frame. TAPSE was measured as the maximum value of this displacement curve. RV e' was calculated as the second global peak of the time-derivative of this curve.

The test set, comprising 840 annotations for each TV insertion point, from 28 patients, was evaluated against manual annotation. The spatial annotation error was measured with: (i) the Euclidean distance, which measured in millimeters the distance error between the ground-truth and predicted annotations, (ii) the angular distance, which measured in degrees the inner intersection angle of the ground-truth and automated planes defined by both TV points, and (iii) the trajectory jitter, which evaluated the difference ratio between the actual point

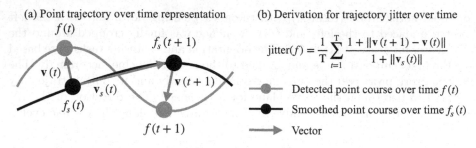

(a) Point trajectory over time representation

(b) Derivation for trajectory jitter over time

$$\text{jitter}(f) = \frac{1}{T} \sum_{t=1}^{T} \frac{1 + \|\mathbf{v}(t+1) - \mathbf{v}(t)\|}{1 + \|\mathbf{v}_s(t)\|}$$

— Detected point course over time $f(t)$

— Smoothed point course over time $f_s(t)$

→ Vector

Fig. 2. Point trajectory jitter illustration (a) and derivation (b). The jitter for two successive points over time t of the trajectory with T temporal frames was measured as the difference between their vectors $\mathbf{v}(t)$ and $\mathbf{v}(t+1)$ from $f(t)$ to its smoothed contour $f_s(t)$, after a low-pass filter and normalized by the distance, magnitude of $\mathbf{v}_s(t)$, between the two corresponding points on $f_s(t)$. The trajectory jitter was evaluated by averaging every two points jitter along the trajectory $f_s(t)$ and was performed for every annotated point course.

course over time and its smoothed version, representing a more physiological pattern, as illustrated in Fig. 2. Clinical metric (TAPSE and RV e') comparisons were performed using linear regression analysis, Bland-Altman plots, the intra-class correlation coefficient (ICC), and the coefficient of variance (CoV) between the automated and manual measures. For inter-observer variability analysis, observer 2, a researcher of 7 years of cardiac MRI experience, performed manual annotations on the same test set and the same evaluation was assessed. The threshold for statistical significance was considered for $p < 0.05$.

Additional Stage Influence. This evaluation was performed for the results of using only the first stage (stage 1), both stages (stage $1 + 2$), and three successive iterations (stage $1 + 2 + 2$, stage $1 + 2 + 2 + 2$, and stage $1 + 2 + 2 + 2 + 2$), to show the influence of the additional stage (see red box in Fig. 1(b)).

3   Experiments and Results

3.1   Implementation

TVnet was implemented as a plug-in in the medical image analysis software Segment v3.1 R8109 [15] (http://segment.heiberg.se) and uploaded to https://github.com/ra-gonzales/TVnet, which is freely available for research purposes. Training time took 22 h for each network. For the dual-stage pipeline, on the GPU, testing time took 1.8 s per patient with an additional 0.8 s for every iteration, whereas on a CPU, it took 3.8 s per patient with an additional 1.6 s for every iteration, compared to an average manual annotation time from 4 to 20 min. Additional movie files, in Supplementary Material, demonstrate automated TV tracking across a cardiac cycle.

Table 1. Automated spatial annotation accuracy (n = 840) of tricuspid valve tracking.

Stage	Lateral point error (mm)	Septal point error (mm)	Angular error (°)	TAPSE error (mm)
1	4.03 ± 3.13	3.98 ± 3.16	4.40 ± 3.91	0.32 ± 2.86
1 + 2	2.61 ± 2.01	2.61 ± 1.59	3.82 ± 3.36	0.29 ± 1.86
1 + 2 + 2	2.43 ± 1.92	2.45 ± 1.48	3.78 ± 3.34	0.13 ± 1.90
1 + 2 + 2 + 2	2.41 ± 1.97	2.43 ± 1.50	3.87 ± 3.38	−0.02 ± 1.80
1 + 2 + 2 + 2 + 2	2.38 ± 1.93	2.48 ± 1.51	3.95 ± 3.37	0.07 ± 2.03
Observer 2	2.18 ± 1.78	3.67 ± 3.49	4.83 ± 6.01	−0.20 ± 2.30

The mean ± standard deviation are reported for Euclidean (for lateral and septal points) and angular distances and tricuspid annular plane systolic excursion (TAPSE) errors between detected and ground-truth manual annotations by observer 1.

3.2 Annotation Accuracy

The reproducibility analysis of the TVnet annotations and the clinical metrics is reported in Table 1. The proposed framework improved the accuracy after every iteration converging on stage $1 + 2 + 2$, with a marginal improvement afterward. With this chosen model, the method achieved a mean Euclidean distance error of 2.44 ± 1.71 mm, an angular distance error of $3.78 \pm 3.34°$, and a TAPSE error of 0.13 ± 1.90 mm with an ICC = 0.94. The mean Euclidean distance error with standard deviation across folds was 2.79 ± 0.39 mm. The inter-observer variability for manual annotation achieved 2.92 ± 2.87 mm, $4.83 \pm 6.01°$, -0.20 ± 2.30 mm, and ICC = 0.92, respectively, with a marked discordance in tracing the septal point. The trajectory jitter with the framework was 0.70 ± 0.10 whereas the trajectory jitter of observer 2 was 0.81 ± 0.13. Compared to the difference between automated and ground truth annotations, the second observer achieved a stronger agreement with the lateral point annotation but weaker agreement with the septal point annotation, whereas the automated annotations exhibited less jitter and achieved a better agreement with the clinical metrics.

The automatically derived annotations yielded average values of 19.9 ± 5.5 mm for TAPSE and 7.8 ± 3.3 cm/s for RV e'. These were comparable with manually derived average values of these clinical metrics from both observers (observer 1: 19.8 ± 5.7 mm and 6.8 ± 2.8 cm/s and observer 2: 19.6 ± 5.8 mm and 9.2 ± 4.1 cm/s respectively for TAPSE and RV e'). The regression and Bland-Altman plots for TAPSE and RV e' between the automated (stage $1 + 2 + 2$) and manual measurements are presented in Fig. 3(a), where an excellent correlation and agreement were observed for each parameter. Figure 3(b) shows the corresponding inter-observer variability of these metrics, demonstrating similar excellent agreement for TAPSE, but less agreement for RV e'. This is due to the sensitivity of RV e' to jitter, as RV e' requires a time derivative. All reported correlation values were significant ($p < 0.0001$).

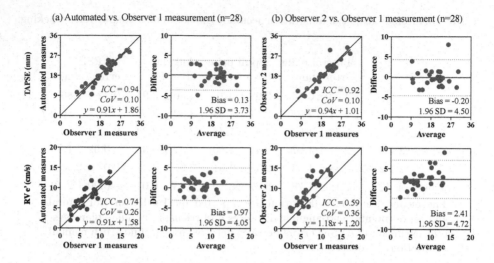

Fig. 3. Correlation and bias of tricuspid valve MRI-derived clinical measures between an expert manual annotation (observer 1) and (a) the automated method, and (b) annotation by observer 2, on the testing set of 28 patients. The first column of each analysis shows the regression plots whereas the second shows the Bland-Altman plots of the tricuspid annular plane systolic excursion (TAPSE), in brown, and the right ventricular early diastolic velocity (RV e'), in blue. In each scatter plot the red line denotes the regression line and the black line denotes the identity line, whereas in each Bland-Altman plot, the red line denotes the mean difference (bias) and the two light dotted lines denote ±1.96 standard deviations from the mean. (Color figure online)

4 Discussion and Conclusion

In this work, we proposed a dual-stage residual learning framework, TVnet, for time-resolved annotation of the TV in four-chamber views from standard long-axis cine MRI images. The proposed method was fast, fully automated, and showed excellent agreement with manual annotation by expert readers in terms of valve points positioning as well as with the subsequently extracted RV function parameters. Tedious manual labor is not needed, reducing the processing time from 4–20 min to 2 s. This enables fully automated, accurate, fast, and reproducible assessment of RV function in clinical routine. Moreover, this method can be applied retrospectively to any four-chamber image acquisition, which is routinely acquired in a standard cardiovascular MRI exam.

The initial step, used to standardize the images with a linear transformation, substantially reduced the spatial annotation and clinical-metric errors. It further exploited the utility of artificial neural networks in readjusting the result from a baseline model using the TV anatomical position and reducing the image variability. The low Euclidean distance and angular orientation errors displayed consistency in the tracking. The former assured a point placement around the ground-truth annotation, whereas the latter assured a concordant TV plane displacement, which for the RV is known to vary greatly throughout the cardiac cycle. The high parity with human-level performance may be explained with the low trajectory jitter of the model, as the automated point prediction was smoother compared to a second observer, which is vital in calculating TV velocity. Future work includes validation on a multi-vendor, multi-center population and expansion to other cardiac landmark annotations.

Acknowledgement. The authors gratefully acknowledge funding from NHLBI R01HL144706, RAG acknowledges Magnus Caspersen, MSc for his guidance in deep learning, and DCP acknowledges James W. Goldfarb, PhD for his ideas on the utilization of deep learning for cine valve-tracking, many years ago.

References

1. Dimopoulos, K., et al.: Echocardiographic screening for pulmonary hypertension in congenital heart disease: JACC review topic of the week. J. Am. Coll. Cardiol. **72**(22), 2778–2788 (2018)
2. Amsallem, M., Mercier, O., Kobayashi, Y., Moneghetti, K., Haddad, F.: Forgotten no more: a focused update on the right ventricle in cardiovascular disease. JACC: Heart Failure **6**(11), 891–903 (2018)
3. D'Andrea, A., et al.: The impact of age and gender on right ventricular diastolic function among healthy adults. J. Cardiol. **70**(4), 387–395 (2017)
4. Ho, C.Y., Solomon, S.D.: A clinician's guide to tissue doppler imaging. Circulation **113**(10), e396–e398 (2006)
5. Abraham, T.P., Dimaano, V.L., Liang, H.Y.: Role of tissue doppler and strain echocardiography in current clinical practice. Circulation **116**(22), 2597–2609 (2007)
6. Valsangiacomo Buechel, E.R., Mertens, L.L.: Imaging the right heart: the use of integrated multimodality imaging. Eur. Heart J. **33**(8), 949–960 (2012)
7. Carlsson, M., Ugander, M., Mosén, H., Buhre, T., Arheden, H.: Atrioventricular plane displacement is the major contributor to left ventricular pumping in healthy adults, athletes, and patients with dilated cardiomyopathy. Am. J. Physiol. Heart Circulatory Physiol. **292**(3), H1452–H1459 (2007)
8. Seemann, F., et al.: Valvular imaging in the era of feature-tracking: a slice-following cardiac MR sequence to measure mitral flow. J. Magn. Reson. Imag. **51**(5), 1412–1421 (2020)
9. Caudron, J., Fares, J., Vivier, P.H., Lefebvre, V., Petitjean, C., Dacher, J.N.: Diagnostic accuracy and variability of three semi-quantitative methods for assessing right ventricular systolic function from cardiac MRI in patients with acquired heart disease. Eur. Radiol. **21**(10), 2111–2120 (2011)
10. Leng, S., et al.: Three-dimensional tricuspid annular motion analysis from cardiac magnetic resonance feature-tracking. Ann. Biomed. Eng. **44**(12), 3522–3538 (2016)
11. Seemann, F., et al.: Time-resolved tracking of the atrioventricular plane displacement in cardiovascular magnetic resonance (CMR) images. BMC Med. Imag. **17**(1), 19 (2017)
12. Hu, C., et al.: T1-refBlochi: high resolution 3D post-contrast T1 myocardial mapping based on a single 3D late gadolinium enhancement volume, Bloch equations, and a reference T1. J. Cardiovascular Magn. Reson. **19**(63), 1–17 (2017)

13. Seemann, F., et al.: Assessment of diastolic function and atrial remodeling by MRI-validation and correlation with echocardiography and filling pressure. Physiol. Rep. **6**(17), e13828 (2018)
14. Gonzales, R.A., et al.: Automated left atrial time-resolved segmentation in MRI long-axis cine images using active contours. BMC Med. Imag. **21**(1), 1–12 (2021)
15. Heiberg, E., Sjögren, J., Ugander, M., Carlsson, M., Engblom, H., Arheden, H.: Design and validation of segment-freely available software for cardiovascular image analysis. BMC Med. Imag. **10**(1), 1 (2010)
16. He, K., Zhang, X., Ren, S., Sun, J.: Deep residual learning for image recognition. In: Proceedings of the IEEE Conference on Computer Vision and Pattern Recognition, pp. 770–778 (2016)
17. Deng, J., Dong, W., Socher, R., Li, L.J., Li, K., Fei-Fei, L.: ImageNet: a large-scale hierarchical image database. In: 2009 IEEE Conference on Computer Vision and Pattern Recognition, pp. 248–255. IEEE (2009)
18. Kingma, D.P., Ba, J.L.: Adam: a method for stochastic gradient descent. In: ICLR: International Conference on Learning Representations (2015)

Clinical Applications - Vascular

Critical Applications - Vascular

Deep Open Snake Tracker for Vessel Tracing

Li Chen(✉) 🆔, Wenjin Liu, Niranjan Balu, Mahmud Mossa-Basha,
Thomas S. Hatsukami, Jenq-Neng Hwang, and Chun Yuan

University of Washington, Seattle, WA 98109, USA
cluw@uw.edu

Abstract. Vessel tracing by modeling vascular structures in 3D medical images with centerlines and radii can provide useful information for vascular health. Existing algorithms have been developed but there are certain persistent problems such as incomplete or inaccurate vessel tracing, especially in complicated vascular beds like the intracranial arteries. We propose here a deep learning based open curve active contour model (DOST) to trace vessels in 3D images. Initial curves were proposed from a centerline segmentation neural network. Then data-driven machine knowledge was used to predict the stretching direction and vessel radius of the initial curve, while the active contour model (as human knowledge) maintained smoothness and intensity fitness of curves. Finally, considering the non-loop topology of most vasculatures, individually traced vessels were connected into a tree topology by applying a minimum spanning tree algorithm on a global connection graph. We evaluated DOST on a Time-of-Flight (TOF) MRA intracranial artery dataset and demonstrated its superior performance over existing segmentation-based and tracking-based vessel tracing methods. In addition, DOST showed strong adaptability on different imaging modalities (CTA, MR T1 SPACE) and vascular beds (coronary arteries).

Keywords: Vascular tracing · Active contour model · Snake · Artery modeling · Vascular tree · Vessel tracker

1 Introduction

Cardiovascular disease is one of the leading causes of death and disability worldwide, primarily via coronary artery disease and stroke [1]. Medical imaging such as computed tomography angiography (CTA) and magnetic resonance angiography (MRA) allows visualization of vasculatures. Through 3D vascular map construction, topological and morphometric information can be quantified, providing vascular information for clinical diagnosis and vascular health assessment, including the presence of stenosis, occlusion, collateral arteries, and distribution [2–6].

Electronic supplementary material The online version of this chapter (https://doi.org/10.1007/978-3-030-87231-1_56) contains supplementary material, which is available to authorized users.

© Springer Nature Switzerland AG 2021
M. de Bruijne et al. (Eds.): MICCAI 2021, LNCS 12906, pp. 579–589, 2021.
https://doi.org/10.1007/978-3-030-87231-1_56

A critical and challenging step for vascular map construction is accurate and automated vessel tracing, which is converting original vascular images into a vascular network with topological representation of branches, including centerlines and radii. In general, two main approaches exist for automated vessel tracing. **1) Segmentation-based approaches.** Voxels belonging to the vascular region are segmented, then the vessel skeletonization method identifies the centerline through iterative thinning. Finally, radii along the centerlines are estimated. Methods belonging to this approach mainly differ in the segmentation algorithms, which are comprehensively discussed in review papers [7–9]. Recently, convolutional neural network (CNN) based vascular segmentation has become the predominant method, such as using the patch origin encoded Y-Net [10] and distance transformed segmentation [11]. For segmentation-based approaches, there is no guarantee for the smoothness and continuity of vessels after skeletonization. Moreover, two nearby vessels close to each other might be traced as one large vessel. **2) Tracking-based approaches.** Initial seeds are identified as the starting points from the vascular images, then the artery centerline and radius are directly identified from vascular images from seed points through iteratively stretching both ends of the trace during tracking. It is critical to predict the correct direction for stretching, which can either be from a human designed vesselness enhancement filter (for example, the first principal direction of the Jacobian matrix of images [12]), or from a neural network used in CNN Tracker and DCAT [13, 14]. The tracking-based approach is sensitive to initial seed placement. Improper seed selection is likely to cause tracing leakage into the background or result in an incomplete vascular tree. Traces might also be rough or zigzagged due to a lack of smoothness constraints.

In this study, we proposed a deep open active contour (snake) tracker (DOST), a hybrid method taking advantage of both segmentation and tracking approaches for robust vessel tracking. More importantly, DOST, by merging deep learning predictions into the traditional open snake algorithm, combines human knowledge (a snake algorithm to ensure the smoothness and intensity fitness of traces) and data-driven machine knowledge (deep learning to ensure reliable and robust vascular tracing) to construct a topologically correct vascular tree.

The main innovations and contributions of our work include: 1) DOST, proposed to solve automated vessel tracing; 2) by benchmarking with an intracranial MRA dataset, we achieved the state-of-the-art performance on vessel tracing; 3) DOST is applicable to multiple vascular beds and imaging modalities; 4) new evaluation metrics were used to evaluate multi-vessel connection accuracy of complicated vascular structures, a supplement to existing overlap-based metrics.

2   Methods

2.1   Deep Open Curve Snake

Active contour model, also called snake [15], was first introduced as an algorithm for contour delineation on noisy 2D images. Initial contours are provided by users, then contours are refined through iterative minimization of the energy function, considering the external energy for fitting image intensity and the internal energy conserving contour smoothness. Wang et al. [12, 16] extended snake to open curve snake (OCS) for 3D tubular neural fibers tracing by optimizing on open traces (initialized from seed points) instead of closed contours. However, three problems limit the use of OCS on more complicated vessel tracing: 1) seed initialization and initial stretching direction are sensitive to noise; 2) stretching directions are decided purely by the local image gradients, which is sensitive to noise and not applicable to vessel bifurcation/branches; 3) each trace is independently identified without global constraints. Connection mistakes such as loops might appear in the vascular tree (Fig. 1 (a)).

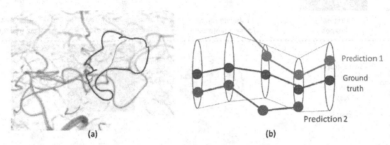

(a)   (b)

Fig. 1. (a) An example of loop (red) due to errors in tracing intracranial arteries using the OCS method. (b) An example when overlap measures are high but traced by multiple traces. (Color figure online)

Deep open snake tracker (DOST) is designed to improve OCS with deep learning and global tree structure constraints. DOST includes key steps: curve proposal from centerline segmentation, deep snake tracing, and global tree construction (Fig. 2).

The tracing target for each scan in our task is a constructed snake list $T = \{t_i, i = 1, 2, \ldots, N\}$ with N traces. Ground truth labels are annotated with hat symbols, for example, $\hat{T}$.

Each snake is a list of points with 3D position $p_{i,j} = (x_{i,j}, y_{i,j}, z_{i,j})$ along with the corresponding radius of lumen area $r_{i,j}$.

$$t_i = (P_i, R_i) = (\{p_{i,j}, j = 1, 2, \ldots, n\}, \{r_{i,j}, j = 1, 2, \ldots, n\}) \tag{1}$$

2.2 Curve Proposal from Centerline Segmentation

Instead of seed points [12, 13], DOST initializes from *vessel curves* predicted from the centerline segmentation, which better utilizes vascular structures from segmentation-based tracking methods to avoid initial stretching errors. A 3D patch-based encoder-decoder centerline segmentation network (2 blocks in encoder/decoder, each with 2 3D convolutional layers + RELU followed by 3D max pooling/up sampling layers) was used for segmentation on the vascular images. To separate the nearby vessels, the centerline distance transform [11] was used to map the vascular regions with continuous values according to the distance to the centerline. Each voxel at $p = [x, y, z]$ within the vascular labels was transformed into a labeled distance map d, where centerlines voxels had highest values.

$$d[x, y, z] = \max_i \max_j \left(\frac{\max(0, r_{i,j} - \|p_{i,j} - p\|)}{r_{i,j}} \right) \tag{2}$$

582   L. Chen et al.

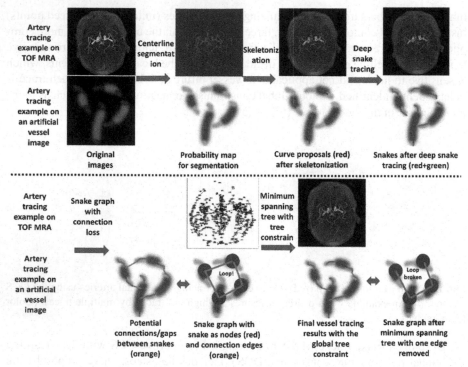

Fig. 2. Workflow of DOST. Centerlines for arteries were identified by a centerline segmentation CNN and skeletonized into pieces of 1-voxel thin vessel curves, which were used as the initial traces for deep snake tracing. CNN predicted the stretching directions and radii for both ends of the snake while trace smoothness and fitness to image intensities were maintained. Then a graph model was used to select and construct a topologically correct vascular tree.

Considering the majority of d is zeros. The L2 loss for training the segmentation network is masked by the non-zero regions in d. $1(\cdot)$ is an indication function.

$$Loss_{seg} = \left\| 1\big(d[x,y,z] \neq 0\big)\big(d[x,y,z] - \hat{d}[x,y,z]\big) \right\|_2 \tag{3}$$

For each dataset, an individual segmentation network was trained with initial weights from a meta-segmentation model [17] trained with our previous MRA data [6, 10, 18] (no overlap with datasets used in this study) so that the adaptation was efficient.

Threshold after segmentation was chosen (from the validation set) to make binary predictions. Zhang's skeletonization algorithm [19] was applied to identify initial curves (grouped if voxels were 26-connected) for tracing.

2.3 Deep Snake Tracing

An initial curve $c(s)$ was represented as a parametric open curve model $(x(s), y(s), z(s))$, $s \in [0, 1]$. The snake energy used in DOST E_{DOST} was a combination of the internal energy E_{int} and the external energy of E_{ext}

$$E_{DOST} = \int_0^1 E_{int}(c(s)) + E_{ext}(c(s)) \, ds \tag{4}$$

$$E_{int}(c(s)) = \alpha(s)|c_s(s)|^2 + \beta(s)|c_{ss}(s)|^2 \tag{5}$$

"Elasticity" $\alpha(s)$ and "stiffness" $\beta(s)$ of the snake were set to be zero at $s = 0$ or 1 to allow snake stretching. c_s and c_{ss} indicate first and second order derivatives.

$$E_{ext}(c(s)) = -I(x(s), y(s), z(s)) + E_{str}(c(s)) \tag{6}$$

I is the image intensity. E_{str} is the energy for stretching directions at both ends of the snake. $E_{str}(c(s)) = 0$ for points other than both ends.

Different from the OCS, the stretching directions at the end of the snake were predicted from a deep neural network following the settings from the CNN tracker [13]. The network structure has 6 blocks followed by a convolution layer with $D + 1$ dimensions ($D = 500$), each block was a 3D convolutional layer + batch normalization + RELU. A 3D image patch $P(c(s))$ with the size of 19 [3] (large enough to cover the vessel diameter) was extracted centered at $c(s)$, $s = 0, 1$ for prediction of the 1-dimensional radius $r(P(c(s)))$ and D-dimensional stretching direction $v_{m=1,2,...,D}(P(c(s)))$ by the network. The predicted stretching magnitudes $\{k_m\}$ has D dimensions indicating evenly distributed 3D unit directions v_m. The training targets for each patch were generated from semi-automatedly traced arteries using iCafe [20].

$$\nabla E_{str}(c(s)) = - \begin{cases} \max_m(sign(-c_s(s) \cdot v_m) \cdot k_m), & s = 0 \\ \max_m(sign(c_s(s) \cdot v_m) \cdot k_m), & s = 1 \\ 0, & 0 < s < 1 \end{cases} \tag{7}$$

The directions of $-c_s(s)|s = 0$ and $c_s(s)|s = 1$ pointed outward from the curve, indicating the correct stretching directions for v_m to stretch the snake.

Snakes were traced independently. In each iteration for a snake tracing, the snake stretched at both ends with the step size of $\gamma = 0.5r(P(c(s)))$ with minimized snake energy, then resampled evenly for the next iteration.

Snake stretching was terminated either when a snake end point reached another traced snake or the predicted direction v_m can no longer predict a confident direction indicated by the normalized entropy [13].

$$H = \frac{\sum_m -k_m \cdot log_2(k_m)}{log_2(D)} \tag{8}$$

2.4 Global Tree Construction

Deep snake tracing can reliably trace individual arteries. However, traces were not connected with each other through merging or branching to form a topologically meaningful vascular tree. In addition, the global constraint on tree structure (no loop) can be used to fix connection errors in tracing. On rare occasions when loops naturally exist, such as in individuals with collateral arteries or with a complete circle of Willis, a manual extra step of loop reconnection can be added.

The vascular tree was constructed using an undirected *snake graph*, in which vertices indicated snakes, and edges between vertices indicated the connection loss $Loss_{con}(i, j)$ for the snake pair i, j (through merging or branching).

The connection loss $Loss_{con}(i, j)$ was based on intensities of point list $t_{i,j}$ from two snakes $c_i \cup c_j$ and their gap g_i (minimum distance from one point of the snake to any point on the other snake). Foreground intensities for the snake pair were estimated with normal distributions N_{i_f, δ_f}. i_f, δ_f were mean and standard deviation of intensity along $t_{i,j}$. Background intensities for the snake pair were sampled from points at twice the radius around $t_{i,j}$ with normal distributions N_{i_b, δ_b}. i_b, δ_b were mean and standard deviation of background intensities. Mean intensity along g_i was i_g.

$$Loss_{con}(i, j) = \frac{N_{i_f, \delta_f}(i_g)}{N_{i_f, \delta_f}(i_g) + N_{i_b, \delta_b}(i_g)} \tag{9}$$

When $Loss_{con}(i, j)$ was below the threshold of 0.05 or the gap was above the maximum distance for connection consideration (10 mm), edges were removed from the graph. Parameters were empirically chosen for the best vascular tree construction.

Kruskal's algorithm [21] was used for minimum spanning tree (MST) construction from the snake graph with connection losses. Edges in MST were used to fill the gaps for snakes to construct a whole vascular tree.

3   Experimental Settings and Results

3.1   Datasets

We used the BRAVE dataset [5, 22] with 167 TOF MRA from elderly hypertensive subjects to evaluate DOST comparing with state-of-the-art methods. To evaluate the robustness of DOST on more vascular beds and imaging modalities, we used 1) the Rotterdam Coronary Artery Challenge (CAT08): eight CTA scans for coronary artery tracing [23], and 2) Harborview dataset: clinical scans with MR vessel wall imaging using T1 SPACE (black blood) from 15 patients with history of stroke. Detailed dataset properties are described in Table 3 (supplementary material). Ethics approval was waived due to the retrospective study design.

Ground truth for CAT08 was provided. For other datasets, arteries were automatically generated by the OCS and manually corrected in an analysis tool of iCafe [20, 24].

3.2 Evaluation Metrics

We evaluated the tracing accuracy following the metrics used in the CAT08 Challenge [23], including Average inside (AI) and Overlap (OV). However, AI and OV mainly evaluate centerline overlap for a single artery and cannot reflect the multi-vessel connection accuracy. For example, a vessel matched by two traces (Fig. 1 (b)) has 100% OV but gets no penalties on additional broken predictions.

Thus following multiple object tracking tasks [25], we adopted three 3D multi-vessel connection accuracy metrics: ID switch (IDS), multiple object tracking accuracy (MOTA) and IDF1 metrics to evaluate connection errors in vessel tracing. After $\{t_i\}$ were matched ($min_{i,j}(\|\boldsymbol{p}_{i,j} - \hat{\boldsymbol{p}}_{i,j}\| < \hat{r}_{i,j})$) with $\{\hat{t}_i\}$, for each $\hat{t}_i$, ID switch IDS_i was defined as the additional count of sources from the predicted $\{t_i\}$. TP was the number of $\{\hat{t}_i\}$ having matching t_i, FN was the number of $\{\hat{t}_i\}$ having no matching t_i, and FP was the number of $\{t_i\}$ having no matching $\hat{t}_i$. T was the number of points in $\{\hat{t}_i\}$. $MOTA$ penalizes FP, FN as well as IDS, leading to the best possible value of 1. $IDF1$ is the F1 score for artery matching ranging from 0 to 1, with lower score either by larger FP or FN.

$$IDS = \sum_i IDS_i \tag{10}$$

$$MOTA = 1 - \frac{FN + FP + IDS}{T} \tag{11}$$

$$IDF1 = \frac{2 \cdot TP}{2 \cdot TP + FP + FN} \tag{12}$$

3.3 Evaluation on BRAVE

Comprehensive quantitative comparisons with DOST were made with traditional and deep learning methods. For segmentation-based approaches, Frangi vesselness filter [26], U-Net [27] and Deep Distance Transform (DDT) [11] were selected. For tracking-based approaches, open curve snake (OCS) [16], CNN-tracker [13] and Discriminative Coronary Artery Tracking (DCAT) [14] were selected. Results were shown in Table 1.

All experiments were implemented in PyTorch 1.4.0 and TensorFlow 1.15.0 on a workstation with an Intel Xeon E5-1650 v4 CPU and a NVIDIA Titan V GPU.

DOST showed higher performance than most methods. Note that the ground truth was generated based on OCS, which was a natural bias. An example of artery tracing on a MRA data is shown in Fig. 3.

3.4 Ablation Study

We tested DOST performance on the BRAVE dataset without key modules of deep snake tracing, and global tree construction. Results are shown in Table 2. Deep snake tracing improved overall performance, and global tree constraint mainly improved multi-vessel connection accuracy but with minor drop of OV and AI.

Table 1. Quantitative comparison results for intracranial artery tracing

Tracing approach	Model name	OV↑	AI↓	MOTA↑	IDF1↑	IDS↓
Traditional segmentation	Frangi	0.617	0.956	0.238	0.621	343.9
Deep learning segmentation	U-Net	0.662	0.724	0.300	0.696	398.3
Deep learning segmentation	DDT	0.683	0.703	0.281	0.712	423.0
Traditional tracking	OCS*	0.672	**0.356**	**0.372**	0.694	**74.8**
Deep learning tracking	CNN tracker	0.562	0.860	−0.312	0.595	108.5
Deep learning tracking	DCAT	0.564	0.943	−0.241	0.601	137.8
Hybrid	DOST (our)	**0.732**	0.592	0.318	**0.731**	104.1

*Ground truth was modified manually based on OCS results

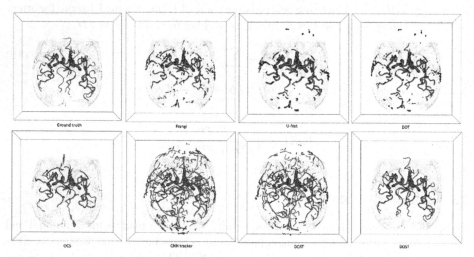

Fig. 3. An example of artery tracing for comparison. DOST did not have the problem in segmentation-based method for broken arteries, and it considered global tree structures in tracing, thus avoiding loops or many noise branches.

Table 2. Ablation study of DOST modules evaluated on BRAVE dataset.

Modules	OV↑	AI↓	MOTA↑	IDF1↑	IDS↓
Initial curve	0.655	0.781	0.274	0.702	467.0
Initial curve + deep snake tracing	**0.719**	**0.651**	0.292	0.732	323.8
DOST (Initial curve + deep snake tracing + global tree)	0.709	0.655	**0.339**	**0.738**	**120.0**

3.5 Adaptability of DOST on Other Datasets

On the CAT08 dataset, DOST achieved OV of 99.6% and AI of 0.22, which were comparable with CNN tracker (OV: 98.7% and AI: 0.18). Tracing results are shown in Fig. 4 (supplementary material).

On the Harborview dataset, DOST achieved OV of 91.7%, AI of 0.69, MOTA of 0.843, IDF1 of 0.925 and IDS of 7.4. For black blood MRI, traditional methods did not work, but DOST was feasible after retraining the centerline segmentation using a small training set (5 cases). An example is shown in Fig. 5 (supplementary material).

4 Discussions and Conclusion

A deep learning based open curve snake model (DOST) was developed and evaluated. DOST combines deep learning-based direction prediction/radius estimation and the classic parametric curve modeling. It allows data driven machine learning knowledge to complement human prior knowledge on the structure of vessels (smoothness and stretchiness) and the topology of the vasculature, so that DOST out-performed existing models with either human or machine knowledge. In addition, DOST, as an adaptive hybrid (segmentation and tracking-based) tracing method, is able to identify complete vascular trees from multiple vascular beds and modalities.

DOST is not purely post processing/smoothing on CNN based tracking results. The snake constraint was applied during the trace stretching and the tracing was initiated from segmentation-based curve proposal instead of seed points.

The main limitation of DOST is its requirement of supervised training to get the initial curve proposals and predict stretching directions. The training labels requires detailed semi-automated artery tracings [20]. However, we have demonstrated DOST works even with 6 cases in CAT08 and 5 cases in Harborview dataset.

With the combination of artery labeling [28], vessel wall segmentation [29], feature quantification [20] and visualization [30], a complete workflow for 3D vascular map construction and analysis can be automated, which will greatly benefit quantitative vascular research and has much potential on clinical diagnosis on vascular diseases.

Acknowledgement. This work was supported by National Institute of Health under grant R01-NS092207. We are grateful for the collaborators who provided the datasets for this study, including the BRAVE investigators, Harborview Medical Center, and the public data from Erasmus MC, Rotterdam. We gratefully acknowledge the support of NVIDIA Corporation for donating the Titan GPU.

References

1. Callaway, C.W., Carson, A.P., Chamberlain, A.M., et al.: Heart disease and stroke statistics—2020 uspdate a report from the American heart association (2020). https://doi.org/10.1161/CIR.0000000000000757
2. Hameeteman, K., Zuluaga, M.A., Freiman, M., et al.: Evaluation framework for carotid bifurcation lumen segmentation and stenosis grading. Med. Image Anal. **15**(4), 477–488 (2011). https://doi.org/10.1016/j.media.2011.02.004

3. Han, K., Chen, L., Geleri, D.B., Mossa-basha, M., Hatsukami, T., Yuan, C.: Deep-learning based significant stenosis detection from multiplanar reformatted Images of traced Intracranial arteries. In: American Society of Neuroradiology 58th Annual Meeting (2020). https://doi.org/10.1002/mrm.26961

4. Chen, Z., Chen, L., Shirakawa, M., et al.: Intracranial vascular feature changes in time of flight MR angiography in patients undergoing carotid revascularization surgery. Magn. Reson. Imaging **75**(August 2020), 45–50 (2021). https://doi.org/10.1016/j.mri.2020.10.004

5. Liu, W., et al.: Uncontrolled hypertension associates with subclinical cerebrovascular health globally: a multimodal imaging study. Eur. Radiol. **31**(4), 2233–2241 (2020). https://doi.org/10.1007/s00330-020-07218-5

6. Chen, L., Sun, J., Hippe, D.S., et al.: Quantitative assessment of the intracranial vasculature in an older adult population using iCafe (intracranial artery feature extraction). Neurobiol. Aging **79**, 59–65 (2019). https://doi.org/10.1016/j.neurobiolaging.2019.02.027

7. Lesage, D., Angelini, E.D., Bloch, I., Funka-Lea, G.: A review of 3D vessel lumen segmentation techniques: models, features and extraction schemes. Med. Image Anal. **13**(6), 819–845 (2009). https://doi.org/10.1016/j.media.2009.07.011

8. Bibiloni, P., Massanet, S.: A survey on curvilinear object segmentation in multiple applications. Pattern Recognit. **60**, 949–970 (2016). https://doi.org/10.1016/j.patcog.2016.07.023

9. Zhao, F., Chen, Y., Hou, Y., He, X.: Segmentation of blood vessels using rule-based and machine-learning-based methods: a review. Multimedia Syst. **25**(2), 109–118 (2017). https://doi.org/10.1007/s00530-017-0580-7

10. Chen, L., Xie, Y., Sun, J., et al.: 3D intracranial artery segmentation using a convolutional autoencoder. In: 2017 IEEE International Conference on Bioinformatics and Biomedicine (BIBM) 3D. IEEE (2017). https://doi.org/10.1109/BIBM.2017.8217741

11. Wang, Y., Wei, X., Liu, F., Chen, J., Zhou, Y., Shen, W.: Deep distance transform for tubular structure segmentation in CT scans, 3833–3842 (2020)

12. Wang, Y., Narayanaswamy, A., Tsai, C.L., Roysam, B.: A broadly applicable 3-D neuron tracing method based on open-curve snake. Neuroinformatics **9**(2–3), 193–217 (2011). https://doi.org/10.1007/s12021-011-9110-5

13. Wolterink, J.M., Hamersvelt, R.W., Viergever, M.A., Leiner, T., Išgum, I.: Coronary artery centerline extraction in cardiac CT angiography using a CNN-based orientation classifier. Med. Image Anal. **51**, 46–60 (2019). https://doi.org/10.1016/j.media.2018.10.005

14. Yang, H., Chen, J., Chi, Y., Xie, X., Hua, X.: Discriminative coronary artery tracking via 3D CNN in cardiac CT angiography. In: Shen, D., et al. (eds.) MICCAI 2019. LNCS, vol. 11765, pp. 468–476. Springer, Cham (2019). https://doi.org/10.1007/978-3-030-32245-8_52

15. Kass, M., Witkin, A., Terzopoulos, D.: Snakes: active contour models. Int. J. Comput. Vis. **1**(4), 321–331 (1988). https://doi.org/10.1007/BF00133570

16. Wang, Y., Narayanaswamy, A., Roysam, B.: Novel 4-D open-curve active contour and curve completion approach for automated tree structure extraction. In: Proceedings of IEEE Computer and Social Conference on Computer and Vision Pattern Recognition, pp. 1105–1112 (Published online 2011). https://doi.org/10.1109/CVPR.2011.5995620

17. Liu, Q., Dou, Q., Heng, P.-A.: Shape-aware meta-learning for generalizing prostate MRI segmentation to unseen domains. In: Martel, A.L., et al. (eds.) MICCAI 2020. LNCS, vol. 12262, pp. 475–485. Springer, Cham (2020). https://doi.org/10.1007/978-3-030-59713-9_46

18. Chen, L., Dager, S.R., Shaw, D.W.W., et al.: A novel algorithm for refining cerebral vascular measurements in infants and adults. J. Neurosci. Methods. **340**(April), 108751 (2020). https://doi.org/10.1016/j.jneumeth.2020.108751

19. Zhang, T.Y., Suen, C.Y.: A fast parallel algorithm for thinning digital patterns. Commun. ACM. **27**(3), 236–239 (1984). https://doi.org/10.1145/357994.358023

20. Kruskal, J.B.: On the shortest spanning subtree of a graph and the traveling salesman problem. Proc. Am. Math. Soc. **7**(1), 48 (1956). https://doi.org/10.1090/S0002-9939-1956-0078686-7

21. Liu, W., Chen, Z., Ortega, D., et al.: Arterial elasticity, endothelial function and intracranial vascular health: a multimodal MRI study. J. Cereb. Blood Flow Metab. 0271678X2095695 (Published online 20 October 2020). https://doi.org/10.1177/0271678X20956950

22. Schaap, M., Metz, C.T., van Walsum, T., et al.: Standardized evaluation methodology and reference database for evaluating coronary artery centerline extraction algorithms. Med. Image Anal. **13**(5), 701–714 (2009). https://doi.org/10.1016/j.media.2009.06.003

23. Chen, L., Mossa-Basha, M., Balu, N., et al.: Development of a quantitative intracranial vascular features extraction tool on 3D MRA using semiautomated open-curve active contour vessel tracing. Magn. Reson. Med. **79**(6), 3229–3238 (2018). https://doi.org/10.1002/mrm. 26961

24. Chen, L., Mossa-Basha, M., Sun, J., et al.: Quantification of morphometry and intensity features of intracranial arteries from 3D TOF MRA using the intracranial artery feature extraction (iCafe): a reproducibility study. Magn Reson Imaging. **2019**(57), 293–302 (2018). https://doi.org/10.1016/j.mri.2018.12.007

25. Bernardin, K., Stiefelhagen, R.: Evaluating multiple object tracking performance: the CLEAR MOT metrics. EURASIP J. Image Video Process. **2008**(1), 1 (2008). https://doi.org/10.1155/ 2008/246309

26. Frangi, A.F., Niessen, W.J., Vincken, K.L., Viergever, M.A.: Multiscale vessel enhancement filtering. In: Wells, W.M., Colchester, A., Delp, S. (eds.) MICCAI 1998. LNCS, vol. 1496, pp. 130–137. Springer, Heidelberg (1998). https://doi.org/10.1007/BFb0056195, https://doi. org/10.1016/j.media.2004.08.001

27. Ronneberger, O., Fischer, P., Brox, T.: U-Net: convolutional networks for biomedical image segmentation. In: Navab, N., Hornegger, J., Wells, W.M., Frangi, A.F. (eds.) MICCAI 2015. LNCS, vol. 9351, pp. 234–241. Springer, Cham (2015). https://doi.org/10.1007/978-3-319- 24574-4_28

28. Chen, L., Hatsukami, T., Hwang, J.-N., Yuan, C.: Automated intracranial artery labeling using a graph neural network and hierarchical refinement. In: Martel, A.L., et al. (eds.) MICCAI 2020. LNCS, vol. 12266, pp. 76–85. Springer, Cham (2020). https://doi.org/10.1007/978-3- 030-59725-2_8

29. Chen, L., Sun, J., Canton, G., et al.: Automated artery localization and vessel wall segmentation using tracklet refinement and polar conversion. IEEE Access. **8**, 1 (2020). https://doi. org/10.1109/access.2020.3040616

30. Chen, L., Geleri, D.B., Sun, J., et al.: Multi-planar, multi-contrast and multi-timepoint analysis tool (MOCHA) for intracranial vessel wall imaging review. In: Proceedings of the Annual Meeting of the International Society for Magnetic Resonance in Medicine, 2020 (2020). https://doi.org/10.1002/mrm.24254.6

MASC-Units:Training Oriented Filters for Segmenting Curvilinear Structures

Zewen Liu[✉] and Timothy Cootes

Division of Informatics, Imaging and Data Science, The University of Manchester,
Stopford Building, Manchester M13 9PT, UK
zewen.liu@manchester.ac.uk

Abstract. Many medical and biological applications involve analysing vessel-like structures. Such structures often have no preferred direction and a range of possible scales. We take advantage of this self-similarity by demonstrating a CNN based segmentation system that requires far fewer parameters than conventional approaches. We introduce the Multi Angle and Scale Convolutional Unit (MASC) with a novel training approach called Response Shaping. In particular, by reflecting and rotating a single oriented kernel we can generate four versions at different angles. We show how two basis kernels can lead to the equivalent of eight orientations. This introduces a degree of orientation invariance by construction. We use Gabor functions to guide the training of the kernels, and demonstrate that the resulting kernels generally form rotated versions of the same pattern. Invariance to scale can be added using a pyramid pooling layer. A simple model containing a sequence of five such blocks was tested on CHASE-DB1 dataset, and achieved better performance comparing to the benchmark with only 0.6% of the parameters and 25% of the training examples. The resulting model is fast to compute, converges more rapidly and requires fewer examples to achieve a given performance than more general techniques such as U-Net.

Keywords: Steerable filter · Image segmentation

1 Introduction

Many CNN-based approaches have been proposed for solving image segmentation tasks in medicine and biology. Such models are usually huge, with millions of parameters. However some tasks involve segmenting structures with a large degree of local self similarity. For instance, many curvilinear structures (such as blood vessels or neuron fibrils) can be thought of as being constructed from rotated and scaled versions of a small number of cannonical templates. We propose a system which explicitly re-uses kernels at multiple scales and orientations and is trained in a way that encourages invariance to rotation. This leads to a model with less than 1% of the parameters of more general approaches, such as U-Net, yet which gives performance at least as good.

M. de Bruijne et al. (Eds.): MICCAI 2021, LNCS 12906, pp. 590–599, 2021.
https://doi.org/10.1007/978-3-030-87231-1_57

We introduce a novel structure, called Multi-Angle Convolution (MAC) Unit, which is designed to encourage rotation invariance.

Unlike many steerable networks [2,6,15], a MAC Unit kernel is not manipulated by combining basis filters, but learns rotated versions during training.

The MAC Unit involves applying a set of filters, corresponding to different orientations, to a patch to compute a vector of outputs. The shape of these outputs (response vs orientation) is compared against an expected shape for each orientation in order to select the best angle. The output is then given by a weighted sum of the individual outputs, with the weights being angle specific (full details are given below). The advantage of this approach is that rather than just choose the best response, all filters contribute to the final output, which makes training more stable.

Since this approach involves training filters to achieve a particular shape of output responses, we call it "Response Shaping".

We can make the unit robust to scale changes by using a pyramid representation [5]. A MAC Unit is applied to different downsampled versions of the input, the results are upsampled and a max operation performed to select the best fitting scale.

In the following we describe the approach in detail, and demonstrate the approach on the the CHASE-DB1 dataset [13]. We show that we can achieve equivalent performance to the benchmark U-Net with only 0.6% of the parameters and 25% of the training samples.

2   Related Works

Steerable filters have been popular in image analysis for tasks such as edge detection and texture feature extraction [4]. A core property is rotational invariance, as the filters are evenly-distributed across orientations. Typically steerable filters require only a few parameters. One of the most popular is the Gabor filter [12]. They combine a cosine wavelet function with a 2D Gaussian.

Recently encouraging results have been achieved by combining steerable filters with convolutional networks. Optimisation can lead to steerable models that are more flexible, and have less filter redundancy [6]. In [2], Cohen and Welling designed an operator that can rotate a kernel by a chosen angle. The work had lower error rate comparing to the state-of-arts at the time. Worral et al. [16] proposed a new approach of compositing steerable filters using some atomic filters. Based on their results, Weiler et al. [15] proposed a model of arbitrary directions, with the corresponding combining process being trainable. However, these methods are computationally expensive as the kernel size is constrained by the number of predefined circular harmonic patterns. To bring more variance to the kernel pattern, large kernels are preferred but this brings increased computation. Ghosh and Gupta [6] proposed a generative kernel which is also scalable. This enables filters to detect a pattern at different scales. However, the kennel pattern is still not as flexible as desired and can be hard to optimize. In [1], Bekkers et al. enabled CNNs to deal with rotation and translation effects via bi-linear interpolation. In other works, the idea of using combined kernel is investigated. For example, in [9,10], the authors combined a bank of generative Gabor filters with some random initialized kernel. The resulting assembled neuron showed good performance on MNIST dataset.

Most steerable CNN models are based on a steering operator and the rotation basis is pre-defined to some degree. This is theoretically favored as it gives perfect rotation, but could put implicit constraints on training.

Our proposed MAC Unit variants achieve approximate rotation invariance by response shaping. This enables the model to be fully trainable, and also can be applied to any number of directions.

Unlike conventional matched filters for detecting particular signals in radar and images, both the optimal template and the response shape is unknown for MASC model before training.

In recent projects, multi-scale pooling has been shown to be beneficial when extracting features. Studies such as [7] suggested using a pyramid representation in analyzing multi-scale information. [5] introduced cross connections between different scale layers, and found they bring improvement to pose detection and image segmentation. Using pyramid pooling is much more computational efficient than the filter-scaling strategy used by the previous generative-based steerable models. For efficiency we use the invertible bottleneck [14]. Its narrow-wide-narrow shape can achieve comparable accuracy with fewer parameters.

3   Methods

Limitations of conventional steerable convolutional models include their computational expense, their inflexiblity/indifferentiablity in pattern generation and direction selection, and most are not fully trainable. Our method abandoned operator-based rotating, instead we use response shaping to solve these problems.

3.1   Rotatable MAC Unit and Response Shaping

A simple approach to achieving approximate rotational invariance would be to construct a set of n filters which are rotated versions of the basis kernel. Each would be convolved with the target image, and the strongest response taken at each pixel. In CNN terms, this is equivalent to performing a maximum operator over channels, where each channel is the output of one oriented filter.

Let $\mathbf{w}_i$ be a vector containing all the elements of the filter at orientation i (of n), which has been normalised so that $|\mathbf{w}_i| = 1$. If $\mathbf{x}$ is the vector of the input patch, then the response of the filter to the patch is given by $v_i = \frac{\mathbf{w}_i \cdot \mathbf{x}}{|\mathbf{x}|}$. The output of the simple approach would then be $\max_i(\frac{\mathbf{w}_i \cdot \mathbf{x}}{|\mathbf{x}|})$.

Let $M_{ij} = \mathbf{w}_i \cdot \mathbf{w}_j$ be the response of filter i to an image patch equal to filter j. Typically this will have a strong response when i is near j, and weak response when they are more mismatched. Let $\mathbf{m}_i = (M_{i1}|...|M_{in})^T$. The elements of $\mathbf{m}_i$ give the shape of the response of filter i to the other filters. See, for instance, Fig. 2.

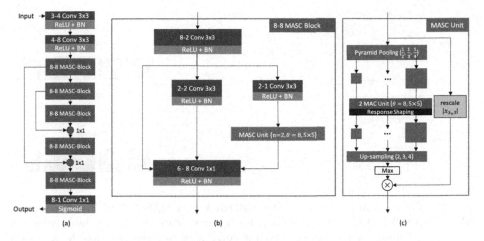

Fig. 1. (a) Model MASC-5-2-8, with 5 stacking MASC Blocks, each with 2 parallel MAC Units of 8 5 × 5 directional kernels; (b) 8-8 MASC Block, with 8 input and 8 output channels and a thin bottleneck; (c) MASC Unit, including 2 MAC Units applied on pyramid representations, where only the maximum is taken from across the scale channels. MASC is a normalized calculation, the result value will be rescaled with a 3 × 3 L2pooling node

In order to use information from all filters, the Multi-Angle Convolutional Unit performs the following operations to get its output for a patch $\mathbf{x}$;

$$v_i = \frac{\mathbf{w}_i \cdot \mathbf{x}}{|\mathbf{x}|} \ \forall i \tag{1}$$

$$R(\mathbf{x}) = \max_i \left(\frac{\mathbf{v} \cdot \mathbf{m}_i}{|\mathbf{m}_i|} \right) \text{ where } \mathbf{v} = (v_1|...|v_n)^T \tag{2}$$

Thus it is comparing the shape of $\mathbf{v}$ with that of each $\mathbf{m}_i$, and choosing the one with the largest similarity - the orientation used is that which provides an output most similar to the shape given by comparing each angle filter with all the others (rather than the strongest response from the individual filters).

If we initialise each $\mathbf{w}_i$ as oriented filters, such as Gabor kernels at equally spaced angles, they will have a particular shape of output when compared to one another.

During training we modify the filters, but seek to retain the relationship between them by encouraging the output of all of them to follow a particular shape using (2).

The first step uses normalised cross correlation, with the output dependent only on the angle between $\mathbf{x}$ and $\mathbf{w}_i$. This would discard the overall intensity, and would tend to exaggerate noise in near flat regions. In order to retain some information about the overall intensity we rescaled the output using

$$\text{MAC Unit}(\mathbf{x}) = |\mathbf{x}_{3\times3}|R(\mathbf{x}) \tag{3}$$

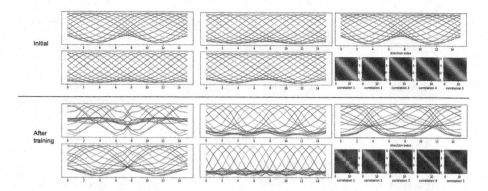

Fig. 2. The correlation curves of the first block in a MASC5-5-16 9×9 model, in which 5 parallel MAC units are used. Each curve represents a row in M which is illustrated bottom right. It shows that the correlation between two close kernels with similar directions is strong. This pattern is retained after training.

where $\mathbf{x}_{3\times3}$ is a vector containing the pixel information in the central 3×3 part of the input patch.

Instead of taking the maximum signal from $\mathbf{v}$ like [6,11,16], we combine all the signals in a way that is dependent on the estimate of the best angle. Extracting information by simple maximum selection can lead to poor training for directional kernels, kernels for rare directions can be underfitted. One the advantage of response shaping is that it includes all the directional parameters in the forward propagation. Also, using an n-length directional response vector better depicts the input in the feature space than a single maximum scalar.

During training the values of M_{ij} are updated before every forward propagation iteration.

Though MASC is not inspired by the idea of 'matched filters', they shared some common characteristics.

The MASC approach can also be thought of as a variant of an 'attention' mechanism. The directional response vector V can be regarded as the query vector, and the optimal response matrix M can be treated as the keys. Comparing to ViT (vision transformer) [3] on ImageNet, a medical task typically does not contain arbitrary objects, and useful features tend to be local. Unlike the common self-attention unit, here, the encoder for query and key are the same, and the model has n candidate keys for each query. Also, it does not refer to context information to compute keys but uses a learned bank of patterns. Since the mechanism is different from the 'self-attention', in the paper, we named it as response shaping.

3.2   Filter Re-use

Suppose that we have a $k \times k$ (k odd) filter which gives a strong response to structures at orientation θ, with $0 < \theta < \pi/2$. If we rotate the kernel 90°, we get a new filter which would give a strong response at angle $\pi/2 + \theta$. If we transpose the filter (reflect in the line $y = x$), then the new filter gives a strong response at angle $\pi/2 - \theta$. Thus a single set of filter weights can be re-used in four filters at angles $\theta, \pi/2 - \theta, \pi/2 + \theta, \pi - \theta$. So we need only two basis filters, at angles $\pi/16, 3\pi/16$ to generate a set of 8 filters equally spread over all angles ($\theta_i = (1+2i)\pi/16$). In general, for filters at $4a$ orientations we need a basis filters at angles $\theta_j = (1 + aj)\pi/(8a), j = 0..a - 1$. This enables a significant reduction in the total number of parameters in the model.

3.3 Initialization Strategy

We initialise the basis filters using Gabor filters at the appropriate angles, but with the other parameters (amplitude, σ, aspect ratio) chosen randomly.

We have tried two approaches to train the filters.

1. Initialise the basis filters with Gabor filters and optimise;
2. Use a Gabor function to weight the filter elements.

In the second case the basis filter, W_i, is given by the per-element (Hadamard) product of a Gabor kernel, G, with a filter B_i, so $W_i = G \circ B_i$. During optimisation we vary the parameters defining the Gabor kernel and the elements of B_i. We find that this is often more stable than simply initialising with a Gabor.

3.4 Multi-scale Processing with Pyramids

In the works of [17], multi-scale pooling has been proven effective in against scale variance and signal discontinuity. Here, instead of using different filters, we apply the same MAC Unit on each pyramid representation of the input. The outputs are rescaled via bilinear interpolation. Then only the maximum signal among scale channels is taken. The scale ratio selection depends on the nature of task. In our model, ratios of $(\frac{1}{2}, \frac{1}{3}, \frac{1}{4})$ are used, see Fig. 1(c).

The method of pyramid pooling takes less computation than using scaled kernels [6], as the kernel is applied on smaller feature maps. We use max pooling during the downsampling, for efficiency. We call the resulting network a Multi-Angle and Scale Convolutional (MASC) Unit, see Fig. 1(c).

We create a MASC-Block, as shown in Fig. 1(b). The input is to two nodes, a standard Conv layer and a MASC Unit. The combined feature map is concatenated with the residuals. A summarization layer combines the information.

More than one MAC Unit can be used in a MASC Unit. They are managed in parallel when processing the pooled features, and their results are concatenated. In experiments, we found the marginal profit of using more MAC Units tapers off gradually, and the speed of convergence depends on initialization. The channel numbers are arranged as wide-narrow-wide-narrow with a thin bottleneck.

By stacking 5 MASC blocks, each includes 2 independent 8-directional MAC Unit arranged in parallel (same input), a MASC-5-2-8 model is illustrated in Fig. 1(a). Here, a single MASC Unit can work like 24 filters (in 8 directions and 3 scales).

4   Experiments

In this paper, we tested the MASC-5-2-8 model on retina vessel images in the
CHASE-DB1 dataset [13] (see Fig. 2(b)). The Ground truth of 1stHO is used.

The dataset contains 30 color images. The first 20 were taken as training
set, and the last 10 as test set as stated in [8]. Then 8000 50×50 patches were
randomly cropped from the training images.

The MASC-5-2-8 model was initialized as described above. An Adam optimizer was used with plateau learning rate scheduler(start = 0.01, ratio =
0.5, patience = 10). The initialization noise was picked from range $(0, 0.1] \times$
Gabor amplitude. In pilot experiments, we found setting Gabor pattern to be
more peaked (high γ, e.g. 15, exaggerates the difference among directions) can
bring significant improvements in training, but the optimum choise may vary for
different tasks.

The results are summarised in Table 1. The MASC model with only 3,292
parameters outperforms the UNet benchmark almost on all columns, and is
generally similar to other benchmarks despite having so few parameters. On
the metric such as averaged specificity and F1 score, our method achieves best
results.

Ablation Experiments. As a baseline we replaced all MASC blocks with 8-
8 convolutional layers with 3×3 kernels (using more parameters) - "MASC-
replaced" in the table. This demonstrates the benefits of the MASC over simpler convolutions. The MASC5-2-8-init represents the case in which we initialize
kernels W_i with Gabor distribution plus noise, rather than using a Hadamard
product $(W_i = G \circ B_i)$ during optimisation. It is not as stable as when using
the product, occasionally failing to converge to a sensible form. MASC5-2-8-
w/o-rescale is a version without the intensity rescaling (Eq. 3) - the rescaling
helps.

In model MASC5-2-8-max, the response shaping is replaced by a max function. This can be regarded as a special case of response shaping, where all directional kernels are mutually orthogonal and the correlations are equal to 0 except
to itself. In this case, the max of combined responses is equivalent to the max
of single-direction response. In the experiment, we found the performance of
MASC5-2-8-max is close to the full version but is less stable when training.

We also explored how well the model performance varied with the size of
the training set. Figure 3(a) shows how the Area Under the ROC Curve (AUC)
declines much more slowly as the training set shrinks, compared to a U-Net.
The MASC model can achieve good results even with relatively small numbers
of training examples.

In Fig. 3(c) and (d), some intermediate outputs from MASC Units with 16
directions are illustrated, together with the 9×9 kernel patterns and the maximum index map generated by Eq. (2). The response shaping approach has
encouraged the kernels to represent rotated versions of the same pattern, which
can also be read from the index map. For example, in the index map, the horizontal vessels are mostly identified by the yellow response shape. The indices for

Table 1. Comparison with other methods on CHASE_DB1 (*results obtained from [8])

Methods	F1	Se	Sp	Acc	AUC
U-Net	79.2	79.1	97.4	95.4	95.3
Residual U-Net*	78.0	77.3	98.2	95.5	97.8
Reccurent U-Net*	78.1	74.6	**98.4**	96.2	98.0
R2U-Net*	79.3	77.6	98.2	**96.3**	**98.1**
LadderNet*	79.0	78.6	98.0	96.2	97.7
VesselNet*	79.1	78.2	98.1	96.2	97.6
MASC5-2-8-w/o-rescale	79.8	80.3	97.4	95.4	97.4
MASC5-2-8-max	79.7	78.6	97.7	95.5	97.2
MASC5-2-8-init	79.1	76.7	97.9	95.5	97.3
MASC-replaced	76.4	72.1	97.9	95.1	97.1
MASC5-2-8	**80.5**	**81.3**	97.4	95.5	97.6

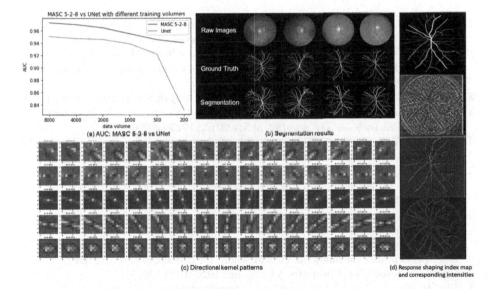

Fig. 3. (a) AUC comparison between MASC5-2-8 and UNet with different training set sizes; (b) Segmentation results; (c) Illustration of MASC5-5-16-9x9 directional kernels, MASC with response shaping can achieve a pseudo-steerable effect via gradient decent; (d) The maximum index map generated by Eq. (2), the colors represent the chosen indices. From top to bottom, i. ground truth, ii. response shaping index map, iii. masked index map by ground truth, iv. corresponding shaped response intensity map. The index map shows that structures with particular directions are picked out with specific index values.

the background area are generally random. The intensity map shows the vessel area has a higher shaped response value.

5   Conclusion

We have introduced novel model for analysing curvilinear structures which are composed of self similar elements at arbitrary orientation and scale. The system learns a set of filters which can be transformed easily to produce responses at a range of angles. We show how this can be extended to include a range of scales. The resulting model is very parameter efficient.

On the task of retina vessel segmentation the model achieves accuracy equivalent to of the benchmark U-Net model with only 0.6% of the parameters and 25% of the training set. It is thus potentially very useful where limited numbers of training examples are available. Though this work focuses on retinal images, we have also applied it successfully to tracking growing axons in microscopy images. In future work we will extend it to 3D volume data.

References

1. Bekkers, E.J., Lafarge, M.W., Veta, M., Eppenhof, K.A.J., Pluim, J.P.W., Duits, R.: Roto-translation covariant convolutional networks for medical image analysis. In: Frangi, A.F., Schnabel, J.A., Davatzikos, C., Alberola-López, C., Fichtinger, G. (eds.) MICCAI 2018. LNCS, vol. 11070, pp. 440–448. Springer, Cham (2018). https://doi.org/10.1007/978-3-030-00928-1_50
2. Cohen, T.S., Welling, M.: Steerable CNNs. arXiv preprint arXiv:1612.08498 (2016)
3. Dosovitskiy, A., et al.: An image is worth 16x16 words: transformers for image recognition at scale. arXiv preprint arXiv:2010.11929 (2020)
4. Freeman, W.T., Adelson, E.H., et al.: The design and use of steerable filters. IEEE Trans. Pattern Anal. Mach. Intell. **13**(9), 891–906 (1991)
5. Ghiasi, G., Lin, T.Y., Le, Q.V.: NAS-FPN: learning scalable feature pyramid architecture for object detection. In: Proceedings of the IEEE/CVF Conference on Computer Vision and Pattern Recognition, pp. 7036–7045 (2019)
6. Ghosh, R., Gupta, A.K.: Scale steerable filters for locally scale-invariant convolutional neural networks. arXiv preprint arXiv:1906.03861 (2019)
7. He, K., Zhang, X., Ren, S., Sun, J.: Delving deep into rectifiers: surpassing human-level performance on imagenet classification. In: Proceedings of the IEEE International Conference on Computer Vision, pp. 1026–1034 (2015)
8. Liu, B., Gu, L., Lu, F.: Unsupervised ensemble strategy for retinal vessel segmentation. In: Shen, D., et al. (eds.) MICCAI 2019. LNCS, vol. 11764, pp. 111–119. Springer, Cham (2019). https://doi.org/10.1007/978-3-030-32239-7_13
9. Liu, Z., Cootes, T., Ballestrem, C.: An end to end system for measuring axon growth. In: Liu, M., Yan, P., Lian, C., Cao, X. (eds.) MLMI 2020. LNCS, vol. 12436, pp. 455–464. Springer, Cham (2020). https://doi.org/10.1007/978-3-030-59861-7_46
10. Luan, S., Chen, C., Zhang, B., Han, J., Liu, J.: Gabor convolutional networks. IEEE Trans. Image Process. **27**(9), 4357–4366 (2018)

11. Marcos, D., Volpi, M., Tuia, D.: Learning rotation invariant convolutional filters for texture classification. In: 2016 23rd International Conference on Pattern Recognition (ICPR), pp. 2012–2017. IEEE (2016)
12. Mehrotra, R., Namuduri, K.R., Ranganathan, N.: Gabor filter-based edge detection. Pattern Recogn. **25**(12), 1479–1494 (1992)
13. Owen, C.G., et al.: Measuring retinal vessel tortuosity in 10-year-old children: validation of the computer-assisted image analysis of the retina (CAIAR) program. Invest. Ophthalmol. Vis. Sci. **50**(5), 2004–2010 (2009)
14. Sandler, M., Howard, A., Zhu, M., Zhmoginov, A., Chen, L.C.: MobileNetV2: inverted residuals and linear bottlenecks. In: Proceedings of the IEEE Conference on Computer Vision And Pattern Recognition, pp. 4510–4520 (2018)
15. Weiler, M., Hamprecht, F.A., Storath, M.: Learning steerable filters for rotation equivariant CNNs. In: Proceedings of the IEEE Conference on Computer Vision and Pattern Recognition, pp. 849–858 (2018)
16. Worrall, D.E., Garbin, S.J., Turmukhambetov, D., Brostow, G.J.: Harmonic networks: deep translation and rotation equivariance. In: Proceedings of the IEEE Conference on Computer Vision and Pattern Recognition, pp. 5028–5037 (2017)
17. Zhao, H., Shi, J., Qi, X., Wang, X., Jia, J.: Pyramid scene parsing network. In: Proceedings of the IEEE Conference on Computer Vision and Pattern Recognition, pp. 2881–2890 (2017)

Vessel Width Estimation via Convolutional Regression

Rui-Qi Li[1,2], Gui-Bin Bian[1,2], Xiao-Hu Zhou[1,2], Xiaoliang Xie[1,2],
Zhen-Liang Ni[1,2], Yan-Jie Zhou[1,2], Yuhan Wang[1], and Zengguang Hou[1,2,3,4](✉)

[1] State Key Laboratory of Management and Control for Complex Systems,
Institute of Automation, Chinese Academy of Sciences, Beijing 100190, China
zengguang.hou@ia.ac.cn
[2] School of Artificial Intelligence, University of Chinese Academy of Sciences,
Beijing 100049, China
[3] Center for Excellence in Brain Science and Technology, CAS, Beijing 100190, China
[4] CAS-MUST Joint Laboratory of Intelligence Science and Technology,
Institute of Systems Engineering, Macau University of Science and Technology,
Macau 999078, China

Abstract. Vessel width estimation has a wide range of applications in disease diagnosis and treatment. In this paper, vessel width estimation is cast as a regression problem, and a novel Convolutional Neural Network (CNN) based method is proposed for vessel width estimation. In our CNN-based method, the idea of divide-and-conquer is introduced to solve the challenge of imbalanced training samples. Besides, in order to solve the shortage of training samples required by CNN, a vessel width label generation method is proposed to generate width labels from vessel segmentation labels. In the experiments, we apply our vessel width label generation method and CNN-based width estimation method to two tasks which are retinal vessel width estimation and coronary artery width estimation. Experimental results show that our width label generation method can generate sufficiently realistic width labels using accurate segmentation labels. Also, our CNN-based method can solve the challenge of imbalanced training samples, achieving state-of-the-art performance with less inference time.

Keywords: Vessel · Width estimation · Deep learning

1 Introduction

The vessel width estimation plays a weighty role in both disease diagnosis and clinical treatment. For retinal vessel, studies have shown that a decreased ratio of arterial to venous retinal vessel width forms an independent risk factor for stroke, myocardial infarct as well as eye disease [1]. For coronary artery, during the implementation of percutaneous coronary intervention, surgeons need to measure the width of the coronary artery to determine the type of the stent. Therefore, vessel width estimation has a wide application prospect.

© Springer Nature Switzerland AG 2021
M. de Bruijne et al. (Eds.): MICCAI 2021, LNCS 12906, pp. 600–610, 2021.
https://doi.org/10.1007/978-3-030-87231-1_58

Existing vessel width estimation algorithms fall into two categories: semi-automatic [2,3,10] and fully automatic [4–9,11]. The semi-automatic algorithms require the user to nominate the profile of the vessel, and then locate the positions of two vessel edges by analyzing the changes of the pixel intensity on the profile. The vessel width on this profile is the distance of two vessel edges. Since the user is required to nominate the profile, the quality of the estimation result largely depends on the accuracy of the profile given by the user. Instead of requiring the user to nominate a profile, the fully automatic algorithms can estimate the widths of all the vessel segments in the image. Fully automatic algorithms first segment all vessels, then extract the centerline of the segmented vessels and compute the profiles of all vessels. Finally, vessel widths are calculated by a semi-automatic method with the calculated profiles.

Semi-automatic algorithms require the user to nominate the vessel profile and are not intelligent. Although automatic algorithms are intelligent, all the operations in automatic algorithms (extracting the centerline, calculating the vessel profiles and predicting the width of all the profiles) require repeated processing of all the image pixels, resulting in the processing time of an image requiring tens of seconds or even minutes. Therefore, an automatic and fast vessel width estimation algorithm needs to be proposed.

Up to now, many methods based on Convolutional Neural Network (CNN) have been used in many fields of medical image processing, but as far as we know, the CNN-based method has not been applied to the vessel width estimation task. The main reason is the lack of datasets that can be used for training. Besides, how to use CNNs to predict the vessel widths is also a problem to be solved. In this paper, we address the above difficulties by proposing a vessel width label generation method and a novel CNN-based vessel width estimation method. Compared with the existing automatic algorithms, our CNN-based method can achieve the same or even better results in precision, and far exceeds the existing methods in speed.

Our contributions are as follow: (1) The width estimation problem is transformed into a pixel-level width regression problem, which brings a new idea to solve this problem. (2) To address the lack of width labels for training, a vessel width label generation method is proposed. (3) A CNN-based method for automatic vessel width estimation is proposed, which can solve the uneven width distribution in training samples. To the best of our knowledge, our method is the first work that uses a deep learning model for vessel width estimation.

2   Method

2.1   Vessel Width Label Generation Method

To address the lack of width labels for training, a method that generates vessel width labels using the vessel segmentation labels is proposed. This method has similar procedure with existing automatic width estimation algorithms. However, since the difference of algorithm objective, our method has different processing in some steps compared with existing width estimation algorithms. Here, taking the coronary artery as an example, we will describe our method. We only detail the special parts of our method, and briefly cover the parts that are the same as the existing width estimation algorithms. The processing steps are as follow:

(1) Extract the coronary artery centerline in the segmentation labels. The thinning algorithm we use is the method proposed in [12].
(2) Remove the intersection points and bifurcation points in the coronary artery's centerline. After these points are removed, the coronary artery tree is cut into artery segments, as shown in Fig. 1(c). In our implementation, artery segments with length less than 10 pixels will be then removed.
(3) Calculate the profile of each pixel on the remaining centerline. Here, we apply the method proposed in [8], that is, using several adjacent centerline pixels on both sides of the target pixel and applying principal component analysis on these pixels. A profile result is shown as the blue line in Fig. 1(d).
(4) Find two artery edge points on each profile and calculate the artery width. Since the accurate segmentation labels have been given, the boundaries between vessel and non-vessel pixels in segmentation labels are the edges of the artery. Therefore, the two edge points are the intersections of the profile and two segmentation boundaries. In our implementation, we proceed from the centerline pixel (red pixel in Fig. 1(d)) along the profile to both sides at a certain step size (0.1 pixel we use), and calculate the intensity of the current coordinate using bilinear interpolation at each step. When the intensity value is less than 0.5, the current coordinate is the edge point. As shown in Fig. 1(d), two yellow cross points are the two edge points calculated this way. The artery width on this profile is the distance between two edge points.
(5) Generate the final labels. For training and for test, the generated labels are different. For test, two edge points' coordinates are regarded as a test sample, just like REVIEW dataset [13]. For training, to train a pixel-level width estimation algorithm, we assign the width value to all the pixels that belong to both the profile and the coronary artery, as shown in Fig. 1(e). There are two purposes to do this. First and most intuitively, this way can enlarge training samples, especially the training samples of thick vessels. Second, in application, we hope to obtain uniform width estimation results

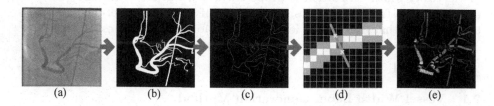

(a)   (b)   (c)   (d)   (e)

Fig. 1. (a) The raw image. (b) Corresponding segmentation label. (c) Coronary artery centerline without intersection and bifurcation points. (d) White pixels are centerline pixels, gray pixels are artery pixels. Blue line is the calculated profile of red pixel, and yellow points are two edge points on the profile. (e) Generated labels for training. (Color figure online)

regardless of which pixel is selected among all the pixels belonging to the same profile. In this generating way, many pixels have no width values, and these pixels will not be used in training. Besides, there will be pixels that belong to multiple profiles and have multiple width values. In this case, we will assign the mean width value to these pixels.

It is worth noting that the labels generated by this method inevitably have very few error labels. For deep learning methods, very few error training labels can not affect the final result. However, for the test labels, you must manually check and eliminate all error labels.

2.2 Vessel Width Estimation Network

In this paper, vessel width estimation is cast as a pixel-level width regression problem, which means to predict a width value for each image pixel, and a CNN-based method is proposed to regress the pixel-level width by extracting the local image features. The vessel width can range from one pixel to dozens of pixels in the images, and the width distribution is often uneven. In general, thin vessels make up the majority of vessels, which will cause the width estimation algorithms to pay too much attention to thin vessels instead of thick vessels. However, from the point of applications, it is usually the thick vessels that need to be measured, while the width of the thin vessels is not important. Therefore, the imbalanced distribution is the main challenge in vessel width regression.

Inspired by the divide-and-conquer idea used in other numerical regressions [14,15], we partition the entire width range into many sub-ranges, and train a local regressor for each sub-range. Each local regressor only uses pixels whose width label belongs to its own sub-range in training to ensure its good performance in its sub-range, regardless of its performance in other sub-ranges. At the same time, a classifier is trained taking each local regressor as a category. The classifier is used for selecting appropriate regressors for different pixels by judging their width sub-ranges. The classifier is trained using the one-hot labels. The advantages of our method include: 1. In each sub-range, the distribution of vessel widths is relatively even. Therefore, the challenge of imbalanced distribution has been addressed. 2. With the partition of the range, the difficulty of each regression task is reduced, which helps to obtain finer regression results.

To implement the regressor and classifier, a network is proposed in Fig. 2. This network can actually be viewed as an U-net [16] with two decoding branches. Compared with U-net, we reduce the depth of the network, and replace the deconvolutional layer with bilinear upsampling layer to reduce the model parameters. As shown in Fig. 2, both the regression branch and the classification branch output the results with the size of $H \times W \times N$. N is the number of sub-ranges, also the number of local regressors. Each output channel of the regression branch represents the estimated widths of each local regressor for all pixels. For each pixel, N local regressors output N width predictions $(1 \times 1 \times N)$. The classification branch outputs the probability $(1 \times 1 \times N)$ that each pixel belongs to N

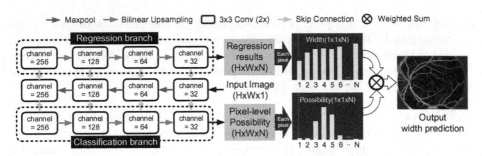

Fig. 2. Schematic diagram of our CNN-based method. N is the number of sub-ranges.

regressors. Instead of using the predicted width of the local regressor with maximum probability, the output width of each pixel is the weighted sum of the predicted widths of all regressors taking the probability as weights. The use of weighted sum is beneficial to obtain smoother width estimation results.

For the loss function, mean-square error loss is used for the regression, and softmax cross-entropy loss is utilized for the classification.

$$
\begin{aligned}
Loss &= \lambda L_{reg} + L_{cls} \\
&= \frac{\sum_{i=1}^{P}[\lambda(W(i) - W^*(i))^2 - \sum_{j=1}^{N} p(j)log\ q(j)]\delta(v_i = 1)}{\sum_{i=1}^{P} \delta(v_i = 1)}
\end{aligned} \quad (1)
$$

where P is the number of pixels, N is the number of sub-ranges. v_i is the trainable flag for pixels, and only the pixels have width labels are used in training. W and W^* are the predict width and ground-truth width respectively. $q(j)$ and $p(j)$ are the output and ground-truth probabilities, λ is the weight set for L_{reg}.

3   Dataset

3.1   Retinal Vessel Dataset for Width Estimation

There is an open dataset for evaluation of the width estimation task of retinal vessel: REVIEW [13]. The REVIEW dataset includes four image sets and we only use three of them those are challenging (KPIS is not used since it only contains 3 vessel segments): (1) High Resolution Image Set (HRIS); (2) Vascular Disease Image Set (VDIS); (3) Central Light Reflex Image Set (CLRIS). These image sets include 14 images with 190 vessel segments, and contain 4902 manually marked profiles in total. These profiles are marked by three observers, with the mean value used as the ground-truth width. Since the dataset was established in 2008, before the boost of deep learning methods, this dataset is aimed at the unsupervised retinal vessel width estimation methods, and only contains the test samples. Two samples are shown in Fig. 3(d–e).

In order to obtain training samples for our CNN-based method, we use the proposed width label generation method to generate. The utilized segmentation

Fig. 3. (a) A sample in DRIVE dataset. (b) Segmentation label of (a). (c) Generated training label of (a). (d–e) Two samples in REVIEW dataset. The green part is the vessel segments with manually marked profiles. (Color figure online)

dataset is the most commonly used dataset: DRIVE [17]. DRIVE dataset contains a total of 40 images with precision pixel-level segmentation annotations, of which 20 images for training and 20 images for test. Each image has a resolution of 565×584. An image sample is shown in Fig. 3(a). The segmentation label and generated vessel width labels are shown in Fig. 3(b–c). Due to the existing of REVIEW (test samples), all 40 images of DRIVE dataset are used as the training samples of our proposed CNN-based method.

3.2 Coronary Artery Dataset for Width Estimation

Unlike retinal vessels, there is no public width estimation dataset for coronary arteries, or even a public segmentation dataset. Therefore, a coronary artery segmentation dataset established by us is used to generate the required training and test samples. Our segmentation dataset includes 745 images selected from 25 independent Digital Substraction Angiography (DSA) continuous sequences. Among them, 568 images from 13 DSA sequences are used as the training set, and 177 images from 12 DSA sequences are used as the test set. The DSA sequences are generated by Siemens Artis zee III ceiling, and are based on a flat panel detector. Each image has a resolution of 512×512. Using our proposed width label generation method, training samples and test samples for width estimation will be generated by the training samples and test samples of segmentation dataset respectively.

4   Experiment

4.1   Retinal Vessel Width Estimation

Our CNN-based method is compared with existing methods on the REVIEW dataset to verify the superiority of our CNN-based width estimation method and the effectiveness of the width label generation method.

Implementation Details: Similar to other methods [6,8], only the green channel of the retinal image is used as input, since the green channel shows the highest contrast between the blood vessels and background. Since the training set and the test set are from two independent datasets with a large scale difference, in order to ensure the consistency of vessel width distributions of the two sets, the resolutions of all images should be unified. The resolution of the DRIVE images is doubled to 1130×1168. For the REVIEW dataset, the images of HDIS are resized to 1792×1232, the images of CLRIS are resized to 1440×960, and the images of VDIS remain unchanged (1360×1024).

Due to the limited training data, we adopt a commonly used trick in retinal vessel segmentation: image cropping. During the training, the image patches of 288×288 are randomly cropped from the training images as the network input. Except random image cropping, random flip, random grayscale adjustment between -20 to 20, and random contrast ratio between 0.8 to 1.2 are also adopted for image augmentation. The retinal vessel width ranges from 0 to 20 pixels. We set the sub-range to be 4 (5 sub-ranges in all). λ in loss function is 5. The network is implemented using Tensorflow 1.10. For optimization, Adam optimizer [18] is applied with the batch size of 4. We use an initial learning rate of 0.0005, and the initial learning rate is multiplied by 0.8 every 1000 steps to avoid overfitting. Training takes about 3 h on an NVIDIA Titan XP for 2000 epochs.

Metric: Since REVIEW dataset is used for evaluation, the evaluation metrics designed by REVIEW dataset is also used. We report the success rate (SR) and the standard deviation of the width error (σ_E) as evaluation metrics. The success rate is the percentage of the test samples that the algorithm returns a meaningful width value. The σ_E is the metric proposed by the authors of REVIEW dataset. They argue that σ_E is more suitable to evaluate the performance of the width estimation algorithm, while the mean width error is incompetent in evaluation.

For our CNN-based method, these metrics are calculated in the following way: Firstly, since REVIEW dataset provides two edge points on the profiles, the center point coordinates and corresponding ground-truth vessel widths are calculated. Secondly, since CNN-based method will predict a width value for all pixels, the width corresponding to center point coordinate is the predicted width. Since the coordinates are usually not integers, bilinear interpolation is used to get the width predictions. Finally, the success rate and σ_E are calculated. Since each center point must have a return value, the value greater than zero is regarded as a meaningful measurement to calculate the success rate.

Experimental Results: Our CNN-based method is compared with some fully automatic and semi-automatic methods, including: HHFW, 1-D gaussian, 2-D gaussian, ESP, graph-based method, 3D model. The results are shown in Table 1. We can see that all success rates of our method are 100%. In terms of the σ_E, our method achieves state-of-the-art results on two of three sets: CLRIS and HRIS. It is worth noting that these results are obtained when the training data and the test data are from different datasets. We believe that the result can be better if the training data and the test data come from the same dataset. In addition, the experimental results also prove that our width label generation method can generate sufficiently realistic width labels for training our CNN-based method. If the training samples generated are not ideal enough, our CNN-based method cannot achieve such width estimation accuracy in the condition of cross-datasets.

Table 1. Results on REVIEW dataset

Method	HRIS		CLRIS		VDIS	
	SR (%)	σ_E	SR (%)	σ_E	SR (%)	σ_E
HHFW [2]	88.3	0.926	0	–	78.4	0.879
1D Gaussian [5]	99.6	0.896	98.6	4.137	99.9	2.11
2D Gaussian [6]	98.9	0.703	26.7	6.019	77.2	1.328
ESP [7]	99.7	0.42	93	1.469	99.6	**0.766**
Graph-based [8]	100	0.567	94.1	1.78	96	1.43
3D model [10]	99.4	0.65	98	1.56	97.8	1.14
Ours	**100**	**0.41**	**100**	**1.33**	**100**	1.41

In fact, the advantage of the CNN-based method lies not only in its estimation accuracy, but also in inference speed. Due to the complex image processing, existing automatic width estimation methods may take tens of seconds to process an image. While the CNN-based method only needs 30 ms (on an NVIDIA Titan XP GPU) to process an image with a resolution of 512×512, which can even achieve real-time performance.

4.2 Coronary Artery Width Estimation

Since there is no public dataset for coronary artery, no method can report its performance on this task, thus we cannot compare our method with other methods. However, as mentioned in dataset section, we use a coronary artery segmentation dataset to generate the training data and test data, thus we can run ablations to validate whether the idea of divide-and-conquer proposed in our CNN-based method can help obtain better vessel width estimation results.

For comparison, we respectively remove each branch of our CNN-based method to form two methods without the divide-and-conquer idea: (1) Regression Only: After removing the classification branch, the regression branch directly regresses the vessel width on full width range using a single regressor (instead of N regressors). (2) Classification Only: After removing the regression branch, we discretize the width values into 30 categories (1 to 30 pixels) and use the classification branch to classify the width value of all pixels.

Implementation Details: The input size of CNN-based method is 512×512. The coronary artery width ranges from 0 to 30 pixels. We set the sub-range to be 3, so there are 10 sub-ranges in all. We use an initial learning rate of 0.001, and the initial learning rate is multiplied by 0.8 every 1000 steps to avoid overfitting. Training takes about 2.5 h on an NVIDIA Titan XP for 200 epochs. Other details are the same as the experiments of retinal vessels.

Metric: Since the regression way can guarantee a 100% success rate, we only report the σ_E as the evaluation metric. Since the algorithm is expected to perform well over all widths, we will report the σ_E over different width ranges.

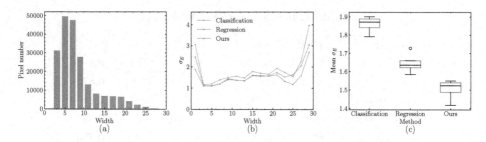

Fig. 4. Results: (a) Width distribution of samples. (b) The performance of three models at different width sub-range. (c) The box diagram of mean σ_E of three models

Specifically, we divide the range of 0 to 30 into 15 sub-ranges, and report the σ_E on all sub-ranges.

Experimental Results: The width distribution of our dataset is statistically analyzed in Fig. 4(a). It's clear that the width distribution is imbalanced. As shown in Fig. 4(b), this imbalance leads the width estimation algorithms to pay more attention to the sub-ranges with more samples. No matter the classification only or regression only, their performance on the thin vessels with more samples are significantly better than those on the thick vessels with fewer samples. However, after introducing the idea of divide-and-conquer, our proposed method maintains the performance on thin vessels, and significantly improves the performance on thick vessels, effectively alleviating the problem of sample imbalance. We also calculate the mean σ_E under different width ranges in multiple trials and draw a box diagram. As shown in Fig. 4(c), our method is significantly better than the comparison methods.

5   Conclusion

This paper proposes a CNN-based method for vessel width estimation, which is a brand new idea compared with the existing methods. In order to solve the most important challenge (lack of training data) when using CNN-based method, a method which can generate width labels using segmentation labels is also proposed. In order to solve the inevitable sample imbalance in the generated width labels, the idea of divide-and-conquer is introduced into our CNN-based method. In the experiments, we apply our method to retinal vessels, indicating that our method can achieve the state-of-the-art performance with less inference time. In addition, we apply our method to coronary arteries, indicating that the proposed divide-and-conquer method can alleviate the sample imbalance.

Acknowledgments. This work was supported in part by the National Natural Science Foundation of China under Grant 62073325, Grant 62003343, Grant U1913601, and Grant U20A20224; in part by the National Key Research and Development Program of China under Grant 2019YFB1311700; in part by the Youth Innovation Promotion Association of CAS under Grant 2020140; and in part by the Strategic Priority Research Program of CAS under Grant XDB32040000.

References

1. Klein, R., Klein, B.E., Moss, S.E.: The relation of systemic hypertension to changes in the retinal vasculature: the Beaver Dam Eye Study. Trans. Am. Ophthalmol. Soc. **95**, 329–350 (1997)
2. Brinchmann-Hansen, O., Heier, H.: Theoretical relations between light streak characteristics and optical properties of retinal vessels. Acta Ophthalmol. **64**, 33–37 (1986)
3. Rezaeian, M., Butlin, M., Golzan, S.M., Graham, S.L., Avolio, A.P.: A novel method for retinal vessel segmentation and diameter measurement using high speed video. In: 41st Annual International Conference of the IEEE Engineering in Medicine and Biology Society (EMBC), pp. 2781–2784 (2019)
4. Sun, G., Liu, X., Wang, S., Gao, L., Liu, M.: Width measurement for pathological vessels in retinal images using centerline correction and k-means clustering. Measurement **139**, 185–195 (2019)
5. Zhou, L., Rzeszotarski, M.S., Singerman, L.J., Chokreff, J.M.: The detection and quantification of retinopathy using digital angiograms. IEEE Trans. Med. Imaging **13**(4), 619–626 (1994)
6. Lowell, J., Hunter, A., Steel, D., Basu, A., Ryder, R., Kennedy, R.L.: Measurement of retinal vessel widths from fundus images based on 2-D modeling. IEEE Trans. Med. Imaging **23**(10), 1196–1204 (2004)
7. Al-Diri, B., Hunter, A., Steel, D.: An active contour model for segmenting and measuring retinal vessels. IEEE Trans. Med. Imaging **28**(9), 1488–1497 (2009)
8. Xu, X., et al.: Vessel boundary delineation on fundus images using graph-based approach. IEEE Trans. Med. Imaging **30**(6), 1184–1191 (2011)
9. Li, Q., You, J., Zhang, D.: Vessel segmentation and width estimation in retinal images using multiscale production of matched filter responses. Expert Syst. Appl. **39**(9), 7600–7610 (2012)
10. Aliahmad, B., Kumar, D.K.: Adaptive Higuchi's dimension-based retinal vessel diameter measurement. In: 38th Annual International Conference of the IEEE Engineering in Medicine and Biology Society (EMBC), pp. 1308–1311 (2016)
11. Huang, F., Dashtbozorg, B., Yeung, A.K.S., Zhang, J., Berendschot, T.T.J.M., ter Haar Romeny, B.M.: A comparative study towards the establishment of an automatic retinal vessel width measurement technique. In: Cardoso, M.J., et al. (eds.) FIFI/OMIA -2017. LNCS, vol. 10554, pp. 227–234. Springer, Cham (2017). https://doi.org/10.1007/978-3-319-67561-9_26
12. Zhang, T.Y., Suen, C.Y.: A fast parallel algorithm for thinning digital patterns. Commun. ACM **27**(3), 236–239 (1984)
13. Al-Diri, B., Hunter, A., Steel, D., Habib, M., Hudaib, T., Berry, S.: REVIEW - a reference data set for retinal vessel profiles. In: 30th Annual International Conference of the IEEE Engineering in Medicine and Biology Society (EMBC), pp. 2262–2265 (2008)
14. Li, W., Lu, J., Feng, J., Xu, C., Zhou, J., Tian, Q.: BridgeNet: a continuity-aware probabilistic network for age estimation. In: Proceedings of 2019 IEEE/CVF Conference on Computer Vision and Pattern Recognition (CVPR), pp. 1145–1154 (2019)
15. Chen, S., Zhang, C., Dong, M., Le, J., Rao, M.: Using ranking-CNN for age estimation. In: Proceedings of 30th IEEE/CVF Conference on Computer Vision and Pattern Recognition (CVPR), pp. 742–751 (2017)

16. Ronneberger, O., Fischer, P., Brox, T.: U-Net: convolutional networks for biomedical image segmentation. In: Navab, N., Hornegger, J., Wells, W.M., Frangi, A.F. (eds.) MICCAI 2015. LNCS, vol. 9351, pp. 234–241. Springer, Cham (2015). https://doi.org/10.1007/978-3-319-24574-4_28
17. Staal, J., Abramoff, M.D., Niemeijer, M., Viergever, M.A., van Ginneken, B.: Ridge-based vessel segmentation in color images of the retina. IEEE Trans. Med. Imaging **23**(4), 501–509 (2004)
18. Kingma, D.P., Ba, J.L.: Adam: a method for stochastic optimization. In: Proceedings on 3rd International Conference on Learning Representations (ICLR), pp. 1–15 (2015)

Renal Cell Carcinoma Classification from Vascular Morphology

Rudan Xiao[1], Eric Debreuve[1], Damien Ambrosetti[2], and Xavier Descombes[1(✉)]

[1] Université Côte d'Azur, Inria, CNRS, I3S, Nice, France
xavier.descombes@inria.fr
[2] Hôpital Pasteur, CHU, Nice, France

Abstract. Renal Cell Carcinoma (RCC) is one of the most common malignancies, and pathological diagnosis is the gold standard for RCC diagnostic method. Recognizing the type of RCC tumor and the possibility of cell migration highly depends on the geometric and topological properties of the vascular network. Motivated by the diagnosis pipeline, we explore whether the vascular network visible in RCC histopathological images is sufficient to characterize the RCC subtype. To achieve this, we firstly build a new vascular network-based RCC histopathological image dataset of 7 patients, namely VRCC200, with 200 well-labeled vascular network annotations. Based on these vascular networks of RCC histopathological images, we propose new hand-crafted features, namely skeleton features and lattice features. These features well represent the geometric and topological properties of the vascular networks of RCC histopathological images. Then we build strong benchmark results with various algorithms (both traditional and deep learning models) on the VRCC200 dataset. The result of skeleton and lattice features can outperform popular deep learning models. Finally, we further prove the robustness and advantage of proposed features on an additional database VRCC60 of 20 patients, with 60 vascular annotated images. All of the results of our experiments prove that the vascular network structure of RCC is one of the most important biomarkers for RCC diagnosis.

Keywords: RCC histopathological image dataset · Vascular network · Skeleton features · Lattice features · RCC classification

1 Introduction

RCC is a highly malignant tumor in the urinary system. 90% of kidney cancers are RCC [18], which is mainly divided into clear cell RCC [25] (ccRCC) accounting for 75% of RCC and papillary RCC (pRCC) accounting for 10% of RCC. RCC classification is a challenging task. Cell morphology, tumor architecture, phenotype and genetics data are mainly used to define the tumor subtype. Most of the current classification research is focused on the search for biological biomarkers, aiming to define the RCC subtype and also predict the behavior of the tumor [8,9,15]. However, vasculature also plays an important role [26] in histopathological diagnosis

© Springer Nature Switzerland AG 2021
M. de Bruijne et al. (Eds.): MICCAI 2021, LNCS 12906, pp. 611–621, 2021.
https://doi.org/10.1007/978-3-030-87231-1_59

and is a characteristic diagnostic feature for ccRCC [28]. The vascular structure of these two kinds of RCC is different and may lead to an accurate diagnosis. For example, ccRCC is characterized by a fishnet-like vascular architecture while the pRCC has a tree-like structure [37]. As shown in Fig. 1, the ccRCC vascular networks are denser and contain more junctions but fewer end branches.

In this paper, we explore the importance of the vascular network in the RCC diagnosis. Since there is no public RCC dataset with vascular network annotations, we build the **VRCC200** dataset with vascular annotations. VRCC200 is extracted from our larger RCC histopathological image dataset (BigRCC) randomly, which is labeled with RCC categories. The larger dataset contains the data of 158 Whole Slide Images (WSIs), coming from 68 patients, which can be cropped into 39986 image patches of ccRCC and 18254 image patches of pRCC respectively. VRCC200 has 200 vascular network segment images of ccRCC and pRCC, coming from 7 patients (3 for ccRCC and 4 for pRCC).

To further explore the potential of traditional algorithms, we propose two sets of features, called "**skeleton features**" and "**lattice features**", which are extracted from the vascular network. Specifically, we compute the skeleton of the vascular network. The skeleton is a structure that embeds the topological and geometrical properties of the vascular network, as shown in Fig. 1. It is composed of 3 types of elements: junction, non-end branch and end branch. Firstly, we define some meaningful features from the skeleton to form the skeleton features, containing number of end branches (NE), small NE, long NE, and density, etc., as shown in Table 2. Then, we perform a series of operations on the skeleton to obtain the lattice spatial map. The lattice features that represent the regions between vessels are extracted from this spatial map. Lattice features include a set of features, such as the mean area, median area, etc. in Table 2.

Finally, we build solid benchmark results of traditional and deep learning methods on the VRCC200 dataset. The results of traditional algorithms with our skeleton and lattice features can outperform the results of popular deep learning models (Graph Convolutional Network (GCN) [36], Convolutional Neural Network (CNN) [16,30]). Then we test these different methods on a new vascular annotated dataset, namely **VRCC60**, which contains 60 vascular annotated image patches, coming from 20 patients (10 for ccRCC and 10 for pRCC). Our skeleton and lattice features still perform best. This indicates our features are robust and can embed sufficient information to characterize RCC subtypes. Also, we show the first work using GCN [6,36] with vascular graph features [11] for the RCC histopathological image classification.

Our contribution can be summarized as follows:

- We are the first work to investigate the importance of geometric and topological properties of the vascular network for RCCs classification.
- We proposed two sets of hand-crafted features, "skeleton and lattice features" to represent the vascular network, which is extracted from the vascular network segmentation images.
- We build new vascular annotated datasets, "VRCC200" and "VRCC60", for RCCs histopathological image classification.

– We build benchmark results on the VRCC200 and test on the VRCC60, showing that our proposed features based traditional classifier can provide best results at small datasets, and can classify ccRCC and pRCC robustly.

2   Related Works

2.1   Histopathological Images Dataset

Classification from histopathological images plays a key role in computer-aided diagnosis or prognosis. There are some public histopathological images datasets for the classification tasks, such as BreaKHis [33] and BACH [1]for breast cancer, LC25000 [3] with five classes of lung tissue, and DigestPath [23] for colon cancer, etc. However, RCC histopathological image datasets are rare, and no public with vascular annotation. As far as we know, we are the first to perform RCC classification on the vascular network annotated histopathological image dataset.

2.2   Histopathological Images Classification

In the histopathological image classification task, Deep learning-based methods are the most widely used method. Wei et al. [34] developed a BiCNN model to classify breast cancer histopathological images. Deepak et al. [14] classify cancers using GCNs by modeling a tissue section as a multi-attributed multi-relational spatial graph of its constituent cells. In this paper, we classify topological information of RCC vascular networks by GCNs. Although the deep learning methods have achieved great progress for medical imaging tasks, traditional machine learning algorithms are still crucial due to interpretability. Shweta et al. [29] using kernelized weighted extreme learning machine to classify breast cancer histopathology image. Abhinav et al. [21] propose a novel CoMHisP framework based on a fuzzy support vector machine with within-class density information (FSVM-WD) for histopathological image classification.

2.3   Hand-Crafted Features

Hand-crafted features perform an irreplaceable role in medical classification tasks due to low data requirements and easy to train. Li et al. [24] use support vector machines (HC-SVM) to obtain hand-crafted features and their performance can comparable to CNN in colon histology images. Jeena et al. [19] proposed hand-crafted texture features for stroke diagnosis. Zubiolo et al. [37] use hand-crafted features (number of end branches (NE) & junctions (NJ), the length of end branches (LE) & non-end branches (LJ) and their ratios) to represent the vascular network of RCC. In this paper, we not only proposed a much bigger dataset, but also new hand-crafted features (skeleton and lattice features) of RCC and perform a series of classification tasks as opposed to the former study [37].

3 Dataset

3.1 Dataset Building

The original WSIs are Hematoxylin Eosin-stained, scanned by SCN400 scanner at 40x magnification ($\sim$ 60000 × 60000 Pixels).

Tumor and non-tumor (necrosis, fiber, normal) areas were annotated using the open-source software ASAP and stored in XML format file. Then we cut the annotated areas of WSIs into smaller image patches (2000 × 2000 Pixels) to form the BigRCC dataset. The statistical distribution of the BigRCC dataset is shown in Table 1.

Table 1. The number of image patches of each category in BigRCC.

BigRCC dataset	Necrosis	Fiber	Normal	Tumor	Total
ccRCC	3324	1941	7459	27287	39986
pRCC	1602	920	2105	13637	18254

3.2 VRCC200

We have annotated the vascular of 200 tumor image patches from BigRCC to construct the VRCC200 dataset. We use "ImageJ" software to annotate the vascular networks and checked them by medical specialists. As shown in Fig. 1, the vascular structure of ccRCC is like a "fishnet", while pRCC looks like a "tree". To describe the vascular network, we consider *junctions, end branches* and *non-end branches* (branches between two junctions).

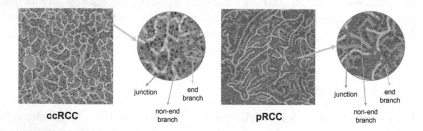

Fig. 1. The vascular network examples of ccRCC and pRCC images.

4 Vascular Network Feature

4.1 Hand-Crafted Features

Skeleton Features. As shown in Fig. 1, the fishnet has more junctions, more non-end branches, the tree has more end branches. The graph visualization on Fig. 3 also supports our assumption. We use NE, small NE (less than nuclear size ×10), long NE (more than nuclear size ×10), NJ, LE, LJ, density as the

Table 2. The explanation of each skeleton and lattice feature.

Skeleton features	Note
NE	the **N**umber of **E**nd branches
LE	average **L**ength of the **E**nd branches
Small NE	NE that LE less than nuclear size ×10
Long NE	NE that LE more than nuclear size ×10
NJ	**N**umber of **J**unctions
LJ	Average **L**ength of the non-End branches
Density	Sum of skeleton pixels
NE/NJ	NE/NJ Ratio
LE/LJ	LE/LJ Ratio
NE/(LJ+LE)	NE/(LJ+LE) Ratio
NJ/(LJ+LE)	NJ/(LJ+LE) Ratio
LJ/(LJ+LE)	LJ/(LJ+LE) Ratio
Lattice features	Note
Mean area	Mean of all lattice areas
Median area	Median of all lattice areas
Mean perimeter	Mean of all lattice perimeters
Median perimeter	Median of all lattice perimeters
Mean eccentricity	Mean of all lattice eccentricities
Mmedian eccentricity	Mean of all lattice eccentricities

basic features of the skeleton. Moreover, we also consider NE/NJ, NE/(LJ+LE), NJ/(LJ+LE), LE/LJ, LJ/(LJ+LE) to make the skeleton features robust and comprehensive. The details are shown on Table 2.

Lattice Features. Based on the different structures of vascular networks of ccRCC and pRCC, we further propose a new set of features "lattice features". But they aren't extracted from the skeleton directly. As shown in Fig. 2, we firstly compute the watershed [32] on the vascular skeleton to obtain closed areas. Then we use the minima imposition method [2,27] to solve the problem of local minimum and modify the distance transformation to obtain the lattice spatial map. Finally, we remove surrounding lattice cells which do not include complete vascular information. Next, we define and extract 6 features from the lattice map. including mean area, median area, mean perimeter, median perimeter, mean eccentricity, and median eccentricity. The details are shown in Table 2.

4.2 Deep Learning Feature

Deep Features. To further evaluate if the vascular network can be used alone for RCCs classification, we consider both raw images and vascular segmentation

Fig. 2. The pipeline of obtaining the lattice features.

images as inputs, then learn these deep features with deep learning models as the baseline experiments, such as LeNet [35], AlexNet [20], VggNet [30] and ResNet [16].

Graph Features. Because the vascular network is a kind of graph-like structure, we transform the vascular network into graph with SKL-Graph [11], as shown in Fig. 3. Then we feed the adjacency matrix of the graph into GCN. Although we only consider topological information as the branch length is not coded in the adjacency matrix, this feature is more explainable than the deep features systematically learned during network optimization.

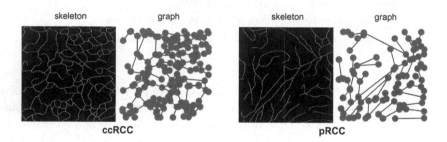

Fig. 3. The graph features of ccRCC and pRCC, red points represent end points of the vascular network, green points represent junctions of the vascular network.

5 Experiments

5.1 Skeleton Features and Lattice Features Analysis

Non-parametric Tests. Since the proposed features are not normally distributed, we use 3 common non-parametric tests methods [17] (Mann-Whitney U Test, Moses extreme reactions Test and Kolmogorov Smirnov Z Test) to calculate the statistical significance between the ccRCC and pRCC features. As shown in Table 3, there are only 5 features whose P-value is larger than 0.05 for at least one statistical test (*italic font*). Finally, we choose the 13 features (**bold font**) that are significant in all 3 tests. These 13 features are more suitable for classifying the ccRCC and pRCC images.

Table 3. P-value of non-parametric tests of every feature.

Features	Mann-Whitney U Test	Moses extreme reactions Test	Kolmogorov-Smirnov Z Test
NE	1.73E−19	1.23E−13	7.01E−16
Small NE	*2.09E−01*	1.39E−04	*8.37E−02*
Long NE	2.89E−21	0.00E+00	4.57E−18
NJ	1.17E−07	2.28E−03	1.60E−05
LE	1.16E−13	7.34E−03	3.46E−14
LJ	2.50E−24	8.41E−19	5.26E−23
Density	7.93E−29	0.00E+00	1.98E−27
NE/NJ	4.76E−28	2.83E−21	5.25E−26
LE/LJ	4.63E−02	*1.00E+00*	2.61E−02
NE/(LJ+LE)	2.60E−02	*9.88E−01*	3.45E−03
NJ/(LJ+LE)	1.59E−28	0.00E+00	3.93E−27
LJ/(LJ+LE)	4.44E−02	*1.00E+00*	2.70E−02
Mean area	3.96E−29	0.00E+00	3.96E−29
Median area	2.65E−22	5.91E−20	5.25E−26
Mean perimeter	4.44E−08	*3.18E−01*	1.26E−07
Median perimeter	2.45E−23	5.91E−20	5.25E−26
Mean eccentricity	3.83E−03	6.00E−06	4.30E−05
Median eccentricity	4.87E−02	1.39E−04	1.60E−03

Performance on Traditional Algorithms. As shown in Table 4, We compare the result of skeleton and lattice features with the baseline features: NE, NJ, LE, LJ, NE/NJ and LE/LJ [37]. Our proposed features, particularly filtered by non-parametric tests (filtered Skeleton & Lattice) achieve higher accuracy on almost all traditional algorithm models (Adaboost [12], DecisionTree [5], Gradient Boosting Tree [13], KNN [10], Logistic Regression [35], Random Forest [4] and two kinds of SVM [7,31]). This proved that our proposed features are more robust and efficient.

Table 4. The accuracy results of 3 feature sets on different algorithms.

Methods	Baseline skeleton	Full skeleton & Lattice	Filtered skeleton & Lattice
Adaboost [12]	**0.975**	0.940	0.940
Decision Tree [5]	0.955	**0.965**	**0.965**
Gradient Boosting Tree [13]	0.955	**0.965**	**0.965**
KNN [10]	0.905	0.910	**0.935**
Logistic Regression [35]	0.970	**0.985**	**0.985**
Random Forest [4]	0.960	0.960	**0.965**
SVM RBF [7]	0.965	0.980	**0.985**
SVM Sigmoid [31]	0.790	**0.965**	**0.965**

5.2   Vascular-Based RCC Classification Benchmark

In this section, we conduct benchmark experiments on the VRCC200 dataset with traditional and deep learning models. The accuracy results of the validation are the average of the "leave one out" cross-validation. There are 7 patients' data of VRCC200, we take 6 patients for training and 1 for testing each time. To check the robustness of these models, we annotate dataset VRCC60 with more patients for testing. VRCC60 dataset has 60 image patches, coming from 20 patients.

Table 5. The benchmark results on the VRCC200 and VRCC60.

Methods	Input	Feature	Acc(val)	Acc(test)
Traditional Algorithm				
Adaboost [12]	Segment	filtered Skeleton & Lattice	**0.940**	0.935
Decision Tree [5]	Segment	filtered Skeleton & Lattice	**0.965**	0.935
Gradient Boosting Tree [13]	Segment	filtered Skeleton & Lattice	**0.965**	**0.965**
KNN [10]	Segment	filtered Skeleton & Lattice	**0.935**	0.915
Logistic Regression [35]	Segment	filtered Skeleton & Lattice	**0.985**	0.915
Random Forest [4]	Segment	filtered Skeleton & Lattice	0.965	**0.985**
SVM RBF [7]	Segment	filtered Skeleton & Lattice	**0.985**	0.915
SVM Sigmoid [31]	Segment	filtered Skeleton & Lattice	**0.965**	0.865
Deep learning				
LeNet [22]	Raw	Deep	0.683	**0.750**
AlexNet [20]	Raw	Deep	0.690	**0.767**
VGG-16 [30]	Raw	Deep	**0.819**	0.783
ResNet-18 [16]	Raw	Deep	0.847	**0.883**
LeNet [22]	Segment	Deep	**0.876**	0.750
AlexNet[20]	Segment	Deep	**0.866**	0.800
VGG-16 [30]	Segment	Deep	0.857	**0.883**
ResNet-18 [16]	Segment	Deep	0.904	**0.950**
GCN+SAGPoolg [36]	Segment	Graph	**0.910**	0.833
GCN+SAGPoolh [6]	Segment	Graph	**0.845**	0.817

Validation Results on VRCC200. As shown in Table 5, segment images are vascular network segment masks. Acc (val) is the accuracy result of "leave one out" cross-validation on VRCC200. Acc (test) is the accuracy result of testing on the VRCC60.

The results of the traditional methods range from 94% to 98.5%, which is higher than most others. This demonstrates that our proposed skeleton and lattice features are efficient and robust.

For the results of deep learning baselines, the accuracy with vascular segmentation input is much better than raw image input. We argue that a bigger dataset is needed to train raw images, but vascular segmentation can work well on a small dataset and proved that we can just use the vascular network to do classification.

The results of the two GCN models are lower than some vascular segment input based deep learning models (85.7% VS 84.5%) and traditional algorithm models (93.5% VS 84.5%). Maybe because the graph features only contain the topological information of vascular, whereas the other features contain both topological and geometrical information.

Testing Results on VRCC60. Compared with the validation results on VRCC200, the accuracy values on VRCC60 (75.0% to 98.5%) similar to the result of VRCC200 (68.3% to 98.5%). Moreover, skeleton and lattice features based models still have the best performers (86.5% to 98.5%). This demonstrates that our proposed skeleton and lattice features are robust and efficient on the more patients' dataset (20 patients).

6 Conclusion

To demonstrate the importance of the vascular network structure of RCC, we build the VRCC200 and VRCC60 datasets with accurate vascular annotations. In this way, we can characterize ccRCC and pRCC from the topological and geometrical information of RCC vascular networks, by our proposed "skeleton and lattice features". The model performance with our features is much better than the "graph features" (only topological information) and the baseline deep learning features. It proved that the vascular network can be used alone for RCC classification and sufficient to define a tumor subtype. Also, our features provide a better explanation for the RCC classification task and perform well on the more challenging dataset (more patients).

References

1. Aresta, G., et al.: Bach: grand challenge on breast cancer histology images. Med. Image Anal. **56**, 122–139 (2019)
2. Beucher, S.: Segmentation d'images et morphologie mathématique. Ph.D. thesis, Ecole Nationale Supérieure des Mines de Paris (1990)
3. Borkowski, A.A., Bui, M.M., Thomas, L.B., Wilson, C.P., DeLand, L.A., Mastorides, S.M.: Lung and colon cancer histopathological image dataset (lc25000). arXiv (2019)
4. Breiman, L.: Random forests. Mach. Learn. **45**(1), 5–32 (2001)
5. Breiman, L., Friedman, J.H., Olshen, R.A., Stone, C.J.: Classification and Regression Trees. CRC Press, Boca Raton (1984)
6. Cangea, C., Veličković, P., Jovanović, N., Kipf, T., Liò, P.: Towards sparse hierarchical graph classifiers. arXiv (2018)
7. Cao, H., Naito, T., Ninomiya, Y.: Approximate RBF kernel SVM and its applications in pedestrian classification. In: MLVMA Workshop (2008)
8. Cheng, J., et al.: Computational analysis of pathological images enables a better diagnosis of tfe3 xp11. 2 translocation renal cell carcinoma. Nat. Commun. **11**(1), 1–9 (2020)

9. Cheville, J.C., Lohse, C.M., Zincke, H., Weaver, A.L., Blute, M.L.: Comparisons of outcome and prognostic features among histologic subtypes of renal cell carcinoma. Am. J. Surg. Pathol. **27**(5), 612–624 (2003)
10. Cover, T., Hart, P.: Nearest neighbor pattern classification. IEEE Trans. Inf. Theor. **13**(1), 21–27 (1967)
11. Debreuve: https://gitlab.inria.fr/edebreuv/sklgraph (2020)
12. Freund, Y., Schapire, R.E.: A decision-theoretic generalization of on-line learning and an application to boosting. J. Comput. Syst. Sci. **55**(1), 119–139 (1997)
13. Friedman, J.H.: Greedy function approximation: a gradient boosting machine. Ann. Stat. **29**, 1189–1232 (2001)
14. Gadiya, S., Anand, D., Sethi, A.: Histographs: graphs in histopathology. arXiv (2019)
15. Gao, Z., Puttapirat, P., Shi, J., Li, C.: Renal cell carcinoma detection and subtyping with minimal point-based annotation in whole-slide images. In: Martel, A.L., et al. (eds.) MICCAI 2020. LNCS, vol. 12265, pp. 439–448. Springer, Cham (2020). https://doi.org/10.1007/978-3-030-59722-1_42
16. He, K., Zhang, X., Ren, S., Sun, J.: Deep residual learning for image recognition. In: CVPR, pp. 770–778 (2016)
17. Hollander, M., Wolfe, D.A., Chicken, E.: Nonparametric Statistical Methods, vol. 751. John Wiley & Sons, New York (2013)
18. Grimm, M.-O., Doehn, C., Krege, S.: Renal cell carcinoma. Der Urologe **59**(2), 133–134 (2020). https://doi.org/10.1007/s00120-020-01130-y
19. Jeena, R., Shiny, G., Sukesh Kumar, A., Mahadevan, K.: A comparative analysis of stroke diagnosis from retinal images using hand-crafted features and CNN. J. Intell . Fuzzy Syst. (Preprint), 1–9 (2021)
20. Krizhevsky, A., Sutskever, I., Hinton, G.E.: Imagenet classification with deep convolutional neural networks. Adv. Neural Inf. Process. Syst. **25**, 1097–1105 (2012)
21. Kumar, A., Singh, S.K., Saxena, S., Singh, A.K., Shrivastava, S., Lakshmanan, K., Kumar, N., Singh, R.K.: Comhisp: a novel feature extractor for histopathological image classification based on fuzzy SVM with within-class relative density. IEEE Trans. Fuzzy Syst. **29**(1), 103–117 (2020)
22. LeCun, Y., Boser, B., Denker, J.S., Henderson, D., Howard, R.E., Hubbard, W., Jackel, L.D.: Backpropagation applied to handwritten zip code recognition. Neural Comput. **1**(4), 541–551 (1989)
23. Li, J., Yang, S., Huang, X., Da, Q., Yang, X., Hu, Z., Duan, Q., Wang, C., Li, H.: Signet ring cell detection with a semi-supervised learning framework. In: IPMI. pp. 842–854. Springer (2019)
24. Li, W., Manivannan, S., Akbar, S., Zhang, J., Trucco, E., McKenna, S.J.: Gland segmentation in colon histology images using hand-crafted features and convolutional neural networks. In: ISBI, pp. 1405–1408. IEEE (2016)
25. Lopez-Beltran, A., Scarpelli, M., Montironi, R., Kirkali, Z.: 2004 who classification of the renal tumors of the adults. Eur. Urol. **49**(5), 798–805 (2006)
26. Loukas, C.G., Linney, A.: A survey on histological image analysis-based assessment of three major biological factors influencing radiotherapy: proliferation, hypoxia and vasculature. Comput. Methods Prog. Biomed. **74**(3), 183–199 (2004)
27. Meyer, F., Beucher, S.: Morphological segmentation. J. Vis. Commun. Image Repre. **1**(1), 21–46 (1990)
28. Prasad, S.R., Humphrey, P.A., Catena, J.R., Narra, V.R., Srigley, J.R., Cortez, A.D., Dalrymple, N.C., Chintapalli, K.N.: Common and uncommon histologic subtypes of renal cell carcinoma: imaging spectrum with pathologic correlation. Radiographics **26**(6), 1795–1806 (2006)

29. Saxena, S., Shukla, S., Gyanchandani, M.: Breast cancer histopathology image classification using kernelized weighted extreme learning machine. Int. J. Imag. Syst. Technol. **31**(1), 168–179 (2021)
30. Simonyan, K., Zisserman, A.: Very deep convolutional networks for large-scale image recognition. arXiv (2014)
31. Smola, A.J., Bartlett, P., Schölkopf, B., Schuurmans, D.: Probabilities for SV machines (2000)
32. Soille, P., Vincent, L.M.: Determining watersheds in digital pictures via flooding simulations. In: VCIP, vol. 1360, pp. 240–250. International Society for Optics and Photonics (1990)
33. Spanhol, F.A., Oliveira, L.S., Petitjean, C., Heutte, L.: A dataset for breast cancer histopathological image classification. IEEE Trans. Biomed. Eng. **63**(7), 1455–1462 (2015)
34. Wang, C., Shi, J., Zhang, Q., Ying, S.: Histopathological image classification with bilinear convolutional neural networks. In: EMBC, pp. 4050–4053. IEEE (2017)
35. Wright, R.E.: Logistic regression (1995)
36. Zhang, M., Cui, Z., Neumann, M., Chen, Y.: An end-to-end deep learning architecture for graph classification. In: AAAI, vol. 32 (2018)
37. Zubiolo, A., Debreuve, E., Ambrosetti, D., Pognonec, P., Descombes, X.: Is the vascular network discriminant enough to classify renal cell carcinoma? In: CBMI, pp. 1–6. IEEE (2016)

Correction to: TVnet: Automated Time-Resolved Tracking of the Tricuspid Valve Plane in MRI Long-Axis Cine Images with a Dual-Stage Deep Learning Pipeline

Ricardo A. Gonzales⊙, Jérôme Lamy⊙, Felicia Seemann⊙,
Einar Heiberg⊙, John A. Onofrey⊙, and Dana C. Peters⊙

Correction to:
Chapter "TVnet: Automated Time-Resolved Tracking
of the Tricuspid Valve Plane in MRI Long-Axis Cine Images
with a Dual-Stage Deep Learning Pipeline"
in: M. de Bruijne et al. (Eds.): *Medical Image Computing*
and Computer Assisted Intervention – MICCAI 2021,
LNCS 12906, https://doi.org/10.1007/978-3-030-87231-1_55

In a former version of this paper, Reference 12 referred to issue 1 rather than to issue 63, which led to an error in the CrossRef link. This has been corrected.

The updated version of this chapter can be found at
https://doi.org/10.1007/978-3-030-87231-1_55

Author Index

Afacan, Onur   430
Alexander, Daniel C.   44
Ambrosetti, Damien   611
Arvinte, Marius   350
Austin, Thomas   265

Bae, Hyeon-Min   222
Balu, Niranjan   579
Belov, Alcksandr   254
Beqiri, Arian   495
Bi, Shanshan   506
Bian, Gui-Bin   600
Bryan, Robert   265

Carson, Richard E.   34
Chang, Dong-Jin   402
Chang, Wei-Tang   191
Chao, Hanqing   441
Chatterjee, Sudhanya   392
Chellappa, Rama   75
Chen, Hu   243
Chen, Jiawei   107
Chen, Junxiao   24
Chen, Li   307, 579
Chen, Ming-Kai   34
Chcn, Xinjian   372
Chen, Xuejin   119
Chen, Yinran   412
Chen, Yong   161, 191, 329
Chen, Zhijie   329
Chen, Zhongyue   372
Chênes, Christophe   451
Cheng, Kai   119
Cheon, Sojeong   402
Cho, Eun-Seo   181
Chung, Sang Hun   191
Cochet, Hubert   537
Cootes, Timothy   590

Debreuve, Eric   611
Deligiannis, Nikos   421
Descombes, Xavier   611
Ding, Qiaoqiao   286
Ding, Xinghao   97, 150

Dong, Bin   86
Dong, Ming   471
Du, Rose   171
Duncan, James S.   34, 485
Dylov, Dmitry V.   254

Emami, Hajar   471

Fei, Hongwen   506
Feng, Chun-Mei   140, 307
Feng, Ruiming   65
Frangi, Alejandro F.   201
Frisken, Sarah   171
Fu, Huazhu   140, 307
Fu, Jiajun   296

Gao, Hao   286
Gharbia, Omar   361
Gholipour, Ali   430
Glide-Hurst, Carri K.   471
Golby, Alexandra   171
Gonzales, Ricardo A.   567
Guo, Pengfei   13

Habes, Mohamad   265
Haouchine, Nazim   171
Harrison, Josquin   537
Hatsukami, Thomas S.   579
Heckbert, Susan   265
Heiberg, Einar   567
Hou, Benjamin   495
Hou, Zengguang   600
Hu, Chen   382
Hu, Xindi   201
Huang, Hongyu   97
Huang, Meiping   506
Huang, Xiaoqiong   201
Huang, Yongqiang   243
Huang, Yue   97, 150
Huang, Yuhao   201
Huynh, Khoi Minh   191
Hwang, Jenq-Neng   579

Iriart, Xavier   537

Jensen, Paul   265
Ji, Hui   286
Jiang, Shanshan   13
Jiang, Xiajun   361
Joel, Suresh Emmanuel   392
Jung, Euijin   318
Juvekar, Parikshit   171

Kainz, Bernhard   495
Kapur, Tina   171
Kastryulin, Sergey   254
Kellman, Michael   461
Kim, Myeong-Gee   222
Kim, Theodore   485
Kim, Youngmin   222
Kondi, Lisimachos   421
Kwon, Hyuksool   222

Lamy, Jérôme   567
LaViolette, Aaron K.   129
Lee, Deukhee   402
Lee, Yong Oh   402
Lee, Yueh   191
Leeson, Paul   495
Legghe, Benoit   537
Li, Chao   232
Li, Cheng   382
Li, Chenxin   150
Li, Jiang   372
Li, Lei   557
Li, Qian   86
Li, Rui   24
Li, Rui-Qi   600
Li, Xiao-Xin   329
Li, Yang   119
Li, Yuexiang   107
Li, Yuwei   65
Li, Zheng   3
Li, Zhihe   506
Li, Zhiyuan   361
Liang, Liang   485
Liang, Xiaokun   55
Lin, Hongxiang   44
Lin, Xin   150
Liu, Chi   34
Liu, Feng   211
Liu, Hangfan   265
Liu, Huafeng   547
Liu, Jing   412
Liu, Minliang   485

Liu, Qiegen   382
Liu, Wenjin   579
Liu, Xiaoqing   97, 150
Liu, Xiaozhao   65
Liu, Xinwen   211
Liu, Yan   243
Liu, Yilin   161
Liu, Zewen   590
López-Rodríguez, Domingo   340
López-Rubio, Ezequiel   340
Lorenzi, Marco   537
Lou, Xin-Jie   329
Lu, Zhiyang   3
Luna, Miguel   318
Luo, Gongning   516
Luo, Jianwen   412
Luo, Jie   171
Luo, Mingyuan   201
Luo, Xiongbiao   412
Luo, Yanmei   276
Lustig, Michael   461
Lyu, Yuanyuan   86, 296

Ma, Kai   107, 526
Ma, Xinghua   516
Ma, Yiting   119
Marivani, Iman   421
Maza-Quiroga, Rosa   340
McKay, Raymond   485
Mecca, Adam P.   34
Meng, Deyu   107
Missel, Ryan   361
Moon, Leo   129
Mossa-Basha, Mahmud   579
Mu, Lide   547

Nasrallah, Ilya   265
Nejad-Davarani, Siamak P.   471
Nguyen, Thanh D.   232
Ni, Dong   201
Ni, Zhen-Liang   600
Niu, Chuang   441

O'Dell, Ryan S.   34
Oh, SeokHwan   222
Onofrey, John A.   567

Pak, Daniel H.   485
Park, Sang Hyun   318
Patel, Vishal M.   13

Peng, Cheng   75, 296
Peng, Yuanyuan   372
Peters, Dana C.   567

Rashid, Tanweer   265
Ravikumar, Nishant   201
Reynaud, Hadrien   495
Ryu, Hyun   181

Sabuncu, Mert   232
Sabuncu, Mert R.   129
Sandino, Christopher M.   461
Sapp, John L.   361
Schmid, Jérôme   451
Schnabel, Julia A.   557
Seemann, Felicia   567
Sermesant, Maxime   537
Shanbhag, Dattesh Dayanand   392
Shang, Kun   86
Shen, Dinggang   3, 276, 329
Shi, Jun   3
Shi, Yiyu   506
Shin, Changyeop   181
Slator, Paddy J.   44
Spincemaille, Pascal   232
Stadelmann, Joël   254
Suh, Sungho   402
Sui, Yao   430
Sun, Bin   119
Sun, Huaiqiang   243
Sun, Liyan   97, 150
Sun, Wei   485

Tamir, Jonathan I.   350, 461
Tewfik, Ahmed H.   350
Thurnhofer-Hemsi, Karl   340
Toloubidokhti, Maryam   361
Tsiligianni, Evaggelia   421

Valanarasu, Jeya Maria Jose   13
Van Dyck, Christopher H.   34
Vasanawala, Shreyas S.   461
Venkatesan, Ramesh   392
Vishwanath, Sriram   350
Vlontzos, Athanasios   495

Wang, Alan Q.   129
Wang, Ce   86
Wang, Ge   441
Wang, Haifeng   382

Wang, Hong   107
Wang, Jing   211
Wang, Jun   3
Wang, Ke   461
Wang, Kuanquan   516
Wang, Lianyu   372
Wang, Linwei   361
Wang, Meng   372
Wang, Puyang   13
Wang, Rui   34
Wang, Shanshan   382
Wang, Tao   243
Wang, Tianchen   506
Wang, Ting   372
Wang, Tingting   372
Wang, Wei   516
Wang, Yan   276
Wang, Yi   232
Wang, Yuhan   600
Ware, Jeffrey   265
Warfield, Simon K.   430
Wei, Dong   526
Wei, Hongjiang   65
Wei, Jia   24
Wu, Qing   65
Wu, Xi   276

Xia, Wenjun   243
Xiao, Rudan   611
Xie, Xiaoliang   600
Xing, Lei   55
Xiong, Xin   171
Xu, Chris   129
Xu, Lan   65
Xu, Xiaowei   506
Xu, Xuanang   441
Xu, Yong   140, 307

Yan, Pingkun   441
Yan, Yunlu   307
Yang, Junwei   329
Yang, Qing   65
Yang, Xin   201
Yao, Chenpu   372
Yap, Pew-Thian   161, 191
Yoon, Young-Gyu   181
Yu, Boliang   65
Yu, Jingyi   65
Yu, Lequan   55
Yu, Stella X.   461

Yu, Yizhou   97, 150
Yuan, Chun   579
Yuan, Shuhao   140

Zerva, Matina   421
Zhan, Bo   276
Zhang, Haimiao   86, 107
Zhang, Hang   232
Zhang, Jiajin   441
Zhang, Jiawei   506
Zhang, Jinwei   232
Zhang, Kevin   461
Zhang, Xiaoqun   286
Zhang, Yi   243
Zhang, Yunlong   150
Zhang, Yuyao   65
Zhang, Zhicheng   55
Zhao, Wei   55

Zheng, Hairong   382
Zheng, Yefeng   107, 526
Zhou, Bo   34
Zhou, Jiliu   243, 276
Zhou, Jinyuan   13
Zhou, Luping   276
Zhou, S. Kevin   75, 86, 211, 296
Zhou, Xiao-Hu   600
Zhou, Yan-Jie   600
Zhou, Yi   372
Zhou, Yukun   44
Zhu, Weifang   372
Zhuang, Jian   506
Zhuang, Xiahai   557
Zhuang, Yihong   150
Zimmer, Veronika A.   557
Zou, Yuxin   201
Zu, Chen   276

Printed in the United States
by Baker & Taylor Publisher Services